D1571225

# Molecular Complexes

# Molecular Complexes

## *A Lecture and Reprint Volume*

ROBERT S. MULLIKEN
Distinguished Service Professor of Physics and Chemistry
*University of Chicago*
Distinguished Research Professor of Chemical Physics
*Florida State University (Winters)*

WILLIS B. PERSON
Professor of Chemistry
*University of Florida*

Wiley-Interscience, A Division of John Wiley & Sons
New York · London · Sydney · Toronto

10 9 8 7 6 5 4 3 2 1

Library of Congress Catalog Card Number: 71-84970

SBN 471 62370 9

Printed in the United States of America

# Preface

It had been our intention to prepare a substantial book on molecular complexes and their spectra. However, the work went slowly because of the many questions that need answering when one tries to find a valid interpretation of the everincreasing body of experimental information. These questions are especially numerous because of the fascinatingly varied types of complex that are now known, also because of the need of considering both solid-state and solution data and trying to relate them to a theoretical framework that is based primarily on isolated molecules of complex. Vapor-state data to which the theoretical considerations would be directly applicable are unfortunately difficult to obtain, but recently several instructive and intriguing papers on vapor-state complexes have been published.

Rather than wait longer we have decided to release the present small book based on some lectures given at Chicago, supplemented by a reprinting of several relevant journal articles. During much of 1965–66 both of us were in Chicago. In the fall of 1965 a lecture course given by R. S. M. with the cooperation of W. B. P. afforded an opportunity to clarify and to organize some of the subject matter presented here. The lectures approximately corresponding to Chapters 5, 6, 7, and in part 9, 10, and 14 were given by W. B. P., the rest by R. S. M. The lectures were first written up by W. B. P. and then were jointly edited. The content of the original lectures has been rearranged and revised, enlarged, and updated, and includes a considerable amount of previously unpublished material.

However, the whole still forms only an incomplete and in many respects only tentative account, which should nevertheless serve to outline at least some of the major aspects of the subject and to give an impression of some of the unsolved problems. The reprints—even though some of the ideas expressed therein need revision, change of emphasis, or generalization—give additional detail and additional perspective in various directions.

Our main effort is concentrated on 1:1 complexes in solution. We have not attempted to deal seriously with typical solid-state properties but rather to use solid-state data only for whatever light they can throw on individual molecules of complex, in particular on their geometrical configurations and on polarization directions in their electronic spectra. We give only brief consideration to the role of electron donors and acceptors in intramolecular charge transfer, and in complex ions, carbonyls, and other coordination compounds.

We have attempted a reasonably representative critical sampling of topics, including most that we thought important in 1966. However, the immense number of publications has made it impossible not to overlook some important contributions. Our apologies go to those authors whose work has not been properly quoted or has been overlooked.

The main body of the manuscript was first completed in 1966, but the delay in final preparation permitted us to modify some portions of the material to make some use of recent ideas up to April, 1968, and to include also some afterthoughts added in the "Postscript" late in 1968.

Robert S. Mulliken
*Chicago, Illinois*

Willis B. Person
*Gainesville, Florida*

*June 1969*

# *Acknowledgements*

We are greatly indebted to the Office of Naval Research for assistance during the preparation of this book, and W. B. P. is indebted to the National Science Foundation for a senior postdoctoral fellowship during a considerable part of his stay in Chicago (on leave of absence from the University of Iowa).

We are grateful to Prof. Milton Tamres for critical reading and comments on the book.

We thank our coauthors and the journal editors for permission to reproduce the reprints at the end of this book: *The Journal of the American Chemical Society* (R1, R3, R7, R9–11); *The Journal of Physical Chemistry* (R4); *Recueil des travaux chimiques des pays-bas* (R8); *Annual Review of Physical Chemistry* (R12); *Journal of Chemical Physics* (R13, R15); and *Journal de chimie physique* (R14). In addition we are grateful for the cooperation of Academic Press, permitting us to quote freely from our Chapter 10 for *Physical Chemistry,* Vol. III (edited by H. Eyring, D. Henderson, and W. Jost) (Academic, New York, 1969).

We have made much use of the books by G. Briegleb, *Elektronen-Donator-Acceptor-Komplexe* (Springer, Berlin, 1961) and by L. J. Andrews and R. M. Keefer, *Molecular Complexes in Organic Chemistry* (Holden-Day, San Francisco, 1964) and wish to acknowledge our indebtedness to them, especially for their tables of data and their efforts in organizing this mass of material.

Finally, we acknowledge our debt to the following authors for permitting us to use their illustrations, as follows:

O. Hassel and Chr. Rǿmming (University of Oslo), Figs. 5-1, 5-3.
L. Hall (University of Wisconsin, River Falls), Fig. 5-4.
S. C. Wallwork (University of Nottingham), Figs. 5-5 and 5-6.
T. Kubota (Shionogi Research Laboratory, Osaka), Fig. 7-2.
S. Nagakura (University of Tokyo), Fig. 7-3.
N. J. Rose (University of Washington) and R. S. Drago (University of Illinois), Fig. 7-4.
M. Tamres (University of Michigan), Fig. 7-5.
E. Rabinowitch (University of Illinois), Fig. 10-1*a* and *b*.
G. Kortüm (University of Tübingen), Fig. 10-1*c* and *d*.

Permission to use these figures was granted by the editors of the following societies or journals:

The Chemical Society (Figs. 5-1, 5-3).
Acta Crystallographica (Figs. 5-5 and 5-6).
Journal of the American Chemical Society (Figs. 7-2 to 8-5).
Faraday Society (Fig. 10-1*a* and *b*).
Zeitschrift für Naturforschung (Fig. 10-1*b* and *c*).

Robert S. Mulliken
Willis B. Person

# *Contents*

*POSTSCRIPT* 299

Molecular Compounds and Their Spectra. Some General Considerations. 301

*REPRINTS*

R1 Structures of Complexes Formed by Halogen Molecules with Aromatic and with Oxygenated Solvents [*J. Am. Chem. Soc.*, **72**, 610 (1950)]. R. S. Mulliken. 313

R2 The Interaction of Electron Donors and Acceptors (paper 25 of ONR Report on September 1951 Conference on "Quantum-Mechanical Methods in Valence Theory"). R. S. Mulliken. 322

R3 Molecular Compounds and Their Spectra. II. [*J. Am. Chem. Soc.*, **74**, 811 (1952)]. R. S. Mulliken. 331

R4 Molecular Compounds and Their Spectra. III. The Interaction of Electron Donors and Acceptors [*J. Phys. Chem.*, **56**, 801 (1952)]. R. S. Mulliken. 345

R5 Intermolecular Charge-Transfer Forces. (Proceedings of the International Conference on Theoretical Physics, Kyoto and Tokyo, September 1953). R. S. Mulliken. 367

R6 The Interaction of Electron Donor and Acceptor Molecules (Symposium on Molecular Physics at Nikko 1953). R. S. Mulliken. 374

R7 Molecular Compounds and Their Spectra. IV. The Pyridine-Iodine System [*J. Am. Chem. Soc.*, **76**, 3869 (1954)]. C. Reid and R. S. Mulliken. 381

R8 Molecular Complexes and Their Spectra. VI. Some Problems and New Developments [*Rec. trav. chim.*, **75**, 845 (1956)]. R. S. Mulliken. 387

R9 Molecular Complexes and Their Spectra. VII. The Spectrophotometric Study of Molecular Complexes in Solution; Contact Charge Transfer Spectra [*J. Am. Chem. Soc.*, **79**, 4839 (1957)]. L. E. Orgel and R. S. Mulliken. 395

R10 Molecular Complexes and Their Spectra. IX. The Relationship Between the Stability of a Complex and the Intensity of its Charge-Transfer Bands [*J. Am. Chem. Soc.*, **81**, 5037 (1959)]. J. N. Murrell. 403

R11 Molecular Complexes and Their Spectra. XII. Ultraviolet Absorption Spectra Caused by the Interaction of Oxygen with Organic Molecules [*J. Am. Chem. Soc.*, **82**, 5966 (1960)]. H. Tsubomura and R. S. Mulliken. 410

R12 Donor-Acceptor Complexes [*Ann. Rev. Phys. Chem.*, **13**, 107 (1962)]. R. S. Mulliken and W. B. Person. 419

R13 Electron Affinities of Some Halogen Molecules and the Charge-Transfer Frequency [*J. Chem. Phys.*, **38**, 109 (1963)]. W. B. Person. 439

R14 The Interaction of Electron Donors and Acceptors [*J. chim. phys.*, **61**, 20 (1964)]. R. S. Mulliken. 447

R15 Infrared Spectra of Charge-Transfer Complexes. VI. Theory [*J. Chem. Phys.*, **44**, 2161 (1966)]. H. B. Friedrich and W. B. Person. 466

*General References* 477

*Partial Glossary of Symbols* 479

*Index* 487

# Molecular Complexes

CHAPTER 1

# Introduction

## 1-1 SCOPE

In this book the main emphasis is on an approximate theoretical understanding of the structure and spectra of the addition compounds, or molecular complexes, formed by the association of molecules or other entities. A *molecular complex* between two molecules is an association, somewhat stronger than ordinary van der Waals associations, of definite stoichiometry (one-to-one for the cases we shall consider). The partners are very often already *closed-shell* (saturated valence) electronic structures. In loose complexes the identities of the original molecules are to a large extent preserved. The tendency to form complexes occurs when one partner is an *electron acceptor* (Lewis acid) and the other is an *electron donor* (Lewis base). We abbreviate to the term *donor-acceptor complex* to include all such associations and use $D$ for electron donor, $A$ for electron acceptor.

The discussion is at first in terms of applications of the simplest form of an intermolecular charge-transfer resonance theory presented by Mulliken in 1951–1952 (R1–6).[†] As the discussion progresses the need for using the theory in more general form is recognized and applications made. Further developments of the theory (R7–15) and a comparison

[†] The reprints that are included in this volume, numbered 1 to 15 (see Contents) are referred to in the following text as R1, R2, and so on.

with an alternative approach using overall molecular orbitals for the whole complex, are also considered. Many of the original papers to which reference is made during the course of the discussion are reprinted here following the text of the book.

Electron donor-acceptor complexes and their spectra have been the subject of several reviews,† and three books.‡ The most comprehensive book is that by Briegleb (GR1). Another, by Andrews and Keefer (GR2), emphasizes applications to organic chemistry. A chapter in *Physical Organic Chemistry* by Kosower (GR3) also provides a valuable recent survey with emphasis on applications to reaction mechanisms. A recent book by Rose (GR 4) gives additional information, particularly emphasizing chemical properties. A useful condensed version of the material in the present volume is given as Chapter 10 of Volume III of *Physical Chemistry,* an advanced treatise edited by Eyring, Henderson and Jost (GR5).

Most studies of complexes thus far have been made in solution, in solvents that are as inert as possible. We may then assume that the London dispersion attractions, which are important between D and A in the vapor state, are very approximately cancelled by losses of solute-solvent dispersion-force attractions when a complex is formed from free donor and acceptor in solution. Roughly, one donor-solvent plus one acceptor-solvent contact is replaced by one donor-acceptor and one solvent-solvent contact.

The theory of donor-acceptor complexes and their spectra as presented below is a *vapor-state* theory, except for the omission of the London-dispersion attraction terms. We believe that this theory, after small corrections for solvation energies (see Fig. 9-2) should still be essentially valid for solutions in inert solvents (see also GR5). The few studies that have as yet been made on vapor-state complexes are in general agreement with these expectations but show some puzzling features (see Section 7-4).

The reader should refer to the Postscript following Chapter 18 for some further general considerations based on a last-minute fresh look at certain phases of the subject matter of this book, written after the preparation of the text had been concluded.

## 1-2 INTRODUCTORY DISCUSSION

G. N. Lewis [1] explained coordination compounds, or dative compounds

† For references to many of these see R12.

‡ See "General References" at the end of the book. Individual references are given at the end of each chapter.

(e.g., $R_3N \cdot BCl_3$, which can also be considered as an especially stable molecular complex) in terms of a structure with sharing of the electron lone pair of the nitrogen atom between the N and B atoms. By this sharing the N atom as well as the B atom are surrounded by a complete octet of outer-shell electrons. This sharing can be expressed in quantum-theory language by an approximate wave function $\Psi$ that is a combination of two resonance structures (D here is $R_3N$, A is $BCl_3$):

$$\Psi(\mathrm{D} \cdot \mathrm{A}) \simeq a\,\underset{\text{(no-bond)}}{\Psi_0\,(\mathrm{D},\,\mathrm{A})} + b\,\underset{\text{(dative)}}{\Psi_1\,(\mathrm{D}^{+}\text{—}\mathrm{A}^{-})}. \tag{1-1}$$

The dative structure corresponds to an ionic plus a covalent bond and has sometimes been called a semipolar double bond. The interpretation of the N—B dative bond in the complex as given by (1-1) is analogous to the approximate ionic-covalent resonance interpretation of the chemical bond in HCl:

$$\Psi(\mathrm{HCl}) \simeq a\,\underset{\text{(ionic)}}{\Psi_0\,(\mathrm{H}^{+},\,\mathrm{Cl}^{-})} + b\,\underset{\text{(covalent)}}{\Psi_1\,(\mathrm{H}\text{—}\mathrm{Cl})}. \tag{1-2}$$

In both examples $b > a$. The inclusion of the no-bond structure in (1-1) is probably even more important than that of the ionic structure in (1-2), although this fact seems to have been overlooked (or at any rate underemphasized) in pre-quantum discussions of dative compounds. [Even possibly $a > b$ in (1-1).]

Complexes are classified as *strong* or *weak* depending on whether the energy of formation and the equilibrium constant $K$ are large or small:

$$\mathrm{D} + \mathrm{A} \overset{K}{\rightleftharpoons} \mathrm{D} \cdot \mathrm{A}.$$

Increasingly strong donors and/or acceptors form increasingly stable complexes.

Modern interest in complexes was awakened by the observation by H. A. Benesi and J. H. Hildebrand [2] of a new absorption band in a solution of benzene and iodine dissolved in $n$-heptane, a band that did not appear in the spectrum of either component alone. The explanation was provided by Mulliken in 1950 at a symposium on complexes arranged for the American Chemical Society by Hildebrand. This is the basis for the 1952 papers referred to above (R1–6).

Although the new band found by Benesi and Hildebrand is in the ultraviolet region, an analogous band occurs in the visible for many other complexes. To demonstrate this a solution of tetracyanoethylene (TCNE) in methylene dichloride may be added to a series of aromatic hydrocarbons dissolved in methylene dichloride. Benzene gives a yellow solution, xylene an orange, durene a deep red, and hexamethylbenzene a deep purple.

The succession of methylated benzenes used as donors, with TCNE as acceptor, also illustrates *fortification*. The $\pi$-electron molecules, ethylene and benzene, can act either as weak donors or very weak acceptors. Other things being equal, donor ability increases with decreasing ionization potential $I$; and acceptor ability, with increasing electron affinity $E$. Among aromatic hydrocarbons $I$ decreases and $E$ increases with increasing size; graphite, with $I = E$, is the extreme example and is in fact both a good acceptor and a good donor. Now starting with any unsaturated or aromatic hydrocarbon, either its donor or its acceptor capability can be strengthened by the introduction of suitable substituent groups. The weak donor property of the benzene is fortified by adding more and more electron-releasing methyl groups (inductive effect) whereas the four electrophilic CN groups in TCNE greatly fortify the acceptor capability of the ethylene.

The two kinds of molecular complexes discussed above provide examples of $n \cdot v$ (strong) and of $\pi \cdot \pi$ (weak) complexes. The common types of donors and acceptors are listed in Table 1-1.

**TABLE 1-1**

**Common Types of Donors and Acceptors**

| Donor Type | Example | Dative Electron[a] from | Acceptor Type | Example | Dative Electron[a] Goes to |
|---|---|---|---|---|---|
| $n$ | $:NR_3$ | Nonbonding lone pair | $v$ | $BCl_3$ | Vacant orbital |
| $b\pi$ | Benzene | Bonding $\pi$ orbital | $a\pi$ | TCNE | Antibonding $\pi$ orbital |
| | | | $a\sigma$ | $I_2$, HQ[b] | Antibonding $\sigma$ orbital |

[a] "Dative electron" refers to the electron transferred from donor to acceptor.
[b] Molecules such as phenol, water, and other molecules that give hydrogen bonding.

Equation 1-1 implies that the complex is stabilized by resonance between $\Psi_0$ and $\Psi_1$, the forces involved being called *charge-transfer* forces. However, classical electrostatic forces (including induction forces) also contribute to the stability of complexes and may even be of predominant importance for the stability of most hydrogen-bonded complexes and of the weaker of the complexes of the $b\pi \cdot a\sigma$ and the benzene. $I_2$ ($b\pi \cdot a\sigma$) type (Hanna [3], and R14).

The description of the bonding in complex formation has been given above in terms of resonance structures. It is also possible to give a description (see Chapter 12) in terms of whole-molecule molecular orbitals (regarding the complex as a single molecule). The latter type of description is much more useful in the case of a complex ion, such as for example $[Co(NH_3)_6]^{3+}$, which could be regarded as a 6:1 molecular complex.

## 1-3 STRUCTURES AND SPECTRA—A SIMPLE DESCRIPTION

In terms of the resonance structure description of (1-1) the structure of the ground state of any 1:1 complex† is

$$\Psi_N = a\Psi_0\,(D, A) + b\Psi_1\,(D^+—A^-).$$

This function is normalized as follows:

$$\int \Psi_N \Psi_N\, dv = a^2 + b^2 + 2abS_{01} = 1, \tag{1-3}$$

where

$$S_{01} = \int \Psi_0 \Psi_1\, dv, \tag{1-4}$$

with the integration carried over all space, is the overlap (or nonorthogonality) integral between the functions $\Psi_0$ and $\Psi_1$. If the complex is loose, $S_{01}$ is small and

$$a^2 + b^2 \simeq 1.$$

Hence $b^2$ approximately measures the weight of the dative structure or the fraction of an electron transferred from the donor to the acceptor in the ground state. More accurately, note that the term $2abS_{01}$ in (1-3) can easily be as large as $b^2$ or larger. Half of this term can reasonably be assigned to the donor and half to the acceptor, so that the fractions $F_0$ and $F_1$ in the no-bond and dative structures are

$$F_0 = a^2 + abS_{01}, \quad F_1 = b^2 + abS_{01}. \tag{1-5}$$

In loose complexes between closed-shell donors and acceptors $b^2 \ll a^2$. For benzene $\cdot I_2$, $b^2$ is perhaps 0.06 or less; for pyridine $\cdot I_2$,

---

† See Sections 2-5 to 2-7 for details of $\Psi_0$ and $\Psi_1$. Although (1-1) is to be used mainly for discussing complexes in their equilibrium configuration, it is equally valid for the donor-acceptor pair at any distance as they approach each other.

$b^2 \simeq 0.2$; for trimethylamine $\cdot I_2$, $b^2$ may be about 0.4.

If the ground-state structure of a complex (weak or strong) is given by $\Psi_N$, then according to quantum-theory principles there must be also an excited[†] state $\Psi_V$, which can be called a *charge-transfer* (CT) *state*, given by

$$\Psi_V = -b^* \Psi_0 (\mathrm{D, A}) + a^* \Psi_1 (\mathrm{D^+—A^-}). \qquad (1\text{-}6)$$

The coefficients $b^*$ and $a^*$ are determined by the quantum-theory requirement that the excited-state wave function be orthogonal to the ground-state function:

$$\int \Psi_V \Psi_N \, dv = 0.$$

The excited-state function $\Psi_V$ is normalized as follows:

$$\int \Psi_V^2 \, dv = a^{*2} + b^{*2} - 2a^* b^* S_{01} = 1. \qquad (1\text{-}7)$$

This makes $a^* \simeq a$, $b^* \simeq b$. If $S_{01}$ were 0, $a = a^*$ and $b = b^*$ would be true exactly.

Since for loose molecular complexes the ground state is mostly no-bond ($a^2 >> b^2$) and so (because of the orthogonality requirement) the excited state is mostly dative ($a^{*2} >> b^{*2}$), excitation of an electron from $\Psi_N$ to $\Psi_V$ essentially amounts to the transfer of an electron from D to A. Further, the theory shows that spectroscopic absorption from $\Psi_N$ to $\Psi_V$ should occur with generally high intensity. Charge-transfer spectra of this kind were already well known for $Na^+Cl^-$ in the vapor state, for silver chloride (roughly $Ag^+Cl^-$) in the solid state (the photographic process), and for $Cl^-$ in water solution [hence really $Cl^-(H_2O)_n$], to list a few examples in which $Cl^-$ is the electron donor. It can be shown also that a CT absorption is possible for any pair of molecules (or ions or other entities) if in contact, even if they do not form a stable complex.

Not all CT states are related to the ground state in the manner of (1-6). In (1-1) and (1-6) the two functions $\Psi_0$ and $\Psi_1$ must belong to the same group-theoretical symmetry species. But CT states of different symmetry species are in general possible; for such a state we might write

$$\Psi_{V'} = \Psi_1' (\mathrm{D^+—A^-}), \qquad (1\text{-}8)$$

where of course the nature of $\Psi_1'$ must be specified further if we wish to consider actual examples. A discussion of some examples is given in Section 13-3 (after which reference to Section 11-2 and perhaps also

[†] The notation ($N$ and $V$) for the ground and excited states of complexes is chosen by analogy with the $N$ and $V$ states of diatomic molecules. (See Mulliken [4].)

Section 14-1 is instructive).

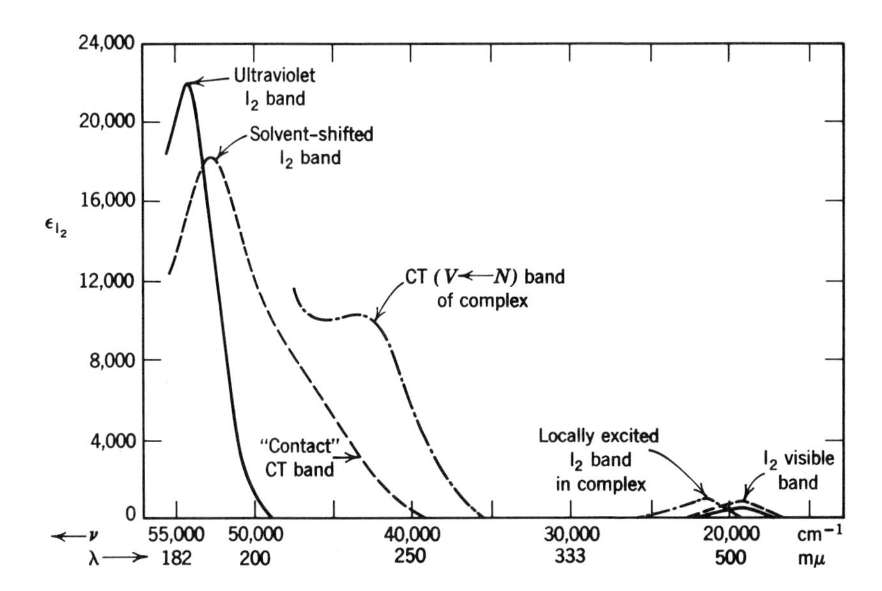


**Figure 1-1** The apparent molar absorptivity of $I_2$ vapor and of $I_2$ and EtOH · $I_2$ in *n*-heptane. Here —— is for $I_2$ vapor, --- for $I_2$ in *n*-heptane, and -·-·- for $I_2$ in *n*-heptane with 3.4 *M* ethyl alcohol.

However, most of the theoretical discussions in the following text deal with CT states only of the type $V$ of (1-6) and CT spectra involving such states, and ignores states of the type $V'$ of (1-8) and corresponding CT spectra. The reader should keep these omissions in mind.

Complexes have been studied mostly in solution, but some studies have been made of solids and, in the case of $n \cdot v$ compounds and also recently [5] of a few complexes of weaker types, in the vapor state. Complexes in solids, even when of 1:1 stoichiometry, do not always occur in the form of pairwise units. Studies in the vapor are difficult because $K$ is small and interference from overlapping spectra of the uncomplexed components is often severe. The latter difficulties also occur for solution studies but are less troublesome because $K$ is larger.

The complete absorption spectrum of a complex consists of absorption to the following:

1. *Locally excited states* (states of D or of A, more or less but usually not greatly modified in the complex).
2. *Charge-transfer states* [$\Psi_V$, as in (1-3), and other CT states involving excited dative structures, for example $\Psi(D^{+*}—A^-)$, $\Psi(D^+—A^{-*})$].

Figure 1-1 shows the change that occurs in the spectrum of $I_2$, first when it is dissolved in *n*-heptane, and then when ethanol is added [6]. (Ethanol is transparent up to 220 m$\mu$.) The peak of the $C_2H_5OH \cdot I_2$ charge-transfer band is marked in the figure, and the position of the shifted visible absorption band of $I_2$ in the complex (a transition to a locally excited state) is also indicated. The contact CT band, which appears as a long wavelength shoulder on the ultraviolet iodine band when $I_2$ is dissolved in heptane, is discussed in Section 14-3.

## 1–4 REFERENCES

[1] G. N. Lewis, *Valence and the Structure of Atoms and Molecules,* Reinhold, New York, 1923.
[2] H. A. Benesi and J. H. Hildebrand, *J. Am. Chem. Soc.*, **71**, 2703 (1949); ibid., **70**, 3978 (1948).
[3] M. W. Hanna, *J. Am. Chem. Soc.*, **90**, 285 (1968) and references 27–29 given there.
[4] R. S. Mulliken, *J. Chem. Phys.*, **7**, 20 (1939); *Phys. Rev.*, **50**, 1017, 1028 (1936); ibid., **51**, 310 (1937).
[5] M. Kroll, *J. Am. Chem. Soc.*, **90**, 1097 (1968) and references given there.
[6] L. Julien, Ph.D. thesis, University of Iowa, 1966.

CHAPTER 2

# Simplified Resonance-Structure Theory

In this chapter we shall go through in some detail the derivation of the theory presented in R3, in an extremely simplified form assuming only one CT state and only two resonance structures ($\Psi_0$ and $\Psi_1$).

## 2-1 GENERAL REMARKS ABOUT $\Psi_N$ AND $\Psi_V$

Let us begin with the Schrödinger equation

$$\mathcal{H}\Psi = W\Psi, \quad \text{or} \quad (\mathcal{H} - W)\Psi = 0. \tag{2-1}$$

This equation holds accurately if $\mathcal{H}$ is the correct Hamiltonian operator for the system and if $W$ and $\Psi$ are an *exact* energy eigenvalue and an exact eigenfunction, respectively. In principle we could compute $W$s and $\Psi$s for a complex, but in practice this is very difficult at present (see, however, Sections 12-2, 12-8, and 15-1). Instead we here assume reasonable forms for $\Psi$, then compute $W$ and other properties (e.g., $W \simeq \int \Psi^{\dagger}\mathcal{H}\Psi\, dv$), and afterwards judge the reasonableness of our assumption by seeing how well the results fit the body of experimental data. For donor-acceptor complexes we have already introduced in Chapter 1 [cf. (1-1) and (1-6)] the expressions

$$\Psi_N = a\Psi_0\,(\mathrm{D, A}) + b\Psi_1\,(\mathrm{D^+—A^-}), \tag{2-2}$$

$$\Psi_V = a^*\Psi_1\,(\mathrm{D^+—A^-}) - b^*\Psi_0\,(\mathrm{D, A}), \tag{2-3}$$

which should be rather accurate (at least in the case of $\Psi_N$) for loose complexes but only approximate for strongly bound complexes.

A very important point is that $\Psi_0$ and $\Psi_1$, and therefore of course $\Psi_N$ and $\Psi_V$, have to belong to the same group-theoretical symmetry species, in terms of the symmetry of the complex. (See Herzberg [1] for symmetry species of diatomic and of polyatomic molecules; or see Jaffé and Orchin [2] or Cotton [3].) Symmetry species of atomic states are indicated by symbols such as ${}^1S$, ${}^1P$, and ${}^2P_{3/2}$; for atomic orbitals the species symbols are $s$, $p$, $d$, etc. For states of homopolar diatomic molecules the species symbols are ${}^1\Sigma_g^+$, ${}^1\Sigma_u^-$, ${}^3\Pi_g$, etc.; symbols for homopolar diatomic orbital species are $\sigma_g$, $\sigma_u$, $\pi_g$, etc. For polyatomic molecules there are many different kinds of symmetry, each with several symmetry species. Later on we shall consider individual donors and acceptors, looking at their symmetry and symmetry species in detail.

Equations 2-2 and 2-3 can be made formally accurate by writing

$$\Psi_N = c_0\Psi_0 + c_1\Psi_1 + c\Omega, \tag{2-4}$$

$$\Psi_V = c_0^*\Psi_0 + c_1^*\Psi_1 + c^*\Omega^*. \tag{2-5}$$

Here $c\Omega$ or $c^*\Omega^*$ contains whatever else has to be included along with $\Psi_0$ and $\Psi_1$ to obtain exact eigenfunctions. The terms $\Omega$ and $\Omega^*$ are assumed to be normalized, as are $\Psi_0$ and $\Psi_1$. Note also that if $\Psi_N$ and $\Psi_V$ are always assumed normalized, the coefficients of $\Psi_0$ and $\Psi_1$ in (2-4) and (2-5) are somewhat different from those in 2-2 and 2-3. (*Approximately*: $c_0 = a$, $c_1 = b$, $c_0^* = -b^*$, $c_1^* = a^*$.) In correspondence to (1-3) and (1-7) we now have

$$\begin{aligned} c_0^2 + c_1^2 + 2c_0c_1S_{01} + c^2 &= 1, \\ c_0^{*2} + c_1^{*2} + 2c_0^*c_1^*S_{01} + (c^*)^2 &= 1. \end{aligned} \tag{2-6}$$

We have noted before that $\Psi_0$ and $\Psi_1$ are not orthogonal. However, instead of (2-4) we *could* write

$$\Psi_N = c_0'\Psi_0 + c_1'\Psi_1' + c\Omega,$$

where $\Psi_1'$ is a linear combination of $\Psi_0$ and $\Psi_1$ such that $\Psi_1'$ is orthogonal to $\Psi_0$; then $(c_1')^2 + (c_2')^2 + c^2 = 1$ if $\Psi_N$ is normalized; but we shall not actually use this expression for $\Psi_N$. The orthogonal residue $\Omega$ is orthogonal to $\Psi_0$ and $\Psi_1$ (and $\Psi_1'$).

## 2-2 THE ENERGIES OF THE NORMAL AND CHARGE-TRANSFER STATES; $h\nu_{CT}$

Let us now go back to the Schrödinger equation (2-1) and substitute our exact expression (2-4) for $\Psi_N$ to obtain the exact equation

$$(\mathcal{H} - W)(c_0\Psi_0 + c_1\Psi_1 + c\Omega) = 0.$$

By multiplying from the left by $\Psi_0^\dagger\, dv$ and integrating over all space we obtain

$$c_0(H_{00} - W) + c_1(H_{01} - S_{01}W) + cH_{0\Omega} = 0. \qquad (2\text{-}7)$$

Here $H_{00} = \int \Psi_0^\dagger \mathcal{H} \Psi_0\, dv$; $H_{01} = \int \Psi_0^\dagger \mathcal{H} \Psi_1\, dv$, $H_{0\Omega} = \int \Psi_0^\dagger \mathcal{H} \Omega\, dv$, and the other terms are as defined earlier.* There is no term containing the factor $cW$ because $\Omega$ is orthogonal to $\Psi_0$.

Multiplying by $\Psi_1^\dagger\, dv$ instead of $\Psi_0^\dagger\, dv$ and integrating, we obtain

$$c_0(H_{10} - S_{01}W) + c_1(H_{11} - W) + c(H_{1\Omega}) = 0. \qquad (2\text{-}8)$$

Now let us make the in general rather severe simplification of ignoring the terms with coefficient $c$ in (2-7) and (2-8) (see Chapter 11 for a consideration of the effects of these terms). Then we have just two homogeneous equations in $c_0$ and $c_1$. These have a solution if, and only if, the determinant of the coefficients is equal to zero. Hence

$$\begin{vmatrix} H_{00} - W & H_{01} - S_{01}W \\ H_{10} - S_{01}W & H_{11} - W \end{vmatrix} = 0. \qquad (2\text{-}9)$$

Expansion of this determinant leads to a quadratic equation in $W$, the two roots of which [see (2-10) below] are $W_N$, the energy of the ground state, and $W_V$, the energy of the CT state. These roots, when substituted into the equations for $c_0$ and $c_1$ [(2-7) and (2-8) after dropping the term $cH_{0\Omega}$ or $cH_{1\Omega}$, give two sets of equations that can be solved to obtain the ratio $c_1/c_0$ or $c_1^*/c_0^*$ by putting $W = W_N$ or $W_V$, respectively [see (2-16) and (2-17) below]. The normalization conditions [(2-6), dropping $c^2$ or $(c^*)^2$] can then be used to fix the absolute values of $c_0$, $c_1$, $c_0^*$, and $c_1^*$, which represent nothing other than $a$, $b$, $-b^*$, and $a^*$ respectively of (2-2) and (2-3).† The solution is somewhat complicated because $S_{01} \neq 0$.

To simplify the notation from here on we write $W_0$, $W_{01}$, and $W_1$ instead of $H_{00}$, $H_{01}$, or $H_{10}$ ($H_{01}$ and $H_{10}$ are equal here), and $H_{11}$.

---

* Note that the complex conjugate of $\Psi_0$ (called $\Psi_0^\dagger$) should be used in the manner illustrated in (2-7) if $\Psi_0$ is complex. However, we shall usually be dealing only with real functions and shall normally omit the symbol for the complex conjugate, taking it as understood if needed.

† When $\Omega$ and $\Omega^*$ are dropped $W$, $c_0$, and $c_1$ of both (2-7) and (2-8) can be considered to belong to *either* $\Psi_N$ or $\Psi_V$. The particular values for these two states, obtained via solution of (2-9), we then call $W_N$, $c_0$, $c_1$ (or $W_N$, $a$, $b$) and $W_V$, $c_0^*$, $c_1^*$ (or $W_V$, $-b^*$, and $a^*$).

The exact solutions of the quadratic equation (2-9) for $W$ are then

$$W(1-S_{01}^2) = (1/2)(W_0 + W_1) - S_{01}W_{01} \pm \sqrt{(\Delta/2)^2 + \beta_0\beta_1}. \quad (2\text{-}10)$$

Here

$$\beta_0 \equiv W_{01} - W_0 S_{01} \quad (\beta_0 < 0), \quad (2\text{-}11)$$

$$\beta_1 \equiv W_{01} - W_1 S_{01} \quad (\beta_1 < 0), \quad (2\text{-}12)$$

and

$$\Delta \equiv W_1 - W_0 \quad (\Delta > 0 \text{ usually but sometimes } \Delta < 0), \quad (2\text{-}13)$$

$$\beta_1 - \beta_0 = -S_{01}\Delta. \quad (2\text{-}14)$$

Thus $|\beta_1| > |\beta_0|$ if $\Delta > 0$ and $S_{01} \neq 0$. If $S_{01}$ were zero, the result would be much simpler ($\beta_0 = \beta_1$, etc.). (See Section 3-5 for a detailed examination of the magnitudes of $\beta_0$ and $\beta_1$.)

The lower energy root is always called $W_N$ (normal state), the upper is $W_V$, even if $\Delta < 0$. The frequency of the CT band is

$$h\nu_{CT} = W_V - W_N = \frac{2\sqrt{(\Delta/2)^2 + \beta_0\beta_1}}{1 - S_{01}^2}. \quad (2\text{-}15)$$

Substituting $W_N$ [from (2-10)] for $W$ in either (2-7) or (2-8) and neglecting $\Omega$ we may solve for the ratio $b/a$ (or $\rho$) for the ground-state wavefunction, $\Psi_N$. Thus

$$\rho \equiv \frac{b}{a} = c_1/c_0 = -\frac{(W_0 - W_N)}{(W_{01} - S_{01}W_N)} = -\frac{(W_{01} - S_{01}W_N)}{(W_1 - W_N)}. \quad (2\text{-}16)$$

Similarly for $\Psi_V$, again using (2-10),

$$\rho^* \equiv b^*/a^* = -c_0^*/c_1^* = \frac{(W_{01} - S_{01}W_V)}{(W_0 - W_V)} = \frac{(W_1 - W_V)}{(W_{01} - S_{01}W_V)}. \quad (2\text{-}17)$$

We may now write the wavefunctions as

$$\Psi_N = \frac{(\Psi_0 + \rho\Psi_1)}{\sqrt{1 + 2\rho S_{01} + \rho^2}} \quad (2\text{-}18)$$

and

$$\Psi_V = \frac{(\Psi_1 - \rho^*\Psi_0)}{\sqrt{1 - 2\rho^* S_{01} + (\rho^*)^2}} \quad (2\text{-}19)$$

It is possible for $W_1$ and $W_0$ to be approximately equal ($\Delta \simeq 0$). Then, or in fact whenever $\beta_0\beta_1$ is not small compared with $(\Delta/2)^2$, the exact

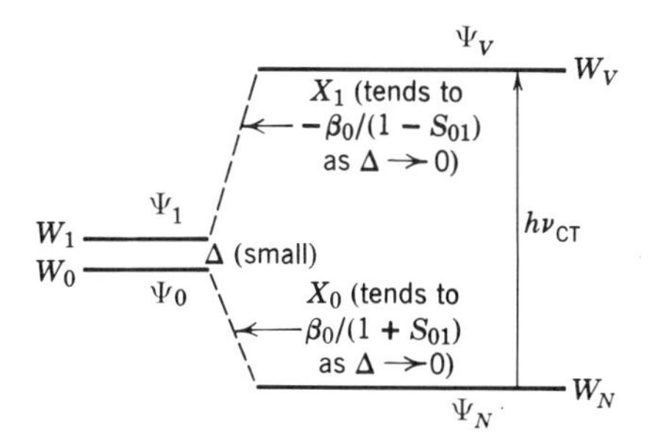


**Figure 2-1** Energy diagram for small $\Delta$; for $\Delta = 0$ in (2-10) $a = b$ in (2-2) $a^* = b^*$ in (2-3), and $\beta_0 = \beta_1$. Here $X_1 = W_V - W_1$ and $X_0 = W_N - W_0$; these terms are more completely defined in Fig. 9-2. (Figure from GR5.).

equation (2-10) must be used to compute $W_N$ and $W_V$ from $W_1$ and $W_0$. We find the result as shown in Fig. 2-1 (note that $W_{01} < 0$).†

## 2-3 APPROXIMATE ENERGIES; PERTURBATION THEORY

If $(\Delta/2)^2 \gg \beta_0\beta_1$, and if $S_{01}^2 \ll 1$ as is expected to be more or less true for weak complexes, the exact expressions above reduce to the following:
From (2-9)

$$W_N = W_0 - \frac{\beta_0^2}{\Delta} + \text{small correction terms,} \tag{2-20}$$

and

$$W_V = W_1 + \frac{\beta_1^2}{\Delta} + \text{small correction terms.} \tag{2-21}$$

† Alternative relations between resonance energy or CT frequency and the coefficients $\rho$ are given by other authors. See, for example, S. Kobinata and S. Nagakura, *J. Am. Chem. Soc.*, **88**, 3905 (1966). Note, however, that Eqs. 11 and 14 of that reference contain typographical errors (Prof. S. Nagakura, private communication). In the former the expression for $X_1$ should be

$$X_1 = \frac{-k^*\beta_1}{1 - k^*S};$$

in the latter

$$h\nu_{CT} = W_E - W_N = -\beta_0\,\frac{1 + 2kS + k^2}{k(1 + kS)\,(1 - S^2)}.$$

(See Kobinata and Nagakura for definitions.)

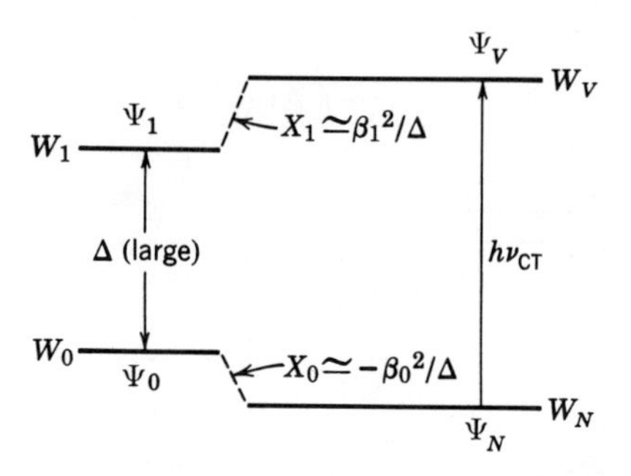


**Figure 2-2** Energy diagram to illustrate perturbation-theory equations (2-20) through (2-22). (Figure from GR5.)

Subtracting (2-20) from (2-21) gives

$$h\nu_{CT} = \Delta + \frac{\beta_0^2 + \beta_1^2}{\Delta} + \cdots. \tag{2-22}$$

In a similar degree of approximation we find from (2-16), (2-17), (2-11), and (2-12) that

$$\rho \equiv \frac{b}{a} = \frac{-\beta_0}{\Delta} \quad (\rho > 0 \text{ if } \Delta > 0). \tag{2-23}$$

and

$$\rho^* \equiv \frac{b^*}{a^*} = \frac{-\beta_1}{\Delta} \quad (\rho^* > 0 \text{ if } \Delta > 0). \tag{2-24}$$

Finally, the wavefunctions are given approximately by (2-18) and (2-19) with $\rho$ and $\rho^*$ from (2-23) and (2-24).†

The results for the energies are illustrated graphically in Fig. 2-2. In contrast with the case in which $W_1 \simeq W_0$, discussed at the end of Section 2-2, we note that $\beta_1^2 > \beta_0^2$ if $\Delta > 0$ [cf. (2-14) and Section 3-6]. Furthermore $\Delta$ is much larger here than either resonance term.

† In R3 it is stated that $b^* \simeq b$ for loose complexes; actually this is not quite correct. (However, it is true that $a^* \simeq a$ since these two coefficients are both approximately equal to 1 for loose complexes). In fact [see (2-14), (2-23), and (2-24)].

$$b^* - b \simeq \rho^* - \rho = \frac{-(\beta_1 - \beta_0)}{\Delta} = S_{01}.$$

Finally, $W_1$ may be lower in energy than $W_0$. Then $\Delta < 0$, $b > a$ in $\Psi_N$, $\beta_0^2 > \beta_1^2$, and the energy diagram is

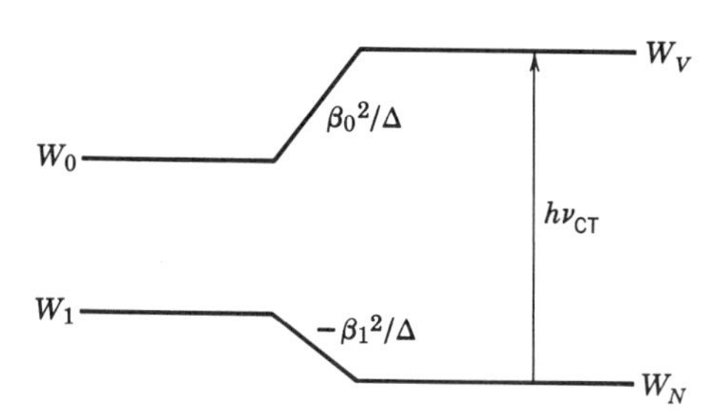


In Chapter 9 we show how to estimate $\Delta$ from the ionization potential $I_D$ of the donor and the electron affinity $E_A$ of the acceptor, together with some estimate of the change in energy as D and A (for $W_0$) or $D^+$ and $A^-$ (for $W_1$) are brought together from complete separation to the equilibrium separation found in the complex. Having made some estimate of $\Delta$ and also for $S_{01}$ we can then—for the case of loose complexes, for which the present approximate theory is reasonably good—use the observed CT frequency, for example, with (2-22) and (2-14), to evaluate $\beta_0$. (Regarding $\beta_0$ see also Section 3-6.) From $\Delta$ and $\beta_0$ we can calculate $\rho$ from (2-23) and hence estimate other properties of the complex. However, all this is distinctly unreliable, especially since the estimation of $S_{01}$ is still largely a matter of guesswork.

All the above approximate theory of Section 2-3 can also be obtained from a second-order perturbation-theory treatment. This treatment takes us well into Section II of R3, which should be consulted for further details.

## 2-4 DIPOLE AND TRANSITION MOMENTS

From the expressions for the approximate wavefunctions of the complex various properties can be computed. Thus the *dipole moment* in the ground state is given by

$$\mu_N = \int \Psi_N \, \mu_{op} \, \Psi_N \, dv = a^2\mu_0 + b^2\mu_1 + 2ab\mu_{01}, \qquad (2\text{-}25)$$

where $\mu_{op}$ (see Section 3-1), $\mu_0$, and so on are vectors.†

† In this book we shall use bold-face type to denote vectors. The same symbol ($\mu_0$, for example) in ordinary Roman or italic type refers to the *magnitude* of that vector.

Here $\mu_0 = \int \Psi_0 \mu_{op} \Psi_0 \, dv$, and so on, and $\mu_0$ is the vector sum of $\mu_D$ and $\mu_A$. If D and A have zero dipole moments, it is natural to suppose that $\mu_0 = 0$; however, Hanna has shown [4] that classical inductive effects included in $\Psi_0$ of (2-2) should even then give rise to a small $\mu_0$. If the complex is weak so that $b$ is not large, $b^2$ is small, and $a$ is near 1, both the terms $2ab\mu_{01}$ and $b^2 \mu_1$ can nevertheless be important since $\mu_{01}$ is approximately proportional to $S_{01}$, which may be far from negligible (see Section 3-5 and Fig. 3-4), whereas $\mu_1$, the dipole moment of $\Psi_1$, is now large. The major contribution to $\mu_1$ is from the charge separation; that is, $\mu_1 \simeq e\mathbf{R}$, where $\mathbf{R}$ is the average distance between the charges on $D^+$ and $A^-$ (more exactly, $\mu_1 = \mu_0 + e\mathbf{R}$).

$$\underset{R}{\underline{D^+\text{—}A^-}}$$

As a specific example consider the benzene · TCNE complex. Let us suppose $\mu_0 = 0$; $R$ is approximately the sum of the van der Waals distances = 1.7 + 1.7 = 3.4 Å, assuming here that the plane of the TCNE molecule is parallel to the ring plane. In order to achieve better overlap the TCNE may be shifted off center with respect to the benzene molecule, increasing $R$ somewhat and causing $\mu_1$ to be directed at an angle away from the perpendicular to the two planes. (See Section 3-3 and Fig. 3-2.) We shall not include this effect here. Then $\mu_1 \simeq 3.4 \times 4.8 \simeq 16$ debyes. We note that $\mu_1$ is generally large so that an appreciable dipole moment is expected for the complex even if the contribution of the dative structure ($b^2$) is small in $\Psi_N$. Thus, if $b^2$ is about 0.1, $\mu_N$ is predicted to be about 1.5 D. In principle the dipole moment of the complex can be used to estimate the value of $b^2$ if allowance is made for the $\mu_{01}$ term in (2-25), and for the polarization dipole term in $\mu_0$. (See Hanna [4].)

Besides $\mu_N$ we can compute the *transition moment* for the CT absorption if $\Psi_N$ and $\Psi_V$ are known:

$$\mu_{VN} = \int \Psi_V \mu_{op} \Psi_N \, dv.$$

The *intensity* of the CT band is proportional to the square, $\mu^2_{VN}$, of the transition moment. However, it is often important, especially in considering intensities, to include $\Omega$ and especially $\Omega^*$ of (2-4) and (2-5) in $\Psi_N$ and $\Psi_V$.

## 2-5 ON THE FORM OF $\Psi_0$ AND THE ENERGIES $W_0$ AND $W_1$

It is important to note that the no-bond structure $\Psi_0$ is defined for a hypothetical state where the D and A molecules are pressed together, with more or less modification of their original geometrical configurations, into the configurations they have in the complex. $\Psi_0$ is a function of the coordinates and spins of the electrons in the complex. It can be expressed as follows:

$$\Psi_0^{[N]}(\mathrm{D, A}) = \mathcal{A}'[\Psi_{\mathrm{mod}}^{[1 \cdots M]}(\mathrm{D})\,\Psi_{\mathrm{mod}}^{[M+1 \cdots N]}(\mathrm{A})]. \qquad (2\text{-}26)$$

Here the donor is assumed to contain $M$ electrons and the acceptor, $N - M$, so that the complex has $N$ electrons, these numbers being indicated by the bracketed superscripts in (2-26). The term $\Psi_{\mathrm{mod}}(\mathrm{D})$ is the electronic wavefunction for the normal state of the donor as it would be if (a) the nuclear framework of the free donor were altered to the configuration that the donor has in the complex, and (b) the donor is subject to exchange repulsion forces and dispersion and classical electrostatic attraction forces (see Section 1-1) due to the presence of the acceptor. The term $\Psi_{\mathrm{mod}}(\mathrm{A})$ has a corresponding meaning. (The subscript "mod" means *modified.*) The function $\Psi_{\mathrm{mod}}(\mathrm{D})$ is antisymmetric with respect to the exchange of any two of its $M$ electrons, whereas $\Psi_{\mathrm{mod}}(\mathrm{A})$ is antisymmetric in its $N - M$ electrons. The modifications make $\Psi_{\mathrm{mod}}(\mathrm{D})$ and $\Psi_{\mathrm{mod}}(\mathrm{A})$ somewhat different, or in some strong complexes considerably different, from the corresponding *free* donor and acceptor functions $\Psi_{\mathrm{D}}$ and $\Psi_{\mathrm{A}}$.

In (2-26) $\Psi_{\mathrm{mod}}(\mathrm{D})$, like any closed-shell singlet-state molecular wavefunction expressed in terms of molecular orbitals (MO's), may be assumed to be expressed in the LC · AS · MSOP form (linear combination of antisymmetrized molecular-spinorbital-product functions), consisting of one main term [i.e., an SCF (selfconsistent-field) wavefunction that corresponds to the ordinary MO electron-configuration description], plus a series of terms with small coefficients (configuration mixing, or configuration interaction, terms). For the free donor these added terms make $\Psi_{\mathrm{D}}$ *exact* by taking care of electron correlation. However, the individual terms and their coefficients are of course somewhat modified in $\Psi_{\mathrm{mod}}$. The case is similar for $\Psi_{\mathrm{mod}}(\mathrm{A})$.

The supplementary antisymmetrizer $\mathcal{A}'$ in (2-26) is an operator that makes $\Psi_0(\mathrm{D, A})$ antisymmetric in *all* the $N$ electrons of the complex (see Section 2-6), as is required by the Pauli exclusion principle. Associated with the overall antisymmetrization is an exchange or closed-shell repulsion between D and A, which increases as they are brought closer and closer together; this repulsion is of the same nature as that between two helium atoms when forced together.

As D and A are brought closer together $W_0$ and $W_1$ change somewhat, the changes being expressible as a function of a suitably defined distance $R_{\mathrm{DA}}$ between D and A, of the orientation of D with respect to A,

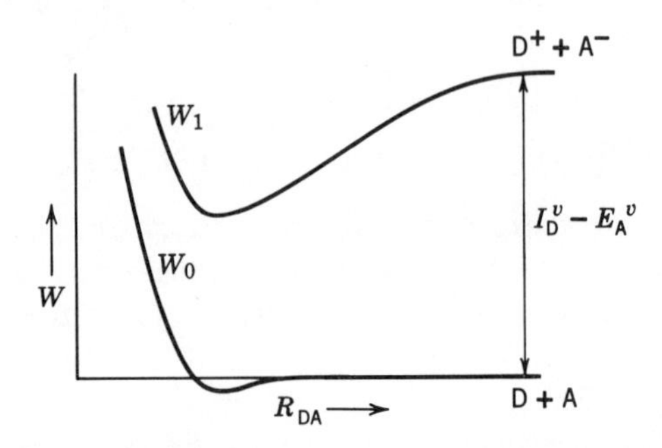


**Figure 2-3** Schematic diagram indicating energies of the no-bond and dative states as functions of the D—A separation in the complex.

and of changes in conformation of D and A from their forms in the gas phase. For a vapor-state complex $W_0$ contains all the classical electrostatic (including induction) attractions and the London-dispersion-attraction terms as well as the closed-shell-repulsion terms and the nuclear configurational modification energies. In solution the London-dispersion terms are omitted because they are approximately cancelled out (see Section 1-1).

If we suppose D and A to be modified to the configurations they have in the complex and oriented so that they line up as they do in the complex, we can plot $W_0$ and $W_1$ as functions of $R_{DA}$ as in Figure 2-3. Here $I^v_D$ is the ionization potential for the free donor *constrained* to remain after ionization in the conformation it has in its neutral normal state, and $E^v_A$ is the electron affinity of the similarly *constrained* acceptor. (The superscript $v$ refers to a "vertical" process.)

## 2-6 A FOOTNOTE: ANTISYMMETRIZATION OF SPINORBITAL-PRODUCT FUNCTIONS

The antisymmetrizer $\mathcal{A}$ for an $N$-electron spinorbital-product wavefunction is

$$\mathcal{A} = \sqrt{\frac{1}{N!}}\, \Sigma\, (-1)^p\, P, \tag{2-27}$$

where $P$ is a permutation operator that interchanges electrons from one orbital to another. The superscript $p$ is even or odd depending on whether

$P$ is an even or odd permutation; that is, whether it can be effected by an even or by an odd number of successive exchanges of two electrons. Thus for a four-electron function $\Psi$

$$\Psi = \lambda_A(1)\,\lambda_B(2)\,\lambda_C(3)\,\lambda_D(4),$$

$$P_0\Psi = \Psi \qquad (p = 0),$$

$$P_{12}\Psi = \lambda_A(2)\,\lambda_B(1)\,\lambda_C(3)\,\lambda_D(4) \quad (p = 1),$$

$$P_{12,34}\Psi \equiv P_{34}P_{12}\Psi = \lambda_A(2)\,\lambda_B(1)\,\lambda_C(4)\,\lambda_D(3) \quad (p = 2),$$

$$P_{142}\Psi \equiv P_{24}P_{12}\Psi = \lambda_A(4)\,\lambda_B(1)\,\lambda_C(3)\,\lambda_D(2) \quad (p = 2),$$

and so on. Here each $\lambda$ is a spinorbital, a product of an orbital $\phi$ and an electron spin function ($\alpha$ or $\beta$); for example, for an atom $\lambda_A$ might be $2p_z\alpha$ or $2p_z\beta$, so $\lambda_A(1)$ would be $2p_z(1)\alpha(1)$ or $2p_z(1)\beta(1)$. In the present resonance-structure theory MSOs (molecular spinorbitals), which are products of MOs and spin functions, will be used. The summation in (2-27) is over the $N!$ possible permutations of $N$ electrons.

In (2-26) $\Psi_{mod}(D)$ was assumed to have already been made antisymmetric in its $M$ electrons; and $\Psi_{mod}(A)$, in its $N - M$ electrons. The further antisymmetrization needed to make the overall function $\Psi_0$ obey Pauli's principle is provided by the supplementary antisymmetrizer $\mathcal{A}'$, given by

$$\mathcal{A}' = \frac{\sqrt{1/N!}}{\sqrt{1/M!}\,\sqrt{1/(N-M)!}}\;\Sigma'\,(-1)^p\,P. \qquad (2\text{-}28)$$

Here $\Sigma'$ is taken not over all possible permutations of the $N$ electrons but only over those that exchange electrons between D and A.

## 2-7 DETAILED FORMS OF $\Psi_0$ AND $\Psi_1$; EVALUATION OF $S_{01}$

The general form of $\Psi_0$(D, A) has been given in (2-26). Now for the dative function $\Psi_1$ we have

$$\Psi_1(\mathrm{D}^+\text{—}\mathrm{A}^-) = \frac{(\Psi_{\mathrm{I}} + \Psi_{\mathrm{II}})}{\sqrt{2(1 + S_{\mathrm{I,II}})}}. \qquad (2\text{-}29)$$

The wavefunctions $\Psi_{\mathrm{I}}$ and $\Psi_{\mathrm{II}}$ are defined by

$$\Psi_{\mathrm{I}} = \mathcal{A}'\,\Psi^{(\alpha)}_{\mathrm{D}+}\,\Psi^{(\beta)}_{\mathrm{A}-} \quad \text{and} \quad \Psi_{\mathrm{II}} = \mathcal{A}'\,\Psi^{(\beta)}_{\mathrm{D}+}\,\Psi^{(\alpha)}_{\mathrm{A}-}, \qquad (2\text{-}30)$$

where each superscript in parentheses denotes the spin of the *odd electron* in $\mathrm{D}^+$ or $\mathrm{A}^-$. Thus transfer of an electron of $\beta$ spin from an $\alpha,\beta$ spin pair on the donor leaves $\Psi^{(\alpha)}_{\mathrm{D}+}$ and produces $\Psi^{(\beta)}_{\mathrm{A}-}$; and vice versa for transfer of an $\alpha$-spin electron.

To see all this more clearly let us write out the spinorbital products. First [neglecting all but the main AS · MSOP term of $\Psi_{mod}(D)$ and $\Psi_{mod}(A)$ in (2-26)] we have

$$\begin{aligned}\Psi_0 &= \mathcal{A}'\Psi_D\Psi_A \simeq \mathcal{A}[\phi_d(1)\alpha(1)\phi_d(2)\beta(2)\phi_d'(3)\alpha(3)\cdots \\ &\quad \phi_a(M+1)\alpha(M+1)\phi_a(M+2)\beta(M+2)\phi_a'(M+3)\alpha(M+3)\cdots].\end{aligned} \tag{2-31}$$

[Note the shift from $\mathcal{A}'$ to $\mathcal{A}$ when we replace the product of AS · MSOP forms of $\Psi_D$ and $\Psi_A$ separately by an MSOP form that includes MSOs of both D and A.] Here $\phi_d$ refers to one of the MOs on the donor, $\phi_d'$ to another, and so on, whereas $\phi_a$ and $\phi_a'$ refer to MOs on the acceptor. In the no-bond structure $\Psi_0$ the neutral donor MO $\phi_d$ that supplies the electron in donor-acceptor action contains two electrons, whereas the acceptor MO into which the electron will go is unocuppied.

Now we look at $\Psi_I$, again considering only the main AS · MSOP terms:

$$\begin{aligned}\Psi_I &= \mathcal{A}'\Psi_{D+}^{(\alpha)}\Psi_{A-}^{(\beta)} \simeq \mathcal{A}[\phi_d(1)\alpha(1)\phi_{a^-}(2)\beta(2)\phi_d'(3)\alpha(3)\cdots \\ &\quad \phi_a(M+1)\alpha(M+1)\phi_a(M+2)\beta(M+2)\phi_a'(M+3)\alpha(M+3)\cdots].\end{aligned} \tag{2-32}$$

This function is just the same as the no-bond AS · MSOP in (2-31) except that electron number 2 has been moved from the spinorbital $\phi_d\beta$ and placed in the spinorbital $\phi_{a^-}\beta$. (We call the acceptor MO that accepts the electron $\phi_{a^-}$ because in $\Psi_1$ it is an MO of the negative ion $A^-$ in the usual case where D and A are neutral molecules.) The term $\Psi_{II}$ is the same as $\Psi_I$ except that now in the MSOP expression electron 1 from $\phi_d$ instead of electron 2 is placed into the orbital $\phi_{a^-}$:

$$\begin{aligned}\Psi_{II} &= \mathcal{A}'\Psi_{D+}^{(\beta)}\Psi_{A-}^{(\alpha)} \simeq \mathcal{A}[\phi_{a^-}(1)\alpha(1)\phi_d(2)\beta(2)\phi_d'(3)\alpha(3)\cdots \\ &\quad \phi_a(M+1)\alpha(M+1)\phi_a(M+2)\beta(M+2)\phi_a'(M+3)\alpha(M+3)\cdots].\end{aligned} \tag{2-33}$$

By substituting (2-32) and (2-33) into the definition for the overlap integral between functions $\Psi_I$ and $\Psi_{II}$ we see† that

$$S_{I,II} = \int \Psi_I\Psi_{II}\,dv \simeq \int \phi_d(1)\phi_{a^-}(2)\phi_{a^-}(1)\phi_d(2)\,dv' = S_{da^-}^2 \tag{2-34}$$

where

$$S_{da^-} = \int \phi_d\phi_{a^-}\,dv. \tag{2-35}$$

† The "integrals" over the spin functions disappear, because $\int\alpha^2\,d\sigma = 1$, $\int\beta^2\,d\sigma = 1$ if $\sigma$ is the spin variable.

Thus the normalization factor in (2-29) takes the form $1/\sqrt{2(1+S_{da^-}^2)}$. The term $S_{da^-}$ can be fairly large even for weak complexes (see Section 3-5).

Finally, let us consider the overlap integral $S_{01}$ of (1-4). By substituting the relations given in (2-31), (2-29), (2-32), (2-33), (2-34), and (2-35) we obtain

$$S_{01} \simeq \frac{\sqrt{2}\, S_{da^-}}{\sqrt{1+S_{da^-}^2}}. \tag{2-36}$$

This is Equation 19 of R3.

(The material given in this section is also discussed in Appendices 1-3 of Murrell [5].)

## 2-8 THE SINGLET AND TRIPLET CHARGE-TRANSFER STATES; COMPARISON WITH HEITLER-LONDON STATES OF THE HYDROGEN MOLECULE

If we consider electrons 1 and 2 only, the dative function $\Psi_1$ of (2-29), after $\Psi_I$ and $\Psi_{II}$ expressions from (2-32) and (2-33) have been inserted, is of the same form as the Heitler-London function for the normal state of the hydrogen molecule. The combination $\Psi_I + \Psi_{II}$ has $m_{s\,1} = +1/2$, $m_{s\,2} = -1/2$, $M_S = 0$ and is a singlet wavefunction ($S = 0$).

Just as for $H_2$ [6], there is also a triplet state, $\Psi_T = \Psi_1'$ ($D^+ \leftrightarrow A^-$), with $S = 1$, in which the spins of the electrons 1 and 2 in the orbitals $\phi_d$ and $\phi_{a^-}$ are parallel instead of paired opposite. In a complex, this state is a CT state. For $M_S = 0$, after using (2-34), its wavefunction is given by

$$\Psi_{T,0} = \frac{(\Psi_I - \Psi_{II})}{\sqrt{2(1-S_{da^-}^2)}}. \tag{2-37}$$

For $M_S = \pm 1$, $\Psi_{T,\pm 1}$ is the same as $\Psi_I$ of (2-32), except that the expression in brackets starts with $\phi_d(1)\alpha(1)\phi_{a^-}(2)\alpha(2)$ for $M_S = +1$, or $\phi_d(1)\beta(1)\phi_{a^-}(2)\beta(2)$ for $M_S = -1$. In the (2-30) type of description these two functions are

$$\Psi_{T,+1} = \mathcal{A}'\Psi_{D^+}^{(\alpha)}\Psi_{A^-}^{(\alpha)} \quad \text{and} \quad \Psi_{T,-1} = \mathcal{A}'\Psi_{D^+}^{(\beta)}\Psi_{A^-}^{(\beta)}.$$

Charge-transfer bands are observed in absorption for the transition from the singlet ground state $\Psi_N$ to the singlet excited state $\Psi_V$, and in fluorescence for the reverse transition, but the triplet CT state $\Psi_T$ also exists, and a phosphorescent emission spectrum from it to $\Psi_N$ occurs. (See Section 9-5.)

## 2-9 REFERENCES

[1] (a) G. Herzberg, *Spectra of Diatomic Molecules,* 2nd ed., Van Nostrand, Princeton, N.J., 1950. (b) G. Herzberg, *Infrared and Raman Spectra of Polyatomic Molecules,* Van Nostrand, Princeton, N.J., 1945; and (c) G. Herzberg, *Electronic Spectra and Electronic Structure of Polyatomic Molecules,* Van Nostrand, Princeton, N.J., 1966.
[2] H. Jaffé and M. Orchin, *Symmetry in Chemistry,* Wiley, New York, 1965.
[3] F. A. Cotton, *Chemical Applications of Group Theory,* Interscience, New York, 1963.
[4] M. W. Hanna, *J. Am. Chem. Soc.,* **90**, 285 (1968).
[5] J. N. Murrell, *The Theory of the Electronic Spectra of Organic Molecules,* Wiley, New York, 1963.
[6] J. N. Murrell, S. F. A. Kettle, and J. M. Tedder, *Valence Theory,* Wiley, London, 1965.

CHAPTER 3

# The Intensity of the Charge-Transfer Transition (Simplified Theory)

## 3-1 THE CHARGE-TRANSFER ABSORPTION; DIPOLE MOMENTS AND TRANSITION MOMENTS

Let us expand the discussion begun in Section 2-4 on the ground-state dipole moment and on the transition moment of the CT absorption. We ought now to recognize that accurate electronic wavefunctions such as $\Psi_N$, $\Psi_V$, or $\Psi_0$ depend on vibrational as well as electronic coordinates, also that for a more detailed understanding than in Section 2-4 we must introduce vibrational wavefunctions for each electronic state. If, as usually, the Born-Oppenheimer approximation is adequate, then

$$\Psi_m^v = \Psi_m (\mathbf{r}_i, \mathbf{R}_k)\, \chi_{v_m}^m (\mathbf{R}_k). \tag{3-1}$$

Here $v$ refers to an appropriate set of vibrational quantum numbers (plus vibrational continuum sometimes) for the $m$th electronic state, with electronic wavefunction $\Psi_m$ and vibrational wavefunctions $\chi_{v_m}^m$. The vectors $\mathbf{r}_i$ and $\mathbf{R}_k$ refer to the locations of the electrons and nuclei respectively.

Consideration of nuclear vibrations is especially important for weak complexes, in which very-low-frequency vibrational modes are expected and the populations of complexes in excited vibrational states of this kind are expected to be large even at room temperature. However, we shall still ignore molecular rotation; the dipole moments we compute correspond to nonrotating but usually vibrating molecules.

The transition moment for a spectroscopic transition between vibrational levels of two electronic states $m$ and $n$ is now given by

$$\mu_{mn}^{v_m, v_n} = \int \Psi_m \, \chi_v^m \, \mu_{op} \, \Psi_n \, \chi_v^n \, d\mathbf{r}_i \, d\mathbf{R}_k, \tag{3-2}$$

where $d\mathbf{r}_i$ and $d\mathbf{R}_k$ are volume elements for the electronic and vibrational coordinates respectively.

The dipole-moment operator has the form

$$\mu_{op} = e \sum_k z_k \, \mathbf{R}_k - e \sum_i \mathbf{r}_i \tag{3-3}$$

Here $Z_k$ is the charge on the $k$th nucleus. The definition makes $\mu_{op}$ independent of the choice of the origin of the coordinate system in case $\Sigma \, Z_k$ is equal to the total number of electrons—that is, if the system is neutral. Note also that the definition is so chosen as to make the $\mu$ vector positive when drawn in the direction from the negative to the positive end of the $\mu_{op}$ or the $\mu_{mn}$ dipole.

By substituting (3-3) into (3-2) we have

$$\mu_{mn}^{v_m, v_n} = \int \chi_{v_m}^m \left[\int \Psi_m \, \mu_{op} \, \Psi_n \, d\mathbf{r}_i\right] \chi_{v_n}^n \, d\mathbf{R}_k \tag{3-4}$$

Integration over the electronic coordinates then gives $\mu_{mn}^{el} \, (\mathbf{R}_k)$:

$$\mu_{mn}^{v_m, v_n} = \int \chi_{v_m}^m \, (\mathbf{R}_k) \, \chi_{v_n}^n \, (\mathbf{R}_k) \, \mu_{mn}^{el} \, (\mathbf{R}_k) \, d\mathbf{R}_k. \tag{3-5}$$

Here $\mu_{mn}^{el}$ is always independent of origin, even for the case of a charged system, as can be seen by considering new coordinates $\mathbf{r}_i'$ such that $\mathbf{r}_i = \mathbf{r}_i' + \mathbf{a}_0$ and $\mathbf{R}_k = \mathbf{R}_k' + \mathbf{a}_0$. After substitution in (3-4) the constant additive terms, when integrated over the electronic coordinates, give zero, since $\Psi_n$ and $\Psi_m$ are orthogonal.

Equation 3-5 must next be integrated over the vibrational coordinates; however, we shall postpone discussion of this step until Chapter 8.

Let us now turn to the dipole moment of the normal state. This is obtained from (3-4) if we put $m = n = N$ and $v_m = v$.* For $m = n$, integration over the electronic coordinates gives us a term that can be written as

$$\mu_N^v = e \left| \sum_k z_k \, \mathbf{R}_k - N \bar{\mathbf{r}}^v \right|.$$

Here $N$ is the total number of electrons and $\bar{\mathbf{r}}^v$ is an average over the electronic positions in $\Psi_N$ for vibrational state $v$; $\bar{\mathbf{r}}^v$ is not expected to vary much with $v$. The term $\Sigma \, z_k \, \mathbf{R}_k$ is just the sum over the average or expectation values for the $k$ nuclear coordinates. The moment $\mu_N^v$ is easily seen to be independent of the coordinate origin if there is no net charge. The result for $\mu_N^v$ given here is essentially the same as that given in Section 2-4; $\mu_N$ there is essentially the same as $\mu_N^v$ except

* For $v_m \neq v_n$, the transition moment for an infrared vibrational transition of state $N$ is obtained. If this is desired, it is convenient first to transform from the $\mathbf{R}_k$ to vibrational normal coordinates.

for the fact that we now note that $\mu_N^v$ varies slightly with $v$.

## 3-2 OSCILLATOR STRENGTHS†

The molar absorptivity or extinction coefficient $\epsilon_\nu$ of any molecular species is defined by Beer's law:

$$I_\nu = I_\nu^0\, 10^{-\epsilon_\nu c l}. \tag{3-6}$$

Here $c$ is the concentration in moles per liter, $l$ is the path length in centimeters, and $I_\nu$ and $I_\nu^0$ are the intensities of the transmitted and incident light of frequency $\nu$. The oscillator strength $f$ for an electronic transition is defined by

$$f = \frac{2.303\, mc^2}{\pi e^2 N_0} \int \epsilon_\nu\, d\nu,$$

or, substituting for the values of the fundamental constants,

$$f = 4.318 \times 10^{-9} \int \epsilon_\nu\, d\nu. \tag{3-7}$$

The integration is made over the entire absorption band or bands that

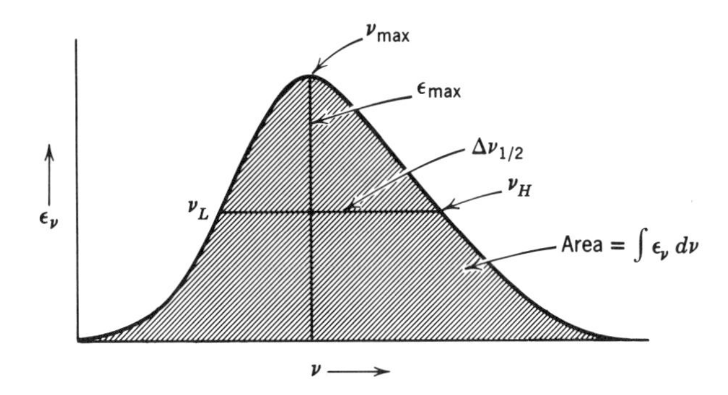


**Figure 3-1** Appearance of absorption transition without vibrational structure. (See also Section 8-4.)

belong to the particular electronic transition (see Fig. 3-1 for the case in which the transition shows no vibrational structure); if $\epsilon_\nu$ is known

† For discussion of material in this and the next section see papers by Mulliken [1,2].

empirically, (3-7) can be used to obtain an empirical value of $f$. For a lazy man's estimate the integral can be fairly well approximated by taking the product $\epsilon_{max}\,\Delta\nu_{1/2}$;† that is,

$$f \simeq 4.32 \times 10^{-9}\,\epsilon_{max}\,\Delta\nu_{1/2}. \tag{3-8}$$

Here $\Delta\nu_{1/2}$ is the width in $cm^{-1}$ of the band between the two frequencies at which $\epsilon = (1/2)\,\epsilon_{max}$.

The theoretically computed oscillator strength is related to the magnitude of the theoretical electronic transition dipole $\mu_{mn}^{el}$ in esu · cm, here abbreviated to $\mu_{mn}$, by

$$f_{mn} = \left(\frac{8\pi^2 mc}{3h}\right)\bar{\nu}\left(\frac{\mu_{mn}^2}{e^2}\right), \tag{3-9}$$

or

$$\mu_{mn}\ \text{(debyes)} = 0.0958\left[\frac{\int \epsilon\, d\nu}{\bar{\nu}}\right]^{1/2} \simeq 0.0958\left[\frac{\epsilon_{max}\,\Delta\nu_{1/2}}{\bar{\nu}}\right]^{1/2}$$

Here $\bar{\nu}$ is the average wavenumber in $cm^{-1}$ (it is approximately $\nu_{max}$, the wavenumber of the maximum absorption). Since $\mu_{mn}$ is a vector,

$$\mu_{mn}^2 = |\mu_{mn}|^2 = (\mu_x^{mn})^2 + (\mu_y^{mn})^2 + (\mu_z^{mn})^2,$$

where $\mu_x^{mn}$ is the component of the transition dipole along the molecular $x$-axis and so on.

This discussion applies to the vapor state. In solution the oscillator strength computed from the vapor-state theory by (3-9) should, according to solution theory (see Section 7-4), be multiplied by a factor that depends on the dielectric constant (or polarizability) of the medium before comparing it with the experimental value from (3-7) [see (7-12)]. Experimentally [3], however, the observed intensity changes from vapor to solution do not agree well with those predicted from just this one effect, so that for the present we shall recognize the existence of this effect without trying to define its magnitude, and thus use (3-9) for solution data as well. (See further discussion in Section 7-4, in which we shall see that the resulting error in $f$ from solution data might be an overestimate by about a factor of 2 or so.)

---

† The accuracy of this approximation depends on the band shape. (Cf. Fig. 3-1.) If the shape were triangular, the approximation would be perfect; if the shape is Gaussian, the integral can be evaluated exactly and is $(\sqrt{\pi/2 \ln 2})\,\epsilon_{max}\,\Delta\nu_{1/2}$ (or $= 1.05\,\epsilon_{max}\,\Delta\nu_{1/2}$); if the shape is Lorentzian, the integral can be shown to be $(\pi/2)\,\epsilon_{max}\,\Delta\nu_{1/2} = 1.57\,\epsilon_{max}\,\Delta\nu_{1/2}$. The band shapes actually found are usually not symmetrical (see Section 8-4) and do not fit any of these shapes exactly but are usually more nearly Gaussian than Lorentzian.

The theoretical expression in (3-9) is reliable only for transition moments computed from *exact* wavefunctions $\Psi_m^{el}$ and $\Psi_n^{el}$. There are also two other ways of defining $f$ theoretically. Equation 3-9 relates $f$ to the dipole moment, another definition relates it to the dipole velocity, and a third to the dipole acceleration. If exact wavefunctions are used, identical results from the three definitions are obtained for $f$. If approximate functions are used, the results are of course inaccurate; furthermore, the results from the three definitions cannot be expected to agree. In that case experience suggests that the definition in terms of velocities is likely to be the most accurate. However, we use here the definition in terms of dipole moments because the calculations are easier.

If a given set of wavefunctions gives nearly the same value of $f$ for all three definitions, it indicates (but does not conclusively prove) that these functions are good approximations. (See Ehrenson and Phillipson [4] for a calculation that illustrates the application of the above discussion for a strong vacuum-ultraviolet $H_2$ transition. Also see Hameka [5] for some criticism emphasizing that with approximate functions the dipole-moment calculation is not the best way of estimating $f$.)

## 3-3 INTENSITY AND POLARIZATION OF THE CHARGE-TRANSFER ABSORPTION BAND

Now let us apply all these ideas to the $V \leftarrow N$ transition, using the approximate wavefunctions of (2-2) and (2-3) and Section 2-7. The result for the transition moment, from here on assumed to be integrated over the complete band or bands that belong to the electronic transition, is as follows (see Eq. 13 of R3):

$$\mu_{VN}^{el} = (a^* b\mu_1^{el} - ab^* \mu_0^{el}) + (aa^* - bb^*)\, \mu_{01}^{el} + \cdots. \tag{3-10}$$

(The dots indicate correction terms resulting from the inaccuracies of the $N$ and $V$ wavefunctions. These terms we shall ignore for the present, but they may often be important.)

We note that $\mu_{VN}$ (from here on we drop the superscript "el") is a vector whose direction within the complex is determined by the specific wavefunctions $\Psi_N$ and $\Psi_V$ (approximately by the MOs $\phi_d$ and $\phi_{a^-}$ defined in Section 2-7). If we can place the complex in a fixed orientation (e.g., in a crystal), we may expect to be able to measure experimentally the polarization of the CT absorption spectrum and to check the prediction for the orientation of $\mu_{VN}$. In a $b\pi$-donor, $a\pi$-acceptor complex we would expect the direction of $\mu_{VN}$ to be as shown in Fig. 3-2 for the (probably usual) case where the donor is not symmetrically located with respect to

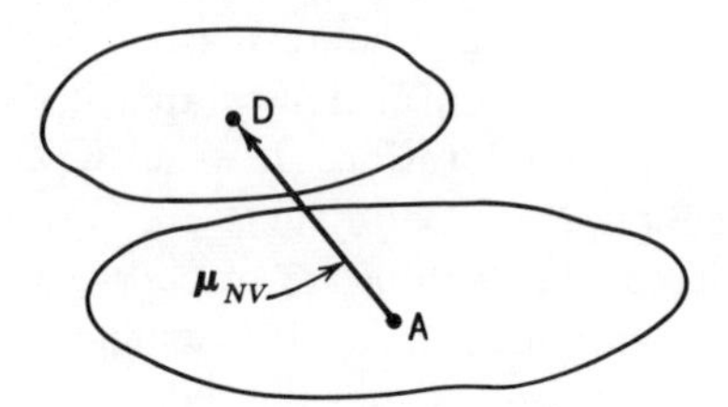


**Figure 3-2** Transition dipole for the CT band of a $b\pi \cdot a\pi$ complex, illustrating the deviation of $\mu_{VN}$ from the perpendicular orientation.

the acceptor, yet their planes are parallel. (See Chapter 5 for further discussion.)

Equation 3-10 can be transformed (See Eq. 17 of R3) to give

$$\mu_{VN} = a^*b\,(\mu_1 - \mu_0) + (aa^* - bb^*)\,(\mu_{0\,1} - S_{0\,1}\mu_0). \qquad (3\text{-}11)$$

In (3-11) we can show (by consideration of what would happen on substituting $\mathbf{r}_i'$ and $\mathbf{R}_k'$ for $\mathbf{r}_i$ and $\mathbf{R}_k$ as discussed above) that each of the terms, and thus the whole $\mu_{VN}$, is independent of the choice of origin for the electronic coordinates. The first term in (3-11) is the same as if a fraction $a^*b$ of one electron had been transferred from $\phi_d$ to $\phi_{a^-}$. We do not need to know accurately $\mu_1$ or $\mu_0$ individually; the difference is simply

$$\mu_1 - \mu_0 = e(\bar{\mathbf{r}}_d - \bar{\mathbf{r}}_{a^-}), \qquad (3\text{-}12)$$

assuming that all the other electrons are unaffected when we change from $\Psi_0$ to $\Psi_1$. This assumption is not quite accurate since the removal of an electron from D and its transfer to A must cause some modification in the MOs occupied by the other electrons. However, the effects of these changes are usually not large. In (3-12) $\bar{\mathbf{r}}_d$ and $\bar{\mathbf{r}}_{a^-}$ are the average positions of the donated electron in $\phi_d$ on D before transfer and in $\phi_{a^-}$ on $A^-$ after transfer, respectively.

## 3-4 MAGNITUDES OF TERMS CONTRIBUTING TO $\mu_{VN}$

It is usually assumed, as stated in R3, that the first term in (3-11), the one in $\mu_1 - \mu_0$, is the more important. However, this assumption may not always be correct. Let us examine the second term in some detail. First, in loose complexes the factor $(aa^* - bb^*)$ is not much less than $aa^*$,

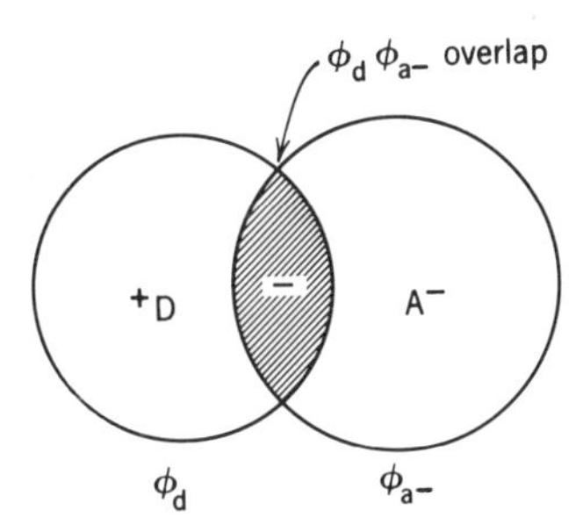


**Figure 3-3** The overlap charge distribution $\phi_d \phi_{a-}$ of (3-13).

which is near 1. Using the approximate wavefunctions $\Psi_0$ and $\Psi_1$ of Section 2-7 we find (see R3, footnote 19) very nearly

$$\mu_{01} - S_{01}\mu_0 = -eS_{01} \int \left( \frac{\phi_d \phi_{a-}}{S_{da-}} - \phi_d^2 \right) \mathbf{r}\, d\mathbf{r}. \tag{3-13}$$

The term in parentheses in (3-13) when multiplied by $-e$ represents a dipole type of electric-charge distribution, and the integral gives the dipole moment of this distribution. The dipole consists of a positive charge of magnitude $S_{01}e$ centered on the donor and a negative charge $-S_{01}e$ centered in the overlap region. All this is illustrated schematically in Fig. 3-3.
Another way of describing this term is [see Eq. 19 of R3 and (2-36)]:

$$\mu_{01} - S_{01}\mu_0 = eS_{01}(\bar{\mathbf{r}}_d - \bar{\mathbf{r}}_{da-}) = \sqrt{2}eS_{da-} \frac{\bar{\mathbf{r}}_d - \bar{\mathbf{r}}_{da-}}{\sqrt{1 + S^2_{da-}}}, \tag{3-14}$$

where $\bar{\mathbf{r}}_{da-}$ is the average position of the overlap change $\phi_d \phi_{a-}$.

Consideration of two limiting cases throws light on the importance of this term: (1) the case of small overlap of D with A, for a very weak complex, or a mere contact, and (2) the case of strong overlap of D with A, for a strong complex. We note that $\phi_d$ is of about the same size as the donor molecule, and $\phi_a$ (the highest filled orbital in the free acceptor; *not* $\phi_{a-}$) is of the same size as A. For very weak interaction, the molecules are at about the van der Waals distance, and $\int \phi_d \phi_a \, dv$ should be very small. However, the acceptor orbital $\phi_{a-}$ is generally expected

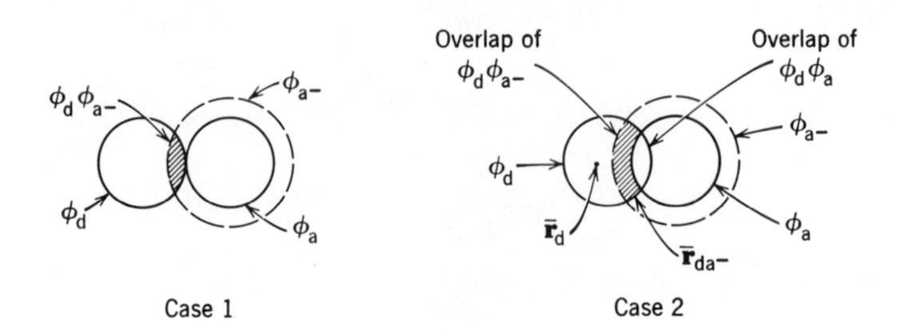


**Figure 3-4** Comparison of the overlap charge contribution to $\mu_{VN}$ for weak and strong complexes. Case 1: weak complex; case 2: strong complex.

to be larger than A, since it is an antibonding orbital of A, and since the net negative charge encourages expansion. Hence the term $\int \phi_d \phi_{a-}\, dv$ may still be fairly large. The effect, which is probably considerably more important for $I_2$ as acceptor than for a $\pi$ acceptor, is illustrated in Fig. 3-4.

It is important to note that if $\int \phi_d \phi_a\, dv$ is small, $b$ should also be small, as Murrell has shown. (See Section 3-5.) The first term in (3-11) should then also be small. However, the second term, which depends on $\int \phi_d \phi_{a-}\, dv$, is not necessarily small† even when the first term is. Hence it seems that in weak complexes the second term might be large enough to give some considerable intensity to the CT band even when the first term would indicate very little intensity. Experimentally a probable example of this situation is found in the observation of a new band in the absorption spectrum of iodine when dissolved in $n$-heptane or similar solvents that are such weak donors that there is no evidence for the existence of a complex with iodine. This new band is the *contact* CT band indicated in Fig. 1-1. (Some of these topics are discussed at length by Murrell [6] and in R10.)

To summarize, the first term in (3-11) is a dipole term that is approximately equal to $eb(\bar{r}_d - \bar{r}_{a-})$. For a $\phi_d$ that is symmetrical about the center of the donor $\bar{r}_d$ would be at the center of the D molecule; similarly

† However, it is worth noting that in $b\pi \cdot a\pi$ complexes the exchange repulsion term is quite large, so that D and A do not come very close together. Also, $\phi_{a-}$ for these acceptors is probably not much larger than $\phi_a$. Hence both terms in (3-11) tend to be relatively small in $b\pi \cdot a\pi$ complexes. This fact affords an explanation of the relatively small and variable intensities of the CT bands in these complexes. In $n \cdot \sigma$ complexes $\phi_{a-}$ is much larger than $\phi_a$, corresponding to case 2.

$\bar{\mathbf{r}}_{a-}$ would be at the center of the A molecule if that is centrosymmetric. Hence for centrosymmetric D and A this term would be equal to the dipole moment that would be generated by transferring a charge $-eb$ from the donor to the acceptor.

The second term in (3-11) is equal to the dipole moment generated when a charge $-eS_{01}$ ($\simeq -\sqrt{2}eS_{da-}$) is transferred from the center of the donor molecule to the average position of the overlap charge (given by $\bar{\mathbf{r}}_{da-}$), near the center of the doubly cross-hatched region in Fig. 3-4. If the average position of this overlap charge is at the center of the D–A bond, then

$$\mu_{01} = (1/2)\, S_{01} + S_{01}\mu_0.$$

In case 2 of Fig. 3-4, in which $b$ is substantial, probably the first term in (3-11) is usually larger than the second term. In case 1 $b$ should be very small, making the first term of (3-11) small, so that the second term, which may still be appreciable, may be the predominant one. The latter case may well often explain the rather considerable intensities of CT absorption bands for weak complexes, although Murrell has advocated a different explanation. (See Section 14-3 but see also Section 3-5.)

## 3-5 MAGNITUDES OF THE RESONANCE INTEGRALS $\beta_0$ AND $\beta_1$

Now let us look at certain expressions for $\beta_0$ and $\beta_1$ [see (2-11) and (2-12)] given by Murrell [6]. These will be of especial interest later on but are discussed here because they are closely related to the second term in (3-11), as given by (3-14) and Fig. 3-4. In [6] Murrell discusses only a one-electron case, which applies strictly only to the case of an odd-electron donor (e.g., a sodium atom); in R10 he discusses more general cases. The results for $\mu_{VN}$ are almost the same for a two-electron discussion as for the one-electron problem except for a readily understandable factor of 2 in $f$ (or $\sqrt{2}$ in $\mu$). (Actually Murrell considered a four-electron case with two $\phi_d$ electrons and two $\phi_a$ electrons as in Section 3-4 and Fig. 3-4. As a first approximation we may ignore the "kickback" on $\phi_a$ resulting from the advent in $\Psi_1$ of an electron into $\phi_{a-}$ and assume that the acceptor (and donor) MOs that are not directly involved in the donor-acceptor action are the same in $\Psi_1$ as in $\Psi_0$.)

Murrell's approximate expression for the interaction integral $\beta_0$ for the usual case (two electrons initially in $\phi_d$) can be written as

$$\beta_0 = \sqrt{2}\, S_{da-} \int \left( \frac{\phi_d \phi_{a-}}{S_{da-}} - \phi_d^2 \right) V_A \, dv. \tag{3-15}$$

All these terms have been defined previously (Section 3-4) except $V_A$. This symbol represents $-e$ times the electrostatic potential function corresponding to the field of the neutral acceptor molecule. The integral then represents the potential energy of two electrons, each distributed in space in the manner indicated by the expression in parentheses, in the field of the neutral acceptor. This energy is evidently very small unless the overlap charge function $-e\phi_d\phi_{a-}$ appreciably penetrates the acceptor, which is true only if the overlap $\phi_d\phi_a$ of the orbitals of the *neutral* donor and acceptor is appreciable, as in case 2 of Fig. 3-4.

In case 1 of Fig. 3-4 the overlap distribution function $\phi_d\phi_{a-}$, and all the more so the distribution $-S_{da-}\phi_d^2$, lies almost entirely outside A, hence mainly outside the region where $V_A$ has appreciable values. Consequently $\beta_0$ as given by (3-15) must be very small in case 1. We may expect this case to be applicable for distances between donor and acceptor that are equal to or greater than the sum of van der Waals radii. However, contrary to Murrell's conclusion, $\mu_{01} - S_{01}\mu_0$ in (3-11) could still be very appreciable since the integral defining it [see (3-13)] involves $\mathbf{r}$ instead of $V_A$, and we have already seen (cf. Fig. 3-4) that this can be very appreciable even in case 1. For the interaction integral $\beta_1$ [see (2-12)] Murrell gives

$$\beta_1 = \sqrt{2}\, S_{da-} \int \left(\frac{\phi_d\phi_{a-}}{S_{da-}} - \phi_{a-}^2\right) V_{D+}\, dv. \tag{3-16}$$

Here $V_{D+}$ is the potential energy function for an electron in the field of the donor positive ion, $D^+$. Since even at large distances from the donor $V_{D+}$ falls off only as $1/R$, where $R$ is the intermolecular D–A distance, $-\beta_1$ should be much larger than $-\beta_0$; this is qualitatively consistent with the relationship $\beta_1 - \beta_0 = S_{01}\Delta$ [(2-14)], which must always hold. However, it is not clear how well the approximations made by Murrell in obtaining (3-16) and (3-17) permit $\beta_1 - \beta_0$ to satisfy quantitatively this necessary relation.

## 3-6 REFERENCES


[1] R. S. Mulliken, *J. Chem. Phys.*, **7**, 14 (1939).

[2] R. S. Mulliken and C. A. Rieke, *Reports on Progress in Physics of the Physical Society* (London), **8**, 231 (1941).

[3] L. E. Jacobs and J. R. Platt, *J. Chem. Phys.*, **16**, 1137 (1950).

[4] S. Ehrenson and P. Phillipson, *J. Chem. Phys.*, **34**, 1224 (1961).

[5] H. F. Hameka and L. Goodman, *J. Chem. Phys.*, **42**, 2305 (1965).

[6] J. N. Murrell, *Quart. Rev.*, **15**, 191 (1961).

*CHAPTER 4*

# *Classification*

## 4-1 CLASSIFICATION OF DONORS AND ACCEPTORS

Donors and acceptors are classified according to their structure and function as follows (see Tables I and II in R14, and the detailed discussion in R4):

| Donors | | | Acceptors | | |
|---|---|---|---|---|---|
| Number of Electrons | Functional Type | Structure Type | Number of Electrons | Functional Type | Structure Type |
| Odd | Radical | R | Odd | Radical | Q |
| Even | Increvalent | $n$ | Even | Increvalent | $v$ |
| Even | Sacrificial | $b\sigma$ | Even | Sacrificial | $a\sigma$ |
| Even | Sacrificial | $b\pi$ | Even | Sacrificial | $a\pi$ |

In connection with this classification it should be kept in mind that in some cases the same molecule can function either as a donor or as an acceptor according to circumstances. An example is water, which is an $n$ donor but can also act as an $a\sigma$ acceptor. Furthermore, in a large molecule simultaneous functioning at one site as a donor and at another as an acceptor is possible.

The "complexes" R—Q between radical donors and acceptors are mostly just compounds with ordinary chemical bonds, but there are many possibilities of complexes between radicals or radical ions (e.g., aromatic anions) and closed-shell ions or molecules. Complexes between radical ions, such as $(TCNQ)_2^{2-}$, have recently been studied extensively [1]. Such dimers between $TCNQ^-$ acting as an R$\pi$-donor to another $TCNQ^-$ molecule acting as a Q$\pi$-acceptor force us to recognize radical donors and acceptors as important classes. Further consideration has shown that it is important to subdivide this classification further into R$\sigma$- and R$\pi$-donors, and Q$\sigma$- and Q$\pi$-acceptors.

The case of an *increvalent* donor combined with an increvalent acceptor can be illustrated by the $n \cdot v$ addition compound $R_3N \cdot BR_3$ (R = alkyl or other group). Here the nitrogen valence *increases* from 3 to 4, and so does the boron valence, in the dative structure $\Psi_1$. We note that the structure of the complex, as given by (1-1), should have a compromise geometry; that is, since the electronic structure is neither purely no-bond nor purely dative, the nuclear configuration also should be a compromise between those appropriate to the two extreme electronic structures. This expectation is confirmed. (See Section 5-2.)

Consider now $R_3N \cdot I_2$ as an example of an $n \cdot a\sigma$ complex. Here the $I_2^-$ in the dative function has added an antibonding electron (in a $\sigma_u$ MO). Now $I_2^-$ has approximately a one-half bond (like $He_2^+$). Thus $I_2$ is a "sacrificial" $\sigma$ acceptor; the bond between the atoms is weakened if it accepts an electron, as it does to the extent that $\Psi_1$ is represented in $\Psi_N$. The larger the mixing coefficient $b$ in (1-1), the further apart we should expect the two iodine atoms to be in the complex as a result of the compromise geometry that predicts that the I—I distance in the complex should be intermediate between that for $I_2$ and that for $I_2^-$. This expectation is confirmed. (See Section 5-3.)

The strongest complexes tend to be those in which both D and A are increvalent—and the weakest, those in which both D and A are sacrificial. Nevertheless there are many examples of stable $\pi \cdot \pi$ and $\pi \cdot \sigma$ complexes, in which both D and A are sacrificial (but the stability may be due to a considerable extent to classical interactions), whereas there are many examples of weak complexes of other types (in particular of the types $b\pi \cdot v$ and $n \cdot a\pi$, but also $n \cdot a\sigma$ and even $n \cdot v$ types).

In order to understand why $\pi$-donors and $\pi$-acceptors both are sacrificial consider the $\pi$-orbitals of benzene shown in Fig. 4-1. The six $\pi$-electrons in benzene are distributed with two in each of the three bonding MOs. Thus if one is removed to form $Bz^+$ in donor action, the bonding is weakened. The lowest unfilled $\pi$-orbital in Fig. 4-1 is antibonding; thus if an electron is added to form $Bz^-$ in acceptor action, this also

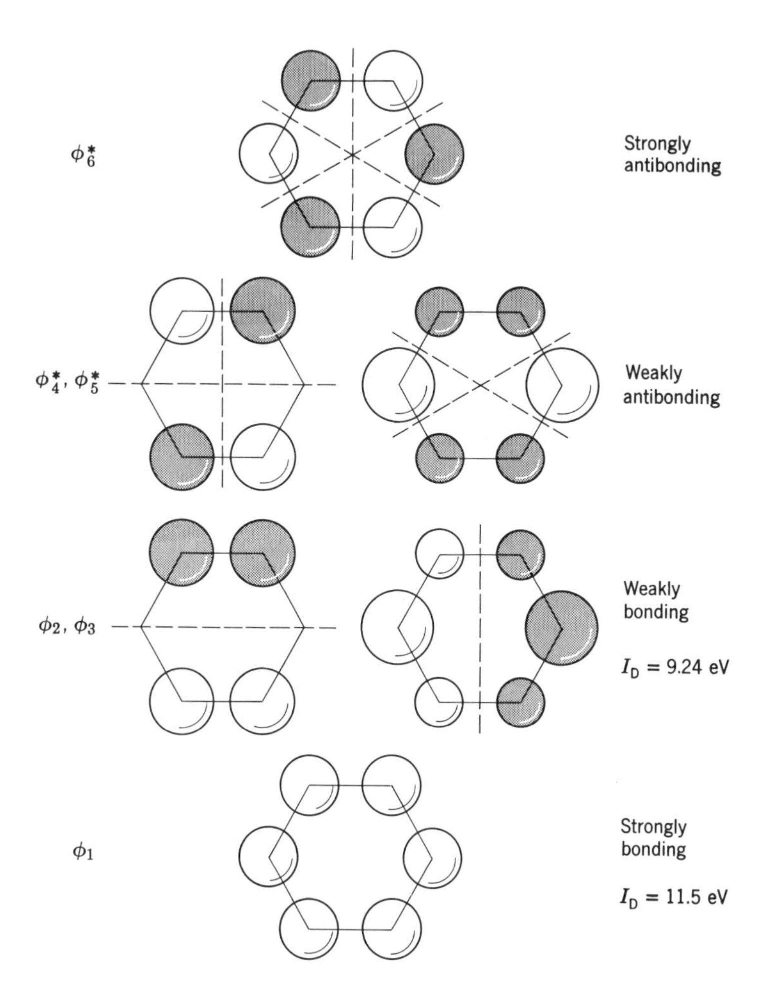


**Figure 4-1** Top view of $\pi$ molecular orbitals of benzene and their ionization energies. For a simple discussion and further pictures, see Gray [2]. The ionization energy for the degenerate pair $\phi_2,\phi_3$ is well established. That for $\phi_1$ is 11.5 eV according to El-Sayed, Kasha, and Tanaka [3]; however, theoretical calculations indicate that it should be about 14 or 15 eV (see [4] for a recent calculation), and it seems possible that 11.5 eV may represent a $\sigma$-type ionization. Here an open circle ○ means a $2p_z$ orbital perpendicular to the plane of paper with negative lobe above the plane; a solid circle ● means a $2p_z$ orbital with positive lobe above the plane. In order to indicate the electron density the radii of the circles are proportional to the two-thirds powers of the coefficients of the AOs in the Hückel expressions for the wavefunctions. Dashed lines indicate nodal planes.

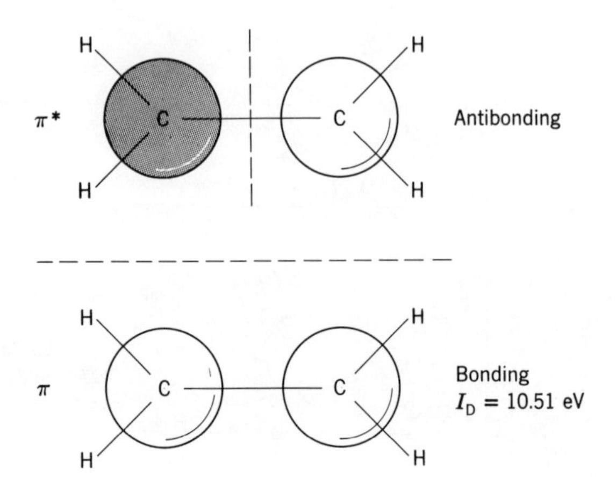


**Figure 4-2** The $\pi$ molecular orbitals of ethylene.

decreases the bonding relative to Bz. Thus Bz is sacrificial as a donor, and it is again sacrificial as an acceptor.

If we compare benzene with ethylene, we note that the latter has only two $\pi$-orbitals,—the bonding MO ($\pi$) filled and the antibonding MO ($\pi^*$) empty,—as in Fig. 4-2.

The sacrificial action for ethylene acting as a donor or acceptor must be much greater than that for benzene since it involves 1 electron out of 4 binding electrons (2 $\sigma$ and 2 $\pi$), whereas for benzene the sacrificial action involves only 1 out of 12 binding electrons (6 $\sigma$ and 6 $\pi$).

The $Ag^+$-ethylene complex is believed to be a two-way CT complex—(1-1) with a term $c\Psi_2$ ($D^-$—$A^+$) added (see Chapter 18)—with the ethylene being sacrificial both ways. Hence the C=C stretching vibration frequency in the complex is expected to be smaller than that in free ethylene; the same is true for ethylene derivatives. As an example, $\nu_{C=C}$, the frequency for the stretching vibration of the double bond, comes at 1653 $cm^{-1}$ in cyclohexene, but at 1584 $cm^{-1}$ in the Raman spectrum of the $Ag^+$ complex [5]. However, this decrease is only about 4% of the C=C frequency and is thus not really a very large effect.

Besides the aromatic and olefinic $\pi$-donors and acceptors, acetylene and its derivatives show marked $\pi$-donor ability.

## 4-2 INTRAMOLECULAR DATIVE ACTION AND FORTIFICATION

Now consider the odd-odd, R—Q interaction between atoms, using three examples—$H_2$, HCl, and NaCl—and treating them like complexes. For $H_2$ we need $\Psi_0$(H—H), $\Psi_1(H^+,H^-)$, and $\Psi_2(H^-,H^+)$. Here a *covalent* structure appears as $\Psi_0$ instead of the no-bond structure of even-even complexes, and *ionic* structures appear as $\Psi_1$ instead of the dative structures of even-even complexes. In the three molecules under consideration we have the following approximate structures for the normal state:

$$H_2\text{:}\ \Psi_N = \underset{\text{(covalent)}}{a\Psi_0} + \underset{\text{(ionic)}}{b(\Psi_1 + \Psi_2)} \quad (a > b) \text{ (two-way CT action)}$$

$$HCl\text{:}\ \Psi_N = a\Psi_0 + b\Psi_1(H^+,Cl^-) + c\Psi_2(H^-,Cl^+) \quad (b > c) \text{ (two-way but less so)}$$

$$NaCl\text{:}\ \Psi_N = a\Psi_0 + b\Psi_1(Na^+,Cl^-) \quad (a \ll b) \text{ (essentially one-way).}$$

Here the hydrogen or sodium atom is D and the H or Cl atom is A, and we regard HCl or NaCl as having a covalent $\sigma$-bond modified by $\sigma$-electron charge transfer mostly from D to A.

Consider now the difference between ammonia and trimethylamine as electron donors. The latter is a much stronger donor because of $\sigma$-bond charge transfer from $CH_3$ to N. We may speak of *fortification* by $CH_3$ groups.

But first consider the $\sigma$-bond charge transfer from H to N (like that from H to Cl in HCl) that has already taken place in $NH_3$ to give the charge separations indicated in Fig. 4-3. As a result of the negative charge on N provided by this action the lone pair of electrons has a lower ionization potential than for a free N atom. Hence the energy difference $\Delta$ (2-13) is lower and the resonance energy $\beta_0^2/\Delta$ is larger, so $NH_3$ can more readily act as an $n$ donor than does the N atom. Furthermore, $\phi_d$ (the N-atom lone-pair orbital) is larger because of the negative charge; also, because of hybridization, it protrudes out along the symmetry axis away from the H atom. Both these effects favor a large $S_{da^-}$ and so a large $-\beta_0$ (see Section 3-5) and $\beta_0^2/\Delta$.

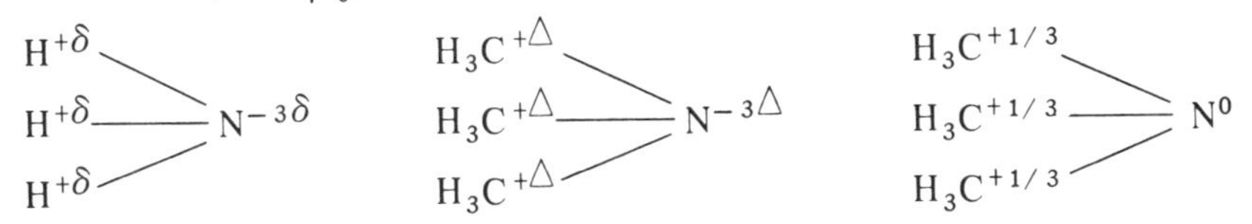


**Figure 4-3** Charge distribution in $NH_3$, $(CH_3)_3N$, and $(CH_3)_3N^+$ ($\Delta > \delta$).

Fig. 4-3 also shows the charge distribution in trimethylamine, which undoubtedly has a larger positive charge on each methyl group than that on the H atoms in $NH_3$. Population analyses of theoretically computed wavefunctions for $NH_3$ and for $(CH_3)_3N$ indicate [6] that in Fig. 4-3 $\delta$ may be about 0.25 $e$[†] and $\Delta$ about 0.33 $e$. Hence the N atom in $(CH_3)_3N$ is even more negative than that in $NH_3$, and trimethylamine is a better donor for complexing in inert solvents.

Thus, although the donor action of alkylamines at first sight might seem to be localized on the lone pair, that is only part of the story. It really depends very much on the cooperation of the rest of the molecule.

The positive ion $[(CH_3)_3N]^+$ may have a charge distribution that is somewhat like that indicated on the right side of Fig. 4-3, in which the N atom is far from being $N^+$.

Now consider, as another example, benzene and some of its derivatives. Here the charge distribution[‡] is

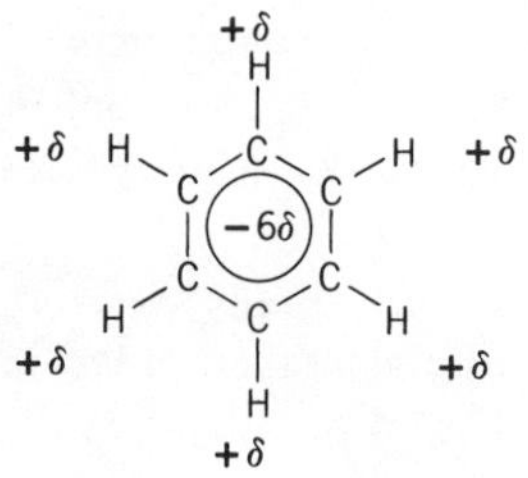

As in HCl and $NH_3$, there is $\sigma$-bond charge transfer, this time from H to C. Again as in Fig. 4-3, if we replace H by $CH_3$, we can transfer a larger charge to the ring, thus lowering the ionization potential and increasing the donor ability.

Aromatic compounds can be fortified as acceptors too. Let us, however, first consider an example in which there is fortification of two opposite kinds at once, leading to a very small net effect. If H in benzene is replaced by Cl, charge in the $\sigma$-bond is transferred away from the ring to the Cl atom (this phenomenon is usually called an inductive effect)

[†] For $NH_3$ Palke and Lipscomb [6b] find that $\delta = +0.15\ e$, whereas Peters [6a] found that $\delta$ ranged from 0.13 to 0.25 $e$, and Body [6c] found $\delta$ to be 0.17 $e$, the result depending on which approximate wavefunctions are used.

[‡] Here $\delta$ is of course not the same as in Fig. 4-3.

tending to leave a positive charge on the ring. However, although the Cl atom is in this way a $\sigma$-bond acceptor in an R—Q type of donor-acceptor action, it is also an $n\pi$-donor in a donor-acceptor action like that in (1-1). (The symbol $n\pi$ means that the donor action is from the "$\pi$-island" lone pair in the $p\pi$-orbital of Cl, which is one of the two $p$-orbitals with axes that are perpendicular to the C—Cl bond.) As the overall result of this two-way *intramolecular* charge transfer action (R$\sigma$,Q$\sigma$ one way; $n\pi,a\pi$ the other) there is a small net transfer of negative charge from the ring to the Cl radical, increasing $I_D$ slightly and resulting in a slightly weaker donor than for benzene itself.

The $n\pi,a\pi$ transfer in chlorobenzene just described can also be called *dative conjugation*. In valence-bond language this is represented in terms of resonance structures

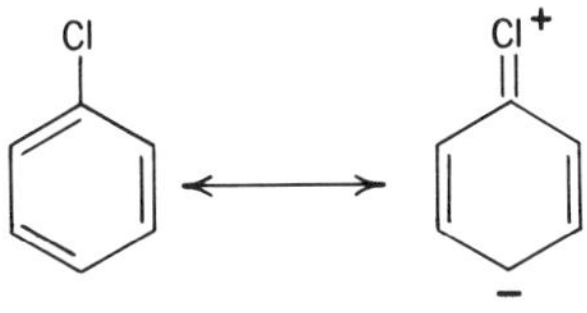

Cl Cl$^+$

(and two others obtained by putting the transferred electron on the *ortho* carbon atoms). The $\pi$-transfer here is increvalent ($n\pi$) for Cl but sacrificial ($a\pi$) for benzene, the overall $\pi$-transfer effect being isovalent. The term "isovalent" refers to the fact that there is an equal number (three) of double bonds in both structures. A similar discussion is applicable to fluorobenzene.

Although in chlorobenzene the two opposite directions of fortification (R$\sigma$,Q$\sigma$ and $n\pi,a\pi$) almost cancel, there are other cases in which benzene can be strongly fortified as either an acceptor or as a donor. The nitro group ($NO_2$) acts as an $a\pi$-acceptor, and in trinitrobenzene the intramolecular effect of the nitro groups

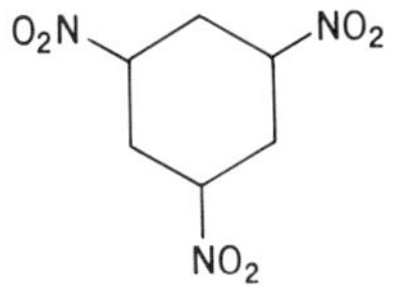

$O_2N$ $NO_2$

$NO_2$

has converted benzene into a fairly strong intermolecular $a\pi$-acceptor.

On the other hand introduction of substituent groups such as OH, $OCH_3$, $NH_2$, or $NR_2$ causes strong fortification of benzene as an intermolecular donor. (See Section 18-3 for further discussion of the anilines.) In these cases the $n\pi,a\pi$ intramolecular donor action from the $\pi$-island lone pair of the oxygen or nitrogen atom to the ring definitely predominates over the R,Q intramolecular $\sigma$-bond acceptor action of the substituent groups.

## 4-3 TYPES OF COMPLEXES

Examples of important types of combinations of donors and acceptors are listed in Table 4-1. (See R4 and R14 for more complete discussion.)

**TABLE 4-1**

**Combinations of Donors and Acceptors**

| Donor Type | Acceptor Type: Q | $v$ | $a\sigma$ | $a\pi$ |
|---|---|---|---|---|
| R | Compounds and $(TCNQ^-)_2$ | Reaction intermediate | Reaction intermediate | Reaction intermediate |
| $n$ | Reaction intermediate | $R_3N \cdot BR_3$ | $R_3N \cdot I_2$ | Often two-way |
| $b\sigma$ | Reaction intermediate | $RX \cdot AlX_3$ | Contact | Contact |
| $b\pi$ | Reaction intermediate | Often two-way | $Bz \cdot I_2$ | $HMB_Z$ · chloranil |

It is not always obvious just how a given electron donor or acceptor may function in a particular complex. Experimental information that is helpful in deciding this question is often obtained from X-ray diffraction studies of the geometrical configurations of solid complexes. (See Chapter 5.) The observed orientation often makes it possible to decide which orbitals of D and A are being used in CT stabilization of the complex. Experimental confirmation of the possible sacrificial nature of the donor-acceptor action may be found in the lengthening of bonds in D or A. Changes in vibrational spectra of D and A can also provide information about the symmetry of the complex, or about the strength of D—A bonding, or about changes in the internal bonding in D or A. (See the discussion in Section 4-1 on the cyclohexene $\cdot$ $Ag^+$ complex.)

From such experimental information guided by the theory we can obtain a fairly good understanding of the complexes that have been thoroughly studied. In general, insofar as the configuration of a complex is determined by CT forces, the theory (Section 2-3) says that the configuration should be that which causes $\beta_0^2/\Delta$ to be maximum (2-20). We note further that $\beta_0^2$ is approximately proportional to $S_{da-}^2$. [See (3-15).] Minimum $\Delta$ [see (2-13)] is obtained if $\phi_d$ is the most easily ionized MO of D and $\phi_{a-}$ is the acceptor orbital of largest electron affinity so that this choice of $\phi_d$ and $\phi_{a-}$ tends to be favored. However, other choices may be preferred if $\beta_0^2$ is then sufficiently larger. The configuration is also affected, and in some cases largely controlled, by electrostatic attraction and exchange-repulsion terms contained in $W_0$. (See Figs. 9-2 and 9-3.)

In some complexes two isomeric forms are possible, or at least conceivable, that differ strongly in the extent of charge transfer and often at the same time in nuclear configuration. Usually only one of the two forms is stable in any one environment, but a change of environment (e.g., from a nonpolar to a polar solvent) may cause a change from one to the other form. Such isomeric forms are called *outer complexes* (usually loosely bound and with little charge transfer) and *inner complexes* (usually with more or less complete charge transfer). They will be discussed, along with *middle complexes*, in Chapters 15 and 16. (See also R2 and R4.) Some middle complexes belong to the category of *symmetrized complexes*, of which $I_3^-$ and $HF_2^-$ are simple examples. Those examples are linear molecules, $(III)^-$ and $(FHF)^-$, which can be thought of as $n \cdot a\sigma$ complexes of $I^-$ with $I_2$ or of $F^-$ with HF, respectively, but symmetrized. (See R4, Section 12-9, and Chapter 15.)

## 4-4 REFERENCES


[1] R. H. Boyd and W. D. Phillips, *J. Chem. Phys.*, **43**, 2927 (1965).
[2] H. B. Gray, *Electrons and Chemical Bonding*, Benjamin, New York, 1964.
[3] M. F. El-Sayed, M. Kasha, and Y. Tanaka, *J. Chem. Phys.*, **34**, 334 (1961).
[4] M. D. Newton, F. P. Boer, and W. N. Lipscomb, *J. Am. Chem. Soc.*, **88**, 2367 (1966).
[5] H. J. Taufen, M. J. Murray, and F. F. Cleveland, *J. Am. Chem. Soc.*, **63**, 3500 (1941).
[6] (a) D. Peters, *J. Chem. Phys.*, **36**, 2743 (1962) and *J. Chem. Soc.*, 4017 (1963); (b) W. E. Palke and W. N. Lipscomb, *J. Am. Chem. Soc.*, **88**, 2384 (1966); (c) R. Body, D. S. McClure, and E. Clement, *J. Chem. Phys.*, **49**, 4916.

CHAPTER 5

# *The Geometrical Configurations of Complexes*

Let us now look at some experimental results from studies of complexes. Most of these results can be understood in terms of the simplified theory presented in Chapter 2; some of them, however, indicate the need for the more general theory presented in Chapter 11. First let us consider the geometrical configurations of solid complexes as determined in X-ray diffraction studies of their crystals. In so doing it is necessary to recognize that the configurations found in crystals cannot be relied on to be the same as for individual 1:1 complexes in solution, except probably in especially stable complexes whose crystals are built up out of individual 1:1 units. In weaker complexes the structures typically consist of chains or sheets in which D and A molecules alternate in a regular way (*N:N* complexes).

## 5-1 X-RAY DIFFRACTION STUDIES

What features should we look for in examining the results from crystal-structure studies? The following are of special interest:

1. General configuration (e.g., is the axial or resting model found for benzene-halogen complexes?)
2. D—A distances compared with van der Waals distances and covalent-bond distances.

3. Changes in the internal structure of D and of A.

In examining available results it is useful to refer to Table 4-1. Of the possibilities listed there we shall examine results for $n \cdot v$, $n \cdot a\sigma$, $b\pi \cdot a\sigma$, and $b\pi \cdot a\pi$ complexes. Some information is also available for two-way complexes such as, for example, $Ag^+$ (or other metal-ion) complexes with benzene or ethylene derivatives. However, few X-ray studies have been made of the other kinds of complexes listed in the table, and we shall not examine two-way complexes here. (For surveys of this material see GR1 and GR2.)

## 5-2 RESULTS FOR $n \cdot v$ COMPLEXES

In addition compounds between $n$-donors and $BF_3$ we expect the limiting dative structure ($R_3N^+$—$B^-F_3$) to be ethanelike. We are thus especially interested in the N—B distance and in the FBF angle. The latter should serve as a measure of the extent to which the $BF_3$ in the complex has gone over toward tetrahedral geometry from the planar geometry of free $BF_3$. Some results of X-ray studies are shown in Table 5-1. Since the structures here consist of 1:1 units they can probably be taken as approximately valid for individual 1:1 complexes.

We note that the N—B distance approaches the covalent bond distance, and the FBF angle becomes tetrahedral (109° 28') for the stronger complexes. We note with interest that the B–F distance increases with donors of increasing donor strength, indicating that the "vacant" acceptor orbital in free $BF_3$ is not entirely vacant but has some bonding character which is progressively lost as the no-bond structure in (1-1) gives way to the dative structure during complex formation. Another way to state this is that the formation of the complex introduces competition with the intramolecular dative $\pi$ bonding ($F \rightarrow B$) which is present in planar $BF_3$. Some relevant calculations have been made by Cotton and Leto (See Section 18-4 for this and further discussion of $BX_3$).

## 5-3 RESULTS FOR $n \cdot a\sigma$ COMPLEXES

Most of the experimental results on crystalline $n \cdot a\sigma$ complexes with halogen molecules as $a\sigma$-acceptors have been obtained by Hassel and his coworkers at the University of Oslo. (See his review [4].)

Experimental error is an important fact of life for all X-ray studies, but it is especially important with complexes. We are particularly interested in complexes in comparing distances—for example, in comparing

**TABLE 5-1**

**Some Structural Information about $n \cdot v$ Complexes with $BF_3$ as the $v$-Acceptor (All Distances in Å)[a]**

| $r_i$ or Angle | $BF_3$ | $CH_3CN \cdot BF_3$ | $H_3N \cdot BF_3$ | $(CH_3)_3N \cdot BF_3$ | $BF_4^-$ | $r_{vdW}$[b] | $r_{cov}$[c] |
|---|---|---|---|---|---|---|---|
| N—B | – | 1.63 | 1.60 | 1.58 | – | 3.0 | 1.54 |
| B—F | 1.30 | 1.33 | 1.38 | 1.39 | 1.43 | – | 1.40 |
| < FBF | 120° | 114° | 111° | 107° | 109° | – | – |
| < FBN | – | 103° | 107° | 112° | – | – | – |

[a] Data from [1] and from L. E. Sutton (ed.), *Tables of Interatomic Distances and Configuration in Molecules and Ions*, Special Publication No. 11, The Chemical Society, London, 1958.
[b] Estimated van der Waals distance; see Pauling [2].
[c] Estimated; see footnote a and also Pauling [2].

the I—I distance in the complex with that for the free molecule. Here we need to know whether two experimentally measured distances, $r_1$ and $r_2$, are truly different. If the difference $|r_1 - r_2|$ is equal to $\sigma$, where $\sigma$ is the standard deviation in the measurement of $r$, there is a 68% probability that $r_1$ and $r_2$ are *really* different. To be quite sure they are different we should require that the difference $|r_1 - r_2| \geq 3\sigma$; then the probability that they are really different is better than 99%.

Unfortunately, standard deviations are often not listed in reports on X-ray studies of complexes. Hence Hassel's own evaluation [4] is of considerable interest: "It should be emphasized that...[these studies]... involve single-crystal work at well below room temperature, in some cases even at very low temperatures, which may mean that it is very difficult to obtain crystals comparable in quality to those easily obtainable from crystals melting above room temperature. ...[and so on, pointing out the considerable difficulty in handling crystals and apologizing for the resulting poorer quality of the data]... The accuracy actually obtained in two-dimensional X-ray studies may in general be expected to be of the order 0.02—0.03 Å for the halogen-halogen distance, and better than 0.1 Å in the case of the halogen-donor distances." (Note that halogen atoms scatter X-rays much better than carbon or nitrogen atoms, which makes it harder to locate the lighter atoms accurately.)

With this experimental uncertainty in mind we now examine the results shown in Tables 5-2 and 5-3 for $n \cdot a\sigma$ complexes. In all cases the D···X—Y grouping is linear, as would be expected for maximum $S_{da^-}$ [see (2-35)] if the XY molecule acting as an electron acceptor uses its molecular orbital of maximum electron affinity, which is of the antibonding $\sigma$ type.

In the case of the strong amine-iodine complexes in Table 5-2, whose crystals are built up from well-defined 1:1 units, we can be reasonably sure that the configurations are not radically different than those of the individual 1:1 complexes. In these complexes the D—A distance is rather close to that for a covalent bond, suggesting a covalent D—A bond order in the complex of perhaps two-thirds. This small D—A distance is direct evidence of an important resonance interaction ($\beta_0^2/\Delta$) in the structure of the complex since classical electrostatic plus van der Waals forces alone would result in a distance between molecules that would not be very different from the van der Waals distance. Note that the N—I distance is nearly the same in all the amine complexes listed, indicating similar strength of CT interaction in all.

In the column labeled $\delta$ we can see the "expansion" of the D—X "bond" relative to the covalent distance. In Table 5-3 this expansion increases fairly regularly for the weaker $I_2$ complexes from the strong

**TABLE 5-2**

**X-Ray Data for Halogen Complexes with Strong *n*-Donors[a] (All Distances in Å)**

| Complex[b] | [c]$\delta$ | $(D—X)_{obs}$ | D—X (sum of covalent radii) | D—X (sum of vdW radii) | $(X—Y)_{obs}$ | $(X—Y)_{free}$ |
|---|---|---|---|---|---|---|
| $(CH_3)_3N \cdot I_2$ | 0.24 | 2.27 | 2.03 | 3.65 | 2.84 | 2.67 |
| $(CH_3)_3N \cdot ICl$ | 0.27 | 2.30 | ” | ” | 2.52 | 2.32 |
| Py · ICl | 0.23 | 2.26 | ” | ” | 2.51 | ” |
| $\gamma$-Pic · $I_2$ | 0.28 | 2.31 | ” | ” | 2.83 | 2.67 |
| HMT · $2Br_2$ | 0.32 | 2.16 | 1.84 | 3.45 | 2.43 | 2.28 |
| $Py_2I^+$ | 0.13 | 2.16 | 2.03 | 3.65 | — | — |
| $I_3^-$ (sym)[d] | 0.23 | 2.90 | 2.67 | 4.30 | 2.90 | 2.67 |
| $Cs^+I_3^-$ (asym)[d] | 0.36 | 3.03[e] | 2.67 | 4.30 | 2.83[e] | 2.67 |
| $Br_3^-$ | 0.26 | 2.54 | 2.28 | 3.90 | 2.54 | 2.28 |
| $ICl_2^-$ | 0.23 | 2.55 | 2.32 | 3.95 | 2.55 | 2.32 |
| $(I_2^-)_{est}$ | | | | | (3.15) | 2.67 |
| $(Br_2^-)_{est} = (ICl^-)_{est}$ | | | | | (2.80) | 2.28, 2.32 |

[a] Data from GR2, Chapter III., and from the review by E. H. Wiebanga, E. E. Havinga, and K. H. Boswijk, *Adv. Inorg. Radiochem.*, **3**, 133 (1961) for the polyhalide ions except for $ICl_2^-$, which is from G. J. Visser and A. Vos, *Acta Cryst.*, **17**, 1336 (1964); estimates for $XY^-$ are from R13.

[b] Py = pyridine; $\gamma$-Pic = 4-methyl pyridine (= $\gamma$-picoline); HMT = Hexamethylenetetramine.

[c] $\delta$ means $(D—X)_{obs}$ − (sum of D and X covalent radii.)

[d] $I_3^-$ in most crystals is somewhat *asymmetrical* and nonlinear, but in $(C_6H_5)_4As^+I_3^-$, and we think in solution, it is linear and symmetrical.

[e] The two different I—I distances here may be interpreted as approximately $I^- \overset{3.03}{\cdots} I \overset{2.83}{—} I$.

**TABLE 5-3**

**X-Ray Data from Halogen Complexes with Weaker *n*-Donors[a,b]**

| Complex | $\delta$[c] | $(D—X)_{obs}$ | D—X (sum of covalent radii) | D—X (sum of vdW radii) | $X—Y_{obs}$ | $X—Y_{free}$ |
|---|---|---|---|---|---|---|
| Diselenane · $I_2$ | 0.33 | 2.83 | 2.50 | 4.15 | 2.87 | 2.67 |
| Dithiane · $I_2$ | 0.50 | 2.87 | 2.37 | 4.00 | 2.79 | 2.67 |
| Benzyl sulfide · $I_2$ | 0.47 | 2.84 | 2.37 | 4.00 | 2.81 | 2.67 |
| Dioxane · 2 ICl | 0.58 | 2.57 | 1.99 | 3.55 | 2.33 | 2.32 |
| Dioxane · $Br_2$ | 0.91 | 2.71 | 1.80 | 3.35 | 2.31 | 2.28 |
| Dioxane · $Cl_2$ | 1.02 | 2.67 | 1.65 | 3.20 | 2.02 | 1.99 |

[a] Data from references cited in footnote a, Table 5-2.
[b] All distances in this table are in Å.
[c] The term $\delta$ means $(D—X)_{obs}$ − (sum of D and X covalent radii.)

*n*-donor diselenane through dithiane to the relatively weak dioxane. Even in the dioxane complexes, however, the D—X distance is much smaller than the sum of van der Waals radii.

Table 5-2 includes for comparison some data for the planar, centrosymmetric $Py_2I^+$ ion, where the N—I bond, if the ion is symmetrical, should have a Coulson bond order of $1/\sqrt{2}$. If symmetrical, this ion is expected to have a structure given by equal resonance between two structures like

$$\overset{+}{N}\text{—I} \cdots N$$

Another item of interest is the substantial increase in the X—Y distance as the complex forms. This observation is direct experimental confirmation that the acceptor is sacrificial and accepts into its antibonding $\sigma$ MO. We note that the X—Y bond length is about one-third of the way between that for free XY and for the $XY^-$ ion in the case of the amine-XY complexes in which the weight of the dative term in $\Psi_N$ may perhaps be not very far below 50%. Thus in this respect the complex exhibits the compromise geometry that we expected.

In this connection the results for the trihalide ions in Table 5-2 are of considerable interest. These ions may be understood as *symmetrized* CT complexes between a halide ion as donor and a halogen molecule as acceptor. (See Sections 4-3 and 12-9.) In some crystals, especially with small positive cations, the structures of the trihalide ions are not symmetrical. Presumably this asymmetrical structure occurs because of polarization by the positive ion. (Another possibility is that it may result from a weakening of the halide ion's donor action toward the halogen molecule through competition for the halide ion by the cation.) Presumably when isolated in solution the ions are linear and symmetrical. Although a simpler description of the structures of these symmetrical ions is given in terms of whole-complex MOs (see Section 12-9), it is interesting in such cases (e.g., in $I_3^-$) to note the strong sacrificial effect in the I—I bond, as expected for a strong resonance interaction between $I^-$ as D and $I_2$ as A.

## 5-4 SPECTROSCOPIC AND CHEMICAL ORBITALS IN MOLECULES

At this point it is useful to introduce the concept of a chemical orbital as distinguished from a spectroscopic orbital. The orbital that is associated with the actual minimum ionization potential of an atom or molecule—or *equally* with any *higher* actual ionization potential—may be called a *spectroscopic orbital* because, when an electron is either excited spectroscopically or removed, it is out of such an orbital that it comes. Spectroscopic MOs are in general nonlocalized; that is, they spread more or less over all the atoms in a molecule, although in special cases they can be almost completely localized on particular atoms.

On the other hand it is convenient in discussing chemical bonding to replace spectroscopic MOs by bond MOs, each *localized* in a particular chemical bond [5,6], together with lone-pair AOs on the atoms†; Mulliken has called these *chemical orbitals* [7]. Chemical orbitals can be constructed as hybrids (i.e., linear combinations) of spectroscopic MOs or AOs. Of principal interest for the present discussion are hybrids of certain spectroscopic MOs in *n*-donors that have approximately the form of lone-pair AOs. A more complete discussion is given in Section 11-3.

## 5-5 COMPLEXES OF OXYGEN LONE-PAIR AND HOMOLOGOUS DONORS WITH HALOGEN-MOLECULE ACCEPTORS

Let us now turn to a further discussion of the dioxane · halogen complexes and the acetone · $Br_2$ complex. Their structures as reported by Hassel [4] are sketched in Fig. 5-1. Fig. 5-1*a* shows one of the dioxane · $Br_2$ chains which form this crystal. Fig. 5-1*b* shows part of the sheets forming the acetone · $Br_2$ crystal.

Noteworthy in both these cases is the fact that the distances are the same to both neighbors of a given $Br_2$ molecule. This situation contrasts with that found for the stronger complexes, which are definitely 1:1 in the crystal. Here the structures of the crystals are strongly suggestive but not at all conclusive as to the configurations in individual 1:1 complexes. Hassel called the D · · $Br_2$ · · D bonding in these weaker complexes *halogen bridging*.

Of special interest is the bond angle between the O—Br bond and the other bonds in the donor molecule. The most easily ionized electrons in an ether molecule are in the oxygen atom 2*p* AO with its axis of symmetry (*x*-axis) perpendicular to the R—O—R plane. This is a lone-pair AO,

† *Some* spectroscopic MOs are almost identical with lone-pair AOs, but in general lone-pair AOs are too localized to be spectroscopic MOs.

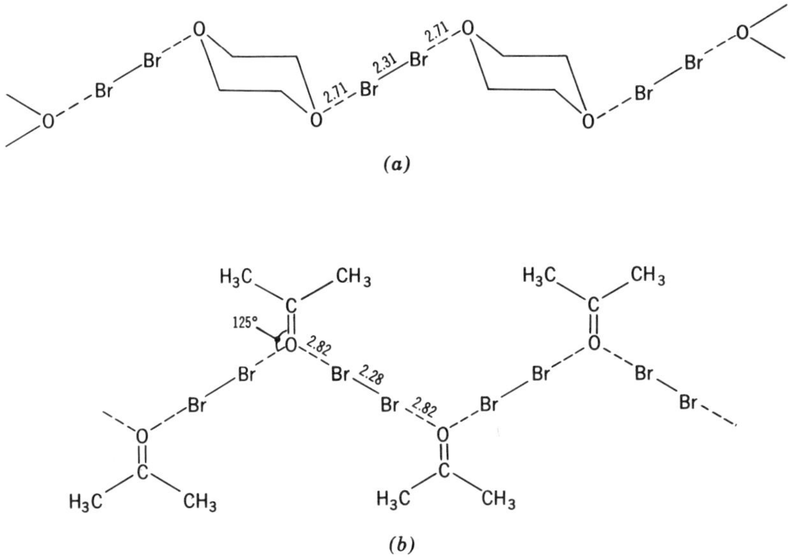

(a)

(b)

**Figure 5-1** Donor-acceptor arrangements in crystal structures of (*a*) dioxane · $Br_2$ and (*b*) acetone · $Br_2$ complexes. (Adapted from Hassel and Romming [4].)

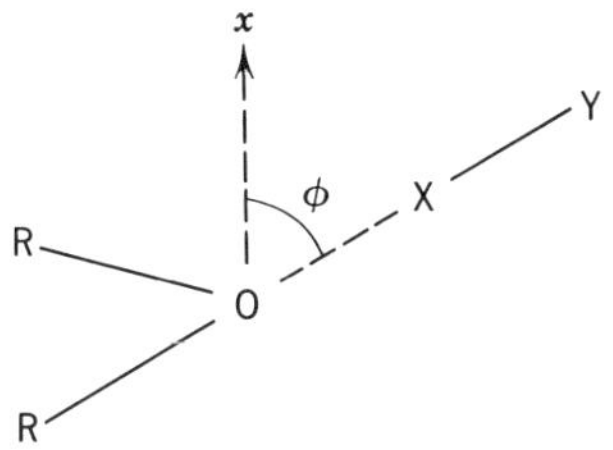

**Figure 5-2** Diagram showing how $\phi$ of Table 5-4 is defined for ether-halogen complexes; the axis O$x$ is perpendicular to the R—O—R plane.

which is at the same time a spectroscopic MO. If this MO were being used as the donor orbital in $\Psi_1$ of (1-1), the O · · Br—Br bond should be perpendicular to the R—O—R plane, along the direction O$x$ in Fig. 5-2. The actual experimental results for some etherlike donors complexed with halogens are shown in Table 5-4.

Also for the crystalline acetone · $Br_2$ complex (Fig. 5-1*b*), the C—O—Br

**TABLE 5-4**

**Angles Found in Complexes with Ethers and Thioethers[a]**

| Complex | $\phi$[b] | < C—O—X[c] | < C—O—C[c] |
|---|---|---|---|
| Dioxane · $Br_2$ | 53° | 116° | 116° |
| Dioxane · $Cl_2$ | 53° | 118° | 111° |
| Dioxane · $H_2SO_4$ | 49° | 116° | 113° |
| Benzyl sulfide · $I_2$ | 15° | 100° | 93° |
| Dithiane · $I_2$ | 16° | 100° | 101° |
| Diselenane · $I_2$ | 17.5° | 101° | 101° |

[a] Computed from data given by Hassel, [4], and by G. Y. Chao and J. D. McCullough, *Acta Cryst.*, **13**, 727 (1960), and ibid., **14**, 940 (1961).
[b] See Fig. 5-2.
[c] Or < C—S—X or < C—S—C etc.

angle is 125° instead of the 90° value that would be expected if the spectroscopic orbital with the lowest ionization potential were used for bonding. In acetone this would be an MO that is roughly an oxygen-atom lone-pair† $2p$ AO with its axis at the plane of the paper and perpendicular to the C═O bond.

We note in Table 5-4 that in the thioether complexes the halogen molecule is approximately perpendicular to the C—S—C plane ($\phi$ of Fig. 5-2 small), indicating that mainly the $2p$ lone-pair spectroscopic donor MO is actually used in the bonding in these complexes. However, in the ethers, the angle $\phi$ differs considerably from 0°. This indicates that here a hybrid, essentially lone-pair, chemical orbital of the oxygen* is used to bond the halogen in the complex, in spite of the larger $\Delta$ as compared with use of the most easily ionized MO alone. Apparently the increase in $S_{da-}$ and concomitant increase in $\beta_0$ overpowers the cost in stability due to the increase in $\Delta$ in the expression $\beta_0^2/\Delta$ [cf. (2-20)]. This result indicates the need to modify the simplified theory of Chapter 2 so that a dative function $\Psi_1$ may be introduced in which $D^+$ is not in its lowest energy configuration. (See Section 11-3 for further discussion.)

† This so-called $n$-orbital is really not quite a lone-pair oxygen AO. Strictly, it is an MO that is somewhat antibonding, though *largely* concentrated on the oxygen atom.
* That is, a hybrid of the first and second spectroscopic MO's, of which the first is almost exactly, and the second to a rough approximation, just an oxygen atom AO.

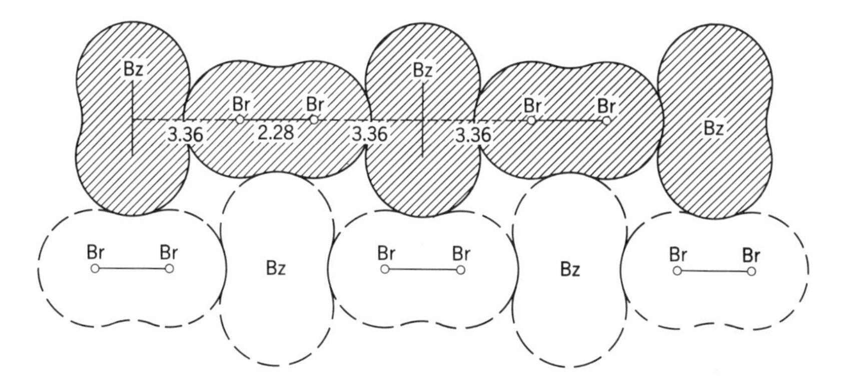


**Figure 5-3** The structure of the benzene $\cdot$ $Br_2$ crystal (schematic). The molecule chain shown with dashed lines is above the other chain by one-half a unit cell dimension. (This figure is adaped from Fig. 6 of Hassel and Rømming [4].)

## 5-6 RESULTS FOR $b\pi \cdot a\sigma$ COMPLEXES

The only examples of this type for which crystal structures are available are the 1:1 benzene complexes with $Br_2$ and with $Cl_2$ [4]. The structures of both crystals are similar and consist of chains of alternating $Br_2$ and benzene molecules, as indicated in Fig. 5-3. Each $Br_2$ molecule is aligned along the sixfold axis of the neighboring benzene molecules, perpendicular to the ring planes of the latter. Three *neighboring* chains are arranged more or less symmetrically around each chain; one of these is indicated in the figure. The chains are staggered, and the whole forms a reasonably compact structure of slightly positive disks and slightly negative cylinders.

The Bz $\cdot\cdot$ Br distances (3.36 Å) are all equal, indicating halogen bridges again for this weak complex.[†] Note that the Bz $\cdot\cdot$ Br distance

[†] However, a study of the infrared spectrum [W. B. Person, C. F. Cook, and H. B. Friedrich, *J. Chem. Phys.*, **46**, 2521 (1967)] indicates that centers of symmetry at the $Br_2$ and Bz molecules are not present in a crystal formed at $-200°$C by deposit from the vapor. Hence it is suggested that these halogen bridge structures found in the X-ray study may be *average* structures resulting from a crystal of random chains containing 1:1 complexes with alternating long and short Bz $\cdot\cdot$ Br distances. Alternatively, Hassel's results may be for a different phase that forms at high temperatures ($-40°$C) in contrast with the low-temperature ($-200°$C) phase of the infrared study.

is smaller than the sum of van der Waals radii (3.65 Å). Because of the uncertainty in the latter we cannot be quite sure whether or not this shorter distance is significant evidence of an important resonance interaction, but it seems very likely that it is.

Because of the possible complications of neighbor interactions and because of the *N:N* instead of 1:1 nature of the halogen bridge structure (if it is real), we cannot be sure that the axial structure found here in the crystal for the $Bz \cdot Br_2$ complex also exists for an isolated $Bz \cdot Br_2$ complex in solution. However, in the absence of information to the contrary this seems to be the most likely configuration for this 1:1 complex and for the analogous $Bz \cdot I_2$ complex. It would then follow that the donor MO out of which an electron has been transferred in the dative function $\Psi_1$ in (2-2) is not the most easily ionized $\pi$ MO ($\phi_2$ or $\phi_3$ from Fig. 4-1) but is the deeper lying $\pi$ MO, $\phi_1$; this problem will be further discussed in Section 11-2.

## 5-7 RESULTS FOR $b\pi \cdot a\pi$ COMPLEXES

In considering these complexes the summary articles by Wallwork [8] and by Boeyans and Herbstein [9] are most helpful. When contrasted with the usual structures for aromatic crystals (indicated in Fig. 5-4 for benzene [10]), the most striking results of the studies of the complexes is the parallel-plane (or stacked) configuration of alternate D and A molecules found in the latter.

Since aromatic hydrocarbons have both $\pi$-donor and $\pi$-acceptor properties, it is reasonable to ask whether CT forces may have some importance for the structures of aromatic hydrocarbon crystals. No conclusive answer is available, but it seems likely that such CT forces may (at least usually) be too weak to be decisive in determining configurations. Let us consider the CT interaction of two benzene molecules, one above the other with their planes parallel. If these were to form a complex, this would be an example of a *self-complex* and would involve two-way CT action; that is, two $\Psi_1$ terms ($Bz^+$—$Bz^-$ and $Bz^-$—$Bz^+$) would be required, with equal coefficients. For CT stabilization it is necessary to have a good $\beta_0$ for stabilization of $\Psi_0$ by these $\Psi_1$ terms, and as has been seen earlier [see (3-15)] $\beta_0$ is approximately proportional to $S_{da^-}$. Now if the two benzene molecules are placed with their sixfold axes coincident, $S_{da^-}$ is readily seen to be zero (see Fig. 4-1 for the benzene $\pi$ MOs) for transfer of an electron from any occupied $\pi$ MO of the donor benzene to any unoccupied $\pi$ MO of the acceptor benzene in either of the $\Psi_1$ terms. However, with a suitable lateral shift of one benzene, keeping

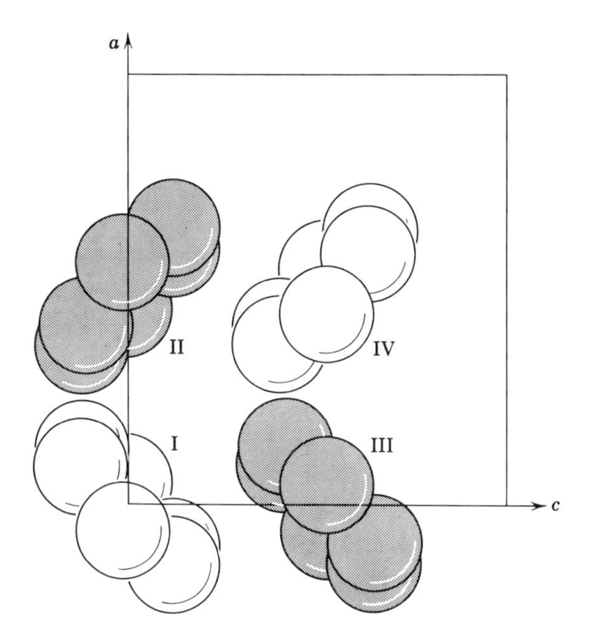


**Figure 5-4** Representation of the orthorhombic unit cell of benzene viewed along the *b*-axis. The two shaded molecules are (1/2) *b* above the unshaded molecules. This edge-on view of the crystal shows how the neighboring molecules in a given plane (for example, I and IV) are almost perpendicular to each other in contrast to the parallel stacks that are found in crystals of complexes. (From L. Hall, Ph.D. thesis, University of Iowa, 1961; based on data by Cox et al. [10].)

its plane parallel to that of the other, positions with $S_{da^-} \neq 0$ can be found, so that CT stabilization of a $Bz_2$ self-complex with parallel-plane geometry is in principle possible. However, benzene is relatively weak both as a donor ($I_D = 9.24$ eV) and as an acceptor ($E_A < 0$), making $I_D - E_A > 10$ eV; thus [(2-20) and (2-13)] CT forces between benzene molecules are probably unimportant. In larger aromatic hydrocarbons $I_D - E_A$ is smaller; the most favorable case among aromatic molecules for self-complexing is that of graphite, in which $I_D - E_A = 0$.

Let us now turn to more typical complexes in which one molecule is definitely a donor and the other is definitely an acceptor. An example is the complex between anthracene (An) and *s*-trinitrobenzene (TNB) [11]. A stack of alternating An—TNB molecules along the *c*-axis for the solid

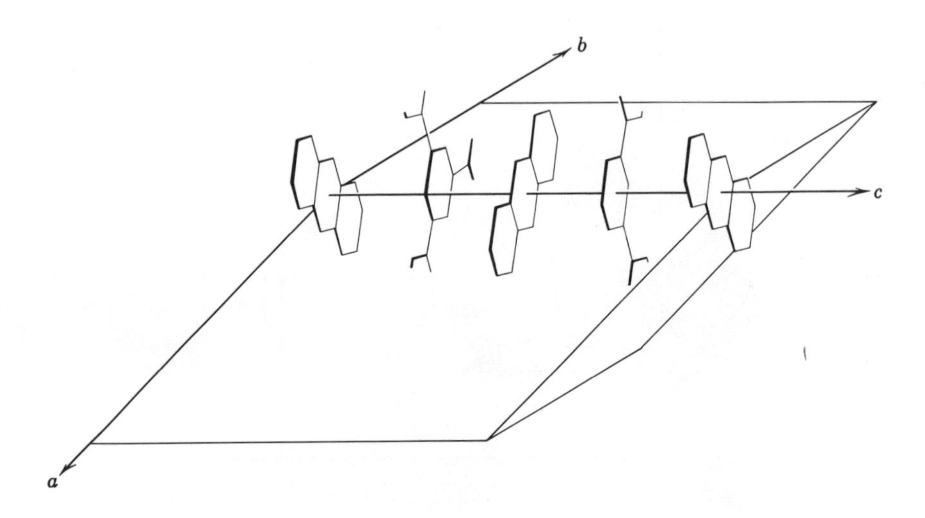


**Figure 5-5** Part of the cell contents of the anthracene-*s*-trinitrobenzene complex, showing the stacking of molecules along the *c*-axis. There is a similar column of molecules generated from this one by the *C* face centering. (From Brown, Wallwork, and Wilson [11].)

complex is shown in Fig. 5-5. The molecules are tilted slightly with respect to the stack axis, with the plane of the TNB molecule parallel to the *b*-axis whereas that of An makes an angle of 8° with the *b*-axis. The closest intermolecular contacts between the An and TNB molecules within the stack are shown in Fig. 5-6. We note in Fig. 5-5 that the *c*-axis passes approximately through the center of the anthracene molecule, but it passes through one carbon of the TNB ring, so that the TNB is shifted off center, as expected for nonzero overlap between D and A.

Although the numbers shown in Fig. 5-6 are for C—C′ distances and are not exactly average interplane separations, they do appear to be significantly shorter than the van der Waals separation (3.40 Å). These distances are fairly typical of the values reported for $b\pi \cdot a\pi$ complexes. Careful experimental work is required to show that the interplane separations within a stack are indeed shorter than 3.40 Å, but there are now several examples of complexes in which shortenings of this order (0.1 to

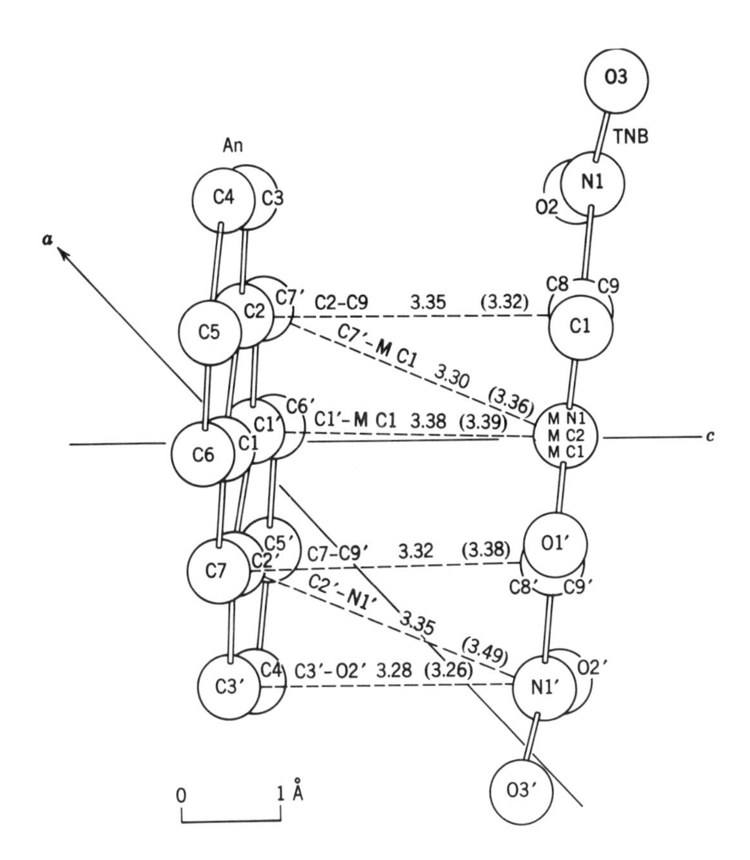


**Figure 5-6** Closest intermolecular contacts between anthracene and *s*-trinitrobenzene molecules at −100°C. (Room-temperature values in parentheses.) The projection is as in Fig. 5-5. (From Brown, Wallwork, and Wilson [11].)

0.2 Å) have been observed. (See Boeyens and Herbstein [9].)

In general the D—A distances in $b\pi \cdot a\pi$ complexes appear to be only slightly shorter than van der Waals contacts, in contrast with complexes involving $a\sigma$-acceptors. This behavior of $b\pi \cdot a\pi$ complexes is reasonably explicable as due to strong exchange repulsions between the electrons in the other $\pi$-orbitals that are not involved in the D—A resonance stabilization. Hence the CT stabilization, which is relatively weak because both the donor and the acceptor actions are sacrificial, is not sufficient to pull the molecules close enough together to result in big decreases from the van der Waals distances. However, the interaction is readily apparent in that the molecules do form stacks of alternating pairs of parallel molecules, with normal van der Waals distances *between* stacks but with slightly shorter D—A interplane distances *within* the stacks.

## 5-8 REFERENCES

[1] J. L. Hoard, S. Geller, and T. B. Owen, *Acta Cryst.*, **4**, 405 (1951).
[2] L. Pauling, *The Nature of the Chemical Bond*, Cornell University Press, Ithaca, N.Y., 1960.
[3] F. A. Cotton and J. R. Leto, *J. Chem. Phys.*, **30**, 993 (1959).
[4] O. Hassel and C. Rømming, *Quart. Rev.*, **16**, 1 (1962).
[5] D. Peters, *J. Chem. Soc.*, 2003, 2015, and 4017 (1963).
[6] C. Edmiston and K. Ruedenberg, *J. Phys. Chem.*, **68**, 1628 (1964).
[7] R. S. Mulliken, *Science*, **157**, 13 (1967).
[8] S. C. Wallwork, *J. Chem. Soc.*, 494 (1961).
[9] J. C. A. Boeyens and F. H. Herbstein, *J. Phys. Chem.*, **69**, 2160 (1965).
[10] E. G. Cox, D. W. J. Cruikshank, and J. A. Smith, *Proc. Roy. Soc.* (London), **A247**, 1 (1958).
[11] D. S. Brown, S. C. Wallwork, and A. Wilson, *Acta Cryst.*, **17**, 168 (1964).

CHAPTER 6

# *More Ground-State Properties; Vibrational Spectra*

## 6-1 GENERAL

In addition to the results from X-ray diffraction studies information bearing on the nature of the bonding in complexes may be obtained from the study of their vibrational (infrared and Raman) spectra. Some expected effects are the following:

1. New motions must appear in the complex in which D vibrates against A. Counting degrees of freedom, 3 translations are lost from D and also from A, and either 2 (if linear) or 3 rotations from each, making a total of 10, 11, or 12 degrees of freedom lost on complex formation. The complex, however (if nonlinear), now has six translational and rotational degrees of freedom. Hence, for example, in $R_3N \cdot I_2$ (with D nonlinear but A linear) there are five new vibrational degrees of freedom, distributed as follows in this $C_{3v}$ complex: one D—A stretch, a doubly degenerate rocking vibration of D about the A-axis, and a doubly degenerate motion of D moving perpendicular to the A-axis, giving three new vibrational frequencies. Thus in the strong $(CH_3)_3N \cdot I_2$ complex, the N—I stretching vibration has been observed [1] at 145 $cm^{-1}$; the two bending vibrations are expected to be lower in frequency, perhaps from 50 to 100 $cm^{-1}$ or less, comparable to the ordinary lattice vibrations of molecular crystals. The D—A stretching frequencies in weaker complexes should also be lower than 150 $cm^{-1}$, approaching ordinary lattice frequencies.

With the new far-infrared instruments now available we may expect a rapid expansion in the quantity of the now (fall 1965) very sparse data on these new vibrations.†

2. The frequencies and intensities of vibrational bands belonging to the D and A molecules in the complex are expected to change to some extent, including appearance of absorption bands of D or A in the complex that are forbidden in the free molecules. If the D—A interaction is very strong, the vibrational spectrum of the complex may be so different from the sum of the separate components as to make it difficult to correlate the bands of the complex with those of the parent components. In weak complexes, however, the correlations are easier. Obviously the assignment of the spectrum of a large complexed donor molecule is much harder than is the assignment of the spectrum of a diatomic or triatomic complexed acceptor. Hence we shall concentrate on the latter to illustrate the process and the kind of information obtained.

Most of the first studies of the infrared spectra of complexes were of the changes in the internal vibrations of the donor [2,3]. Polyatomic donors have many vibrations, and even slight shifts in frequency and moderate changes in intensity can create a fairly complicated problem of interpretation. Furthermore, complexes in solution are often prepared by adding a large excess of donor so that the spectrum of complexed donor is obscured by that of excess free donor; for example, $I_2$ is not very soluble and its complexes are often weak; hence in order to complex the $I_2$ in a solution a large excess of donor may be needed.

For these reasons substantial progress in understanding the vibrational spectra of complexes was made only after studies were begun of the diatomic halogen acceptors. The first interpretable dramatic change in the spectrum of a molecule participating in a complex was the report in 1955 by Collin and D'Or [4] of the appearance of the Cl—Cl stretching vibration in the infrared spectrum of $Cl_2$ dissolved in benzene. Since $Cl_2$ is symmetrical, absorption of infrared light is forbidden for the isolated molecule; the appearance of this vibration with moderate intensity (later found to be 153 cm/mmole) in the solution of the $Bz \cdot Cl_2$ complex was widely interpreted as evidence for an asymmetrical location of the $Cl_2$ molecule in the complex [5]. The shift in frequency (530 $cm^{-1}$ in the complex compared to 557 $cm^{-1}$ in the gas phase‡ and 545 $cm^{-1}$ from the

---

† Even in 1969 as this is edited again the data on the D—A vibrations are still sparse. However, see R. F. Lake and H. W. Thompson, *Proc. Roy. Soc.* (London), **A297**, 440 (1967) and J. Yarwood and W. B. Person, *J. Am. Chem. Soc.*, **90**, 594 (1968), and ibid., **90**, 3930 (1968) for some recent studies of D—A vibrations of complexes with halogens.

‡ Although the infrared transition is forbidden in the gas phase, the vibrational frequency is known in that phase from studies of the electronic spectrum. Of course the transition is allowed in the Raman spectrum for symmetric $Cl_2$.

Raman spectrum of a solution in $CCl_4$) suggests that the Cl—Cl bond in the complex is weaker than that in the free molecule, supporting the idea that $Cl_2$ is acting as a sacrificial $a\sigma$-acceptor.

The first systematic study of halogen acceptors was made by Person, Humphrey, Deskin, and Popov [6] for ICl complexes with several donors. They found that the ICl stretching frequency decreases and its intensity increases in different complexes as the strength of the donor increases. The frequency changes from 375 $cm^{-1}$ for uncomplexed ICl dissolved in $CCl_4$, to 355 $cm^{-1}$ for ICl complexed with the weak donor benzene, on to 290 $cm^{-1}$ for ICl complexed with the strong donor pyridine, whereas the intensity of the I—Cl vibration for the complex increases from 1050 to 1900 to 12,000 cm/mmole, respectively. Further studies of other halogen complexes have supported the hypothesis that these changes are typical and are due to the weakening of the ICl bond by the $a\sigma$-acceptor action of the ICl.

For ICN complexes Person, Humphrey, and Popov [7] later showed that the I—C stretching vibration in the complexed ICN also decreases in frequency and increases in intensity as the strength of the donor increases, whereas the frequency and intensity of the bending vibration and the frequency, at least, of the C—N stretching vibration do not change appreciably or in any systematic way. This behavior is consistent with the idea that the I—C bond acts as a sacrificial $a\sigma$-acceptor when ICN forms complexes. The behavior observed for these two $a\sigma$-acceptors is quantitatively similar to the change in infrared spectrum found for the O—H bond in R—OH when hydrogen-bonded complexes are formed, suggesting that the changes in the infrared spectra observed for the latter are also due to $a\sigma$ sacrificial action in the O—H acceptor bond; and similarly for other hydrogen-bonded complexes.

The removal of forbiddenness in the halogen vibration on complexing with benzene is no longer interpreted as evidence for an asymmetrical location of the halogen molecule in the complex. What is probably essentially the correct explanation for the intensity changes was given by Ferguson and Matsen [8–10]. This explanation accounts for the intensity increase as a vibronic effect that can occur regardless of the configuration of the halogen in the complex. The subject has recently been reviewed by Friedrich and Person (R15); we shall outline here their presentation first for the frequency shifts and then for the intensity changes.

## 6-2 FREQUENCY SHIFTS

In order to predict the frequency shifts resulting from complexing we need to know the change of the potential surface as a function of the normal coordinates of the complex in the normal ($N$) state: $W_N(Q_1, Q_2, Q_3 \cdots Q_{3N-6})$, as compared to the combined potential surface for the free D and A molecules.

As D and A are brought together to form the complex the potential energies for motions in certain directions on $W_0$ are modified by the proximity of the partner. Since the interaction is relatively weak, the vibrations in general are not strongly affected, and the potential surface of the complex, $W_N$, bears strong resemblance to that ($W_0$) of the pair of isolated molecules.

If the potential surface along a particular coordinate $Q_i$ changes as a result of the interaction between D and A, we can say something about *how* it changes. We have seen that the electronic function $\Psi_N$ for the complex in its ground state can be described approximately by modifying the wavefunction $\Psi_0$ for the isolated molecules by mixing in a little of the dative wavefunction $\Psi_1$. Similarly the motion described by $Q_i$ on the potential surface $W_N$ should differ from that for the corresponding motion in the free molecule in such a way that it becomes a little more like this motion in the dative structure $D^+—A^-$. The magnitude of the change depends on the extent of modification from $W_0$ as a result of the charge transfer action. It seems plausible to assume that we can measure the latter by $(b^2 + abS_{01})$, the weight of the dative structure in $\Psi_N$ [cf. (1-5)].

Friedrich and Person (R15) have attempted to justify this assumption; it has not been easy to do so quantitatively. The assumption is based on the principle of compromise geometry, which states that the properties of the ground electronic state are expected to be weighted averages of the properties of the two principal resonance structures: the no-bond structure $\Psi_0$ and the dative structure $\Psi_1$. This assumption has been used successfully before—for example, to predict that the $k_{C—C}$ force constant for stretching a C—C bond in benzene is approximately the average of the force constant for a single bond and of that for a double bond (since the C—C bond in benzene is expected to have about 50% single-bond and 50% double-bond character).

In order to clarify the nature of this assumption further let us look now at the cross section through the $W_N$ potential surface obtained when all the normal coordinates are fixed except one, $Q_i$. This cross section should be approximately a quadratic function of $Q_i$. As already noted, this potential surface cross section for most of the vibrational coordinates should not be very different from that corresponding to the cross section through $W_0$ (or $W_1$ for that matter), and there should be little change in the frequency on complexing. However, for stretching the sacrificial bond in a sacrificial $a\sigma$-acceptor (e.g., the I—Cl or I—C bond in a molecule such as ICl or ICN) a considerable decrease may be

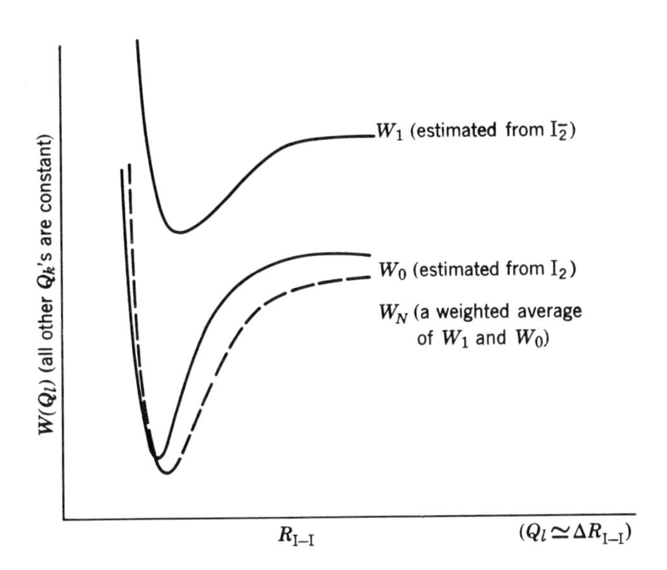


**Figure 6-1** Schematic potential curve cross sections for the I—I stretching vibration. $W_0$ is the potential curve for this bond in $\Psi_0$; $W_1$ is the potential curve for stretching the I—I bond in $\Psi_1$ estimated to be similar to that for free $I_2^-$. The potential-energy curve for free $I_2^-$ is lower than the curve for free $I_2$ (see Fig. 1 of R13), but the cross section through the potential surface for the dative state $\Psi_1$ is of course higher than that for the potential surface for $\Psi_0$. $W_N$ is a schematic reproduction of the potential curve for the I—I bond in the complex; it is taken to be mostly $W_0$, as modified a little toward $W_1$ by mixing a small amount of $\Psi_1$ with $\Psi_0$.

expected in the associated force constant for the complex as compared with the free molecule, because of the decrease in this force constant in the $A^-$ ion in the dative function, which is mixing in with $\Psi_0$ more importantly to form $\Psi_N$. This change is expected because this bond changes from a whole bond in the free molecule toward what is expected to be about a half-bond in $A^-$.

As an example consider an $I_2$ complex in which one $Q_i$ can be identified very nearly with the change in the I—I distance ($R_{I—I}$). The change in $W$(I—I) on going from $I_2$ as in $W_0$ to $I_2^-$ as in $W_1$ can be estimated from a comparison of the potential curve for $I_2^-$ (see R2 and R13) with that of free $I_2$. These potential curve cross sections for the I—I bond are shown in Fig. 6-1 for both $W_1$ and $W_0$.

The force constant for the vibration $Q_l$ in state $N$ is

$$k_i^N = \left(\frac{\partial^2 W_N}{\partial Q_i^2}\right)_{\text{eq}}.$$

Here the subscript eq means "evaluated at the equilibrium configuration of the complex." Now referring to Fig. 6-1 we see that, if $W_N$ is taken to be a weighted average of $W_0$ and $W_1$, then the second derivative at its minimum ($k_i^N$) may also reasonably be given by the weighted average of $k_l^0$, the force constant for the $l$th bond in $\Psi_0$, and $k_i^1$, the corresponding force constant in the dative structure $\Psi_1$. Hence

$$k_i^N \simeq (a^2 + abS_{01})k_i^0 + (b^2 + abS_{01})k_i^1. \tag{6-1}$$

Rearranging this equation with the help of the normalization condition $a^2 + b^2 + 2abS_{01} = 1$, we obtain

$$\frac{\Delta k}{k} \equiv \left(\frac{k_i^0 - k_i^N}{k_i^0}\right) = \left(1 - \frac{k_i^1}{k_i^0}\right)(b^2 + abS_{01}). \tag{6-2}$$

According to (6-2) no change in force constant from free molecule to complex is expected if $k_i^1 = k_i^0$, but if $k_i^1 << k_i^0$, then $\Delta k/k$ is equal to the weight of the dative structure $(b^2 + abS_{01})$ in the ground state of the complex.†

The I—I stretching-force constant for $I_2^-$ has been estimated (see R13) to be about one-fourth that for $I_2$; hence for the I—I bond in an $I_2$ complex

$$(b^2 + abS_{01}) \simeq \frac{4}{3}\frac{\Delta k}{k},$$

or, even more approximately,

$$(b^2 + abS_{01}) \simeq \frac{\Delta k}{k}. \tag{6-3}$$

For small frequency shifts in weak complexes,‡

$$\frac{2\Delta\nu}{\nu} \equiv \frac{2(\nu_0 - \nu)}{\nu_0} \simeq \frac{\Delta k}{k}, \tag{6-4}$$

† See K. Szczepaniak and A. Tramer, *J. Phys. Chem.* **71**, 3035 (1967), who give a treatment for hydrogen-bonded complexes [following P. G. Puranik and V. Kumar, *Proc. Indian Acad. Sci.*, **A58**, 29, 327 (1963)] which leads to an equation for the shift of the O–H stretching frequency that is analogous to those given by (6-2) to (6-5). Their treatment is based on the Gordy relationship between force constant and bond order. [See W. Gordy, *J. Chem. Phys.*, **14**, 305 (1946).]

‡ Since $\nu^2 = (k/m_r)/4\pi^2c^2$, $2\nu\, d\nu = (dk/m_r)/4\pi^2c^2$. Dividing the second by the first equation gives (6-4).

so that for the I—I vibration in weak complexes (6-2) reduces approximately to[†]

$$(b^2 + abS_{01}) \simeq \frac{8}{3} \frac{\Delta\nu_{\text{I—I}}}{\nu_{\text{I—I}}}, \tag{6-5}$$

## 6-3 THE WEIGHT OF THE DATIVE STRUCTURE

We find in (6-5) or in the more approximate (6-3) a relationship between a relatively easy to measure experimental quantity (here the frequency shift of the I—I stretching frequency) and the weight of the dative structure in the ground state of the complex. The relationship in (6-3) is expected to be valid for any bond for which the force constant in the dative state is expected to be much less than that for the free molecule. (In practice this means for the stretching of a $\sigma$-bond of an *aσ*-acceptor molecule.) Large frequency shifts are not expected for any of the other vibrations of the molecules participating in a complex.

It is important to note that the force constant $k_i^0$ for the no-bond structure may not be identical to the force constant for the free molecule. Thus Hanna and Williams [11] have argued that an electrostatic interaction between D (such as benzene) and A (such as $I_2$) may cause a decrease in the I—I stretching frequency. This effect would appear in the treatment described here as a difference between $k^0_{\text{I—I}}$ and the force constant for the *free* $I_2$ molecule.

We now note that the $l$th vibrational frequency of A (or D) is expected to *increase* as a result of an interaction between D and A which does *not* cause the internal force constant associated with this $l$th vibration to change. As an example, consider the effect on the X—Y stretching frequency, $\nu_{\text{XY}}$, ($= \nu_i$) of a diatomic molecule occurring as a result of an increasingly strong interaction with an atom, Z:

† We stress that (6-5) or the even more approximate relation form (6-3) that $(b^2 + abS_{01}) \simeq 2\Delta\nu/\nu$ applies only to the very restricted class of weak $I_2$ complexes. For stronger complexes $\Delta\nu/\nu$ is not simply related to $\Delta k/k$, especially if the D—I and I—I vibrations mix. In the case of stronger complexes the interaction constant $k_{12}$ that relates the D—I and the I—I stretching coordinates must also be known, so that three parameters—namely, $k_{\text{I—I}}$, $k_{\text{D—I}}$, and $k_{12}$—determine two experimental frequencies, $\nu_{\text{I—I}}$ and $\nu_{\text{D—I}}$. Furthermore, in strong complexes these motions may mix with other low-frequency motions of the donor (or acceptor), for example, with the $CH_3$ bending motions of $(CH_3)_3N \cdot I_2$. These complications require a much more complete knowledge of those complexes than is currently available; a more complete normal-coordinate treatment to obtain better estimates of the force constants for use with (6-3) might then be justified.

$$\mathrm{X}\overset{k^0_{\mathrm{X-Y}}}{\text{———}}\mathrm{Y}\overset{k_{\mathrm{YZ}}}{\text{- - -}}\mathrm{Z}.$$

Let us suppose that the interaction constant between X—Y and Y—Z bonds, $k_{12} \simeq 0$, and that the interaction causes $k_{YZ}$ to rise from zero to a small finite value but does not change the X—Y force constant, $k^0_{X-Y}$. We can easily show by a simple perturbation treatment that the X—Y stretching frequency—which is the higher-frequency stretching vibration of the new triatomic molecule (X—Y—Z)—increases. The result of this treatment is that the frequency of this X—Y stretching motion is given by:

$$\lambda_{XY} = \lambda^0_{XY} + \frac{\mu_y^2 k^0_{XY} k_{YZ}}{\lambda^0_{XY}}.$$

Here $\lambda^0_{XY} = 4\pi^2 c^2 (\nu^0_{XY})^2$, where $\nu^0_{XY}$ is the frequency in the isolated molecule; $\lambda_{XY}$ is the corresponding expression involving that frequency in (X—Y - - - Z); $\mu_y$ is the reciprocal of the mass of the Y atom; and $k^0_{XY}$ and $k_{YZ}$ are the two force constants defined earlier. Since all of the terms causing $\lambda_{XY}$ to differ from $\lambda^0_{XY}$ are positive,† these considerations show that in our complex the *l*th frequency should *increase* slightly in the no-bond structure unless an unexpectedly large electrostatic interaction cancels this expected increase. Hence the observed *decrease* in frequency for the I—Cl vibration, for example, yields a fairly strong argument that $k^N_{I-Cl}$ is lower than the force constant for the free molecule, probably because of the sacrificial *aσ* action.

Now let us examine the values of $\Delta k/k$ and the values deduced therefrom for $(b^2 + abS_{01})$ and $b$ for some halogen complexes of interest (see Table 6-1). It is difficult to test definitively the validity of the assumptions made above except by examining the results for consistency. The values of $b$ in the table are in general reasonable and not seriously inconsistent with the expectation that ICl should form stronger complexes than the homopolar halogens. The value of $(b^2 + abS_{01})$ for the strong complex $ICl_2^-$ is also in reasonable agreement with the expected value of 0.50.

For $I_2$ complexes the Table 6-1 values of $b^2 + abS_{01}$ can be checked by comparing them with values obtained from the dipole moments of the complexes. To obtain the latter the measured dipole moment ($\mu_{exp}$) of the complex is used with (2-25), together with an estimate for $\mu_1$, to obtain the value of $(b^2 + abS_{01})$.‡ Equation 2-25 reads

†If $k_{X-Y}$ in the complex (X—Y - - - Z) *decreases* from $k^0_{X-Y}$ as a result of the interaction, the expression for $\lambda_{XY}$ should be modified by replacing $\lambda^0_{XY}$ by a new lower value characteristic of the lower $k_{XY}$.

‡ But this method of estimating $(b^2 + abS_{01})$ assumes that $\mu_0$ is zero for the no-bond structure $\Psi_0$. See the discussion in Section 2-4, in the "Postscript", and in M. W. Hanna, *J. Am. Chem. Soc.*, **90**, 285 (1968).

## TABLE 6-1

**Estimates of $b^2 + abS_{01}$ from (6-3)[a]**

| Complex | $\nu_0(cm^{-1})$ | $\Delta\nu(cm^{-1})$ | $b^2 + abS_{01}$ [b] | $a$ [c] | $b$ [c] |
|---|---|---|---|---|---|
| $Bz \cdot I_2$ | 207 | 2[d] | 0.02[e] | 0.99 | 0.10 |
| $Bz \cdot Br_2$ | 312 | 7 | 0.04 | 0.97 | 0.17 |
| $Bz \cdot Cl_2$ | 541 | 16 | 0.06 | 0.96 | 0.20 |
| $Bz \cdot ICl$ | 375 | 20 | 0.11 | 0.93 | 0.28 |
| $Tol \cdot ICl$ | 375 | 19 | 0.10 | 0.94 | 0.27 |
| p-Xy $\cdot$ ICl | 375 | 21 | 0.11 | 0.93 | 0.28 |
| $Py \cdot ICl$ | 375 | 83 | 0.30[e] | 0.76 | 0.41 |
| $ICl_2^-$ | 375 | 108 | 0.57[f] | — | — |
| $Py \cdot I_2$ | 207 | 24 | 0.29[e] | 0.78 | 0.43 |
| $Me_3N \cdot I_2$ | 207 | 22 | { 0.61[g]<br>0.41[e] | — | — |

[a] From Friedrich and Person (R15), except as indicated.
[b] All values from (6-3) except for strong complexes in which the N—I and I—I stretching vibrational frequencies mix, so that $\Delta k/k$ has to be calculated from an approximate normal coordinate analysis, where an additional uncertainty arises (see footnote e). Note that the values in this column would be larger (by 8/3) if (6-5) had been used to estimate $b^2 + abS_{01}$.
[c] Estimated from the values of $b^2 + abS_{01}$, using (1-3), with the assumption that $S_{01} \simeq 0.1$ for the $b\pi \cdot a\sigma$ complexes and $S_{01} \simeq 0.4$ for $n \cdot a\sigma$ complexes.
[d] From the infrared spectrum. However, the frequency shift of $I_2$ in benzene is 6 $cm^{-1}$, as measured in the Raman spectrum, [H. Stammreich, R. Forneris and Y. Tavares, *Spectrochim. Acta.* **17**, 1173 (1961).] Hence, we believe the estimate for $b^2 + abS_{01}$ of $Bz \cdot I_2$ is a lower limit, and it is possibly the same as for $Bz \cdot Br_2$ and $Bz \cdot Cl_2$, within the experimental error.
[e] From (6-3) with force constants from J. Yarwood and W. B. Person, *J. Am. Chem. Soc.*, **90**, 594 (1968); *ibid.*, **90**, 3930 (1968).
[f] From (6-3) with force constants given by Maki and Forneris [12]. These are uncertain because of uncertainty in the interaction constant, $k_{12}$, relating the I—I stretch to the N—I stretch.
[g] From the $k$'s given by Yada, Tanaka, and Nagakura [1].

**TABLE 6-2**

**Comparison of $b^2 + abS_{01}$ Values from Infrared Frequency Shifts with Those from Dipole Moments**

| Complex | $\mu_1$ | $b^2 + abS_{01}$ | |
|---|---|---|---|
| | | [a]from $\Delta\nu/\nu$ | [b]from $\mu$ |
| $Bz \cdot I_2$ | 24.0 | 0.02[c] | [d]0.075 |
| $Py \cdot I_2$ | 17.7 | 0.29 | [d]0.25 |
| $Et_3N \cdot I_2$ | 17.7 | (0.4) | [e]0.28–0.35 |
| $Me_3N \cdot I_2$ | 17.7 | [f]0.41 | [e]0.33 |

[a] From Table 6-1.
[b] See also estimates for related aliphatic amine $\cdot I_2$ complexes given by S. Kobinata and S. Nagakura, *J. Am. Chem. Soc.*, **88**, 3905 (1966).
[c] This value is a lower limit. See footnotes *b* and *d* of Table 6-1.
[d] The value of $\mu_{exp} = 1.80$ D for $Bz \cdot I_2$, 4.5 D for $Py \cdot I_2$, from Kortüm and Walz [13]. Mulliken had earlier estimated $\mu$ for $Bz \cdot I_2$ as $\sim 0.72$ (R3), but this estimate was based on data [14] from iodine dissolved in pure benzene and so is probably not very reliable. However there is some question about the reliability of the value for $Bz \cdot I_2$ reported by Kortüm and Walz [13], since they used a larger value for $K_x$ of that complex than is presently accepted. See also the footnotes on pp. 67 and 68,
[e] Lower limit from $\mu = 5.6$ D, as reported for triethylamine in *n*-heptane by P. Boule, *J. Am. Chem. Soc.*, **90**, 517 (1968). Also, see A. J. Hamilton and L. E. Sutton, *Chem. Comm.*, 460 (1968), who report $\mu = 6.9$ for $Et_3N \cdot I_2$ and $\mu = 6.5$ for $Me_3N \cdot I_2$. Here $a^2 \mu_0$ is taken to be 0.6, based on $\mu$ for $R_3N$ from A. L. McClellan, *Tables of Experimental Dipole Moments* (Freeman, San Francisco, 1963).
[f] This value seems the more probable of the two listed in Table 6-1.

$$\mu_{exp} \simeq \mu_N = a^2\mu_0 + b^2\mu_1 + 2ab\mu_{01}.$$

From Section 3-4 or from footnote 21 of R3 we may estimate $2\mu_{01}$ to be approximately equal to $S_{01}\mu_1$, whereas from $\mu_1 \simeq eR_{D-A}$, with $R_{D-A} \simeq 5.0$ Å for $Bz \cdot I_2$,† $\mu_1$ for that complex is estimated as about 24 D. The value of $\mu_1$ for the amine $\cdot I_2$ complexes is estimated similarly and listed

†This value for $R_{D-A}$ is appropriate for the axial model of the $Bz \cdot I_2$ complex. For the resting model $R_{D-A} \simeq 3.4$ (see Section 9-2), and $\mu_1 \simeq 16$, so that $b^2 + abS_{01}$ is about 0.11 from $\mu = 1.80$ (see Table 6-2).

in Table 6-2. To the few comparisons in Table 6-2 we should perhaps also add the values of $b^2 + abS_{01}$ computed from the force constants for the symmetrical trihalide ions [12,16]; these agree with the expected value (0.5) about as well as is shown for $ICl_2^-$ in Table 6-1. On the whole the comparison gives empirical support to the relationship between $(b^2 + abS_{01})$ and the frequency shifts for $a\sigma$-acceptors.

## 6-4 INTENSITY CHANGES

As a result of the lower symmetry of the donor in the complex some vibrations that are forbidden in the free donor are expected to mix somewhat in the complexed donor with allowed vibrations of the free donor. These frequencies should then appear in the spectrum of the complex with appreciable intensity. For such changes due to mixing of normal coordinates in the complex the sum of intensities for the vibrations that mix should be very nearly the same as the intensity sum for these bands in the free molecule.

In addition there may be changes in intensities because the vibrations of D or A may have different intensities for $\Psi_0$ and $\Psi_1$ because of the different electronic distributions in D or A in these two structures. The intensities in the complex for such bands might then perhaps be weighted averages of those for the same bands in the no-bond and the dative structures. Lack of general information about vibrational intensities in $\Psi_1$ prevents further pursuit of this idea.

There is one effect, however, that can result in a major change in intensity on complexing and that was first suggested by Ferguson and Matsen [8–10]. This effect is a vibronic contribution to the infrared intensities resulting from the fact that $\Psi_N^{el}$ is in general a function of nuclear configuration; hence it varies during vibrations. The resulting shift of electronic charge with configuration makes an important contribution to the intensity of certain vibrations.

Before reviewing this effect, however, it may be worthwhile noting just what intensity changes are observed for the benzene $\cdot I_2$ complex as a specific example. We have already stated that only large effects can be seen, because the large excess of *uncomplexed* benzene in solutions of this complex interferes with seeing the spectrum of the *complexed* benzene. In all the observed vibrational spectrum of the complex in solution (from 4000 to 150 $cm^{-1}$) only three changes are found:

1. The $a_{1g}$ vibration at about 990 $cm^{-1}$ (inactive in the free molecule) appears weakly; Ferguson and Chang estimate that its intensity in the complex is 190 cm/mmole [17].

2. The only $e_{1g}$ vibration of benzene, at about 850 $cm^{-1}$ and inactive in the free molecule, appears in the complex with an estimated intensity of 640 cm/mmole [17].

3. The inactive I—I stretching vibration appears at 205 $cm^{-1}$ in the complex with an estimated intensity of 160 cm/mmole [18]. We note especially that the second $a_{1g}$ vibration of benzene, at 3060 $cm^{-1}$, cannot be observed in the spectrum of the complex because of the overlap with the allowed C—H stretching frequencies. There are apparently no appreciable frequency shifts on complexing, which indicates that relatively little mixing of normal coordinates occurs in the complexed benzene. The heavy mass of the iodine atoms and the resulting low amplitude of motion in the I—I stretching vibration means that the small observed intensity (or small value of $\partial\mu/\partial Q_{I—I}$) still corresponds to a considerable dipole-moment change, $\partial\mu/\partial R_{I—I}$ [since $\partial\mu/\partial R_{I—I} \simeq \sqrt{m_r}(\partial\mu/\partial Q_{I—I})$, as we shall see later]. We may keep these observations in mind as we search for reasons why any intensification is found.†

Now let us go on to review the vibronic contribution to the infrared intensity changes observed on complexing. (See also R15.) The basic change from the usual intensity theory (such as that reviewed in Chapter 3) is that we must remember that the electronic wavefunction changes when the nuclear configuration changes. Specifically, for donor-acceptor complexes the coefficients $a$ and $b$ of (2-2) are functions of the nuclear coordinates $Q$:

$$\Psi_N^{el}(Q,\mathbf{r}_i) = a(Q)\Psi_0(Q,\mathbf{r}_i) + b(Q)\Psi_1(Q,\mathbf{r}_i). \tag{6-6}$$

Here the nuclear configuration, denoted by $Q$, refers to the entire set of normal coordinates for the complex, whereas $\mathbf{r}_i$ refers to the electronic positions. The transition moment for the transitions between the ground and first excited vibrational levels in the same electronic state is given by equation (3-5), with $m = n$:

$$\mu_N^{0,1} = \int \chi_0^N(Q_i)\, \chi_1^N(Q_i)\, \mu_N(Q_i)\, dQ_i \tag{6-7}$$

Here the $i$th vibrational coordinate is written as $Q_i$. The operator $\mu_N^{el}$ is just the dipole moment integral defined by equation (2-25), using the wavefunction of 6-6, to show specifically the dependence of the coefficients on $Q_i$. Friedrich and Person (R15) have defined a term $\mathbf{M}_{0,1}$, which

---

† Another indication of the kind and magnitude of the changes that are due to complexing can be seen in the spectrum of solid benzene · $Br_2$. [See W. B. Person, C. F. Cook, and H. B. Friedrich, *J. Chem. Phys.*, **46**, 2521 (1967).] Several additional small changes in the benzene vibrations are found in the spectrum of this solid, but its structure in the solid is probably different than that for the 1:1 complex in solution (see Section 5-6).

corresponds to the dipole moment derivative, $\partial\mu/\partial Q_i$, in the usual treatment of intensities. Using harmonic oscillator functions for the $\chi_i^N(Q_i)$, they show that integration of (6-7) gives:

$$\left.\begin{aligned} \mu_N^{0,1} &= \mathbf{M}_{0,1}\sqrt{\frac{b}{8\pi^2 c\omega_i}} \\ \text{with} \\ \mathbf{M}_{01} &= \frac{\partial\mu_N}{\partial Q_i} + [2a\frac{\partial a}{\partial Q_i}\mu_0 + 2b\frac{\partial b}{\partial Q_i}\mu_1 + (2a\frac{\partial b}{\partial Q_i} + 2b\frac{\partial a}{\partial Q_i})\,\mu_{01}]. \end{aligned}\right\} \quad (6\text{-}8)$$

This derivative assumes that $a$ and $b$ do not depend strongly on any of the nuclear coordinates except $Q_i$ (which might for example,be the I—I stretching coordinate in benzene $\cdot\, I_2$). In (6-8) the first term $(\partial\mu_N/\partial Q_i)$ is the one responsible for the vibrational intensity as usually described; the remaining terms are the vibronic contributions that arise because of the changes with $Q_i$ of the coefficients $a$ and $b$ in (6-6). The values of $a$ and $b$ in (6-8) are those at equilibrium.

In (6-8), $\omega_i$ is the wavenumber of the $i$th vibration and $\partial\mu/\partial Q_i$ is essentially the intrinsic contribution to the intensity from the free molecules to which $Q_i$ approximately belongs.†

Utilizing the relationships $\mu_{01} \simeq \frac{1}{2}S_{01}\,\mu_1 + S_{01}\,\mu_0$ (See Section 3-4 and Eq. 19, and footnote 21 of R3) and $a^2 + b^2 + 2abS_{01} = 1$, given earlier, we may rearrange (6-8) to obtain

$$\begin{aligned} \mathbf{M}_{01} = \frac{\partial\mu_N}{\partial Q_i} &+ [\frac{S_{01}a^2 + 2ab + S_{01}b^2}{a + S_{01}b}]\,[\frac{\partial b}{\partial Q_i}]\,[\mu_1 - \mu_0] \\ &+ [\frac{S_{01}}{a + S_{01}b}]\,[\frac{\partial b}{\partial Q_i}]\,[a^2 - b^2]\,\mu_0 \end{aligned} \quad (6\text{-}9)$$

For weak complexes $S_{01} \simeq 0$, and we see that (6-9) reduces to:

$$\mathbf{M}_{01} \simeq \frac{\partial\mu_N}{\partial Q_i} + 2b\frac{\partial b}{\partial Q_i}[\mu_1 - \mu_0]. \quad (6\text{-}10a)$$

Furthermore, if $S_{01} \simeq 0$, $\mu_1 - \mu_0 \simeq \mu_{VN}/ab$ [from equation (3-11)]. Hence,

$$\mathbf{M}_{01} \simeq \frac{\partial\mu_N}{\partial Q_i} + \frac{2}{a}\frac{\partial b}{\partial Q_i}\mu_{VN}, \quad (6\text{-}10b)$$

† More accurately, this quantity is $a^2(\partial\mu_0/\partial Q_i)_{a,b} + b^2(\partial\mu_1/\partial Q_i)_{a,b} + 2abS_{01}\,\partial\mu_{01}/\partial Q_i$. If, for example, $Q_i$ is essentially $R_{I-Cl}$, $\partial\mu_0/\partial Q_i$ and $\partial\mu_1/\partial Q_i$ would refer to the dipole derivatives within ICl in $\Psi_0$ and within $ICl^-$ in $\Psi_1$.

which is the equation derived by Ferguson and Matsen [8].

Now why should $b$ depend on $Q_i$ so that $\partial b/\partial Q_i$ is not zero? There are several possible reasons. Using (2-23) for weak complexes, namely $\rho \equiv b/a = -\beta_0/\Delta$, we have

$$\frac{\partial b}{\partial Q_i} = \frac{b}{a}\left(\frac{\partial a}{\partial Q_i}\right) - \frac{a}{\Delta}\left(\frac{\partial \beta_0}{\partial Q_i}\right) - \frac{b}{\Delta}\left(\frac{\partial \Delta}{\partial Q_i}\right). \tag{6-11}$$

Rearranging and using the normalization condition, we obtain

$$\frac{1}{a^2}\left(\frac{\partial b}{\partial Q_i}\right) = -\frac{a}{\Delta}\left(\frac{\partial \beta_0}{\partial Q_i}\right) - \frac{b}{\Delta}\left(\frac{\Delta \partial}{\partial Q_i}\right). \tag{6-12}$$

Hence the condition that $\partial b/\partial Q_i \neq 0$ requires that either $\partial\beta_0/\partial Q_i$ or $\partial\Delta/\partial Q_i$ (or both) be nonzero. Let us consider each possibility.

Friedrich and Person (R15) have argued that $\partial S_{\mathrm{da}^-}/\partial Q_i$ and hence $\partial\beta_0/\partial Q_i$ should not be expected to be much different from zero except for the new D—A vibrations. When free (D or A) vibrates, its center of mass does not change; if these normal coordinates are not changed very much in weak complexes, then there may be little relative motion of D against A and so $S_{\mathrm{da}^-}$ may not be a function of $Q_i$. Furthermore, it might be expected that vibrations of D that tend to decrease $R_{\mathrm{DA}}$ may merely push A away from D instead of increasing $S_{\mathrm{da}^-}$ as they would for a fixed D—A distance, since the repulsion energy between D and A rises rapidly as $R_{\mathrm{DA}}$ decreases.

On the other hand Ferguson [10] has attributed the intensification of the $e_{1g}$ benzene vibration at 850 cm$^{-1}$ to the change in overlap between D and A during this vibration. This vibration results in a rotation of the benzene plane against the iodine axis and is one of the few vibrations of the complexed benzene for which $S_{\mathrm{da}^-}$ may change importantly. (One might expect the $a_{2u}$ vibration at 680 cm$^{-1}$ to be another, but no evidence has been put forward for an intensity change in this strongly allowed benzene vibration.) Since there seems to be no reason for $\Delta$ to be a function of $Q_i$ for the $e_{1g}$ vibration, we must accept Ferguson's hypothesis and recognize the importance of $\partial S_{\mathrm{da}^-}/\partial Q_i$ and hence $\partial\beta_0/\partial Q_i$ for enhancement of this vibration.†

Three terms contribute to $\Delta$; that is,

$$\Delta = I_{\mathrm{D}}^{\mathrm{V}} - E_{\mathrm{A}}^{\mathrm{V}} + (G_1 - G_0). \tag{6-13}$$

† In an independent treatment of this vibronic intensity effect—specifically applied to hydrogen-bonded complexes—K. Szczepaniak and A. Tramer, *J. Phys. Chem.*, **71**, 3035 (1967) give equations for the change in overlap integral with $Q_i$. They find that this change in $S$ is important in determining the intensity changes in hydrogen bonding.

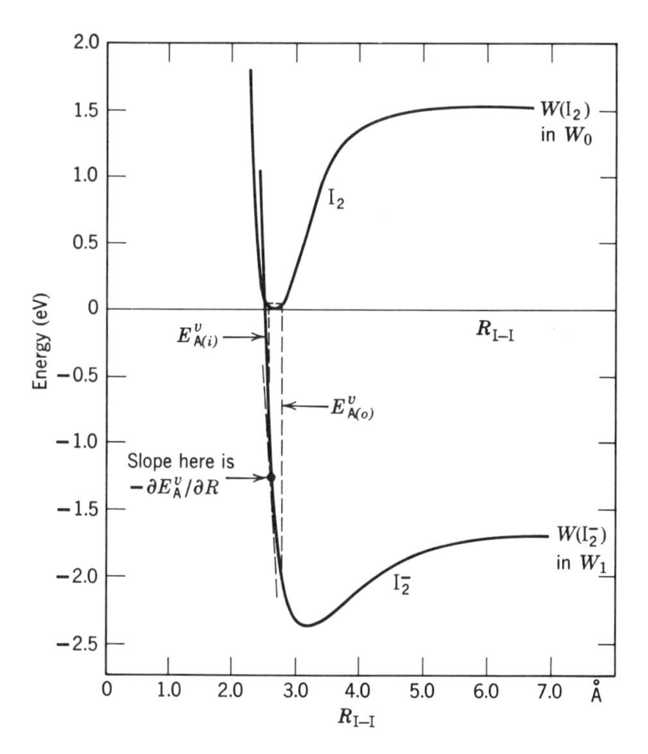


**Figure 6-2** Schematic potential curves for $I_2$ and $I_2^-$, showing the change in vertical electron affinity as the I—I bond is stretched. (See Fig. 1 of R13.)

These quantities are described in detail in Chapter 9. We merely state here that we expect $G_1 - G_0$ to vary with $Q_i$ only for the same reasons that $S_{da^-}$ varies with $Q_i$; hence we argue that $G_1 - G_0$ does not vary significantly except possibly for the $e_{1g}$ and $a_{2u}$ vibrations of benzene and for the D—A vibrations (in $Bz \cdot X_2$ complexes).

The quantity $\partial I^v_D/\partial Q_i$ can differ from zero only for totally symmetric vibrations of the donor since these are the only ones that vary the size of the donor molecule and hence affect its ionization potential.[†] The

[†] However, it is conceivable that electronic rearrangements associated with other kinds of vibrations (e.g., an out-of-plane bending vibration of benzene) could change the frontier orbital so that its ionization energy changes. Probably the statement here concerning the changes in the $a_{1g}$ benzene vibrations is valid but the indication that other vibrations do not affect $I^V_D$ is oversimplified.

observed enhancement of the intensities of the $a_{1g}$ vibrations of benzene in the complex can thus be accounted for qualitatively in terms of $\partial I_D^v/\partial Q_i$.

Finally consider $\partial E_A^v/\partial Q_i$. The vertical electron affinity of $I_2$ is expected to vary strongly as the I—I distance changes in the vibration (and a similar effect is expected in analogous vibrations of other $a\sigma$-acceptors). We refer again to the potential curves for $I_2$ and $I_2^-$ redrawn in Fig. 6-2 in order to show $\partial E_A^v/\partial Q$. Here we have indicated the value $E_{A(i)}^v$ of the vertical electron affinity at the inner classical turning point of a vibration, and also its value, $E_{A(o)}^v$, at the outer turning point. The change in vertical electron affinity is thus the negative slope of the tangent to the lower potential curve evaluated at the I—I equilibrium distance for free $I_2$. As Fig. 6-2 shows, this derivative may be large. However, we note that $E_A^v$ is expected to depend so strongly on $R_{I-I}$ only because the $I_2^-$ ion has its equilibrium configuration at such a large $R_{I-I}$ value, so that the vertical energy change as an electron is added to $I_2$ brings us (in Fig. 6-2) to the $I_2^-$ potential curve on the steep repulsive side of that curve.†

Let us assume that the term $\partial E_A^v/\partial Q_i$, where $Q_i$ is essentially $R_{X-Y}$, is the only term that contributes significantly to $\partial b/\partial Q_i$ for the X—Y stretching vibration of an XY sacrificial $a\sigma$-acceptor. Then substitution of (6-12) and (6-13) into (6-11) gives

$$\frac{\partial b}{\partial Q_i} = \frac{a^2 b}{\Delta}\frac{\partial E_A^v}{\partial Q_i} = \frac{a^2 b}{\sqrt{m_r}\,\Delta}\frac{\partial E_A^v}{\partial R_i}. \tag{6-14}$$

Substitution of (6-14) into (6-10a) gives the approximate equation:

$$M_{0,1} \simeq \frac{\partial \mu_N}{\partial Q_i} + \frac{2a^2b^2}{\sqrt{m_r}\,\Delta}\frac{\partial E_A^v}{\partial R_i}|\mu_1 - \mu_0| \equiv \frac{\partial \mu_N}{\partial Q_i} + M_d \tag{6-15}$$

(This identity defines the *delocalization moment*, $M_d$.)
We have written (6-15) in terms of the magnitudes of the vectors of (6-10). Here $m_r$ is the reduced mass in amu for the X—Y stretching vibration. The term $(\partial\mu_N/\partial Q_i)$ is approximately equal to $(1.1936\,\sqrt{B}/\sqrt{N_0}) \times 10^{10}$ D/Å, where $B$ is the intensity in cm/mmole of the band in an inert solvent, and $N_0$ is Avogadro's number.

† This situation, which is characteristic for $I_2$ complexes, may be expected to be more or less characteristic of the bond stretching vibration of all $a\sigma$-acceptors. However, it should not be assumed that $E_A^v$ will be so strongly dependent on $Q_i$ for all acceptors; for example, in $a\pi$ acceptors the acceptor orbital may not be very strongly antibonding, so that $\partial E_A^v/\partial Q_i$ may be much smaller than for $a\sigma$ acceptors. Similar remarks hold for other vibrations of a polyatomic acceptor; for example, $\partial E_A^v/\partial Q_k$ is expected to be small in ICN complexes for the case that $Q_k$ is the CN stretching vibration of ICN.

An even better estimate for the intensity can be obtained by substituting the value of $\partial b/\partial Q_i$ obtained using (6-14) back into the more exact equation (6-9). The delocalization moment $M_d$ is then defined as the collection of all terms containing $\partial b/\partial Q_i$ in (6-9).

We may now estimate $M_d$ using equation (6-9) with the parameters evaluated as described below. First, the reduced mass $m_r$ appears because we have assumed that $Q_i = \sqrt{m_r}(\Delta R_i)$ where $\Delta R_i$ is the change in the bond length of the $i$th bond (e.g., the I—I bond in $I_2$ complexes). It may be simpler, as discussed below, to define $M_d'$ $(= \sqrt{m_r}\, M_d)$ and use $\partial \mu_N/\partial R_i$ and $\partial b/\partial R_i$ in (6-9). However, let us compute here $M_d$ in debyes/[Å(amu)$^{1/2}$]. The values of $a$ and $b$ needed in (6-9) can be estimated from the shift in frequency of the $i$th vibrational band, as described in Section 6-2, and tabulated in Table 6-1. The value of $S_{01}$ is estimated to be about 0.1 for weak $b\pi \cdot a\sigma$ complexes, and about 0.4 for the $n \cdot a\sigma$ complexes, (See Chapters 9 and 10.) Values for $\Delta$ in eV [equal to $W_1 - W_0$, see (2-13)] are approximately equal to $h\nu_{CT}$ for typical weak complexes; for stronger $n \cdot a\sigma$ complexes, (6-13) is not exact, and estimation of $\partial b/\partial Q_i$ from (6-14) is not strictly correct. We use it here, however, for the Py $\cdot$ $I_2$ complex to show the quality of agreement obtained between calculated and experimental intensity values even for that strong complex.

In previous papers (R15, for example) the value of $|\mu_1 - \mu_0|$ has been estimated to be equal to $\mu_{VN}/ab$. However, this estimate may be badly in error for benzene $\cdot$ halogen and analogous complexes, if axial geometry for these is correct. (See Chapters 5 and 11.) In that case, it is better to estimate $|\mu_1 - \mu_0|$ as described above in the discussion of dipole moments, using, for example, the values from Table 6-2. If the axial configuration is, in fact, correct for the benzene $\cdot$ halogen type of complexes, the CT transition is not to the dative state which mixes to stabilize the ground state; $\Delta$ is not equal to $h\nu_{CT}$. For $b\pi \cdot X_2$ complexes with axial configuration, we have assumed $\Delta$ (in $eV$) $\simeq h\nu_{CT} + 2$. (See Section 10-3.)

The values of $\partial E_A^v/\partial R$ (in $eV$/Å) can be obtained for halogen complexes for the X—Y stretch from the potential curves estimated by Person (R13), but are, of course uncertain. The sensitivity of $\partial E_A^v/\partial R$ to the estimates of the parameters determining the potential curve for $I_2^-$ was examined by Friedrich and Person (R15) and found to vary by a factor of 2 for quite reasonable changes in the values of the parameters determining the $I_2^-$ curve. However, using their preferred values (e.g., $\partial E_A^v/\partial R = +4.25$ $eV$/Å, for $I_2^-$) we find the values for the vibronic term $M_d$ to be as shown in Table 6-3 for some halogen complexes. For the most part, we see that the estimated values of $M_{0,1}$ in Table 6-3 agree with the experimental

**TABLE 6-3**

**Comparison of the estimated $M_{0,1}$ for the X—Y stretching vibrations in halogen complexes with the observed values.[a,b]**

| Complex | $(\partial \mu_N/\partial Q_i)$ | Calculated[c] | | Observed | Calc.[d] | Obs. |
|---|---|---|---|---|---|---|
| | | $M_d$ | $M_{0,1}$ | $M_{0,1}$ | $M'_d$ | $M'_d$ |
| $Bz \cdot I_2$ | 0 | 0.06 | 0.06 | 0.19 | 0.5 | 1.5 |
| $Bz \cdot Br_2$ | 0 | 0.17 | 0.17 | 0.31 | 1.0 | 1.9 |
| $Bz \cdot Cl_2$ | 0 | 0.38 | 0.38 | 0.19 | 1.6 | 0.8 |
| $Bz \cdot ICl$ | 0.51 | 0.57 | 1.08[e] | 0.67 | 3.0 | 1.0 |
| $Tol \cdot ICl$ | 0.51 | 0.53 | 1.04[e] | 0.71 | 2.9 | 1.2 |
| $p$-Xy · ICl | 0.51 | 0.58 | 1.09[e] | 0.60 | 3.1 | 0.7 |
| $Py \cdot I_2$ | 0 | 0.97 | 0.97 | 0.72 | 7.7 | 9.5 ± 0.5[f] |

[a] Units are (D/[Å (amu)$^{1/2}$].

[b] The observed values in this table are computed from R15, with corrections, except for $Bz \cdot I_2$ and $Py \cdot I_2$, which are taken from J. Yarwood and W. B. Person, *J. Am. Chem. Soc.*, **90**, 594 (1968).

[c] Calculated from (6-9) and (6-14), by using parameters evaluated as described in the text. We have taken $a$ and $b$ from Table 6-1; $\partial E_A^v/\partial R$ from R15; $\Delta$ from R15, with $\Delta = h\nu_{CT} + 2$, except for $Py \cdot I_2$ where $\Delta = 9.24 - 7.3$ $eV$, from Table 10-1; $|\mu_1 - \mu_0|$ is taken to be 24 D. for $Bz \cdot I_2$, 22 D. for $b\pi \cdot XY$ complexes, and 17 D. for $Py \cdot I_2$; and $S_{01} \simeq 0.1$ for $b\pi \cdot X—Y$ complexes and 0.4 for $Py \cdot I_2$. Values of $\sqrt{m_r}$ are 7.97, 6.32, 4.21, and 5.27 amu for $I_2$, $Br_2$, $Cl_2$, and ICl complexes, respectively.

[d] Defined by (6-16), with $M'_d = \sqrt{m_r}\, M_d$ for diatomic molecules, but see text for polyatomic molecules.

[e] If $(\partial \mu_N/\partial Q_i)$ and $M_d$ have opposite signs, these numbers would be about 0.1 D/Å (amu)$^{1/2}$ in obvious disagreement with the experimental observation of an increase in intensity from the free molecule.

[f] This uncertainty is due to the lack of certainty in the values of the normal coordinate coefficients (the $L_{ij}^{-1}$ values). The value here (9.5 ± 0.5) was computed by Yarwood and Person [18] and it includes the range of $M'_d$ values corresponding to normal coordinates arising from reasonable assumptions about the value of the interaction force constant, $k_{12}$.

values within a factor of 3, which may be within the uncertainty of estimating $\partial E_A^v/\partial R$.

We note that the entries in Table 6-3 are different from the calculated values of $M_{0,1}$ in R15 and in Yarwood and Person [18]. These differences are caused by several factors; the earlier papers made very questionable[†] use of equation (6-10b) to estimate $M_d$, occasionally made incorrect conversion from $\partial\mu/\partial Q$ to $\partial\mu/\partial R$, used higher estimates of $b$ for some of the $b\pi \cdot X_2$ complexes (which we now believe to have been in error), and applied (6-14) erroneously to the strong pyridine $\cdot I_2$ complex. (However, we note that application of this equation to that complex does lead to a predicted value of $M_{0,1}$ in Table 6-3 which is in very good but probably fortuitous agreement with the experimental value.)

The differences from the previous treatments (R15 and [18]) are such that they have reduced the extent of agreement between the experimental intensities and the calculated ones that was believed to be indicated by tables corresponding to Table 6-3. In particular, the low intensity calculated for the $Bz \cdot I_2$ complex in Table 6-3 suggests that perhaps the correction term suggested by Hanna and Williams [11], which will appear here as a correction to $\partial\mu_N/\partial Q_i$, is indeed required.[‡] However, for the stronger complexes, the calculated intensity is too large, as would be expected if $\partial E_A^v/\partial R$ is overestimated.

Thus, although the agreement between calculated and experimental values in Table 6-3 is not convincing evidence that the theoretical discussion presented above is quantitatively correct, we believe the consistency is sufficient to establish the general validity of many of the ideas involved, with directions indicated for further modification of these ideas as needed. At least a major portion of the observed intensification of the X—Y vibrations must come from the vibronic mechanism presented here.[††]

Let us now consider for just a moment longer the question of conversion from normal coordinate, $Q_i$, to stretching coordinate, $\Delta R_{I-I}$. In the preceeding treatment we have assumed (in evaluating $\partial E_A^v/\partial Q_i$) that

---

[†] Questionable, because its use does not apply to the axial model for $Bz \cdot I_2$, although it does apply if the resting model is the correct configuration for this complex.

[‡] However, since $\partial\mu_N/\partial Q_i$ is determined for example, for ICl complexes, from the intensity observed for $\nu_{I-Cl}$ in an "inert" solvent, it is possible that some of the electrostatic contribution suggested by Hanna and Williams [11] has already been included empirically. Also this low value for $M_{0,1}$ in Table 6-3 for $Bz \cdot I_2$ is based on the value of $b^2 + abS_{01}$ from Table 6-1, which is probably a lower limit. See footnotes b and d of Table 6-1.

[††] In this discussion we have not tried to include the effect of solvent either on the intensities ($M_{0,1}$) or on the parameters of the theory (such as $\partial E_A^v/\partial R$). We do not believe that this neglect introduces an error that is any larger than those due to the other approximations in this treatment.

$Q_i = \sqrt{m_r}\,(\Delta R_{\mathrm{I—I}})$. If some mixing of coordinates occurs so that for example, $Q_i = L_{11}^{-1}\,(\Delta R_{\mathrm{I—I}}) + L_{11}^{-1}\,(\Delta R_{\mathrm{DI}})$,† some modification of this treatment is necessary. The easiest way to effect this is to modify (6-9) or (6-15) so as to define $\partial\mu/\partial R_{\mathrm{I—I}}$ for the bond in the complex by

$$\partial\mu/\partial R_{\mathrm{I—I}} = \left[\frac{\partial\mu_N}{\partial R}\right]_0 + 2a^2b^2\left[\frac{(\partial E_{\mathrm{A}}^{\mathrm{v}}/\partial R)}{\Delta}\right][\mu_1 - \mu_0] = (\partial\mu/\partial R)_0 + M_d'. \quad (6\text{-}16)$$

Here $R$ has replaced $R_{\mathrm{I—I}}$ for convenience.

The value of $\partial\mu/\partial R_{\mathrm{I—I}}$ is obtained from the experimental intensity by using

$$\frac{\partial\mu}{\partial R_{\mathrm{I—I}}} = \sum_i L_{ij}^{-1}\,\frac{\partial\mu}{\partial Q_i},$$

with a corresponding equation defining $(\partial\mu_N/\partial R)_0$ for the uncomplexed A (or D) vibration. Here the values of $\partial\mu/\partial Q_i$ are obtained from the experimental intensity values ($B_i$) as outlined above. In this procedure, then, we must know the normal coordinate transformation (the values of $L_{ij}^{-1}$); and so the analysis becomes more complicated for strong complexes, just as it did for the force constants.

The values of $M_d'$ have also been included in Table 6-3. For some purposes it is desirable to look at these values, which do not include the mass-dependence as do the $M_{0,1}$ values. For example, the observation of almost identical intensities (values of B) for the X—X stretching vibrations of $\mathrm{Bz\cdot I_2}$ and $\mathrm{Bz\cdot Cl_2}$ complexes corresponds to a much larger observed vibronic effect for the former than for the latter complex, as indicated correctly by the larger observed value for $M_d'$.

Finally, note that the mechanism for intensification does not depend on the orientation of A with respect to D; $\partial E_{\mathrm{A}}^{\mathrm{v}}/\partial R$ is predicted to be different from zero whether the $\mathrm{I_2}$ is parallel or perpendicular to the benzene ring. Hence, contrary to the original naive interpretation of the intensification, the observed intensification does not provide us with any information about orientation.

## 6-5 REFERENCES


[1] H. Yada, J. Tanaka, and S. Nagakura, *J. Mol. Spectry.*, **9**, 461 (1962).

[2] D. L. Glusker, H. W. Thompson, and R. S. Mulliken, *J. Chem. Phys.*, **21**, 1407 (1953).


† Here $L_{11_j}^{-1}$ is one of the coefficients of the normal coordinate transformation matrix ($\mathbf{Q} = \mathbf{L}^{-1}\mathbf{R}$). See E. B. Wilson, Jr., J. C. Decius, and P. C. Cross, *Molecular Vibrations*, McGraw-Hill, New York, 1955.

[3] D. L. Glusker and H. W. Thompson, *J. Chem. Soc.*, 471 (1955).
[4] J. Collin and L. D'Or, *J. Chem. Phys.*, **23**, 397 (1955).
[5] For example, see R. S. Mulliken, *J. Chem. Phys.*, **23**, 397 (1955).
[6] W. B. Person, R. E. Humphrey, W. A. Deskin, and A. I. Popov, *J. Am. Chem. Soc.*, **80**, 2049 (1958).
[7] W. B. Person, R. E. Humphrey, and A. I. Popov, *J. Am. Chem. Soc.*, **81**, 273 (1959).
[8] E. E. Ferguson and F. A. Matsen, *J. Chem. Phys.*, **29**, 105 (1958).
[9] E. E. Ferguson and F. A. Matsen, *J. Am. Chem. Soc.*, **82**, 3268 (1960).
[10] E. E. Ferguson, *J. chim. phys.*, **61**, 257 (1964).
[11] M. W. Hanna and D. E. Williams, *J. Am. Chem. Soc.*, **90**, 5358 (1968).
[12] A. G. Maki and R. Forneris, *Spectrochim. Acta*, **23A**, 867 (1967).
[13] G. Kortüm and H. Walz, *Z. Electrochem.*, **57**, 73 (1953).
[14] F. Fairbrother, *J. Chem. Soc.*, 1051 (1948).
[15] K. Toyoda and W. B. Person, *J. Am. Chem. Soc.*, **88**, 1629 (1966).
[16] W. B. Person, G. R. Anderson, J. N. Fordemwalt, H. Stammreich, and R. Forneris, *J. Chem. Phys.*, **35**, 908 (1961).
[17] E. E. Ferguson and I. Y. Chang, *J. Chem. Phys.*, **34**, 628 (1961).
[18] Estimated from the results of E. K. Plyler and R. S. Mulliken, *J. Am. Chem. Soc.*, **81**, 823 (1959); see also the measured value given by J. Yarwood and W. B. Person, *J. Am. Chem. Soc.*, **90**, 594 (1968).

CHAPTER 7

# *Experimental Determination of K and $\epsilon$; Gas Phase Studies and Their Comparison with Solution Data*

## 7-1 SPECTROPHOTOMETRIC METHODS

Now let us consider some practical questions that concern the methods of obtaining formation constants ($K$) and other thermodynamic properties of complexes and the related problem of obtaining molar absorptivities $\epsilon$ and hence the intensities of CT and other bands. (For a general reference, see Briegleb [1].)

The formation constant is defined for the reaction

$$D + A \rightleftharpoons C,$$

by the expression†

$$K = \frac{c_C}{c_D c_A}. \tag{7-1}$$

Here, for example, $c_D$ is the concentration (in moles per liter) of the donor that exists in the solution *at equilibrium*. If we take solutions with a fixed total concentration $c_A^0$ of A but with increasing concentrations of D, (7-1) implies that the concentration $c_C$ of the complex C increases as shown in Fig. 7-1. (The absorption observed may be that of

---

† $K$ can be defined in terms of equilibrium concentrations in moles per liter, in which case we shall call it $K_c$; or in terms of concentrations in mole-fraction units, in which case we call it $K_x$. (See Section 7-2.)

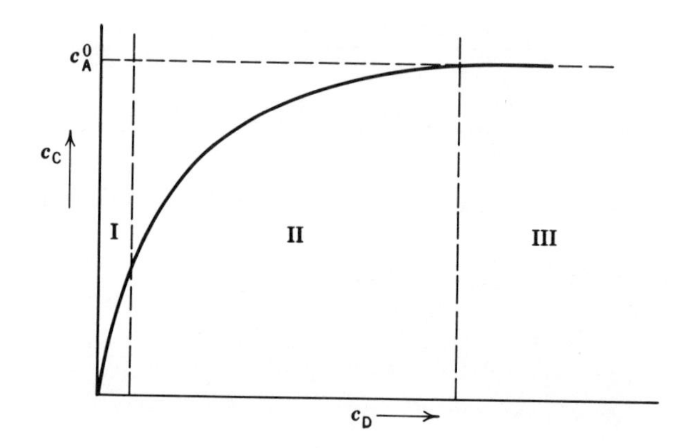


**Figure 7-1** The concentration of complex as a function of donor concentration for a fixed total acceptor concentration $c_A^0$. Region I: $c_C$ is approximately a linear function of $c_D$. Region III: Saturation has been reached, and $c_C$ is constant and equal to $c_A^0$.

the CT band or it could be absorption for a locally excited band of D or A, shifted in the complex.) Since the *absorbance* $A_C$ due to C in a region in which the complex absorbs is given by Beer's law,

$$A_C \equiv \log \frac{I_0}{I} = \epsilon c_C l.$$

Here $\epsilon$ is the molar absorptivity of C and $l$ is the length in centimeters of the absorbing path. The value of $A_C$ must increase as $c_D$ increases, as is seen from Fig. 7-1. Depending on how large $K$ is, $c_C$ at *maximum possible* donor concentration (namely, pure-liquid donor) may closely approach $c_A^0$ (region III of Fig. 7-1, $K > 1.0$) or be limited to region II ($K \simeq 0.1$) or even to region I ($K < 0.01$). (In the latter two cases the $c_D$ scale in Fig. 7-1 would of course be modified.)

As a specific example, the visible spectrum [2] of the complex between pyridine-*N*-oxide and $I_2$ is shown in Fig. 7-2. This figure illustrates the increasing absorbance near 450 m$\mu$ of the complexed $I_2$ and the decreasing absorbance near 520 m$\mu$ due to the familiar locally excited band of uncomplexed $I_2$ as the concentration of donor is increased, with an

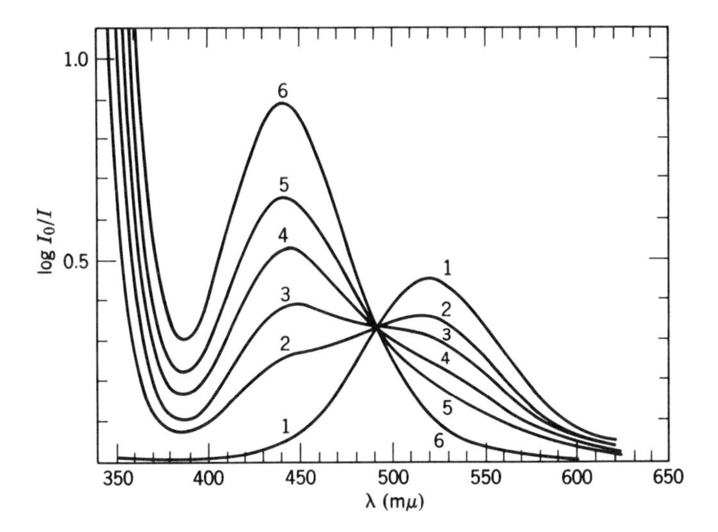


**Figure 7-2** The visible absorption spectrum of pyridine-*N*-oxide · iodine in carbon tetrachloride (23°C, 5-cm cell). Curve 1 is for iodine ($9.550 \times 10^{-5}$ *M*). The concentrations of pyridine-*N*-oxide are $4.210 \times 10^{-3}$ *M* for curve 2, $8.421 \times 10^{-3}$ *M* for curve 3, $16.84 \times 10^{-3}$ *M* for curve 4, and $33.68 \times 10^{-3}$ *M* for curve 5; curve 6 is a calculated curve for the absorption that is due solely to the complexed iodine molecule. (From T. Kubota [2].)

*isosbestic point* at 490 mμ. The existence of this point confirms the assumption that only two species in the solution (complexed $I_2$ and uncomplexed $I_2$) absorb in this region of the spectrum. (See [3] for a more general discussion.)

Figure 7-2 also illustrates the problem of overlapping absorption of two species, which often occurs. In principle a correction is made for the absorbance due to the uncomplexed $I_2$ in order that the absorbance due to the complex alone may be obtained (illustrated by curve 6 in Fig. 7-2). In practice the details of this correction may be somewhat troublesome (see Briegleb [1]); for example, we determine $\epsilon_A$ for the acceptor alone in a solvent and compute the correction $A_A$ to be subtracted from the total absorbance $A_T$ to obtain $A_C$, the absorbance of a D · A complex C. In so doing we usually assume that the absorption for uncomplexed A does not change when the solvent is changed by adding D. This assumption is at best questionable in the case of weak complexes, for which

$c_D$ has to be made rather large.

From the corrected absorbance $A_C$ at a given frequency $\nu$ for a series of concentrations $c_D$, the values of $K$ and $\epsilon_\nu$ can be obtained using the method[†] of Benesi and Hildebrand [4]. The Benesi-Hildebrand analysis starts from the assumption that only one equilibrium exists in the solution and that the constant is defined as in (7-1).[‡] Then, using a zero superscript to denote the total concentration ($c_D^0 = c_D + c_C$, etc.), we have

$$\frac{1}{K} = \frac{(c_D^0 - c_C)(c_A^0 - c_C)}{c_C} = \left(\frac{c_D^0 c_A^0}{c_C} - c_D^0\right) - c_A^0 + c_C. \quad (7\text{-}2)$$

Under the usual conditions ($K$ small, A relatively insoluble) $c_D^0$ is very much greater than $c_A^0$, in order to form enough complex: $c_D^0 >> c_A^0 > c_C$.[††] From Beer's law $c_C = A_C/\epsilon_\nu l$; hence to a good approximation

$$\frac{1}{K} = \frac{\epsilon_\nu l\, c_D^0 c_A^0}{A_C} - c_D^0. \quad (7\text{-}2a)$$

Dividing by $c_D^0 \epsilon_\nu$ and rearranging, we obtain the Benesi-Hildebrand equation

$$\frac{l c_A^0}{A_C} = \frac{1}{K\epsilon_\nu} \frac{1}{c_D^0} + \frac{1}{\epsilon_\nu}. \quad (7\text{-}3)$$

Here in a given experiment we know $l$, $c_A^0$, and $c_D^0$; $A_C$ is measured for a series of solutions with varying $c_D^0$, and the results are plotted as shown in Fig. 7-3. From the slope and intercept the values of $K$ and $\epsilon$ can be obtained.

A number of modifications of the foregoing analysis have been developed, each usually being designated by the name of the person who proposed it; for example, if $\epsilon_A$ is of the order of magnitude of $\epsilon_C$, a corrected analysis is necessary. Various modifications have been summarized by Briegleb [1].

---

[†] This method was developed for use with data from the Beckman DU spectrometer in which the spectrum is obtained point by point at each frequency. The continuous spectra given routinely by more modern instruments contain somewhat more information ($A_C$ as a continuous function of $\nu$) than was readily available in the older studies. This additional information is utilized to some extent in the Liptay method of analysis (see below) but the optimum scheme of analysis has apparently not yet been found.

[‡] Since Benesi and Hildebrand measured concentrations in mole fractions instead of moles per liter, the actual Benesi-Hildebrand equation contains $x_D$ instead of $c_D^0$.

[††] However, note that this approximation fails for the case of strong complexes, when $c_A^0 \simeq c_C$, so that $1/K$ depends on a small difference between the term in parentheses and the other (small) term ($c_A^0 - c_C$). In such cases no term can be dropped. See Tamres [5].

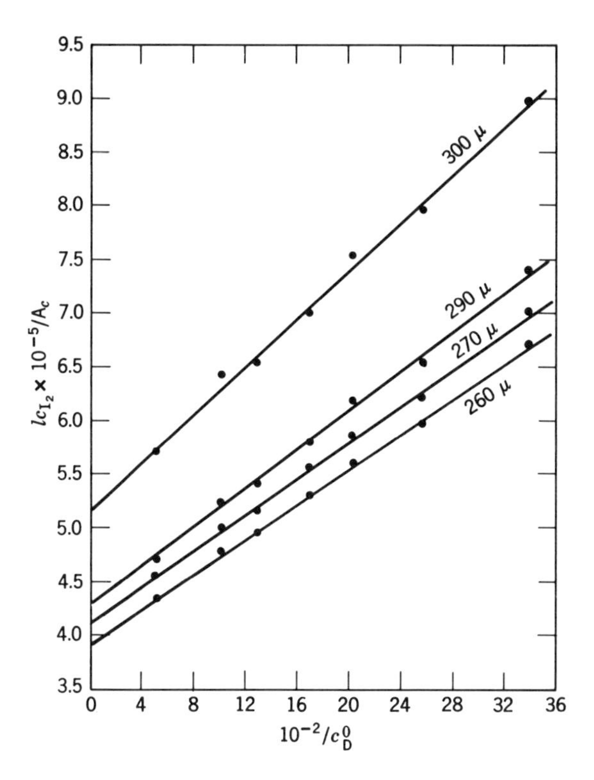


**Figure 7-3** Illustration of the use of the Benesi-Hildebrand equation to obtain $K$ and $\epsilon_\nu$ at four wavelengths for the triethylamine · $I_2$ complex. [From S. Nagakura, *J. Am. Chem. Soc.*, **80**, 520 (1958)]. The intercept of each line is $1/\epsilon_\nu$; the slope is $(1/K\epsilon_\nu)$.

One of the most important is the Scott equation [6], obtained from (7-2*a*) by division by $\epsilon_\nu$:

$$\frac{lc_D^0 c_A^0}{A_C} = \frac{1}{\epsilon_\nu} c_D^0 + \frac{1}{K\epsilon_\nu}. \tag{7-4}$$

For practical reasons (the points are weighted directly according to $c_D^0$ instead of by $1/c_D^0$) this form of the equation is preferable to (7-3).

We must always remember that the points that determine the straight line in either (7-3) or (7-4) are experimentally measured—and there is always some error in their measurement. Hence the straight line has

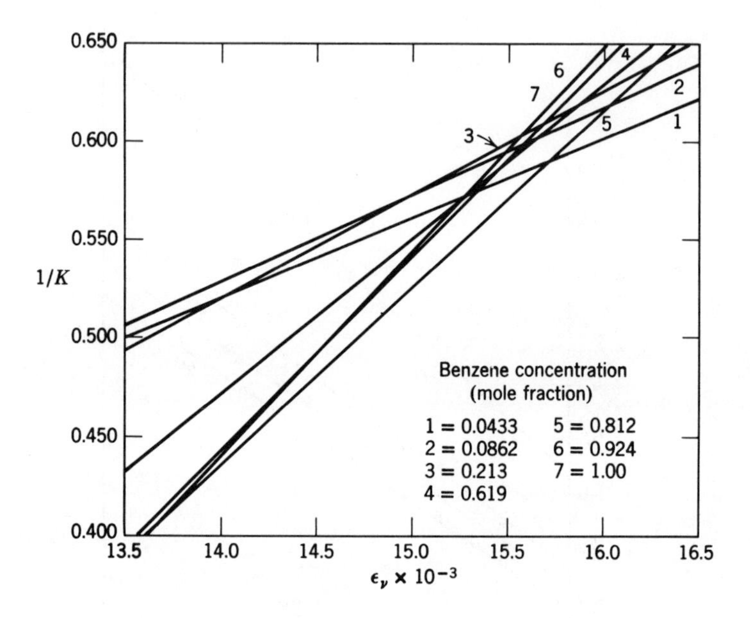


**Figure 7-4** Illustration of the Rose-Drago procedure to obtain $K$ and $\epsilon_\nu$. (From Rose and Drago [7].) The numbers refer to different solutions; there is one line from each solution. (Note that solutions No. 1 and 5 give lines that are apparently inconsistent with the other data and so should probably be eliminated to obtain the best $K$ and $\epsilon$ values.)

some uncertainty in its slope and intercept, leading to some uncertainty in $K$ and $\epsilon_\nu$. In some cases the uncertainty in each may be as large as a factor of 10. Here it is important to note that the product $K\epsilon_\nu$ is usually known more accurately than either $K$ or $\epsilon_\nu$ separately.

One procedure that has the merit of showing directly the scatter in the data is that of Rose and Drago [7]. They start with (7-2$a$) and apply it to a series of solutions each characterized by a particular value of $c_D^0$ (and/or $c_A^0$). For each such solution all the quantities in (7-2$a$) are known except $K$ and $\epsilon_\nu$. They assume a series of trial values for $\epsilon_\nu$ for this particular solution and then calculate corresponding values $1/K$ from (7-2$a$). These are then plotted against $\epsilon_\nu$ to give one of the straight lines of Fig. 7-4. Repeating this procedure for a solution of different concentration gives another straight line with a slightly different slope;

the intersection of these two lines in Fig. 7-4 gives values of $K$ and $\epsilon$ that are consistent for the two solutions. Repeating this procedure for several solutions gives a set of intersecting lines such as that shown in Fig. 7-4. The method makes it possible to eliminate any one bad measurement, which would give bad intersections with *all* the other lines from the other solutions. The procedure has special merit in showing graphically the scatter in the data in terms of a range of intersecting points in Fig. 7-4. The entire procedure can now be repeated at several frequencies to give a good average value for $K$ since if what is present is really a 1:1 complex, $K$ should be independent of frequency.

Finally we mention the Liptay method [8], which is described in detail by Briegleb [1] and illustrated nicely in a step-by-step application to a set of data on pp. 212–219 in GR1. In this procedure absorbance data at several different frequencies $i$ are first combined in the following way: First for a particular solution of concentrations $c_D^0$ and $c_A^0$ the absorbance $A_i$ of the complex at frequency $i$ is compared with the absorbance $A_m$ at some suitably chosen reference frequency $m$ (e.g., at the absorption maximum) and the ratio $A_i/A_m$ is called $\zeta_i$. Values of $\zeta_i$ for all values of $c_D^0$ (and/or $c_A^0$) are then averaged to obtain $\overline{\zeta_i}$ ($\overline{\zeta_i} = \overline{A_i/A_m}$). The merit of this procedure is that the true *relative absorbance* $\zeta_i$ of each solution, if accurately measured, must be the same as that from any other solution made with different $c_D^0$ (and/or $c_A^0$) if only a 1:1 complex is present. Trends in the data for different solutions due to the formation of 2:1 complexes (or other causes) show up easily in the $\zeta$-matrix

Next for a solution of concentration $c_A^0$ and $c_D^0$ we *compute* $A_m^{(\zeta_i)}(c^0)$ (equal to $A_i(c^0)/\overline{\zeta_i}$)†, the absorbance at the reference frequency $m$, using the observed absorbance $A_i(c^0)$ at each frequency $i$ and the average $\overline{\zeta_i}$ value to get a set of values $A_m^{(\zeta_i)}(c^0)$. If the data were perfect, all computed $A_m^{(\zeta_i)}(c^0)$ values would equal the measured absorbance $A_m(c^0)$ for this particular solution; the scatter or trends with frequency measure the error. Now the values of $A_m^{(\zeta_i)}(c^0)$ for each solution are averaged over all frequencies $i$ to obtain $\overline{A_m}(c^0)$, the average absorbance for this solution at frequency $m$. These averaged values of $\overline{A_m}(c^0)$ for a number of different values of $c^0$ are then used in the Scott equation (7-4) to obtain $K$ and $\epsilon$. This procedure has the advantages of using better the data for $A_C$ as a function of frequency and of displaying the data in $\zeta$-matrices and absorbance matrices to show readily the scatter or trends in the absorbance data with concentration or wavenumber in order to suggest errors in the analysis or complications due to additional ($m$:$n$) complexing

† $A_i(c^0)$ means the absorbance at frequency $i$ for a solution of concentrations $c_D^0$ and $c_A^0$.

or reactions. This method is highly recommended, especially when computers are available.

Often the absorbance data for a series of solutions are read at each of several frequencies, and the values of $K$ and $\epsilon_\nu$ are determined by independent analysis at each frequency using (7-3) and (7-4). If the absorbance is due to 1:1 complexes, the values of $K$ so determined should be the same at every frequency. There has been some concern with the fact that $K$s determined at different frequencies sometimes appear to vary with the frequency; for example, Johnson and Bowen [9] have recently reexamined this question for TCNE complexes. (See also [10] for an earlier study recognizing 2:1 complexes.) Johnson and Bowen suggest that such trends in $K$ computed for 1:1 complexes may be due to the unrecognized presence of some 2:1 complexes. However, as Briegleb [1] notes, they may also often be a result of failure of the assumption that the molar absorptivities ($\epsilon_A$, $\epsilon_C$, or $\epsilon_D$) are the same in the pure inert solvent as in the mixed solvent containing excess donor—or the trends may even be the result of experimental errors.

In general we must maintain a healthy skepticism toward many of the results quoted in the literature for $K$ and $\epsilon$, especially those that are given without an error analysis. The review by Ross [11] (which, however, is rather unfair as a book review) emphasizes this attitude and lists a number of references that present the arguments of the skeptics.

## 7-2 EMPIRICAL RESULTS; THE MAGNITUDES OF $K$ AND $\epsilon$

Let us now take a quick look at some actual values of $K$ and $\epsilon$. The most complete survey is presented by Briegleb, GR1, in Chapters V and IX. He lists $K$ values (along with other thermodynamic functions) on pp. 120–130, and $\lambda_{max}$ and $\epsilon_{max}$ values for the charge-transfer band on pp. 30–37. He has grouped existing data by acceptor. By far the most extensive data are for complexes of $I_2$.

Before we examine the magnitudes of these quantities one possible source of confusion should be mentioned. As we have seen, $K$ in (7-1) was defined by using concentrations in moles per liter and so was called $K_c$; however, mole-fraction units, $x_A = n_A/(n_A + n_D + n_C + n_S)$ can be used instead of $c_A$, and so on, to give $K_x$. Here $n_A$ is the number of moles of A in the solution at equilibrium, and so on; S is the solvent. Under usual experimental conditions for weak complexes (D in large excess) the denominator is approximately $n_D + n_S$, and the relationship between $K_c$ and $K_x$ is then

**TABLE 7-1**

**Simplified Summary of Data for Some Classes of Complexes at 20°C[a]**

| Type of complex | Examples[b] | $K_c$ (1/mole) | $-\Delta H$ (kcal/mole) | $\epsilon_{max}$ | $\Delta\nu_{(1/2)}$(cm$^{-1}$) |
|---|---|---|---|---|---|
| $b\pi \cdot a\sigma$ | $Ar \cdot I_2$ | 0.1–2.0 | 1–4 | 5000–15,000 | ~ 5000 |
| $n \cdot a\sigma$ | $Am \cdot I_2$, $R_2O \cdot I_2$, $R_2S \cdot I_2$ | 0.5–7,500 | 4–13 | 3000–30,000 | 5000–8000 |
| $b\pi \cdot a\pi$ | Ar · TCNE | 3–200 | 2–7 | | |
| | Ar · Chl | 0.3–10 | 1.5–5.5 | 500–5000 | 4000–6000 |
| | Ar · TNB | 0.2–12 | 0.5–3.0 | | |

[a] See GR1, pp. 28–47, 120–130.

[b] Abbreviations: Ar = aromatic hydrocarbon, Am = aliphatic amine, $R_2O$ = ether or alcohol, TCNE = tetracyanoethylene, Chl = chloranil, TNB = trinitrobenzene.

$$K_x = K_c \left(\frac{n_S + n_D}{n_S\overline{V}_S + n_D\overline{V}_D}\right) \simeq \frac{K_c}{\overline{V}_S}. \tag{7-5}$$

Here $\overline{V}_S$ and $\overline{V}_D$ are the molar volumes in liters of the solvent and of the donor, respectively. For many solvents $\overline{V}_S \simeq 0.1$, so that $K_x$ is often about 10 times as large as $K_c$.

In Table 7-1 we summarize Briegleb's data (GR1) for a wide range of complexes in a schematic view of this mass of information. The scatter of values for any one complex as measured by different workers is often as much as ±50% of the value of $K$ or sometimes even worse. Some of this scatter can be attributed to the fact that the measurements were made in different solvents; considerable solvent effects can occur [12].

In order to get some feeling for the values of the constants in Table 7-1 we note that the concentration of benzene in the pure liquid state is 11.3 $M$. The value of $K_c$ is 0.15 for the benzene · $I_2$ complex (GR1); hence the ratio of $c_C/c_{I_2}$ in pure liquid benzene is 1.7. In other words, even when dissolved in pure benzene, apparently only 63% of the iodine is complexed. Examination of Fig. 7-1 reveals that the concentration of the complex varies most sensitively with $c_D$ when about 10 to 90% of A is complexed. In fact, if $K_c < 0.01$ so that the system never moves out of Region I for acceptor dissolved in pure liquid donor, $K$ and $\epsilon$ cannot be determined separately. Thus for weak complexes, such as Bz · $I_2$, the determination of $K_c$ and $\epsilon$ is made using a range of solutions in which the ratio of D to solvent may vary from 0.1 to 10, so that the nature of the solvent changes drastically, possibly resulting in changes in the properties (e.g., in $\epsilon$) being measured. Furthermore, the limited solubilities of many donors in inert solvents result in an upper limit to the value of [D] which can be achieved, so that the experimental conditions fall in Region I of Fig. 7-1, even for complexes with larger $K_c$ than Bz · $I_2$, and the accurate determination of $K_c$ and $\epsilon$ separately may not be possible. These observations suggest that we cannot reliably measure $K$ values that are much less than that for benzene · $I_2$ ($K_x = 1.5$ or $K_c = 0.15$) and that the values for $K_x < 1.0$ listed in Briegleb's Table 60 are probably not reliable. (See Person [13] for further discussion.)

Table 7-1 shows that $\epsilon_{max}$ for weak $b\pi$ complexes of $I_2$ ranges from about 5000 to 15,000 and even as high as 30,000 in the strong amine complexes. In the weak $b\pi \cdot a\pi$ complexes—such as those with trinitrobenzene, chloranil, or TCNE as acceptors—$\epsilon_{max}$ values are smaller, in the neighborhood of 3000.

There is some indication that the half-intensity width of the CT band increases as $\nu_{CT}$ increases (see GR1, pp. 45–47), but for a surprisingly large number of complexes the width is approximately constant at about 5000 cm$^{-1}$. Hence oscillator strengths ($f$ values) are roughly proportional to $\epsilon_{max}$.

## 7-3 OTHER THERMODYNAMIC PROPERTIES

If $K$ can be measured, the other thermodynamic properties that are associated with complex formation can also be obtained by using well-known thermodynamic relations [14]. The free-energy change at temperature $T$ for converting 1 mole of D and 1 mole of A in their standard states to 1 mole of complex in its standard state is

$$\Delta F^0 = -RT \ln \frac{a_C}{a_A a_D} = -RT \ln K_a. \tag{7-6}$$

Here $K_a$ is the *thermodynamic* equilibrium constant expressed in terms of *activities* (based either on molar concentrations or on mole fractions). The thermodynamic equilibrium constant is related to $K$ (either $K_x$ or $K_c$, depending on how the activity coefficients are defined) by

$$K_a = K \left(\frac{\gamma_C}{\gamma_A \gamma_D}\right)_i = K\Gamma_i \quad (i = x \text{ or } c). \tag{7-7}$$

If $K_x$ is used in (7-7), the standard states of D, A, and C are conveniently chosen to be pure liquids; if $K_c$ is used, the standard states are approximately the 1-$M$ solutions [14]. As the actual concentrations that are used in measuring $K$ depart from standard-state conditions, we may expect the $\gamma$s to differ from unity so that $\Gamma$ may be significantly different from unity. Its numerical value obviously depends on how the standard state is chosen. For badly nonideal solutions $\Gamma$ itself is a function of concentration. However, for a first approximation to the thermodynamic properties of solutions $\Gamma_c$ is taken to be unity.

We can obtain $\Delta H^0$ from measurements of $K$ at more than one temperature by using the van't Hoff equation [14]

$$\ln K_i = -\frac{\Delta H^0}{R}\left(\frac{1}{T}\right) + \frac{\Delta S_i^0}{R} \quad (i = x \text{ or } c). \tag{7-8}$$

Assuming that $\Delta H^0$ is constant over the temperature range involved, a plot of ln $K$ against $1/T$ should then be a straight line whose slope gives $\Delta H^0$ and whose intercept is $\Delta S_i^0/R$. In this way the enthalpy change $\Delta H^0$ and the standard entropy change $\Delta S^0$ for complex formation can be obtained. (Values for several representative complexes are given in GR1, pp. 120–130, and are summarized here in Table 7-1.) The molarity and mole-fraction enthalpies are related by $\Delta H_x^0 = \Delta H_c^0 + aRT^2$, where $a$ is the thermal-expansion coefficient of the solvent if the solutions are ideal. (See GR2.) The term that involves $a$ is small and thus usually can be neglected.

From (7-8) the value of $\Delta S_i^0$ is numerically different depending on whether we use $K_x$ or $K_c$ because of the different choice for the standard states.

At this point it is worth noting that the difficulty in separating $K$ and $\epsilon$, described above, may cause a considerable scatter in the $K$ values that are obtained at various temperatures. Hence it may be desirable to obtain $\Delta H^0$ and $\Delta S^0$ for weak complexes from a plot of the logarithm of the more accurately determined *product* $K\epsilon$ against $1/T$ rather than from (7-8). The molar absorptivity of any complex is expected to change with temperature, but only by a very small amount over the usual temperature range ($\sim$ 50°C). However, we should then if possible verify this approximate constancy of $\epsilon$, since if, for example, 2:1 as well as 1:1 complexes are present in important amounts, $\epsilon$ would appear to change with temperature.

The term "stability of the complex" (or "strength of the donor or acceptor") usually refers to the magnitude of $K$ (or $\Delta F^0$). A donor is "strong" if it forms a complex with a large $K$. However, as we shall see below, the enthalpy of formation $\Delta H^0$ is more closely related to the stabilization by the charge-transfer forces and the concepts of strength based on the theory.

We note that as the complex becomes more stable (i.e., $\Delta H^0$ becomes more negative) the decrease in entropy due to the loss of freedom as D and A combine to form C also tends to become larger. In fact it has been found empirically that $-\Delta S^0$ is a linear function of $-\Delta H^0$ for most complexes. (See GR1, pp. 140–143, and Fig. 7-5.) These two functions have opposing effects on the value of $\Delta F^0$ so that as the charge-transfer resonance interaction causes $\Delta H^0$ to become more negative the corresponding decrease in entropy prevents $\Delta F^0$ from becoming negative as rapidly as does $\Delta H^0$. However, the change in enthalpy predominates at room temperature so that the stability of the complex as measured by $K$ nearly always goes qualitatively in the same way as $\Delta H^0$. (See Person [15] for further discussion of the linear relation between $\Delta S^0$ and $\Delta H^0$.)

## 7-4 COMPARISON BETWEEN RESULTS IN SOLUTION AND IN THE GAS PHASE

Since the theory has been developed for the interaction between isolated D and A molecules, it applies strictly only to the vapor phase. We should then ask how the results that are obtained from solution studies can be related to those for the gas phase. This question has not been an easy one to answer in spite of several recent attempts. (See Kroll [16], Tamres and Goodenow [17] and Rice [18] among other discussions.)

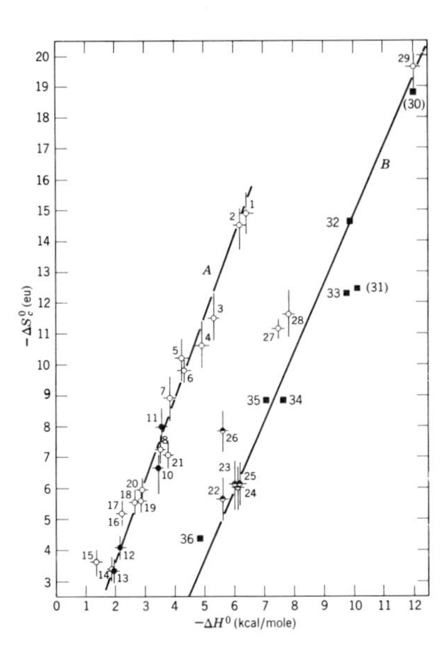


**Figure 7-5** Variation of entropy of formation, $\Delta S^0$, with enthalpy of formation, $\Delta H^0$, for iodine complexes in various solvents ($H$ = *n*-heptane, $C = CCl_4$). Ethers (○), alcohol (●), methylpyridine *N*-oxides (◉), alkylbenzenes (△), pyridine (□), amines (■): (1) trimethylene oxide ($H$); (2) 2-methyltetrahydrofuran ($H$); (3) tetrahydrofuran ($H$); (4) tetrahydropyran ($H$); (5) ethyl ether ($H$); (6) ethyl ether ($C$); (7) propylene oxide ($H$); (8) 1,4-dioxane (*n*-hexane); (9) 1,4-dioxane ($C$); (10) *t*-butyl alcohol ($C$); (11) ethyl alcohol (*n*-hexane); (12) ethyl alcohol ($C$); (13) methyl alcohol ($C$); (14) hexaethylbenzene ($C$); (15) benzene ($C$); (16) *p*-xylene ($C$); (17) *sym*-tri-*t*-butylbenzene ($C$); (18) *sym*-triethylbenzene ($C$); (19) durene ($C$); (20) mesitylene ($C$); (21) hexamethylbenzene ($C$); (22) pyridine-*N*-oxide ($C$); (23) 2-picoline *N*-oxide ($C$); (24) 3-picoline *N*-oxide ($C$); (25) 4-picoline *N*-oxide ($C$); (26) 2,6-lutidine *N*-oxide ($C$); (27) pyridine ($C$); (28) pyridine ($H$); (29) triethylamine ($H$); (30) $Me_3N$; (31) piperidine; (32) $Et_2NH$; (33) $Me_2NH$; (34) $EtNH_2$; (35) $MeNH_2$. (Me = methyl, Et = ethyl.) Dotted lines represent estimated error limits in those cases where none were specified.

Drawing modified (to include the data for the amines[g]) from Fig. 1 from M. Tamres and Sr. M. Brandon, *J. Am. Chem. Soc.*, **82**, 2134 (1960). Here $\Delta S^0$ is calculated for standard states of 1-*M* solutions ($\Delta S_c^0$). The equation for curve *A* is $-\Delta S_c^0 = -3.75\ \Delta H^0 + 2.50$; the equation for curve *B*, through the amine data, is $-\Delta S_c^0 = -2.34\ \Delta H^0 + 7.9$. References for the data are from Tamres and Brandon except for H. Yada, J. Tanaka, and S. Nagakura, *Bull. Chem. Soc. Japan*, **33**, 1660 (1960), on the amine $I_2$ complexes.

For the following discussion let us attempt to predict the changes that are expected in the measured thermodynamic properties ($K$, $\Delta H^0$, etc.) for complex formation on changing from the gas phase to solution in an "inert" solvent, on the assumption that the CT and electrostatic interactions that stabilize the complex are essentially the same in these two phases. A comparison of these predicted changes with the experimental results then should enable us to reach conclusions about the validity of this assumption.

For this purpose it is convenient to construct a thermodynamic cycle, as shown in Fig. 7-6 for $\Delta H$. Here the actual reaction measured in

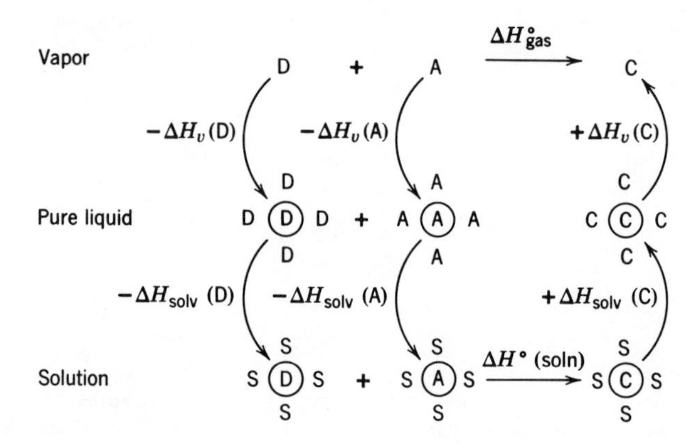


**Figure 7-6** The relationship between the thermodynamic properties (here $\Delta H^0$) of complexes measured in solution and those measured in the vapor phase. Here $\Delta H_v$(D) is the enthalpy of vaporization of 1 mole of pure liquid D, and so on. For an ideal solution $\Delta H_{\mathrm{solv}}$ (D), the solvation molal enthalpy change from pure liquid D to solution in S, is zero; for a regular solution it is estimated from the solubility parameters [18] $\delta_{\mathrm{D}}$ for D and $\delta_1$ for S to be: $\Delta H_{\mathrm{solv}}(\mathrm{D}) = \overline{V}_{\mathrm{D}}\phi_1^2 \times (\delta_{\mathrm{D}} - \delta_1)^2$. Here $\overline{V}_{\mathrm{D}}$ is the molar volume of pure liquid D and $\phi_1$ is the volume fraction of the solvent. Similar expressions can be written for the $\Delta H_{\mathrm{solv}}$ for A and C; $\delta_c$, the solubility parameter for C, may be estimated to be the geometric mean of $\delta_{\mathrm{D}}$ and $\delta_{\mathrm{A}}$.

solution is indicated on the lower line as occurring between solvated D and A molecules to form a solvated complex with a measured standard enthalpy change $\Delta H^0_{\mathrm{soln}}$. The corresponding reaction in the gas phase is indicated along the top line between free D and A molecules to form a free complex molecule C, with standard enthalpy change $\Delta H^0_{\mathrm{gas}}$.

In considering the relationship between $\Delta H^0_{soln}$ and $\Delta H^0_{gas}$, it is convenient to break down the change from solution to gas into two parts: first, the enthalpy change from solution to pure liquid, and then from pure liquid to vapor. The first part of the change ($\Delta H_{solv}$) can be estimated from "regular" solution theory, using the *solubility parameters* $\delta_i$ of Hildebrand and Scott [19]. The second part ($\Delta H_v$) consists of the known heats of vaporization of D and A and the estimated heat of vaporization of C.

Summing over the various terms in Fig. 7-6 we find

$$\Delta H^0_{gas} = \Delta H^0_{soln} - \Delta H_v(\mathrm{A}) - \Delta H_v(\mathrm{D}) + \Delta H_v(\mathrm{C}) + \Delta(\Delta H_{solv}). \quad (7\text{-}10)$$

If the solution is "ideal" in D, then the solvation energy of D, $\Delta H_{solv}(\mathrm{D})$, is zero. If the solution is ideal in all components, then $\Delta(\Delta H_{solv}) = 0$. If, furthermore, the energies of vaporization of the pure liquids A, D, and C are all of about the same order of magnitude (say about 5 to 8 kcal/mole), then the enthalpy change in the gas phase is expected to be more negative than that in the solution by approximately the enthalpy of vaporization of 1 mole of liquid (or about 5 to 8 kcal/mole).

Another way of looking at this relation is to note that the ordinary London dispersion attractive energy between D and A is approximately cancelled in the measurement in solution since in that reaction one D—S and one A—S contact are replaced by one D—A plus one S—S contact. However, the enthalpy change that is measured in the gas phase includes this D—A London dispersion contribution, which is expected to be about −5 kcal/mole in a typical case.

If the solution is nonideal, "regular" solution theory [19] can be used to estimate the heats of solvation. The expression for these heats in terms of solubility parameters is given in the caption for Fig. 7-6. Evaluating $\Delta(\Delta H_{solv})$ for the reaction when D is benzene, A is $I_2$, and the solvent is *n*-heptane, using the solubility parameters from Hildebrand and Scott [19], with $\delta_C$ taken to be the geometric mean of $\delta_{I_2}$ and $\delta_{Bz}$, leads to an estimate of $\Delta(\Delta H_{solv})$ of about +1.2 kcal/mole. Hence from this estimate for the contribution from the nonideality of this solution we estimate that $\Delta H^0_{gas}$ should be more negative than $\Delta H^0_{soln}$ by about 3.5 to 6.5 kcal/mole, depending on the heat of vaporization.

The theory refers to the *internal energy* change $\Delta W^0_f$ for complex formation. In turn $\Delta W^0_f$ is related to $\Delta H^0_{gas}$ by the following relation [$\Delta n$ = −1 for the equilibrium defined in (7-1)]:

$$\Delta W^0_f = \Delta H^0 - \Delta(PV) = \Delta H^0 - RT\Delta n = \Delta H^0_{gas} + RT. \quad (7\text{-}11)$$

Recently there have been several attempts [16,17,20,21] to measure

**TABLE 7-2**

**Comparison of Gas-Phase and Solution Properties of Complexes[a]**

| Property | p-Xylene · TCNE | | $Et_2S \cdot I_2$ | |
|---|---|---|---|---|
| | Gas | Solution ($CH_2Cl_2$) | Gas | Solution (n-heptane) |
| $\epsilon_{max}$ | [a]800 ± 250 | [b]2770 | [a]3500 ± 1000<br>[c]8700 ± 1640 | [c]24,800 ± 1050 |
| $K_c$ (298°K) | [a]280 ± 100 | [b]7.20 | [a]750 ± 250<br>[c]285 | [c]200 |
| $\Delta H$ (kcal/mole) | [a]−8.1 ± 0.5 | [b]−3.37 | [a]−8.3 ± 0.5<br>[c]−8.4 ± 0.4 | [c]−8.9 ± 0.6 |
| $\Delta S$ (cal/mole-deg) | [a]−15.5 ± 2.0 | [b]−7.38 | [a]−14.2 ± 1.9<br>[c]−14.8 ± 1.0 | [c]−19.4 ± 2.0 |

[a] From results tabulated by Kroll [16].

[b] Results from Merrifield and Phillips [26].

[c] Results by Goodenow and Tamres (vapor) [17] or Tamres and Searles (solution) [17], using $K_c = 200$ at 298°K in n-heptane solution.

thermodynamic properties for complexes in the gas phase by using spectrophotometric methods. The studies of the $(C_2H_5)_2S \cdot I_2$ complex [16,17] and of the TCNE complexes [16] may be accurate enough to provide a useful comparison between gas-phase and solution results. This comparison is given in Table 7-2, which is taken from the more extensive tables of Kroll [16]. We must note that there is almost a 100° temperature difference between the gas and solution measurements because a higher temperature was needed to get enough vapor pressure in the gas phase; hence for the comparison the gas-phase results have to be extrapolated to the temperature of the solution experiment, with considerable likelihood of error in the extrapolated value of $K_c$ [16].

Looking first at the $\Delta H$ values we see that the results for the TCNE complexes (represented here by *p*-xylene · TCNE) fit with the predictions of (7-10), whereas those for the stronger complex between diethyl sulfide and $I_2$ do not. At present there is no quantitative explanation for these differences. It is possible that solvation energies $[\Delta(\Delta H_{solv})]$ contribute more than the 1.2 kcal/mole estimated for $Bz \cdot I_2$, but it seems more likely that the differences from the predictions of (7-10) are due to a failure with some complexes (here $Et_2S \cdot I_2$ and its analogues) of the assumption that the CT and electrostatic stabilization energies are the same in the gas phase as in solution (see Kroll [16] and also others [17,21]).

Let us also examine the apparent change in $\epsilon_{max}$ reported in Table 7-2 and emphasized also in the other two studies of gas-phase properties of complexes. As mentioned in Section 3-2, some increase in intensity on going from gas to solution may be expected because of the change in the effective electric field of the light wave in the polarizable medium. This effect was predicted in simple form by Chako [22], and the model has been improved by Buckingham [23]. While it has not been convincingly proven that their equations explain all the observed intensity changes from the gas phase to the "ideal" liquid phase [24], we may tentatively accept them and estimate the change in intensity expected on going from gas to solution because of the "effective field". We can use the Chako equation for this rough estimate:

$$\frac{\epsilon_i}{\epsilon_g} = \left[\frac{n_0^2 + 2}{3}\right]^2 \frac{1}{n_0}. \tag{7-12}$$

Here $\epsilon_i$ is the total integrated absorbance and $n_0$ is the index of refraction of the pure liquid complex. (In solution a more complicated equation should properly be used. However, the simple Chako equation is expected to give at least the correct order of magnitude for this effect.) For $n_0 \simeq 1.5$, a typical value, (7-12) predicts that $\epsilon_i/\epsilon_g = 1.33$. Hence a small portion of the observed increase in intensity from gas to solution that is reported in Table 7-2 can apparently be attributed to the field effect, but most of it must probably be attributed to some other effect. The large increase in the intensity in solution is consistent with the idea [16,17,21] suggested above to account for the enthalpy discrepancy that the complex in solution is stabilized more by the CT resonance than it is in the vapor phase.

Rice [18] has suggested that the observed differences between the gas and solution properties (e.g., the intensity of the CT absorption) may be evidence for a much smaller degree of CT resonance stabilization of the gas phase complex. He suggests a model for the gas phase complexes in which there is very little preferential orientation. (He attributes orientation to the requirements of maximum overlap for CT resonance stabilization.) However, we believe that the true explanation for the puzzling intensity behavior in the gas phase must be somewhat more complicated than that suggested by Rice [18].

Another possible explanation for the apparent difference in the $\epsilon$-values from gas to solution may lie in the neglect of solvation effects in the Scott analysis of the solution studies. It has been suggested [25] that this error in treating solution data may lead to an overestimation of $\epsilon$ by a factor of 10 to 20. We do not think that the error in the analysis of the solution data is quite so large; we shall investigate this effect in the following treatment.

Basically this error is due to the assumption that the solutions are ideal and to the neglect of the concentration dependence of $\Gamma_i$ in (7-7) in the analysis of the data by the Scott equation [(7-4)]. This error has been recognized and discussed by many authors, including Carter, Murrell, and Rosch [25], Merrifield and Phillips [26], Corkill, Foster, and Hammick [27], Trotter and Hanna [28], Tamres [5], and Scott [6]. In our treatment we shall estimate $\Gamma_i$ as a function of donor concentration by using the "regular" solution theory [19] and then show that an analysis of the data by the Scott Eq. (7-4) that neglects this variation in $\Gamma_i$ leads to an error in $\epsilon$ that can be estimated. Although this treatment differs in its details and in the magnitude of the estimated errors from the other treatments [5,6,25–27], it is basically equivalent to them and includes most of them as special cases.

If $\Gamma_i$ is a linear function of the donor concentration; namely,

$$\Gamma_x = k_0 \,(1 + bx_\mathrm{D}), \tag{7-13}$$

the Scott or Benesi-Hildebrand methods of analyzing the data do not give true values of $K$ and $\epsilon$. For weak complexes such a large amount of donor is required to form a reasonable amount of complex that the nature of the solvent may be changed during a set of experiments with increasing $x_D$ so that $\Gamma_x$ is expected to be *some* function of $x_D$, although perhaps not *linear* as in (7-13). However, $\Gamma_x$ calculated as a function of $x_D$ from the solubility parameters [19] does take the form indicated in (7-13) as we see below.

Using (7-1), (7-7), and (7-13), we rederive the Scott Eq. (7-4):

$$\frac{l x_D c_A^0}{A_C} = \frac{k_0}{K_a \epsilon_\nu} + \left(\frac{k_0}{K_a} b + 1\right) \frac{1}{\epsilon_\nu} x_D = \frac{1}{K_x^0 \epsilon_\nu} + \left(\frac{b}{K_x^0} + 1\right) \frac{1}{\epsilon_\nu} x_D. \quad (7\text{-}14)$$

Here $K_x^0$ is the limiting mole-fraction equilibrium constant measured in very dilute solutions ($x_D \to 0$). If we use the Scott procedure (7-4) without realizing that $\Gamma_x$ depends on $x_D$ [(7-13)] to obtain $K$ and $\epsilon$ from experimental data, we shall find an *apparent* $\epsilon_{app}$ and $K_{app}$ from the slope and intercept that are related to the *true* $\epsilon_{true}$ and $K_x^0$ by

$$\epsilon_{app} = \frac{1}{\text{slope}} = \epsilon_{true} \left[\frac{1}{(b/K_x^0) + 1}\right] = \epsilon \frac{K_x^0}{K_{app}}, \quad (7\text{-}15)$$

$$K_{app} = \frac{\text{slope}}{\text{intercept}} = K_x^0 + b.$$

Using solubility parameters [19] as described above to estimate the dependence of $\Gamma_i$ on $x_D$, we find for benzene plus iodine in $n$-heptane that

$$\Gamma_x = 0.26\ (1 - 0.077\ x_D). \quad (7\text{-}16)$$

Here we have used what seems to be the most reasonable value, $\delta_{I_2} = 13.6$, for the solubility parameter of $I_2$ in $n$-heptane and we have estimated $\delta_C$ from the geometric mean of $\delta_{I_2}$ and $\delta_{Bz}$. From (7-16) $K_x^0 = 1.23$ (compared with $K_{app} = 1.15$ from [4]) and $\epsilon_{app} = 1.07\,\epsilon_{true}$ so that the neglect of the solvent dependence of $\Gamma_x$ has led to an overestimate of $\epsilon$ in solution by only about 7% and an underestimate of $K$ by about the same amount.

However, if the solubility parameter of $I_2$ is taken as $\delta_{I_2} = 14.1$, which is the value that is ordinarily used [19] for most solutions,† the situation is much worse, with

$$\Gamma_x = 0.14\ (1 + 0.71\ x_D); \quad (7\text{-}16a)$$

† The solubility parameter measured for $I_2$ depends partly on the solvent. The value measured in $n$-heptane is 13.6, whereas the average value for a large number of solvents is 14.1. (See J. H. Hildebrand and R. L. Scott, *The Solubility of Nonelectrolytes*, 3rd ed., Dover, New York, 1964, Chapter 17.)

from this $K_x^0 = 0.45$, with $\epsilon_{app} = 0.392\epsilon_{true}$. These two estimates of the error in $\epsilon$ suggest that the neglect of solvent imperfection in the Scott analysis of the data may lead to errors as large as a factor of 2 in $\epsilon$. However, the consistency of results found in different solvents suggests that the errors due to neglect of solvent imperfection are probably at most no larger than a factor of 2. Thus the change in $\epsilon$ from gas to solution that is indicated in Table 7-2 must have some other explanation.

## 7-5 REFERENCES


[1] G. Briegleb, *Elektronen-Donor-Acceptor-Komplexe*, Springer Verlag, Berlin, 1961, Chapter 12.
[2] T. Kubota, *J. Am. Chem. Soc.*, **87**, 458 (1965).
[3] M. D. Cohen and E. Fischer, *J. Chem. Soc.*, 3044 (1962).
[4] H. A. Benesi and J. H. Hildebrand, *J. Am. Chem. Soc.*, **71**, 2703 (1949).
[5] M. Tamres, *J. Phys. Chem.*, **65**, 654 (1961).
[6] R. L. Scott, *Rec. trav. chim.*, **75**, 787 (1956).
[7] N. J. Rose and R. S. Drago, *J. Am. Chem. Soc.*, **81**, 6138 (1959).
[8] W. Liptay, *Z. Elektrochem.*, **65**, 375 (1961).
[9] G. D. Johnson and R. E. Bowen, *J. Am. Chem. Soc.*, **87**, 1655 (1965).
[10] J. Landauer and H. McConnell, *J. Am. Chem. Soc.*, **74**, 1221 (1952).
[11] S. Ross, *J. Am. Chem. Soc.*, **87**, 3032 (1965).
[12] R. S. Drago, T. F. Bolles, and R. J. Niedzielski, *J. Am. Chem. Soc.*, **88**, 2717 (1966). See also C. C. Thompson, Jr., and P. A. D. DeMaine, *J. Phys. Chem.*, **69**, 2766 (1965).
[13] W. B. Person, *J. Am. Chem. Soc.*, **87**, 167 (1965).
[14] G. N. Lewis and M. Randall, *Thermodynamics*, 2nd ed. (revised by K. S. Pitzer and L. Brewer), McGraw-Hill, New York, 1961.
[15] W. B. Person, *J. Am. Chem. Soc.*, **84**, 536 (1962).
[16] M. Kroll, *J. Am. Chem. Soc.*, **90**, 1101 (1968).
[17] (a) J. M. Goodenow and M. Tamres, *J. Chem. Phys.*, **43**, 3393 (1965); (b) M. Tamres and J. M. Goodenow, *J. Phys. Chem.*, **71**, 1982 (1967). [Also Tamres and Searles, *J. Phys. Chem.*, **66**, 1099 (1962).]
[18] O. K. Rice, *Int. J. Quantum Chem.*, **II S**, 219 (1968).
[19] J. H. Hildebrand and R. L. Scott, *Regular Solutions*, Prentice-Hall, Englewood Cliffs, N.J., 1962.
[20] F. T. Lang and R. L. Strong, *J. Am. Chem. Soc.*, **87**, 2345 (1965).
[21] (a) J. Prochorow, *J. Chem. Phys.*, **43**, 3394 (1965); (b) J. Prochorow and A. Tramer, *J. Chem. Phys.*, **44**, 4545 (1966).
[22] N. Q. Chako, *J. Chem. Phys.*, **2**, 644 (1934); also see S. R. Polo and M. K. Wilson, *J. Chem. Phys.*, **23**, 2376 (1955).
[23] A. D. Buckingham, *Proc. Roy. Soc.* (London), **255**, 32 (1962);
[24] For example, see L. E. Jacobs and J. R. Platt, *J. Chem. Phys.*, **16**, 1137 (1948).
[25] S. Carter, J. N. Murrell, and E. J. Rosch, *J. Chem. Soc.*, 2048 (1965).
[26] R. E. Merrifield and W. D. Phillips, *J. Am. Chem. Soc.*, **80**, 2779 (1958).
[27] J. M. Corkill, R. Foster, and D. L. Hammick, *J. Chem. Soc.*, 1202 (1955).
[28] P. J. Trotter and M. W. Hanna, *J. Am. Chem. Soc.*, **88**, 3724 (1966).

CHAPTER 8

# *The Franck-Condon Principle and the Width and Shape of the Charge-Transfer Band*

## 8-1 THE FRANCK-CONDON PRINCIPLE

Let us review the Franck-Condon principle [1] for diatomic and polyatomic molecules as preparation for some discussion of the bandwidth and the shape of the spectrum of the CT transition. In Fig. 1-1 this spectrum in absorption is shown as a broad and structureless band. Typical values for $\Delta\nu_{(1/2)}$, its width at half maximum, are about 5000 $cm^{-1}$ [see Section (7-2)]. The band is usually not symmetrical but is usually broader on the high-frequency side. (See GR1, pp. 45–47.)

According to the Franck-Condon principle, when light is absorbed causing a molecule to change from one electronic state to another, the *most probable* occurence is that the nuclear configuration does not change (that is, $\Delta\mathbf{R}_k = 0$) and the nuclear momenta also do not change ($\Delta\mathbf{P}_k = 0$). The principle can be described as a *verticality* principle (Franck) partially relaxed by *quantum-mechanical spread* (Condon). The verticality principle states that the positions and momenta of the nuclei do not change during the electronic transition. On a potential energy diagram this process is represented by a vertical line, of length equal to the energy separation between points of identical nuclear configuration on the two potential surfaces; also, equal kinetic energies (often this is just zero-point kinetic energy of the lower state) are assumed. Actually the absorption of light occurs over a range of frequencies (Figs. 8-4 and 3-1); this spread is explained by the quantum-mechanical Heisenberg uncertainty principle, which allows transitions with $\Delta\mathbf{R}_k \neq 0$ and $\Delta\mathbf{P}_k \neq 0$ but with fairly rapidly decreasing intensity for increasing $\Delta\mathbf{R}_k$ and $\Delta\mathbf{P}_k$. The Franck-Condon principle is discussed here in terms of the complete wavefunctions of the two electronic states in order to see in more detail the origin of the quantum-mechanical spread.

## 8-2 DIATOMIC MOLECULES

For a diatomic molecule there is only one vibrational coordinate $R$, equal to the displacement $r - r_e$ of the internuclear distance from its equilibrium value $r_e$, and in (3-5), $d\mathbf{R} = dR = dr$. Using the harmonic approximation [$W = (1/2)kR^2$, where $k$ is the force constant (Chapter 6)] for the potential energy,

$$E_v = hc\omega_e (v + 1/2); \qquad \omega_e = \frac{1}{2\pi c}\sqrt{k/m_r}\ \text{cm}^{-1}, \tag{8-1}$$

$m_r$, the reduced mass, is defined for a diatomic molecule by

$$\frac{1}{m_r} = \frac{1}{m_A} + \frac{1}{m_B}. \tag{8-2}$$

The vibrational wavefunctions are the well-known Hermite orthogonal functions [2], depending as follows on the vibrational quantum number $v$:

$$\chi_v = N_v e^{-(1/2)aR^2} H_v(\sqrt{a}R). \tag{8-3}$$

Here $a = 4\pi^2\mu c\omega_e/h$, $H_v(\sqrt{a}R)$ is the Hermite polynomial of $v$th degree, and the normalization factor is given by

$$N_v^2 = (1/2^v v!)\sqrt{a/\pi}.$$

Now $\mu_{mn}^{el}(R)$ is a slowly varying function of $R$ and is often assumed to be a constant with the same value as for $r = r_e$ $(R = 0)$. With this assumption (3-5) becomes

$$\mu_{mn} = \mu_{mn}^{el}(r_e)\int \chi_v^m \chi_v^n \, dR. \tag{8-4}$$

(For accurate calculations the variation of $\mu_{mn}^{el}$ with $R$ must be taken into account.) We now consider the transition moment in absorption, starting from the $v = 0$ level in the electronic state $m$; for example, in the normal state $N$. Ordinarily, except for very heavy molecules, the population in higher vibrational states of state $N$ is small at room temperature; since the population in vibrational state $v$ is proportional to $e^{-hc\omega_e v/kT}$, the population at ordinary temperatures for $v > 0$ is negligible for $\omega_e$ greater than about 600 $\text{cm}^{-1}$ or about $3kT$. For $v_m = 0$ in

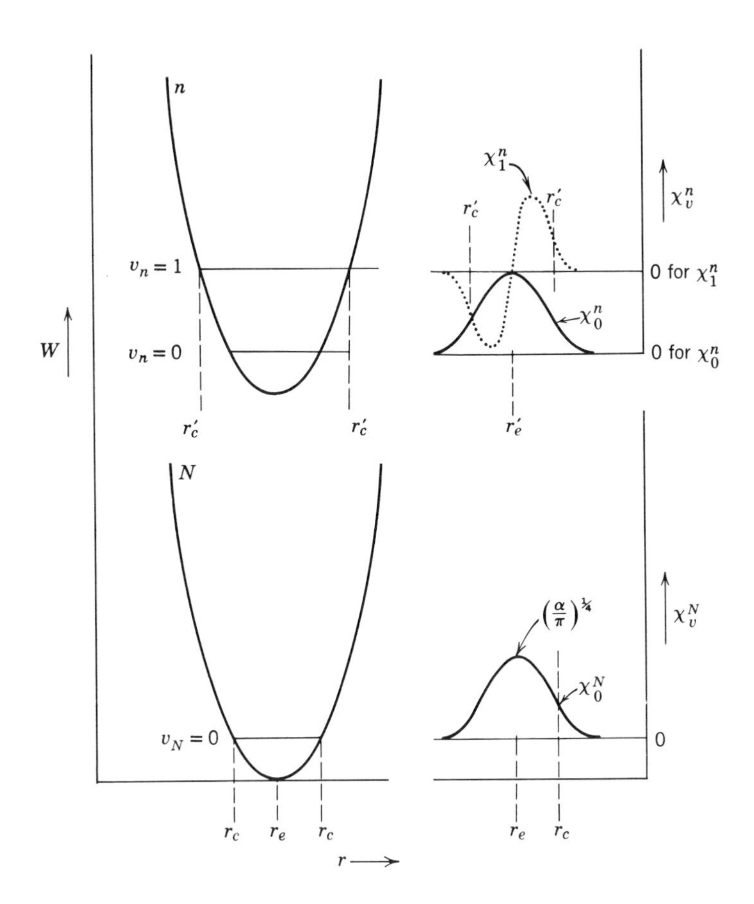


**Figure 8-1** Case 1. Potential curves identical in $r_e$ and $k$ for electronic states $n$ and $N$. Vibrational wavefunctions are sketched on the right.

(8-4), $\chi_v^m = \chi_0^m = (\alpha/\pi)^{(1/4)} e^{-(1/2)\alpha R^2}$ from (8-3). We can then compute the transition moment for any vibrational transition by multiplying the intrinsic electronic moment $\mu_{mn}^{el}(r_e)$ by a factor that is just the overlap integral between $\Psi_0^{vib}$ of the ground electronic state and the vibrational wavefunction of the excited electronic state.

Let us consider a few specific cases.

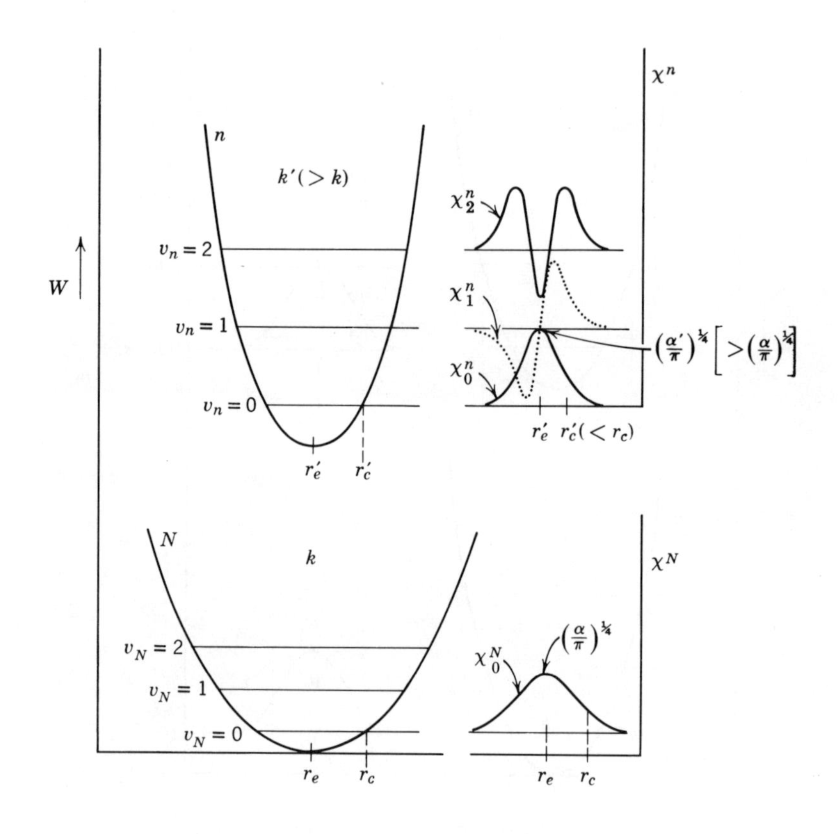


**Figure 8-2** Case 2. Potential curves for identical $r_e$ values but different $k$ values for electronic states $n$ and $N$. Vibrational wavefunctions are shown on the right.

### Case 1

The potential curve in the upper electronic state $n$ is identical (same $r_e$ and $k$) with that of the ground state $N$. (See Fig. 8-1.) Always the classical turning point $r_c$ for a vibration corresponds to a point of inflection for $\chi_v$; this relation is illustrated in Fig. 8-1. Inspection of Fig. 8-1 shows that $\int \chi_0^N \chi_0^n \, dR \neq 0$, but $\int \chi_0^N \chi_1^n \, dR = 0$. In fact, because of the orthogonality of the vibrational functions belonging to any one potential curve and because of the identity of the curves $\chi_v^N$ and $\chi_v^n$ in this

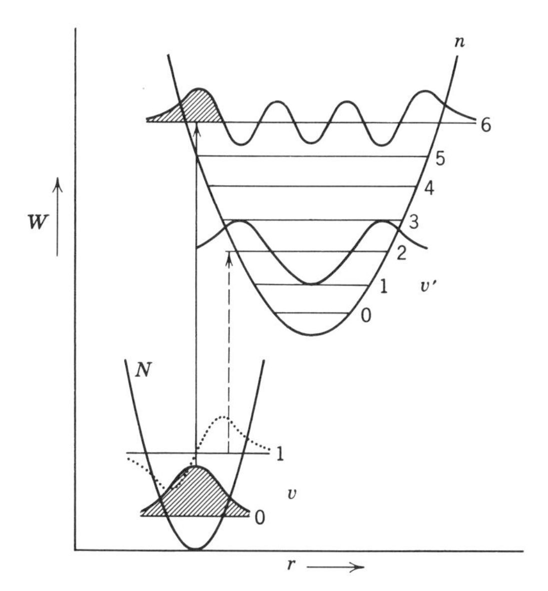


**Figure 8-3** Case 3. Potential curves for both $r_e$ and $k$ different for electronic states $n$ and $N$. (In diatomic molecules increased $r_e$ always goes with decreased $k$.)

case, *only* the $0 \leftarrow 0$ transition occurs, and there is just one sharp absorption band, thus no quantum-mechanical spread.

### Case 2

The upper curve has the same equilibrium value of $r$ as the lower curve, but the force constant is quite different. (See Fig. 8-2.) Although cases approximating to this are unknown for diatomic molecules, a very similar case *does* occur for polyatomic molecules (see below). Inspection of Fig. 8-2 shows that vibrational transitions can occur from $v_N = 0$ to any *even* value of $v_n$, but not to odd values. Of these the $0 \leftarrow 0$ transition is always the strongest, very much so unless the force constants of states

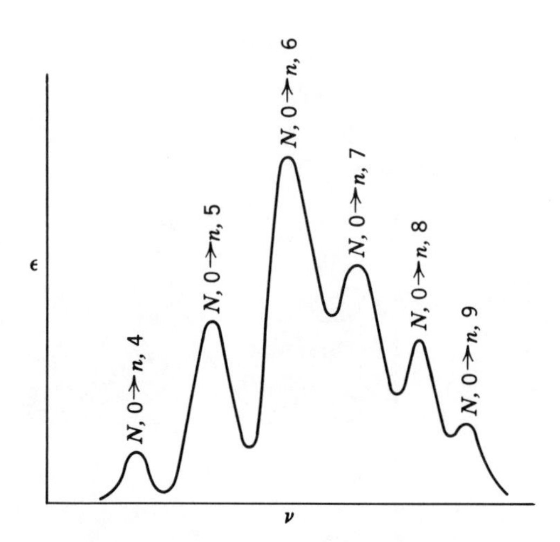


**Figure 8-4** Absorption band shape for transition to a typical bound excited state, such as in Fig. 8-3; see Fig. 3-1 for the case of transition to an unbound excited state, as in Fig. 8-5. The intensity variation illustrates the Franck-Condon principle.

$N$ and $n$ differ greatly—as can be seen by consideration of Fig. 8-2 or variations of it with varying degrees of inequality of the force constants.

## Case 3

The upper curve has both $r_e$ and $k$ different in electronic states $N$ and $n$ (Fig. 8-3). This case is fairly typical for transitions between valence-shell states. In Fig. 8-3 the overlap $\int \chi_6^n \chi_0^N \, dR$ that gives the largest nonzero value is between the two shaded lobes; we see that the most probable transition is to the level whose left-hand, relatively large, wave-function loop most nearly lines up with $\chi_0^N$. (The first and relatively large maximum of each $\chi_v$ occurs a little to the right of the left-hand turning point of the classical vibration; the first point of inflection occurs at the left-hand turning point.) Transitions to other levels occur with lower probability as the vibrational wavefunctions in the upper level shift to the right or left. The absorption spectrum of the electronic transition then has the appearance shown in Fig. 8-4. If the vibrational

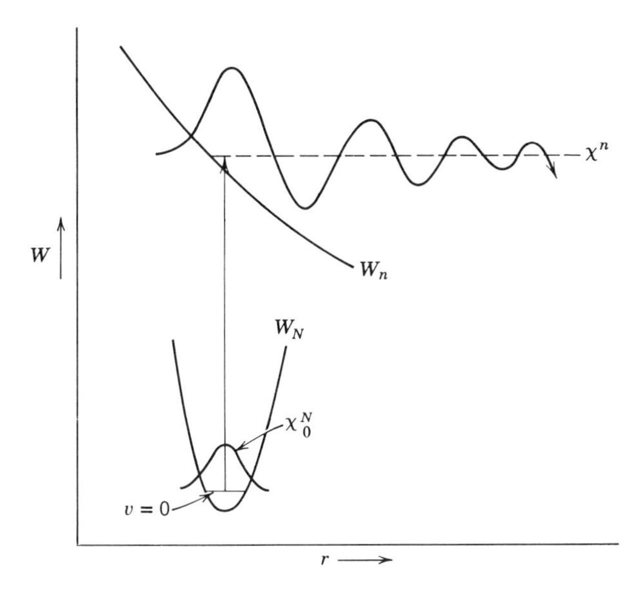


**Figure 8-5** Case 4. Potential curves for the case in which the upper state is not bound.

levels are sufficiently closely spaced and/or if a number of vibrational levels of the normal state are populated, overlapping of the separate vibrational bands may cause the overall picture to be smeared out to give an apparently smooth band shape like that in Fig. 3-1.

The effect of transitions from excited normal-state vibrational levels is also shown for $\chi_1^N$ in Fig. 8-3 by the dashed lines. As the temperature increases molecules in the lowest vibrational state are more and more excited into $\chi_1^N$ and higher vibrational levels. Transitions from this level to the upper electronic state make the electronic absorption band less intense in the middle and more intense in the wings.

### Case 4

The upper state is not bound. This case is shown in Fig. 8-5. The wavefunction indicated for the upper state is a free-particle function with energy chosen so as to give maximum overlap with $\chi_0^N$. This overlap is proportional to the transition moment for absorption at the frequency indicated by the vertical arrow. Obviously at higher energies and higher absorption frequencies the first peak of the $\chi^n$ wavefunction moves continuously to the left, decreasing its overlap with $\chi_0^N$. Similarly the overlap between $\chi^n$ and $\chi_0^N$ decreases continuously as the frequency of the transition is decreased. The absorption spectrum is then expected to resemble approximately that shown in Fig. 3-1. In this case the quantum-mechanical spread is quite clearly due to the zero-point spread in $\chi_0$.

The half-intensity width $\Delta\nu_{(1/2)}$ of Fig. 3-1 is evidently governed by the steepness of $W_n$ near $R = 0$. It can also be seen from Fig. 8-3 or even more from Fig. 8-5 that as a result of the *increasing* steepness of $W_n(R)$ as $r$ decreases, the spread of the absorption band on the two sides of $\nu_{max}$ should be unequal, being greater on the high-frequency side.

## 8-3 POLYATOMIC MOLECULES

For a polyatomic molecule with $N$ nuclei the discussion is similar to that in Section 8-2 except that there are $3N - 6$ degrees of freedom (or $3N - 5$ for linear molecules) so that potential curves are replaced by multidimensional potential surfaces. Instead of the single displacement coordinate $R$ we use the $3N - 6$ mass-weighted displacement coordinates, or normal coordinates, $Q_i$, which correspond to the actual fundamental modes of vibration. For each normal coordinate there is a corresponding fundamental frequency $\omega_i$, and in harmonic approximation

$$W^{vib} = hc \sum_{i=1}^{3N-6} \omega_i (v_i + 1/2).$$

The wavefunctions $\chi^m_{v_i}$ are again the Hermite orthogonal functions of (8-4) except that $\sqrt{a}R$ is replaced by $\sqrt{a_i}Q_i$, and the total vibrational function is the product of the individual vibrational functions. Then, again neglecting the dependence of $\mu^{el}_{mn}$ on the $Q$s, the transition-moment integral is

$$\mu^{v_m, v_n}_{mn} = \mu^{el}_{mn}(r_{1e}, r_{2e}, \cdots) \int \chi^m \chi^n \, dQ_1 dQ_2 \cdots, \tag{8-5}$$

where $\chi^n = [\chi^n_{v_1}(Q^n_1)\, \chi^n_{v_2}(Q^n_2) \cdots] + [\chi_{anh}(Q^n_1, Q^n_2 \cdots)]$, with a similar expression for $\chi^m$. Here $\chi_{anh}$ is a correction function that is needed because the vibrations are not exactly harmonic. It is often ignored for the sake of simplicity or because data to determine it are lacking.

In general the equilibrium configuration is different for the two electronic states $m$ and $n$. If so, a vertical absorption transition from $m$ to $n$ in general leaves the excited molecule in an excited vibrational state. For each vibrational degree of freedom with respect to which there is departure from equilibrium in state $n$ a series ("progression") of bands can appear, corresponding to the quantum-mechanical spread in the diatomic case 3 (and/or 4).

If the molecule in the excited state has the same equilibrium *symmetry* as in the ground state but different interatomic distances, vertical excitation would lead only to *totally symmetric* vibrations of the excited molecule, and only totally symmetric vibrations of the excited molecule would appear in progressions. Because of the quantum-mechanical spread, however, non-totally-symmetric vibrations can sometimes be excited weakly, but only by even numbers of quanta (see below).

We can illustrate by considering excitation of the carbon dioxide molecule to an excited state in which the molecule is still linear and symmetrical but with increased equilibrium separations. The normal vibrations of carbon dioxide are

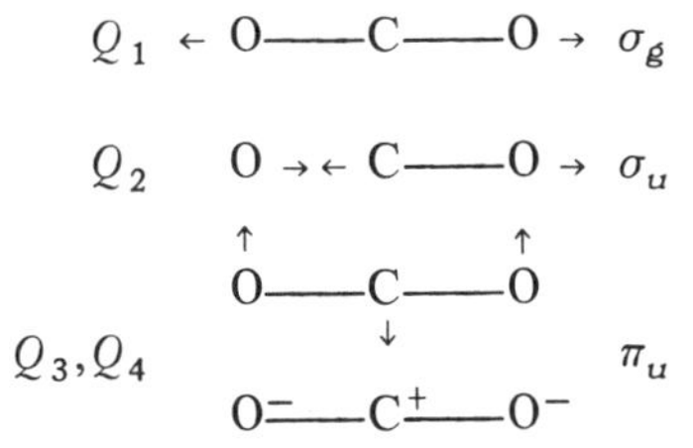

$Q_1$ ← O——C——O → $\sigma_g$

$Q_2$ O →← C——O → $\sigma_u$

$Q_3, Q_4$ (↑O——C↓——O↑; $O^=$——$C^{\pm}$——$O^-$) $\pi_u$

For a vertical transition from $m$ to $n$ the carbon dioxide molecule would be left with the oxygen atoms in the positions indicated by the dashed circles in the diagram below. If the equilibrium C—O distance in state $n$ is as indicated by the solid circles, the molecule can reach equilibrium by expanding according to vibrational mode $Q_1'$ of the excited molecule:

$n$ $CO_2^*$ O——(O)——C——(O)——O

↑ $h\nu$ ↑ $h\nu$

$m$ $CO_2$ O—C—O

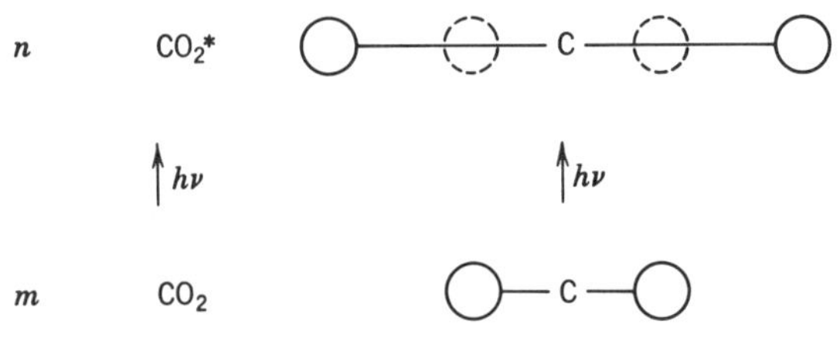

Then because of quantum-mechanical spread, this electronic absorption transition should exhibit a vibrational progression in $\omega_1'$.

As another example consider the ethylene molecule, which has three totally symmetric vibrations as follows:†

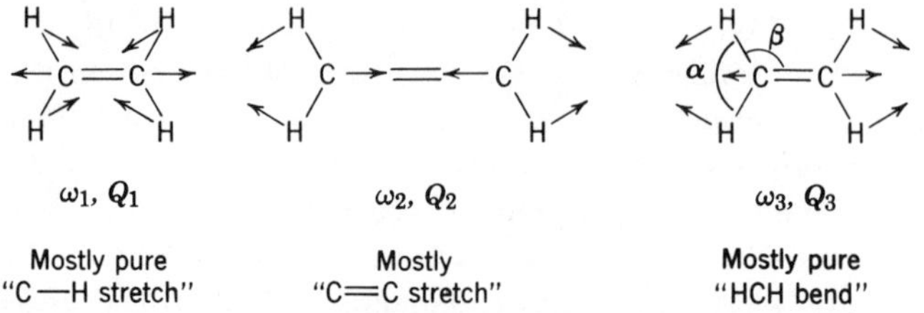

$\omega_1$, $Q_1$ Mostly pure "C—H stretch"

$\omega_2$, $Q_2$ Mostly "C=C stretch"

$\omega_3$, $Q_3$ Mostly pure "HCH bend"

As stated before, if the excited state when in equilibrium has the same symmetry as the ground state but different dimensions, the absorption spectrum may be expected to show bands that correspond to excitation of one or more of these vibrations.

In the picture for ethylene the normal coordinates that describe the forms of the vibrations are linear combinations of the following symmetry coordinates: $R_{C=C}$; $r_1 + r_2 + r_3 + r_4$ where $r_1$ is the change in the C—H distance to hydrogen atom 1, etc.; and $\Delta\alpha_1 + \Delta\alpha_2 - \Delta\beta_1 - \Delta\beta_2 - \Delta\beta_3 - \Delta\beta_4$. In the ground state of ethylene the $Q$s are mostly pure stretch or bend; in the excited molecule the weights of the symmetry coordinates that contribute to the different $Q$s may very well change considerably from those in the ground state. Then the normal coordinates $Q_i^n$ used in the integral in (8-5) and (8-3) to define the excited vibrational wavefunction $\chi_{v_i}^n$ are different from those, $Q_i^m$, that define the lower state function $\chi_{v_i}^m$. Hence we must know the relationship between $Q_i^n$ and $Q_i^m$ and then write $\chi_v^n$, using (8-3), in terms of the ground-state coordinates $Q_i^m$. The resulting expression for $\chi_v^n$ may then be written in terms of the three symmetric wavefunctions by using ground-state normal coordinates (or else vice versa) before evaluating the integral in (8-4). The result is that there may be progressions in all three symmetric vibrations even though the distortion is along only *one* of the symmetric normal coordinates of the ground state.

In addition to these large effects that involve the totally symmetric vibrations there is some quantum-mechanical spread involving those non-totally-symmetric vibrations for which the values of $\omega_i$ are different in

† Based on Fig. 44 of G. Herzberg, *Electronic Spectra and Electronic Structure of Polyatomic Molecules*, Van Nostrand, Princeton, N.J., 1966.

the excited state from those in the ground state. Consider, for example, an electronic transition in carbon dioxide without change of equilibrium symmetry and suppose that the antisymmetrical ($\sigma_u$) vibration has a frequency $\omega_2'$ in the excited state that is quite different from $\omega_2$ in the ground state. Here the normal coordinate is of unchanged form on electronic excitation since it is the only one of $\sigma_u$ symmetry. Hence its behavior in an electronic transition is just like that for the case 2 diatomic molecule, and there is some quantum-mechanical spread in $\omega_2'$, as compared with the expectation of *no change* in v that we would obtain from the verticality principle alone. Just as in the diatomic case 2, only transitions in which the change from $v$ to $v'$ is an even number have nonzero overlap integrals in (8-5); also just as there, for the case $\omega_2' = \omega_2$, only the $0 \leftarrow 0$ (or $v \leftarrow v$) transition occurs. Similarly other non-totally-symmetric transitions with even values of $\Delta v$ may appear in progressions in the absorption spectrum as a result of this type of quantum-mechanical spread, but usually such progressions are much weaker than those for the totally symmetric vibrations. Even for the most extreme case ($k' \simeq 0$) transitions with $\Delta v > 0$ are weaker than for $\Delta v = 0$ and fall off rapidly in intensity with increasing $\Delta v$.

We emphasize here that the integral in (8-5) can be different from zero only if its integrand is totally symmetrical. For this to be true $\chi_{v_i}^n$ and $\chi_{v_i}^m$ must both belong to the same group-theoretical species. This is the only completely *rigorous* requirement for a nonzero integral. If, for example, $\chi_{v_i}^m$ is totally symmetric, $\chi_{v_i}^n$ must be totally symmetric, which is true for all vibrational states in which any or all totally symmetric vibrations are excited by any number of quanta and any non-totally-symmetric vibrations are excited by any *even* numbers of quanta. The above-stated rigorous selection rule is seen to be more permissive than the approximate rule from the Franck verticality principle, which would say that only totally symmetric vibrations can be excited in an electronic transition.

If the upper electronic state has a symmetry that is *different* from the ground electronic state, the statement of the verticality principle must be cast in a more general form; for example, at equilibrium in one of the lower excited electronic states of carbon dioxide the molecule is known to be bent ($C_{2v}$ symmetry). Hence

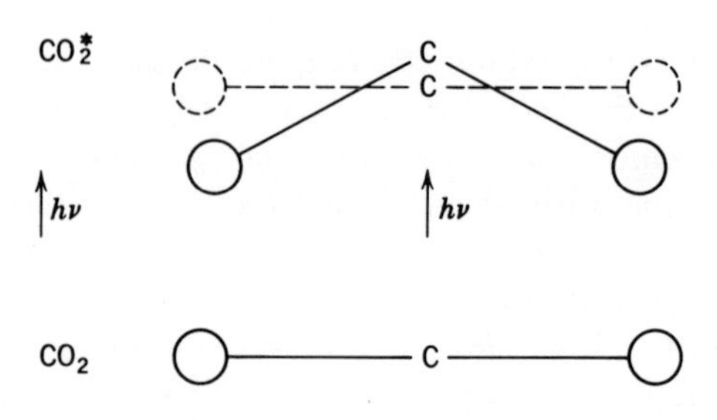


Clearly the verticality principle now predicts the occurrence of a vibrational progression in a bending vibration $\omega_3'$ (and very likely also in $\omega_1'$, since in general the equilibrium C—O bond length changes). The statement that progressions occur mainly in the totally symmetric vibrations still holds true in terms of the upper state equilibrium configuration; the bending vibration for $CO_2^*$ is totally symmetric under the $C_{2v}$ equilibrium symmetry of the upper electronic state.

## 8-4 MOLECULAR COMPLEXES—SHAPE OF THE CHARGE-TRANSFER BAND

In typical band systems (Figs. 8-3 and 8-4) the *total* intensity that is obtained by summing the intensities of the individual bands for transitions to various upper vibrational states is almost the same as the value of the intensity computed for the $0 \leftarrow 0$ transition of Fig. 8-1 for the case in which $\nu[\mu_{VN}^{\text{el}}(R)]^2$ at the equilibrium configuration is the same; that is, the intensity sum (or integral) is the same as the intensity for a single pure electronic transition $n \leftarrow N$. These statements also remain true if the transitions go wholly or partly to the continuous levels above the dissociation limit of the upper electronic state.

In the case of transitions for which $\mu_{mn}^{\text{el}}$ is zero (forbidden transitions) when evaluated at the equilibrium configuration of the nuclear coordinates, the variation of $\mu_{mn}^{\text{el}}$ with nuclear configuration often becomes important. If the value of $\mu_{mn}^{\text{el}}$ differs from zero when the molecule is distorted during a vibration, a symmetry-forbidden electronic transition can in many cases be observed as a result of "vibronic interaction," resulting from the variation of $\mu_{mn}^{\text{el}}$ with $R_k'$. Such vibronic interactions are not believed to be of primary importance in the spectra of complexes, however.

For the charge-transfer band of a molecular complex considerations similar to those for a polyatomic molecule apply. If several of the lower vibrational states are populated (as would be true for the case of complexes at room temperature or even at the temperature of liquid nitrogen, because of their low-frequency vibrations for the motion of D against A), then we expect a set of bands such as the series shown in Fig. 8-4 for *each* of the lower vibrational states. If the vibrational spacings in the normal electronic state are small enough, the absorption as obtained at room temperature is likely to appear continuous as in Fig. 3-1, although for a different reason. In this case the strength of the electronic transition dipole is spread not only over transitions *to* the various charge-transfer-state vibrational levels but also over transitions *from* the various vibrational states of the ground electronic state. Actually most charge-transfer transitions have the appearance of a single continuous band, although $V$-state vibrational structure has been seen in a few cases [3].

Furthermore, both $N$ and $V$ states involved in a CT transition have minima at some value of $R$; in general the minimum in $V$ is at a value of $R$ that is different from the value of $R$ for the minimum in $N$. (See Section 9-1.) Near the minimum the potential curve for the $V$-state as a function of $R$ is roughly parabolic; it can be seen that if $R_e'$ (the value of $R$ for the minimum in the $V$-state) is *either* greater or less than $R_e''$ (the value at the minimum in $N$), then, when quantum-mechanical spread is allowed for, the resulting absorption band $V \leftarrow N$ is asymmetrical with greater breadth on the high-frequency side. Hence, we cannot tell from the asymmetry whether $R_e'$ or $R_e''$ is greater. Only if $R_e' = R_e''$ is the CT band expected to be symmetrical. In fact, as noted in Section 8-1, Briegleb (GR1) has commented on the asymmetry of CT bands. Quite generally CT bands are broader on the high-frequency side with $\nu_H - \nu_L \simeq 2.4\ (\nu_{max} - \nu_L)$ instead of $2\ (\nu_{max} - \nu_L)$ as it would for a symmetrical band. (Here $\nu_L$ is the frequency on the low-frequency side of the band at which the intensity drops to one-half its maximum value; $\nu_H$ is the corresponding value on the high-frequency side of the band.) (See Fig. 3-1.)

In some complexes the overlapping of vibrational frequencies is sufficiently reduced for some vibrational structure to be seen. We may then expect the band shape to be governed by considerations that are similar to those for case 3 diatomic molecules, except that now several symmetric vibrations (including the D—A stretching vibration) may show progressions.

Whereas the half-intensity width of the charge-transfer band is probably determined to some extent by the steepness of the upper potential curve (and hence by the magnitude of $R_e' - R_e''$), it seems likely that the width is governed *primarily* by the population of the molecules in the low-frequency D—A vibrations in the ground electronic state and the corresponding band spread. (See Section 8-2.) Presumably this explanation accounts for the rather strikingly similar values found for $\Delta\nu_{(1/2)}$ of the charge-transfer bands from quite different kinds of complexes; for example, from $b\pi \cdot a\pi$, from $b\pi \cdot a\sigma$, and from $n \cdot a\sigma$ complexes. As we shall see in Chapter 10, however, there is a definite increase in $\Delta\nu_{(1/2)}$ with increasing strength of interaction for the $n \cdot a\sigma$ complexes (such as the amine $\cdot I_2$ complexes), very probably reflecting the changes in the relative positions of the *N* and *V* curves in these stronger complexes.

(Some of the material in this chapter is discussed briefly in footnotes 15 and 17 of R3.)

## 8-5 REFERENCES

[1] G. Herzberg, *Molecular Spectra and Molecular Structure. I. Spectra of Diatomic Molecules,* 2nd ed., Van Nostrand, New York, 1950, pp. 199 ff.
[2] L. Pauling and E. B. Wilson, Jr., *Introduction to Quantum Mechanics,* McGraw-Hill, New York, 1935.
[3] For example, see (a) B. Chakrabarti and S. Basu, *J. chim. phys.,* **63**, 1044 (1966) and (b) S. Saha, A. S. Ghosh, and S. Basu, *J. chim. phys.,* **65**, 673 (1968), which report vibrational fine structure in the CT absorption bands of some $b\pi \cdot a\pi$ complexes.

CHAPTER 9

# *Energy Parameters, Potential Surfaces, and Charge-Transfer Spectra*

Chapter 2 contains some preliminary discussion of the estimation of the energy $W_1$ of the hypothetical pure dative wavefunction $\Psi_1$ and its relation to the energy $W_0$ of the pure no-bond function. We now examine this problem further and then examine the resonance energies that finally lead to the energies $W_N$ and $W_V$ of the normal and CT states. Further illustration of these ideas is provided for $I_2$ complexes in Chapter 10 and more generally in Chapter 14.

## 9-1 POTENTIAL-ENERGY SURFACES; ESTIMATION OF $W_1$

First let us try to estimate the energy of the dative wavefunction $\Psi_1$ in terms of known quantities. These estimates are necessarily very crude because the "known quantities" are not well known; nevertheless, they have some value for the correlation of experimental properties of the complex.

Figure 2-3 showed qualitatively the forms of $W_1$ and $W_0$ as functions of the distance $R_{DA}$ between two convenient points or planes in D and A. That figure is repeated here, with elaborations, as Fig. 9-1. Figure 9-1 shows not only $W_0$ and $W_1$ but also a breakdown of these into sums of their values at $R_{DA} = \infty$ plus the energy changes $G_0$ and $G_1$, respectively, that occur when D and A are brought together in the complex. Finally the resonance energies $-\beta_0^2/\Delta$ and $+\beta_1^2/\Delta$ are added to give $W_N$

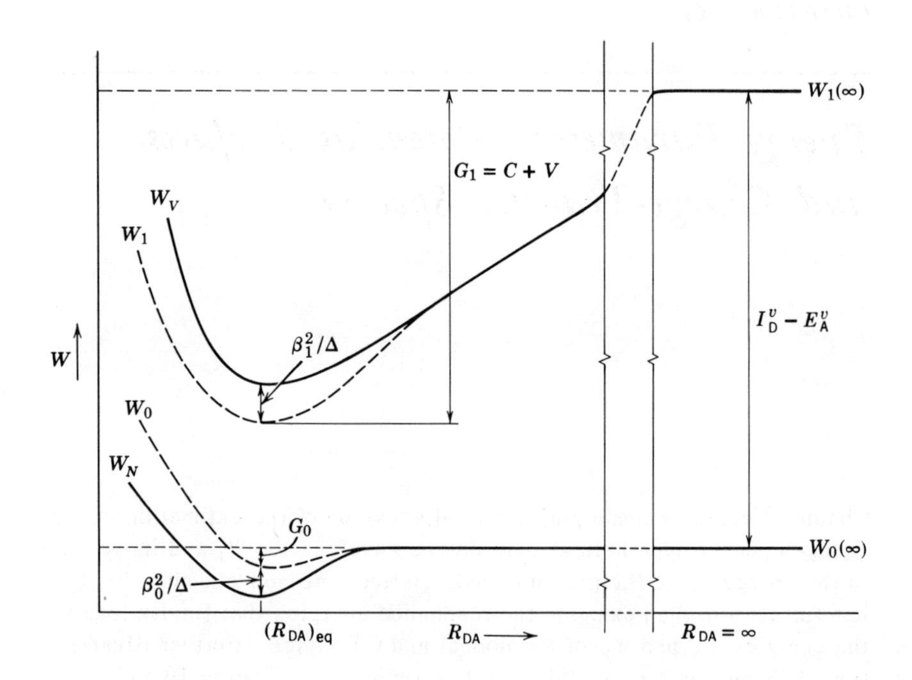


**Figure 9-1** The energies $W_V$ and $W_N$ as a function of the donor-acceptor separation $R_{DA}$.

and $W_V$, respectively. All the energy quantities are functions of $R_{DA}$. *It is assumed* in Fig. 9-1 and following figures that during the variation of $R_{DA}$ *all other* configurational coordinates (those that are internal to D and A as well as those that are external) are continuously adjusted so as to minimize the energy of the complex. In the dative structure $\Psi_1$ the relatively large negative quantity $G_1$ is mostly Coulomb energy that is lost when the $D^+$ and $A^-$ ions are brought together. The energy difference $W_1 - W_0$ for separated D and A ($R_{DA} = \infty$) is just equal to the energy difference $I_D^v - E_A^v$ using *vertical* values of $I_D$ and $E_A$. (See Section 2-4.) *Included* in $G_0$ and $G_1$ are any needed energy contributions that correspond to internal readjustments of D and A to their configurations in the complex. In a weak complex, however, the configurations

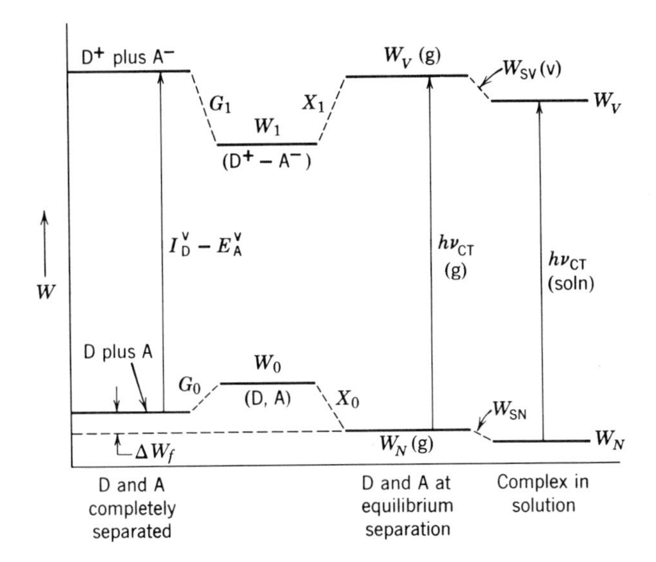


**Figure 9-2** Energy diagram showing a convenient breakdown of terms for $h\nu_{CT}$. $G_0$ (here shown positive) may also be negative, especially in weak complexes. (Figure taken from GR5.)

of D and A are nearly the same as in the free molecules.

In Fig. 9-1 the $G_0$ term is shown as slightly negative. This term represents the interaction energy of donor and acceptor if they were brought together in the no-bond structure $\Psi_0$. In the vapor state London dispersion forces should result in a considerable negative value for $G_0$ if $R_{DA}$ is approximately the sum of van der Waals radii, but in solution these energy contributions are approximately canceled out. (See Section 1-1.) Electrostatic attractions (direct and induced) should also contribute (*both* in vapor and in solution) if D and/or A are ions or have dipole or quadrupole moments. If the minimum energy for the complex in $W_N$ is at smaller D—A distances than the minimum for $W_0$ (i.e., if $R_{DA}$ at the minimum in $W_N$ is less than the sum of van der Waals radii), then $G_0$—at that value of $R_{DA}$—may be *positive,* corresponding to the "steric," or nonbonded, repulsion of two molecules forced together. Generally, $G_0$ should usually be small for loose complexes, but for stronger complexes it may be expected to be rather large and positive.[†]

Fig. 9-2 illustrates the estimation of $W_0$ and $W_1$ at the $R_{DA}$ distance that corresponds to the minimum of $W_N$ in Fig. 9-1. From Fig. 9-2

$$\Delta \equiv W_1 - W_0 = I_D^v - E_A^v + G_1 - G_0. \qquad (9\text{-}1)$$

Furthermore, $\Delta W_f$, the energy of formation of the complex from D and A, is seen to be

$$\Delta W_f = G_0 + X_0. \qquad (9\text{-}2)$$

Here $X_0$ is the resonance energy; in weak complexes $X_0 = -\beta_0^2/\Delta$. [See (2-20).] Finally the Franck-Condon *vertical* frequency $\nu_{max}$ for the CT band (see Fig. 3-2*b*) is

$$h\nu_{CT} = I_D^v - E_A^v + G_1 - G_0 + X_1 - X_0, \qquad (9\text{-}3)$$

which can be equated in general to $h\nu_{CT}$ of (2-15) or, for weak complexes for which it is sufficiently accurate, to (2-22).

The main part of the diagram corresponds to the theory for a complex in the gaseous state,[‡] but at the right the diagram is adapted to solutions by the inclusion of solvation-energy terms. In Fig. 9-2 the solvation energies are shown as rather small, as they should be for nonpolar solvents. For polar solvents they would be much larger for state $N$ and/or state $V$.

Figure 9-3 repeats Fig. 9-2 with some modifications: $G_0$ is shown as the sum of an electrostatic attraction (Attrn) and an exchange-repulsion (Repn) contribution, $G_1$ is shown as the sum of an electrostatic attraction term (Attrn), an exchange-repulsion term (Repn), and a valence term (±Exch). Here "+Exch" leads to $W_1'$ and to the triplet charge-transfer state $T$ and "−Exch" to $W_1$ and to the singlet charge-transfer state $V$. The resonance-energy terms $X_0$ and $X_1$ are now designated "Res." Solvation energies are indicated as in Fig. 9-2.

## 9-2 EVALUATION OF PARAMETERS—COULOMB ENERGY AND $G_1$ IN THE CHARGE-TRANSFER STATE

We are now interested in the values of the parameters that are involved in (9-3) for $h\nu_{CT}$. The ionization energy $I_D$ for the free donor molecule

---

[†] In drawing potential curves such as these one question which has been especially difficult for us to settle concerns the relative positions of the minima of the $V$ and $N$ curves. (See Section 9-4 and Chapter 14.)

[‡] For the gas state $G_0$ would include London-dispersion force attractions, but these should be approximately eliminated (see Section 1-1) in solution.

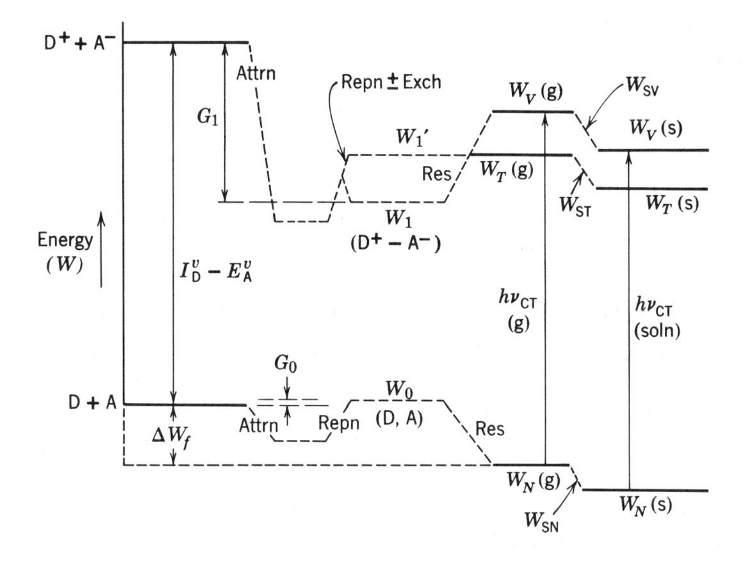


**Figure 9-3** Energy diagram for neutral donor and acceptor. *All* partial as well as total energies (except at extreme left) refer to the same configuration, that of equilibrium in state $N$. In state $N$ examples occur with Res > Attrn, others with Attrn > Res. The possibilities $W_V > W_T$ or $W_T > W_V$ depend on the size of Exch and Res at $W_1$.

is often known experimentally, and $I_D^v$ can then usually be estimated rather accurately. From photoionization experiments Watanabe [1] has obtained reliable values of $I_D$ (usually within $\pm 0.01$ or $0.02$ eV) for many electron donors, and there are many good spectroscopic values from Rydberg series.† The electron affinity $E_A^v$ is harder to obtain. This quantity is not generally available but has been estimated for a number of acceptors. (See Briegleb [2] for a review. Many of the relative values of $E_A$ that he lists—the absolute values are less reliable—are obtained from the CT frequency by using the inverse of the reasoning presented here.) The other quantities, $G_0$ and $G_1$, must be estimated.

For weak complexes $G_0$ should be small and has usually been ignored.

† See Appendix, Section 9-6, for an abridged table of ionization potentials taken from Watanabe, et al. [1]. Note that these are all *minimum* ionization potentials, but that each molecule has a number of higher ionization potentials which in general are also relevant (see Section 11-1). Recently Turner and his coworkers have determined these higher ionization potentials for a number of molecules by photoelectron spectroscopy. (For example, see D. W. Turner, Chap. 3. *Physical Methods in Advanced Inorganic Chemistry,* H. A. O. Hill and P. Day, Eds. Interscience, New York, 1968).

On the other hand $G_1$ is large and negative, mainly because of the large negative term $C$:

$$G_1 = C \pm V \simeq -\frac{q_{D^+} q_{A^-}}{R_{D^+A^-}} \pm V. \tag{9-4}$$

Here $C$ is Coulomb energy plus two other small contributions (additional electrostatic attraction terms—polarization energy—plus exchange repulsion) that we shall neglect, assuming partial cancellation; $\pm V$ is the same as "±Exch" of Fig. 9-3; for the $V$ state it is the energy of formation of the valence bond in the dative function $\Psi_1(D^+—A^-)$; $q_{D^+}$ and $q_{A^-}$ are the charges (usually $\pm e$) on the donor and the acceptor. In weak complexes (large $R_{DA}$) $V$ is very small and can be ignored; in strong complexes it should be appreciable.

Let us now estimate $C$. We note that there is a problem here as to the locations of the positive and negative charges; for example, consider the $Bz \cdot I_2$ complex. Is the center of the negative charge in the center of the $I_2^-$ ion, or is it displaced toward one iodine atom? Also, what configuration has this complex? And can we assume that the positive charge is concentrated at the center of the benzene ring, or should we be more realistic and spread it over the carbon and hydrogen atoms? Since we cannot answer these questions exactly, we can try various assumptions and see how they affect the value of $C$.

In Table 9-1 $C$ is tabulated as a function of $R_{D^+A^-}$, assuming point charges on D and A. As an example for the estimation of $R_{D^+A^-}$ let us first consider the axial model of the $Bz \cdot I_2$ complex (symmetry axis of $I_2$ coincident with the sixfold axis of Bz). Here the D—A distance should be a little smaller than the distance from the positive charge at the center of the benzene ring to the negative charge at the center of the I—I bond computed as follows:

$$r_{vdW}(Bz) + r_{vdW}(I) + r_{cov}(I) = 1.70 + 2.15 + 1.33 = 5.18\ \text{Å}.$$

Here $r_{vdW}$ is the van der Waals distance [3]. It is reasonable to expect a shortening of a few tenths of an angstrom in the complex, to about 5.0 Å, due to the net attraction between D and A.

In amine complexes the positive charge might, for example, be supposed to be centered either on the nitrogen ($H_3N^+—I_2^-$) or distributed over the hydrogens ($H_3^{+(1/3)}N—I_2^-$); values of $C$ for both possibilities are listed in Table 9-1. As we have seen in Section 4-2, molecular-orbital calculations indicate a charge of about $-0.45e$ on the nitrogen in $NH_3$ and about $+0.15e$ on each hydrogen [4a-c]; removal of an electron from nitrogen without other changes would leave about $+0.25e$ on the nitrogen and on each hydrogen. However, we might expect most of

**TABLE 9-1**

**The Coulomb Energy $C$ in $W_1$ for Various $D^+$—$A^-$ Distances**

| $R_{D^+A^-}$ (Å) | $-C$ (eV) | Probable examples |
|---|---|---|
| 5.5 | 2.7 | |
| 5.0 | 2.9 | $Bz^+$—$I_2^-$ (axial model) |
| 5.0 | 2.3–2.7 | $Bz^+$—$I_2^-$ (axial model) (charges spread)[a] |
| 4.3 | 3.3 | $(H_3C^{+(1/3)})_3N$—$I_2^-$ |
| 4.0 | 3.6 | $H_3^{+(1/3)}N$—$I_2^-$ |
| 3.7 | 3.9 | $H_3N^+$—$I_2^-$; $Bz^+$—$I_2^-$ (resting model) |
| 3.4 | 4.2[b] | $b\pi \cdot a\pi$ complexes (charges in center of $D^+$ and $A^-$) |
| 3.4 | 3.3–3.6 | $b\pi \cdot a\pi$ complexes (charges spread)[a] |

[a] Here, for example, the plus and minus charges are spread equally over the six carbon atoms in the benzene ring of the aromatic donor or aromatic acceptor. The range covers various possible spreads of the plus charges. See text.
[b] Here the plus and the minus charges in $D^+$ and $A^-$, respectively, are assumed to be concentrated in the centers of the donor and acceptor planes, which are located in the face-to-face, centered position with respect to each other.

the positive charge on the nitrogen atom to readjust to the H atoms, as we have assumed in the calculation for $H_3^{+(1/3)}N$—$I_2^-$.

Almost all attempts in the past to estimate these parameters (see, for example, R13) have calculated $-C$ by assuming a point positive charge centered on $D^+$ and a point negative charge centered on $A^-$. In such a model $-C$ is the same for all complexes between aromatic molecules (e.g., benzene, naphthalene, and anthracene) and $I_2$ or $a\pi$-acceptors. However, it would be more realistic to assume that the positive charge on the donor is spread out over the carbon atoms in the ring (or in part over the hydrogen atoms or the $CH_3$ groups in methylated ring compounds). When the positive charge is divided equally among the six carbon atoms in benzene, with the negative charge still considered to be centered in the $I_2$ bond, the Coulomb energy at $R_{D^+A^-} = 5.0$ Å is computed to be 2.7 eV, as shown in Table 9-1. The most important point of this calculation, however, is that the stabilization energy depends on the size of the donor molecule: if the positive charge is placed on six centers that are 2.8 Å apart (a crude model representing naphthalene) instead of 1.4 Å as for benzene, then $-C$ is only 2.3 eV.

This effect is even more pronounced for $b\pi \cdot a\pi$ complexes. Dividing both the positive and negative charges equally among the six ring atoms of $D^+$ and of $A^-$ in, for example, benzene · tetracyanobenzene (TCNB), while leaving $R_{D^+A^-}$ unchanged, changes $-C$ from 4.2 eV for the point-charge model to 3.6 eV. (See Table 9-1.) If the six charges are placed 2.8 Å apart, $-C$ drops to 3.3 eV. If the center of the TCNB is shifted so that it is not directly over the center of the benzene, $-C$ drops from 3.6 to about 3.3 eV for a 1.2-Å shift of the centers.

Iwata, Tanaka, and Nagakura [5] have evaluated by semiempirical methods the Coulomb integrals $C_{ij} = \int \phi_{i\mathrm{d}}(1)\, \phi_{i\mathrm{d}}(1) [e^2/r_{12}]\, \phi_{j\mathrm{a}}(2)\, \phi_{j\mathrm{a}}(2)\, d\tau$ for the benzene · TCNB case. Here $\phi_{j\mathrm{a}}$ is the $j$th molecular orbital for the acceptor (TCNB) and $\phi_{i\mathrm{d}}$ is the $i$th molecular orbital for the donor (benzene), obtained from Hückel-type treatments of TCNB and benzene separately. This integral, $C_{ij}$, is then the Coulomb energy of interaction. They found that $C_{ij} \simeq 3.3$ eV, with a range of ±0.1 eV for various acceptor orbitals and assumed orientations of the complexed molecules. Their procedure is to be recommended for future attempts to estimate the Coulomb energy, although the estimates described above assuming spread-out charges agree moderately well with this more exact estimate.

Hence we conclude that $-C$ is most probably about 3.3 to 3.6 eV for $b\pi \cdot a\pi$ complexes, varying with the sizes of D and A. We do not at this time believe that $-C$ should vary much over a set of amine donors ($R_3N$) as R changes, for example, in size from H to $C_2H_5$, so that a constant value of $-C$ of about 3.6 or 3.9 eV should be a good estimate for amine complexes of all kinds. As we shall see later (Sections 9-3 and 10-2) the possible variation of $-C$ with size of the donor has important and hitherto unrecognized consequences in the interpretation of the relationship between $h\nu_{CT}$ and $I_D$.

We note that $-C = 3.3 \pm 0.5$ eV includes most of the cases listed in Table 9-1; indeed, this range of $-C$ values includes most of the possibilities that we may expect to encounter, and in the absence of specific knowledge of configurations and detailed charge distributions we can assume that $-C \simeq 3.3$ eV without risk of serious error. For particular types of complexes with known configurations we can of course refine the estimates.

Let us now consider the $V$ term in (9-4). For weak complexes this term can be neglected. For amine $\cdot I_2$ complexes the N—I distance in state $N$ suggests a bond order of about two-thirds (see Chapter 5), and so $V$ for $\Psi_1$ in the dative state (at the N—I distance found in state $N$) may be about two-thirds of the bond energy for an N—I bond. The N—I bond strength $D_{N—I}$ can be estimated [3] from the N—N and I—I bond energies and the electronegativity difference of nitrogen and iodine:

$$D_{N—I} \simeq (D_{I—I} \times D_{N—N})^{(1/2)} + \Delta = (36 \times 38)^{(1/2)} + 30(0.5)^2 =$$

$$= 44 \text{ kcal/mole} = 1.9 \text{ eV}.$$

Hence $-V_{N—I} \simeq (2/3)D_{NI} = 1.2$ eV. For $b\pi \cdot a\pi$ complexes, in which $D^+$ and $A^-$ in $\Psi_1$ of the dative state are separated by nearly van der Waals distances (3.4 Å) as compared with 1.54 Å for typical C—C bonds, $V$ can be assumed to be negligible. Similarly for the aromatic $\cdot I_2$ complexes $V$ can probably be assumed to be negligible.

Using the estimates for $C$ from Table 9-1 and the above estimates for $V$, we arrive at the $G_1$ estimates listed in Table 9-2.

**TABLE 9-2**

**Estimates of $G_1$ for Some Typical Complexes**

| Complex[a] | $-G_1$ (eV) |
|---|---|
| $Ar \cdot I_2$ (axial model; point charges) | 3.0 |
| $Ar \cdot I_2$ (axial model; charges spread) | 2.4–2.8[b] |
| $Ar \cdot I_2$ (resting model) | 4.0 |
| $b\pi \cdot a\pi$ (point charges) | 4.3 |
| $b\pi \cdot a\pi$ (charges spread) | 3.4–3.7[b] |
| Amine $\cdot I_2$ | 4.8 |

*Note:* All values are approximately ± 0.3 eV.

[a] "Ar" refers to benzene and its alkyl derivatives.

[b] The value to be chosen for these cases depends on the sizes of D and A.

## 9-3 RELATIONSHIP BETWEEN $h\nu_{CT}$ AND $I_D$; EMPIRICAL EVALUATION OF $\beta_0$ AND $\Delta$

For weak complexes Fig. 9-2 and (2-22) (recall that $\Delta \equiv W_1 - W_0$) give

$$h\nu_{CT} = \Delta + \frac{\beta_0^2 + \beta_1^2}{\Delta}. \tag{9-5}$$

For a series of weak complexes between a set of donor molecules and a single acceptor $E_A^v$ is constant, and it has usually been assumed that $G_1$ is roughly constant, although we have just seen in Section 9-2 that $-C$ (hence $G_1$) should vary appreciably with the size of D, for aromatic donors at least. For similar donors $G_0$ may probably be expected to be fairly constant and usually near zero for weak complexes. Also, for weak complexes of a single acceptor with similar donors we may perhaps expect $\beta_0$ and the overlap integral $S_{01}$ to be relatively constant and small. Given $\beta_0$, $S_{01}$, and $\Delta$, $\beta_1$ is calculated from (2-14).

Combining (9-1) and (9-5) for a set of closely related weak complexes of a single acceptor and assuming constant $-C$ values, the following approximate relation is obtained:

$$h\nu_{CT} \simeq \Delta + \frac{C_2}{\Delta} = I_D - C_1 + \frac{C_2}{I_D - C_1}, \tag{9-6}$$

where

$$C_1 = I_D - \Delta = (I_D - I_D^v) + E_A^v - G_1 + G_0, \tag{9-7}$$

and

$$C_2 = \beta_0^2 + \beta_1^2. \tag{9-8}$$

Note that (9-7) involves explicitly the difference between the *vertical* ionization potential $I_D^v$, which is needed by the theory, and the adiabatic ionization potential from experiment, $I_D$. (The latter is the energy difference between the donor ion in its equilibrium configuration and the donor molecule in its equilibrium configuration.) For most molecules (e.g., benzene) the equilibrium geometry of the ion is not expected to be very different from that of the molecule, so that $I_D - I_D^v \simeq 0$. However, for example, for amines the equilibrium configuration of the ion is believed to be planar rather than pyramidal as for the molecule. In such molecules $I_D - I_D^v$ may be appreciable. Later (Section 10-3) we estimate that $I_D - I_D^v$ may range from 0 to $-0.5$ eV for the amines.

Briegleb [6] gives some examples of plots of experimental values of $\nu_{CT}$ against $I_D$ for sets of complexes with various acceptors, as well as values of $C_1$ and $C_2$ obtained by fitting (9-6) to the data. (See Table 9-3.)

**TABLE 9-3**

**Constants of (9-6) for Complexes of $b\pi$-Donors with Several Acceptors**[a]

| | Acceptor | | | |
|---|---|---|---|---|
| | Tetracyanoethylene | Chloranil | Trinitrobenzene | Iodine |
| $C_1$ (eV) | 6.10 | 5.70 | 5.00 | 5.2 |
| $C_2$ (eV) | 0.54 | 0.44 | 0.70 | 1.5 |

[a] Values are from GR1, Table 31, p. 171. The empirical values for iodine complexes listed here give a curve from (9-6) that does not fit the data as well as does a curve using other values. Furthermore, the naive interpretation of $C_1$ and $C_2$ in terms of (9-7) and (9-8) is meaningless if the $b\pi \cdot I_2$ complexes conform to the "axial" model rather than the "resting" model, as is probable. (See discussion of $Bz \cdot I_2$ in Section 10-3.)

It has been found empirically that $h\nu_{CT}$ data from complexes of a series of chemically related donors with a single acceptor, when plotted against $I_D$, fit well on a straight line. (See McConnell, Ham, and Platt [7], also R12 and GR1, Section III.) Since (9-6) clearly has quadratic dependence on $I_D$, it may seem strange that such good linear relations are found empirically. The explanation may be that (9-6) and (2-15) predict such small deviations from linearity over the experimentally feasible range of $I_D$ that the points appear to fall on a straight line within normal limits of experimental error. This topic has been discussed in some detail in R12; Fig. 2 of R12 and Fig. 10-2 from Chapter 10 are especially relevant. Further discussion of the curve-fitting process is given in Section 10-3, where it is applied in a detailed analysis of data for $I_2$ complexes.

Returning now to the evaluation of parameters, we should expect to be able for weak complexes to compare the empirical $C_1$ values of Table 9-3 (from $h\nu_{CT}$ versus $I_D$) with the values predicted from (9-7) by using the parameters discussed above. For $b\pi \cdot a\pi$ complexes, choosing $G_1 = -3.5$ eV from Table 9-2 and $G_0 = 0$, we obtain

$$C_1 \simeq E_A - G_1 + G_0 \simeq E_A + 3.5 \text{ eV}.$$

If we compare these values of $C_1$ with the empirical values in Table 9-3, we see that electron affinities for the $\pi$-acceptors in the range from 1.5 eV for trinitrobenzene to 2.6 eV for tetracyanoethylene are consistent with the empirical value of $C_1$. These values seem possible although they are higher than the values given by Briegleb [2] by about 1 eV because of the smaller value of $-C$ used here. However, we emphasize the uncertainty in $G_1$ for these complexes.

For the complexes involving iodine, $E_A^v$ has been estimated to be 1.7 eV. (See R13 for a review.) This subject is discussed in detail in Chapter 10, where we shall see that the agreement between the empirical values of $C_1$ and the parameters given here is reasonably good. For the strong amine $\cdot I_2$ complexes (9-5) and (9-6) are no longer applicable, and the expression given by (2-15) must be used instead, since the ratio of $\beta_0\beta_1$ to $(\Delta/2)^2$ is too large for validity of the second-order perturbation treatment. (See Chapter 10.)

The values of $\beta_0$ that are obtained from the empirical constant $C_2$ of (9-6) in the case of the weaker complexes are of some interest. For weak complexes $S_{01} \simeq 0.1$ is plausible (whereas $S_{01}$ might be 0.3 or more for more closely bound $n \cdot a\sigma$ complexes), and $\beta_0$ can then be obtained from $C_2$:

$$C_2 = \beta_0^2 + \beta_1^2 = 2\beta_0^2 - 2\beta_0 S_{01}\Delta + (S_{01}\Delta)^2. \tag{9-9}$$

Equation 9-9 is derived from (2-14). The value of $\Delta$ varies from case to case, but let us use an average value to compute $\beta_0$ from the empirical value of $C_2$. A reasonable average value for $\Delta$ for the $b\pi \cdot a\pi$ complexes of Table 9-3 is about 3 to 4 eV—say about 4 eV (except for chloranil complexes, where 3 eV might be better).

By using the empirical values of $C_2$ from Table 9-3 and (2-14), we find the $\beta_0$ and $\beta_1$ values listed in Table 9-4.

However, it is important to note that we could also interpret $C_2$ by a model with *no* charge-transfer resonance energy ($\beta_0 = 0$) if we allow (cf. Section 9-2) for the variation of $-C$ with size.[†] We do not believe that this extreme model, advocated by Dewar,[‡] is correct, but it is clear that some allowance should be made for variation of $-C$ with size, leading to quite small $-\beta$ values, even smaller than those in Table 9-4. Since both donor and acceptor are sacrificial in the $b\pi \cdot a\pi$ and $b\pi \cdot a\sigma$ complexes (for the $b\pi \cdot I_2$ complexes see Section 10-3), it is reasonable that $-\beta_0$ values and charge transfer in state $N$ may be especially small for them. That they are however appreciable is indicated by unpublished preliminary calculations by Friedrich[††] using the whole-complex-MO method of

---

[†] If we consider just the points for benzene and the methylated benzenes in Figs. 31 and 32 of GR1, the scatter of the data about the calculated line with $\beta_0 = 0$ is large enough so that the line apparently fits the data within its experimental uncertainty, even without allowance for a size effect (which, however, would probably be small in any case for this restricted set of data).

[‡] See M. J. S. Dewar and C. C. Thompson, Jr., *Tetrahedron*, Suppl. 7, 97 (1966).

[††] Private communication (1968) from H. B. Friedrich, University of Iowa.

**TABLE 9-4**

**Parameter Values for Some $b\pi \cdot a\pi$ Complexes[a]**

| Type of Complex | $b\pi \cdot a\pi$ | $b\pi \cdot a\pi$ | $b\pi \cdot a\pi$ |
|---|---|---|---|
| Acceptor | TCNE | Chloranil | TNB |
| $G_0$ (eV, assumed) | 0 | 0 | 0 |
| $S_{01}$ (assumed) | 0.1 | 0.1 | 0.1 |
| $\beta_0$ (eV) | –0.28 | –0.30 | –0.36 |
| $\beta_1$ (eV) | –0.68 | –0.60 | –0.76 |

[a] These values should not be taken as absolutely reliable.

Chapter 12. These yield about 3% charge transfer (i.e., $b^2 \simeq 0.03$, which would correspond to $C_2$ of about 1.1 (eV)$^2$ and $\beta_0$ of about –0.5 eV if $S_{01} = 0.1$) in the case of Bz · TCNE.

In a series of complexes in which $R_{DA}$ becomes increasingly less than the van der Waals distance, $S_{01}$ should increase. As stated in Chapter 2, $-\beta_0$ should in general increase as $S_{01}$ increases; and resonance energy, since it increases as $\beta_0^2$ for weak complexes [cf. (2-19)], should then go roughly as $S_{01}^2$. [See (2-36) and (3-15) for further insight.] On the other hand the nonbonded repulsion energies should also increase as overlap increases—and in fact approximately proportionally to the squares of relevant overlap integrals [8,9] (but note that we must consider here for the repulsion *all* overlap integrals between D and A, and not just those involved in the CT action).

Figure 9-4 is introduced here to help clarify ideas about overlap integrals. It shows the estimated overlap integral $S_{da^-}$ for a nitrogen atom orbital of $sp^3$ type with a $5p\sigma$ iodine-atom orbital as a function of the N—I distance. From this figure, for example, approximate values of $S_{da^-}$ for amine · $I_2$ complexes can be estimated if $R_{DA}$ is known. (See Chapter 5.) Further illustration of these ideas is given for iodine complexes in Chapter 10.

## 9-4 POTENTIAL SURFACES FOR $W_N$ AND $W_V$

Now let us examine the potential surfaces in some detail for the simplest case of even-electron neutral donor and acceptor and assuming only one CT state. (For further detail see R14, especially Figs. 1 to 4.

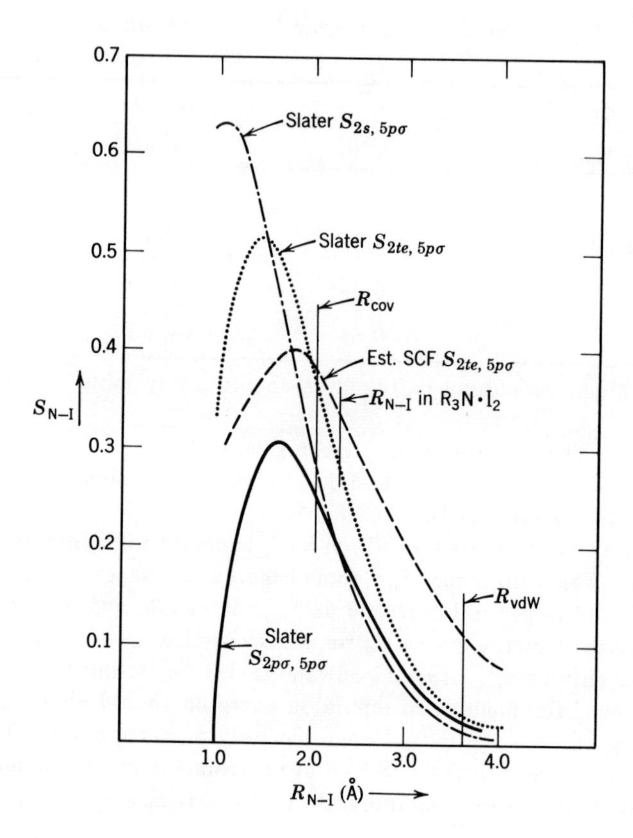


**Figure 9-4** The overlap integral ($\simeq S_{da^-}$) between nitrogen-atom AOs and the $5p\sigma$ AO of the iodine atom as a function of the N—I distance. The dotted line gives the overlap integral between a tetrahedral ($2te$) nitrogen-atom AO and the iodine-atom $5p\sigma$ AO, computed by using Slater AOs. By analogy with a comparison between overlap integrals computed for carbon by using Slater AOs and SCF AOs (for example, see R. S. Mulliken, *Record of Chemical Progress,* **13**, 67 (1952) and also [8]), we estimate that the overlap integral based on SCF AOs for nitrogen and for iodine would be that shown here by the dashed line. This line represents our best estimate of the overlap integral to be used for amine $\cdot I_2$ complexes. The covalent and van der Waals N—I distances [4] are shown by vertical lines; the N—I distance that is found in many complexes is indicated by the shortest vertical line. (See Section 13-1 for definitions of Slater AOs.)

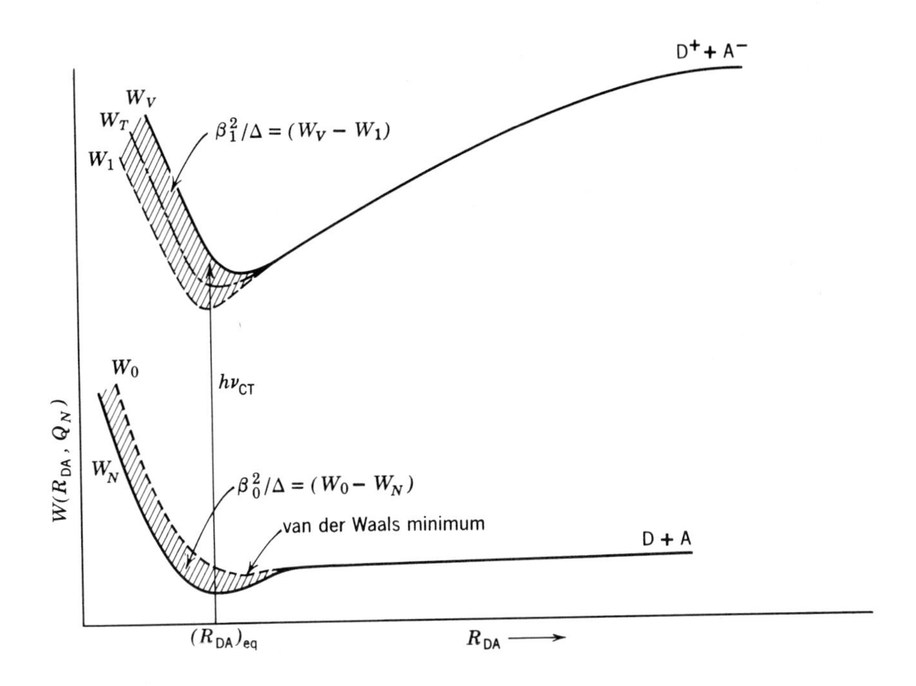


**Figure 9-5** Schematic potential curves for a weak complex.

We shall go into this subject much more carefully in Chapter 14.)

First we emphasize again that $W_N$, $W_V$, $W_0$, and $W_1$ all are expected to depend to some extent on all the coordinates that describe the configuration of the complex; hence they really should be represented by multidimensional potential-energy surfaces. The usual potential curves discussed here represent only cross sections through such surfaces, with all the configuration coordinates fixed except $R_{DA}$. However, the other coordinates (collectively called $Q$) are still there and should not be forgotten. In Figs. 1, 3, and 4 of R14 the potential curves $W_N$ shown for the $N$-state of the complex assume that all the coordinates other than $R_{DA}$ are being adjusted continuously as $R_{DA}$ varies so as to keep the energy as low as possible for state $N$; the minimum in the $W_N(R_{DA})$ curve is at the equilibrium energy of the complex. As a typical example of a strong complex, consider an $R_3N \cdot I_2$ complex. Here $W_N$ depends strongly not only on the N—I distance ($R_{DA}$) but also on the I—I distance, probably on the N—I—I angle, and possibly on other $Q$s; but of all the $Q$s, probably the variations of only one or two, notably $R_{DA}$ and the I—I distance, need be considered in determining $W_N$.

Figure 9-5 shows a more or less typical set of potential curves expected for the interaction of two neutral molecules in a weak complex.[†] Figure 9-5 displays a number of interesting points. The magnitudes of the resonance interactions ($-\beta_0^2/\Delta$ and $+\beta_1^2/\Delta$) are very definitely functions of $R_{DA}$; they are shown in Fig. 9-5 by the shaded portion. We see that they are zero at large $R_{DA}$, and become important only when the overlap between $\phi_d$ and $\phi_{a^-}$ becomes important at $R_{DA}$ values near the van der Waals distance and smaller. (With regard to this overlap and its relation to $\beta_0$ see Sections 3-5 and 3-6.) As a result of this interaction, the $R_{min}$ for $W_V$ is greater than for $W_1$, and $R_{min}$ for $W_N$ is less than that for $W_0$. We recall here that $|\beta_1| > |\beta_0|$ always, at any one $R_{DA}$ value.

The triplet charge-transfer state $W_T$ (see Section 2-7 and Fig. 9-3) is also shown. If there were no valence pairing of the $\phi_d$ and $\phi_{a^-}$ electrons in $\Psi_1$, $W_T$ and $W_1$ would coincide. The valence interaction per se splits this degeneracy, pushing $W_T$ up and $W_1$ down, just as in the analogous case of the Heitler-London theory for the two hydrogen atoms in $H_2$. (See Section 2-7 and Fig. 9-3.) However, the interaction of $W_0$ with $W_1$ now pushes the latter up, so that $W_V$ may come out higher than $W_T$, which does not interact with $W_0$; or, depending on the relative strengths of the two types of interaction, $W_V$ can be below $W_T$. The question as to which of these cases is more likely, or under what circumstances, has not yet been studied carefully. The matter is often further complicated by the interaction of either or both $W_T$ and $W_V$ with locally excited states of D or A; this sort of interaction for $W_V$ is discussed in Chapter 11. Finally, we see that it seems as though the value of $R_{DA}$ for the minimum of $W_V(R_{DA},Q)$ may often be rather close to the value for $W_N$.

In Fig. 9-5 the $W_1$ and $W_V$ curves as well as the $W_0$ and $W_N$ curves are supposed to be drawn with values (call them $Q_N$) for all the configuration coordinates $Q$ such that the energy as a function of $R_{DA}$ is a minimum in state $N$. The reason for doing this is that according to the Franck-Condon principle the intensity maximum of the charge-transfer band, to which $h\nu_{CT}$ in Fig. 9-5 refers, corresponds to a *vertical* transition, which would mean that not only $R_{DA}$ is kept at its state-$N$ equilibrium value (vertical line in Fig. 9-1) but that all the other coordinates also are kept

---

[†] In R14 the $W_1$ curves were drawn with their minima at larger $R_{DA}$ distances than for $W_0$. It is not definitely known where these minima should come, but it seems more likely that they should be at *smaller* $R_{DA}$ values for $W_1$ than for $W_0$.

at their state-$N$ equilibrium values. Thus, for example, in Fig. 9-5 as applied to the $Bz \cdot I_2$ complex the I—I distance in $W_1$ when $R_{DA}$ has its state-$N$ equilibrium value would not be the equilibrium I—I distance for the dative function $\Psi_1$ (about 3.15 Å) but rather the distance (about 2.7 Å) that it has at equilibrium in state $N$ of the complex. The parameters $W_1$ and $W_V$ in Fig. 9-5 would attain lower minima if we allowed the $Q$s to readjust to equilibrium.

As we discussed in connection with the Franck-Condon considerations in Section 8-4, one interpretation of the large half-intensity widths $\Delta\nu_{(1/2)}$ that are observed for charge-transfer bands is that the equilibrium configuration in the $V$ state is considerably different from that in the $N$ state. It seems probable from the considerations in this chapter, however, that $R_{DA}$ at equilibrium is often about the same for both states;† if so, $\Delta\nu_{(1/2)}$ may be attributable to changes in the equilibrium values of the other $Q$s (such as the I—I distance) from state $N$ to state $V$. In Chapter 10 (see Table 10-2), we shall find that $\Delta\nu_{(1/2)}$ increases strongly from ammonia to the trialkylamines; this change can be reasonably explained by the increasing difference in the equilibrium value of $R_{DA}$ in states $N$ and $V$, with the larger value in the $V$ state. However, it seems difficult to account for the magnitude of $\Delta\nu_{(1/2)}$ for $b\pi \cdot a\pi$ complexes in this same way.

## 9-5 PHOTOCHEMISTRY—FLUORESCENCE AND PHOSPHORESCENCE

Any changes from the vertical to the equilibrium values of the configuration coordinates of the charge-transfer state $V$ have interesting implications for the photochemistry of complexes. After absorption of light the excited complex may be expected to dissipate its excess configurational energy and to move toward the equilibrium configuration of all the $Q$s; for example, let us consider the pyridinium halides (which will be discussed in detail in Section 13-4). Here $\Psi_0$ is the ion pair with the $I^-$ perhaps located above the ring near the nitrogen atom. The vertically

---

† For strong complexes, such as amine $\cdot I_2$ complexes, we should expect that $R_e'$ (the value of $R_{DA}$ for which $W_V$ is a minimum) is greater than $R_e''$ (the value for which $W_N$ is a minimum), since the resonance interaction at small $R_{DA}$ values between $W_1$ and $W_0$ tends to push the former curve up and increase $R_e'$. (See Section 14-5.) In weak complexes we might expect $R_e'$ (which is here nearly the minimum for $W_1$ also) to be a bit less than $R_e''$ because of polarization and bonding in the dative state. However, the high nonbonded repulsion that is expected to prevent close approach of donor and acceptor for $b\pi \cdot a\pi$ complexes would probably result in $R_e' \simeq R_e''$. All these considerations apply of course to the $Q_N$ configuration. For $Q_V$ we may expect $R_e'$ to be slightly smaller.

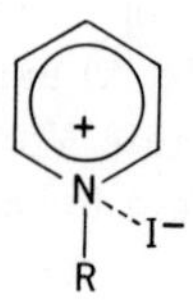


obtained CT state would then comprise the pyridinyl radical with an iodine atom above it at about van der Waals distance and therefore only slightly bonded to it. However, at equilibrium the CT state structure might rearrange to give, for example, the definitely bonded form

Fluorescence is another possibility, but the time required for fluorescence (about $10^{-9}$ sec) is much longer than that for configurational readjustment (about $10^{-12}$ sec). Charge-transfer fluorescence does occur for some $b\pi \cdot a\pi$ complexes [10]; in these complexes the change in equilibrium configuration from the $N$- to the $V$-state is probably much less drastic than in some other cases.

If we wish to consider the fluorescence process, we should redraw the potential curves of Fig. 9-5 to correspond to those that exist when the configuration coordinates have readjusted to the equilibrium configuration for the $V$-state. For simplicity, however, let us still plot $W_V$ as a function of the same $R_{\mathrm{DA}}$ as before, but with $Q$s whose values minimize the energy of the $V$-state at each $R_{\mathrm{DA}}$ value and which can be called $Q_V$. (See Fig. 9-6.)

Figure 9-6 shows two potential curves for each state—one for the coordinate values $Q_N$ and one corresponding to the set of values $Q_V$; for example, in $\mathrm{Bz} \cdot \mathrm{I_2}$, among the $Q_V$, the I—I distance would correspond to the equilibrium value expected in state $V$. However, as discussed above, the equilibrium values of *most* $Q$s probably do not change much

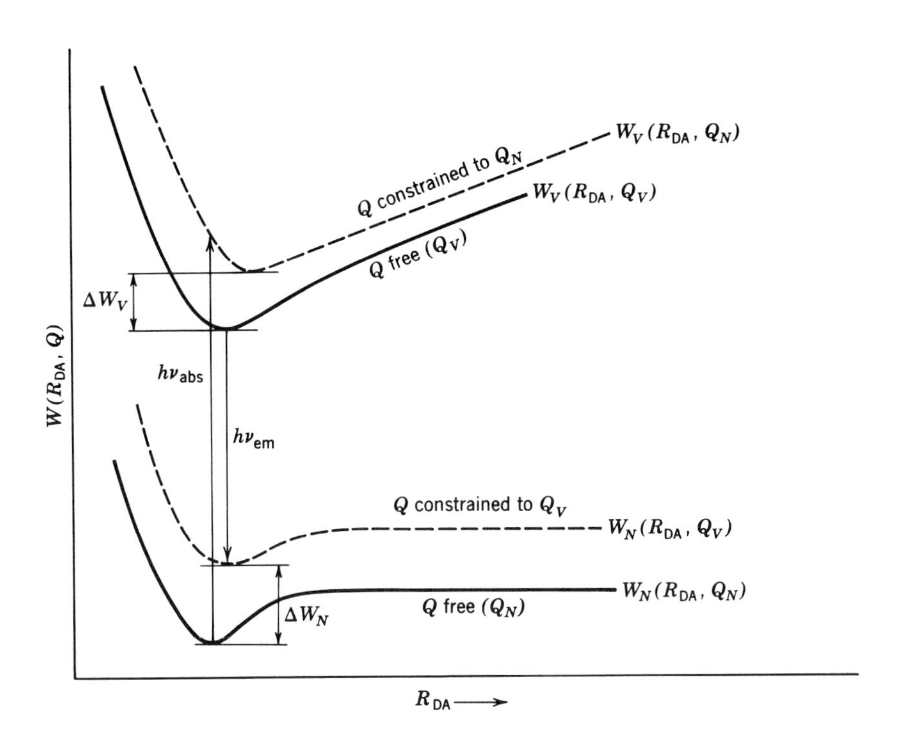


**Figure 9-6** Schematic potential curves illustrating the readjustment of $Q$ before fluorescence. (This figure is essentially the same as Fig. 2 of R14 except that the predicted value of $R_{\min}$ for $W_V$ differs.)

on going from the $N$- to the $V$-state. For iodine complexes we expect the changes to occur mainly in the D—I distances ($R_{DA}$) and the I—I distances; for example, in the $V$-state of the $R_3N \cdot I_2$ complexes we expect changes in the N—I and I—I distances and possibly in the two intermolecular bending coordinates (see Chapter 6), so that state $V$ still resembles state $N$ but approaches closer to $R_3N^+$—I · · $I^-$, with shortened N—I distance and lengthened I—I distance.

In Fig. 9-6 fluorescence is expected usually to occur from the equilibrium configuration of the $V$-state. Thus $\nu_{max}$ for the fluorescent light is expected to be lower than $\nu_{max}$ in absorption. Since the most probable fluorescence process is vertical according to the Franck-Condon principle, $\nu_{max}$ corresponds to a transition to the $W_N$ curve for a configuration $R_V,Q_V$ constrained to be the same as for equilibrium in the $V$-state. If fluorescence were always observed after CT absorption, it would be possible to learn something about the relative shapes of the $W_N$ and $W_V$ curves. Unfortunately, CT fluorescence has been seen for only a relatively few $b\pi \cdot a\pi$ complexes [10].

From Fig. 9-6 the difference in $\nu_{max}$ for absorption and fluorescence is

$$h(\nu_{abs} - \nu_{em}) = \Delta W_V + \Delta W_N.$$

Let us assume that the only $Q$ that changes much for the $N \rightarrow V$ transition for iodine complexes is the I—I distance. Let us further assume that the relative potential curves for this coordinate in the no-bond ($\simeq \Psi_N$) and dative ($\simeq \Psi_V$) functions are approximately those shown in Fig. 6-1. If so, then $\Delta W_V$ is the energy given up when the I—I distance relaxes from that in the free iodine molecule to that in $I_2^-$. From Fig. 1 of R13 (or Fig. 6-1) this energy is about 2.5 − 1.7 = 0.8 eV. Also, $\Delta W_N$ is the energy given up when the $I_2$ distance in $W_N$ is constrained to the equilibrium value for $I_2^-$ or about 0.7 eV. Therefore we very tentatively predict (for weak $I_2$ complexes)

$$h(\nu_{abs} - \nu_{em}) = 0.8 + 0.7 = 1.5 \text{ eV},$$

or

$$\nu_{abs} - \nu_{em} \simeq 12{,}000 \text{ cm}^{-1}.$$

Although absorption to the triplet CT state (Section 2-8 and Fig. 9-4) is expected to be too weak to be observed, phosphorescent emission from it to state $N$ should be observable in favorable cases. After absorption to its CT state (or to a locally excited singlet state) a complex may lose energy by crossing to a locally excited triplet state, or to its triplet CT state if that is lower in energy than the locally excited triplet. Then if the $T$-state is the lowest energy triplet state and if it lies below the $V$-state, phosphorescent emission from it is possible. In the observed cases [11] ($b\pi$-donors, e.g., methylated benzenes, with tetracyanobenzene or tetrachlorophthalic anhydride as acceptor) these conditions are fulfilled. Phosphorescence has also been observed for several $b\pi \cdot a\pi$ complexes with a $^3$LE (locally excited triplet) state, usually of the *donor* (see GR1), as the emitting state. In cases in which phosphorescent emission from $^3$CT states has been observed, the $^3$CT wavefunction is to some extent (varying from 4 to 96%) mixed with a $^3$LE state with the *acceptor* excited.

## 9-6 APPENDIX

**TABLE 9-5**

**Minimum Ionization Potentials for Some Possible Donors[a]**

| Compound | Ionization Potential (eV) | Compound | Ionization Potential (eV) |
|---|---|---|---|
| Methane | 12.98 | Formaldehyde | 10.87 |
| Ethane | 11.65 | Acetone | 9.60 |
| Propane | 11.07 | Formic acid | 11.05 |
| *n*-Butane | 10.63 | Methyl formate | 10.81 |
| *n*-Pentane | 10.35 | Acetaldehyde | 10.21 |
| *n*-Hexane | 10.16 | Acetic acid | 10.37 |
| *n*-Heptane | 10.08 | Methyl acetate | 10.27 |
| Cyclopropane | 11.06 | | |
| Cyclopentane | 10.53 | Ammonia | 10.15 |
| Cyclohexane | 9.88 | Methylamine | 8.97 |
| | | Dimethylamine | 8.24 |
| Methyl chloride | 11.28 | Trimethylamine | 7.82 |
| Carbon tetrachloride | 11.47 | Ethylamine | 8.86 |
| Ethyl chloride | 10.98 | Diethylamine | 8.01 |
| Methyl bromide | 10.53 | Triethylamine | 7.50 |
| Methyl iodide | 9.54 | | |
| Ethyl iodide | 9.33 | Formamide | 10.25 |
| Chlorotrifluoromethane | 12.91 | Dimethylformamide | 9.12 |
| | | Acetamide | 9.77 |
| Water | 12.59 | Dimethylacetamide | 8.81 |
| Methanol | 10.85 | | |
| Ethanol | 10.48 | Hydrogen cyanide | 13.91 |
| Dimethyl ether | 9.96 | Acetonitrile | 12.22 |
| Diethyl ether | 9.53 | Propionitrile ($C_2H_5CN$) | 11.84 |
| Ethylene oxide | 10.56 | | |
| Tetrahydrofuran | 9.42 | Ethylene | 10.51 |
| Tetrahydropyran | 9.25 | Propylene | 9.73 |
| *p*-Dioxane | 9.13 | Vinyl chloride | 10.00 |
| Furan | 8.89 | Acetylene | 11.41 |
| | | Methylacetylene | 10.36 |
| Hydrogen sulfide | 10.46 | | |
| Methyl mercaptan ($CH_3SH$) | 9.44 | Benzene | 9.24 |
| Dimethyl sulfide | 8.68 | Toluene | 8.82 |
| Dimethyl disulfide $(CH_3)_2S_2$ | 8.48 | Ethyl benzene | 8.77 |
| Ethyl mercaptan ($C_2H_5SH$) | 9.28 | *o,m*-Xylene | 8.58 |
| Diethyl sulfide | 8.43 | *p*-Xylene | 8.48 |
| Diethyl disulfide $(C_2H_5)_2S_2$ | 8.27 | Mesitylene | 8.40 |
| Thiophene | 8.86 | Durene | 8.02 |
| | | | |
| | | Naphthalene | 8.12 |
| | | Biphenyl | 8.27 |

**TABLE 9-5 continued**

| Compound | Ionization Potential (eV) | Compound | Ionization Potential (eV) |
|---|---|---|---|
| Phenol | 8.50 | Benzotrifluoride | 9.68 |
| Benzaldehyde | 9.50 | Nitrobenzene | 9.92 |
| Aniline | 7.70 | Anisole | 8.22 |
| Fluorobenzene | 9.20 | Benzonitrile | 9.70 |
| Chlorobenzene | 9.07 | | |
| *m*-Dichlorobenzene | 9.12 | Pyridine | 9.24 |
| Bromobenzene | 8.98 | 2,6-Lutidine | 8.85 |
| Iodobenzene | 8.73 | 4-Picoline | 9.04 |

[a] Values taken from K. Watanabe, T. Nakayama, and J. Mottl [1], except for pyridine (from M. F. A. El-Sayed, M. Kasha, and Y. Tanaka, *J. Chem. Phys.*, **34**, 334 (1961); and for dimethyl ether, diethyl ether, tetrahydrofuran, and tetrahydropyran, which are from G. J. Hernandez, *J. Chem. Phys.*, **38**, 1644 (1963); ibid., **39**, 1355 (1963); and ibid., **38**, 2233 (1963), respectively. Thus all these values are from spectroscopic analyses of Rydberg series or from photoionization experiments. Again, we should like to draw attention to Turner's recent elegant results from photoelectron spectroscopy. (See the footnote on p. 119.)

## 9-7 REFERENCES


[1] K. Watanabe, T. Nakayama, and J. Mottl, *J. Quant. Spectrosc. Radiat. Transfer,* **2**, 369 (1962). See also V. I. Vedeneyev, L. V. Gurvich, V. N. Kondrat'yev, V. A. Medvedev, and Ye. L. Frankevich, *Bond Energies, Ionization Potentials and Electron Affinities,* English ed. (translated by Scripta Technica Ltd.) Arnold, London, 1966.

[2] G. Briegleb, *Angewandte Chemie* (Int. Ed.), **3**, 617 (1964).

[3] L. Pauling, *The Nature of the Chemical Bond,* 3rd ed., Cornell University Press, Ithaca, N.Y., 1960.

[4] (a) D. Peters, *J. Chem. Phys.*, **36**, 2743 (1962); *J. Chem. Soc.*, 4017 (1963); (b) W. E. Palke and W. N. Lipscomb, *J. Am. Chem. Soc.*, **88**, 2384 (1966); (c) R. Body, D. S. McClure, and E. Clement, *J. Chem. Phys.*, **49**, 4916.

[5] S. Iwata, J. Tanaka, and S. Nagakura, *J. Am. Chem. Soc.*, **88**, 894 (1966).

[6] G. Briegleb, *Elektronen-Donor-Acceptor-Komplexe,* Springer Verlag, Berlin, 1961, pp. 74–82.

[7] H. McConnell, J. S. Ham, and J. R. Platt, *J. Chem. Phys.*, **21**, 66 (1953).

[8] R. S. Mulliken, *J. Am. Chem. Soc.*, **72**, 4493 (1950).

[9] R. S. Mulliken, *J. Phys. Chem.*, **56**, 295 (1952).

[10] G. Briegleb, *Elektronen-Donor-Acceptor-Komplexe,* Springer Verlag, Berlin, 1961, pp. 82–88.

[11] S. Iwata, J. Tanaka, and S. Nagakura, *J. Chem. Phys.*, **47**, 2203 (1967). H. Hayashi, S. Iwata, and S. Nagakura, *J. Chem. Phys.*, **50**, 993 (1969). H. Beens and A. Weller, paper 15 presented at the International Conference on Molecular Luminescence at Loyola University, August 21, 1968 (triplet charge-transfer states formed by interaction of $b\pi$-donors with *excited* $a\pi$-acceptors in a manner that is similar to that of excimer formation).

CHAPTER 10

# *Survey of Experimental Results on Iodine Complexes, with Application of the Simplified Theory*

Now let us illustrate the application of some of the ideas given in the previous chapters to some experimental results from studies of iodine complexes. We choose to look at these results in detail because iodine is one of the simpler acceptor molecules whose complexes have been studied rather thoroughly.

## 10-1 THE ELECTRONIC STRUCTURE AND SPECTRUM OF $I_2$

Before examining the results of spectroscopic studies of complexes of iodine let us review what is known about iodine itself. (See R1.) The state and the electronic configuration for the outermost electrons in the ground state of the iodine molecule are given by

$$\cdots(\sigma_g 5p)^2(\pi_u 5p)^4(\pi_g 5p)^4, \quad {}^1\Sigma_g^+. \tag{10-1}$$

(See [1] for a discussion of the notation, also see Section 11-1.) The lowest energy vacant MO is the antibonding MO $\sigma_u 5p$; numerous excited states must arise when one or two electrons are excited from the occupied MOs in (10-1) to the $\sigma_u 5p$ MO. Almost all of these states have predicted energies within 6 eV of the ground level. The iodine molecule might then be expected to absorb light almost continuously in the spectral region from the visible out to 2000 Å because of transitions to these states. Actually, because of selection rules only a few of these

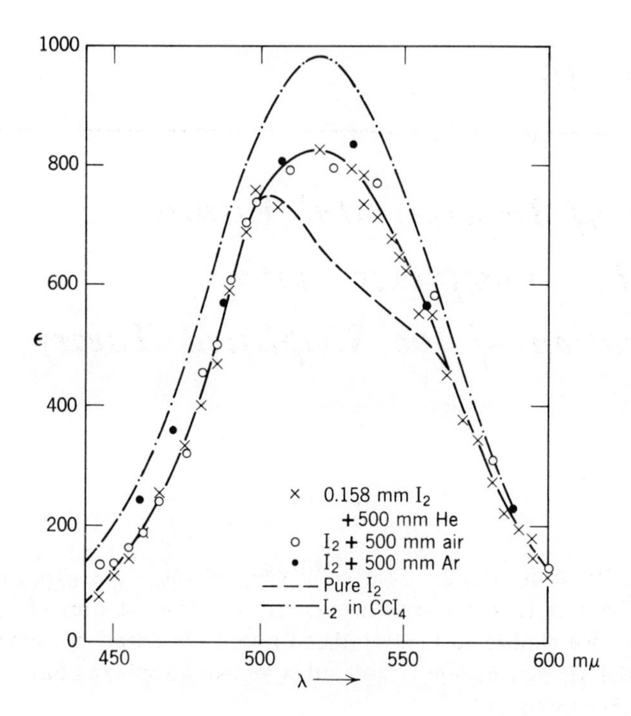


**Figure 10-1** Some low-resolution spectral studies of $I_2$ absorption. (*a*) The "limiting" extinction curve of $I_2$.

transitions are observed with appreciable intensity.

The lowest energy excited configuration and states of the iodine molecule are

$$\cdots(\sigma_g 5p)^2(\pi_u 5p)^4(\pi_g 5p)^3(\sigma_u 5p), \quad {}^3\Pi_{2u}, {}^3\Pi_{1u}, {}^3\Pi_{0^-u}, {}^3\Pi_{0^+u}, {}^1\Pi_u. \quad (10\text{-}2)$$

In iodine the strongest transition to this group of levels is to the ${}^3\Pi_{0^+u}$ state; this, together with weaker absorption to the ${}^1\Pi_u$ level, gives rise to the well-known visible absorption near 5200 Å. The doubly anomalous transition to the ${}^3\Pi_{0^+u}$ (singlet to triplet, which is normally forbidden, and, still more anomalous, ${}^1\Sigma$ to ${}^3\Pi_0$) is explained in terms of the strong intraatomic spin-orbit coupling in the heavy iodine atoms [2], which is believed to cause mixing into the ${}^3\Pi_{0^+u}$ of the ${}^1\Sigma_u^+$ (*V*-state) of (10-3),

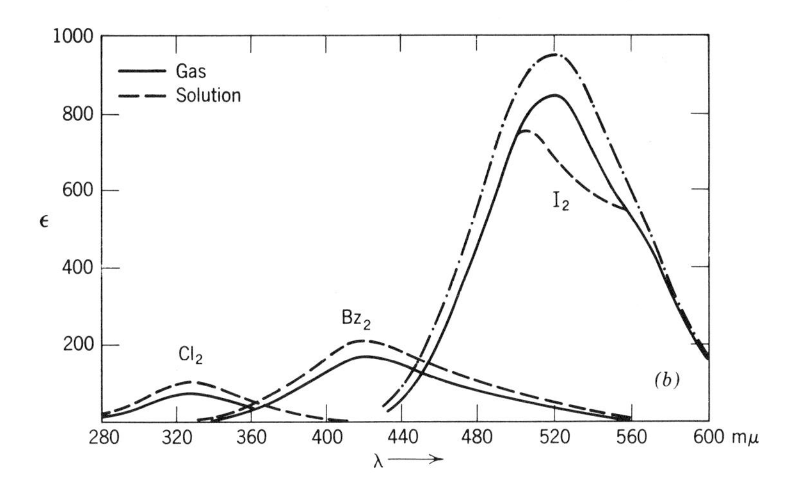


**Figure 10-1** Some low-resolution spectral studies of $I_2$ absorption. (*b*) The extinction curves of the three halogens in gas and solution; both Fig. 10-1 (*a*) and (*b*) are from E. Rabinowitch and W. C. Wood, *Trans. Faraday Soc.*, **32**, 540(1936).

to which absorption is very strong (see below). It is of interest that, like the $V \leftarrow N$, the $^3\Pi_{0^+_u} \leftarrow N$ transition should be polarized along the internuclear axis ("parallel" polarization).

Although the spectrum of $I_2$ has been the subject of many detailed investigations, we need consider here only its general features as obtained with low-resolution spectrometers (see Figs. 10-1). In Fig. 10-1*c* we note that $I_2$ vapor does absorb over the entire visible and ultraviolet region but at most wavelengths the absorption is very weak. The major feature between 7000 and 2300 Å is the visible "band" with maximum near 5200 Å.

In the vapor spectrum this band really consists of very many individual lines, grouped in numerous vibrational bands, from 6000 to 5000 Å, at which point continuous absorption sets in and continues to shorter wavelengths, corresponding to dissociation of the excited molecule into atoms. In the vapor the apparent absorption intensity (Fig. 10-1*a*) in the fine-structure region (between 5000 and 5500 Å) is misleading when observed with a low-resolution spectrometer, because then the absorption is not

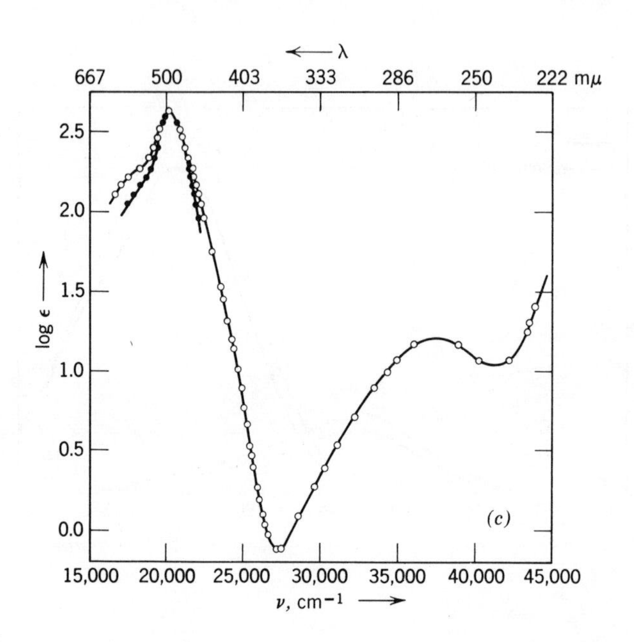


**Figure 10-1** Some low-resolution spectral studies of $I_2$ absorption. (*c*) Absorption spectrum of iodine vapor (•, 80°C; o, 340°C).

proportional to the absorption coefficient where the latter is large. Rabinowitch and Wood [4] showed that when iodine vapor was pressurized by the addition of a neutral gas (helium or argon) to wipe out fine structure, the somewhat peculiar apparent band shape for the unpressurized vapor changed into a smooth band† (Fig. 10-1*a*), which reflects the true (but smoothed out) $\epsilon_\nu$ distribution. The apparent maximum absorption shifts from 5000 Å in unpressurized iodine vapor to 5200 Å in the pressurized gas. There is no further shift in position when the spectrum is observed in solution in an inert solvent.

In Fig. 10-1*c* we note that after falling to a minimum near 3700 Å the intensity of absorption increases again toward shorter wavelengths. Fig. 1-1 continues the spectrum which in Fig. 10-1 is cut off at 2200 Å, and shows very strong absorption near 2000 Å with maximum, in the vapor,

† An alternative, but unacceptable, explanation has been given for this change by Goy and Pritchard [5]. See also Ogryzlo and Thomas [6].

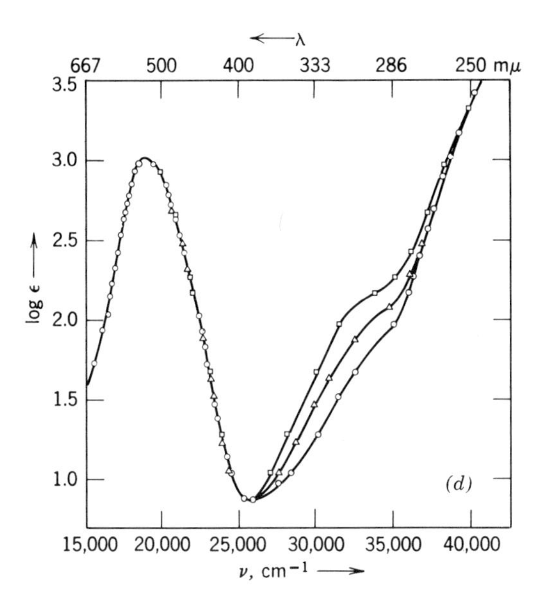


**Figure 10-1** Some low-resolution spectral studies of $I_2$ absorption. (*d*) Absorption spectrum of the "violet" solution of iodine in cyclohexane ○: $c = 3.97 \times 10^{-3}$ *M*; Δ: $c = 2.02 \times 10^{-2}$ *M*; □: $c = 3.97 \times 10^{-2}$ *M*. Both Figs. 10-1 (*c*) and (*d*) are from G. Kortüm and G. Friedheim, *Z. Naturforschung*, **2a**, 20 (1947).

at 1800 Å. This absorption is believed to correspond to a transition to the excited state

$$\cdots(\sigma_g 5p)(\pi_u 5p)^4(\pi_g 5p)^4(\sigma_u 5p), \quad {}^1\Sigma_u^+. \tag{10-3}$$

It is identified with the $V \leftarrow N$ transition [7] of $I_2$, an intramolecular two-way CT transition that is rather similar to an intermolecular CT transition. This $V \leftarrow N$ band has $\epsilon_{max} \simeq 22{,}000$ and $f \simeq 0.4$ in the vapor state.† It shows what appears to be a normal solvent red shift [8] when observed in solution in *n*-heptane (see Fig. 1-1). As mentioned in Chapter 1, Fig. 1-1 also illustrates the appearance, as a shoulder near 2300 Å, of a "contact charge-transfer band" for iodine dissolved in *n*-heptane.

Now turning to Figs. 10-1*c* and 10-1*d* we note that the weak absorption that is observed near 2700 Å, both in iodine vapor and in iodine solutions, does not obey Beer's law. This absorption is explained as being due to

† These values are from L. Julien and W. B. Person, *J. Phys. Chem.*, **72**, 3059 (1968).

absorption in part by $I_2$ and in part by the iodine dimer, $I_4$. In solution, deMaine [9] and Keefer and Allen [10] have been able to separate the absorptions that are attributable to $I_2$ and to $I_4$. They have also obtained the heat of formation for $I_4$ and find it to be of the same order of magnitude as for the benzene $\cdot I_2$ complex. McConnell has pointed out [11] that on the basis of the relationship between frequency and ionization potential in Chapter 9 (the ionization potential of $I_2$ is 9.28 V) the frequency of the absorption corresponds to that predicted for a CT band of $I_4$ regarded as an $I_2 \cdot I_2$ complex stabilized by two-way charge transfer. However, it seems likely that cross-bonding between the two $I_2$ molecules may be important as well as two-way charge transfer in stabilizing $I_4$. The presence of this dimer usually does not interfere in studies of iodine complexes since the $I_2$ concentration in such studies is usually so small ($< 10^{-3}$ $M$) that the amount of dimer is negligible.

A weak shoulder is observed on the long wavelength side of the visible $I_2$ band, near 6900 Å. This corresponds to absorption to the $^3\Pi_{1u}$ state of equation (10-2). A transition to the related $^3\Pi_{2u}$ state, were it not forbidden, should give a continuous spectrum with a Franck-Condon maximum near 8000 Å [2]. A diligent search of the spectrum of iodine in *n*-heptane (and also that of the pyridine $\cdot I_2$ complex) in this region by Peters [12] failed to indicate any absorption due to this expectedly strongly forbidden transition.

There are a number of interesting and unanswered questions concerning possible changes from the electronic spectrum of free $I_2$ to the spectrum of the corresponding "locally-excited" $I_2$ transitions (cf. Section 1-3) that should occur when the iodine molecule has entered a complex; for example, we might expect that the lowered symmetry of the $I_2$ in the complex could result in the appearance of various locally excited transitions whose analogues are forbidden in the free molecule. Such transitions have, however, as yet not been found.

The "blue shift" that is observed for the visible band on complexing (see below) suggests that a similar shift might be expected for the strong $V \leftarrow N$ transition near 1800 Å (perhaps opposed, however, by a normal solvent shift). More information about this locally excited transition in complexed $I_2$ would be desirable in view of its close relationship to the visible iodine absorption as discussed above.

## 10-2 THE ELECTRONIC STRUCTURE AND SPECTRUM OF $I_2^-$

The potential function of the ground state of $I_2$ is probably one of the best known of all potential curves [13]. In contrast, very little is known experimentally about the radical-ion $I_2^-$. Because of the appearance of $I_2^-$ in $\Psi_1$ of (2-2) for iodine complexes, its properties are of special importance. This subject has been discussed in a footnote by Mulliken (footnote 25 of R1) that was later expanded by Person (see R13). The electronic spectrum of $I_2^-$, formed by X-ray irradiation of potassium iodide crystals and trapped at low temperatures in the crystal lattice, has been studied by Delbecq, Hayes, and Yuster [14]. These studies provide information (see R13) about excited states of $I_2^-$.

The ground-state structure of $I_2^-$ is

$$\cdots(\sigma_g 5p)^2(\pi_u 5p)^4(\pi_g 5p)^4(\sigma_u 5p), \quad {}^2\Sigma_u^+. \tag{10-4}$$

The electron in the antibonding $\sigma_u$ orbital is expected to weaken the I—I bond to give a dissociation energy about one-half that for $I_2$ ($D_{I_2}$ = 1.54 eV; $D_{I_2^-} \simeq 0.7$ eV), a longer equilibrium distance ($r_{I_2}$ = 2.67 Å; $r_{I_2^-} \simeq 3.15 \pm 0.1$ Å), and a smaller vibrational frequency ($\omega_{I_2}$ = 215; $\omega_{I_2^-} \simeq 115 \pm 50$ cm$^{-1}$) (see R13). Person suggested that the properties of the $I_2^-$ ion might be expected to be similar to the known properties of iodine in its ${}^3\Pi_{0^+u}$ state; the estimates above are derived mainly from that comparison. These predictions have already been used in discussing the configurations and the vibrational spectra of iodine complexes in Chapters 5 and 6.

As noted in Chapter 6 the electron affinity of $I_2$ as a function of $R_{I—I}$ would be given by a knowledge of the potential curve of $I_2^-$ relative to that for $I_2$; an estimated curve deduced by Person is given as Fig. 6-2 and as Fig. 1 in R13 (see also R2). On the whole we can feel moderately confident about our knowledge of $I_2^-$, even though direct experimental information is lacking.

## 10-3 THE FREQUENCIES OF CHARGE-TRANSFER BANDS

The frequencies of the charge-transfer bands of some complexes of $I_2$ are plotted against the minimum ionization potentials of the donors (see Chapters 2 and 9) in Fig. 10-2. The ionization potentials are mostly from Watanabe [15]; the charge-transfer frequencies are from the literature (see GR1) without regard to solvent and thus they may shift a bit because of varying solvent effects.

Figure 10-2 indicates three different families of complexes with $I_2$: (*a*) those with $b\pi$-donors, (*b*) those with aliphatic amine (*n*) donors, and (*c*) those with alcohol and ether (*n*) donors. Before the publication of the data of Yada, Tanaka, and Nagakura [16] on the aliphatic amine complexes the need for separate treatment of different donor types was not recognized,

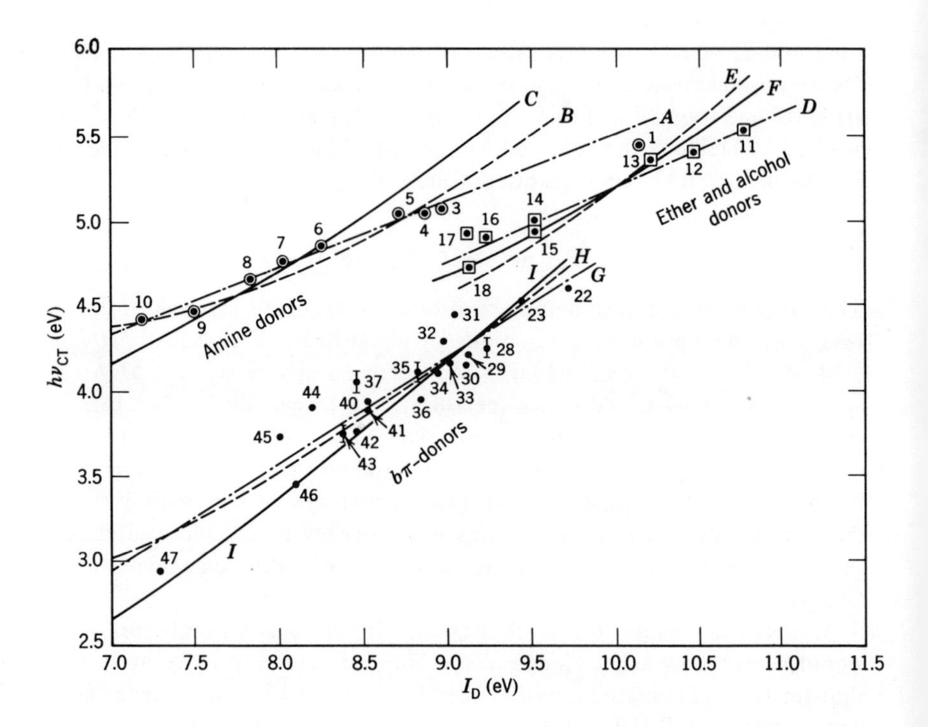


**Figure 10-2** Frequencies of the charge-transfer bands for some $I_2$ complexes, plotted against the minimum ionization potentials of the donors. The latter are photoionization potentials taken from Watanabe [15] where possible. The vertical lines indicate the scatter of data, often for different solvents, to give *some* idea of experimental uncertainty in the charge-transfer frequency.

The donors are 1. $NH_3$; 3. $MeNH_2$; 4. $EtNH_2$; 5. *n*-$BuNH_2$; 6. $Me_2NH$; 7. $Et_2NH$; 8. $Me_3N$; 9. $Et_3N$; 10. (*n*-Pr)$_3$N; 11. MeOH; 12. EtOH; 13. propylene oxide; 14. tetrahydrofuran; 15. $Et_2O$; 16. tetrahydropyran; 17. dimethylformamide; 18. dioxane; 22. propylene; 23. hexene-1; 28. benzene; 29. *cis*-butene; 30. *trans*-butene; 31. 1,3-butadiene; 32. bromobenzene; 33. cyclopentene; 34. cyclohexene; 35. toluene; 36. isoprene; 37. *p*-xylene; 40. *o*-xylene; 41. *m*-xylene; 42. styrene; 43. mesitylene; 44- anisole; 45. durene; 46. naphthalene; 47. anthracene, with estimated $I_D$.

Frequency data from the summary by Briegleb (GR1) except for the alcohols (From L. Julien, Ph.D. thesis, University of Iowa, 1966.) (Me = methyl; Et = ethyl; *n*-Bu = *n*-butyl; *n*-Pr = *n*-propyl.)

and, for example, Briegleb (GR1) fitted all the data then available with one curve of the form of (9-6). This curve is shown in Fig. 10-2 by the solid line through the data for the $b\pi$-donors marked $I$, with $C_1 = 5.2$ eV, $C_2 = 1.5$ $(\text{eV})^2$.

Yada, Tanaka, and Nagakura recognized that a more general equation, of the form of (2-15) must be used to fit their data; they fitted separate curves to the data for primary, secondary, and tertiary amines. With somewhat poorer fit one single curve was fitted to all the amine data by Mulliken and Person (R12 and R13). The curve that they drew was similar to the one shown here and marked $C$; the parameter values that they listed ($C_1 = 6.9$, $-\beta_0 = 2.0$ eV, and $S_{01} = 0.3$) are slightly different from the values given here in Table 10-1.†

To illustrate the difficulty of finding unique curves to fit the data, we show in Fig. 10-2 three curves through each set of data. The full line ($I$ for $b\pi$-donors, $F$ for the alcohol and ether donors, and $C$ for aliphatic amine donors) represents one curve through each set of data. The dashed, curved lines ($B$, $E$, $H$) for the most part fit the data somewhat better but with parameters that seem rather less acceptable. All of these curves have been drawn to fit equations of the form of (2-15), although for $b\pi$-donors this amounts to the same thing as a fitting to (9-6), to which (2-15) reduces in the case of very weak complexes. In addition to the curves we show a dashed straight line ($A$, $D$, $G$) through each set of data (see below).

Before attempting to use (2-15) we note that it expresses a relation between $h\nu_{CT}$ and $I_D$ in terms of the three independent parameters $S_{01}$, $C_1$, and $\beta_0$, each of which can be expected to be roughly constant for each type of donor; namely, $\Delta$ of (2-15) is identical by definition with $I_D - C_1$ (Eq. 9-7), and $\beta_1 = \beta_0 - S_{01}\Delta$ (Eq. 2-14). About all we can do for the parameter $S_{01}$ is to assume a reasonable value. The other two ($\beta_0$ and $C_1$) can then be determined by a trial-and-error procedure such that the resulting curve of the form of (2-15) fits the $h\nu_{CT}$, $I_D$ data points within their experimental uncertainty. As we shall see, this empirical procedure is not very satisfactory because several curves with different parameter values can be found (as in Fig. 10-2) that fit the data about equally well. We can require that the parameter values be "reasonable" so that, for example, the empirical $C_1$ agrees with the value estimated from (9-7) as described in Chapter 9. Further, $S_{01}$ is reasonably expected to be between 0.4 and 0.5 for strong aliphatic amine $\cdot I_2$ complexes,‡ or between 0.05 and 0.15 for the weak $b\pi \cdot I_2$ complexes.

† These differences are partly the result of an arithmetic error by Persons in the earlier papers.

‡ From Fig. 9-4, $S_{da^-} = 0.33$ for $(CH_3)_3N \cdot I_2$; with $S_{01} = \sqrt{2}S_{da^-}/\sqrt{1+S_{da^-}^2}$ [see (2-36)], this gives $S_{01} = 0.45$.

We may also require the empirical values of $\beta_0$ to be approximately proportional to $S_{01}$ (Section 3-5), adding an additional constraint that helps to determine what empirical values of the parameters are reasonable. In this way (and as described in more detail below) we have obtained the values of the parameters listed in Table 10-1.

Each of the straight lines ($A$, $D$, $G$) in Fig. 10-2, of the form

$$h\nu_{\mathrm{CT}} = mI_{\mathrm{D}} + n,$$

provides a better empirical fit to each set of data than either curved line. The values of $m$ are all less than 1.0 ($m = 0.62$, 0.40, and 0.38, with $n = -1.40$, $+1.20$, or $+1.70$ eV, for the complexes with $b\pi$-donors, alcohol and ether donors, or aliphatic amine donors, respectively). We earlier (R12, GR5) interpreted these small values for $m$ as evidence that CT resonance energy is important, causing the data to be fitted by a flat curve (see below) that is approximately a straight line with slope less than 1.0.

However, the *change in size* of the donor as $I_{\mathrm{D}}$ decreases (especially for the $b\pi$-donors) and the subsequent change in $-C$ (see Section 9-2) could also cause this slope to be less than 1. In fact we can estimate the change in Coulomb stabilization from $\mathrm{Bz} \cdot \mathrm{I_2}$ ($-C = 2.7$ eV, charges spread) to naphthalene $\cdot \mathrm{I_2}$ ($-C = 2.3$ eV, charges spread); the resulting calculated straight line with $C_2 = 0 = \beta_0$ for $b\pi \cdot \mathrm{I_2}$ complexes is almost exactly the dashed straight line shown here. Nevertheless we believe that $C_2 > 0$ but think there is *also* some decrease in $-C$ with increasing $I_{\mathrm{D}}$. Here, however, it is necessary to point out that for $\mathrm{Bz} \cdot \mathrm{I_2}$ and related $b\pi \cdot a\sigma$ complexes the forgoing considerations based on the use of (2-15) are valid only if the resting model is correct for them. If the axial model is correct, as is rather probable, a modified discussion is necessary [see below, in the paragraphs including (10-9)], although the fact that changing $-C$ can decrease the slope is still relevant.

Let us now consider the process of choosing the parameters of (2-15) to fit the data in Fig. 10-2, ignoring the possibility of changing values of $-C$. Let us define $y = h\nu_{\mathrm{CT}}$, and $x = \Delta = I_{\mathrm{D}} - C_1$, assuming that $C_1$ is constant for all iodine complexes with a given type of donor. Then (2-15) becomes

$$y = \frac{2[(x/2)^2 + \beta_0\beta_1]^{(1/2)}}{1 - S_{01}^2}. \tag{10-5}$$

Substituting the value of $\beta_1$ from (2-14), replacing $\Delta$ by $x$, and squaring both sides of (10-5), we obtain

$$y^2 (1 - S_{01}^2)^2 - x^2 + 4\beta_0 S_{01} x = 4\beta_0^2.$$

Defining $x' = x - 2\beta_0 S_{01}$ and rearranging, we have

**TABLE 10-1**

**Likely Values of the Parameters of (2-15) for Some $I_2$ Complexes[a]**

| Complex with | $S_{01}$ | $C_1$ (eV) | $-\beta_0$ (eV) |
|---|---|---|---|
| $\pi$-Donors: | | | |
| Curve *I* | <0.1 | 5.3 | 1.0 |
| Curve *H* | <0.1 | 6.2 | 1.4 |
| Summary | 0–0.1 | 5.0–6.2 | 0.7–1.4 |
| "Reasonable" values[b] | 0–0.15 | 4.7[c]–5.7[d] | 0.5–1.0 |
| Ethers and alcohols: | | | |
| Curve *F* | 0.3 | 8.7 | 2.0 |
| Curve *E* | 0.3 | 8.1 | 1.8 |
| Summary | 0.3–0.4 | 7.5–9.0 | 1.8–2.2 |
| "Reasonable" values[b] | 0.3–0.5 | 6.5–8.0 | 1.5–3[e] |
| Aliphatic amine donors: | | | |
| Curve *C* | 0.40 | 7.3 | 1.8 |
| Curve *B* | 0.40 | 8.2 | 2.0 |
| Summary | 0.3–0.45 | 6.5–8.7 | 1.8–2.2 |
| "Reasonable" values[b] | 0.4–0.5 | 6.5–7.5 | 1.7–3[e] |

[a] $S_{01}$ values are assumed. Parameters given in the table are related to quantities in (2-15) as follows: $C_1 = I_D - \Delta$ [by definition, see (9-7)]; $\beta_1 = \beta_0 - S_{01}\Delta$ [see (2-14)].
[b] Based on considerations given in Chapter 9.
[c] Estimated for axial model (Chapter 9).
[d] Estimated for resting model; the values of $C_1$ that are obtained from fitting curves *H* and *I* to the data implicitly assume this model.
[e] Based on a "reasonable" ratio of $S_{01}$ for these complexes to $S_{01}$ for $\pi$-donor $\cdot I_2$ complexes ($\simeq$ 3) assuming $-\beta_0 \simeq 0.5$ to 1.0 for the latter.

$$y^2\left[\frac{(1 - S_{01}^2)}{4\beta_0^2}\right] - x'^2\left[\frac{1}{4\beta_0^2(1 - S_{01}^2)}\right] = 1. \qquad (10\text{-}6)$$

This is the standard equation for a hyperbola centered on the $y$-axis with

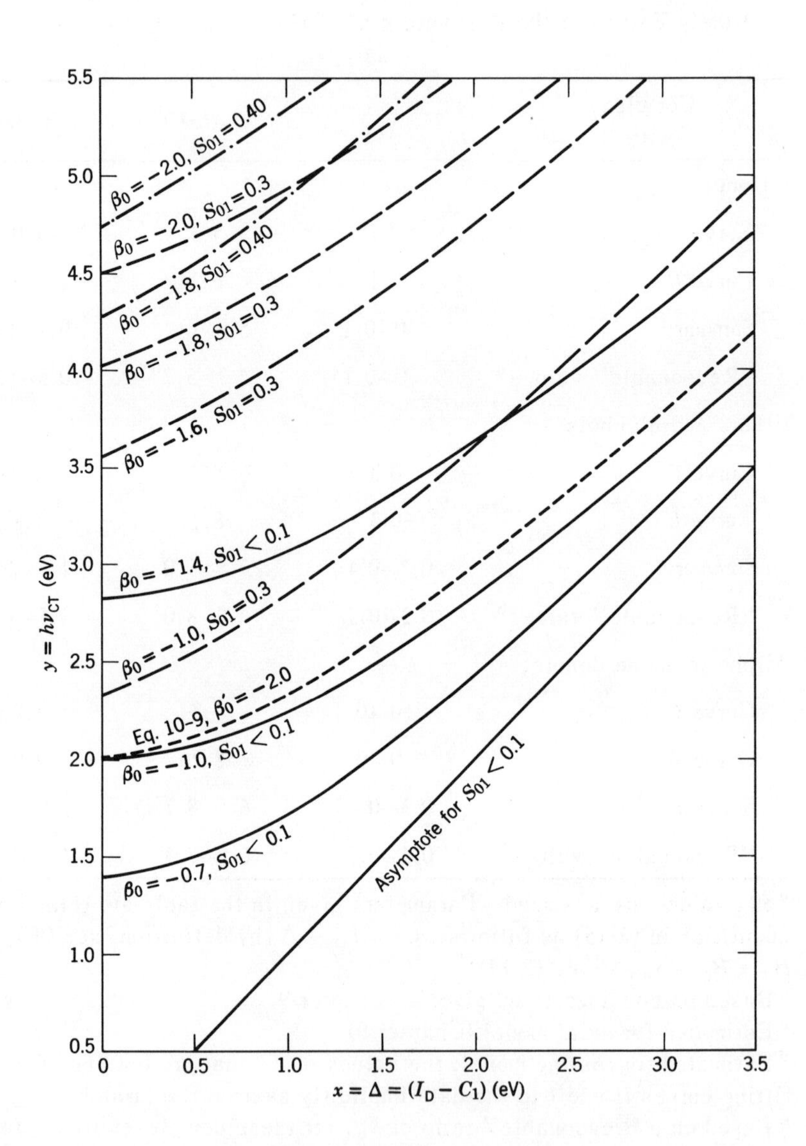


**Figure 10-3** Graph showing curves of (10-6) plotted for several different parameter values. Curves for $S_{01} < 0.1$ are —; those for $S_{01} = 0.3$ are — — —; those for $S_{01} = 0.40$ are —·—·—. A curve for (10-9) is - - - -.

vertex at $x' = x - 2\beta_0 S_{01} = 0$, $y = -2\beta_0/\sqrt{1 - S_{01}^2}$, and with asymptotic straight lines $y = \pm x/(1 - S_{01}^2)$.

Members of this family of hyperbolas are plotted in Fig. 10-3, giving flat curves, appearing almost linear over large ranges (2 to 3 eV) of $x(\equiv \Delta \equiv I_D - C_1)$. The slopes of these curves $\{dy/dx = x'/(1 - S_{01}^2) \times \times [4\beta_0^2 (1 - S_{01}^2) + x'^2]^{(1/2)}\}$ approach unity for small $S_{01}$ and $-\beta_0$ or $1/(1 - S_{01}^2)$ for very large values of $\Delta$. As $S_{01}$ and $-\beta_0$ increase for a fixed value of $\Delta$, the curves become flatter and shallower and are displaced toward the left as $-2\beta_0 S_{01}$ becomes larger, but the curvature is never very great. We see that the slopes of the curves approach zero as $x'$ tends to zero; hence we must choose curves with $\Delta$ fairly small in order to fit the data of Fig. 10-2.

We now illustrate the choice of parameters ($\beta_0$, $S_{01}$, and $C_1$) for a curve of the type of (10-6) to fit a set of data such as that in Fig. 10-2 for the $I_2$ complexes with one donor type. We do so by superimposing the set of curves shown in Fig. 10-3 on the graph of the data. The superposition is made so that the ordinates ($y \equiv h\nu_{CT}$) coincide on the two figures, but one figure is moved with respect to the other along the abscissa to determine possible values of $C_1$ (with "reasonable" values of $\beta_0$ and $S_{01}$) by determining the values of $x$ ($\equiv\Delta$) for which curves from Fig. 10-3 fit the data points of Fig. 10-2. Several possible curves from Fig. 10-3 will be found to fit the data almost equally well (for example, *H* and *I* in Fig. 10-2). One then tries to choose a curve that has parameters that appear to be more reasonable than the others; for example, curve *H* seems to us to give less reasonable parameters than does curve *I*, in that $\beta_0$ and $C_1$ are larger for curve *H* than might reasonably be expected for these weak complexes. The procedure is repeated for the other sets of data in Fig. 10-2 to give the parameters in Table 10-1.

Even though this procedure may be quite unsatisfying, still the parameters of (10-6) are fixed, within admittedly rather large limits, to values near those that we list in Table 10-1. These values may be changed to some extent, but we wish to emphasize that there *are* bounds for them, as indicated in the summary Table 10-1.

One argument against large values of $-\beta_0$ in the case of the weak complexes of $I_2$ with $b\pi$-donors, if the resting model is correct, is that the CT resonance-stabilization energy ($-\beta_0^2/\Delta \simeq -0.25$ eV for $Bz \cdot I_2$ with $\beta_0 = -1.0$ eV) is rather too large in comparison with the observed energies of formation ($\simeq -0.1$ eV for these weak complexes).† However, from Table 6-1 we see that values of $b/a$ ($\simeq -\beta_0/\Delta$) that are listed there

† The same difficulty occurs with the axial model (see below), which leads to an even larger value of the resonance-stabilization energy.

are of the order of 0.2 for these $b\pi \cdot I_2$ complexes, in reasonable agreement with $-\beta_0$ of about 0.8 eV. It is our present belief that a value of $-\beta_0$ from 0.5 to 1.0 eV with $S_{01} \simeq 0.1$ and $C_1$ approximately constant and equal to about 5.0 eV is a not unreasonable set of parameters for the $b\pi \cdot I_2$ complexes if the resting model is correct. Some change in $-C$ with the size of the donor must occur so that better parameters[†] should result if $-C$ (hence $C_1$) is allowed to vary with the donor size as discussed above. This whole subject is currently (1968) very much in flux, so we shall not attempt to discuss it further here.

In general it appears that the energy of formation $\Delta W_f$ of complexes is determined by a balance of three rather large terms: the CT resonance attraction energy, the electrostatic attraction energy, and the repulsion energy, which is especially large for $n \cdot a\sigma$ complexes. If the repulsion energy is ignored, we might conclude in the case of weak complexes that either the resonance energy *or* the electrostatic energy explains the stability of the complex. We believe that this neglect of repulsion by some authors has led to an underestimation of the importance of the charge-transfer forces stabilizing the hydrogen bond, which represents another example of $n \cdot a\sigma$ complexes very similar to the iodine complexes of $n$-donors. At the same time the importance of the charge-transfer attraction in weak complexes such as benzene $\cdot I_2$ may have been overestimated as a result of neglect of the electrostatic contribution [17].

On the other hand we do not see any reason to question the rather large values of $-\beta_0$ found for the amine $\cdot I_2$ complexes. The large CT resonance stabilization is expected to be partially cancelled by repulsion energy (estimated as about 0.5 to 1.0 eV) in these strong complexes; such estimates for the energy of formation are in reasonable agreement with experiment. It is of some interest to note that $b^2 + abS_{01}$ is calculated from the parameters of Table 10-1 ($C_1 = 7.3$, $\beta_0 = -1.8$, $S_{01} = 0.40$) for $(CH_3)_3N \cdot I_2$ to be 0.41. This value is in good agreement[‡] with the empirical value 0.33 from dipole moments listed in Table 6-2. For some further discussion of $(CH_3)_3N \cdot I_2$, with potential curves, see Section 14-5.

We next ask whether $C_1$ for the amine $\cdot I_2$ complexes could be smaller, as Table 10-1 suggests, than for the alcohol and ether complexes. One possible reason is that the change from the pyramidal amine to the planar ion requires a correction [see (9-7)] from the tabulated $I_D$ values of Section 9-6 to the vertical ionization potential $I_D^V$; we estimate that this

---

[†] In particular, allowing $-C$ to vary slightly (as discussed in Chapter 9) would permit use of a curve of the form of (10-6) with a smaller value of $-\beta_0$.

[‡] The value computed with the set of parameters from curve $B$ is 0.57, in worse agreement with the value from the dipole moment.

correction contributes from 0 to −0.5 eV to $C_1$ for amine complexes but not for alcohols and ethers.

For the axial model of the $B_z \cdot I_2$ complex neither (9-6) nor (2-15) is applicable, and we must proceed to the more complicated treatment outlined in Section 11-2. As we see there,

$$h\nu_{CT} = W'_V - W_N \simeq W_1 - W_0 + \frac{\beta_0'^2}{\Delta'}. \tag{10-7}$$

Here $W_1 - W_0$ is the energy difference between the ground state and the dative state to which the transition occurs ($W'_V$ in Fig. 11-3), $\Delta'$ is the difference in energy between $W_0$ and the CT state that stabilizes the complex ($W_V$ in Section 11-2), and $\beta_0'$ is the resonance integral between these two interacting states. We can write $W_1 - W_0 = \Delta = I_D - C_1$, where these terms have their usual meaning, and $\Delta' = I'_D - C'_1$, where $I'_D$ is not the frontier ionization potential of the molecule but, for example, in benzene is the ionization potential of the $a_{2u}$ $\pi$ orbital $\phi_1$ of Fig. 4-1, and so is about 2 eV above the frontier ionization potential. These inner ionization potentials of a series of related $\pi$-donors might be expected to vary more or less in the same way as the frontier ionization potentials. Also, we may expect $C'_1$ to be about the same as $C_1$. Hence we may estimate $\Delta'$ to be

$$\Delta' \simeq \Delta + 2 = I_D - C_1 + 2 \text{ eV}. \tag{10-8}$$

Thus for the axial benzene $\cdot I_2$ complex

$$h\nu_{CT} \simeq \Delta + \frac{\beta_0'^2}{(\Delta + 2)}. \tag{10-9}$$

This curve is also hyperbolic and is shown in Fig. 10-3 for the case when $\beta_0 = -2.0$ eV. We can see that this curve, with $C_1 = 5.5$ eV, is identical with curve *I* in Fig. 10-2. This value of $C_1$ is in good agreement with the value that is estimated in Table 10-1 (5.7 eV) from the considerations given in Chapter 9 for the *axial* model of the $Bz \cdot I_2$ complex. Since the empirical values for $C_1$ are consistent with values estimated for either the axial or the resting model, we cannot eliminate one model on the basis of inconsistency. However, the high value of $-\beta_0'$ and the resulting relatively large resonance-stabilization energy, $-\beta_0'^2/\Delta'$($\simeq$ 0.6 eV) is a weak argument against the axial model, although other strong arguments favor it.

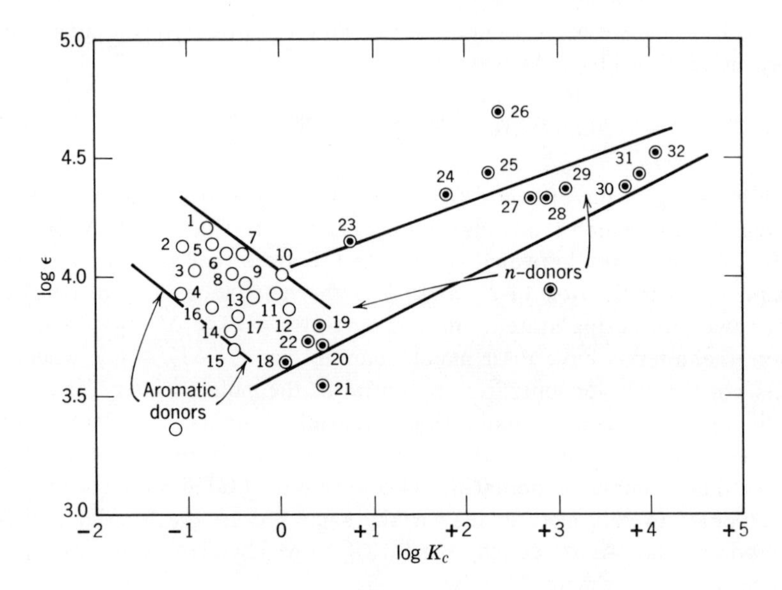


**Figure 10-4** Attempted correlation between intensity (log ε) of the charge-transfer band of some $I_2$ complexes and the strength measured by the formation constant (log $K_c$). Data from Briegleb (GR1), (a) from J. Peters and W. B. Person, *J. Am. Chem. Soc.*, **86**, 10 (1964), and (b) from Yada, Tanaka, and Nagakura, [16]. Key to donors: 1. benzene; 2. chlorobenzene; 3. bromobenzene; 4. *o*-dichlorobenzene; 5. toluene; 6. *o*-xylene; 7. *m*-xylene; 8. *p*-xylene; 9. mesitylene; 10. durene; 11. pentamethylbenzene; 12. hexamethylbenzene; 13. triethylbenzene; 14. tri-*t*-butylbenzene; 15. styrene; 16. naphthalene[a]; 17. phenanthrene[a]; 18. dioxane; 19. tetrahydropyran; 20. tetrahydrofuran; 21. 2-methyltetrahydrofuran; 22. diethylether; 23. diethyldisulfide; 24. $NH_3^b$; 25. diethylsulfide; 26. pyridine; 27. methylamine[b]; 28. ethylamine[b]; 29. tri-*n*-propylamine[b]; 30. triethylamine[b]; 31. dimethylamine[b]; 32. trimethylamine[b].

## 10-4 THE INTENSITIES OF CHARGE-TRANSFER BANDS

The values of $\beta_0$ and $\Delta$ that are obtained from the study of CT frequencies imply estimates of $b$ and $a$ using appropriate equations from Chapter 2 as discussed in the preceding section. From these or from the values in Tables 6-1 or 6-2 we should be able to use (3-11), with an estimate for $\mu_1$ as described in (3-12), in order to predict the intensity of the CT band.

One of the purposes of the survey in this chapter is to illustrate that predictions such as these are on the whole consistent with the experimental results.

Before looking at quantitative predictions for intensities, let us examine the more qualitative expectations. First of all (see Chapter 3) as the extent of charge transfer [measured by $(b^2 + abS_{01})$] increases the intensity of the transition should increase. At the same time the strength of the interaction, as measured by $-\Delta H^0$ or $-\Delta F^0$, is expected to increase. Hence some correlation is expected between the intensity of the CT band and $\Delta F^0$ or $\Delta H^0$. To explore this expectation log $\epsilon$ for the CT band has been plotted in Fig. 10-4 against log $K$ (proportional to $-\Delta F^0$).

The $n$-donor data in Fig. 10-4 show general agreement with the expectation, but the data from $b\pi$-donor complexes trend in the opposite direction. This apparent contradiction with expectations for $b\pi$-donor complexes has led to a considerable amount of discussion [18,19] (see also R9 and R10). One qualifying fact that has already been noted in Chapter 7 is the difficulty in obtaining accurate intensities. However, the trend toward decreasing intensity with increasing strength for aromatic donors is probably qualitatively real. The magnitude of the trend is hardly as great as might be implied by the extensive literature on it; most of the apparent intensities vary only by a factor of 2, and $\epsilon$ is still large even for the least intense of these transitions.

One possible explanation for the apparent decrease in intensity in Fig. 10-4 with increasing strength of the $b\pi$-donor is that given by Orgel and Mulliken (see R9); namely, that a large part of the intensity in the case of the weakest complexes (e.g., $Bz \cdot I_2$) comes from *contact* CT absorption by molecules that are not actually bound in a complex and makes the apparent intensity greater in these than in the stronger complexes. Studies of this absorption at high temperatures, at which essentially all complexes would be dissociated, might clarify this matter.

A more likely explanation is connected with the fact that the intensity theory that is developed in Chapter 3 is *not applicable* if the correct configuration for aromatic-hydrocarbon complexes of iodine is the axial model, as is probable. The transition moment in this event is of an entirely different character (see Section 11-2), and it is difficult to predict what to expect for the CT band intensity.

As indicated in Section 3-3, the intensity of a CT transition is best measured by its oscillator strength $f$ or by its transition dipole moment $\mu_{VN}$. Table 10-2 presents some values of these quantities for a few representative complexes, mostly of $n$-donors. (It should be kept in mind that $f$ is proportional to $\mu_{VN}^2$, and to frequency.) Unfortunately, over-

**TABLE 10-2**

**The Charge-Transfer Band and Derived Data for Some Iodine Complexes**

| Donor | $K_c$ (20°C) | $-\Delta H$ (kcal/mole) | $\lambda_{max}$ (m$\mu$) | $\epsilon_{max}$ | $\Delta\nu_{(1/2)}$ (cm$^{-1}$) | $f_{mn}$ | $\mu_{mn}$ (D) |
|---|---|---|---|---|---|---|---|
| Benzene[a] | 0.15 | 1.4 | 292 | 16,000 | 5100 | 0.38 | 4.64 |
| Naphthalene[a] | 0.26 | 1.8 | 360 | 7,250 | 4700 | 0.21 | 4.00 |
| Methanol[b] | 0.23 | 3.5 | 232 | 13,700 | 5700 | 0.34 | 4.1 |
| Ethanol[b] | 0.26 | 4.5 | 230 | 12,700 | 6800 | 0.37 | 4.3 |
| Diethylether[c] | 0.97 | 4.3 | 249 | 5,700 | 6900 | 0.170 | 2.99 |
| Diethyl disulfide[c] | 5.62 | 4.62 | 304 | 15,000 | 7200 | 0.466 | 5.49 |
| Diethyl sulfide[c] | 210 | 7.82 | 302 | 29,800 | 5400 | 0.695 | 6.68 |
| Ammonia[d] | 67 | 4.8 | 229 | 23,400 | 4100 | 0.42 | 4.53 |
| Methylamine[d] | 530 | 7.1 | 245 | 21,200 | 6400 | 0.59 | 5.65 |
| Dimethylamine[d] | 6800 | 9.8 | 256 | 26,800 | 6450 | 0.75 | 6.39 |
| Diethylamine[d] | 6320 | 12.0 | 278 | 25,600 | 8100 | 0.90 | 7.28 |
| Trimethylamine[d] | 12,100 | 12.1 | 266 | 31,300 | 8100 | 1.10 | 7.87 |
| Pyridine[c] | 269 | 7.8 | 235 | [e]50,000 | [e]5200 | [e]1.12 | 7.48 |

[a] Data from summary by J. Peters and W. B. Person, *J. Am. Chem. Soc.*, **86**, 10 (1964).
[b] Data from L. Julien, Ph.D. thesis, University of Iowa, 1966.
[c] Data from summary by Tsubomura and Lang [20].
[d] Data from Yada, Tanaka, and Nagakura [16], supplemented by private communication from Prof. Nagakura.
[e] See text, top of p. 156.

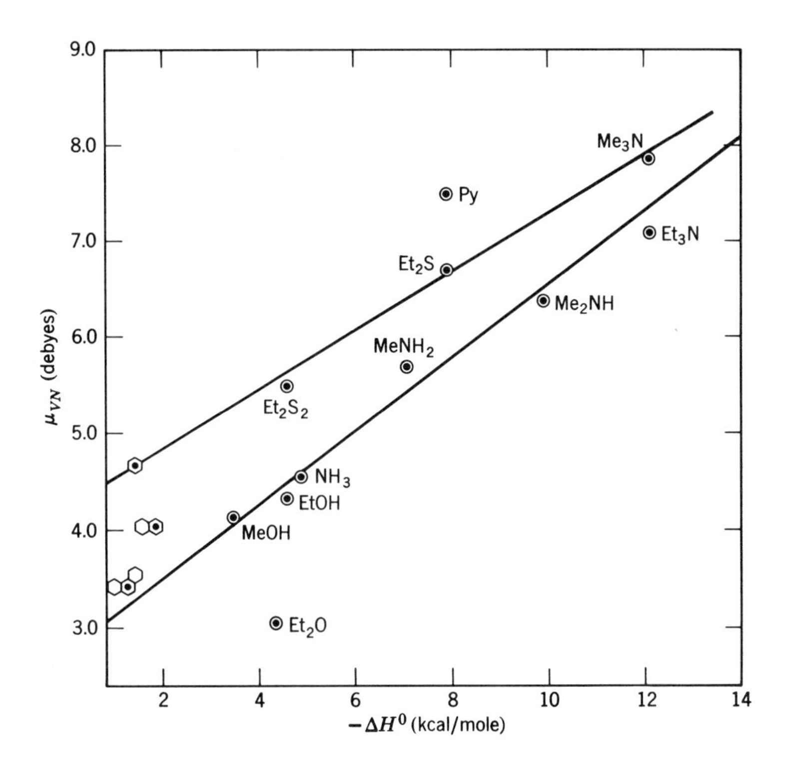


**Figure 10-5** Correlation between the transition dipole $\mu_{VN}$ for the charge-transfer band of some $I_2$ complexes and the strength of the complex, measured by $-\Delta H$. (See Table 10.2.)

lapping bands or general laziness have resulted in a sparsity of *accurate* data. Most of the values in Table 10-2 are estimated from $\epsilon_{max}$ and $\Delta\nu_{(1/2)}$ by using (3-8) and (3-9). The $\mu_{VN}$ data are plotted against $-\Delta H$ in Fig. 10-5. The anomalous decrease in intensity with aromatic donors is still evident in this figure, but it is apparent that for the most part the transition-dipole moment for the CT band increases steadily with increasing strength of the complex for iodine complexes with *n*-donors.

However, Fig. 10-5 shows an apparently anomalously low intensity for the complex with diethyl ether. This anomaly is characteristic of the CT bands of all ether complexes with $I_2$ and is not understood. The exceptionally high intensity for the pyridine $\cdot I_2$ complex can be ascribed to the superposition, on the parallel-polarized transition due to n-donor functioning of the pyridine, of a perpendicularly-polarized CT transition corresponding to a CT state in which an electron is transferred out of a MO of the pyridine [20a].

Consider now the quantitative use of (3-11), together with (3-14), to predict the intensity of the CT band. In order to do so we may estimate values of the CT coefficients $a (\simeq a^*)$ and $b (\simeq b^* - S_{01})$ from Tables 6-1 and 6-2, using $\mu_1$ from the latter and other parameters from Table 10-1. In (3-14) we assume that the overlap charge is approximately in the center of the bond so that $e(r_d - r_{da^-})$ is approximately $(1/2)\mu_1$. Hence (3-11) is

$$\mu_{VN} \simeq a^* b(\mu_1 - \mu_0) + (1/2)(aa^* - bb^*) S_{01}\mu_1. \qquad (10\text{-}10)$$

For benzene $\cdot I_2$ (assuming that the complex does not have axial geometry, so that the theory applies) $b \simeq 0.2$, $S_{01} \simeq 0.1$, $b^* \simeq 0.3$, $a^* \simeq a \simeq 1$, and $\mu_1 \simeq 16$. Thus, $\mu_{VN}$ from (10-10) is estimated to be $3.6 + 0.75 \simeq 4.4$ debyes, in excellent but probably fortuitous agreement with the value of 4.6 debyes from Table 10-2. For the trimethylamine $\cdot I_2$ case, with the parameters from curve $C$ of Table 10-1, $a = 0.67$, $b = 0.52$, $a^* = 0.62$, and $b^* = 0.58$; hence if $\mu_1 \simeq 18$, $\mu_{VN} = 5.8 + 0.4 = 6.2$ debyes, as against the experimental value of 7.8 debyes. (See Table 10-2). Other examples give about the same quality of agreement.

## 10-5 THE BLUE SHIFT OF THE $I_2$ VISIBLE SPECTRUM

In addition to the appearance of a CT band, the electronic spectra of iodine complexes are characterized by a blue shift of the locally excited $I_2$ visible band (at 5200 Å in the vapor) and usually also by an increase in its intensity.† This blue shift is illustrated in Fig. 10-6 for $I_2$ with a series of donors of increasing strength in order to dramatize the increasing blue shift and intensification with increasing donor strength. This change is often large enough to make possible spectrophotometric determinations of thermodynamic constants of complexes for which the CT band is not observed either because it is too far in the ultraviolet (< 220 mμ, as in complexes of halogens with acetonitrile) or because it is overlapped by a donor band (as, for example, in complexes of amides).

† Sometimes there is some overlap with the CT band and consequent difficulty in obtaining information about the shifted visible band.

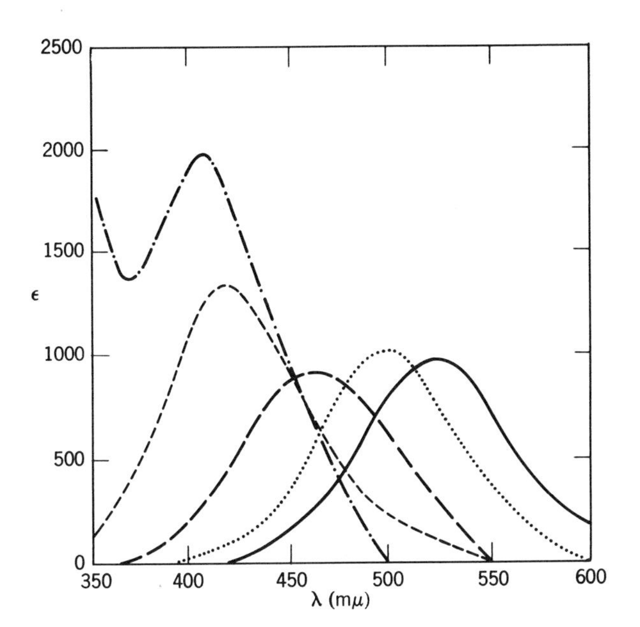


**Figure 10-6** The spectrum of the locally excited visible $I_2$ band in various complexes, illustrating the blue shift. Here —— is $I_2$ in *n*-heptane[a], . . . . . is $I_2$ in benzene[b], – – – – is $I_2$ with diethylether in $CCl_4$[c], xxx is $I_2$ with pyridine in *n*-heptane[a], and -·-·-·- is $I_2$ with triethylamine in *n*-heptane[d].

[a] Redrawn from Reid and Mulliken (R7).
[b] From H. A. Benesi and J. H. Hildebrand, *J. Am. Chem. Soc.*, **71**, 2703 (1949).
[c] From P. A. D. de Maine, *J. Chem. Phys.*, **26**, 1192 (1957).
[d] From S. Nagakura, *J. Am. Chem. Soc.*, **80**, 520 (1958).

Such complexes can be studied spectrophotometrically by the same techniques as are used for the CT band (see Chapter 7), but the measurements are less sensitive because of the relatively lower intensity of the visible band ($\epsilon_{max} \simeq 1000$). However, if the blue shift is large enough, it is possible to make rather satisfactory measurements by using higher concentrations.

A rough overall summary of changes in the visible band of $I_2$ for several types of complexes is presented in Table 10-3. We see there that rather large shifts occur in wavelength and in intensity for complexes

## TABLE 10-3

**Frequency and Intensity Changes for the Visible Band of $I_2$ in Complexes with Various Donors[a]**

| Donor Type | System | $\lambda_{max}$ (m$\mu$) | $\epsilon_{max}$ |
|---|---|---|---|
| None | $I_2$ in heptane | 520 | 900 |
| $b\pi$ | Benzene and substituted benzenes, polynuclear aromatics, ethylenes | 490–510 | 900–1200 |
| $n$ | Donors at an oxygen atom | 435–470 | [b]700–1300 |
| | Donors at a sulfur atom | 400–460 | 1300–4000 |
| | Donors at a nitrogen atom | 410–430 | 1000–2600 |

[a] Data from J. Walkley, D. N. Glew, and J. H. Hildebrand, *J. Chem. Phys.*, **33**, 621 (1960); from Tsubomura and Lang [20]; from Yada, Tanaka, and Nagakura [16]; from Brandon, Tamres, and Searles [24]; and Lang [21].
[b] The low-intensity values are for $H_2O$ and alcohols, which are known to react with $I_2$, thus perhaps giving a spuriously low value for $\epsilon_{max}$. However, the decrease in intensity is reproducible and may indeed be real.

with the stronger $n$-donors. There seems to be a definite correlation between the strength of the complex and the wavelength and $\epsilon_{max}$ for this band [21].

Ham [22] was the first to point out that the blue shift can be correlated rather well with the heat of formation. This correlation has been accepted by most workers since then [21,23,24]. It is illustrated in Fig. 10-7 for a number of iodine complexes. Here the change in energy $\Delta W_{vis}$ ($= h\nu_{complex} - h\nu_{I_2}$) for $\nu_{max}$ of the visible $I_2$ band is plotted against the enthalpy of formation ($-\Delta H^0$) for a number of complexes. The plot is seen to be linear with a slope of about 1.5, showing that the increase in transition energy exceeds the energy of formation of the complex by half.

Mulliken has pointed out (see R8) that the $\sigma_u$ antibonding MO, which contains the electron excited by the absorption of light in the visible band, must be larger than the outer occupied MOs in the normal state of $I_2$. Then when the iodine molecule, paired off with a close partner in a complex, is excited by absorption of visible light ($\sigma_u \leftarrow \pi_g$), its suddenly swollen size increases the repulsion energy between it and the donor,

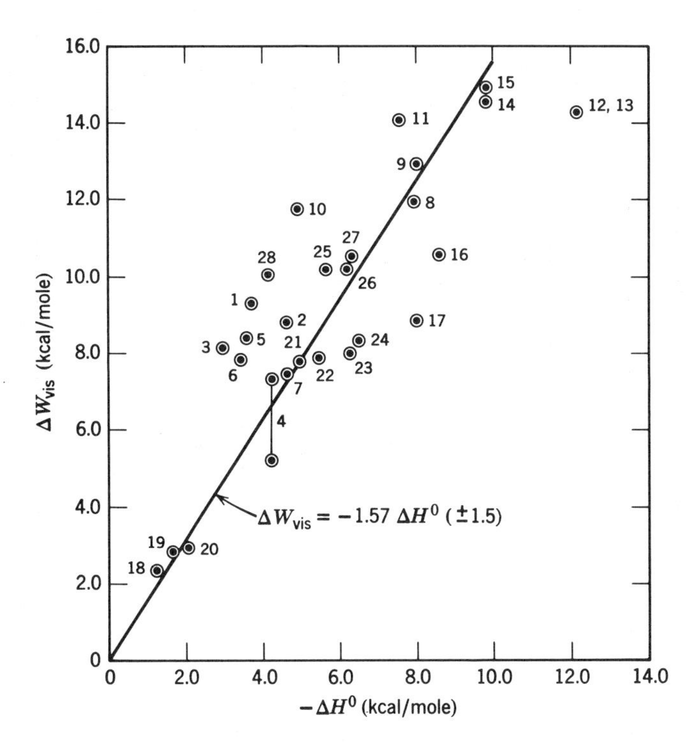


**Figure 10-7** Correlation between the blue shift of the visible $I_2$ band ($\Delta W_{vis} = h\nu_{comp} - h\nu_{free\ I_2}$) and the enthalpy of formation, $-\Delta H^0$, for some $I_2$ complexes. Data from (a) L. Julien, Ph.D. thesis, University of Iowa, 1966; (b) Tsubomura and Lang [20]; (c) Yada, Tanaka, and Nagakura [16]; (d) Sr. M. Brandon, M Tamres and S. Searles [24]; (e) J. S. Ham, *J. Chem. Phys.*, **20**, 1170 (1952). 1. MeOH[a]. 2. EtOH[a]; 3. tri-*n*-butylphosphate[b]; 4. $Et_2O$[d]; 5. dioxane[b]; 6. *t*-butylalcohol[e]; 7. $Et_2S_2$[b]; 8. $Et_2S$[b]; 9. pyridine[d]; 10. $NH_3$[c]; 11. $EtNH_2$[c]; 12. $Et_3N$[c]; 13. $Me_3N$[c]; 14. $Me_2NH$[c]; 15. $Et_2NH$[c]; 16. $Me_2Se$[b]; 17. $Me_2S$[b]; 18. benzene[d]; 19. toluene[d]; 20. *o*-xylene[d]; 21. tetrahydropyran[d]; 22. tetrahydrofuran[d]; 23. 2-methyltetrahydrofuran[d]; 24. trimethylene oxide[d]; 25. pyridine-*N*-oxide[d]; 26. 3-picoline-*N*-oxide[d]; 27. 4-picoline-*N*-oxide[d]; 28. *N,N*-dimethylformamide[b]. (Me = methyl, Et = ethyl.)

almost as if it had suddenly collided with the latter. This repulsion energy, which should be greater the more intimate the complex, is added to the usual energy of the excited iodine molecule, tending to give a blue

**TABLE 10-4**

**The Visible $I_2$ Band and Derived Data for Some Complexes of *n*-Donors with $I_2$ [a]**

| Donor | $K_c$ (20°C) | $-\Delta H$ (kcal-mole) | $\lambda_{max}$ (mμ) | $\epsilon_{max}$ | $\Delta\nu_{(1/2)}$ (cm$^{-1}$) | $f_{mn}$ | $\mu_{mn}$ |
|---|---|---|---|---|---|---|---|
| *n*-Heptane | – | – | 522 | 897 | 3200 | 0.0124 | 1.17 |
| Methanol | 0.47 | 3.5 | 440 | 973 | 4250 | 0.0179 | 1.29 |
| Ethanol | 0.45 | 4.5 | 443 | 1082 | 4500 | 0.0210 | 1.41 |
| *Tri-n*-butyl phosphate | 21 | 2.9 | 456 | 1280 | 4200 | 0.0232 | 1.50 |
| *N,N*-dimethylformamide | 1.1 | 4.0 | 441 | 1189 | 4300 | 0.0221 | 1.44 |
| Diethyl ether | 1.0 | 4.2 | 465 | 920 | 4100 | 0.016 | 1.26 |
| Diethyl disulfide | 5.6 | 4.6 | 460 | 1369 | 4100 | 0.0242 | 1.54 |
| Diethyl sulfide | 210 | 7.8 | 435 | 1960 | 4200 | 0.0352 | 1.80 |
| Pyridine | 269 | 7.8 | 422 | 1320 | 4300 | 0.0245 | 1.48 |
| Dimethyl selenide | 661 | 8.5 | 430 | 2600 | 4500 | 0.05 | 2 |
| Triethylamine | 6460 | 12.0 (12.9) [b] | 414 | 2030 | 5700 | 0.0501 | 2.10 |

[a] Data from summary by Tsubomura and Lang [20], except for the modified value of $-\Delta H$ for the alcohols, which comes from L. Julien, Ph.D. thesis, University of Iowa, 1966.

[b] M. Tamres, [*J. Phys. Chem.*, **65**, 654 (1961)] has re-evaluated the data from Nagakura [*J. Am. Chem. Soc.*, **80**, 520 (1958)] to obtain this higher value for $-\Delta H$.

shift in the absorption frequency. Of course $-\Delta W_f$ of the complex also contributes to the blue shift, but its effect would tend to be canceled or perhaps more than canceled by the $-\Delta W_f$ that would be expected for complex formation of the donor with the $^3\Pi_{0^+u}$ excited iodine molecule if it were not for the repulsion effect just discussed. Potential-energy diagrams that illustrate this repulsion effect are given in Chapter 14 (Figs. 14-1 and 14-4).

The characteristic intensification that is found for the shifted visible band (Table 10-3) is presumably to be explained by increased mixing of the $^3\Pi_{0^+u}$ state with either the CT state or with the $^1\Sigma_u^+$ state of $I_2$, as a result of the shift in position of the $^3\Pi_{0^+u}$ level to higher energies. The spectral properties of this band are summarized in Table 10-4. The correlation between the transition dipole $\mu_{mn}$ and the strength of the complex is shown in Fig. 10-8. The intensification is not strikingly large, but it does appear to be significant.

All the data that are described above are for complexes in solution. Recent examination [25] of the spectrum of the strong $(C_2H_5)_2S \cdot I_2$ complex in the *vapor* state shows no blue shift in the visible band of the complex. (See Section 7-4.) No good explanation is available for this difference from solution behavior. Perhaps the geometry of the complex is different in the vapor state from that in solution.

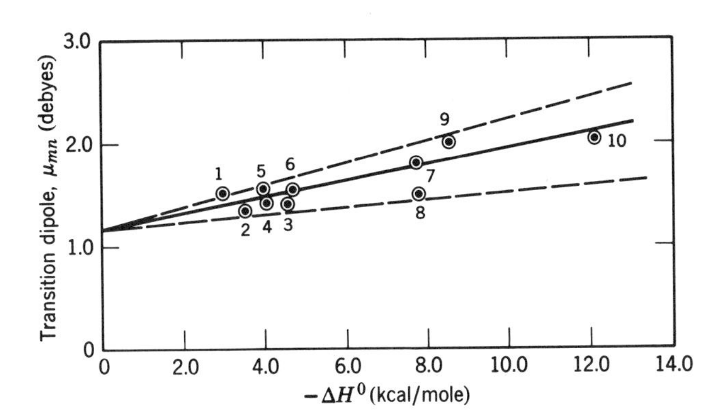


**Figure 10-8** Correlation between the intensity ($\mu_{mn}$) of the locally excited visible $I_2$ band in some $I_2$ complexes with the strength of the interaction, measured by $-\Delta H^0$. The dashed lines indicate the uncertainties in the correlation. The data are from Table 10-4. The donors are 1. tri-*n*-butyl phosphate; 2. MeOH; 3. EtOH; 4. $Et_2O$; 5. *N,N*-dimethyl formamide; 6. $Et_2S_2$; 7. $Et_2S$; 8. pyridine; 9. $Me_2Se$; 10. $Et_3N$. (Me = methyl, Et = ethyl.)

## 10-6 REFERENCES

[1] J. N. Murrell, S. F. A. Kettle, and J. M. Tedder, *Valence Theory*, Wiley, London, 1965.
[2] R. S. Mulliken, *Phys. Rev.*, **46**, 549 (1934); **57**, 500 (1940); also *J. Chem. Phys.*, **8**, 234, 382 (1940).
[3] G. Kortüm and G. Friedheim, *Z. Naturforschung*, **2a**, 20 (1947).
[4] E. Rabinowitch and W. C. Wood, *Trans. Faraday Soc.*, **32**, 540 (1936).
[5] C. A. Goy and H. O. Pritchard, *J. Mol. Spectry.*, **12**, 38 (1964).
[6] E. A. Ogryzlo and G. E. Thomas, *J. Mol. Spectry.*, **17**, 198 (1965).
[7] R. S. Mulliken, *J. Chem. Phys.*, **7**, 20 (1937).
[8] See references 21 and 22 of Chapter 7.
[9] P. A. D. deMaine, *J. Chem. Phys.*, **24**, 1091 (1957). M. M. deMaine, P. A. D. deMaine, and G. E. McAlonie, *J. Mol. Spectry*, **4**, 271 (1960).
[10] R. M. Keefer and T. L. Allen, *J. Chem. Phys.*, **25**, 1059 (1956).
[11] H. McConnell, *J. Chem. Phys.*, **22**, 760 (1954).
[12] J. Peters, unpublished results at Laboratory of Molecular Structure and Spectra, University of Chicago.
[13] R. D. Verma, *J. Chem. Phys.*, **32**, 738 (1960); see also, S. Waissman, J. T. Vanderslice, and R. Battino, *J. Chem. Phys.*, **39**, 2226 (1963).
[14] C. J. Delbecq, W. Hayes, and P. H. Yuster, *Phys. Rev.*, **121**, 1043 (1961).
[15] Watanabe et al. See footnote [a] of Table 9-5.
[16] H. Yada, J. Tanaka, and S. Nagakura, *Bull. Chem. Soc. Japan*, **33**, 1660 (1960).
[17] M. Hanna, *J. Am. Chem. Soc.*, **90**, 285 (1968).
[18] J. N. Murrell, *Quart. Revs.*, **15**, 191 (1961).
[19] S. P. McGlynn, *Chem. Revs.*, **58**, 1113 (1958).
[20] H. Tsubomura and R. Lang, *J. Am. Chem. Soc.*, **83**, 2085 (1961).
[20a] R. S. Mulliken, *J. Am. Chem. Soc.*, **91**, 1237 (1969).
[21] R. Lang, *J. Am. Chem. Soc.*, **84**, 1185 (1962).
[22] J. Ham, *J. Am. Chem. Soc.*, **76**, 3875 (1954).
[23] T. Kubota, *J. Am. Chem. Soc.*, **87**, 458 (1965).
[24] Sr. M. Brandon, M. Tamres, and S. Searles, Jr., *J. Am. Chem. Soc.*, **82**, 2129 (1960).
[25] For example, see M. Kroll, *J. Am. Chem. Soc.*, **90**, 1097 (1968).

*CHAPTER* **11**

# *Generalized Resonance-Structure Theory*

## 11-1 A MORE GENERAL TREATMENT

In the discussion of experimental results in preceding chapters it became apparent that the simplified intermolecular charge-transfer resonance-structure theory developed in Chapter 2 is often inadequate. Here we consider briefly the direction in which this theory should be generalized. In Chapter 2 approximate wavefunctions of the form $a\Psi_0 + b\Psi_1$ for state $N$, and $a^*\Psi_1 - b^*\Psi_0$ for a charge-transfer excited state called $V$, where $\Psi_0$ and $\Psi_1$ are no-bond and dative functions, respectively, were used. A $2 \times 2$ secular equation was then set up and solved for the energies $W_N$ and $W_V$ (Eq. 2-10), and for weak complexes these energies were approximated by expressions (Eqs. 2-21, 2-22), which are the same as those of second-order perturbation theory. Because it is not feasible to give general formulas that correspond to a generalized ($N \times N$, where $N$ is moderately large) secular equation, we shall be content here with generalized second-order-perturbation expressions, which should be satisfactory at least for weak complexes.

Equations 2-2, 2-3, 2-11, 2-12, 2-13, 2-20, 2-21, and 2-23 can rather obviously be generalized as follows for the $m$th state of a complex:

$$\Psi_m^{\mathrm{el}} = a_m(\Psi_m^0 + \sum_{\substack{n \neq m \\ m \geq 0}}' \rho_{mn}\Psi_n^0); \qquad \rho_{mn} \equiv \frac{-\beta_m^n}{\Delta_{mn}},$$

$$W_m = W_m^0 - \sum_{n \neq m}' \frac{(\beta_m^n)^2}{\Delta_{mn}}, \tag{11-1}$$

$$\Delta_{mn} = W_n^0 - W_m^0; \qquad \beta_m^n = W_{mn}^0 - S_{mn}^0 W_m^0.$$

The prime in $\Sigma'$ in (11-1) means that only the $\Psi_n^0$ terms that are *of the same group-theoretical symmetry species* as $\Psi_m^0$ are to be included in the summation.† This restriction implies further that there is a separate set of Eqs. 11-1 for each different symmetry species of the complex and that, among others, there are separate sets for singlet and triplet states. In general the $\Psi^0$s for any one species include the following:

1. No-bond functions, all or most of which are locally excited states $\Psi^0(D, A^*)$, $\Psi^0(D^*, A)$, and in principle also $\Psi^0(D^*, A^*)$; if the symmetry species is that of $\Psi_N$, the unexcited no-bond function $\Psi_0^0(D, A)$ is included here.
2. A variety of dative functions $\Psi^0(D^+—A^-)$, $\Psi^0(D^+—A^{-*})$, $\Psi^0(D^{+*}—A^-)$, $\Psi^0(D^{+**}—A^-)$, $\Psi_0(D^-—A^+)$, and so on.

Equations 11-1 are obviously applicable to state $N$ as a special case ($\Psi_0^{\mathrm{el}}$ is $\Psi_N^{\mathrm{el}}$) and also to all excited states of the complex. Equations 11-1 for $W_m$ have the virtue that for $m = N$ they allow for stabilization of state $N$ by more than one dative function and for excited states allow for possible mixing of different charge-transfer excited states among themselves and with locally-excited states. Although in principle (11-1) should include a vast number of excited and continuum states of the D,A pair, in practice we can include only the more important states of no more than moderately high energy. [For an example see (13-3) and (13-4) for lithium fluoride considered as a complex of $F^-$ and $Li^+$.]

As already mentioned, an accurate treatment would call for the solution of a many-dimensional secular equation. However, if $\Delta_{mn}$ is small and $W_{mn}^0$ is not very small for a pair of important excited $W^0$s (say $W_1^0$ and $W_2^0$ belonging to $\Psi_1^0$ and $\Psi_2^0$), then it is useful first to solve a $2 \times 2$ secular equation that is similar to (2-9) but involves $W_1^0$, $W_2^0$, $W_{12}^0$, and $S_{12}$ to obtain new first-order energies $W_1^1$ and $W_2^1$.‡ After this the second-order-perturbation equations (11-1) can be used to deal with the further interactions, but with $W_1^1$, $W_2^1$ substituted for $W_1^0$ and $W_2^0$, and correspond-

† The $\Psi_n^0$ terms of other species, if included, will be found to have zero coefficients.

‡ These correspond to new zero-order wavefunctions $\Psi_1^1 = a\Psi_1^0 + b\Psi_2^0$ and $\Psi_2^1 = a^*\Psi_2^0 - b^*\Psi_1^0$, obtainable in a manner analogous to those in (2-16) to (2-19).

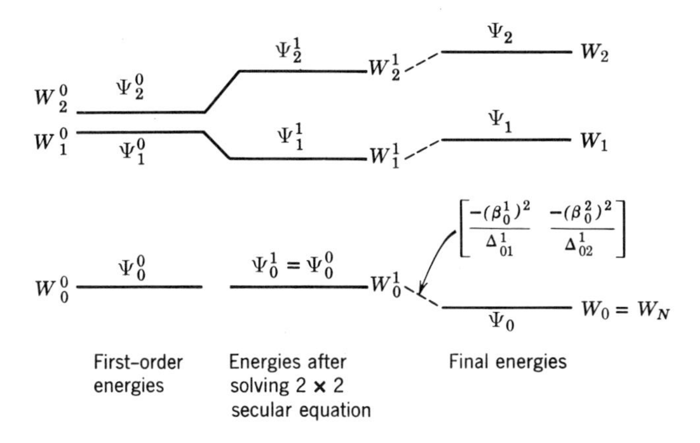


**Figure 11-1** Energy-level diagram illustrating the energies for a system in which there is a near degeneracy between two first-order energies $W_1^0$ and $W_2^0$ in (11-1). (Here $\Delta_{nm}^1 = W_n^1 - W_m^1$.)

ing changes in the $\Delta$s, $S$s, and $\beta$s. Figure 11-1 illustrates the situation just described.

Now let us turn to $\Psi_N$ and consider the following several special cases for the application of (11-1):

1. $\Psi_0^0$ may be stabilized mostly by interaction with *one* dative structure, perhaps $\Psi_1^0(D^+—A^-)$. Here the treatment is just that given in Chapter 2; the energy levels are illustrated in Fig. 2-2, in Chapter 9, and again below.

$W_V(\Psi_V)$

$W_1(\Psi_1^0)$

$W_0^0(\Psi_0^0)$

$W_N(\Psi_N)$

Note, however, that $\Psi_1^0$ often corresponds to a locally excited state, so that the lowest energy dative function may be not $\Psi_1^0$ but $\Psi_k^0$ with $k > 1$; for example, in iodine complexes the lowest energy excited $\Psi^0$s are functions with local excitation in the $I_2$.

2. In connection with the requirement that only $\Psi^0$s of the same symmetry species can mix, two comments are desirable. (a) Usually complexes are necessarily less symmetrical than D or A alone, so that the

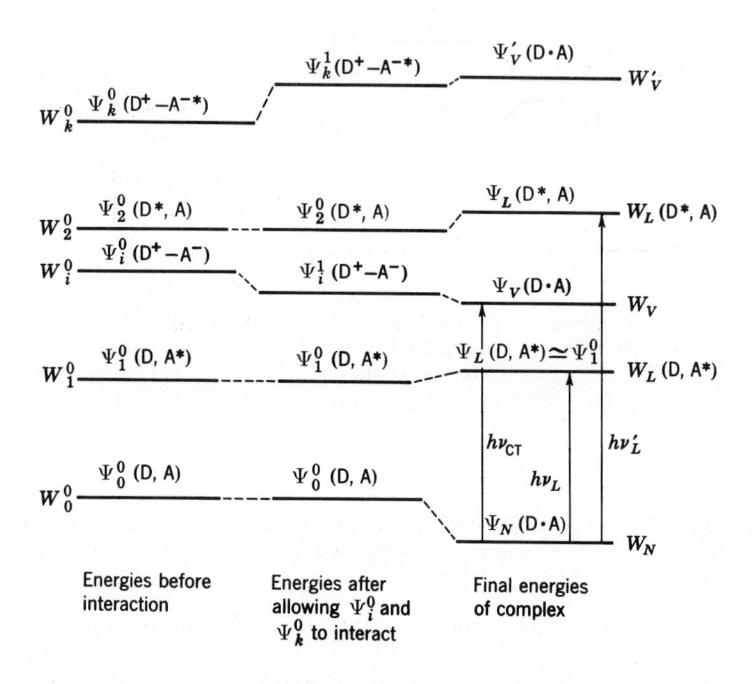


**Figure 11-2** Possible energy-level pattern for the case of two CT wavefunctions ($\Psi_i^0$ and $\Psi_k^0$) and some locally excited functions of the same symmetry as $\Psi_0^0$. Probable example: $(CH_3)_3N \cdot I_2$ complex. (See also Fig. 14-4.)

requirement is less restrictive than we might at first suppose. (b) Even if a fairly high symmetry for state $N$ of a complex were possible, there should be a tendency for the complex to assume a lower symmetry because this would relax the restrictions and permit mixing of more $\Psi^0$s and might give increased stabilization of $\Psi_N$.

3. $\Psi_0^0$ may be stabilized mainly by one dative function as in case 1, but this may be not the lowest energy dative function. This case can occur if the lowest energy dative function is of a different group-theoretical species than $\Psi_0^0$; a probable example is discussed in Section 11-2.

4. $\Psi_0^0$ may be stabilized to an important extent by two (or more) dative functions; for example, by $\Psi_i^0(D^+—A^-)$ and $\Psi_k^0(D^+—A^{-*})$. Including, for generality, possible locally excited $\Psi^0$s, the energy-level pattern is somewhat as shown in Fig. 11-2.

The amine · iodine (and pyridine · iodine) complexes, although they have fitted apparently satisfactorily under the simplified theory in preceding chapters, are examples of cases in which the use of two dative functions would probably represent an appreciable (possibly a considerable) improvement. The two functions in question correspond to structures $D^+—I_2^-$ with normal-state $D^+$ in both and $I_2^-$ in the respective states

$$\cdots(\sigma_g)^2(\pi_u)^4(\pi_g)^4\,\sigma_u,\quad {}^2\Sigma_u^+ \text{ in } \Psi_i^0$$

and

$$\cdots(\sigma_g)(\pi_u)^4(\pi_g)^4\,\sigma_u{}^2,\quad {}^2\Sigma_g^+ \text{ in } \Psi_k^0$$

Here the molecular orbitals $\sigma_g$ and $\sigma_u$ have the respective LCAO forms

$$N(5p\sigma_a \pm 5p\sigma_b),\quad \text{with } N = 2^{-(1/2)}(1 \pm S_{ab})^{-(1/2)} \qquad (11\text{-}2)$$

where $a, b$ refer to the two iodine atoms and the plus sign refers to $\sigma_g$; the minus sign refers to $\sigma_u$.

Both $\Psi_i^0$ and $\Psi_k^0$ have the proper symmetry to mix with the no-bond structure if the configuration of the isolated $R_3N \cdot I_2$ complex has the iodine on the threefold axis, with essentially $C_{3v}$ symmetry, as found from the crystal structure of the solid $(CH_3)_3N$ complex (see Chapter 5):

$H_3C$
$H_3C$—N · I—I
$H_3C$

The two mixtures $\Psi_i^0 \pm \Psi_k^0$ would correspond to $Me_3N^+—I \cdot\cdot I^-$ and $Me_3N^+ \cdot\cdot I^- \cdot\cdot I$, in which the $I_2^-$ is polarized away from (with the minus sign) or toward (with the plus sign) the $N^+$, respectively; the use of $\Psi_1^0$ taken alone corresponds to $Me_3N^+—I_2^-$) with a symmetrical charge distribution in $I_2^-$. (See R14, Section III, for details.) Polarization of the $I_2^-$ as in $Me_3N^+—I \cdot\cdot I^-$ would be favored by greater $N^+—I$ bond energy but $Me_3N^+I^-I$ would have greater Coulomb energy—and perhaps a compromise close to using symmetrical $I_2^-$, hence $\Psi_i^0$ only, may be not far from correct. Furthermore, the resonance terms $(\beta_0^k)^2/\Delta_{0k}$ are of major importance. All in all it does not seem possible at present to decide how important the inclusion of $\Psi_k^0$ is. However, it seems likely that the use of $\Psi_i^0$ only, as in preceding chapters, may be a fairly good approximation for these complexes.

For another informative example see the examination by Kuroda, Ikemoto, and Akamatu [1] of this question for the pyrene · TCNE complex, which concludes that *many* charge-transfer states contribute to the stabilization of that complex.

5. The important $\Psi_k^0$s may include a $D^-—A^+$ type of function in

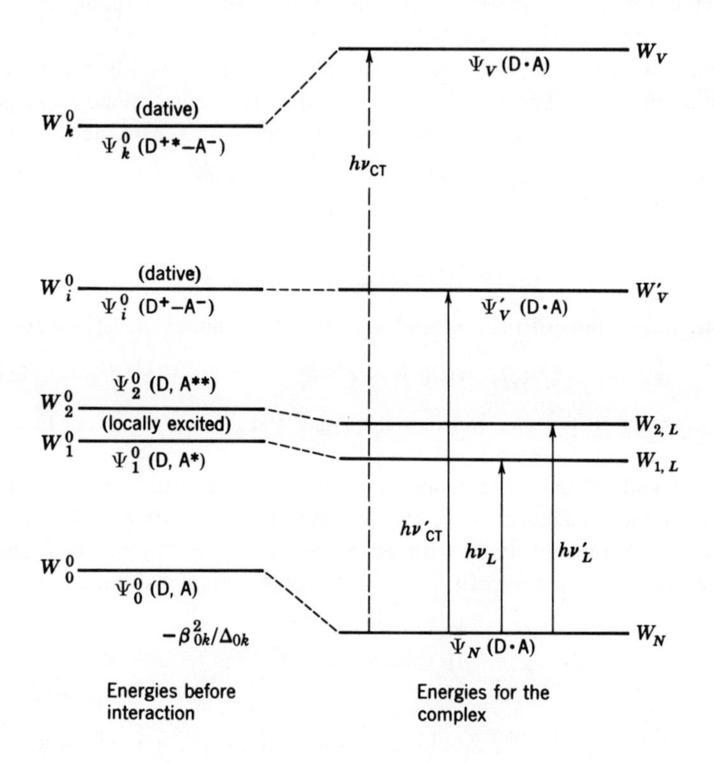


**Figure 11-3** Possible energy-level pattern for the case of two CT states of different symmetry with only the upper one of the proper symmetry to interact with $\Psi_0^0$ to stabilize $\Psi_N$. Example (but see Fig. 14-1 for more quantitative relations): axial model of $Bz \cdot I_2$ complex. The excited states shown are singlet states. In addition there are corresponding triplet states to which extremely weak absorption transitions should occur.

addition to one or more $D^+—A^-$ functions. If so, we speak of the normal state $\Psi_N$ as a two-way charge-transfer complex in which D acts not only as a donor but also somewhat as an acceptor. (See Chapter 17 for details.) An example is provided by the complexes between $Ag^+$ and benzene, as, for example, in the compound $Ag^+BzClO_4^-$, which is known in crystalline form and also in solution in benzene. The following four Ag states are relevant:

$$\text{Ag: } \cdots 4d^{10}5s,\ {}^2S; \quad \text{Ag}^+\text{: } \cdots 4d^{10},\ {}^1S;$$

$$\text{Ag}^{+*}\text{: } \cdots 4d^9 5s,\ {}^3D,\ {}^1D; \quad \text{Ag}^{2+}\text{: } \cdots 4d^9,\ {}^2D.$$

The $Ag^+$ ion should be a $v$-acceptor and the benzene a $b\pi$-donor; hence $\Psi_0^0$ in (11-1) should be $\Psi_0^0(\text{Bz, Ag}^+)$, and the first dative function should be $\Psi_k^0(\text{Bz}^+\text{—Ag})$. (We cannot immediately say that $k = 1$, since possibly a locally excited state $\Psi^0(\text{Bz, Ag}^{+*})$ may have a lower $W^0$.) It is seen also that $Ag^+$ might function as a lone-pair donor, utilizing an electron from the $4d$ shell, whereas benzene could then act as an $a\pi$-acceptor, in $\Psi_l^0(\text{Ag}^{2+}\text{—Bz}^-)$. This is made easier by the fact that the $Ag^+$ is in proximity to a negative ion (e.g., $ClO_4^-$ or $NO_3^-$). Experience suggests that $b\pi \cdot v$ and $n \cdot a\pi$ one-way complexes are not very stable, but that two-way $b\pi \cdot v$ plus $n \cdot a\pi$ action, as in the present example, can result in a stable complex. (See Sections 17-1 to 17-4 and 17-11.)

## 11-2 APPLICATION TO BENZENE · $I_2$

In R3 it was assumed that the simple one-term perturbation theory describes correctly the benzene · $I_2$ complex, but subsequent developments have suggested that this is not true. Let us consider an alternative description involving two charge-transfer functions $\Psi_k^0$ and $\Psi_l^0$, the former, of higher energy, being of the same symmetry as $\Psi_0^0$, the latter, of lower energy, being of different symmetry. The energy levels are illustrated in Fig. 11-3.

How might such a situation arise for the benzene · $I_2$ complex? Let us consider the donor and acceptor orbitals, $\phi_\text{d}$ and $\phi_{\text{a}^-}$. The four most loosely held electrons in benzene are in two degenerate $\pi$ MOs (See Fig. 4-1.) Another pair of electrons occupies the lowest energy MO (See Fig. 4-1.) With a given $\phi_{\text{a}^-}$, two charge-transfer states can then be obtained using $\pi$ MOs of benzene as $\phi_\text{d}$: one of lower energy when the electron is removed from one of the higher benzene $\pi$ MOs (i.e., from $\phi_2$ or $\phi_3$) and one of higher energy when it is removed from the lower energy $\pi$ MO (from $\phi_1$).[†]

Now as discussed in Chapter 10, the valence-electron configuration of the iodine molecule is $\sigma_g{}^2\pi_u{}^4\pi_g{}^4$, and that of the ground state of $I_2^-$ is $\sigma_g{}^2\pi_u{}^4\pi_g{}^4\sigma_u$. This $\sigma_u$ molecular orbital [see (11-2)] is approximately of the LCAO form $\sigma_u = N(5p\sigma_a - 5p\sigma_b)$ and is sketched very schematically in Fig. 11-4.

[†] Higher energy CT states could also be obtained by removing an electron from a benzene MO of $\sigma$-type.

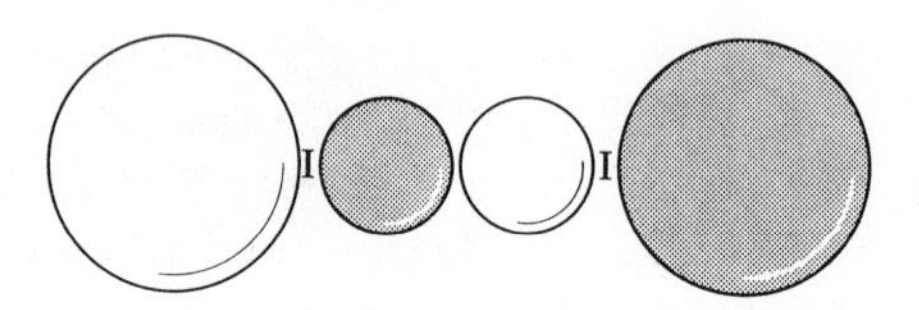


**Figure 11-4** The valence-shell $\sigma_u$ MO of $I_2^-$. (LCAO approximation, schematic.) As usual, the shaded portions represent positive lobes of the orbital.

Within either atom ($a$ or $b$) $5p\sigma$ has three radial nodes and one midplane node. Roughly speaking, $\sigma_u$ resembles $2p\sigma$ since only the outermost loops are important for overlap with $\phi_d$. As compared with the bonding MO $\sigma_g$ [see (11-2)], $\sigma_u$ must be larger because of the node between the atoms in (11-2) and the related form of the normalization factor, and because of the negative charge in $I_2^-$.

We now ask how we may expect the iodine molecule to be oriented toward the benzene when they are brought together to form the complex. According to (11-1) $W_N$ is given by

$$W_0^0 - \frac{(\beta_0^k)^2}{\Delta_{0k}}, \quad \text{or by} \quad W_0^0 - \sum_i{}' \frac{(\beta_0^i)^2}{\Delta_{0i}}$$

if more than one CT function[†] is important for the stabilization. In a consideration of what orientation will minimize $W_N$, small $W_0^0$ and large $(\beta_0^k)^2/\Delta_{0k}$ or $\sum_i{}' (\beta_0^i)^2/\Delta_{0i}$ are desirable; both terms vary with orientation. Let us first consider the $\beta_0^2/\Delta$ terms.

In R3 only one CT state was considered, in which $D^+$ was plausibly assumed to be obtained by removing an electron from a higher energy $\pi$ MO ($\phi_2$ or $\phi_3$). This seemed reasonable because it gives the CT state with the smallest value of $\Delta_{0k}$ and also because the corresponding orientation is especially compact. (See Fig. 11-5.) If this donor MO is used as $\phi_d$, the orientation of the acceptor must be such that the $\sigma_u$ MO belongs, when in the complex, to the same symmetry species. Two (degenerate) possibilities of orientation that satisfy this requirement are indicated in Fig. 11-5. Here the axis of the iodine molecule is parallel to the plane of the benzene molecule and above it. This is the resting model $R$ of R3. The two possibilities sketched correspond to isomeric

[†] Although in principle locally excited states also contribute to $\Psi_N$ in (11-1), this participation is probably small and can be neglected.

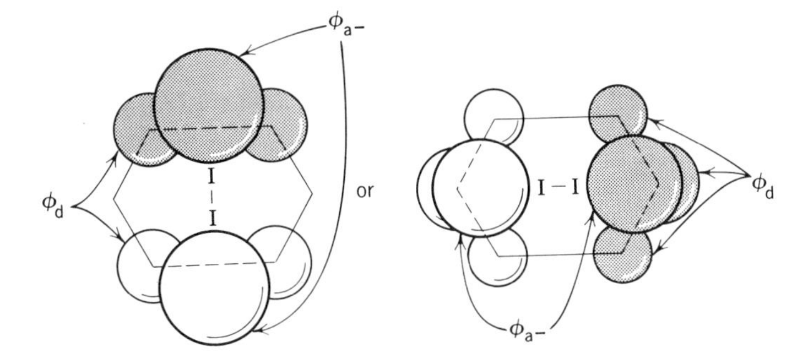


**Figure 11-5** Orientations of $I_2$ and benzene that give nonzero overlap between $\phi_d$ and $\phi_{a-}$, using a higher energy $\pi$ MO of benzene ($\phi_2$ or $\phi_3$) for $\phi_d$. This schematic drawing shows the orbitals viewed from above for the resting model.

forms of the complex of equal energy, but these should be labile since intermediate forms of the degenerate $\pi$ MOs are also equally possible, each with its corresponding rotated orientation of the $I_2$-axis; the energies of all these forms are probably nearly but not quite equal.

After the publication of R3 in 1952 experimental evidence (crystal structures of Bz · $Br_2$ and Bz · $Cl_2$ by Hassel and co-workers [2]—see Section 5-4) appeared which suggests that the correct structure for the equilibrium form of the Bz · $I_2$ complex may be the axial model *A*, with the iodine molecule standing above the plane of the benzene with its axis along the latter's sixfold symmetry axis. Other possibilities for the orientation of the iodine have been discussed in R3. One that seemed (and still seems) reasonable is the oblique model, *O*, which is intermediate between *A* and the *R* models.

In case the *A* model is correct for the isolated complex (which is not certain, as discussed in Chapter 5, since the benzene · $Br_2$ complexes in the crystal are apparently not of 1:1 structure but *n*:*n* chains), the lowest energy CT state that has the same symmetry as $\Psi_0^0$ and so can stabilize the complex is that formed by removing an electron from the *lower* energy $\pi$ MO of benzene, $\phi_1$. For this second dative function $I_D$ is at least 11.6 eV so that $\Delta_{0k}$ is somewhat larger than if $\phi_d$ were $\phi_2$ or $\phi_3$ (with $I_D$ = 9.42 eV), as for the lowest energy dative function. However, the resulting resonance-energy stabilization $(\beta_0^k)^2/\Delta_{0k}$ of the ground state of the axial model using the second dative function could be greater than

$(\beta_0^k)^2/\Delta_{0k}$ for the resting model using the first dative function if $(\beta_0^k)^2$ is sufficiently larger in the former case. Actually approximate computations [3] do indeed indicate that such is the case. Hence CT theory *does* suggest that the axial model is correct and that the energy diagram for benzene $\cdot I_2$ is like that shown in Fig. 11-3. The observed CT band is then to be interpreted as the transition from $W_N$ to the lowest CT state ($V'$ and $h\nu'_{CT}$ in Fig. 11-3), but the stabilization of $W_N$ is probably from the higher CT state ($V$ in Fig. 11-3). This observed CT transition, $h\nu'_{CT}$, is now expected to be polarized perpendicular to the D—A axis, and thus parallel to the benzene plane. (See Section 13-2 for further discussion of perpendicularly polarized CT transitions.) This characteristic provides a possible way for an experimental check on the geometry of the complex.

Thus far we have assumed that the configuration of the complex is governed entirely by a maximization of the CT resonance energy $(\beta_0^k)^2/\Delta_{0k}$ or $\sum_i'(\beta_0^i)^2/\Delta_{0i}$. But $W_N$ also depends on $W_0^0$ and this is expected to differ somewhat for different orientations of the $I_2$. The quantity $W_0^0$ includes whatever classical electrostatic attraction energies may be present, as well as all exchange (steric) repulsion energies between the MO closed shells of electrons of benzene and $I_2$. As noted in Section 9-3, these exchange-repulsion energies should be approximately proportional to the squares of the corresponding overlap integrals. Overlap-integral calculations, as yet unpublished [3], strongly indicate that the exchange repulsions at equally close approach (relative to van der Waals contact) of the benzene and iodine are considerably smaller for the $A$ than for the $R$ model; and that for this reason, together with the resulting larger value of $(\beta_0^k)^2$ when closer approach is thus permitted, model $A$ should be the more stable. Furthermore, this conclusion is reinforced by consideration of the electrostatic stabilization energy [4], which is also expected to be larger for model $A$.

On the other hand, for the oblique model, which seems to be a real possibility, the symmetry is so low that the CT functions that correspond to both the first and the second $\pi$ ionization of benzene can mix with $\Psi_0^0$. The CT absorption would then have both parallel and perpendicular components. Work of Dallinga [5a] on X-ray analysis of the Bz $\cdot I_2$ complex in solutions of iodine in benzene is suggestive of the oblique model. Actually, in view of the small $-\Delta H$ for Bz $\cdot I_2$, the $I_2$ may well at room temperature be moving around vigorously relative to the benzene (the benzene, being lighter, would do most of the moving). In the case of the Mes $\cdot I_2$ complex studied in solutions of iodine in mesitylene, Dallinga [5b] concluded that the $I_2$-axis is parallel to the benzene plane as in the $R$ model but displaced strongly outward from the center toward one methyl group. (See the left-hand part of Fig. 11-5, with the $I_2$ axis displaced toward the right, with its center over the right-hand corner of the ring.) Conceivably such a rather asymmetrical configuration can minimize $W_N$, but no simple qualitative explanation for this possibility is available.

## 11-3 APPLICATION TO ETHER · $I_2$

For the ether molecule the donor MO with lowest $I_D$ is essentially a pure $2p_x$ oxygen atom AO, with its symmetry axis perpendicular to the C—O—C plane, as discussed in Section 5-5. As we saw there, if the $I_2$ orientation were chosen to give maximum overlap between this donor MO and the $\sigma_u$ MO of $I_2$, we should expect the $I_2$ axis to be above the oxygen atom with its axis perpendicular to the C—O—C plane. (Alternatively, it might be on one side of the oxygen atom with the two iodine atoms located symmetrically above and below the C—O—C plane if the repulsion from the other electrons could be ignored.) (See R1.) Since the iodine molecule is found experimentally (see Section 5-5) to be above the oxygen atom with its axis making an angle of about 50° with the C—O—C plane, we conclude that the donor MO is, roughly, a chemical hybrid oxygen orbital. A more accurate description is that *two* CT functions are involved in stabilizing the ground state—not just the one obtained by donation from the $(2p_x)_O$ MO, but also one obtained by donation from a more tightly bound donor MO: approximately, a lone-pair oxygen atom AO that is mainly $2s$ but partly $2p\sigma$, with axis lying in the C—C plane and bisecting the C—O—C angle. This higher energy CT function would favor an orientation of the I—I axis in the plane. Presumably use of this second function increases $S_{da^-}$ and with it $|\beta_0|$ so much that for the higher energy donor MO $\beta_0^2/\Delta_{01}$ is increased even though $\Delta_{01}$ is larger than for a pure $(2p_x)_O$ donor orbital. The combined stabilization by the two CT functions could explain the observed intermediate orientation of the I—I axis (cf. Table 5-4). The energy-level scheme for the ether · $I_2$ complexes, then, is similar to that sketched in Fig. 11-2.

## 11-4 THE INTENSITY OF THE CHARGE-TRANSFER BAND; MIXING OF CHARGE-TRANSFER WAVEFUNCTIONS WITH LOCALLY EXCITED FUNCTIONS

Let us consider the wavefunction of the charge-transfer state under the assumption that one higher energy locally excited donor state mixes somewhat with it.[†] Then [see (11-1)]

$$\Psi_V = a^*[\Psi_1^0(D^+—A^-) + \rho_{10}\Psi_0^0(D, A) + \rho_{12}\Psi_2^0(D^*, A) + \cdots]. \quad (11\text{-}3)$$

The corresponding perturbed locally excited wavefunction is

$$\Psi_L = a'[\Psi_2^0(D^*, A) + \rho_{21}\Psi_1^0 + \cdots]. \quad (11\text{-}4)$$

The ground-state wavefunction is

$$\Psi_N = a(\Psi_0^0 + \rho_{01}\Psi_1^0 + \cdots). \quad (11\text{-}5)$$

The energy levels and absorption transitions that correspond to the wavefunctions of (11-3) to (11-5) have been shown in Figs. 11-2 or 11-3. In Fig. 11-3 the energy of the pure locally excited function is lower than that of the dative function, whereas in Fig. 11-2 an example is shown for the case when the energy of the locally excited state $\Psi_2^0$ is higher than that of the first dative state.[‡]

If the mixings in (11-3) and (11-4) are not large, the transition dipole for the CT band becomes [cf. (3-11)]

$$\mu_{VN} = a^*b(\mu_1 - \mu_0) + (aa^* - bb^*)(\mu_{01} - S_{01}\mu_0) + aa^*\rho_{12}\mu_{D^*D} + \cdots, \quad (11\text{-}6)$$

where $\mu_{D^*D}$ is the transition dipole for the transition between D and D* in the isolated donor molecule; note that, if the energy diagram is as shown in Fig. 11-3, $\rho_{12}$ is negative.[†] Hence the intensity of the CT absorption is decreased, for this example, if $\mu_{D^*D}$ and $\mu_{VN}^0$ are in the same direction.

The transition to the locally excited state can also occur, with transition dipole

$$\mu_{NL} = aa'(\mu_{D^*D} + \rho_{21}\mu_{01} + \cdots). \quad (11\text{-}7)$$

The signs of the $\mu$ terms in (11-6) and (11-7) are to be determined by choosing the signs of the wavefunctions so that the overlap integrals $S_{01}$ and $S_{12}$ are positive. The total intensity sum for the two transitions in the complex should be the same as the sum of the two intensities computed without mixing. Thus for the case examined above for a strong CT and weak local excitation transition, and $\rho_{21}$ $(= -\rho_{12})$ positive,[‡] the local transition is strengthened and the CT band weakened by the mixing.

---

[†] If the interactions between $\Psi_1^0$ and $\Psi_2^0$ are strong, the above perturbation-theory equation must be replaced by solutions of a secular equation (cf. Fig. 11-1).

[‡] If the zero-order dative function is higher in energy than the zero-order locally excited function, each 1 in (11-3) through (11-6) must be replaced by 2, and vice versa, and $\rho_{12}$ becomes $\rho_{21}$, and vice versa. And since $\rho_{21} = -\rho_{12}$, the sign of the coefficient of $\mu_{D^*,D}$ in (11-6) is changed from negative to positive [$\rho_{01}$ and $\rho_{12}$ are negative because $\beta_0^1$ and $\beta_1^2$ are negative while $\Delta_{01}$ and $\Delta_{12}$ are positive in (11-1)].

The relative importance of the first two of the three terms in (11-6) has been discussed in Section 3-5, where it was pointed out that the first term is expected to be predominant in strong complexes, whereas the second term should become important in weak complexes or especially for contact CT spectra. However, in these cases the third term can be important (or even predominant, as indicated by Murrell [6], at least sometimes). Obviously the relative importance of the third term depends greatly on the strength of interaction of charge-transfer and locally excited states, and on the magnitude of $\mu_{D*D}$ (or $\mu_{A*A}$) for the locally excited transition. In the case of weak complexes irregularities of intensity as compared with the simplified theory can perhaps be accounted for in part by the third term in (11-6). However, in strong complexes with large $f$ values (see Table 10-2 for some examples) the third term (contrary to some suggestions by Murrell [6]) can scarcely be of major importance.

## 11-5 REFERENCES


[1] H. Kuroda, I. Ikemoto, H. Akamatu, *Bull. Chem. Soc. Japan*, **39**, 1842 (1966).
[2] O. Hassel and Chr. Rømming, *Quart. Revs.*, **16**, 1 (1962).
[3] Unpublished result of Y. N. Chiu and W. B. Person. Earlier [*Progr. Theor. Phys.*, **22**, 313 (1959)] S. Aono had argued for the axial model on the basis that the $S_{da^-}$ overlap integral, hence $\beta_0$, is very small (0.006) for the $R$ model, but reasonably large (0.11) for the $A$ model, but his way of estimating $S_{da^-}$ for the $R$ model is unconvincing.
[4] M. W. Hanna, *J. Am. Chem. Soc.*, **90**, 285 (1968).
[5] (a) G. Dallinga (abstract only) in *Acta Cryst.*, **7**, 665 (1954). (b) G. Dallinga, reported at the Third International Congress of the International Union of Crystallography, Paris 1954 (see Abstracts, p. 52).
[6] J. N. Murrell: R10 and [6] of Chapter 3.

CHAPTER 12

# *Whole-Complex-Molecular-Orbital Descriptions of Complexes and Comparison with Resonance-Structure Descriptions*

## 12-1 REVIEW OF RESONANCE-STRUCTURE DESCRIPTION

Thus far a resonance-structure description has been used for the wavefunctions of the states of a complex, based on the use of MOs of D and A separately, with charge transfer from the one to the other to give dative resonance structures. The present chapter presents a description using only nonlocalized MOs of the complex as a whole (in other words, DA MOs) and then gives a comparison of the two methods. First, however, let us review certain details of the simplest version of the resonance-structure description, previously discussed in Sections 2-5 through 2-8 but now presented a little differently. (See GR5 for a somewhat modified and slightly expanded summary of the material given in Sections 12-1 to 12-5.)

It is useful here to recall two ways of writing the Heitler-London approximate wavefunction $\Psi_{H\text{-}L}$ for the normal state of $H_2$ (here $a$ and $b$ refer to $1s$ AOs on the two atoms):[†]

$$\Psi_{H\text{-}L} = \left(\frac{\mathcal{A}}{\sqrt{2(1+S_{ab}^2)}}\right)[\phi_a(1)\,\alpha(1)\,\phi_b(2)\,\beta(2) + \phi_b(1)\,\alpha(1)\,\phi_a(2)\,\beta(2)]$$
$$= \left(\frac{1}{\sqrt{2}\,\sqrt{1+S_{ab}^2}}\right)[\phi_a(1)\,\phi_b(2) + \phi_b(1)\,\phi_a(2)]\,S_0(1,2), \qquad (12\text{-}1)$$

[†] The antisymmetrizer, $\mathcal{A}$ (see Section 2-6) here is simply $(P_0 - P_{12})/\sqrt{2}$.

where

$$S_0(1, 2) \equiv \frac{1}{\sqrt{2}} [\alpha(1)\, \beta(2) - \beta(1)\, \alpha(2)].$$

$S_0(1, 2)$ is a two-electron singlet-state spin function. (Regarding the corresponding triplet-state functions, see Section 2-8.)

For a D, A complex [see Chapter 2, especially (2-29) to (2-36)],[†][‡] after adding a double-charge-transfer term $\Psi_{11}(D^{2+}, A^{2-})$ that will be needed in making comparisons with the overall MO method, we have

$$\Psi_N = a\Psi_0 + b\Psi_1 + c\Psi_{11} + \cdots; \qquad \Psi_V = a^*\Psi_1 - b^*\Psi_0 + c^*\Psi_{11} + \cdots$$

$$\Psi_0(\mathrm{D, A}) \simeq \mathcal{A}[\phi_d(1)\, \alpha(1)\, \phi_d(2)\, \beta(2)\, \phi_d{}'(3)\, \alpha(3) \cdots \phi_a(M+1)\, \alpha(M+1) \cdots]$$

$$\Psi_1(\mathrm{D^+{-}A^-}) \simeq \frac{\mathfrak{N}\mathcal{A}}{\sqrt{2(1 + S^2_{\mathrm{da}^-})}} \{[\phi_d(1)\, \alpha(1)\, \phi_{a^-}(2)\, \beta(2) + \tag{12-2}$$

$$\phi_{a^-}(1)\, \alpha(1)\, \phi_d(2)\, \beta(2)]\, \phi_d{}'(3)\, \alpha(3) \cdots \phi_a(M+1)\, \alpha(M+1) \cdots\}.$$

The $\Psi_0$, $\Psi_1$, and $\Psi_{11}$ terms can be rewritten so as to separate out a spin factor $S_0(1, 2)$ for electrons 1 and 2, as for $\Psi_{H\text{-}L}$ in the second equation of (12-1):

$$\Psi_0 = \mathcal{A}''[\phi_d(1)\, \phi_d(2)]\, S_0(1, 2)\, \phi_d{}'(3)\, \alpha(3) \cdots \phi_a(M+1)\, \alpha(M+1) \cdots,$$

$$\Psi_1 = \frac{\mathfrak{N}''\mathcal{A}''}{\sqrt{2}\sqrt{1 + S^2_{\mathrm{da}^-}}} [\phi_d(1)\, \phi_{a^-}(2) + \tag{12-3}$$

$$\phi_{a^-}(1)\, \phi_d(2)]\, S_0(1, 2)\, \phi_d{}'(3)\, \alpha(3) \cdots \phi_a(M+1)\, \alpha(M+1) \cdots.$$

In (12-3) each $\Psi$ has first been made antisymmetric in electrons 1 and 2.[††] The supplementary antisymmetrizer $\mathcal{A}''$ (see Section 2-5 for a somewhat similar situation) is so designed as thereafter to make $\Psi_0$ and $\Psi_1$ in (12-3) antisymmetric in *all* the electrons in the complex. The factor $\mathfrak{N}''$

---

[†] Here, as in (2-29) to (2-36), $\Psi_0$ and $\Psi_1$ are being approximated by single-electron-configuration expressions; for *accurate* $\Psi_N$ and $\Psi_V$, configuration mixing should be introduced internally in D and A, and also the generalizations of Chapter 11 should be brought in.

[‡] Recall also that $\phi_d$ and $\phi_a$ in $\Psi_0$ refer to the most easily ionized MOs of the neutral D and A, respectively, and $\phi_d{}'$ to a less easily ionized donor MO, whereas $\phi_{a^-}$ in $\Psi_1$ refers to the acceptor MO of smallest electron affinity *as it exists* in $A^-$. Strictly speaking, $\phi_d$ and $\phi_d{}'$ in $\Psi_1$ should be relabeled as $\phi_{d^+}$ and $\phi_{d'^+}$ to indicate that they are now MOs of a positive ion, and $\phi_a$ should be relabeled to indicate that it is now in a negative ion.

[††] In each of these cases the orbital factor, in brackets, is symmetric in 1 and 2, whereas the spin factor $S_0(1, 2)$ is antisymmetric, so that the product of the two is antisymmetric.

is a normalizing factor that is close to 1.† The term $\Psi_{11}$ is like $\Psi_0$ except that $\phi_{a^{2-}}$ replaces both of the $\phi_d$ MOs.

For the triplet charge-transfer state ($^3$CT) of a charge-transfer complex (see Section 2-8) we have $\Psi_T \simeq \Psi_1'$ ($D^+ \leftrightarrow A^-$), which can be written

$$\Psi_1' = \frac{\mathfrak{N}''\mathfrak{A}''}{\sqrt{2(1 - S^2_{da^-})}} [\phi_d(1)\, \phi_{a^-}(2) - \phi_{a^-}(1)\, \phi_d(2)]\, S_1(1, 2)\, \phi_d{}'(3)\, \alpha(3) \cdots \qquad (12\text{-}3a)$$

$$\phi_a(M+1)\, \alpha(M+1) \cdots,$$

where $S_1(1, 2)$ refers to the $S = 1$ spin functions with $M_S = 1, 0, -1$.

*Note.* Having introduced the foregoing discussion by reference in (12-1) to the Heitler-London function for $\Psi_N$ of $H_2$, which corresponds formally to $\Psi_1$ of a D, A complex, it is desirable to complete the comparison by writing the following improved $\Psi_N$ for $H_2$:

$$\Psi_N \simeq b\Psi_1 + a(\Psi_0 + \Psi_2) \quad (b > a),$$
$$\Psi_0 = \phi_{a^-}(1)\, \phi_{a^-}(2)\, S_0(1, 2); \quad \Psi_2 = \phi_{b^-}(1)\, \phi_{b^-}(2)\, S_0(1, 2). \qquad (12\text{-}4)$$

Here $\Psi_0$ and $\Psi_2$ are ion-pair terms $\Psi_0(H_a^- H_b^+)$ and $\Psi_2(H_a^+ H_b^-)$,‡ which in $H_2$ are obviously of equal importance. One of these, say $\Psi_0$, is however analogous to $\Psi_0$(D, A) of a typical D, A complex between two neutral molecules. The term $\Psi_2$ then corresponds to the structure $\Psi_{11}(D^{2+}, A^{2-})$, which in (12-2) for a D, A complex is ordinarily omitted as unimportant.

Further, it is instructive to write down the wavefunctions of the two excited electronic CT states of $H_2$, which can be approximated as linear combinations of $\Psi_0$, $\Psi_1$, and $\Psi_2$. These are

$$\Psi_V \simeq \frac{\Psi_0 - \Psi_2}{\sqrt{2}\,\sqrt{1 - S^2_{ab}}}, \qquad (12\text{-}5)$$

$$\Psi_Z \simeq a^*(\Psi_0 + \Psi_2) - b^*\Psi_1. \qquad (12\text{-}6)$$

$\Psi_V$, of species $^1\Sigma_u^+$, is the lowest singlet excited state of $H_2$, and the transition $\Psi_V \leftarrow \Psi_N$ has been classed as a (two-way) intramolecular CT transition [1]. The state $\Psi_Z$, somewhat higher in energy, is of species $^1\Sigma_g^+$, so that the transition $\Psi_Z \leftarrow \Psi_N$ is forbidden, since $\Psi_N$ is also of species $^1\Sigma_g^+$. Still further, $\Psi_T$, the lowest triplet state of $H_2$ in the Heitler-London approximation, is given by the second form of (12-1)

† The factor $\mathfrak{N}''$ would be exactly 1 if there were no overlap between any of the donor MOs other than $\phi_d$ and any acceptor MO other than $\phi_{a^-}$.

‡ Although (12-4) calls for $\phi_{a^-}$ and $\phi_{b^-}$ AOs, actually it has been shown that $\Psi_N$ is better approximated by using the same AOs $\phi_a$ and $\phi_b$ as in $\Psi_1$; and note that (at the equilibrium configuration) $\phi_a$ and $\phi_b$ are best taken to correspond to a somewhat higher nuclear charge than for the free hydrogen atom.

if the plus signs in that equation are replaced by minus signs and $S_0(1, 2)$ is replaced by $S_1(1, 2)$.

## 12-2 WHOLE-COMPLEX-MO DESCRIPTION

For a complex, as for a molecule, it is possible to describe the electronic structure approximately by assigning all the electrons to MOs that in principle are nonlocalized (i.e., extend around all the nuclei) but many of which actually are largely localized (near one or more nuclei in the case of a molecule). To obtain an *accurate* electronic wavefunction, the single-electron-configuration molecule or complex wavefunction that has just been described (whose single configuration may be called, following Jørgensen, the *preponderant* configuration) must be supplemented by forming a linear combination with wavefunctions of other configurations; this procedure is called *configuration mixing*, or configuration interaction.

In the case of the normal state of most molecules, a single-electron-configuration MO wavefunction is a fairly good approximation at the equilibrium nuclear configuration, but not when the bonds are stretched in an approach toward dissociation [2]. However, if the system is composed of closed-shell atoms or (almost as well) a closed-shell atom and one atom that lacks only one electron from a closed shell (e.g., $He_2$, $HeH^+$, ArNe, and $He_2^+$ in their normal states), the single-electron-configuration wavefunction remains a fairly good approximation all the way from small interatomic distances until the atoms are completely separated [2]. The closed shells just referred to are of course AO closed shells.

Now it appears that an analogous statement is also valid when two or more molecules are brought together each of which has a closed-shell structure (or a closed-shell-plus-or-minus-one-electron structure) in terms of MOs of the separate molecules. One-to-one complexes that are formed from an electron donor and acceptor conforming to the condition just stated are examples.

Because of the favorable circumstance just considered it should be a good approximation to describe the electronic structure of a typical complex *in its normal state* (and also, we can show, in some of its excited states) in terms of a single-electron configuration in which all the electrons are assigned to MOs of the complex as a whole (*whole-complex*, or fully delocalized, MOs). The statement just made is applicable (perhaps even especially) to weak complexes, whose small binding energies and large intermolecular distances make them otherwise analogous to almost dissociated molecules.

Now, although in principle all whole-complex MOs are nonlocalized (i.e., spread over the whole complex), actually many or most of them may be only slightly different from the MOs of D and A separately. Often the major part of the D, A interaction can be understood in terms of a single pair of electrons (or just one electron in the case of even-odd complexes) occupying a whole-complex MO that forms a bridge between the D and A (bridging MO), whereas the remaining whole-complex MOs remain almost the same as MOs of the free D and A.

However, even for weak complexes or for strong complexes near dissociation (see Section 12-8 for an example) and quite generally if the approach of D and A is close (as in strong complexes and especially in inner complexes—see Chapters 15 and 16), it seems likely that complex formation may often affect additional MOs nearly as much as the bridging MO, and/or there may be more than one bridging MO. For complete understanding all-electron self-consistent field (SCF) MO calculations are needed; results for a simple example ($F^-Li^+$) that can be regarded as a D · A complex are set forth in Chapter 13. Recent all-electron LCAO SCF theoretical computations by Clementi on the system $H_3N \cdot HCl$ (see Section 15-1) are helping to clarify this question. Furthermore, it seems possible also that some increased configuration mixing may be needed in the whole-complex-MO method, but this is as yet an open question.

Just as for the MOs of an individual molecule, the whole-complex MOs of a complex are the orbitals that would satisfy the requirements of an SCF calculation for a single-configuration wavefunction. Besides those on $H_3N \cdot HCl$, LCAO SCF calculations on $H_3N \cdot BH_3$ are available—see Section 12-6—however, until more such theoretical computations are available we must rely largely on qualitative or semiempirical reasoning to determine the approximate forms of whole-complex MOs.

## 12-3 SIMPLIFIED WHOLE-COMPLEX-MO TREATMENT

Let us begin with a simplified whole-complex-MO treatment, in which the CT action is assumed to be concentrated in one pair of electrons in a bridging MO $\phi_{da}$, thus paralleling the simplified resonance-structure treatment. Then

$$\Psi_N = \mathcal{A}[\phi_{da}(1)\,\phi_{da}(2)\,S_0(1,2)]\,\phi_3'(3)\,\alpha(3)\cdots. \tag{12-7}$$

Here $\phi_3'\alpha$ (and the numerous further occupied molecular spinorbitals (MSOs) that are indicated by dots) are MSOs of the entire complex, but they are here assumed to be the same as the corresponding MSOs in (12-3), part of which are localized on the donor and the rest on the acceptor. In LCMO approximation the bridging MO has the form

$$\phi_{da} = m\chi_d + n\chi_a, \tag{12-8}$$

a simple linear combination of a donor MO $\chi_d$ and acceptor MO $\chi_a$, with $m$ and $n$ positive. Here $\chi_d$ is nearly the same as $\phi_d$ of (12-3), and $\chi_a$ is similar to $\phi_{a-}$ of (12-3).† For weak complexes $m >> n$; hence $\phi_{da}$ is mainly a donor MO that has, however, spread out slightly around the acceptor.

The fraction of an electron that is transferred to the acceptor in $\Psi_N$, analogous to $F_1$ in (1-5) in the resonance-structure method, is obtained after first squaring $\phi_{da}$ of (12-8), multiplying by 2 because there are two electrons in $\phi_{da}$, and integrating as follows:

$$2\int \phi_{da}^2\,dv = 2m^2 \int \chi_d^2\,dv + 4mn \int \chi_d\chi_a\,dv + 2n^2 \int \chi_a^2\,dv$$
$$= 2m^2 + 4mnS_{da} + 2n^2.$$

Here $2m^2$, $4mnS_{da}$, and $2n^2$ represent the net electron population on D, the overlap population, and the net population on A when the total population (2) of electrons in $\phi_{da}$ is broken down into fractions [3]. Since the overlap population involves D and A equivalently, half of it can be allocated to A, along with the net population on A, to give the gross population on A. The latter is equal to the desired quantity $F_1$; that is,

$$F_1 = 2(n^2 + mnS_{da}). \tag{12-9}$$

Further, since $m$ and $n$ are both positive, $\phi_{da}$ has the form of a *bonding* MO. The overlap population $4mnS_{da}$ should be roughly proportional to the bond energy [4]. Notably, $4mnS_{da}$ can be of considerable size even when $F_1$ is rather small; for example, suppose that $n = 0.2$ and $S_{da} = 0.1$; then $F_1$ is about 0.12, and $4mnS_{da}$ is about 0.08. For comparison, in a full-strength nonpolar single bond we would have $m = n = 1/\sqrt{2(1 + S_{da}}$, and if, for example, $S_{da} = 0.35$, $4mnS_{da}$ would be approximately 0.5. The bonding effect of the electron pair in $\phi_{da}$ of (12-7), assuming that only these electrons contribute appreciable bonding effects, is the counterpart [4] in the whole-complex-MO description of the resonance energy $-\beta_0^2/\Delta$ [cf. (2-20)] in the simplified resonance-structure method.

In the CT state one electron must be excited from $\phi_{da}$ to an excited MO in which $F_1$ is increased from nearly 0 to nearly 1. For the present simplified model in which changes are confined to the MOs of two electrons‡

---

† Just how close are these resemblances has not yet been clarified; see the remarks in the second preceding paragraph; also the footnote ‡ on p. 178.

‡ Because $\phi_{da}$ and $\phi_{da}^*$ are orthogonal, there is no factor in (12-10) that is similar to $1/\sqrt{1 + S_{da-}^2}$ in (12-3) for $\Psi$.

$$\Psi_V = \frac{\mathcal{A}''}{\sqrt{2}}[\phi_{da}(1)\ \phi_{da}^*(2) + \phi_{da}(2)\ \phi_{da}^*(1)]S_0(1, 2)\ \phi_3'(3)\ \alpha(3) \cdots, \tag{12-10}$$

$$\Psi_T = \frac{\mathcal{A}''}{\sqrt{2}}[\phi_{da}(1)\ \phi_{da}^*(2) - \phi_{da}(2)\ \phi_{da}^*(1)]T_{0\pm 1}(1, 2)\ \phi_3'(3)\ \alpha(3) \cdots.$$

Here $\phi_{da}^*$ is an antibonding MO that in LCMO approximation is constructed from $\chi_d$ and $\chi_a$:

$$\phi_{da}^* = m^*\chi_a - n^*\chi_d; \tag{12-11}$$

$m^*$ and $n^*$ must be so chosen that $\phi_{da}^*$ is orthogonal to $\phi_{da}$; if $m > n$, $m^* > n^*$. In connection with $\Psi_V$ and $\Psi_T$, we should note that in an accurate SCF whole-complex-MO treatment $\phi_{da}$ in (12-10) and in fact *all* the occupied MOs would be a little different than in $\Psi_N$ because of the somewhat altered field that results largely from the transfer of an electron from $\phi_{da}$, which is mostly on D, to $\phi_{da}^*$, which is mostly on A.

## 12-4 COMPARISON BETWEEN SIMPLIFIED WHOLE-COMPLEX-MO AND SIMPLIFIED RESONANCE-STRUCTURE DESCRIPTIONS

In order to compare the simplified whole-complex-MO and resonance-structure approximations for $\Psi_N$ we substitute the LCMO approximation of (12-8) for $\phi_{da}$ into (12-7) and approximate $\phi_3'$ by $\phi_d'$; then

$$\Psi_N \simeq \mathcal{A}''\{[m\chi_d(1) + n\chi_a(1)]\ [m\chi_d(2) + n\chi_a(2)]\}S_0(1, 2)\ \phi_d'(3)\ \alpha(3) \cdots. \tag{12-12}$$

By multiplying out and designating the resulting expression for $\Psi_N$ by $\Psi_N'$ we obtain

$$\Psi_N' \simeq \mathcal{A}''\{m^2\chi_d(1)\ \chi_d(2) + mn[\chi_d(1)\ \chi_a(2) + \chi_d(2)\ \chi_a(1)] + n^2\chi_a(1)\ \chi_a(2)\}\ S_0(1, 2)\ \phi_d'(3)\ \alpha(3) \cdots. \tag{12-13}$$

In comparing (12-13) with $\Psi_N$ of the resonance-structure approximation [see (12-2 and 12-3)] the first term (two electrons in $\chi_d$) must evidently be identified with $a\Psi_0$ of (12-3) and the second with $b\Psi_1$, to the extent that we can identify $\chi_d$ with $\phi_d$ and $\chi_a$ with $\phi_{a^-}$. The third term in (12-13) corresponds to the term $\Psi_{11}(D^{2+}, A^{2-})$ of (12-2). With these identifications,

$$\Psi_N' \simeq m^2\Psi_0(D, A) + mn\sqrt{2(1 + S_{da^-}^2)}\ \ \Psi_1(D^+—A^-) + n^2\Psi_{11}(D^{2+}, A^{2-}), \tag{12-14}$$

in agreement with the first of (12-2) on putting

$$a = m^2, \quad b = mn\sqrt{2(1 + S^2_{\mathrm{da}^-})}, \quad c = n^2.$$

Thus it is seen that for state $N$ the simplified whole-complex-MO description using a single-MO-configuration wavefunction with one LCMO-approximated bridging MO essentially coincides with the simplified resonance-structure description if [as in (12-2)] the added term $c\Psi_{11}(\mathrm{D}^{2+}, \mathrm{A}^{2-})$ is included and if $\chi_{\mathrm{d}}$ and $\chi_{\mathrm{a}}$ can properly be taken the same in (12-8) and (12-11) and if also they can then be identified with $\phi_{\mathrm{d}}$ and $\phi_{\mathrm{a}^-}$. For a loose complex $c$ $(= n^2)$ approaches zero, so that the two approximations for $\Psi_N$ approach identity. Conversely, for strong complexes $c\Psi_{11}$ is no longer negligible; we may then question whether the relatively large weight $(c = n^2)$ of the function $\Psi_{11}(\mathrm{D}^{2+}, \mathrm{A}^{2-})$ is reasonable. In a more general treatment the coefficient $c$ need not be constrained to the value $n^2$ but can be allowed to go free, with a value to be chosen by the variation method. This modification corresponds to introducing some configuration mixing into the description of $\Psi_N$ in (12-13).

For the singlet charge-transfer state ($^1$CT), using the LCMO approximations for $\phi_{\mathrm{da}}$ and $\phi^*_{\mathrm{da}}$, and expanding $\Psi_V$ of (12-10),

$$\Psi'_V \simeq \sqrt{2}\mathcal{A}''\Big\{\frac{(mm^* - nn^*)}{2}[\chi_{\mathrm{d}}(1)\,\chi_{\mathrm{a}}(2) + \chi_{\mathrm{a}}(1)\,\chi_{\mathrm{d}}(2)]$$
$$- mn^*\chi_{\mathrm{d}}(1)\,\chi_{\mathrm{d}}(2) + m^*n\chi_{\mathrm{a}}(1)\,\chi_{\mathrm{a}}(2)\Big\}S_0(1, 2)\,\phi'_{\mathrm{d}}(3)\,\alpha(3)\cdots. \tag{12-15}$$

With the same provisos stated above for $\Psi_N$ about $\chi_{\mathrm{d}}$ and $\chi_{\mathrm{a}}$ and their identification with $\phi_{\mathrm{d}}$ and $\phi_{\mathrm{a}^-}$, this equation becomes

$$\Psi'_V \simeq a^*\Psi_1(\mathrm{D}^+\text{—}\mathrm{A}^-) - b^*\Psi_0(\mathrm{D}, \mathrm{A}) + c^*\Psi_{11}(\mathrm{D}^{2+}, \mathrm{A}^{2-}),$$

with

$$a^* = (mm^* - nn^*)\sqrt{(1 + S^2_{\mathrm{da}^-})}, \quad b^* = mn^*\sqrt{2}, \quad c^* = m^*n\sqrt{2}. \tag{12-16}$$

Again just as for $\Psi_N$ and $\Psi'_N$, the two approximations called $\Psi_V$ and $\Psi'_V$ agree if the $\Psi_{11}$ term is included. Again as in the case of (12-14), (12-16) can be generalized by varying $c^*$. However, (12-16) shows that $c^*/b^*$ necessarily approaches 1 for loose complexes, or for increasing $R_{\mathrm{DA}}$ in dissociation of a complex, if the whole-complex-MO approximation is used. This behavior is incorrect since $c^*/b^*$ must approach zero in (12-2) for $\Psi_V$, which is surely correct as $R_{\mathrm{DA}} \to \infty$; nevertheless the error becomes unimportant for *loose* complexes since $c^*$ and $b^*$ are then both small. Note that in $\Psi_N$ this fault of the whole-complex-MO method for loose complexes is not present (see above).

For the triplet charge-transfer state ($^3$CT) expansion of $\Psi_T$ of (12-10) gives

$$\Psi_T' \simeq \frac{\mathcal{A}''}{\sqrt{2}} (mm^* + nn^*) [\chi_d(1) \chi_a(2) - \chi_d(2) \chi_a(1)] \times \\ S_1(1, 2) \phi_a'(3) \alpha(3) \cdots . \tag{12-15a}$$

With the same provisos as above, this expression, with $mm^* + nn^* = 1/\sqrt{1 - S_{da^-}^2}$, is identical with the resonance-theory expression (12-3*a*) for $\Psi_1'$. Thus for the $^3$CT state the whole-complex-MO and resonance-theory descriptions *coincide*.

This result for the $^3$CT state is precisely analogous to what is found for the diatomic case of the $^3\Sigma_u^+$ state of $H_2$ formed from two normal hydrogen atoms, where the simple MO description coincides with the Heitler-London description at all $R$ values. On the other hand, for the $^1$CT state the analogy is to the $^1\Sigma_g^+$ ground state of $H_2$, where the MO approximation is good only at small $R$ values and fails completely as $R \to \infty$, whereas the Heitler-London description becomes increasingly accurate as $R \to \infty$. Finally, for the ground state $^1N$ of the charge-transfer complex formed from MO-closed-shell donor and acceptor the analogy is to the interaction of two helium atoms, where, as in the $^3$CT state, the MO and valence-bond approximations coincide as $R \to \infty$.

The preceding discussion affords an explanation of a contradiction in the predictions of the whole-complex-MO and resonance-structure approximations for the $^1$CT and $^3$CT states. In both of these approximations energy expressions for these states contain a term $+K$ for the $^1$CT and $-K$ for the $^3$CT state, where $K$ is an exchange integral that is positive in the MO theory (because $\phi_{da}$ and $\phi_{da}^*$ are orthogonal) but is probably negative (as in the Heitler-London theory for $H_2$) in the resonance-structure theory. The apparent anomaly is resolved when we recognize that for the $^1$CT state the MO theory is incorrect for typical complexes. However, as we have seen, the two approximations coincide in the limit as $R \to \infty$ (here $K = 0$).

We are now led to the following conclusions. *In general* the resonance-structure theory is preferable except for small $R$ values such as occur in the very strong classical complexes such as $R_3N \cdot BCl_3$. For weaker complexes the MO theory gives for the $^1$CT state an incorrect (too high) energy, the error diminishing, however, as we go either to very strong or very weak complexes. Nevertheless for the ground state and the $^3$CT state the MO theory is equally as acceptable as the resonance-structure theory at all $R$ values and is more convenient for computations.

Both the simplified whole-complex-MO and the simplified resonance-structure descriptions that are discussed in this section can be called two-electron approximations in the sense that orbital changes are assumed to occur only for two electrons in a comparison of $\Psi_V$ with $\Psi_N$ or of $\Psi_1$

with $\Psi_0$. In fact the comparisons would come out the same as those given above if we had written down only those parts of the $\Psi$s that refer to electrons 1 and 2 and omitted $\mathfrak{A}''$ and all reference to electrons $3 \cdots N$, which then merely contribute to the field in which electrons 1 and 2 move. However, although they are probably moderately satisfactory in some cases, two-electron approximations are seriously inadequate in others. (See, for example, Section 12-8.)

## 12-5 APPLICATIONS TO SPECIFIC EXAMPLES: $R_3N \cdot I_2$

Consider specifically the complex between $R_3N$ and $I_2$. Then the structures to be considered in (12-14) are $\Psi_0(R_3N,I_2)$, $\Psi_1(R_3N^+—I_2^-)$, and $\Psi_{11}(R_3N^{2+},I_2^{2-})$. How reasonable is it to include $\Psi_{11}$, which comes from the whole-complex-MO description [see (12-12) and (12-13)], with nonzero weight in describing $\Psi_N$? We have already emphasized that the entire $R_3N$ molecule takes part in the donor action, although this is centered in the nitrogen atom lone pair. (See Section 4-2.) The R groups feed charge to the nitrogen-atom through the N—R $\sigma$-bonds; therefore even in $R_3N^{2+}$ the actual charge on the nitrogen atom would probably be only about $+1e$. The $I_2^{2-}$ would have electron configuration $[\cdots \sigma_g{}^2 \pi_u{}^4 \pi_g{}^4 \sigma_u{}^2]$; it is therefore like $Xe_2$ and is expected not to be bound. However, there is no objection to mixing a little of this function into the description of $\Psi_N$, especially since the $I_2^{2-}$ in $\Psi_{11}$ is stabilized by the field of the $R_3N^{2+}$.

Ideally, one would like to carry out an all-electron SCF whole-complex-MO calculation. This is too large an undertaking even for $H_3N \cdot I_2$ at present, although in the year after the lectures on which this volume is based were given (1966) Clementi carried through approximate calculations of this kind for $H_3N \cdot HCl$. (See Sections 15-1 and 12-8.) With a reasonably accurate all-electron calculation many of the questions that are associated with the crude two-electron discussion given above should be answered; until then we may use (12-14) qualitatively to suggest the direction (more toward $\Psi_{11}$) in which prediction from the resonance-structure theory should be modified.

In an accurate SCF whole-complex-MO calculation $\phi_{da}$ of (12-8) would be built from donor and acceptor MOs that have been *modified* as compared with those of the free D and A (or $A^-$). The modifications include (a) scaling (i.e., shrinking or expanding the MOs); and (b) polarization of the MOs to give greater bonding or reduced antibonding. Polarization includes the kind of modification that is commonly called hybridization. Also, there would be very appreciable changes in some of the other whole-complex MOs in view of the rather high $-\Delta H$s of the $R_3N \cdot I_2$ complexes.

## 12-6 APPLICATION TO $H_3N \cdot BH_3$

Consider now the $n \cdot v$ complex (or compound) $H_3N \cdot BH_3$ for which the approximate transformed whole-complex-MO equation (12-14) is a linear combination of $\Psi_0(H_3N,BH_3)$, $\Psi_1(H_3N^+—B^-H_3)$, and $\Psi_{11}(H_3N^{2+},B^{2-}H_3)$. Here again the coefficient of $\Psi_{11}$ is intuitively not expected to be large. If the coefficients of $\Psi_0$ and $\Psi_{11}$ were equal, the structure would resemble that of $H_2$ [see (12-4)] except that the covalent bond is between oppositely charged radicals $H_3N^+$ and $B^-H_3$ instead of between neutral atoms. If $H_3N \cdot BH_3$ maximally resembled the isoelectronic compound $H_3C—CH_3$, these coefficients *would* be equal but both rather small compared with that of $\Psi_1$. Already some rough all-electron Hückel-type calculations [5a] are available; they indicate $a > b > c$ in (12-2) for $H_3N \cdot BH_3$ and give interesting indications about the charge distribution in other similar compounds. More recently some SCF-LCAO whole-complex-MO calculations have been made [5b] (see also [5c]); these confirm Hoffmann's results [5a]. A population analysis [5b] gives the following net charges: on boron, $-0.17e$; on nitrogen, $-0.78e$; on each hydrogen attached to boron, $-0.05e$; on each hydrogen attached to nitrogen, $+0.36e$.

With stronger donors and acceptors, as, for example, in $R_3N \cdot BCl_3$, it seems likely that $b > a > c$ (or possibly $b > c > a$?).

## 12-7 APPLICATION TO $R \cdot a\pi$ AND $b\pi \cdot Q$ COMPLEXES

Consider the complex that is formed between Na and naphthalene (Npt). This can be thought of either as a complex between Na and Npt of form $a'\Psi_0(\text{Na,Npt}) + b'\Psi_1(\text{Na}^+,\text{Npt}^-)$ with $b' \gg a'$ or, probably better, as a complex between $\text{Npt}^-$ and $\text{Na}^+$ of the form $a\Psi_0(\text{Npt}^-,\text{Na}^+) + b\Psi_1(\text{Npt,Na})$ with $a \gg b$.* The former viewpoint would classify the system as an $R$-donor $\cdot$ $a\pi$-acceptor complex; the latter as a $b\pi$-donor $\cdot$ $v$-acceptor complex. Experimentally it is found that the system in solution exists almost entirely as the ion pair $\text{Na}^+\text{Npt}^-$, although electron-spin-resonance studies reveal that the electron spends a little time on the Na nucleus [6]. It is of interest that the essential features of systems of this type can be stated by using a *one-electron* approximate $\Psi_N$ description in which, moreover, the *simplified* single-configuration whole-complex-MO wavefunction is *the same* as the simplified resonance-structure wavefunction. The MO to which the one electron is assigned is $(m\chi_d + n\chi_a)$, with

---

* Of course $b' = a$ and $a' = b$.

$m >> n$, where $\chi_d$ is the odd-electron MO on $Npt^-$, and $\chi_a$ is the 3$s$ AO on Na. Expansion of this simplified whole-complex-MO wavefunction gives just $a\Psi_0(Npt^-, Na^+) + b\Psi_1(Npt, Na)$, with $a = m$ and $b = n$. Similarly in the charge-transfer state one electron is transferred in the MO description from $(m\chi_d + n\chi_a)$ to $(m^*\chi_a - n^*\chi_d)$, with $m^* >> n^*$, which on expansion gives precisely the simplified resonance-structure function $a^*\Psi_1 - b^*\Psi_0$, with $a^* = m^*$ and $b^* = n^*$.

Consider also the complex between the iodine atom and an aromatic donor such as benzene ($b\pi \cdot$ Q class). Here

$$\Psi_N = a\Psi_0(Bz, \dot{I}) + b\Psi_1(\dot{B}z^+, I^-).$$

Strong and co-workers [7], and Porter and co-workers [8] have studied the spectra of these complexes. The iodine atoms are formed by irradiating solutions of $I_2$ to dissociate the $I_2$ molecules. A transitory new absorption band that appears has been interpreted as the CT absorption band of the Bz $\cdot$ I complex. As in the $Npt^-Na^+$ case, there is no covalent bond in either $\Psi_0$ or $\Psi_1$, because one partner is even and the other is odd. For this example $\Psi_0$ involves an electron configuration $(\phi_3)^2_{Bz}(5p\sigma)_I$ (plus other filled Bz and I orbitals); $\Psi_1$ is then $(\phi_3)_{Bz}(5p\sigma_I)^2$. (See Fig. 4-1 for $\phi_3$ of benzene.) In contrast to the $Npt^-Na^+$ case, because now the donor is of closed-shell instead of odd-electron type, it is quite obviously necessary to consider this as a *three-electron* problem if we wish to express even the barest essentials about the nature of $\Psi_N$ and $\Psi_V$.

## 12-8 APPLICATION TO $H_3N \cdot HCl$

Now consider a complex of $NH_3$ and HCl. Such a complex is not known experimentally,† although we might presume that it exists in small amounts in equilibrium with the $NH_3$ and HCl that are usually believed to constitute the gas phase that is obtained when the ionic solid $NH_4^+Cl^-$ is vaporized. In any event $H_3N \cdot HCl$ would seem to be a useful model for a hydrogen-bonded complex, with

$$\Psi_N = a\Psi_0(H_3N \cdot HCl) + b\Psi_1[H_3N^+—(HCl)^-], \quad \Psi_V = a^*\Psi_1 - b^*\Psi_0$$

in the resonance-structure description.

Thus far in the discussion of complexes of a closed-MO-shell donor and acceptor the problem has been treated as one in which only two electrons are involved to an important extent, it being assumed that all others remain in donor and acceptor MOs that are nearly unaffected by

† See Section 15-1 regarding recent work.

complex formation. The terms $\Psi_0$ and $\Psi_1$ are then conveniently written as in Eqs. (12-3), which are antisymmetric in electrons 1 and 2, and then can be made antisymmetric in all the electrons by a supplementary antisymmetrizer $\mathcal{A}''$. In a complex $H_3N \cdot HCl$, however, it appears to be essential to account specifically for a second pair of electrons (namely, the bonding electron pair in HCl) because it would be completely unjustified to assume that the MOs they occupy are the same in $\Psi_1$ as in $\Psi_0$. We then have a four-electron treatment.

To deal with this situation (12-3) can be written in the form

$$\Psi_0 = \mathcal{A}'''[\phi_d(1)\, \phi_d(2)\, S_0(1,2)] \cdots [\phi_a(M+1)\, \phi_a(M+2)\, S_0(M+1, M+2)] \cdots$$

$$\Psi_1 = \frac{\mathcal{N}'''\mathcal{A}'''}{\sqrt{2}\sqrt{1+S^2_{\mathrm{da}-}}}[\phi_d(1)\, \phi_{a-}(2) + \phi_{a-}(1)\, \phi_d(2)]\, S_0(1,2) \cdots \tag{12-17}$$

$$\phi'_a(M+1)\, \phi'_a(M+2)\, S_0(M+1, M+2) \cdots .$$

Here $\phi_d$ is the lone-pair donor MO and $\phi_a$ is the H—Cl bonding MO, itself a $\sigma$ MO approximately of LCAO form

$$\phi_a = \alpha(3p\sigma_{Cl}) + \beta(1s_H), \tag{12-18}$$

with $\alpha > \beta$.[†] Now in $\Psi_1$ the MO $\phi_{a-}$ is an $(HCl)^-$ MO that must be orthogonal to $\phi_a$; hence, if built, like $\phi_a$, from $3p\sigma_{Cl}$ and $1s_H$, it must have the form

$$\phi_{a-} = \alpha^*(1s_H) - \beta^*(3p\sigma_{Cl}), \tag{12-19}$$

with $\alpha^* > \beta^*$. In view of the partial $H^+Cl^-$ polarity of $\phi_a$ $(\alpha > \beta)$ corresponding to the known polarity of the molecule, the MO $\phi_{a-}$ in order to satisfy the orthogonality requirement has to have an opposite $(H^-Cl^+)$ polarity $(\alpha^* > \beta^*)$. A consequence of this orthogonality is that the presence of the electron in the $\phi_{a-}$ MO in $HCl^-$ (with electron configuration $\phi_a^2 \cdots \phi_{a-}$, as compared with just $\phi_a^2 \cdots$ in neutral HCl) must kick back very strongly on the polarity of the $\phi_a$ MO in $HCl^-$, and so in $\Psi_1$, giving it a *modified form* called $\phi'_a$ in (12-17).

To see this note that in neutral HCl, although $\alpha > \beta$ in (12-18), the difference is not large. On the other hand, in the valence-bond description of the structure of $(HCl)^-$,

$$\Psi_{HCl^-} = a\Psi(H, Cl^-) + b\Psi(H^-, Cl),$$

[†] The $3s_{Cl}$ mixes rather strongly with the $3p\sigma_{Cl}$ AO in the $\sigma$ MOs of HCl, as is known from all-electron SCF-MO calculations for HCl, and we really should include it also and use a hybrid $\sigma$ bonding AO of Cl (or better carry out a six-electron treatment; the present treatment is somewhat schematic). The discussion of lithium fluoride in Chapter 13 illustrates some of the complexity that is inherent in an exact all-electron treatment.

everyone will agree that the coefficient $b$ must be relatively small. To match this in the MO description for $HCl^-$ $a$ must there be very much greater than $\beta$; this is necessary in order to get the negative charge concentrated mostly on the Cl. The effect is enhanced by the fact that $a^*$ must then also greatly exceed $\beta^*$. Hence the ratio $a/\beta$ must increase greatly when HCl forms $(HCl)^-$; that is, the $\phi_a$ MO must become much more polar; this change has been signalized in (12-17) by writing $\phi_a'$ in $\Psi_1$ instead of $\phi_a$ as in $\Psi_0$. This behavior, which is in contrast with that for symmetrical $\sigma$-acceptors (e.g., the iodine molecule), in which the contributions from the two polar structures $I \cdot\cdot I^-$ and $I^- \cdot\cdot I$ can perhaps be nearly equal (see Section 11-1), forces us to use a four-electron treatment for the hypothetical amine $\cdot$ HCl complexes.

This discussion illustrates how one can decide whether or not it is necessary to consider the "kickback" of the extra acceptor electron in $\Psi_1$ on the other orbitals of the acceptor, and which ones are affected. Such changes in orbitals not directly concerned with the D—A bonding may be important in understanding the intensity of the CT band.

The above discussion, from the authors' 1965 lectures (see also R14), was based on the assumption that an isolated molecule-pair $NH_3 + HCl$ tends to form a weak hydrogen-bonded complex. Recent (1966) all-electron calculations by Clementi [9] indicate that the interaction actually would result in a very strong complex with a structure that is not far from that of the ion pair $NH_4^+Cl^-$. Nevertheless the above discussion should remain essentially valid (a) for an $NH_3 + HCl$ pair when they have not yet approached each other very closely and (b) for other pairs with weaker acids, such as perhaps $H_3N + HOC_6H_5$, which do actually form hydrogen-bonded complexes in nonpolar environments. For further discussion see Section 15-1.

## 12-9 APPLICATIONS TO SYMMETRIZED COMPLEXES: $I_3^-$ AND $HF_2^-$ AS EXAMPLES

The $HF_2^-$ ion is linear and symmetrical $(F_aHF_c)^-$ in crystals, and the $I_3^-$ ion $(I_aI_bI_c)^-$ is also linear and symmetrical or approximately so, depending on environment (cf. Section 5-3). We can visualize these ions as being formed by the approach, say from the left, of the $n$-donor $F_a^-$ or $I_a^-$ to the $a\sigma$-acceptor $HF_c$ or $I_2$ $(I_bI_c)$, respectively. Equations 2-2 and 2-3 should then be applicable during the earlier stages of the approach. However, as the equilibrium configuration is approached, it is seen that the structure $F_a^- \cdot HF$ or $I_a^- \cdot I_2$ loses any preference over an *equivalent* structure $FH \cdot F_c^-$ or $I_2 \cdot I_c^-$ that could be reached in an approach of the alternative donor-acceptor pair $F_aH$ and $F_c^-$ (or $I_2$ and $I_c^-$) from the *right*. In summary,

$$F_a^- + H—F_c \rightarrow (F_aHF_c)^- \leftarrow F_a—H + F_c^-;$$

$$I_a^- + I_b—I_c \rightarrow (I_aI_bI_c)^- \leftarrow I_a—I_b + I_c^-.$$

Using (2-2) for $\Psi_N$ of $I_3^-$, the left-hand and right-hand approaches give

$$\Psi_N \simeq a\Psi_0(I_a^-, I_2) + b\Psi_1(I_a—I_2^-);\ \Psi_N' \simeq a\Psi_0'(I_2, I_c^-) + b\Psi_1'(I_2^-—I_c), \qquad (12\text{-}20)$$

respectively. However, for the linear symmetrical ion the two expressions must be combined with equal weight. In order to arrive at something sensible it is necessary first to expand the function $\Psi(I_2^-)$ that is implicitly contained in each of the $\Psi_1$s. The MO structure of free $I_2^-$ has been given in (10-4); if $\sigma_u 5p$ there is written in LCAO approximation as $5p\sigma_b - 5p\sigma_c$ or as $5p\sigma_a - 5p\sigma_b$ for the left-hand or right-hand approach, respectively, and the corresponding complete $\Psi$ is expanded, it is found that

$$\Psi_{bc}(I_2^-) = a[\Psi(I_b, I_c^-) + \Psi(I_b^-, I_c)]; \quad \Psi_{ab}(I_2^-) = a[\Psi(I_a, I_b^-) + \Psi(I_a^-, I_b)].$$

Exactly the same expressions are obtained from valence-bond theory directly. When these expressions are inserted into (12-20) the latter become

$$\Psi_N \simeq a\Psi_0(I_a^-, I_2) + ba\Psi_0'(I_2, I_c^-) + ba\Psi(\underbrace{I_a I_b^- I_c}_{|---|})$$

and

$$\Psi_N' \simeq a\Psi_0'(I_2, I_c^-) + ba\Psi_0(I_a^-, I_2) + ba\Psi(\underbrace{I_a I_b^- I_c}_{|---|}).$$

When these expressions for $\Psi_N$ and $\Psi_N'$ are added with equal weight, since they are equivalent and since the resulting $\Psi_N$ must be symmetrical ($^1\Sigma_g^+$ species), the result is

$$\Psi_N \simeq c[\Psi_0(I_a^-, I_2) + \Psi_0'(I_2, I_c^-)] + d\Psi(\underbrace{I_a I_b^- I_c}_{|---|}). \qquad (12\text{-}21)$$

In (12-21) $d/c = 2ba/(a + ba)$. However, this result was based on the assumption that $I_2^-$ in (12-20) is unpolarized, as in free $I_2^-$. If allowance is made for polarization, the ratio $d/c$ is changed. Without trying to follow this idea through, a direct inspection of (12-21) indicates that the coefficient $d$ should be relatively small† since it corresponds to a

† Partial promotion of $I_b^-$ from the closed-shell structure $5p^6$ to $5p^5 5d$, however, would permit $I_b^-$ to be bivalent and to form covalent bonds with both $I_a$ and $I_b$. This possibility favors an increased $d$. However, a final answer must probably await accurate theoretical computations.

"long bond." Each of $\Psi_0$ and $\Psi_0'$ contains a normal bond, and resonance between them gives increased stability according to valence-bond theory. The combination of $\Psi_0$ and $\Psi_0'$ is equivalent to a structure $I^{-(1/2)}$--I--$I^{-(1/2)}$ with the extra-strong *half bond* from the central iodine atom to each of the outer ones.† Equation 12-21 may be called a four-electron resonance-structure description that involves stabilization by CT resonance.‡

The foregoing discussion illustrates that it is possible though awkward to describe a symmetrized CT complex when near its equilibrium configuration from the usual resonance-structure viewpoint of (12-2); this fact remains valid also in the generalized resonance-structure treatment of Chapter 11. However, the whole-molecule-MO description is now especially suitable. Simple LCAO-MO descriptions for $I_3^-$ and $HF_2^-$ were first given by Pimentel [10], and an all-electron approximate SCF-MO calculation for $HF_2^-$ has been published more recently by Clementi and McLean [11]. For linear symmetrical $I_3^-$ in its normal state the MO electron configuration and the state are

$$\cdots \sigma_g^2\, \pi_u^4\, \pi_g^4\, \sigma_u^2\, \pi_u^{*4}, \quad {}^1\Sigma_g^+, \tag{12-22}$$

where in simplest LCAO-MO approximation

$$\sigma_g \simeq N[(1/2)5p\sigma_a + \frac{1}{\sqrt{2}}\, 5p\sigma_b + (1/2)5p\sigma_c)]; \qquad \sigma_u \simeq \frac{5p\sigma_a - 5p\sigma_c}{\sqrt{2}}. \tag{12-23}$$

The coefficients here given in $\sigma_g$ are those that in crude approximation (neglect of overlap and of hybridization effects††) give the strongest possible covalent bonding according to Coulson's bond-order criterion.‡‡ The $\sigma_u$ MO as above approximated is nearly nonbonding. The three $\pi$ MO shells are very nearly equivalent to three localized $\pi$ shells one on

---

† The structure is like that in $XeF_2$, $F^{-(1/2)}$--$Xe^+$--$F^{-(1/2)}$. This description of the bonding in $KrF_2$, $XeF_2$, and $HF_2^-$ including the "long bond" structure has also been given by Coulson. [See C. A. Coulson, *J. Chem. Phys.*, **44**, 468 (1968).]

‡ Four $\sigma$-electrons are actively involved. In either $\Psi_0$ or $\Psi_0'$ there are two in a $5p\sigma$ AO on $I^-$ and two in a $\sigma_g 5p$ MO [cf. (10-1)] in $I_2$. The remaining electrons can be considered, to a not very bad approximation, as occupying closed shells ($5s^2$, $5p\pi^4$, and inner shells), one set on each of the three iodine atoms.

†† An accurate SCF-MO description would show some $5s - 5p\sigma$, $5p\pi - 5d\pi$, and other hybridization effects.

‡‡ The Coulson bond order, which neglects overlap and hence puts $N = 1$, is $2c_a c_b = 2 \times (1/2) \times 1/\sqrt{2} = 1/\sqrt{2}$ for the *ab* link, and the same ($2c_b c_c$) for the *bc* link, or a total of $\sqrt{2}$ for the two links—as compared with $2 \times 1/\sqrt{2} \times 1/\sqrt{2} = 1$ for a single bond in a homopolar molecule (e.g., $H_2$).

each of the three atoms if $d\pi$-hybridization is neglected. Thus the whole-molecule-MO description can be considered as a four-electron one in the same sense as for the resonance-structure description of (12-21).

It can now be shown (cf. Sections 12-3 and 13-2) that expansion of the MO wavefunction for $\Psi_N$, which corresponds to the electron configuration given by (12-22) and (12-23), is identical with the resonance-structure $\Psi_N$ of (12-21) if $d = 0$ in the latter.

In a similar way the resonance-structure description of $\Psi_N$ for $HF_2^-$ is

$$\Psi_N \simeq c\,[\Psi_0(F_a^-, HF_c) + \Psi_0'(F_aH, F_c^-)] + d\Psi(F_aH^-F_c), \qquad (12\text{-}24)$$

with $d$ small, whereas the MO electron configuration and the state are

$$\cdots \sigma_g^2\, \pi_u^4\, \pi_g^{*4}\, \sigma_u^2, \quad {}^1\Sigma_g^+,$$

with, in simplest LCAO MO approximation,[†] (12-25)

$$\sigma_g \simeq a(2p\sigma_a) + b(1s_H) + a(2p\sigma_c); \qquad \sigma_u \simeq \frac{(2p\sigma_a - 2p\sigma_c)}{\sqrt{2}}$$

Some comparisons of $I_3^-$ with $R_3N\cdot I_2$ are now of interest. (For corresponding but somewhat different comparisons of $HF_2^-$ with $H_3N\cdot HCl$ see Section 15-1.) For $R_3N\cdot I_2$ the simple (2-2) expression

$$\Psi_N(R_3N\cdot I_2) \simeq a\Psi_0(R_3N, I_2) + b\Psi_1(R_3N^+{-}I_2^-), \qquad (12\text{-}26)$$

with $b$ in this case being nearly as large as $a$ (see Section 14-5), is probably a fairly good approximation, probably better than in expressions of this simple type in many other cases. (See Chapter 11, especially Section 11-1.) Now let us expand the $\Psi(I_2^-)$ function that is implicitly contained in $\Psi_1$ of (12-26) in the same manner as was done above in going from (12-20) to (12-21). The result is

$$\Psi_N(R_3N\cdot I_2) \simeq a\Psi_0(R_3N, I_b{-}I_c) + c\Psi(R_3N^+{-}I_b, I_c^-) + d\Psi(R_3N^+I_b^-I_c), \qquad (12\text{-}27)$$

(with a dashed bond line drawn beneath linking $I_b$ and $I_c$ in the last term)

with $d = c$ if the $I_2^-$ is unpolarized. Actually more polarization is to be expected, so that $c$ and $d$ should be unequal; how much polarization is not clear; possibly there may be only a little. In any event $c$ is smaller than $a$, but if it were approximately equal to $a$, with $d$ small, (12-27)

---

[†] The accurate SCF-MO description shows some $2s_a - 2p\sigma_a$ and $2s_c - 2p\sigma_c$ hybridization, as well as other modifications. The $\pi$ MOs in LCAO approximation are nearly $(2p\pi_a \pm 2p\pi_c)/\sqrt{2}$; $\pi_u^4\, \pi_g^{*4}$ can, however, very nearly be replaced by the localized shells $2p\pi_a^4\, 2p\pi_c^4$. More accurately, the hydrogen atom contributes a little $2p\pi_H$ to give some bonding in the $\pi_u$ MO.

would closely resemble (12-21) for $I_3^-$ and would correspond roughly to a structure $R_3N^{+(1/2)}$--$I_b$--$I_c^{-(1/2)}$ with a strong half-bond from N to I and another from $I_b$ to $I_c$. It is then probably reasonable to think of the structure of $R_3N \cdot I_2$ as showing a considerable resemblance to that of $I_3^-$ in the way just described. This resemblance receives support from the data in Table 5-2. However, the $R_{N—I}$ and $R_{I—I}$ distances there indicate that $c/a$ is still very considerably less than 1 in (12-27), and other considerations (see Section 11-1) suggest that $d$ may be comparable in magnitude to $c$.

## 12-10 REFERENCES


[1] R. S. Mulliken, *J. Chem. Phys.*, **7**, 24 (1939).
[2] R. S. Mulliken, *J. Am. Chem. Soc.*, **88**, 1849 (1966).
[3] R. S. Mulliken, *J. Chem. Phys.*, **23**, 1833 (1955).
[4] R. S. Mulliken, *J. Chem. Phys.*, **23**, 1841, 2343 (1955).
[5] (a) R. Hoffmann, *J. Chem. Phys.*, **40**, 2474 (1964). (b) M.-Cl. Moireau and A. Veillard, *Theoret. Chim. Acta* (Berlin), **11**, 344 (1968). (c) S. D. Peyerimhoff and R. J. Buenker, *J. Chem. Phys.*, **49**, 312 (1968). (d) D. R. Armstrong and P. G. Perkins, *J Chem Soc.*, 1044 **(1969)**.
[6] H. Nishiguchi, Y. Nakai, K. Nakamura, K. Ishizu, Y. Deguchi, and H. Takaki, *Mol. Physics*, **9**, 153 (1965) and references given there. N. Hirota and R. Kreilick, *J. Am. Chem. Soc.*, **88**, 614 (1966). T. E. Hogen-Esch and J. Smid, ibid., **88**, 307, 318 (1966).
[7] S. J. Rand and R. L. Strong, *J. Am. Chem. Soc.*, **82**, 5 (1960); R. L. Strong, S. J. Rand, and J. A. Britt, ibid., **82**, 5053 (1960); R. L. Strong and J. Perano, ibid., **83**, 2843 (1961).
[8] T. A. Gower and G. Porter, *Proc. Roy. Soc.* (London), **262A**, 476 (1961).
[9] E. Clementi, *J. Chem. Phys.*, **46**, 3851 (1967); ibid., **47**, 2323 (1967); E. Clementi and J. N. Gayles, ibid., **47**, 3837 (1967).
[10] G. C. Pimentel, *J. Chem. Phys.*, **19**, 446 (1951).
[11] E. Clementi and A. D. McLean, *J. Chem. Phys.*, **36**, 745 (1962). Also see A. D. McLean and M. Yoshimine, "Tables of Linear Molecule Wavefunctions," a supplement to "Computation of Molecular Properties and Structure," by A. D. McLean and M. Yoshimine, *IBM Journal of Research and Development*, November 1967.

*CHAPTER* **13**

# *The Lithium Fluoride Molecule and Other Ion-Pair Systems as Charge-Transfer Complexes*

To illustrate further the ideas presented in the two preceding chapters some examples of ion-pair molecules that consist of closed-shell ions[†] and can be looked at as donor-acceptor complexes will now be discussed in terms of both whole-complex-MO and resonance-structure descriptions. Such ion-pair complexes of course differ from typical neutral molecule complexes in the fact that they are predominantly held together by the classical electrostatic forces that are associated with the no-bond term $a\Psi_0(D, A)$ of (2-2) in the intermolecular resonance-structure theory. Here it is seen that in general "no-bond" means *no-covalent-bond.*[†] Further, $b\Psi_1(D^+—A^-)$ is properly described as a "dative" term only if D and A are neutral molecules; in general it might be called a *Heitler-London term*.

A very simple example is the diatomic molecule lithium fluoride, which can be regarded as an $n \cdot v$ complex of $F^-$ as donor with $Li^+$ as acceptor. To be sure, lithium fluoride is an interatomic rather than an intermolecular complex, but an atom (or atom-ion here) is a special case of a molecule, and even the case of lithium fluoride is sufficiently complicated to illustrate important aspects of the general whole-complex-MO theory, the general resonance-structure theory, and the relations between them.

[†] See Section 12-7 for ion-pair complexes composed of one closed-shell and one odd-electron ion. In these both $\Psi_0$ and $\Psi_1$ are no-covalent-bond functions.

## 13-1 MOLECULAR-ORBITAL STRUCTURE OF THE LITHIUM FLUORIDE MOLECULE

The whole-complex MOs of lithium fluoride are of course just diatomic MOs. The preponderant single-electron configuration and state, and the corresponding wavefunction, are

$$1\sigma^2 2\sigma^2 3\sigma^2 4\sigma^2 1\pi^4, \qquad {}^1\Sigma^+$$

and (13-1)

$$\Psi_N \simeq \mathcal{A}[1\sigma(1)\,\alpha(1)\,1\sigma(2)\,\beta(2)\,2\sigma(3)\,\alpha(3)\cdots 1\bar{\pi}(12)\,\beta(12)].$$

The forms of all the MOs, stated in (13-1) in the order of decreasing ionization potential, have been determined rather accurately by McLean [1] in a careful SCF-MO computation.

What do these MOs look like? They are conveniently built up from Slater-type functions (STFs).† An STF is a function that is centered around a particular nucleus (here fluorine or lithium) and conveniently designated by a symbol $nl\lambda^\zeta$. Each STF is of the form

$$nl\lambda^\zeta = Nr^{(n-1)}e^{-r\zeta/a_0}\, Y_{l,\lambda}(\theta, \phi). \tag{13-2}$$

An SCF AO for an *atom* can be approximated as a linear combination of STFs of that atom. Every such STF is like any one term of a hydrogen atom AO but scaled up or down in size depending on the value of the orbital exponent $\zeta$; $n$ is the principal quantum number, $l$ is the orbital-angular-momentum quantum number, and $\lambda$ $(= |m|)$ is the axial-angular-momentum quantum number that is used for diatomic molecules; $Y_{l,\lambda}(\theta, \phi)$ are the standard spherical harmonics, $r$, $\theta$, and $\phi$ being the usual spherical coordinates. An SCF AO usually needs two or three STFs with different $\zeta$-values to give a reasonably good approximation to it [2].

SCF MOs for a diatomic molecule can be constructed as linear combinations of STFs of *two* atoms, most conveniently from STFs with the same $\zeta$-values as were found best for the atoms, here Li and F (or $F^-$). The coefficients of these STFs are then determined by the molecular SCF procedure. McLean used 15 STFs with $\lambda = 0$ ($\sigma$ STFs), from which, through linear combinations with varying coefficients, the forms of all

---

† These are usually called STOs (Slater-type orbitals); however, since in general they are not good AOs themselves but only good building blocks for constructing AOs or MOs, we use the revised designation "STF," which was proposed by Dr. Paul E. Cade and Prof. C. C. J. Roothaan. However, the same set of spectroscopic symbols ($1s$, $2s$, $2p\sigma$, and so on) are used for STFs (see for example Table 13-1) as for complete AOs.

**TABLE 13-1**

**Approximate Self-Consistent-Field Molecular Orbitals for Lithium Fluoride**[a]

| MO | Build-up from STFs | Summary |
|---|---|---|
| $1\sigma$ | $0.89\ 1s_F^{8.86} + 0.04\ 1s_F^{14.41} + 0.08\ 2s_F^{8.07} + \cdots$ | Approximately pure $1s_F$ SCF AO |
| $2\sigma$ | $0.89\ 1s_{Li}^{2.48} + 0.11\ 1s_{Li}^{4.69} + \cdots$ | Approximately pure $1s_{Li}$ SCF AO |
| $3\sigma$ | $(0.52\ 2s_F^{2.00} + 0.56\ 2s_F^{3.27} - 0.08\ 2s_F^{8.07} - 0.21\ 1s_F^{8.86})$ $+ (0.03\ 2s_{Li}^{0.75} - 0.02\ 2s_{Li}^{0.62})$ $+ (0.03\ 2p\sigma_F^{2.36} + 0.03\ 2p\sigma_{Li}^{1.40})$ | Mostly $2s_F$, with a trace of Li and a trace of polarization |
| $4\sigma$ | $(0.01\ 1s_F^{8.86} - 0.13\ 1s_{Li}^{2.48} - 0.05\ 2s_F^{8.07} - 0.04\ 2s_F^{2.00})$ $+ (0.52\ 2p\sigma_F^{2.36} + 0.21\ 2p\sigma_F^{4.74} + 0.35\ 2p\sigma_F^{1.32})$ $+ (0.08\ 2p\sigma_{Li}^{1.40} - 0.08\ 2s_{Li}^{0.75} - 0.02\ 2s_{Li}^{1.77} - 0.06\ 2s_{Li}^{0.62})$ $- (0.02\ 3d\sigma_F^{2.56})$ | Mostly $2p\sigma_F$ but some $2p\sigma_{Li}$ and $2s_{Li}$ |
| $1\pi$ | $0.51\ 2p\pi_F^{2.36} + 0.21\ 2p\pi_F^{1.32} + 0.08\ 2p\pi_{Li}^{0.79}$ $+ 0.02\ 3d\pi_F^{2.16} + 0.03\ 3d\pi_{Li}^{0.88}$ | Mostly $2p\pi_F$, but some $2p\pi_{Li}$ |

[a] From the first situation (basis set X.A) in [1], with smaller coefficients omitted, and the rest rounded off. In a symbol such as $2s_F^{3.27}$ for an STF the 2 is the $n$ of (13-2), the 3.27 is the $\zeta$ of that equation; $s$, $p$, $d$ refer in the usual way to $l = 0, 1, 2$, and $\sigma$ and $\pi$ to $\lambda = 0$ or 1.

the four $\sigma$ MOs of (13-1) were built up. Table 13-1 gives approximate values for the most important coefficients.[†] Table 13-1 also shows how the $\pi$ MO can be constructed from five STFs.

The $1\sigma$ MO is very nearly just a $1s_{\mathrm{F}}$ AO, built up mainly from three $s$-type fluorine-atom STFs, and the $2\sigma$ MO is nearly just a $1s_{\mathrm{Li}}$ AO, built up mainly from two $1s$ lithium STFs. The $3\sigma$ MO is not very different from $2s_{\mathrm{F}}$, but the $4\sigma$ MO is a weakly bridging MO that corresponds to $\phi_{\mathrm{da}}$ in (12-8), with a $2p\sigma_{\mathrm{F}}$ AO (built from three $2p\sigma_{\mathrm{F}}$ STFs) that is somewhat mixed with a little $2s_{\mathrm{F}}$ playing the role of $\chi_{\mathrm{d}}$,—and a mixture of $1s_{\mathrm{Li}}$, $2s_{\mathrm{Li}}$, and $2p\sigma_{\mathrm{Li}}$ playing the role of $\chi_{\mathrm{a}}$. *Approximately*, then, the $\sigma$-MO structure is that of electron pairs in three unchanged orbitals ($1s_{\mathrm{F}}$, $1s_{\mathrm{Li}}$, and $2s_{\mathrm{F}}$) of the donor $\mathrm{F}^-$ and the acceptor $\mathrm{Li}^+$, and one pair in a bridging $\sigma$-orbital ($4\sigma$) that displays partial charge transfer from $\mathrm{F}^-$ onto $\mathrm{Li}^+$; the amount of transfer could be easily calculated by a generalization of (12-9).

Further examination of Table 13-1 shows that not only the $4\sigma$ but also the $1\pi$ MO (or rather the *two* $1\pi$ MOs since the $\pi$-type is twofold degenerate) are bridging MOs, the $\pi$ MOs being responsible for appreciable additional charge transfer from the donor to the acceptor.

In the simplified discussion of Section 12-2 the bridging MO $\phi_{\mathrm{da}}$ in (12-8) was identified as a bonding MO whose bonding characteristics formed a counterpart to the resonance energy $\beta_0^2/\Delta$ of the simplified resonance-structure theory. In lithium fluoride the main stabilization energy in state $N$ is provided by the Coulomb attraction between the ions $\mathrm{Li}^+$ and $\mathrm{F}^-$.[‡] However, this is supplemented by the bonding characteristics of the $1\pi$ MO, which is precisely of the type $m\chi_{\mathrm{d}} + n\chi_{\mathrm{a}}$ of $\phi_{\mathrm{da}}$ in (12-8).[††] The $4\sigma$ bridging MO, on the other hand, is apparently nearly nonbonding or somewhat antibonding, since the $\chi_{\mathrm{Li}}$ in it is of the form of a $2s$-$2p\sigma$ hybrid AO that is directed away from the fluorine and so overlaps the slightly modified $2p\sigma$ $\chi_{\mathrm{F}}$ AO relatively little—that little, however, apparently being antibonding.

---

[†] For present discussion purposes the smaller coefficients and the *exact* values of the larger ones are of no importance. The values in Table 13-1 are from McLean's 1963 paper. Improved values could be given, but the differences are not important for discussion purposes.

[‡] In typical complexes between *neutral* molecules, which are the type to which major emphasis is devoted in this book, it is of course the $V$-state that is mainly stabilized by Coulomb attraction, the resonance energy $+\beta_1^2/\Delta$ making a smaller but destabilizing contribution.

[††] From a perhaps more usual viewpoint, the bonding effect of the $1\pi$-electrons may be regarded as polarization energy that corresponds to polarization of the $2p\pi$ AOs of $\mathrm{F}^-$ by $\mathrm{Li}^+$.

## 13-2 RESONANCE-STRUCTURE DESCRIPTION OF LITHIUM FLUORIDE

Following the procedure used in Section 12-4 we can substitute the LCAO form[†] $m\chi_\mathrm{d} + n\chi_\mathrm{a}$ into an equation such as (12-12) or (12-17), expand, and come out with an expression of the resonance-structure type [compare (12-14) with (12-12)]. In the present example all but the $4\sigma$ and $1\pi$ MOs can be considered to be essentially localized, and if the procedure that has just been indicated is suitably generalized,[‡] there results[††]

$$\Psi'_N = a\Psi_0(\mathrm{F}^-, \mathrm{Li}^+) + b_\sigma\Psi_1^\sigma(\mathrm{Li—F}) + b_\pi[\Psi_1^{\pi_x}(\mathrm{Li}\overset{\pi_x}{—}\mathrm{F}) + \Psi_1^{\pi_y}(\mathrm{Li}\overset{\pi_y}{—}\mathrm{F})] + \cdots, \tag{13-3}$$

with $a$ roughly 1 and $b_\sigma$ and $b_\pi$ roughly, by inspection [cf. (12-14)] from the coefficients in Table 13-1, both about 0.08. Here each of the $\Psi_1$ terms is like the $\Psi_1$ of (12-2), with just one electron of $\mathrm{D}^+$ (i.e., the fluorine atom) paired off with one electron of $\mathrm{A}^-$ (i.e., the lithium atom). In $\Psi_1^\sigma$, $\Psi_1^{\pi_x}$, and $\Psi_1^{\pi_y}$ a $\sigma$-electron, a $\pi_x$-electron, and a $\pi_y$-electron, have been respectively transferred from $\mathrm{F}^-$ to an orbital of similar type on $\mathrm{Li}^+$, and the remaining, now odd, donor electron from a pair that was originally on $\mathrm{F}^-$ is Heitler-London coupled (as always in $\Psi_1$) to the odd electron that is now on the acceptor.

Equation 13-3 [or equally (13-1), with the MOs of Table 13-1] says that the structure of diatomic lithium fluoride is primarily ionic but to *some* extent covalent, though with the covalent bonding essentially of the $\pi$-type rather than of the $\sigma$-type as conventionally assumed.[‡‡] This statement is completely equivalent to saying that ionic binding in diatomic lithium fluoride ($\Psi_0$) is to some extent assisted by charge-transfer binding of the same sort as that between two neutral closed-MO-shell molecules—one a donor and the other an acceptor—as discussed in Chapter 2. However, (13-3) corresponds to the *generalized* resonance-

[†] Here of course $\chi_\mathrm{d}$ and $\chi_\mathrm{a}$ have each been presented in the form of a linear combination of STFs, but that should be regarded as merely a matter of convenience.

[‡] Expanison of a six-electron expression
$\mathcal{A}''''[4\sigma(1)\ 4\sigma(2)\ 1\pi(3)\ 1\pi(4)\ 1\bar{\pi}(5)\ 1\bar{\pi}(6)\ S_0(1,2)\ S_0(3,4)\ S_0(5,6)]$,
omitting the other electrons, leads to the desired results.

[††] There are additional terms of several varieties of the charge-distribution types $\Psi_{11}(\mathrm{F^+Li^-})$ and $\Psi_{111}(\mathrm{F^{2+}Li^{2-}})$. These can be neglected here as being unimportant. In (13-3) the $x$- and $y$-directions are taken perpendicular to the molecular axis, which is taken as $z$-axis.

[‡‡] In view of the probably slightly antibonding nature of the $4\sigma$ MO in lithium fluoride, the "bond" in $\Psi_1^\sigma$ of (13-3) is apparently only formal. (Wavefunctions of the Heitler-London form are usually, but not necessarily, bonding.)

structure theory of Chapter 11, in which more than one charge-transfer structure $\Psi_1$ is of comparable importance in $\Psi_N$.

Instead of (13-3) the expansion of (13-1) in terms of resonance structures can be carried out in more detail, breaking up $\Psi_1^\sigma$ into terms that are based on unhybridized $2s$ and $2p\sigma$ AOs, as follows:

$$\begin{aligned}\Psi_N &= a\Psi_0[\mathrm{F}^-(1s^2 2s^2 2p\sigma^2 2p\pi^4),\ \mathrm{Li}^+(1s^2)]\\ &+ b_{\sigma s}\Psi_1(\sigma_{\mathrm{F}}^{-1}\ s_{\mathrm{Li}}) + b_{\sigma\sigma}\Psi_1(\sigma_{\mathrm{F}}^{-1}\ \sigma_{\mathrm{Li}})\\ &+ b_{\pi\pi}\Psi_{1\pi}\left(\frac{\pi_{x\mathrm{F}}^{-1}\ \pi_{x\mathrm{Li}} + \pi_{y\mathrm{F}}^{-1}\ \pi_{y\mathrm{Li}}}{\sqrt{2}}\right) + b_{ss}\Psi_1(s_{\mathrm{F}}^{-1}\ s_{\mathrm{Li}})\\ &+ b_{s\sigma}\Psi_1(s_{\mathrm{F}}^{-1}\ \sigma_{\mathrm{Li}}) + \cdots.\end{aligned} \tag{13-4}$$

Here a symbol such as, for example, $\Psi_1(\sigma_{\mathrm{F}}^{-1},\ s_{\mathrm{Li}})$ means the function that is obtained from $\Psi_0$ by removing one electron from the $2p\sigma$ AO of $\mathrm{F}^-$ and putting it into the $2s$ AO of $\mathrm{Li}^+$. The coefficients $b_{\sigma s}$, etc., are all small, as is seen from Table 12-1, but all of comparable magnitude, although $b_{\sigma\sigma}$ and $b_{\pi\pi}$ are larger than the others. They are determined by the requirement that $\Psi_N$ in (13-4) be identical with the MO function (13-1). No attempt is made to give values of the various coefficients, since the present discussion is purely illustrative.

The important point here is that *several* dative structures must be included in $\Psi_N$ to represent the accurate wavefunction. The symmetry species of $\Psi_N$ is $^1\Sigma^+$; hence every one of the contributing resonance functions in (13-4), including $\Psi_{1\pi}$, is constructed to have $^1\Sigma^+$ symmetry. However, all these correction terms are small; in other words, $\Psi_0$ alone is a rather good approximation, just as in any weak complex.

## 13-3 CHARGE-TRANSFER STATES AND CHARGE-TRANSFER SPECTRA OF LITHIUM FLUORIDE

Now consider the excited states of lithium fluoride, *all* of which turn out to be CT states.† These states can be predicted from a series of two-electron approximations. Consider first the states that are obtained on transferring an electron out of the $2p\sigma$-orbital of $\mathrm{F}^-$ into a valence-shell orbital of Li. There are two such states with $^1\Sigma^+$ symmetry:

$$\begin{aligned}\Psi_V^{\sigma s},\ {}^1\Sigma^+ &= a^*\Psi_1(\sigma_{\mathrm{F}}^{-1}\ s_{\mathrm{Li}}) - b_{\sigma s}^*\Psi_0 + \cdots,\\ & \qquad\qquad (a^* \gg b^*)\\ \Psi_V^{\sigma\sigma},\ {}^1\Sigma^+ &= a^*\Psi_1(\sigma_{\mathrm{F}}^{-1}\ \sigma_{\mathrm{Li}}) - b_{\sigma\sigma}^*\Psi_0 + \cdots.\end{aligned} \tag{13-5}$$

† For a survey of structure and spectra of alkali halides and related molecules, see [3].

We could now compute $\mu_{VN}$ for transitions from the ground state to each of these two CT states in essentially the same way as for a complex with only one CT state. (See Chapter 3.) It is easily seen that both these transitions are polarized parallel to the lithium fluoride axis ($z$-direction). One more parallel-polarized absorption band is expected at fairly low frequencies; namely, that for the transition to the $^1\Sigma^+$ state $\Psi_V^{\pi\pi}$:

$$\Psi_V^{\pi\pi},\ ^1\Sigma^+ = a^*\Psi_1(\pi_{\mathrm{F}}^{-1},\ \pi_{\mathrm{Li}}) - b_{\pi\pi}^*\Psi_0 + \cdots. \tag{13-6}$$

Many other CT states exist. These include the triplet states ($^3\Sigma^+$) that correspond to all the singlet states described above. Also there are several $^1\Pi$ states:

$$\Psi_V(\sigma_{\mathrm{F}}^{-1}2p\pi_{\mathrm{Li}});\quad \Psi_V(s_{\mathrm{F}}^{-1}2p\pi_{\mathrm{Li}});\quad \Psi_V(\pi_{\mathrm{F}}^{-1}2s_{\mathrm{Li}});\quad \Psi_V(\pi_{\mathrm{F}}^{-1}2p\sigma_{\mathrm{Li}}). \tag{13-7}$$

Transitions from $\Psi_N$ to these states would be polarized along the $x$- and $y$-directions perpendicular to the lithium fluoride axis, unlike the typical CT bands of Chapter 3. However, if the axial model of the $\mathrm{Bz \cdot I_2}$ and related complexes is correct, the lowest CT state of these is of a type that is analogous to the $^1\Pi$ states of (13-7) and their well-known CT bands must then be of the perpendicularly polarized type. Like the lowest CT state of $\mathrm{Bz \cdot I_2}$ if the axial model is correct, none of the $^1\Pi$ functions of (13-7) can mix with $\Psi_0$ in the ground-state wavefunction $\Psi_N$. Of course $^3\Pi$ charge-transfer states that correspond to the $^1\Pi$ states of (13-7) also exist.

It is of interest to see how the transition moment from $\Psi_N$ to one of these $^1\Pi$ $\Psi_V$ states is computed. For $\Psi_V(2p\sigma_{\mathrm{F}}^{-1}2p\pi_{\mathrm{Li}})$ we have

$$\mu_{VN} = \int \Psi_V \mu_{\mathrm{op}}^{\mathrm{el}} \Psi_N \, d\tau \simeq \sqrt{2} \int 2p\sigma_{\mathrm{F}}(-e\mathbf{r})\, 2p\pi_{\mathrm{Li}}\, dv, \tag{13-8}$$

where the first integral in (13-8) is a 12-electron one and the second a 1-electron integral; $d\tau = dv_1 dv_2 \cdots dv_{12}$. The approximate reduction to a one-electron integral is obtained by putting

$$\Psi_N \simeq \Psi_0 \simeq \mathcal{A}''[2p\sigma_{\mathrm{F}}(1)\ 2p\sigma_{\mathrm{F}}(2)\ S_0(1,2)\cdots],$$

$$\Psi_V \simeq \left(\frac{\mathcal{A}''}{\sqrt{2}}\right)\{[2p\sigma_{\mathrm{F}}(1)\ 2p\pi_{\mathrm{Li}}(2) + 2p\pi_{\mathrm{Li}}(1)\ 2p\sigma_{\mathrm{F}}(2)]\, S_0(1,2)\cdots\},$$

$$\mu_{\mathrm{op}}^{\mathrm{el}} = -e(\mathbf{r}_1 + \mathbf{r}_2 + \cdots + \mathbf{r}_{12}).$$

It can be shown that $\int \Psi_V\, \mu_{\mathrm{op}}^{\mathrm{el}}\, \Psi_N\, dv$—if $\Psi_N$ and $\Psi_V$ are approximated as just indicated—is the same as if only two electrons, here called 1 and 2, were present. The result given in (13-8) is then easily obtained. Now $\mathbf{r}$ in the second integral in (13-8) has $x$, $y$, and $z$ components, whereas $2p\pi_{\mathrm{Li}}$ can be either one of $2p\pi x_{\mathrm{Li}}$ or $2p\pi y_{\mathrm{Li}}$. Suppose it is $2p\pi y_{\mathrm{Li}}$.

The integral then becomes $\sqrt{2}\int 2p\sigma_F(-e\mathbf{y})\,2p\pi y_{Li}\,dv$, since only the component $-e\mathbf{y}$ from $\mu_{op}$ has the proper symmetry. This can be seen from the following pictorial representation of the integrand:

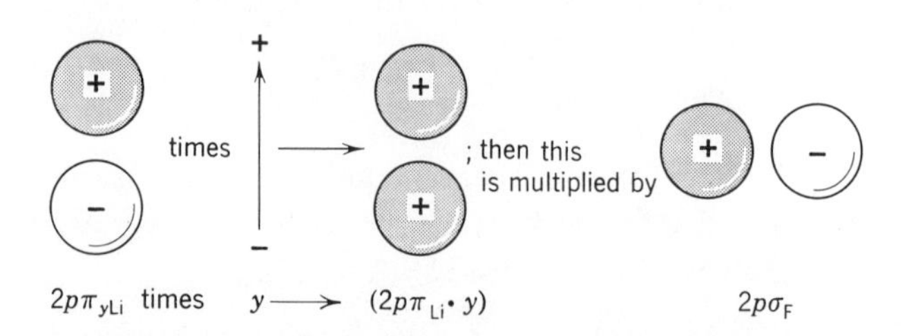


Obviously the product $2p\pi_{yLi}\cdot y$ has the proper symmetry so that the overlap with $2p\sigma_F$ (noting that the overlap is mostly just with the near end of $2p\sigma_F$) is not zero, and the transition is $y$-polarized. Similarly $\int 2p\sigma_F\,\mu_x 2p\pi_{yLi}\,dv = 0$, but $\int 2p\sigma_F\,\mu_x 2p\pi_{xLi}\,dv \neq 0$; and so on.

Let us now examine the description of the excited states of lithium fluoride by using MOs as in (13-1), first recalling that $1\sigma \simeq 1s_F$, $2\sigma \simeq 1s_{Li}$, $3\sigma \simeq 2s_F$, $4\sigma \simeq 2p\sigma_F$, and $1\pi \simeq 2p\pi_F$. Continuing with excited MOs of lithium fluoride in the order of expected increasing energy, we have $5\sigma \simeq 2s_{Li}$, $6\sigma \simeq 2p\sigma_{Li}$, and $2\pi \simeq 2p\pi_{Li}$. Transitions from the ground state to states with an electron in any one of these latter three orbitals are thus CT transitions. The following three transitions can occur with parallel polarization:

$$\Psi_V(4\sigma^{-1}\,5\sigma),\ {}^1\Sigma^+ \leftarrow \Psi_N,\ {}^1\Sigma^+,$$

$$\Psi_V(4\sigma^{-1}\,6\sigma),\ {}^1\Sigma^+ \leftarrow \Psi_N,\ {}^1\Sigma^+,$$

$$\Psi_V(1\pi^{-1}\,2\pi),\ {}^1\Sigma^+ \leftarrow \Psi_N,\ {}^1\Sigma^+.$$

In addition to the ${}^1\Sigma^+$ state that arises from the $1\pi^{-1}\,2\pi$ configuration, there are also ${}^3\Sigma^+$, ${}^3\Delta$, ${}^1\Delta$, and ${}^1\Sigma^-$ charge-transfer states, but transitions from $\Psi_N$ to all these are forbidden. Transitions with perpendicular polarization corresponding to the perpendicular-type charge-transfer bands of the resonance-structure discussion can also occur.

Consider now, using the MO-type wavefunctions, the transition-moment integral for the transition $\Psi_V(4\sigma^{-1}5\sigma,\ {}^1\Sigma^+) \leftarrow \Psi_N$ in two-electron approximation. The transition is $z$-polarized, so we need only the $z$-component of $\mu_{op}$, and find

$$\mu_{VN} = \int \Psi_N \mu_{\rm op}^{\rm el} \Psi_V \, d\tau \simeq -e \int [4\sigma(1)\, 4\sigma(2)]\,[z(1) + z(2)] \times \left[\frac{4\sigma'(1)\, 5\sigma(2) + 5\sigma(1)\, 4\sigma'(2)}{\sqrt{2}}\right] dv_1 dv_2 = -e\sqrt{2}\, S_{4\sigma,4\sigma'} \int 4\sigma\, z\, 5\sigma\, dv \tag{13-9}$$

The calculation here illustrates how when two electrons are available to make a given orbital jump, $\mu$ reduces to a one-electron integral times a factor $\sqrt{2}$ that, when squared, as is required for the *intensity*, gives a factor of 2. This same factor appeared for the same reason in the second form of the integral in (13-8).

Approximate expressions for the mutually orthogonal MOs $4\sigma$ and $5\sigma$ are

$$4\sigma = N(2p\sigma_{\rm F} + \lambda 2s_{\rm Li} + \cdots); \quad 5\sigma = N'(2s_{\rm Li} - \gamma 2p\sigma_{\rm F} + \cdots). \tag{13-10}$$

Neglecting the ellipses in (13-10) for simplicity, since use of the simplified expressions will suffice to illustrate the major points involved, the orthogonality condition is

$$\int 4\sigma\, 5\sigma\, dv = 0 = \lambda - \gamma + S_{s\sigma}(1 - \gamma\lambda), \tag{13-11}$$

where $S_{s\sigma} = \int 2s_{\rm Li} 2p\sigma_{\rm F}\, dv$.
Now

$$\mu_{VN} = -e\sqrt{2}\, NN'[(1 - \lambda\gamma) \int 2s_{\rm Li}\, z\, 2p\sigma_{\rm F}\, dv - \gamma \int 2p\sigma_{\rm F}\, z\, 2p\sigma_{\rm F}\, dv + \lambda \int 2s_{\rm Li}\, z\, 2s_{\rm Li}\, dv]. \tag{13-12}$$

Each of the three integrals defines an average position $\bar{z}$ along the lithium fluoride axis for the electron in one or another orbital function; we then have

$$\mu_{VN} = -e\sqrt{2}\, NN'[(1 - \lambda\gamma)\, S_{s\sigma}\, \bar{z}_{s\sigma} + \lambda \bar{z}_{\rm Li} - \gamma \bar{z}_{\rm F}]. \tag{13-13}$$

By use of the orthogonality relation (13-11) we can transform (13-13) into

$$\mu_{VN}^{(\rm MO)} = -e\sqrt{2}\, NN'[\lambda(\bar{z}_{\rm Li} - \bar{z}_{\rm F}) + (1 - \lambda\gamma)\, S_{s\sigma}(\bar{z}_{s\sigma} - \bar{z}_{\rm F})] \simeq -e\sqrt{2}\, [\lambda(\bar{z}_{\rm Li} - \bar{z}_{\rm F}) + S_{s\sigma}(\bar{z}_{s\sigma} - \bar{z}_{\rm F})]. \tag{13-14}$$

```
          Li              F
z ← · · · * ——— · ——— * · · ·
          ↑       ↑       ↑
        z̄_Li    z̄_sσ    z̄_F
```

Equation 13-14 is of the same form as the corresponding expression†

† Equation 3-11 after making substitutions as given by (3-12) and (3-14); or Equation 20 of R2.

from the resonance-structure method; namely,

$$\mu_{VN}^{(RS)} = -e[a^*b(\bar{z}_{Li} - \bar{z}_F) + (aa^* - bb^*) S_{01}(\bar{z}_{s\sigma} - \bar{z}_F], \qquad (13\text{-}15)$$

where $S_{01} \simeq \sqrt{2}\, S_{da^-}/\sqrt{1 + S_{da^-}^2}$, with $\phi_d = 2p\sigma_{F^-}$ and $\phi_{a^-} = 2s_{Li}$. The two expressions would be identical if

$$a^*b = \sqrt{2}\, NN'\lambda \text{ and } (aa^* - bb^*) = NN' \sqrt{1 + S_{s\sigma}^2}\,(1 - \lambda\gamma). \qquad (13\text{-}16)$$

Actually, they *should* be not quite identical because the resonance-structure and MO approximations are not identical, although they may often be nearly so for loose complexes. In the present example the problem has been simplified to a two-electron one involving just two whole-complex MOs, $4\sigma$ and $5\sigma$, which are related in the manner of (13-10) to two *localized* donor and acceptor MOs (really AOs—$2p\sigma_F$ and $2s_{Li}$) of the resonance-structure method, thus fulfilling the requirements of the simplified two-electron theories of Chapters 2 and 12.

There are just a few more things to be said about the lithium fluoride "complex." Some potential curves for lithium fluoride are shown schematically in Fig. 13-1 [3].

Although diatomic lithium fluoride in its equilibrium configuration exists almost entirely as $Li^+F^-$, it dissociates to neutral atoms (Li, $^2S$ + F, $^2P$). The dashed lines in Fig. 13-1 from the atoms and from the ions are shown as crossing; actually the state of $^1\Sigma^+$ symmetry from Li + F mixes with the $^1\Sigma^+$ state from $Li^+ + F^-$ to give the solid lines shown in the figure.

Starting from separated atoms—Li, $^2S$ and F, $^2P$—the $^2P$ level of the fluorine atom splits in the field of the lithium atom to give two states: $p\sigma p\pi^4$, $^2\Sigma^+$, and $p\sigma^2 p\pi^3$, $^2\Pi$. When these combine with the $^2S$ state of lithium, four molecular states result: $^1\Pi$, $^3\Pi$, $^1\Sigma^+$, and $^3\Sigma^+$. For all of these states the electron configuration differs from that of $Li^+F^-$ by the transfer of one electron from $F^-$ to $Li^+$. Hence these states are all CT states. Transitions from the normal state are allowed, however, only to the singlet levels; the $^1\Sigma^+ \leftarrow {}^1\Sigma^+$ transitions are polarized parallel to the axis, whereas the $^1\Pi \leftarrow {}^1\Sigma^+$ transitions have perpendicular polarization.

As indicated in Fig. 13-1 higher CT states are also possible; these dissociate to give a lithium atom in its $^2P$ first excited state. Interaction of this with the $^2P$ fluorine atom gives rise to a number of singlet and triplet molecular states of types $\Sigma^+$, $\Sigma^-$, $\Pi$, and $\Delta$. Transitions to some of them are forbidden; those to $^1\Sigma^+$ and $^1\Pi$ states give rise to additional CT bands with parallel and perpendicular polarization, respectively. Some of these states are the same as the CT states described above in this section. The spectrum of lithium fluoride itself is little if at all known, but analogous spectra of other alkali halides in the vapor

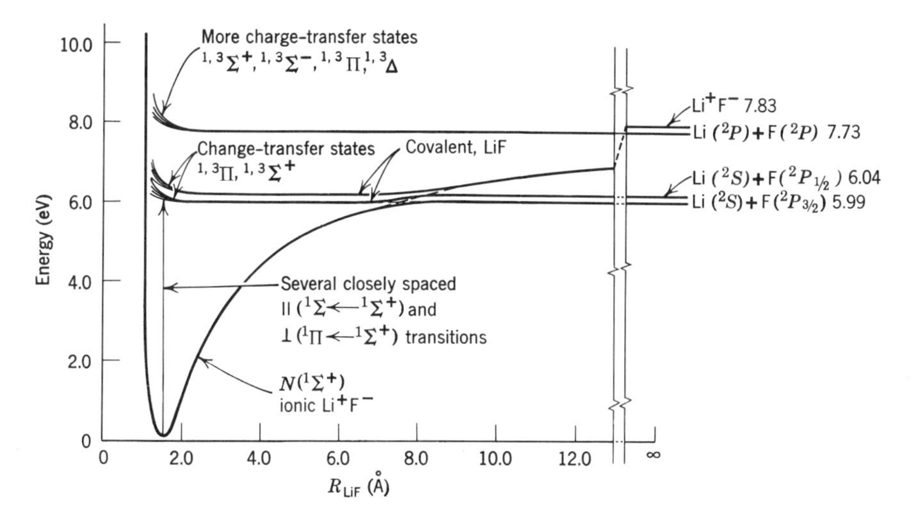


**Figure 13-1** Schematic potential curves for lithium fluoride. (Based on Fig. 3 of [3], using $D_e = 5.99$ eV to Li + F atoms[a], $I_{Li} = 5.39$ eV[b], and $E_F = 3.448$ eV[c].) (a) L. Brewer and E. Brackett, *Chem. Rev.*, **61**, 425 (1961); (b) C. E. Moore, *Atomic Energy Levels*, Vol. I, NBS Circular No. 467; (c) R. S. Berry and C. W. Reimann, *J. Chem. Phys.*, **38**, 1540 (1963).

state provide many examples of interatomic (or intermolecular) CT spectra that are analogous to those just described for lithium fluoride [3].

## 13-4 PYRIDINIUM HALIDES AND OTHER ION-PAIR COMPLEXES

Other examples of ion-pair complexes that are more or less closely analogous to lithium fluoride occur with various organic ions. One example is 1-methylpyridinium iodide (see top page 206) regarded as an $n \cdot a\pi$ complex of the donor $I^-$ and the acceptor $C_5H_5N^+CH_3$.† In the

† This ion pair can also be considered as the inner complex (see Chapter 16) corresponding to the outer complex of pyridine as $n$-donor with $CH_3I$ as a weak $a\sigma$-acceptor. ($CH_3I$ can be classified as an $a\sigma$-acceptor that is analogous to hydrogen iodide and to the iodine molecule, even though perhaps its outer complex with pyridine may be of negligible strength; that is, $K$ and $-\Delta H$ may be so small as to be negligible.)

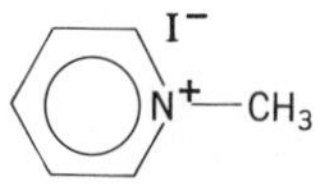

$I^-$

$N^+—CH_3$

spectrum of each of several alkylpyridinium iodides in water and other polar solvents Kosower found a new band that is due neither to $I^-$ nor to the $Py^+R$ ion and that has an intensity that increased as the product of the (equal) $I^-$ and $Py^+R$ concentrations—a fact that is consistent with its attribution to ion-pairs $I^-Py^+R$ [4]. He attributed this band to a CT transition analogous to a CT band of, for example, $Li^+I^-$. Let us compare the following:

$$\begin{aligned}
\Psi_N(I^- \cdot Li^+) &= a\Psi_0(I^-, Li^+) + b\Psi_1(I—Li), \\
\Psi_N(I^- \cdot Py^+R) &= a\Psi_0(I^-, Py^+R) + b\Psi_1(I—PyR), \qquad (13\text{-}17) \\
\Psi_V(I^- \cdot Py^+R) &= a^*\Psi_1(I—PyR) - b^*\Psi_0(I^-, Py^+R).
\end{aligned}$$

For lithium iodide, like lithium fluoride in Fig. 13-1, the covalent bonding in $\Psi_1$ is very weak; this may also be true here for the RPy—I bond. In $\Psi_1$ the electron transferred from $I^-$ has gone into the first, antibonding, $\pi$-acceptor orbital of $Py^+R$ to give the R-pyridinyl radical. In this connection we must of course keep in mind that the peak frequency $\nu_{max}$ of the charge-transfer band is governed by the Franck-Condon principle (see Chapter 8), while in general the *equilibrium* configuration of the $V$-state may differ greatly from that of the $N$-state.

The following questions arise immediately:

1. Where is the $I^-$ located in the $PyR^+I^-$ ion pair? It should be located as near to the centroid of positive charge as closed-shell (steric) repulsions permit. Presumably the positive charge is largely on the nitrogen atom, although it must to some extent be spread over the ring. Hence we may reasonably expect that the $I^-$ is located above the plane of the ring near the nitrogen atom.

2. What about solvation? We might expect polar solvent molecules to be attached to the ions by ion-dipole and CT forces, so that the ions of an $I^-PyR^+$ pair might be separated from each other by a layer or two of solvent molecules. However, the ions that give rise to a CT spectrum involving a transfer of an electron from $I^-$ to $Py^+R$ *must* be in close

contact with each other, with no solvent between; that is, they must form contact, or intimate, ion pairs [5]. Kosower also showed that stabilization of the ion pairs by the solvent led to characteristic variations in the position of the CT band, depending on the polarity of the solvent [5]. With a more polar solvent the energy (of state $N$) of the ion pair is lower, and $h\nu_{CT}$ is increased. Incidentally, this points to the fact that the solvent should often be considered to some extent as part of a complex.

Later Kosower found a second CT band in the spectrum of one of these ion pairs, at shorter wavelengths [6]. We may understand this second band if the I—PyR bond in the $V$-state *for a vertical transition* is so weak that the iodine in this state is nearly a free atom. Then the transition $\Psi_V \leftarrow \Psi_N$ can result in an iodine atom in either its $^2P_{(3/2)}$ ground state or in the $^2P_{(1/2)}$ state, which for a free atom is about 7600 $cm^{-1}$ higher in energy. The second band in various alkylpyridinium iodides has a tendency to overlap the $\pi,\pi^*$ absorption of the ion, making observation difficult experimentally. However, Kosower has reported observations for several substituted pyridinium ions that do show double bands of roughly equal intensity, separated by 7600 $cm^{-1}$ within experimental error. These observations apparently establish the interpretation of the spectrum, described above, including the weakness of the RPy—I bond in the $V$-state for the vertical transition that corresponds to $h\nu_{max}$.

It is of some interest that the intensities for the RPy—I charge-transfer bands are rather low; namely, $\epsilon_{max}$ for the $^2P_{(3/2)}$ band in $CH_3Py^+ \cdot I^-$ in chloroform is about 1200 [4c]. This indicates that $b$ in (13-17) is small for $RPy^+I^-$ [cf. (3-11)], as is no doubt also true for diatomic $Li^+I^-$ (see the discussion of lithium fluoride above).†

In the *equilibrium* configuration of the $V$-state of RPy—I it seems likely that the iodine atom may have moved from a position nearly above the nitrogen atom to some other location on the ring. From the major resonance structures of the RPy radical it appears that the carbon atoms in positions 2 and 6 (adjacent to the nitrogen atom) tend to possess the odd valence electron. Hence it seems quite possible that the equilibrium configuration of the CT state may be

† Kosower and co-workers assumed that the association of $RPy^+$—$I^-$ into intimate ion pairs is virtually complete in chloroform, so that $\epsilon$ based on the total concentration of $RPy^+ \cdot I^-$ in that solvent should be the true intensity of the charge-transfer band [5]. They explained the observed decrease in the apparent $\epsilon$ for this band in other more polar solvents [5,7] as due to dissociation of the intimate ion pairs. However, if the intimate-ion-pair concentration *is* less than the total $RPy^+ \cdot I^-$ concentration in chloroform, then the apparent low intensity of this CT band may result from an overestimate of the intimate-ion-pair concentration.

Kosower states that this form is unstable and reverts to the ion pair; also, various pyridinyl free radicals can be prepared under suitable reducing conditions [8].

If one tries to make $RPy^+OH^-$ instead of $RPy^+I^-$, the 2-substituted compound is formed immediately (without light absorption). Briegleb [9] has investigated a number of compounds of the form

In some cases—for example, if $Q^-$ is $ClO_4^-$—there is neither any CT spectrum nor does the $ClO_4^-$ attach to the ring. Some show CT spectra, like the iodide. In still others Q adds at once to form a bond at the 2 or 6 carbon atom [9].

Other examples have been found of CT spectra of ion pairs with $I^-$ as one partner. These include $(Bu)_4N^+I^-$ (here Bu means the *n*-butyl radical) [10] and $Co(NH_3)_6^{3+}\,(I^-)_3$ [11]. The colors of tropylium halides, for example (see below) can probably also be attributed to CT bands [4b,12].

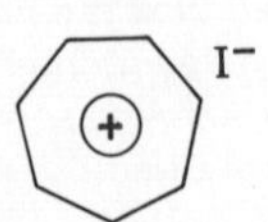

Kosower and Ramsey [13] have shown that the *internally* ion-paired molecule pyridinium cyclopentadienide (see below) has what may be

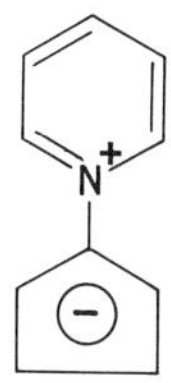

called an *intramolecular* CT band. Interesting examples of ion-pair *intermolecular* CT spectra have been found for various ion pairs consisting of

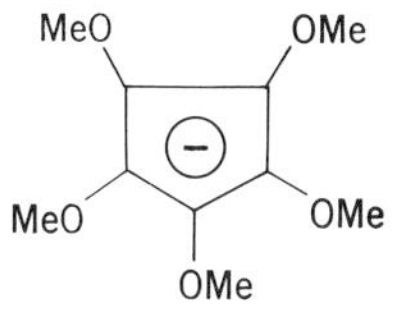

the pentamethoxycyclopentadienyl anion (see above) as donor with various organic cations as acceptors, among them the tropylium ion, the methylpyridinium ion, and the trimethylpyrilium ion [14].

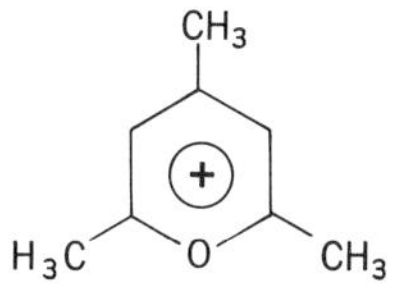

These are all $b\pi$, $a\pi$ complexes. Related to these, although they are not ion pairs, are the $b\pi$, $a\pi$ complexes of various aromatic hydrocarbons as donors with the tropylium ion as acceptor; these show a CT band in the visible or near ultraviolet [12]. (See Chapter 15 for further discussion of ion pairs.)

## 13-5 INTERIONIC DISTANCES IN ION-PAIR COMPLEXES

As remarked above, ion pairs that are in contact (contact, or intimate, ion pairs) can be regarded as donor-acceptor complexes and possess CT spectra. In such ion pairs the distance of approach of the ions may be considerably less than the sum of the crystal radii of the two ions; the latter are known to be approximately equal to van der Waals radii [15]. Thus for the lithium fluoride crystal the closest Li—F distances are 2.01 Å [15], whereas in the free vapor molecule the bond length is 1.56 Å [16]. The decrease is attributed to mutual polarization that is present in the free molecule but pretty well cancels out in the crystal because of the symmetrical environment around each ion [17]. For other alkali halides the vapor-molecule bond lengths mostly range from 0.4 to 0.6 Å less than the crystal-radius sums, more or less independently of the sizes of the ions [18]. The larger ions are more polarizable, but their larger size makes the Coulomb fields acting on them weaker, and the two effects apparently roughly compensate. Somewhat similar relations might be expected in ion-pair complexes such as $RPy^{+}I^{-}$ of Section 13-4; but, when both ions are large molecule-ions such as those that are discussed in the later paragraphs of Section 13-4, it would seem that polarization effects should be smaller so that the distances may be not a great deal less than the sum of van der Waals radii. The CT spectrum of such an ion pair would then be almost like a contact CT spectrum.

## 13-6 REFERENCES

[1] A. D. McLean, *J. Chem. Phys.*, **39**, 2653 (1963). More accurate wave functions are given by A. D. McLean and M. Yoshimina in a Supplement to a Nov. 1967 paper in IBM *J. Res. and Development.*

[2] R. S. Mulliken, C. A. Rieke, D. Orloff, and H. Orloff, *J. Chem. Phys.*, **17**, 1248 (1949), Sec. II.

[3] R. S. Mulliken, *Phys. Rev.*, **51**, 310 (1937).

[4] (a) E. M. Kosower, *J. Am. Chem. Soc.*, **77**, 3883 (1955); (b) E. M. Kosower and P. E. Klinedinst, Jr., ibid., **78**, 3493 (1956); (c) E. M. Kosower and J. A. Skorcz, ibid., **82**, 2188 (1960).

[5] E. M. Kosower, *J. Am. Chem. Soc.*, **78**, 5700 (1956) and subsequently; see especially *J. Am. Chem. Soc.*, **80**, 325 (1958).

[6] E. M. Kosower, J. A. Skorcz, W. M. Schwarz, Jr., and J. W. Patton, *J. Am. Chem. Soc.*, **82**, 2188 (1960).

[7] E. M. Kosower and J. G. Burbach, *J. Am. Chem. Soc.*, **78**, 5838 (1956).
[8] E. M. Kosower and E. J. Poziomek, *J. Am. Chem. Soc.*, **86**, 5515 (1964).
[9] G. Briegleb, W. Jung, and W. Herre, *Z. physik. Chem.*, Neue Folge **38**, 253 (1963).
[10] T. R. Griffiths and M. C. R. Symons, *Mol. Phys.*, **3**, 90 (1960).
[11] M. Linhard and M. Weigel, *Z. anorg. Chem.*, **266**, 49 (1951); M. G. Evans and G. H. Nancollas, *Trans. Faraday Soc.*, **49**, 363 (1953).
[12] M. Feldman and S. Winstein, *J. Am. Chem. Soc.*, **83**, 3338 (1961).
[13] E. M. Kosower and B. G. Ramsey, *J. Am. Chem. Soc.*, **81**, 856 (1959).
[14] E. LeGoff and R. B. LaCourt, *J. Am. Chem. Soc.*, **85**, 1354 (1963).
[15] L. Pauling, *The Nature of the Chemical Bond*, 3rd ed., Cornell University Press, Ithaca, N.Y., 1960.
[16] L. Wharton, W. Klemperer, L. P. Gold, R. Strauch, J. J. Gallagher, and V. E. Derr, *J. Chem. Phys.*, **38**, 1203 (1963).
[17] A. Honig, M. Mandel, M. L. Stitch, and C. H. Townes, *Phys. Rev.*, **96**, 629 (1954).
[18] *Tables of Interatomic Distances and Configuration in Molecules and Ions*, L. E. Sutton, ed., The Chemical Society (London), 1958, 1965 (supplement); R. K. Bauer and H. Lew, *Can. J. Phys.*, **41**, 1461 (1963).

*CHAPTER* *14*

# *Potential-Energy Curves for Specific Cases*

Let us continue the discussion of the potential surfaces of complexes that was begun in Chapter 9, turning from general discussion to the consideration of a few typical specific examples in some detail. We shall use the resonance-structure method, in most cases in its simplified form. (See also R14, in which the discussion up through the first half of p. 35 is probably correct; beyond that point, some of the ideas have since evolved further.)

## 14-1 BENZENE · $I_2$

The benzene · $I_2$ complex has already been discussed in several places, most particularly in Section 11-2, which will serve as an introduction to the present section. Assuming axial geometry, the symmetry of the complex is $C_{6v}$. All the vibration modes and the vibrational and electronic states should then be classified according to this symmetry.

We shall designate the two singlet charge-transfer states that are considered in Section 11-2 as $\Psi_V[Bz^+(\phi_1^{-1}) I_2^-(\sigma_u)]$ and as $\Psi_V'[Bz^+(\phi_2^{-1} \text{ or } \phi_3^{-1}) I_2^-(\sigma_u)]$, using the notation for the benzene MOs given in Fig. 4-1. Under $C_{6v}$, $\Psi_V$ is a $^1A_1$ state since the $\sigma_u$ MO of $I_2^-$

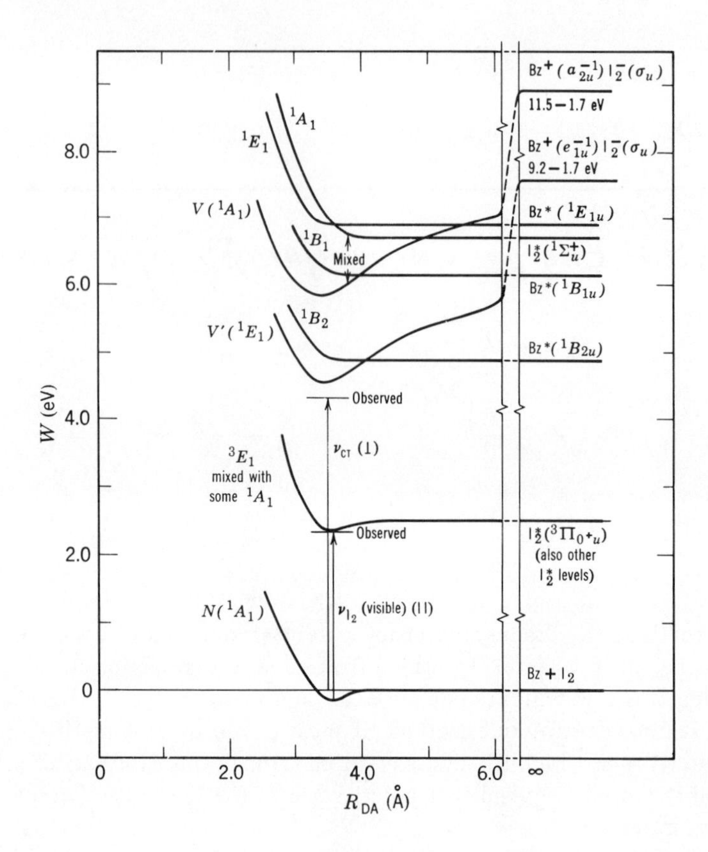


**Figure 14-1** Potential-energy curves for the *axial model* of the benzene • $I_2$ complex. The $I_2$ locally excited $^3\Pi_{0^+u}$ and $^1\Sigma_u^+$ levels are those of (10-2) and (10-3). The benzene locally excited levels (and others) are all of the $(\phi_{2\ 3})^{-1}\phi_{4,5}^*$ configuration (see Fig. 4-1 for the $\phi$s and $\phi^*$s). Although the $^3\Pi_u$ state classifies as $^3E_1$ under $C_{6v}$ symmetry, the $^3\Pi_{0^+u}$ level behaves like $^1A_1$ and the (main part of) the visible band should be parallel polarized. (Figure from GR5.)

and the $\phi_1$ MO of benzene are both of $a_1$ species under $C_{6v}$.† The other

† In $Bz^+$ the symmetry of the state $\phi^{-1}$ that remains after removing one electron from the $\phi_1^2$ shell is determined by the remaining electron and so is $^2A_1$. The $I_2^-(\sigma_u)$ state is also of $^2A_1$ symmetry. The combined state can be either $^1A_1$ (the charge-transfer state $V$) or $^3A_1$ (the corresponding triplet state $T$; see Section 2-8). Similarly the $\Psi_{V'}$ singlet state (with $Bz^+$ in a $^2E_1$ state) is accompanied by a $^3E_1$ charge-transfer state $T'$.

charge-transfer state $\Psi_{V'}$ is of $^1E_1$ species by similar reasoning. For each of the two singlet CT states there must also exist a triplet CT state of the same electron configuration. We shall call these $T'$ and $T$. The former is of species $^3A_1$ and the latter $^3E_1$.

The energies as functions of $R_{DA}$ with all other configurational coordinates kept at $Q_N$ (see Section 9-4 and Fig. 9-4) are shown in Fig. 14-1. On the right side of the figure ($R_{DA} = \infty$) are shown the energies of the CT states computed from $I_D^v - E_A^v$, using the $I_D$ values from Fig. 4-1 and using $E_{I_2}^v = 1.7$ eV. Also shown, from the known ultraviolet spectra of benzene and of iodine, are some of the locally excited levels. In addition a great many more locally excited states of iodine are known to exist (see Section 10-1), and there are the triplet levels of benzene that correspond to the singlet levels shown. Local transitions to these other states are very weak, however.

As the benzene and iodine approach each other, the normal state and the locally excited states are not expected to change in energy until $R_{DA}$ reaches approximately the van der Waals distance. At that point the energy curves of these states are expected to rise because of the steric, or closed-shell, electron-exchange-repulsion term in $G_0$ (see Chapter 9); but those of proper symmetry may be stabilized, as is state $N$, by a resonance-energy term $X_0$ because of interaction with an appropriate CT state.

For both of the CT states the Coulomb attraction (somewhat assisted by mutual polarization of $Bz^+$ and $I_2^-$) must cause the energy to drop as the D—A distance decreases until approximately van der Waals distances or somewhat less are reached. In the case of $\Psi_V$ and $\Psi_T$ the Heitler-London valence-attraction term would tend to cause the $V$-state curve to continue dropping until $R_{DA}$ is slightly smaller and equally to cause the $T$-state curve (not shown in Fig. 14-1) to rise; the $V$-state curve, however, should also be pushed up more or less by the resonance-interaction term $X_1$ (see Chapter 9). In the case of $\Psi_{V'}$ and $\Psi_{T'}$ there is no resonance term or valence-attraction term for $V'$, but there is an atomlike exchange term that tends to make the $T'$ ($^3E_1$) curve (also not shown in Fig. 14-1) fall and the $V'$ ($^1E_1$) curve rise a little. States of like symmetry in Fig. 14-1 can mix more or less to cause some further shiftings.

The foregoing considerations have permitted drawing the curves shown in Fig. 14-1 with some confidence. In the figure the energy change that corresponds to the observed charge-transfer transition is seen to fall somewhat short of the calculated minimum of the $V'$-curve. In this and subsequent figures in the present chapter the curves that are shown are calculated from the simple considerations given above; their positions can then be *compared* with those based on observation; the calculations have *not* been adjusted to fit the observations. The estimated curve for the first charge-transfer state $V'$, of symmetry $^1E_1$, is apparently a little higher than it should be on the basis of the experimental value of the CT frequency (see $h\nu_{CT}$ on the figure), but in view of various uncertainties (including that in $E^v$ for $I_2$, and doubts as to the correctness of the axial model), the agreement is good.

## 14-2 BENZENE · TRINITROBENZENE

As an example of a $b\pi \cdot a\pi$ complex consider the benzene · trinitrobenzene complex. This is apparently quite as stable as the $Bz \cdot I_2$ complex.† In the normal state of typical $b\pi \cdot a\pi$ complexes the equilibrium perpendicular distance of approach between planes, which we may take as $R_{DA}$, is scarcely less than the van der Waals distance. This fact may plausibly be understood if the steric repulsions of the $\pi$ electron shells that are not involved in the CT resonance stabilization of state $N$ are responsible for keeping the D and A separated. If so, these same steric repulsions probably prevent $R_{DA}$ from getting much smaller for the $W_1$ than for the $W_0$ curve, and it then seems probable that the minimum in the $W_1$ curve also comes at not very much less than the van der Waals distance.

The simplest assumption (though very possibly incorrect) for the symmetry of the Bz · TNB complex is that it is $C_{3v}$ with the centers of both Bz and TNB on the axis of symmetry. With this assumption the foregoing considerations and those of Chapter 9 lead to the potential curves for Bz · TNB shown in Fig. 14-2. In arriving at these the estimation of the Coulomb energy in $W_1$ has been made by using the "charges spread" model of Chapter 9, which may not be very good. The value of $I_D$ is 9.24 eV, and $E_A^v$ has been taken as 1.5 eV, somewhat larger than the value of 0.7 eV given by Briegleb [1]. (See Chapter 9.)

As in Fig. 14-1 for $Bz \cdot I_2$, the curve for the CT state crosses those of several locally excited states of benzene. In addition we would expect several locally excited TNB transitions to occur in the same region of the spectrum as the CT band. These include strong absorption to the benzene-like $^1E$ and to intramolecular CT states of TNB with energies near 5 eV which may mix strongly with the CT state of the complex.

† See Briegleb (GR1), Tables 56 and 60, for $K$ and $-\Delta H$ values of Bz · TNB and $Bz \cdot I_2$ in carbon tetrachloride. A comparison of the $C_2$ values (Table 9-3) and resulting $\beta_0$ values (Table 9-4) suggests that the CT resonance stabilization is expected to be in the reverse order. However, we have already noted in Section 10-3 that the stability of these weak complexes is the result of electrostatic and CT stabilization, together with repulsion terms. For these two complexes there may be considerable electrostatic stabilization.

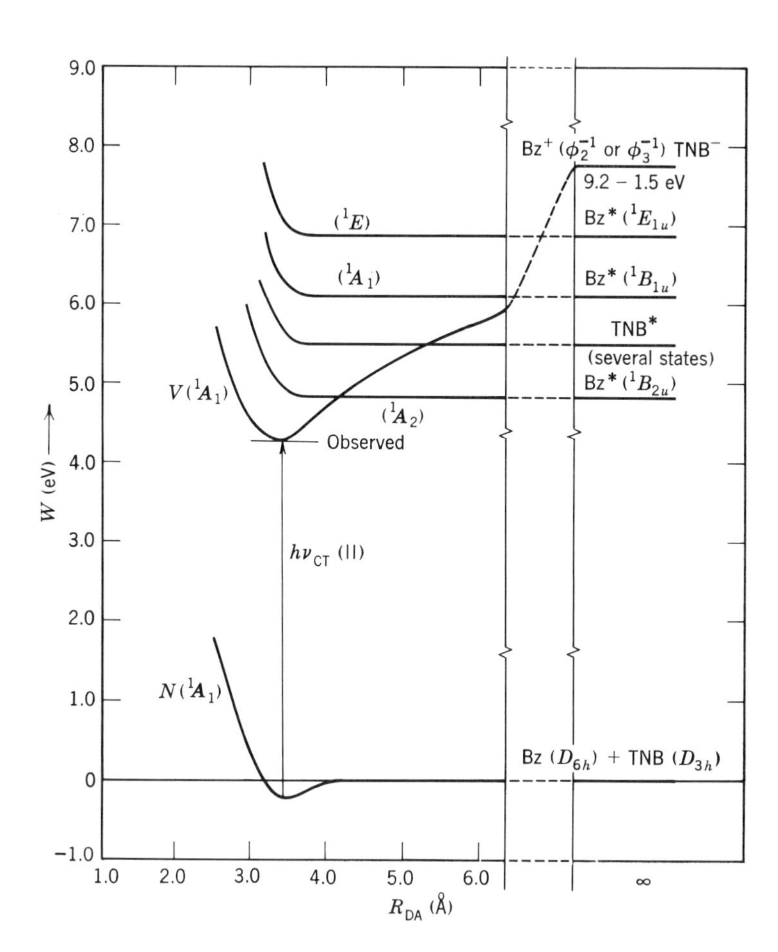


**Figure 14-2** Potential curves for benzene · TNB assuming $C_{3v}$ symmetry. Several additional (triplet and singlet) states that are not shown exist in the same energy range.

## 14-3 CONTACT CHARGE-TRANSFER SPECTRA: IODINE IN ALKANES

Consider now the case in which the resonance interaction is practically

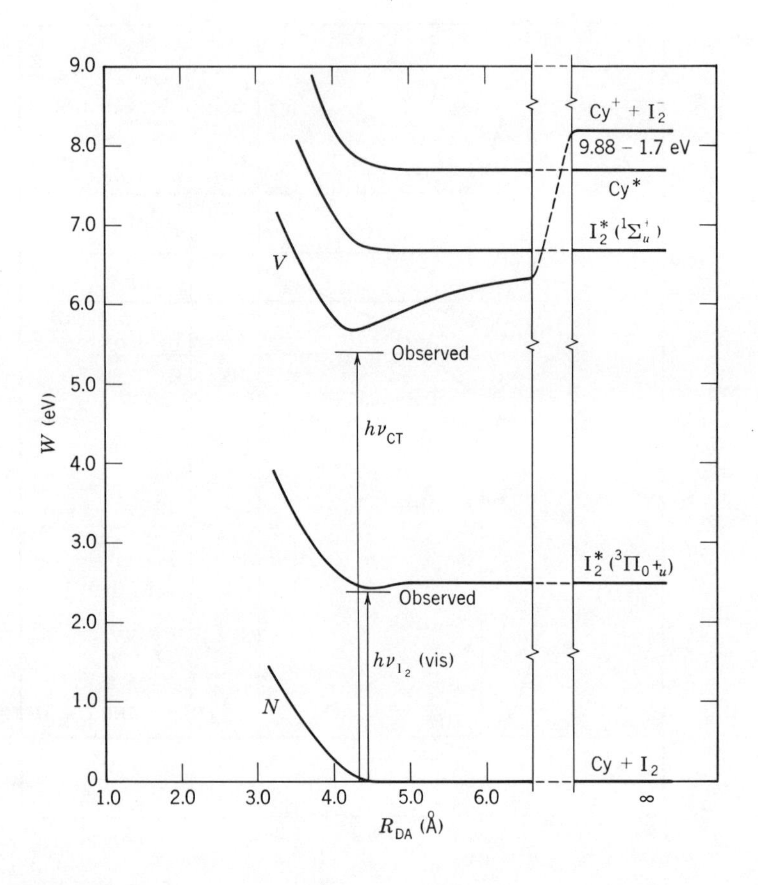


**Figure 14-3** Potential curves for cyclohexane (Cy) plus $I_2$.

zero. Here CT states must nevertheless exist and must be just as real and stable as in the case of stable complexes.† Hence CT absorption bands, here called contact CT bands, should be possible. Iodine dissolved in cyclohexane is thought to provide an example. (See Fig. 1-1.)

Potential curves for this case are shown in Fig. 14-3, assuming that

† The situation is not unlike that of two helium atoms. Although there is no stable ground state, the $He_2$ molecule has a great number of well-known excited states.

the van der Waals distance between a carbon atom of cyclohexane and the nearest iodine atom of $I_2$ is 4.4 Å and assuming $V = 0$ in the CT state. Here we have used values of $I_D = 9.88$ and $E_A = 1.7$ eV. In the absence here of a well-defined equilibrium distance for the complex we might anticipate a rather large width in the contact CT band resulting from CT absorption by molecules in varying degrees of closeness of contact. It is necessary that $\phi_d$ and $\phi_{a^-}$ overlap, which should be possible even at larger than van der Waals distances. (See Section 3-5.) Also, the existence of various relative orientations and various conformations of the cyclohexane might help increase the spread. Empirically, however, $\Delta\nu_{(1/2)}$ is found [2] to be about 7500 $cm^{-1}$, which is relatively large yet only moderately larger than the average for CT bands of $I_2$ complexes. (See Table 10-2.)

Figure 14-3 indicates that some mixing of the wavefunction $\Psi_V$ may occur with those of near-by locally excited states of cyclohexane or of $I_2$. Some mixing of the $V$-state with the locally excited $^1\Sigma_u^+$ state of $I_2$ may perhaps result in enough stabilization of the $V$-state to bring it down to a height in agreement with the observed frequency of the contact CT band.

The transition dipole moment is given[†] by

$$\mu_{VN} = ea^*b(\bar{\mathbf{r}}_D - \bar{\mathbf{r}}_A) + e(aa^* - bb^*)(\bar{\mathbf{r}}_D - \bar{\mathbf{r}}_{DA})S_{01} + c\mu_{DD^*} + d\mu_{AA^*}. \tag{14-1}$$

If there is no resonance interaction, $b = 0$; when the pair is in close contact presumably there may be an instantaneous small value of $b > 0$; but further, as discussed in some detail in Section 3-4, $S_{01} \simeq \sqrt{2}\, S_{da^-}$ need not be zero even if $b = 0$. Hence the second term in (14-1) may well largely account for the intensity of the contact absorption band. Also, as urged by Murrell (see R10), the mixing with the locally excited donor and/or acceptor states (indicated in (14-1) by the third and/or fourth terms) may contribute appreciably to the intensity.

Experimentally, attempts to study the contact CT absorption band for $I_2$ in $n$-heptane or cyclohexane can be complicated by the contact absorption band of oxygen in $n$-heptane. If the oxygen that is normally present as a result of contact with air is removed from $n$-heptane, one sees a more distinct peak for the contact CT band of $I_2$ in $n$-heptane than that shown in Fig. 1-1.[‡]

---

[†] See Section 11-4 and in particular (11-6) after making the substitutions given by (3-12) and (3-14).

[‡] E. C. Lim, Loyola University, private communication.

## 14-4 CONTACT CHARGE-TRANSFER SPECTRA AND SINGLET-TRIPLET ABSORPTION ENHANCEMENT BY OXYGEN

Almost any donor solvent gives a contact charge-transfer band when oxygen is dissolved in it (see R11). Alkanes, water, alcohols, ethers, aniline, benzene, aliphatic amines—all these give contact CT spectra with oxygen yet there is no evidence that oxygen forms a stable complex with any of these. The ability of oxygen to produce contact CT spectra equally with the strongest and weakest electron donors yet not to form complexes with any of them is very puzzling.

For the $Bz + O_2$ contact pair, with benzene and oxygen in their normal states of respective symmetries $^1A_{1g}$ and $^3\Sigma_g^-$, the combined wavefunction is of triplet character. Now if the benzene is excited to, for example, its $^3B_{2u}$ state, singlet, triplet, and quintet combined states result. The energies of these states would be appreciably different in a reasonably strong complex, but in the very weak $Bz \cdot O_2$ complex or contact pair they may be essentially degenerate. Nevertheless from the point of view of the contact pair the forbidden transition that we call $^3B_{2u} \leftarrow {}^1A_{1g}$ in pure benzene can now in part be considered as a triplet $\leftarrow$ triplet transition and therefore to some extent allowed.

After charge transfer, since $Bz^+(^2E_1)$ and $O_2^-(^2\Pi_g)$ each have an odd electron, the pair give rise to a singlet charge-transfer state $V$ and a triplet state $T$ of just the familiar kind (cf. Chapter 2), except that now it is in the $T$-state instead of in the $V$-state that interaction between $\Psi_0$ and $\Psi_1$ occurs:

$$\Psi_T = a^*\Psi_1'(D^+ \leftrightarrow O_2^-) - b^*\Psi_0(D, O_2); \qquad \Psi_V = \Psi_1(D^+—O_2^-). \qquad (14\text{-}2)$$

Furthermore, it is well known that the presence of dissolved oxygen greatly enhances the intensities of singlet-triplet absorption transitions; for example, the $^3B_{2u} \leftarrow {}^1A_{1g}$ absorption transition in benzene is found to occur weakly in liquid benzene saturated with oxygen even though in the absence of oxygen no absorption at all is detectable even with a long liquid path. This phenomenon is probably related to the contact CT phenomenon (see R11) through intermediation of the triplet CT state. (For additional discussion of contact pairs and contact CT spectra see GR5.)

## 14-5 TRIMETHYLAMINE $\cdot$ $I_2$

Let us now consider $Me_3N \cdot I_2$ (Fig. 14-4), using the parameters from

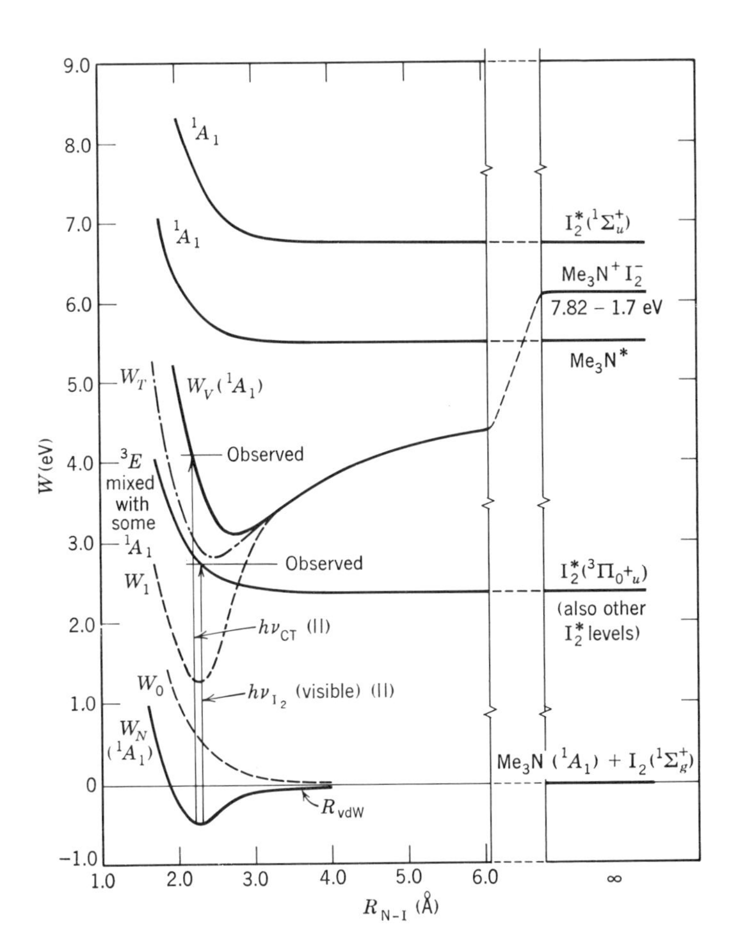


**Figure 14-4** Potential curves for $Me_3N \cdot I_2$, assuming symmetry essentially $C_{3v}$ in interaction region of complex. (This figure has been modified from Fig. 7 of GR5 because of the different values used here (from Section 10-3) for the parameters: $C_1$ especially.)

Table 10-1† and (2-15) for $h\nu_{CT}$. (Here see also the discussion of

† With $\Delta = I_D - C_1$ (Eq. 9-7), $C_1 = 7.3$, and $I_D = 7.82$ for $Me_3N$, $\Delta(\equiv W_1 - W_0)$ is found to be 0.52 eV. For Fig. 14-4 we assume $S_{01} = 0.40$, $\beta_0 = -1.8$ eV, and $\beta_1 = -2.01$ eV. [Cf. (2-14).]

$(CH_3)_3N \cdot I_2$ in Section 10-3 that follows Table 10-1.) The symmetry in the D, A interaction region is essentially $C_{3v}$. (See Section 5-3.) The unperturbed $W_1$ and $W_0$ curves are now predicted to come quite close to each other[†] but to interact strongly to give widely separated $W_N$ and $W_V$ curves, and thus a high charge-transfer band frequency $\nu_{CT}$. The $W_T$ curve is expected, like $W_V$ but for different reasons, to be higher than $W_1$; the unusually large valence-interaction term $V$ in $G_1$ (see Section 9-2) pushes $W_1$ down strongly at $R_{NI}$ less than the van der Waals value from the value of $W_1$ which results from the stabilization by the Coulomb term ($-C$) alone and pushes $W_T$ (the same as $W_1'$—see Section 2-8) up even more strongly (compare the Heitler-London $^1\Sigma_g^+$ and $^3\Sigma_u^+$ states from two normal hydrogen atoms). The charge-transfer resonance between $W_1$ and $W_0$ results in $W_V$ being pushed up again—in Fig. 14-4 it is shown as pushed considerably higher than the $W_T$ curve.

One expected consequence of the large resonance interaction that is believed to exist for these amine complexes (but not for weaker complexes, such as $b\pi \cdot a\sigma$ or $b\pi \cdot a\pi$) is that the equilibrium value $R_{NI}$ for the $V$-state is pushed outward, so that it is larger than for the normal state (see Fig. 14-4). Further, this effect should increase as $I_D$ drops in the series of increasingly strong amine · iodine complexes $H_3N \cdot I_2$, $H_2NMe \cdot I_2$, $HNMe_2 \cdot I_2$, and $Me_3N \cdot I_2$; that is, $R_{NI}$ should decrease for state $N$ and increase for state $V$ within this series. Correspondingly, the $h\nu_{CT}$ *vertical* transition should intersect the $V$-state curve, on the left of its minimum, at an increasingly steep angle. This increasing steepness should lead to increasing quantum-mechanical spread (see Chapter 8), consequently a greater half-width $\Delta\nu_{(1/2)}$ for the charge-transfer band. Exactly this change is observed; $\Delta\nu_{(1/2)}$ increases from 4100 $cm^{-1}$ for $H_3N \cdot I_2$ to 8100 $cm^{-1}$ for $(CH_3)_3N \cdot I_2$. (See Table 10-2.) These observations confirm the supposition that the equilibrium $R_{NI}$ is indeed larger for the $V$ than for the $N$ state—and increasingly so as the strength of the complex increases since, if the reverse relation were true for the $R_{NI}$ terms, $\Delta\nu_{(1/2)}$ should get narrower instead of wider in going from $H_3N \cdot I_2$ to $Me_3N \cdot I_2$.[‡]

It is interesting next to note that the rather large blue shift that is observed for the visible band of iodine in that locally excited transition of the $Me_3N \cdot I_2$ complex seems to be very reasonably explained by a superposition of intra-$I_2^*$ repulsion created by close proximity of D and $A^*$,

[†] Thus coming close to case II [Fig. 3 of R14, also Fig. 14-6*a*].

[‡] Of course this discussion does not take into account the potential energy of these two states as a function of $R_{I—I}$, nor does it properly consider the contribution to $\Delta\nu_{(1/2)}$ from the distribution of the complexes among the low-frequency intermolecular vibrational levels, as discussed in Section 8-4.

plus the stabilization energy of the ground state, minus a perhaps somewhat equal stabilization of the D, A* state. (See Section 10-5 for a detailed discussion.)

Furthermore, we can predict that in the spectrum of the complex there should probably be a considerable blue shift of the locally excited $Me_3N^* \leftarrow Me_3N$ transition and of the locally excited transition to the $^1\Sigma_u^+$ state of $I_2$. Data on these are not available; however, for the $H_3N^* \leftarrow H_3N$ transition analogous to that just mentioned for $Me_3N$, Dressler and Schnepp [3] found a blue shift of about 1 eV on going from the vapor to the solid; this shift they attribute to an effect in the hydrogen-bonded $NH_3$ crystal that is similar to that discussed above (see also [4]).

## 14-6 ALKALI HALIDES AS ODD-ODD CASE III COMPLEXES

The potential curves that have been considered thus far in this chapter have belonged to case I ($W_1 \gg W_0$) or case II ($W_1 \simeq W_0$) of R14. We now consider some examples of case III ($W_1 < W_0$) of R14 (see Fig. 4 of R14). The simplest example of a case III "complex" is an alkali halide—such as lithium fluoride, which was discussed in Chapter 13, but with $\Psi_0$ and $\Psi_1$ now defined from the viewpoint of the neutral *atoms* (which are odd-electron entities) as D and A (see the first paragraph of Section 4-2). Figure 14-5 is a simplification, presented from this viewpoint, of key aspects of Fig. 13-1. Here the curve for the complex, $W_N$, is nearly pure $W_1$ near equilibrium ($R = R_{eq}$), the wavefunction being largely ionic ($\Psi_1$). At large distances, however, $W_N$ changes to $W_0$, and the structure of the complex finally becomes purely covalent ($\Psi_0$) at very large $R_{MX}$ values. Of especial importance for the potential curves is the crossing point of $W_0$ and $W_1$ at $R_c$. Near the crossing point the wavefunctions are very much mixed ($b \simeq a$), but the mixing probably decreases rapidly on each side of $R_c$. (See Chapter 13 for further discussion.)

It is of some interest to ask whether among the alkali halides there are any examples for which the ionic-potential curve is below the covalent curve at all $R$-values. Since at $R = \infty$ the ionic curve is above the covalent curve by an amount of energy $I_M - E_X$, we require an example where the electron affinity of X is larger than the ionization potential of M. At one time it was thought that cesium fluoride is such an example; more recently the electron affinity of fluorine has been established as 3.45 eV [5], whereas $I_{Cs} = 3.89$ eV. Actually, cesium chloride ($E_{Cl} = 3.63$ eV) comes closest to the case we were looking for.

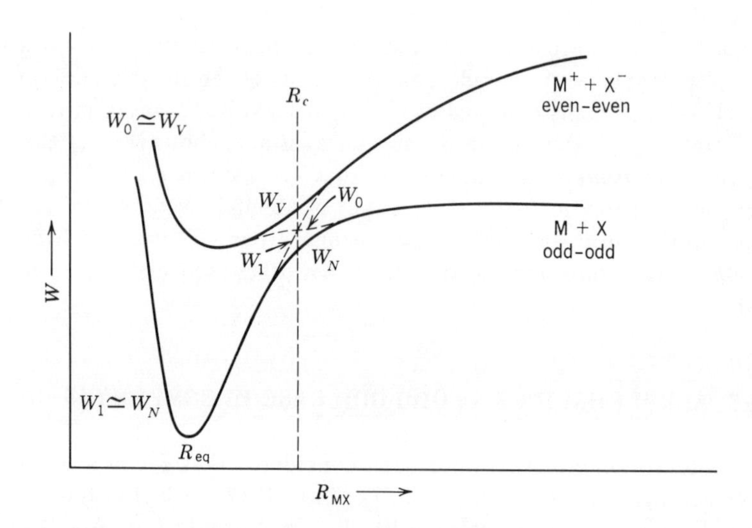


**Figure 14-5** Potential curve for $M^+X^-$ as a case III complex, with $\Psi_N = a\Psi_0(M{-}X) + b\Psi_1(M^+, X^-)$, $b^2 \gg a^2$ at $R_{eq}$.

Very likely FrCl would be an example if it were known.

## 14-7 EVEN-EVEN CASE III COMPLEXES WITH EXTENSIVE CHARGE TRANSFER

Figure 14-6 illustrates two situations that may be expected to occur for $n \cdot v$ complexes, a type that includes some very strong complexes. In strong $n \cdot v$ complexes $b/a$ in (2-2) is expected to be fairly large. However, the two cases $b/a < 1$ and $b/a > 1$ that are possible correspond to case II and case III potential curves, respectively (Fig. 14-6*a* and *b*).[†] If the MO calculations on $H_3N \cdot BH_3$ discussed in Section 12-6 can be taken as a reliable guide, $b/a < 1$ for that complex. On the other hand, it seems likely that for complexes such as $H_3N \cdot BF_3$ and $H_3N \cdot BCl_3$ or especially $R_3N \cdot BF_3$ and $R_3N \cdot BCl_3$[‡] $b/a > 1$.

[†] Fig. 14-6*b* is a simplified redrawing of the left-hand diagram in Fig. 4 of R14.
[‡] In general $BCl_3$ forms somewhat stronger complexes than $BF_3$.

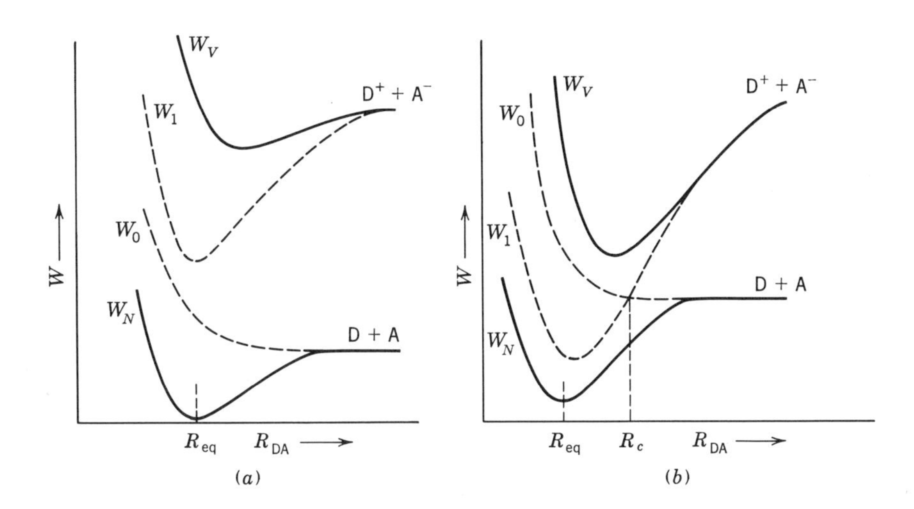


**Figure 14-6** Potential curves for even-even complex with $\Psi_N \simeq a\Psi_0(D, A) + b\Psi_1(D^+—A^-)$. (*a*) Case II: $b < a$ at all $R_{DA}$ values. (*b*) Case III: $b > a$ at $R_{eq}$ and at all $R < R_c$.

In view of the considerable changes in shape (see Table 5-1) that $BR_3$ or $BX_3$ undergoes in forming a complex with $NH_3$ or $NR_3$, it would not be surprising if there were an activation barrier to be surmounted during the approach of the two partners in $n \cdot v$ complexes with a $BR_3$ or $BX_3$ acceptor. Such a barrier would appear in Figs. 14-6*a* and *b* as a small maximum in the $W_N$ curve at fairly large $R_{DA}$ values. Smith and Kistiakowsky [6] report that the barrier, in the case of $H_3N \cdot BF_3$, if it is different from zero, is less than 3 kcal/mole. Furthermore, Bauer and Jones† report that the difference in the energies of activation for the reactions of $NH_3$ and of $N(CH_3)_3$ with $BF_3$ is less than 0.3 kcal, suggesting that the activation energies are zero for both reactions.

Let us now consider the complex between $(t\text{-Bu})_3N$ and $(t\text{-Bu})_3B$, where *t*-Bu stands for the tertiary butyl group. This complex should be like that between $(CH_3)_3N$ and $BCl_3$ except that relatively small substituent groups have been replaced by the bulky *t*-Bu group. This group prevents the

† S. H. Bauer and W. D. Jones, private communication.

nitrogen and boron atoms from approaching each other as closely as in $(CH_3)_3N \cdot BCl_3$, making $b/a$ smaller and $R_{eq}$ larger, and very likely also producing a barrier at larger $R_{DA}$. H. C. Brown has studied the effect of steric hindrance in similar cases in an extensive series of papers beginning in 1942 [7]; see also Taft's review [8]. Apparently compounds that are as strongly hindered as that of $(t\text{-Bu})_3N$ with $(t\text{-Bu})_3B$ have not been studied extensively, but a comparison of the $(CH_3)_3B$ complexes of $t\text{-BuNH}_2$ and of $NH_3$ shows that the strain energy for the former is already about 5 kcal/mole [7]. ($\Delta H$ for dissociation of the $t\text{-BuNH}_2$ complex is 13 kcal/mole, instead of about 19 kcal/mole as otherwise expected.) Following this line of reasoning it is clear that, besides perhaps creating a barrier at large $R_{DA}$, the effect of increasing steric hindrance is to move the $W_1$ curve up, making $W_1 > W_0$ (case II, Fig. 14-7*a*) if this is not already true.

For strong $n \cdot v$ complexes such as that between $H_3N$ and $BCl_3$ no charge-transfer absorption band has been identified, perhaps because it has been pushed into the vacuum ultraviolet and/or perhaps because the structure of the complex has been so modified and the charge-transfer state has been so mixed with locally excited states that it cannot be identified. In fact as yet no bands of $n \cdot v$ complexes that are identifiable as charge-transfer bands have been found, so far as we are aware.

## 14-8 FURTHER CONSIDERATION OF EVEN-EVEN CASE II AND CASE III EXAMPLES

Consider now the potential curves for a series of cases with $W_1$ varying from case II ($W_1 \simeq W_0$) to case III, with $W_1 << W_0$ at small $R$-values, but with weaker interactions between $W_0$ and $W_1$ than in Fig. 14-6. Such a series is shown in four stages in Figs. 14-7 and 14-8.

Figure 14-7*a* shows curves for $W_1 \simeq W_0$ with no crossing. If $W_1$ goes a little lower, it crosses $W_0$, as shown in Fig. 14-7*b*; however, the $W_N$ curve is still not very different from that in Fig. 14-7*a*, differing mainly just in a rather small kink near $R_c$. Nevertheless a marked change in the electronic structure of the complex occurs as $R$ is varied from $R > R_c$ to $R < R_c$.

We now remind ourselves that these curves are cross sections through energy hypersurfaces—actually all the other coordinates $Q$ are changing more or less as $R$ changes, in such a way as to make the energy a minimum at each $R$; that is (see Section 9-4), the curve $W_N$ is drawn for $W_N(R, Q_N)$. Consider as a specific example the hydrogen-bonded complex between an electron donor D and an $a\sigma$-acceptor HX (perhaps phenol).

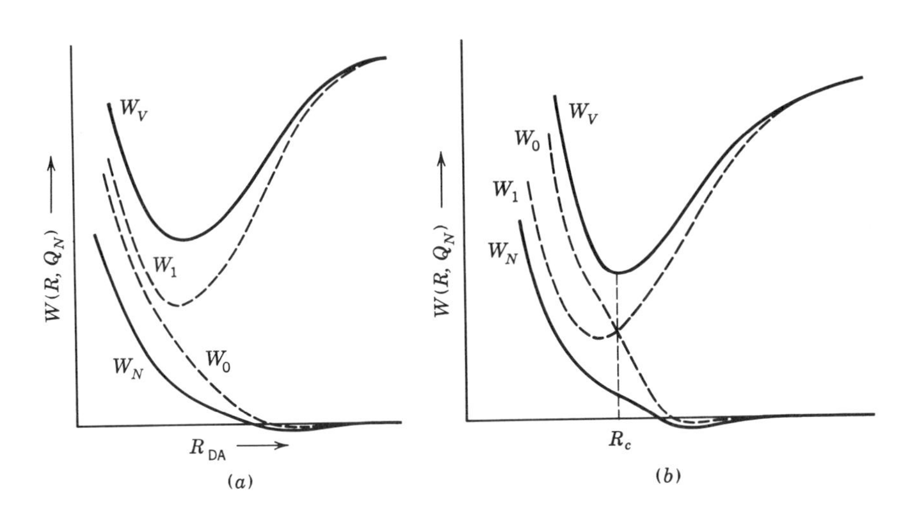


**Figure 14-7** Potential curves illustrating the changes occurring in the transition stages from case II toward case III. System D + HX. $R$ = D—H distance; $Q$ variations are mainly only in H—X distance. (See also Fig. 14-8.) (*a*) Case II ($W_1 \simeq W_0$). (*b*) Incipient case III ($W_1 < W_0$ for $R < R_c$).

In this case it is convenient to take $R_{DH}$ as $R_{DA}$; we then note that among the $Q$ terms there is one, namely $Q_{HX}$, which when $R_{DH}$ varies must also be varied appreciably if $W_N$ is to be kept as low as possible.†

Just to the right of $R_c$, $Q_{HX}$ is only slightly larger than the normal H—X bond length. On crossing to the left of $R_c$ this coordinate stretches considerably, toward the value appropriate for the structure $(D—H)^+X^-$ which should be more or less closely approximated at equilibrium in the pure dative state $\Psi_1$. For a case such as the one shown in Fig. 14-7*b*, although $b > a$ in (2-2) for $\Psi_N$ when $R < R_c$, it is not true that $b^2 >> a^2$: $\Psi_0$ is still considerably mixed into $\Psi_1$, and the ion-pair extreme $DH^+X^-$ is not closely approached.

However, if $W_1$ goes still lower, the situation sketched in Fig. 14-8

† Another possibility would be to use $R_{DX}$ as $R_{DA}$. However, as we decrease $R_{DX}$ smoothly, we may expect that when it reaches a certain value, the H—X bond may stretch and the D—H bond form rather suddenly to give a large change in $R_{DH}$ over a rather small change in $R_{DX}$.

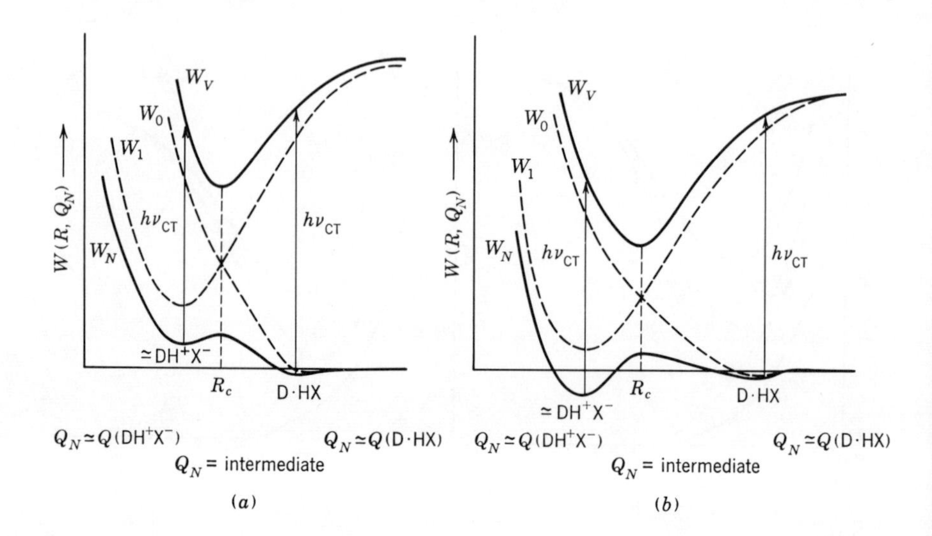


**Figure 14-8** Further stages in the series ranging from case II (see Fig. 14-7) toward strong case III. A still further stage, without activation barrier, is shown in Fig. 14-6. The indicated variation of $Q_N$ with $R$ (i.e. $R_{DH}$) has pulled the minimum of $W_1$ down and to the left, has increased the steepness of rise of $W_0$ toward the left and has raised the left side of the $W_v$ ($Q_N$) curves.

may be reached, with a double minimum in $W_N$. Near the left-hand minimum and at smaller $R$-values the structure may approximate that of the strongly hydrogen-bonded† ion pair $(D—H)^+X^-$; near the right-hand minimum and at larger $R$-values the structure is that of the normal hydrogen-bonded complex D · · H—X, with only relatively small charge transfer. Usually either the left-hand minimum (Fig. 14-8*a*) or the right-hand minimum (Fig. 14-8*b*) is likely to be the stable one, but the existence of two minima with nearly equal energy should sometimes be possible, depending on the relative positions of the $W_1$ and $W_0$ curves. Actually there is evidence from infrared spectra that such double minima of nearly equal depth do sometimes occur for hydrogen-bonded complexes [9].

A further major possibility is that the interaction between $\Psi_0$ and $\Psi_1$ at $R_c$ may be so large as to flatten out the activation barrier between the two minima, leaving only a single minimum at an intermediate position.

† Here the $X^-$ is strongly polarized by the $DH^+$ field.

This case has been discussed in Section 14-7 and depicted in Fig. 14-6*b*.

## 14-9 REFERENCES


[1] G. Briegleb, *Angewandte Chem.* (intnl. ed.), **3**, 617 (1964).
[2] L. Julien and W. B. Person, *J. Phys. Chem.*, **72**, 3059 (1968).
[3] K. Dressler and O. Schnepp, *J. Chem. Phys.*, **33**, 270 (1960).
[4] G. C. Pimentel, *J. Am. Chem. Soc.*, **79**, 3323 (1957).
[5] R. S. Berry and C. W. Reimann, *J. Chem. Phys.*, **38**, 1540 (1963).
[6] F. T. Smith and G. B. Kistiakowsky, *J. Chem. Phys.*, **31**, 621 (1959).
[7] See, for example, the series of papers by H. C. Brown et al., *J. Am. Chem. Soc.*, **75**, 1–28 (1953).
[8] R. W. Taft, Jr., in M. S. Newman's *Steric Effects in Organic Chemistry*, Wiley, New York, 1956.
[9] C. L. Bell and G. M. Barrow, *J. Chem. Phys.*, **31**, 300 (1959) and references given there; K. Bauge and J. W. Smith, *J. Chem. Soc.*, 4244 (1964) and references there.

*CHAPTER* ***15***

# *Inner, Outer, and Middle Complexes; Environmental Cooperative Action*

## 15-1 INNER AND OUTER COMPLEXES

The consideration of case III potential curves in Section 14-8 brings us to the topic of *inner* and *outer complexes*. These were first discussed in R2 and in Sections VIII and IX of R4, to which the reader should refer, noting, however, that since the time when these papers were published our understanding has evolved considerably, in part along lines just discussed in connection with Figs. 14-7 and 14-8. At the time R4 was written, the relation between $W_0$ and $W_1$ curves and inner and outer complexes had not been clarified. Also, a distinction was made in R4 between HQ acceptors (which were called "conditional" acceptors) and other $\sigma$ acceptors such as $I_2$. Hydrogen-bonded complexes were treated in R4 as basically different from donor-acceptor complexes and their stability was attributed essentially to classical electrostatic forces. We are now convinced that ordinary hydrogen-bonded complexes are similar to complexes of other $a\sigma$ acceptors, with differences that are merely matters of degree, not of kind.

With HX acting as an acceptor as in Fig. 14-8, the ordinary hydrogen-bonded complex represented by the outer or right-hand minimum in the potential curve as a function of $R_{DH}$ is called an *outer complex*. The isomeric, strongly hydrogen-bonded, ion-pair form $(D—H)^+X^-$ (the left-hand or inner minimum in Fig. 14-8) is called an *inner complex*. The transformation, occurring as $R_{DH}$ is decreased, from the outer to the inner form of the complex can be called an *even-ionogenic action*, since each of the ions $DH^+$ and $X^-$ that are generated has an even number of electrons.

The HX acceptors were called *dissociative* acceptors in R4 because the inner complex is formed in a "dissociative acceptor action." This description is preferable in the present context to the usual "proton transfer" terminology, for the following reasons. The hydrogen atom in HX is initially far from being a free proton—the bonding is usually much more nearly covalent than ionic. When the donor molecule approaches, a more accurate description of the mechanism of the action between D: and HX than that implied by "proton transfer" is to picture an electron transfer out of the lone pair in D: to the X atom in HX forming $D\cdot^+$ and $X^-$, with simultaneous combination of the $D\cdot^+$ radical-ion with the $\cdot H$ atom to form $D^+$—H.* The advantage of this formulation is that it brings the donor action here and in $n \cdot v$ complexes like $R_3N \cdot BX_3$ under a single unified viewpoint.

In Fig. 14-8*a* and again in Fig. 14-8*b* charge-transfer transitions have been shown at two $R_{DH}$ values. One corresponds to a CT band of the outer complex, in which the electron is transferred in the CT transition from D to H—X, the other to a CT band of the inner complex, in which the electron is transferred from $X^-$ to the $(D—H)^+$ ion.† The CT band in the latter case would be analogous to the observed CT band of methylpyridinium iodide already discussed in Section 13-4; that ion pair (see footnote in Section 13-4) can be considered as an example of an inner complex of the *n* donor pyridine and the *aσ*-acceptor methyl iodide.

The C—I bond in methyl iodide is formally similar to hydrogen iodide and to $I_2$ as an *aσ* acceptor. Both $CH_3I$ and HI are examples of dissociative acceptors; although $I_2$ can also function as a dissociative acceptor, it is much more reluctant to do so, and usually remains at the intermediate state of a merely *sacrificial* acceptor. For HI and $CH_3I$ and the like, the sacrifice goes to completion.‡

---

* The usually fairly small positive charge on the hydrogen atom in HX probably does increase *somewhat* during the course of the transfer, but at no point does it approach +1.

† Actually, no well-defined example of a CT band of a typical hydrogen-bonded complex has yet been found. The reason most probably is that such a band would occur only at wavelengths that are well below 2000 Å, where overlapping locally excited transitions make isolation or identification difficult. [However, see S. Nagakura, *J. chim. phys.*, **61**, 217 (1964).]

‡ However, in the resulting ion pair there is a little back charge transfer (more in some cases). One could then say that the covalent H—I (or $CH_3$—I, or the like) bond is not quite completely broken. The situation is rather similar to that in lithium fluoride. (See Chapter 13.)

For the iodine complexes of triphenylarsine and triphenylstibine, Bhat and Rao [1] have studied the kinetics of transformation of the strong outer complexes to inner complexes:

$$(C_6H_5)_3M + I_2 \rightleftharpoons \underset{\text{outer complex}}{(C_6H_5)_3M \cdot I_2} \xrightarrow{\text{slow}} \underset{\text{inner complex}}{(C_6H_5)_3M^+I\ I^-}$$

$M$ = As or Sb

$$(C_6H_5)_3M^+I\ I^- + I_2 \rightleftharpoons (C_6H_5)_3M^+I\ I_3^-$$

Let us consider some further examples of inner and outer complexes. When dimethyl ether ($Me_2O$) and HCl are mixed in the gas phase, the pressure drops considerably, indicating a strong interaction to give, presumably, the gaseous outer complex $Me_2O \cdot HCl$. In the liquid state, evidence from Raman spectra indicates the presence of the hydrogen-bonded complex when a 1:1 mixture is used, perhaps $Me_2OH^+HCl_2^-$ for a 1:2 mixture, and increasingly conducting species when HCl is in larger excess [2a]. In $Me_2OH^+HCl_2^-$, the extra HCl molecule would act as an auxiliary acceptor helping to stabilize the inner complex. Lawley and Sutton in dielectric constant studies on $Me_2O$ + HCl vapor [2b] found an $Me_2O \cdot HCl$ dipole moment larger by 0.88D than the sum of the moments of $Me_2O$ and HCl; or else a fraction of the associated molecules must have ion-pair structure $Me_2OH^+Cl^-$. They preferred the latter explanation, although one does not expect $Me_2O$ to be so strong a donor. However, this system is apparently an example in which both outer and inner complexes are found, depending on environment.

Now let us consider the interaction between $NH_3$ and HCl (see also Section 12-8). Unlike the $Me_2O$-HCl mixture, when gaseous $NH_3$ and HCl are mixed, solid $NH_4^+Cl^-$ forms immediately as a white cloud. Conversely, when solid $NH_4^+Cl^-$ is sublimed, it is reported in the literature that one gets essentially a gaseous mixture of $NH_3$ and HCl—although it might be expected that a few molecules of outer and/or inner complex would also be present in equilibrium (see R4 for references).

$$NH_3(g) + HCl(g) \rightleftharpoons \begin{bmatrix} NH_3 \cdot HCl \\ \\ [\text{or } NH_4^+Cl^-] \end{bmatrix} \rightleftharpoons [NH_4^+Cl^-]\,(\text{solid}).$$

According to R4 and R14 the potential surface for a single pair of $NH_3$ and HCl molecules is expected to look like that shown in Fig. 14-8*a*, with the inner complex ($NH_4^+Cl^-$) less stable than the outer complex. However, if inner complex $NH_4^+Cl^-$ ion pairs form, they must be expected to interact electrostatically with other ion pairs, resulting in considerable stabilization. Then several of these pairs could cluster together to form solid $NH_4^+Cl^-$ crystals. The electrostatic attraction energies between these different ion pairs give what is called the Madelung energy of the crystal; the crystal may be considered to be a *polymer* of the inner complex, stabilized in that form (instead of in the outer complex) because of the Madelung energy.

However, after the lectures on which the present book is based had been given, Clementi [3a,b] made in 1966 an approximate all-electron SCF whole-complex-MO calculation on the interaction of a single molecule of HCl with one of $NH_3$ at varying distances of approach of the HCl along the threefold axis of the $NH_3$ molecule. Contrary to what was supposed in R4 and R14, the calculated potential-energy curve is not like Fig. 14-8*a*, but of the type of Fig. 14-6*b* (no stable outer complex, but an inner complex of very considerable stability). As we have already noted in Section 12-2, the energy computed by the whole-complex method should be a good approximation for complexes composed of closed-shell molecules all the way from small $R_{DA}$ values out to dissociation. Hence it would seem that the form of the $W(R_{DA}, Q_N)$ curve from Clementi's calculation is very probably at least qualitatively correct. As a matter of fact, one can now see that a curve such as Fig. 14-6*b* should have been *expected* for $NH_3$ + HCl since, for example, ion pairs are known to be formed in the interaction between the relatively weak acid $CH_3COOH$ and the strong base $N(C_2H_5)_3$ in the inert solvent $CCl_4$ [4]. Even with alcohols, which definitely form hydrogen-bonded outer complexes with $N(C_2H_5)_3$, there are indications that a somewhat higher energy minimum exists corresponding to an inner complex [5a], as in Fig. 14-8*a*. Hence the interaction between the strong acid HCl with $NH_3$ should be expected to give the inner complex as a stable form even more readily than do these weaker HQ acids.

In a very interesting paper Bauge and Smith [5b] review earlier evidence by Maryott and others, and present new evidence, concerning the dipole moments of D·HQ pairs in inert solvents. (D is of the type $R_3N$.) They conclude that "the salts of the strongest acids [their dipole moments are of the order of 12D] exist essentially in the form of ion-pairs in which there is probably hydrogen bonding between the cation and anion." Clementi's calculations on $NH_3$ + HCl correspond to this description for the equilibrium state of this pair. Bauge and Smith conclude further "The salts of the weakest acids probably exist as hydrogen-bonded complexes ... The salts of acids with $pK_a$ values, in aqueous solution, between 2 and 4 have dipole moments which are intermediate between the values expected for these extreme states. Preliminary evidence from infrared spectroscopy (see Bell and Barrow [5a]) suggests that in some of these cases the two forms may coexist in equilibrium."

The above comment that in ion pairs (inner complexes) of the type of $R_3NH^+Q^-$ there is hydrogen bonding between the anion (as *n* donor) and the cation (as *a*$\sigma$ acceptor) is of interest. This structure may be characterized as *inner* complex hydrogen bonding, as contrasted with the usual *outer* complex hydrogen bonding. If a hydrogen-bonded outer complex is characterized by $b$ small, $a$ near 1, in (2-2), the corresponding inner complex is expected to have $b$ rather near 1 and $a$ rather small but not close to zero.†

Since no stable hydrogen-bonded outer complex is predicted for the $NH_3 + HCl$ system, the inner complex is directly in equilibrium with $NH_3 + HCl$ as indicated in the equation above for the second route postulated for the intermediate. Calculations by Clementi and Gayles [3c] support this possibility. They indicate that in the hot vapor in equilibrium, a small percentage of the $NH_4^+Cl^-$ inner complex is present in equilibrium with predominant amounts of $NH_3 + HCl$. In recent experiments, Goldfinger and Verhaegen[3d] confirm this prediction.

An apparently very similar case [5c] is that of the anilinium chloride ion pair in the low-dielectric-constant solvent chloroform, which apparently is in equilibrium not with ions (as would be expected in a high-dielectric-constant solvent) nor with the hydrogen-bonded outer complex, but with aniline + HCl.‡

The behavior of $H_2O + HCl$ in the vapor state is more nearly like the ether · HCl system than like the $NH_3 \cdot HCl$ system. However, in solution one finds solvated $H_3O^+$ and $Cl^-$ ions; in the solid a monohydrate of HCl exists. Ferriso and Hornig [5d] summarize evidence supporting an $H_3O^+Cl^-$ structure for the monohydrate and give the infrared spectrum.

Stabilization of the inner complex as a result of the lowering of $W_1$ by the Madelung energy in the formation of ionic crystals is one example of the effect of environment in bringing about even-ionogenic action. Another example is provided by solvent stabilization of the ions when the system is dissolved in a polar solvent. The polar solvent molecules surround each ion or ion-pair in a sheath of molecules held by ion-dipole and CT interactions. They stabilize the inner complex, if it is an ion pair, for example $H_3O^+Cl^-$, but also break it up into separated solvated ions, here $H_3O^+(aq) + Cl^-(aq)$ if the solvent is water (see Section 15-3).

For $b\pi \cdot a\pi$ complexes usually only an outer complex is formed, but in some cases, with the help of a polar solvent or with the assistance of Madelung energy, an ion-pair inner complex or its polymer in the form of an ionic crystal is obtained. In all such cases it is an *odd-odd* ion pair —illustrating *odd-ionogenic* action. In our description of the interaction

---

† Actually *all* ions pairs should show at least some back charge transfer from the anion to the cation, but usually not as much as in the inner hydrogen-bonded case.

‡ Although aniline is a strong $\pi$-donor, it functions toward HCl as an $n$-donor, like an aliphatic amine.

of D with HX, the hydrogen atom with its electron transfers its bond to the D, which at the same time has become $D^+$, having given one of its electrons to the X to form the $X^-$ ion; in this way two even-electron ions are formed. But in $b\pi \cdot a\pi$ complexes no atom is free to migrate with its electron; hence the ions $D^+$ and $A^-$ resulting from the electron transfer from D to A are odd. These ionic $b\pi \cdot a\pi$ inner complexes are important solid organic semiconductors. (See Chapter 16 for further discussion.)

## 15-2 MIDDLE COMPLEXES

The $HCl_2^-$ ion, mentioned above, is probably linear and symmetrical like $HF_2^-$, and is an interesting example of a symmetrized CT complex. One way to understand the structure of this ion (or of $HF_2^-$ or of $I_3^-$, already discussed in Section 12-9) is as a *middle complex*, corresponding to a stable half-way point in a donor-acceptor chemical reaction of an $n$-donor with an $a\sigma$-acceptor:

$$Cl_a^- + HCl_c \rightarrow (Cl_aHCl_c)^- \rightarrow Cl_aH + Cl_c^-;$$

similarly

$$I_a^- + I_2 \rightarrow I_3^- \rightarrow I_2 + I_c^-. \tag{15-1}$$

Another example of a symmetrized complex is $BF_4^-$, regarded as being formed from, or dissociated into, an $n$-donor and a $v$-acceptor, $F^- + BF_3$:

$$F_a^- + BF_3 \rightarrow \left( \text{B with four F} \right) \rightarrow BF_3 + F_i^- \quad (i = b,\ c,\ \text{or } d). \tag{15-2}$$

In many chemical reactions there is no stable minimum along the reaction coordinate and the place of the middle complex in a reaction like those in (15-1) and (15-2) is taken by a configuration at the top of an energy barrier (sometimes regarded as an *activated complex*), as for example in the Finkelstein type of reaction†

$$\underset{A}{HO^- + CH_3I} \rightarrow \underset{B}{\left[ HO\text{---}\overset{\overset{\large H}{|}}{\underset{H\ \ H}{C}}\text{---}I \right]^-} \rightarrow \underset{C}{HOCH_3 + I^-}. \tag{15-3}$$

† Ingold classifies reactions like 15-3, also those like $R_3'N + RX \rightarrow R_3'N^+R\ X^-$ (here viewed as outer complex → inner complex), and many others in which charge transfer accompanies a bond shift, as nucleophilic substitution reactions.

Here in the activated middle complex the $CH_3$ group is presumably planar with the hydrogen atoms forming an equilateral triangle, or nearly so. The $CH_3$ then contains a carbon $2p$ AO that in valence-bond language resonates between bonding to OH (leaving the I as $I^-$) and to I (leaving the OH as $OH^-$); these two forms may be taken as $\Psi_0$ and $\Psi_1$ in our usual donor-acceptor resonance-structure description (2-2). As the reaction proceeds from A through B to C, the coefficients $a$ and $b$ of $\Psi_0$ and $\Psi_1$ change from $a = 1$, $b = 0$ to $a \simeq b$ (at B) to $a = 0$, $b = 1$ (at C). In MO language as applied just to the four active electrons in the middle complex, there are two in an HO—C—I bonding MO [$\simeq b(2p\sigma_O) + a(2p\sigma_C) + c(2p\sigma_I)$, with $a > b \simeq c$] and two in a nearly nonbonding MO [$\simeq (2p\sigma_O - 2p\sigma_I)/\sqrt{2}$] which is approximately half on the OH and half on the I.

## 15-3 ENVIRONMENTAL COOPERATIVE ACTION

Now let us refer again to the potential curves in Fig. 14-8*b*, here redrawn with some changes and additions as Fig. 15-1. This figure indicates the effect of a polar solvent in stabilizing the ionic inner complex by lowering the $W_1$ curve to $W_1'$ so that the energy curve $W_N$ is lowered to $W_N'$. This effect is an example of *environmental cooperative action*. It should be operative for *all* types of complexes, causing an increase in $b/a$ in (2-2) for outer complexes, and increasing the relative stability of ion-pair inner complexes versus outer complexes. (See also Fig. 2 of R4 for a D · HX complex.)

Now let us consider the effect of environmental cooperative action on the complex of a dissociative acceptor. This action may occur in either of at least two ways: (*a*) solvent assistance; (*b*) ionic crystal formation. We may usefully subdivide solvent assistance into action (1) by classical electrostatic forces; or (2) of the solvent as an auxiliary acceptor (or donor).

First consider what happens on the molecular level when an ion, for example $Cl^-$, is solvated [(*a*) above] by a polar solvent, for example water. In the region immediately surrounding the ion, the hydrogen-bonded structure of the solvent is broken up and the water molecules arrange themselves in a sheath around the ion. If the action were purely electrostatic [(1) above], the water molecules would be expected to arrange themselves so that the positive hydrogen atom† gets as close to the $Cl^-$ as possible. However, $Cl^-$ is a strong $n$-donor, and the inter-

† Theoretical calculations indicate that each hydrogen atom in water has a positive charge of about $0.19e$.

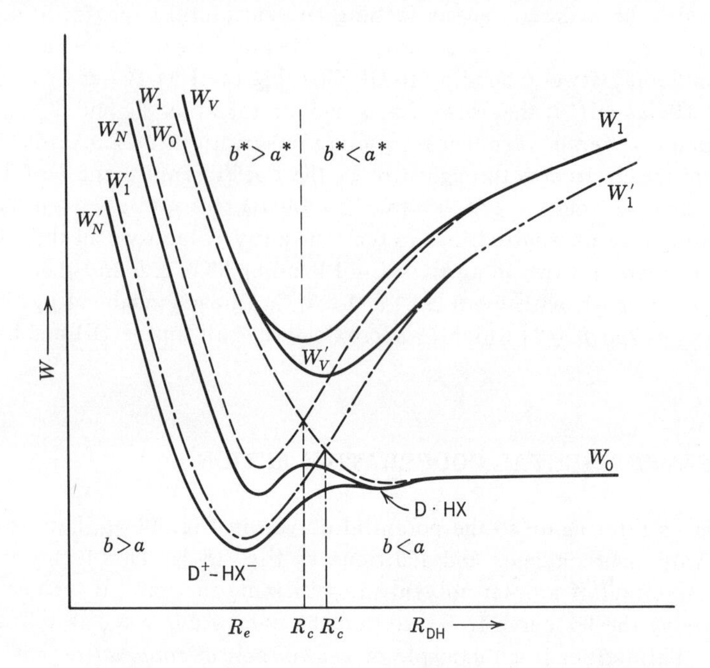


**Figure 15-1** Potential curves for inner and outer complexes, illustrating environmental cooperative action. Here $W_1$ and $W_N$ are the energies of the dative and ground state, respectively, of the isolated complex in the nonpolar solvent; $W_1'$ and $W_N'$ are corresponding energies in a polar solvent, showing the expected stabilization by the latter. The change in solvent is not expected to change the other curves appreciably, except for a change in $W_V$ near $R_c'$ to link it to the $W_1'$ curve rather than to $W_1$.

action with each $H_2O$ should be of more or less typical hydrogen-bonding character, involving charge-transfer [(2) above] as well as classical electrostatic forces. There is interesting infrared spectral evidence on the hydrogen bonding of water to various donors; in the case of $Cl^-$ ions at low concentrations in an inert solvent, it appears that only one hydrogen bond is formed per water molecule [6].

The resulting $H_2O$ sheath around the $Cl^-$ ion, with the negative O atoms on the outside, could then hydrogen bond again to the surrounding water molecules. The ion-dipole and charge-transfer forces between the $Cl^-$ and the inner sheath of $H_2O$ must result in a considerable energy of stabilization of the $Cl^-$ ion; hence if the $Cl^-$ ion is present in the dative state of a complex, $W_1$ is lowered in Fig. 15-1. In evaluating $W_1$, however, the energy involved in the making and breaking of hydrogen bonds of the solvent must also be taken into account.

Rather differently, in the solvation of a positive ion by $H_2O$, one would expect the O atoms, because of their partial negative charges, to arrange themselves next to the ion. In the case of the $NH_4^+$ ion, the solvating water should interact in a strongly hydrogen-bonding fashion. It is of interest that the acceptor MO for an $NH_4^+$ ion surrounded by water molecules would be rather nearly a $3s$ MO of $NH_4$, and the effect would somewhat resemble $n \cdot v$ action between $H_2O$ $n$ donor molecules and $NH_4^+$ as a $v$ acceptor. For the solvation of a positive ion like $Na^+$ by water molecules, there should again be $n \cdot v$-like donor-acceptor action as well as classical electrostatic attraction, but in this case of course no hydrogen bonding; the acceptor orbital would be $3s$ just as for $NH_4^+$.

The environmental cooperative action by ionic crystal formation is believed to be almost purely electrostatic, and results from the increase in Coulomb energy per ion pair in the ionic solid (the Madelung energy) as compared with an *isolated* ion pair, discussed in Section 15-1.

The solvent action discussed above has been in terms of completely separated solvated positive and negative ions. These still exert some attraction for each other through the polar solvent, an attraction that increases with decreasing dielectric constant. On sufficiently close approach the oppositely charged ions may form *solvent-separated ion pairs*, with only one solvent molecule separating them, or even *contact* or *intimate ion pairs* with no solvent between the ions. (The ion-pair inner complexes of Section 15-1 and the ion pairs of Section 13-4 which show CT spectra, are of course contact ion pairs.) Clearly the occurrence of these three possibilities is governed, basically, by the energies of solvation of the ions and the energies of attraction between them. In the contact ion pair, as we have seen in Chapter 13, the attractive energies may include appreciable charge-transfer stabilization. Although relatively little is as yet definitely known, it has been claimed that different spectra have been found for the above-mentioned three stages of ion-pair formation [7].

In cases where Kosower and co-workers have found CT spectra of ion-pair inner complexes, they have also observed large increases in $h\nu_{CT}$ found in going from a less polar to a more polar solvent (see Chapter 13 for references). These shifts are obviously explained by the lowering of $W_1$, hence of $W_N$, through solvation of the ion pair by the polar solvents, while at the same time $W_V$, being largely nonionic like $W_0$, is nearly independent of the solvent.

In passing, one might consider varying $W_1$ continuously by varying the dielectric constant of the medium continuously, using solvent mixtures such as dioxane and water. As indicated above, however, the interaction between solvent and ions must be considered on the molecular level and is more complicated than a simple dielectric effect in a continuous medium. Hence in mixed solvent effects no simple correlation with the dielectric constant need be expected.

## 15-4 INTERACTION BETWEEN HX AND $b\pi$-DONORS

The prototypes of the two major varieties of $b\pi$-donors are ethylene for the unsaturated donors (Un) and benzene for the aromatic donors (Ar). Stronger $b\pi$-donors are derived from these prototypes when they are fortified by electron-donating substituents.

There is good evidence that aromatic donors form moderately strong hydrogen-bonded complexes—for example, with HCl and with alcohols. Benzene and other liquid aromatic hydrocarbons dissolve HCl readily, indicating interaction, and the HX stretching frequency (X = halogen atom) then decreases appreciably. Studies of the solubility of HX in heptane solutions of the aromatic molecule, of varying concentration and at varying temperatures, showed that there is an equilibrium between free HX and HX hydrogen bonded to benzene or other aromatic, and yielded $K$ and $\Delta H$ values [8]. Infrared spectra of solutions of HX and an aromatic molecule both dissolved in an inert solvent are also instructive. Values illustrating the frequency shifts found for HCl and some alcohols in solutions with different donors are presented in Table 15-1. We see there, for example, that $\nu_{HCl}$ decreases by 81 $cm^{-1}$ from the value for HCl alone in carbon tetrachloride for the case of HCl with benzene in carbon tetrachloride; for mesitylene the decrease is 119 $cm^{-1}$ [9]. These shifts are small compared with those for hydrogen bonding in ethers (380 $cm^{-1}$ for HCl in diethyl ether); still they are larger than (and additional to) ordinary "solvent shifts"—such as the decrease of 55 $cm^{-1}$ in $\nu_{HCl}$ observed from gas to solution in carbon tetrachloride. (See [10] for further discussion.)

Even $\nu_{OH}$ in $CH_3OH$ shifts when methyl alcohol is dissolved in aromatics. (See Table 15-1, and also [11].) Basila et al. [12] have examined the infrared spectra of $t$-butyl alcohol in methylated benzene donors. They deduce $\Delta k/k$ values ($k$ = force constant for O—H stretching) ranging from 0.02 for toluene to 0.04 for hexamethylbenzene and discuss these in terms of a small CT resonance interaction (see Chapter 6); classical electrostatic attraction should also contribute to these frequency shifts.

## TABLE 15-1

**Frequency Shifts of the X—H (or X—D) Stretching Vibration for HCl, $CH_3OD$, and *n*-BuOH Solutions in $CCl_4$ with Some Electron Donors**

| Donor | HCl | | $CH_3OD$ | | *n*-Butyl Alcohol | |
|---|---|---|---|---|---|---|
| | $\nu$ | $\Delta\nu_s$ | $\nu$ | $\Delta\nu_s$ | $\nu$ | $\Delta\nu_s$ |
| (None) gas | 2886[a] | – | 2719[c] | – | 3680[f] | – |
| (None) $CCl_4$ | 2831[a] | – | 2689[c] | – | 3640[f] | – |
| Chlorobenzene | 2779[a] | 52 | 2668[c] | 21 | 3616[f] | 24[f] |
| Benzene | 2750[a] | 81 | 2665[c] | 24 | 3606[f] | 34[f] (27)[g] |
| Toluene | 2744[a] | 87 | 2663[c] | 26 | 3600[f] | 40[f] (35)[g] |
| *m*-Xylene | 2723[a] | 108 | 2660[c] | 29 | – | – (40)[g] |
| Nitrobenzene | 2718[b] | 113 | 2653[d] | 36 | – | – |
| Mesitylene | 2712[a] | 119 | 2655[c] | 34 | 3590[f] | 50[f] (49)[g] |
| Diethylether | 2451[b] | 380 | 2593[e] | 96 | – | – |
| *p*-Dioxane | 2469[b] | 362 | 2578[e] | 111 | – | – |

*Note:* The quantity $\Delta\nu_s$ is defined as $\nu_{CCl_4} - \nu_{CCl_4+D}$. It thus represents the *additional* shift in frequency because of the interaction with D. We note that there is *already* a shift from the frequency of the free molecule in the gas phase to that in $CCl_4$ solution of 55 $cm^{-1}$ for HCl, 30 $cm^{-1}$ for $CH_3OD$, and 40 $cm^{-1}$ for *n*-butyl alcohol.

[a] J.-P. Leicknam, J. Lascombe, N. Fuson, and M. L. Josien [9].
[b] W. Gordy and P. C. Martin, *J. Chem. Phys.*, **7**, 99 (1939). The uncertainty in $\nu$ may be as high as $\pm 25$ $cm^{-1}$.
[c] From M. Tamres, *J. Am. Chem. Soc.*, **74**, 3375 (1952), except for gas-phase value which is from G. Herzberg, *Molecular Spectra and Molecular Structure. II. Infrared and Raman Spectra of Polyatomic Molecules*, Van Nostrand, 1945, p. 335.
[d] W. Gordy, *J. Chem. Phys.*, **7**, 93 (1939). The uncertainty in $\nu$ may be as high as $\pm 25$ $cm^{-1}$.
[e] S. Searles and M. Tamres, *J. Am. Chem. Soc.*, **73**, 3704 (1951).
[f] M. L. Josien and G. Sourisseau, *Hydrogen Bonding* (D. Hadzi, ed.), Pergamon, New York, 1959, p. 129.
[g] Values in parentheses are for *t*-butyl alcohol in $CCl_4$ solution, from M. Basila, E. L. Saier, and L. R. Cousins [12].

These examples of hydrogen-bonded interaction between HX and $b\pi$ donors have all been weak outer complexes, even for HCl. This situation is to be contrasted with the behavior of HCl with $n$ donors, where, as we have seen, the tendency to form an inner complex (at least when environmental cooperation is present) is strong. However, under rather extreme conditions of environmental cooperative action, it is possible to obtain inner complexes also for $b\pi$-donors; for example, consider the reactions:

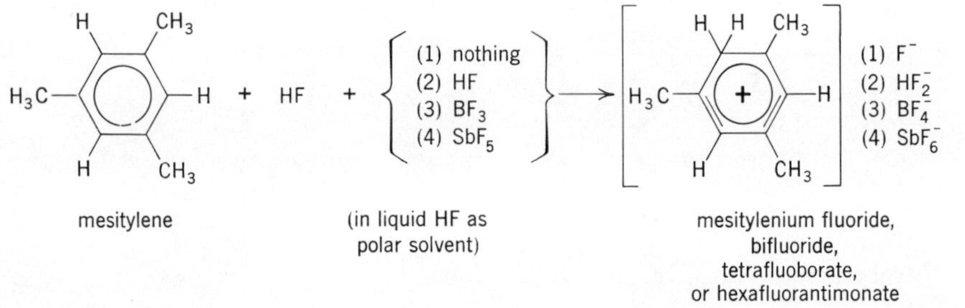

$$\text{mesitylene} + HF + \left\{\begin{array}{l}(1)\ \text{nothing}\\(2)\ HF\\(3)\ BF_3\\(4)\ SbF_5\end{array}\right\} \longrightarrow [\text{mesitylenium}]^{+}\begin{array}{l}(1)\ F^-\\(2)\ HF_2^-\\(3)\ BF_4^-\\(4)\ SbF_6^-\end{array}$$

mesitylene

(in liquid HF as polar solvent)

mesitylenium fluoride, bifluoride, tetrafluoborate, or hexafluorantimonate

Reactions like the above do not proceed far toward completion without environmental cooperation. However, they do go, more or less, in a polar medium, especially if aided by an auxiliary acceptor [HF in (2), or better, $BF_3$ in (3) or $SbF_5$ in (4)] to stabilize the negative ion formed in the reaction. Fortification of benzene by methyl groups greatly increases the ease with which an inner complex can be formed. With benzene itself, the strongest auxiliary acceptors are necessary. McCaulay and Lien [13] show that the formation constant for the HF inner complex with toluene is only about 0.01 that for $p$-xylene, whereas that for hexamethylbenzene is 45,000 times that for $p$-xylene. (See also Ehrenson [14].)

Consider now the ethylenic type of donor. The reaction with HF would yield[†]

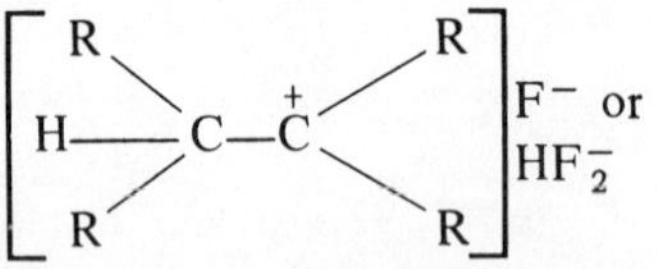

$$\left[\begin{array}{c}R\quad\quad\quad R\\ H{-}C{-}\overset{+}{C}\\ R\quad\quad\quad R\end{array}\right] F^- \text{ or } HF_2^-$$

In thinking about a carbonium ion salt such as this it is useful first to ask about potential curves for dissociation of ethyl chloride, comparing

[†] A bridge structure with the hydrogen midway between the two carbons is conceivable, but on the basis of organic chemical evidence it apparently is less stable

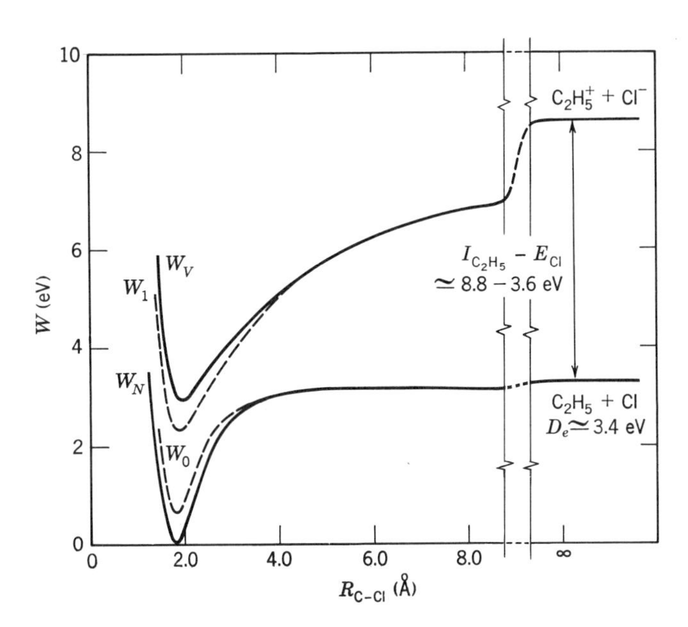


**Figure 15-2** Schematic potential curves for the interaction between $C_2H_5$ and Cl, showing the stabilization of the ground state by ionic-covalent resonance.

them with those for NaCl ($W_1 < W_0$ at $R_e$, see Fig. 14-5) and for HCl ($W_1 > W_0$ at all $R$). Potential curves qualitatively like those for HCl are indicated schematically for $C_2H_5$ + Cl in Fig. 15-2. These show that, although the ground state of $C_2H_5Cl$ is stabilized by ionic-covalent resonance, the C—Cl bond is predominately covalent. This fact suggests that, although an ion-pair inner complex tends to be formed in the interaction of an olefinic compound with HX, the stable state is the simple covalently bonded addition product, unless the ethylene is very strongly fortified (see Section 15-4) and/or given extremely strong environmental cooperation, including an auxiliary acceptor or a highly acidic HX (for example, $HClO_4$).

Hydrogen-bonded outer complexes with olefins as $b\pi$-donors are also known but have not been extensively studied. West has examined the infrared spectra of carbon tetrachloride solutions of the $a\sigma$-acceptor phenol plus various olefins, and for the hydrogen-bonded complexes finds shifts from the sharp OH absorption for phenol at about 3610 $cm^{-1}$ in an inert solvent to a broad absorption for the phenol hydrogen bonded to the olefin. Typical frequency decreases are 62 $cm^{-1}$ for 1-octene, 95 $cm^{-1}$ for cyclohexene, and 113 $cm^{-1}$ for 1-methylcyclohexene [15]. These decreases he finds to be rather larger than with comparably alkylated benzenes as donors; for example, with benzene, the decrease is 47 $cm^{-1}$; with mesitylene, 73 $cm^{-1}$.

The hydrogen-bonded complexes so far discussed are all intermolecular, but intramolecular hydrogen-bonded complexes of various kinds are also familiar [10]. Of particular interest in the present connection are the intramolecular hydrogen bonds between phenolic OH acceptor groups and olefinic $\pi$-electron pairs as donors in compounds of the type of

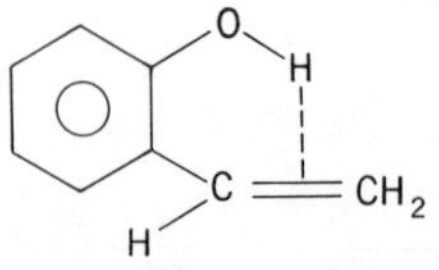

2-allylphenol (see above) [16]. These show shifts for the O—H stretching frequency of the same order of magnitude as those in the intermolecular cases studied by West.

West and Kraihanzel have also made a study of the hydrogen-bonding properties of acetylene derivatives [17]. The 1-alkynes as $b\pi$ electron donors with phenol as $a\sigma$-acceptor show a decrease of 90 $cm^{-1}$ for the O—H stretching band of the phenol, as compared to the phenol frequency alone in the same solvent, indicating that they are somewhat stronger donors than the 1-alkenes or the mono- or dimethyl benzenes. When the terminal hydrogen in a 1-alkyne is replaced by a second alkyl group, decreases of about 135 $cm^{-1}$ are found.

On the other hand, the alkynes can also act as *$a\sigma$-acceptors* in hydrogen bonding. The existence of hydrogen bonding for the acetylenes in basic organic solvents affords an explanation of their marked solubility in these. Further, decreases in the range 60 to 94 $cm^{-1}$ for the CH stretching band near 3315 $cm^{-1}$ of the 1-alkynes and other acetylenes were found when N,N-dimethylformamide was used as electron donor [17].

A great many additional papers have reported shifts in X—H stretching frequencies on hydrogen bonding. (See Pimentel and McClellan [10], for a more complete summary. Also see recent papers by Allerhand and Schleyer [18].)

## 15-5 OTHER ORGANIC IONS

Let us now digress for a moment to consider $(C_6H_5)_3CCl$. Here the C—Cl bond is weak, and if this compound is dissolved in liquid sulfur dioxide, it ionizes in part to give

$$(C_6H_5)_3CCl + SO_2 \underset{SO_2}{\rightleftharpoons} (C_6H_5)_3C^+ + Cl^-SO_2.$$

Conductivity studies of this system were made in a series of papers by Lichtin [19].

For $\phi_3C$—Cl ($\phi = C_6H_5$), $W_1$ is evidently considerably lower relative to $W_0$ than for $C_2H_5Cl$, presumably because the C—Cl bond is itself quite a bit weaker. Hence the environmental cooperative action of the liquid sulfur dioxide on $\phi_3C$—Cl must lower $W_1$ until it crosses $W_0$ so that $W_N$ has largely ion-pair character $(C_6H_5)_3C^+Cl^-$, or perhaps $(C_6H_5)_3C^+(ClSO_2)^-$, and dissociates in part according to the equilibrium expressed in the equation above. Here the solvent sulfur dioxide acts as an auxiliary acceptor. It is, however, not clear whether $Cl^-SO_2$ is an outer complex conforming to (2-2), or more nearly a normal valence compound

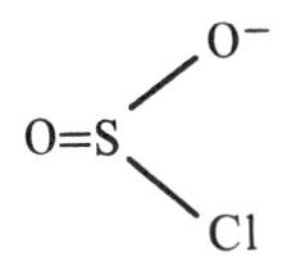


Now let us return to the ethylenic type of donor. Some very interesting compounds with strongly donor-fortified ethylene molecules have been studied by Wizinger; he has summarized his work in an article [20] which is a veritable Thanksgiving feast of information and ideas. He fortified the donor action of the ethylene by replacing part or all of the hydrogens by $p$-$Me_2NC_6H_4$ or by $p$-$MeOC_6H_4$ groups. The resulting donors were so strong that with halogen acceptors (mostly $Br_2$ and $I_2$) the action went beyond the usual limit of outer complex formation, and resulted in interesting inner complexes. One example is shown on the next page. Here inner-complex formation is somewhat assisted by the auxiliary acceptor action of a second $Br_2$ to convert $Br^-$ into $Br_3^-$.

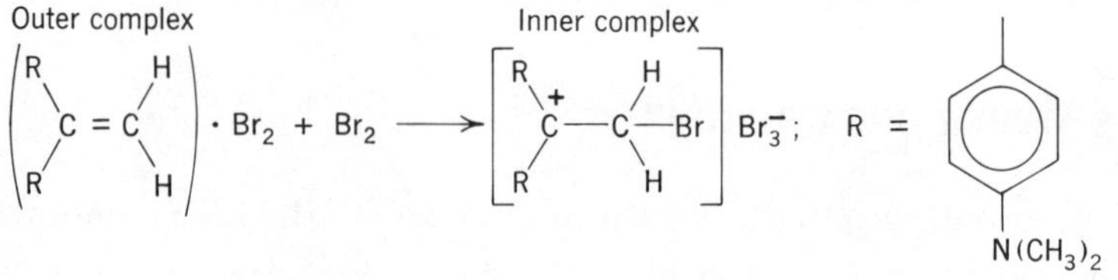

Outer complex — Inner complex

$$\left(\begin{matrix}R & & H\\ & C=C & \\ R & & H\end{matrix}\right)\cdot Br_2 + Br_2 \longrightarrow \left[\begin{matrix}R & & H & \\ & \overset{+}{C}-C-Br & \\ R & & H & \end{matrix}\right] Br_3^-;\quad R = p\text{-}C_6H_4N(CH_3)_2$$

With the tetrasubstituted ethylene as donor, Wizinger reports that he obtains the *dication* salt (like a $Mg^{2+}$ salt):

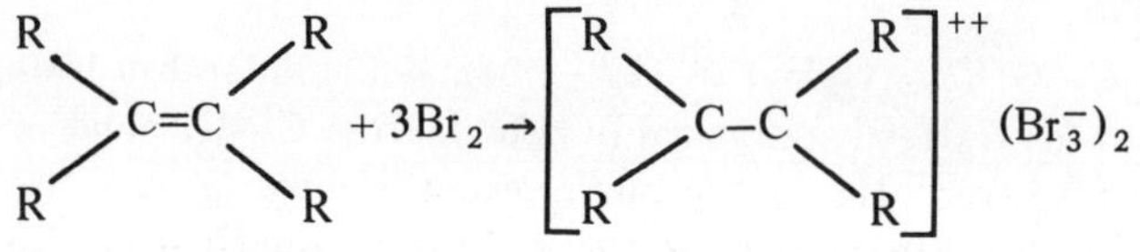

$$R_2C{=}CR_2 + 3Br_2 \rightarrow \left[R_2C{-}CR_2\right]^{++}(Br_3^-)_2$$

R = as above

He called these ionic compounds "carbenium salts." They exist in polar solvents and as crystals.

Wizinger's studies also are intimately connected with the mechanism of the substitution and addition reactions of halogens with unsaturated and aromatic molecules. It is of interest to note here that the normal reaction of halogens with aromatic systems is substitution, for example

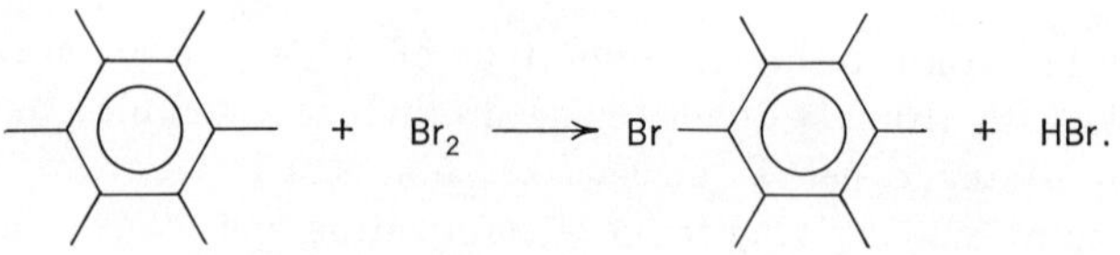

$$C_6R_6 + Br_2 \longrightarrow Br{-}C_6R_5 + HBr.$$

On the other hand, the normal reaction with unsaturated compounds is addition, for example

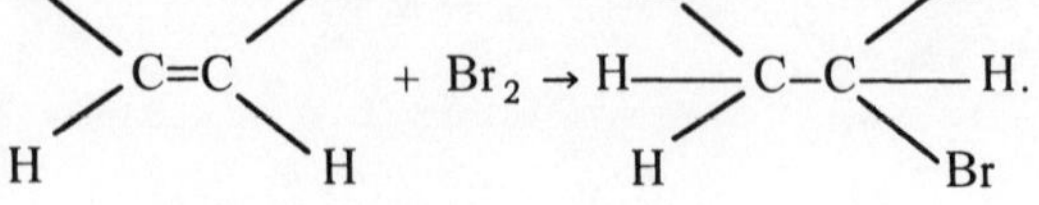

$$H_2C{=}CH_2 + Br_2 \rightarrow H{-}C(Br)(H){-}C(H)(Br){-}H.$$

Further important contributions towards understanding the chemistry of these systems have been made by H. M. Buck [21] (ESR and ultraviolet studies), and by Buckles et al., [22] who have studied the outer complex and the further rearrangement reactions of the monocation, and have verified Wizinger's postulated dication. Some additional examples where $I_2$ functions as a dissociative acceptor are discussed in Section 16-3.

## 15-6 NITRATION REACTIONS

Let us now continue with the discussion of chemical reactions that may involve donor-acceptor action, and consider the nitration of benzene in a strong nitric-sulfuric acid mixture:

$$HONO_2 + Bz \xrightarrow[(H_2SO_4)]{} C_6H_5NO_2 + H_2O$$

As shown by Ingold and co-workers, [5] the nitric acid is converted to solvated $NO_2^+$ by the reaction

$$H_2SO_4 + HONO_2 \rightarrow H_2NO_3^+HSO_4^- \rightarrow (NO_2)^+(HSO_4)^- + H_2O.$$

A plausible cross section of the reaction surface is sketched in Fig. 2 of R6 as a function of a reaction coordinate (see also [23]) which at first can be identified with $R_{DA}$, the distance between $NO_2^+$ and Bz. The first stage is the formation of a $b\pi \cdot v$ outer complex between Bz and $NO_2^+$. As $R_{DA}$ decreases, an inner complex is reached, perhaps over an activation barrier.† The inner complex should be, (see below) analogous to

$$\left[ C_6H_6NO_2 \right]^+ \quad (\text{ring carbon bearing } NO_2 \text{ and } H)$$

† This assumes that the inner complex corresponds to an elevated *minimum* in the potential surface, but obviously there are also other possibilities.

$BzH^+$ discussed in Section 15-4.† This "activated" inner complex $C_6H_6NO_2^+$ is an intermediate in the reaction,‡ acting as a dissociative HQ acceptor by getting together with an $n$-donor—for example, $HSO_4^-$—to yield the products

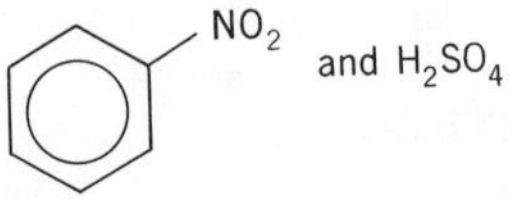

$NO_2$ and $H_2SO_4$

The question of the geometrical structure of the $C_6H_6NO_2^+$ inner complex is of some interest. In the benezenium ion (see below) one hydrogen is be-

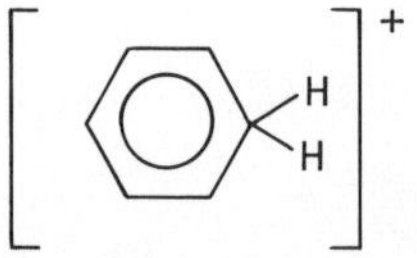

H
H
+

lieved to be above and the other below the plane. Molecular orbitals for this ion can be constructed from the usual $\sigma$- and $\pi$-atomic orbitals of the ring together with two group orbitals of the $H_2$ group, namely a $[\sigma]$, that is to say a quasi-$\sigma$, orbital $(1s_{H_1} + 1s_{H_2})/\sqrt{1 + S_{12}}$ that is to be linked with the $\sigma$-orbital of the adjacent C atom, and a $[\pi]$, that is quasi-$\pi$, group orbital $(1s_{H_1} - 1s_{H_2})/\sqrt{1 - S_{12}}$ which functions like a $\pi$-orbital in linear combinations with the $\pi$-orbitals of the ring carbon atoms in such a way as to yield a partial restoration of the aromatic $\pi$-resonance of neutral benzene. The plus charge (absence of one electron) in the benzenium ion is distributed, probably about equally, over the $H_2$ $[\pi]$ orbital and the two ortho and one para $2p\pi$ orbitals of the ring; this makes the total $\pi$-resonance energy less than if the positive charge were not there but greater than if the positive charge were confined to the ring [24]. This phenomenon is an example of *isovalent hyperconjugation*. The nitrobenzenium ion activated intermediate discussed above should likewise be stabilized by isovalent hyperconjugation [25], and the H and the $NO_2$ groups should be above and below the ring plane. Incidentally, $NO_2^+$,

† The outer and inner complexes can also be considered as $Bz \cdot NO_2^+HSO_4^-$ and $(C_6H_6NO_2)^+(HSO_4)^-$, but the role of the $HSO_4^-$ is a passive one until the final stage of the reaction.

‡ See footnote on page 247.

which is isoelectronic with $CO_2$, is linear—even though neutral $NO_2$ is nonlinear, as is the $NO_2$ group in $HNO_3$, in nitrobenzene, and doubtless also in the intermediate inner complex $C_6H_6NO_2^+$. Thus we see that considerable rearrangements in geometrical structure must occur in the $NO_2$ group at various stages of the interaction reaction.

## 15-7 REFERENCES


[1] S. N. Bhat and C. N. R. Rao, *J. Am. Chem. Soc.* **88**, 3216 (1966).
[2] (a) G. L. Vidale and R. C. Taylor, *J. Am. Chem. Soc.*, **78**, 294 (1956); (b) K. P. Lawley and L. E. Sutton, *Trans. Faraday Soc.*, **59**, 2680 (1963).
[3] (a) E. Clementi, *J. Chem. Phys.*, **46**, 3851 (1967); (b) ibid., **47**, 2323 (1967); (c) E. Clementi and J. N. Gayles, *J. Chem. Phys.*, **47**, 3837 (1967); (d) P. Goldfinger and G. Verhaegen, *J. Chem. Phys.*, **50**, 1467 (1969).
[4] G. M. Barrow and E. A. Yerger, *J. Am. Chem. Soc.*, **76**, 5211 (1954).
[5] (a) C. L. Bell and G. M. Barrow, *J. Chem. Phys.*, **31**, 300 (1959); (b) K. Bauge and J. W. Smith, *J. Chem. Soc.*, 616 (1966); (c) M. M. Davis, *J. Am. Chem. Soc.*, **71**, 3544 (1949); (d) C. C. Ferriso and D. F. Hornig, *J. Chem. Phys.*, **23**, 1464 (1955).
[6] S. C. Mohr, W. D. Wilk, and G. M. Barrow, *J. Am. Chem. Soc.*, **87**, 3048 (1965).
[7] T. R. Griffiths and M. C. R. Symons, *Mol. Phys.*, **3**, 90 (1960); M. J. Blandamer, T. E. Gough, T. R. Griffiths, and M. C. R. Symons, *J. Chem. Phys.*, **38**, 1034 (1963).
[8] For example, see H. C. Brown and J. D. Brady, *J. Am. Chem. Soc.*, **74**, 3570 (1952); H. C. Brown and J. J. Melchiore, ibid., **87**, 5269 (1965) and references given there.
[9] J. P. Leicknam, J. Lascombe, N. Fuan, and M. L. Josien, *Bull. Soc. Chim. France*, 1516 (1959).
[10] G. C. Pimentel and A. L. McClellan, *The Hydrogen Bond*, Freeman, San Francisco, 1960, p. 91.
[11] L. H. Jones and R. M. Badger, *J. Am. Chem. Soc.*, **73**, 3132 (1951).
[12] M. R. Basila, E. L. Saier, and L. R. Cousins, *J. Am. Chem. Soc.*, **87**, 1665 (1965) .
[13] D. A. McCaulay and A. P. Lien, *J. Am. Chem. Soc.*, **73**, 2013 (1951).
[14] S. Ehrenson, *J. Am. Chem. Soc.*, **83**, 4493 (1961).
[15] R. West, *J. Am. Chem. Soc.*, **81**, 1614 (1959).
[16] A. W. Baker and A. T. Shulgin, *J. Am. Chem. Soc.*, **80**, 5358 (1958); ibid., **81**, 1523 (1959); ibid., **81**, 4524 (1959).
[17] R. West and C. S. Kraihanzel, *J. Am. Chem. Soc.*, **83**, 765 (1961).
[18] A. Allerhand and P. v. R. Schleyer, *J. Am. Chem. Soc.*, **85**, 371, 866, 1233, 1715 (1963).
[19] N. N. Lichtin and P. D. Bartlett, *J. Am. Chem. Soc.*, **73**, 5530 (1951); N. N. Lichtin and H. Glazer, ibid., **73**, 5537 (1951); N. N. Lichtin and M. J. Vignale, ibid., **79**, 579 (1957); N. N. Lichtin, P. E. Rowe, and M. S. Puar, ibid., **84**, 4259 (1962).
[20] R. Wizinger, *Chimia*, **7**, 273 (1953).

[21] H. M. Buck, Ph.D. thesis, University of Leyden, 1959.
[22] R. E. Buckles and W. D. Womer, *J. Am. Chem. Soc.*, **80**, 5055 (1958); R. E. Buckles, R. E. Erickson, J. D. Snyder, and W. B. Person, ibid., **82**, 2444 (1960).
[23] S. Nagakura, *Tetrahedron,* **19**, 361 (1963).
[24] N. Muller, L. W. Pickett, and R. S. Mulliken, *J. Am. Chem. Soc.*, **76**, 4770 (1954).
[25] K. Fukui, T. Yonezawa, and C. Nagata, *J. Chem. Phys.*, **27**, 1247 (1957) and subsequent papers.

*CHAPTER* **16**

# *Inner and Outer Complexes with aπ-Acceptors*

## 16-1 $b\pi \cdot a\pi$ INNER COMPLEXES

Most of the familiar $b\pi \cdot a\pi$ complexes are weak outer complexes with potential curves that qualitatively resemble Fig. 14-6*a* but with a shallower minimum in $W_N$ (that is, a considerably smaller $-\Delta H$ and $\Delta E_f$). However, in certain interesting cases with exceptionally small $I_D - E_A$ there apparently exists, at least for solutions in sufficiently polar solvents,† a situation like that shown in Fig. 16-1. Here it is important to keep in mind again that the energies are functions not only of the distance $R_{DA}$ but also to some extent of numerous other coordinates $Q$ (see Chapter 9). Let the values of these coordinates which minimize the energy for the no-bond structure $\Psi_0$ be called $Q_0$. For the ionic (dative) structure $\Psi_1$ a different set of values of the coordinates, say $Q_1$, minimizes the energy. When $W_1$ crosses $W_0$ the nature of $\Psi_N$ for the complex changes from predominantly $\Psi_0$ to predominantly $\Psi_1$. Hence the coordinates $Q_N$ that minimize $W_N$ likewise change from values close to $Q_0$ to values close to $Q_1$. As this change occurs the $W_N$ energy curve that, as in earlier figures, is supposed to be $W_N(R_{DA}, Q_N)$ at all $R_{DA}$ values

† The term "polar solvent" is intentionally somewhat vague. The effects considered in this chapter (see Section 15-3 for some relevant discussion) are to a very considerable effect specific to the particular solvent but are roughly correlated with such quantities as the dielectric constant of the solvent or the dipole moment of its molecules.

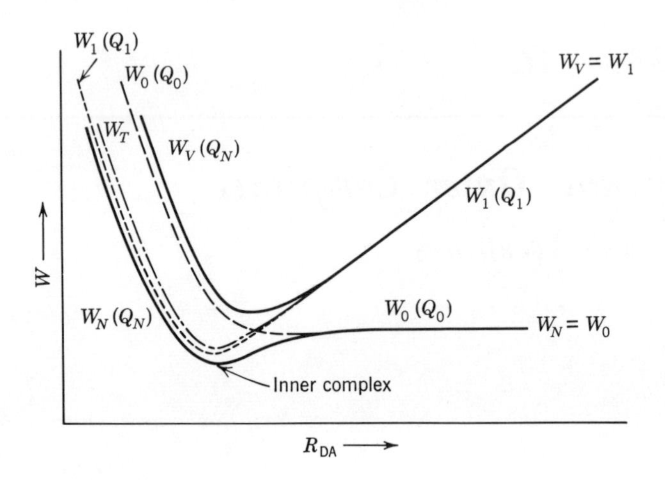


**Figure 16-1** Potential-energy diagram for $b\pi \cdot a\pi$ complexes with small CT interaction and small $I_D - E_A$ in a sufficiently polar solvent. $R_{DA}$ is conveniently taken to be the perpendicular distance between the $b\pi$ and $a\pi$ planes, assumed parallel. Here e.g. $W_1(Q_1)$ means that this curve is the energy of the dative state plotted as a function of $R_{DA}$ but with the other configurational coordinates (including those describing solvation) equal to their values such that $W_1$ is minimized. (See text and Chapter 9.)

(see Chapter 9 for discussion) is expected to change over rather rapidly and smoothly from $W_0(Q_0)$ to $W_1(Q_1)$, which in Fig. 16-1 is shown as somewhat lower in energy.

Here it is necessary to point out that for $W_N$ we must consider the crossing of $W_0(Q_N)$ and $W_1(Q_N)$, *not* that of $W_0(Q_0)$ and $W_1(Q_1)$ which is shown in Fig. 16-1. With the assumption of weak interaction of $\Psi_0$ and $\Psi_1$ made here, however, $W_0(Q_N)$ and $W_1(Q_N)$ would cross not very far above the crossing point of $W_0(Q_0)$ and $W_1(Q_1)$. According to (2-10) [with $\beta_1 = \beta_0$, since $W_1 = W_0$,—see (2-14)], $W_N(R_c, Q_N)$ and $W_V(R_c, Q_N)$ should be given by $W(R_c, Q_N) \pm \beta_0$, where $W(R_c, Q_N) = W_0(R_c, Q_N) = W_1(R_c, Q_N)$. The curve $W_V(Q_N)$ in Fig. 16-1 is drawn in accordance with these considerations; the crossing point of $W_0(Q_N)$ and $W_1(Q_N)$ would then be at an $R_{DA}$ value not far from that for the crossing of $W_0(Q_0)$ and $W_1(Q_1)$, and halfway between the $W_N(Q_N)$ and $W_V(Q_N)$ curves. In Fig. 16-1 it is assumed that $\beta_0$ is small, corresponding to case 1 in Fig. 3-4.

The possibility that $W_1(Q_1)$ can cross $W_0(Q_0)$ occurs only if $I_D - E_A$ is exceptionally small. However, if $W_1$ is not far above $W_0$ in an inert solvent, then under the environmental cooperative action of a sufficiently polar solvent the $W_1$ curve may be lowered to the position shown for $W_1(Q_1)$ in Fig. 16-1 when $Q_N$ changes from $Q_0$ to $Q_1$, and the inner complex ($D^+$—$A^-$) should then be stable. The general nature of these effects, and that of the transition from a $\Psi$ that is mostly $\Psi_0$ to one that is mostly $\Psi_1$ on going from $R > R_c$ to $R < R_c$ is similar to that for $n$, HQ and other $n$, $a\sigma$ complexes (see Figs. 14-8 and 15-1), but there is one main difference; namely, that the inner complex now consists of odd-ion pairs.

As an example consider the complex between the donor tetramethylparaphenylenediamine (TMPPD) and the acceptor chloranil (Chl):

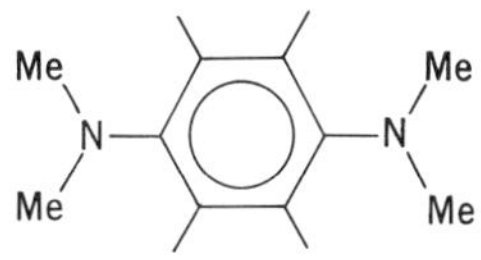

Tetramethylparaphenylenediamine

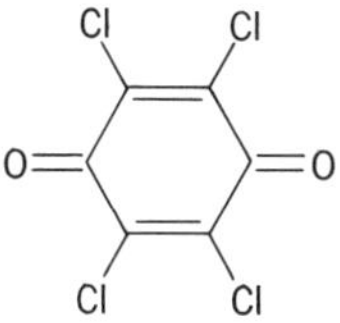

*p*-Chloranil

The ionization potential for TMPPD is low ($I_D \simeq 6.5$ eV; compare 7.70 eV for aniline). The electron affinity of Chl is fairly high; Briegleb [1] gives approximately 1.4 eV. Hence at $R_{DA} = \infty$, $I_D - E_A \simeq 5$ eV. At the equilibrium distance $R_{DA} \simeq 3.3$ Å the Coulomb stabilization energy $-C$ of the dative function should be about 3.5 eV on the point-charge model (see Section 9-2). Hence the minimum of $W_1$ in vapor or an inert solvent would be slightly above $W_0(R_e)$, and after it has been brought lower by a sufficiently polar solvent we can expect that at its minimum $W_N$ is essentially $W_1$ and we expect the ion-pair inner complex to be the stable species in polar solution. Likewise (cf. Section 15-3) we can expect TMPPD · Chl to form ionic crystals.

Experimentally, TMPPD plus chloranil dissolved in the nonpolar solvent cyclohexane forms an outer complex with a normal CT band. In even moderately polar solvents it appears that the ion-pair inner complex $(TMPPD)^+Chl^-$ must be formed, but be extensively dissociated into the odd ions $TMPPD^+$ and $Chl^-$, whose ESR spectra are well known and easily identified [2]. The ion $TMPPD^+$ is the Würster's blue cation and $Chl^-$ is a semiquinone anion. The dimethylaniline-chloranil (TMA · Chl) system, which yields $TMA^+$, the crystal violet cation, shows similar behavior [2]. These and similar studies [3,4,5] (for example, on the complex of TMPPD with TNB [3]) emphasize the importance of ESR in identifying the radical ions.

The *crystals* of TMPPD · Chl show photoconduction, semiconduction, and paramagnetism that increases with increasing temperature and varies with different similar acceptors (for example, *o*-chloranil, bromanil, and iodanil). However, the paramagnetism is much smaller than would be expected if the crystals were composed entirely of free (positive and negative) radical ions.

Kainer and Otting [6] studied the infrared spectra of a number of crystalline $b\pi \cdot a\pi$ complexes. They found that these fall sharply into two classes: either (a) there is very little change in the spectrum from that of the separated components, or else (b) there is a drastic change. They concluded that crystals of the first type are nonionic (corresponding to outer complexes), but those of the second type are ionic (corresponding to inner complexes). Their results seem reasonable: type (a) crystals are formed in cases when $I_D - E_A$ is only moderately small (for example, dimethylaniline · chloranil and hexamethylbenzene · chloranil) and type (b) crystals when $I_D - E_A$ is definitely small (for example, TMPPD · Chl). Matsunaga [7] reports further studies of this type. An examination of the visible electronic absorption spectra of solid TMPPD complexes with a series of $a\pi$-acceptors again shows a sharp division into classes (a) and (b) [8]. With class (a) acceptors a CT band is seen, with class (b) this has disappeared, but two new bands that are attributable presumably to $D^+$ and $A^-$ ions have appeared.

In agreement with the foregoing evidence McConnell et al. [9] have shown that it is *theoretically* reasonable to get *either* $D^+A^-$ *or* $D \cdot A$ crystals, but that one would not expect a mixture of the two forms, as some people earlier had thought to be present in these systems. They reason that the Madelung energy accounts for the stability of the ionic crystal, as discussed in Chapter 15. They also show that the $D^+A^-$ ionic structure is consistent with the observed semiconduction and photoconduction, and they believe that the observed incomplete paramagnetism can also be understood even for a fully ionic crystal.

An interesting question is the following: in cases where in a nonpolar solvent one has an outer complex and has observed its CT spectrum (corresponding essentially to $D^+A^- \leftarrow D,A$), can a reverse CT spectrum (corresponding to $D,A \leftarrow D^+A^-$) be found for the ion-pair inner complex in a polar solvent or in the ionic crystal? Spectra of the latter type are well known for *even-even* ion pairs in solution (see Section 13-4), but now the ion pair is odd-odd. Observations in an interesting paper by Kainer and

Überle on the absorption spectrum of TMPPD · Chl in solution in acetonitrile, a rather strongly polar solvent with a fairly high dielectric constant ($\epsilon = 38$), show characteristic visible absorption bands of $TMPPD^+$ and of $Chl^-$, but also in the near infrared a band whose maximum is somewhat beyond 8000 Å, and whose intensity increases strongly on cooling to −35°C [10]. More recently Foster and Thomson [5] have found the same band with peak at 8430 Å in acetonitrile solution when cooled to −40°C. It seems possible that this band is the $D,A \leftarrow D^+A^-$ reverse CT band of the inner complex $TMPPD^+Chl^-$, present in dissociation equilibrium with its ions.

However, for TMPPD · Chl in cyclohexane solutions Foster and Thomson find a band at 8700 Å that can be identified with very little doubt as the CT band ($D^+A^- \leftarrow D,A$) of the outer complex; no $TMPPD^+$ nor $Chl^-$ absorption is seen in these solutions. Then very likely, after all, the 8430-Å band seen in cold acetonitrile solutions is the outer-complex band, shifted somewhat in position by the change of solvent.† The following explanation now seems very plausible. In polar solvents with high dielectric constants the inner and outer complexes are related qualitatively as in Fig. 14-8*a*, the inner complex being an endothermic tautomer and the outer complex being the more stable. However, the inner complex when formed immediately dissociates almost completely into ions, in a manner similar to that suggested in Fig. 2 of R4; this dissociation of an endothermic complex would be encouraged by the resulting entropy increase. Finally, however, if the temperature is lowered, these endothermic processes are partially reversed, and the complex reverts in part to the lower energy tautomeric form, the outer complex, whose CT band is then seen. An alternative explanation suggested by the observations of Calvin et al.,† would be that the inner complex is the more stable but is reached only somewhat slowly by passage over an activation barrier like that in Fig. 14-8*b*.

The paper of Foster and Thomson contains further interesting evidence on a large number of other $b\pi \cdot a\pi$ complexes. The weaker ones evidently form only outer complexes, whereas the stronger ones give odd ions in polar solvents, implying the presence also in small quantities of a perhaps endothermic inner complex. In any event the evidence that the *crystals* of TMPPD · Chl and similar complexes are composed essentially of ions seems reasonably conclusive.

---

† Calvin et al. [2] report that a CT band at 9240 Å is observed in a freshly prepared acetonitrile solution of TMPPD + Chl in 1:1 proportion, but gradually disappears. However, the ESR spectra of $TMPPD^+ + Chl^-$ are also present in the freshly prepared solution, but the $Chl^-$ intensity decreases and that of $TMPPD^+$ increases with time, indicating the occurrence of further reactions.

Anex and Hill have examined the reflection spectra of TMPPD · Chl crystals in polarized light [11]. In addition to visible and ultraviolet bands polarized parallel to the $TMPPD^+$ and $Chl^-$ planes (which are parallel to each other —see Section 5-7), and which agree with known bands of the two ions, like those seen in $CH_3CN$ solution, they find an infrared band with maximum beyond 10,000 Å and polarized perpendicular to the $TMPPD^+$ and $Chl^-$ planes, as is expected for a CT band (see Sections 3-4 and 5-7). It seems fairly certain that *this* is a reverse CT band $(D,A \leftarrow D^+A^-)$.

Both $D^+$ and $A^-$ are odd-electron systems. Why do they not pair up in the crystal, as in a normal dative state, to form a multiatom covalent bond $(D^+—A^-)$? Before trying to answer this question, let us note that the paramagnetism that is observed for the crystals might be explained if the odd electrons in $D^+A^-$ pairs do interact slightly to form singlet and triplet states, such as one would expect for a $D^+,A^-$ pair with an odd electron on each (see Section 2-8).† If the triplet state is only slightly separated in energy from the ground-state singlet, like $W_T$ in Fig. 16-1, the thermal population of the former could account for the paramagnetism. However, we still have the question of why the interactions of odd π-electron ions to form singlet or triplet states are so small. An answer will be attempted in Section 16-2.

## 16-2 QUINHYDRONE AND RELATED SYSTEMS

For a detailed look at a crystalline $b\pi \cdot a\pi$ complex—one that is not of ionic structure although its donor and acceptor partners are related to others that are—let us examine the 1:1 complex, known as *quinhydrone*, between quinone

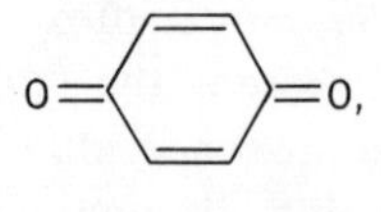

O= =O,

and hydroquinone (see next page). Its geometrical configuration is

† The discussion here in terms of pairwise interactions of course needs generalization when applied to a crystal, but it should be qualitatively valid. Further, the extensive dissociation into odd ions in a polar solvent indicates that the covalent forces are weak also for individual ion pairs.

HO—⟨benzene ring⟩—OH.

described in some detail in Briegleb's book (pp. 177–181) on the basis of studies of related complexes by Wallwork and Harding [12] and of an X-ray diffraction study of quinhydrone itself by Matsuda and co-workers [13]. Papers by Nakamoto [14] and by Anex and Parkhurst [15] are also instructive. The crystal consists of stacks of alternating quinone and hydroquinone molecules whose planes are almost parallel to each other. The centers of the hydroquinone molecules are shifted relative to those in the adjacent quinone molecules by about 2.1 Å along the direction of the C=O groups. Wallwork [12b] attributes the shift to "compromise with hydrogen bond requirements," the parallel planes and the short perpendicular distance (3.16 Å) "being indicative of charge-transfer stabilization by overlap of molecular $\pi$ orbitals." This crystalline complex is believed to be nonionic like an outer complex. It would be interesting to see what the infrared spectrum shows.

Now let us look at the MOs involved in the $b\pi \cdot a\pi$ donor-acceptor interaction. For hydroquinone, if (perhaps wrongly) we assume both O—H bonds to be in the plane of the molecule, the $\pi$ MOs can be constructed as linear combinations of eight $\pi$ (i.e., $2p_x$) AOs—one on each carbon and one on each oxygen atom. Still better, the hydroquinone MOs can be considered to be constructed as linear combinations of the $\pi$ AOs of the two "$\pi$ islands" (cf. Section 4-2), one on each O atom, with the familiar benzene MOs (see Fig. 4-1). The five lowest energy MOs thus obtained are to be occupied by the 10 $\pi$ electrons (originally 2 from each O atom and 1 from each C atom).

Let us now designate the MOs by symbols such as $s_z a_y$, where $s$ means "symmetric" and $a$ means "antisymmetric" with reference to a plane denoted by a subscript ($z$ refers to the $xz$-plane, $y$ to the $xy$-plane, with the $x$-axis perpendicular to the ring plane, and $y$- and $z$-axes chosen as shown in the accompanying figure at top of next page).†

Referring to Fig. 4-1 we see that $\phi_1$ is of $s_z s_y$ type, whereas in the degenerate pair $\phi_2$, $\phi_3$, $\phi_2$ is of $a_z s_y$ type and $\phi_3$ is $s_z a_y$. From the two

† *All* $\pi$ MOs are antisymmetric to the $yz$-plane.

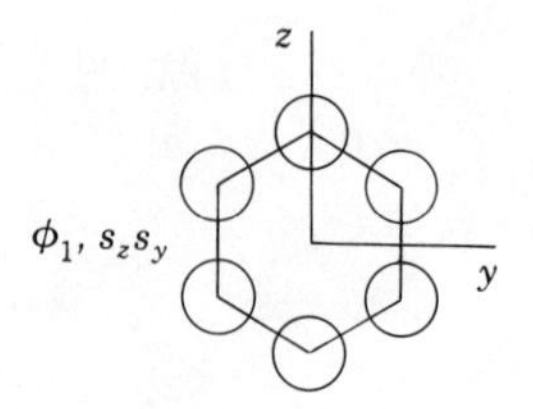


oxygen island AOs $\pi_0$ and $\pi_0'$, which of course are of equal energy in the free molecule, two symmetry orbitals $(\pi_0 + \pi_0')/\sqrt{2}$ and $(\pi_0 - \pi_0')/\sqrt{2}$ are constructed that are respectively $s_z s_y$ and $s_z a_y$ if the long axis of the molecule, passing through the two C—O bonds, is taken as the $z$-axis.

The energy levels and their occupation by electrons are indicated in Fig. 16-2. When the benzene MOs are combined with the $\pi$-island AOs,

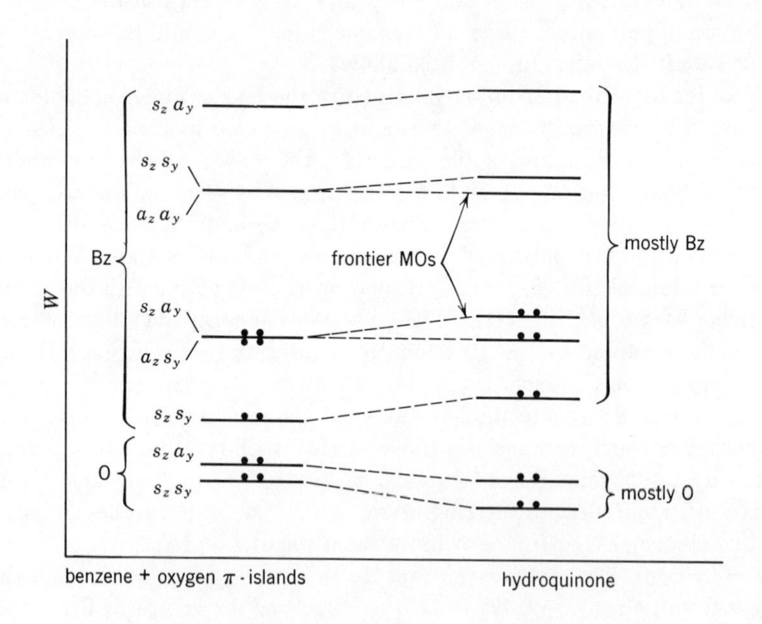


**Figure 16-2** Qualitative energy-level diagram for $\pi$ MOs of hydroquinone.

only orbitals of like symmetry interact. The energies of the resulting MOs of hydroquinone are as indicated qualitatively in Fig. 16-2. The interactions of the ring and O atoms involve a net intramolecular $\pi$-electron charge transfer (cf. Section 4-2) from O to the ring; however, this is realized only when interactions of the lowest two orbitals with the *excited* benzene MOs $\phi_6$ and $\phi_5$, of $s_z s_y$ and $s_z a_y$ symmetry, respectively, are included.[†‡] The highest energy occupied orbital, which is the donor MO in quinhydrone, is called a *frontier orbital*, following Fukui [16]. It is mostly like the benzene MO $\phi_3$ but with some oxygen admixture, perhaps of antibonding character [17]:

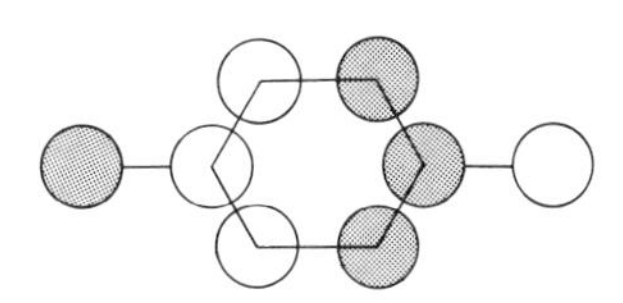

This frontier donor MO is very like the frontier *acceptor* MO on quinone. The MOs of quinone are qualitatively the same as those in Fig. 16-2 for hydroquinone. However, in quinone each oxygen atom has contributed only *one* $\pi$-electron, so that there are only eight $\pi$-electrons. Hence the lowest empty orbital (frontier acceptor MO) of quinone is qualitatively of the same form as the highest filled MO (frontier donor MO) of hydroquinone.[§]

Because of this close similarity of the MOs for this pair of molecules, there should be very good overlap of $\phi_d$ and $\phi_{a-}$ (see Sections 2-7 and 3-4) in the complex to form a very strong *pancake* bond between the molecules in the complex provided only that they come close enough together with corresponding atoms matched up. Why then does not this bonding energy overcome the perhaps conflicting requirements for hydrogen-bond formation and force the molecules in the complex to line up for good $\phi_d$, $\phi_{a-}$ overlap? And in TMPPD · Chl and analogous cases, which except

[†] The simultaneous $\sigma$-bond charge transfer *out* from the ring makes the O atoms more negative and therefore better $\pi$-donors. (This is an example of two-way charge transfer).

[‡] If the O atom $\pi$ AOs interacted only with the *filled* benzene MOs, there would be *no net* $\pi$-transfer. Only by partial migration into *unoccupied* benzene $\pi$ MOs is there a net transfer to the ring.

[§] However, the admixture of benzene MOs, especially $\phi_3$, into the MOs called "mostly O" in Fig. 16-2, is much increased, and corresponding MOs in the two cases are far from identical. The net $\pi$-electron as well as $\sigma$-electron transfer is now doubtless in the direction from the ring to the O atoms.

for the extra $\pi$-electrons on the chlorine atoms in Chl, or the like, is $\pi$-isoelectronic with quinhydrone, why does not a strong $D^+A^-$ pancake bond form between the odd electrons of the $TMPPD^+$ and $Chl^-$, instead of leaving the radical-ions with their odd electrons unpaired or almost unpaired as they have been found to be in many cases, as discussed above for $TMPPD^+Chl^-$?

A plausible answer is that the other $\pi$-electrons and $\sigma$-electrons in filled shells of the two molecules repel each other very strongly as the D—A separation is reduced below the van der Waals distance. A similar effect furnishes a reason for the probable preference for the axial model over the resting model in the $Bz \cdot I_2$ complex. In that case the actual approximate calculations that have been made indicate the importance of $\pi$ closed-shell repulsions (see Section 11-2 and Reference 3 of Chapter 3). On the other hand, the D—A distance (3.16 Å) in hydroquinone and 3.30 Å in $TMPPD^+Chl^-$ is substantially less than the van der Waals distance. How can we understand this?

Adding to the mystery of why ion-radicals of opposite sign with nearly matching MOs do not form pancake bonds is the fact that ion-radicals of *like* sign *do* associate in pairs in solution and in crystals. Of course in these self-complexes there is *exact* matching of the odd-electron MOs, but one would think that this advantage would be outweighed by Coulomb repulsions; for example, $TCNE^-$ dimers with a heat of dimerization of 10 kcal/mole are found in aqueous solution; in this case, to be sure, the Coulomb repulsion would be reduced to a small value by the dielectric effect of the water [18]. Dimerization is also seen in the crystal of a $TCNQ^-$ salt,† [19] and of $TMPPD^+$ salts [20,21].

Dimerization also occurs for the neutral radical *N*-ethyl-phenazyl in

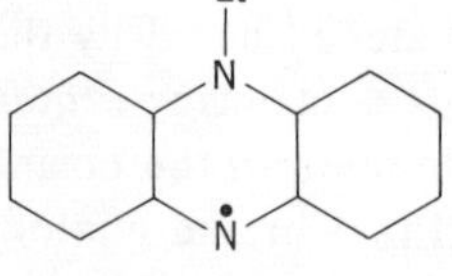

---

† The $TCNQ^{\dot{-}}$ ions are found in plane-to-plane stacks similar to those found for other CT complexes (see Section 5-7) with interplanar spacing of 3.26 Å—significantly less than the van der Waals spacing of 3.4 Å. However, the planes are shifted with respect to each other so that the center of one ring falls not over the center of the neighboring ring but over the center of the quinomethane double bond. The latter position is found by an extended Hückel MO calculation to be an expected stable configuration for a $TCNQ^{\dot{-}}$ self-complex. (Private communication from C. J. Fritchie, Jr., Tulane University, reporting yet unpublished work by C. J. Fritchie, Jr., D. B. Chesnut, and H. E. Simmons.)

crystals and in alcohol-ether solution. In both of the latter Hausser found an absorption band that Hausser and Murrell [22] identified as a CT band of a dimer, for which they believe that a weak pancake bond is present (weak, since the occurrence of paramagnetism at moderate temperatures indicates that $W_T$ is not far above $W_N$ in Fig. 16-1, which should be applicable here). As they note, $W_N$ and $W_T$ here are analogous to the ${}^1\Sigma_g^+$ and ${}^3\Sigma_u^+$ Heitler-London lowest states of $H_2$, and the CT band is analogous to the $V \leftarrow N$ intramolecular two-way CT band of $H_2$ (see Section 12-1).

## 16-3 OTHER π-SYSTEMS THAT YIELD RADICAL-IONS

Consider the interaction between LiBr and tetracyanoethylene (TCNE)

$$\left(\begin{matrix} NC & & CN \\ & C{=}C & \\ NC & & CN \end{matrix}\right).$$

In a polar solvent the $Br^-$ ion acts as an $n$-donor that interacts with the $\pi$-acceptor, TCNE, to give an outer complex [23]. The $Br^-$ ion apparently also forms $n \cdot a\pi$ outer complexes with chloranil, bromanil, and iodanil [23]. On the other hand, although with $I^-$ the CT band of an outer complex with TCNE is seen briefly if the solution is kept very cold (–40°C), a reaction that involves complete charge transfer follows quickly:

$$2\,M^+I^- + 2\,TCNE \rightarrow \underset{\text{outer complex}}{2\,M^+(I^- \cdot TCNE)} \rightarrow \underset{\text{odd-odd}}{2\,[M^+ + \dot{I} \cdot TCNE^{\dot{-}}]} \rightarrow$$

$$\xrightarrow[(M^+I^-)]{} \underset{\text{even-odd}}{2\,M^+(TCNE^{\dot{-}})} + \underset{\text{even-even}}{M^+I_3^-}.$$

Here (in brackets) a high-energy intermediate inner complex has been postulated, somewhat as in Section 15-6 for nitrobenzene, but here of odd-odd type. ($M^+$ is a tetraalkylammonium or similar cation.) The iodide ion reacts in a similar way, with complete charge transfer, with chloranil and its analogues to give $Chl^-$ and so on.

Another system that may be similar, except that it involves a $b\pi$-donor and results in a *positive* radical-ion, may be that between perylene and $I_2$.

$$2\,\text{Per} + I_2 \longrightarrow \left[\begin{array}{c}\text{Per}\cdot I_2\cdot\text{Per}\\ \text{outer complex}\end{array}\right] + 2I_2 \longrightarrow 2\left[\begin{array}{c}(\text{Per}^{\dot{+}})\,I_3^-\\ \text{odd-even}\end{array}\right].$$

perylene
(Per): $b\pi$-donor

The final product may explain the semiconduction found for perylene $\cdot\, I_2$ crystals.

Similarly it seems possible that carotene might react in a similar manner:

$$2 \text{ carotene} + 3I_2 \rightarrow 2(\text{Car}^{\dot{+}})\,I_3^- \quad (?).$$

Lupinski [24a] has obtained evidence that indicates the positive-ion species is the even ion $(\text{Car}\cdot \text{I})^+$; however, Ebrey [24b] has suggested that the strong absorption band cited by Lupinski as evidence for this ion may, in fact, be due to the $\text{Car}^{\dot{+}}$ ion itself.

The examples just given involve large unsaturated molecules with low ionization potentials as donors, and the acceptor action is assisted by the formation of the $I_3^-$ ion, so that $W_1$ may well be lower than $W_0$ at equilibrium.

## 16-4 MEISENHEIMER-TYPE COMPOUNDS

Caldin et al. [25] have studied the kinetics of the following reaction (TNA means 2,4,6-trinitroanisole) in some detail and identified an intermediate stage that they believe to be an outer complex. (See next page.) The inner complex here is called a Meisenheimer compound. The first reaction proceeds rapidly to give a brown solution, followed more slowly by the second reaction to give a purple solution. The resemblance of the Meisenheimer anion to the nitrobenzenium-cation intermediate discussed in Section 15-6 should be noted. Like the latter it should be stabilized by isovalent hyperconjugation.

Servis [26] has published a nuclear-magnetic-resonance study of this reaction in which he concludes that the transient intermediate identified by Ainscough and Caldin [25] as the outer complex is really the following unstable form of the final anion that then rearranges more slowly to give the stable form of inner complex depicted below. However his

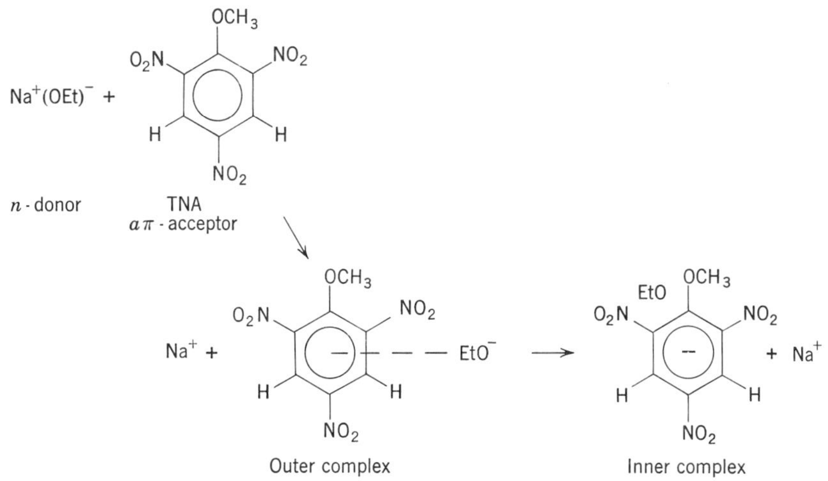

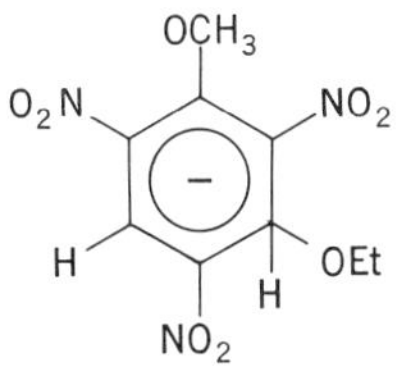

experiments were carried out at room temperature, whereas Ainscough and Caldin stated that the reaction at room temperature proceeds immediately to the inner complex and that in order to observe the outer complex they had to study the reaction at –80°C. Hence it seems possible that Servis has observed still a third step in this interesting reaction.

There are various other examples of this kind of reaction. Thus [27]

$$\underset{n\text{-donor}}{Li^+I^-} + \underset{a\pi\text{-acceptor}}{TNB} \rightarrow \underset{\text{(outer complex only)}}{Li^+ + I^- \cdot TNB}$$

A CT band is observed, whose frequency increases considerably with polarity of the solvent, for reasons discussed near the end of Section 15-3. Bromides and the thiocyanates show behavior like that of the iodides. But

$$\underset{n}{Na^+OH^-} + \underset{a\pi}{TNB} \longrightarrow O_2N\text{-}C_6H_2(NO_2)_2(H)(OH)^- + Na^+,$$

Inner complex

analogous to the reaction of TNA with ethoxide ion. It is now known whether this reaction also proceeds through a distinct intermediate.

For a long time the nature of the interaction of TNB with aliphatic amines has been the cause of much puzzlement [28], but it now seems fairly certain that the reaction in the case of primary and secondary amines is related to that of $Na^+OEt^-$ with TNA and goes as follows:

$$\underset{n}{R_2NH} + \underset{a\pi}{TNB} \longrightarrow \underset{\text{Outer complex}}{R_2NH \cdot TNB} \xrightarrow{(R_2NH)} R_2NH_2^+ \left[O_2N\text{-}C_6H_2(NO_2)_2(H)(NR_2)\right]^-.$$

Salt of Meisenheimer-type inner complex

Here the inner complex that corresponds to $R_2NH \cdot TNB$ would be $H^+Q^-$, where $Q^-$ means the Meisenheimer anion; but actually the $H^+$ is combined with an additional $R_2NH$ to form $R_2NH_2^+$. Charge-transfer spectra that correspond to amine · TNB outer complexes have been obtained in cyclohexane solutions for primary, secondary, and tertiary aliphatic amines

[29,30]. With primary and secondary amines† in polar solvents a following reaction goes very quickly, in most cases in the manner just shown [31]. Briegleb, Liptay, and Cantner [31] in a very thorough study of the reaction of the aliphatic secondary amine piperidine (hexahydropyridine, abbreviated PiH) with TNB gave rather convincing evidence that in acetonitrile solution the interaction proceeds in the manner indicated above.‡ Foster and Mackie [31] conclude that there is an analogous behavior for solutions of substituted TNB in liquid ammonia:

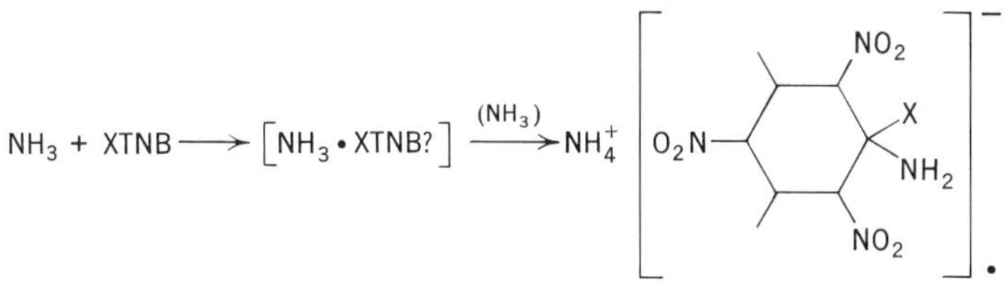

$$NH_3 + XTNB \longrightarrow [NH_3 \cdot XTNB?] \xrightarrow{(NH_3)} NH_4^+ \left[ O_2N\text{–}C_6H_3(NO_2)_2(X)(NH_2) \right]^-.$$

Andrews and Keefer's book (GR2, p. 149), contains further discussion of the reactions of amines.

Briegleb and co-workers [32] have studied the following reaction, which is distinctly related to those just described.

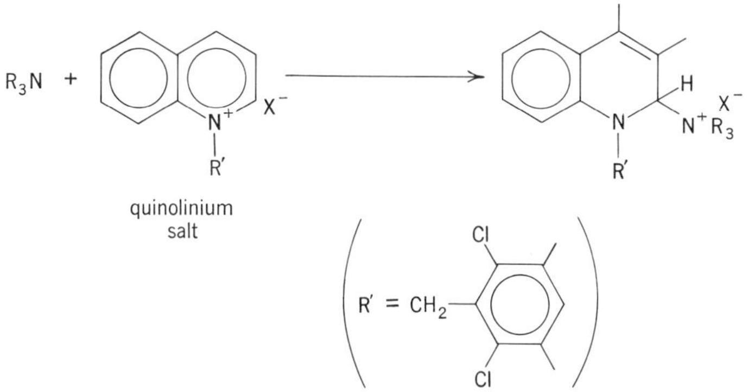

$R_3N$ + quinolinium salt ($N^+$–R′, $X^-$) ⟶ 1-R′-1,2-dihydroquinoline bearing H and $N^+R_3$ at C-2, $X^-$

quinolinium salt

$\left( R' = CH_2\text{–}C_6H_3Cl_2 \right)$

---

† Tertiary amines obviously cannot react this way because of the absence of an H atom attached to N. Aromatic amines form ordinary $b\pi \cdot a\pi$ outer complexes but of 2:1 as well as 1:1 composition.

‡ Additionally, the ion-pair was found to be in equilibrium with its ions $(TNB\text{-}Pi)^-$ and $PiH_2^+$, whereas $PiH_2^+$ with excess PiH was in equilibrium with a hydrogen-bonded combination $Pi_2H_3^+$. Slower secondary reactions also occur.

In this reaction the ring has acquired a negative charge relative to the initial quinolinium ion, and an extra group has become attached to one of the ring atoms, just as in the Meisenheimer anions (or as in the benzenium ion). Hence the product falls in the same category as the Meisenheimer compounds.

## 16-5 CRITIQUE OF THE CONCEPT OF INNER AND OUTER COMPLEXES

The preceding sections in this chapter and in Chapter 15 have illustrated the usefulness of the concept of inner and outer complexes. However, there are many cases in which the concept becomes ambiguous and difficult or impossible to apply. It is the purpose of this section to try to set some boundaries to the worthwhile use of the concept. First let us consider some matters of terminology.

Some equivalent nomenclature is summarized as follows:

| | | | |
|---|---|---|---|
| $\sigma$-complex | intromer | endocomplex | inner complex |
| $\pi$-complex | extromer | exocomplex | outer complex |

Here the left-hand set of names was introduced by H. C. Brown [33] but has the disadvantage of applying only to complexes with $\pi$-donors. The terms *intromer* and *extromer* (or endocomplex and exocomplex) were proposed in lectures by Mulliken for two reasons: (a) the word "complex" is often not very appropriate—for example, the "inner complex" is often really hardly more than a pair of ions in contact, and sometimes the "outer complex" has no stability but is just a pair of reaction products or of reactants (for example, pyridine + $CH_3I$); (b) on the other hand, the relation between outer and inner complex is essentially one of isomerism (or sometimes tautomerism). Nevertheless, the terms "outer complex" and "inner complex" are perhaps best, and are used here.

The usefulness of the concept of inner and outer complexes is perhaps greatest for those cases in which the donor and acceptor are neutral molecules that yield a weak outer complex with small charge transfer and an ion-pair inner complex in which charge transfer has been large, often nearly a whole electron. The inner and outer complexes correspond to two potential minima in energy diagrams such as Figs. 14-8 and 15-1. However, sometimes there is only one minimum, and cases with more than two minima are also to be expected. To clarify matters, it will be useful to look at different D,A and different charge-distribution types separately. Let us begin with types with *even-electron neutral* donors and acceptors.

For many $n \cdot v$ complexes (see Section 14-7 and Fig. 14-6) a single deep minimum with extensive charge transfer [$b$ large, but probably not yet close to 1, in (2-2)] is typical. These may well be called inner complexes. Weak outer complexes of $n \cdot v$ type, although perhaps not yet known, should be possible. The $b\pi \cdot a\pi$ complexes, where both donor and acceptor are sacrificial, are in one respect like the $n \cdot v$ complexes in which both donor and acceptor are increvalent; namely, the donor-acceptor interaction is associative only (that is, there is no *atom* transfer from the one to the other), even in their odd-odd inner complexes. (In the $b\pi \cdot a\pi$ inner complexes the bond breaking that has occurred is of intramolecular $\pi$-bonds.) They differ from $n \cdot v$ complexes in that both outer and inner complexes, in some cases apparently as coexistent tautomers, are well known. Thus for $n \cdot v$ and for $\pi \cdot \pi$ complexes, the concept of inner and outer complexes seem to be clear-cut and useful.

In $n \cdot a\sigma$ and $\pi \cdot a\sigma$ complexes with neutral donors and acceptors the concept of inner and outer complexes again seems to be clear-cut and useful, although now the acceptor functions dissociatively (notably for HQ acceptors) when an inner complex is formed, the bond of one atom (the H atom for HQ acceptors) being transferred to the donor positive ion.

However, if $n$ is an anion in an $n \cdot a\sigma$ pair, with $a\sigma$ still neutral, we find examples such as those in Section 15-2, where the stable form is either a middle complex (in symmetrized CT complexes) or else two separated reactants or reaction products. Here, although weak unsymmetrical complexes that might be called outer complexes are conceivable, they seem not to be known; but if they *were* known, there would be two candidates for the designation "outer complex," corresponding to the left-hand and the right-hand sides of (15-1) to (15-3). In such cases the concept of inner and outer complexes is perhaps not helpful.

On the other hand, in $n \cdot b\pi$ complexes (Section 16-4) both with neutral and with anion $n$-donors it seems useful to talk about inner and outer complexes, with Meisenheimer-type anions or similar structures being regarded as inner complexes. Here and in general it is sometimes arbitrary as to whether one considers an anion alone or an ion-pair that includes an active anion as electron donor; of course if the anion is solvated but not really paired, the solvated anion alone *is* the donor. An analogous remark applies to cations as $v$-acceptors; for example, to $NO_2^+$ in Section 15-6 and to $Ag^+$ in (two-way) complexes (Chapter 17).

There are further ambiguities in the inner-complex, outer-complex concept. One can, for example, regard an ion pair (even including the inner complex of a D,A pair D $\cdot$ HQ or D + MeQ) as an outer complex between the ions, with only small charge transfer relative to the ions and a corresponding covalently bonded structure as inner complex. Or conversely, a covalently bonded structure [for example, $(C_6H_5)_3CCl$] can be regarded as an outer complex and the corresponding ion pair as an inner complex. However, it is probably best not to use the concept of inner and outer complexes in such cases. One final remark, however: in *all* cases, the characteristic *difference* between inner and outer complexes is that of *little* charge transfer in the outer, *large* charge transfer in the inner complex, *relative* to the two molecules or ions that initially one chooses (in case a choice seems open) to regard as D and A.

## 16-6 REFERENCES

[1] G. Briegleb, *Angewandte Chem.* (intnl. ed.), **3**, 617 (1964).
[2] J. W. Eastman, G. Engelsma, and M. Calvin, *J. Am. Chem. Soc.*, **84**, 1339 (1962).
[3] S. Iwata, H. Tsubomura, and S. Nagakura, *Bull. Chem. Soc. Japan*, **37**, 1506 (1964).
[4] W. Liptay, G. Briegleb, and K. Schindler, *Z. Elektrochem.*, **66**, 331 (1962).
[5] R. Foster and T. J. Thomson, *Trans. Faraday Soc.*, **58**, 860 (1962).
[6] H. Kainer and W. Otting, *Chem. Ber.*, **88**, 1921 (1955).
[7] Y. Matsunaga, *J. Chem. Phys.*, **41**, 1609 (1964), and ibid., **42**, 1982 (1965).
[8] R. Foster and T. J. Thomson, *Trans. Faraday Soc.*, **59**, 296 (1963).
[9] H. M. McConnell, B. M. Hoffman, and R. M. Metzger, *Proc. Nat. Acad. Sci.*, **53**, 46 (1965).
[10] H. Kainer and A. Überle, *Chem. Ber.*, **88**, 1147 (1955); see especially Figs. 2 and 3.
[11] B. G. Anex and E. B. Hill, Jr., *J. Am. Chem. Soc.*, **88**, 3648 (1966).
[12] (a) S. C. Wallwork and T. T. Harding, *Acta Cryst.*, **6**, 791 (1953), and *Nature*, **171**, 40 (1953); (b) S. C. Wallwork, *J. Chem. Soc.*, 494 (1961).
[13] H. Matsuda, K. Osaki, and J. Nitta, *Bull. Chem. Soc. Japan*, **31**, 611 (1958).
[14] K. Nakamoto, *J. Am. Chem. Soc.*, **74**, 1739 (1952).
[15] B. G. Anex and L. J. Parkhurst, *J. Am. Chem. Soc.*, **85**, 3301 (1963).
[16] K. Fukui, T. Yonezawa, and C. Nagata, *J. Chem. Phys.*, **27**, 1247 (1957).
[17] S. Nagakura, private communication.
[18] R. H. Boyd and W. D. Phillips, *J. Chem. Phys.*, **43**, 2927 (1965).
[19] (a) P. Arthur, Jr., *Acta Cryst.*, **17**, 1176 (1964); (b) C. J. Fritchie, Jr., *Acta Cryst.*, **20**, 892 (1966); (c) C. J. Fritchie, Jr., and P. Arthur, Jr., *Acta Cryst.*, **21**, 139 (1966).
[20] K. H. Hausser, *Z. Naturforschung*, **11a**, 20 (1956).
[21] H. M. McConnell, in *Molecular Biophysics*, ed. by B. Pullman, Academic, New York, 1965, p. 311, and later papers.
[22] K. H. Hausser and J. N. Murrell, *J. Chem. Phys.*, **27**, 500 (1957).
[23] G. Briegleb, W. Liptay, and R. Fick, *Z. physik. Chem.*, Neue Folge, **33**, 181 (1962); *Z. Elektrochem. Ber. Bunsenges*, **66**, 859 (1962).
[24] (a) J. H. Lupinski, *J. Phys. Chem.*, **67**, 2725 (1963); (b) T. G. Ebrey, *J. Phys. Chem.*, **71**, 1963 (1967).
[25] E. F. Caldin and G. Long, *Proc. Roy. Soc.*, **A228**, 263 (1955); J. B. Ainscough and E. F. Caldin, *J. Chem. Soc.*, 2528 (1956); ibid., 2540 (1956); ibid., 2546 (1956); E. F. Caldin, *J. Chem. Soc.*, 3345 (1959).

[26] K. L. Servis, *J. Am. Chem. Soc.*, **87**, 5495 (1965).
[27] G. Briegleb, W. Liptay, and R. Fick, *Z. Elektrochem. Ber. Bunsenges,* **66**, 851 (1962).
[28] C. R. Allen, A. J. Brook, and E. F. Caldin, *Trans. Faraday Soc.*, **56**, 788 (1960); *J. Chem. Soc.*, 2171 (1961); R. E. Miller and W. F. K. Wynne-Jones, *J. Chem. Soc.*, 2375 (1959), 4886 (1961); R. Foster, *J. Chem. Soc.*, 3508 (1959); R. Foster and R. K. Mackie, *Tetrahedron,* **16**, 119 (1961).
[29] R. Foster and R. K. Mackie, *J. Chem. Soc.*, 3843 (1962).
[30] W. Liptay and N. Tamberg, *Z. Elektrochem. Ber. Bunsenges,* **66**, 59 (1962) (for piperidine as the amine).
[31] G. Briegleb, W. Liptay, and M. Cantner, *Z. physik. Chem.*, Neue Folge, **26**, 55 (1960); R. Foster and R. K. Mackie, *Tetrahedron,* **18**, 161 (1962); M. R. Crampton and V. Gold, *Chem. Commun.*, 549 (1965).
[32] G. Briegleb, W. Liptay, and W. Jung, *Z. Naturforschung,* **19b**, 97 (1964).
[33] H. C. Brown and J. D. Brady, *J. Am. Chem. Soc.*, **74**, 3570 (1952).

*CHAPTER* 17

# *Two-Way Donor-Acceptor Action*

Now let us go on to a different subject: *two-way donor-acceptor complexes*. Various other names have been used to describe the phenomenon involved, including "back donation" and "synergistic action." Most two-way complexes occur in the form of $n$:1 or $m$:$n$ compounds, usually called complex ions or coordination compounds and usually (and most conveniently) discussed in terms of ligand-field theory (see Orgel [1], Ballhausen [2], and Jørgensen [3]), which in its modern form is essentially just the whole-complex-MO theory of Chapter 12. Although this book is primarily limited to 1:1 complexes, some $n$:1 compounds are included in the discussion in Chapters 17 and 18 because they afford particularly good examples of two-way donor-acceptor action.

## 17-1 THE $Ag^+$ ION AS A ONE-WAY ACCEPTOR

When $AgNO_3$, for example, is dissolved in aqueous $NH_3$ solution the complex ion $Ag^+(NH_3)_2$ is formed. This complex must be a 2:1 one-way complex with $Ag^+$ acting as a $v$-acceptor from the $n$-donor, $NH_3$.

$$\Psi = a\Psi_0(Ag^+,2NH_3) + b[\Psi_1(H_3N_a^+\text{—}Ag) + \Psi_1(Ag\text{—}N_b^+H_3)].$$

Each $NH_3$ takes its turn in supplying an electron and bonding to the $Ag^+$. More accurately, the complex in water solution also contains some loosely bound water molecules; although $H_2O$ is a poorer donor for $Ag^+$ than is

$NH_3$, it does participate.

## 17-2 THE $Ag^+$ ION AS A TWO-WAY ACCEPTOR

The $Ag^+$ ion also forms complexes with benzene; when a saturated solution of $AgClO_4$ in water is shaken with benzene the benzene dissolves in the water layer to form the (presumably loosely hydrated) complex ion $Ag^+C_6H_6$. The Raman spectrum of the water layer was studied by Taufen, et al. [4], who found little change in the spectrum of the benzene except that the ring-stretching frequencies decreased, on the average, about 20 $cm^{-1}$, a fact that indicates a sacrificial role of the benzene in the complex.

Silver perchlorate is also very soluble in benzene directly. From the solution the crystalline complex $Ag^+Bz \cdot ClO_4^-$ can be precipatated; its X-ray diffraction pattern has been studied by Smith and Rundle [5]. The $Ag^+$ ion is found to be approximately centered above one of the C—C bonds, and not on the sixfold symmetry axis of the ring as was earlier anticipated. The structure is indicated in Fig. 17-1.

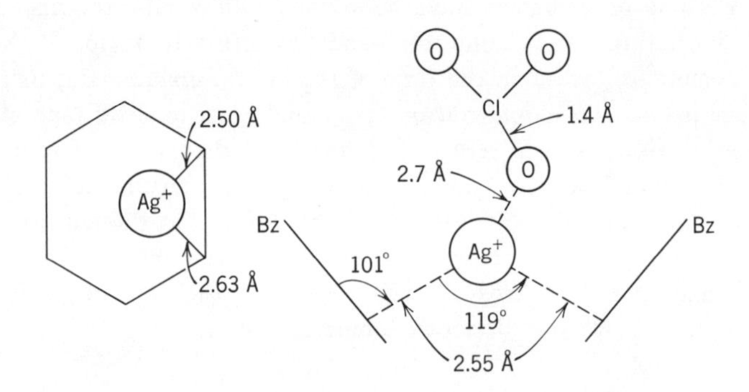


**Figure 17-1** Schematic drawing in projection of the structure of the $Ag^+Bz \cdot ClO_4^-$ crystal.

Each $Ag^+$ is coordinated with two benzenes, and vice versa; thus the solid complex is not a 1:1 complex but is like a 2:1 and a 1:2 complex simultaneously. However, in solution the complex is probably one to one, but the geometry is expected to be similar to that found in the crystal. (See GR2, pp. 62–69.) Presumably one $ClO_4^-$ is attached to each $Ag^+Bz$

in benzene solution, whereas in water solution presumably water molecules are loosely attached to each $Ag^+Bz$.

The crystal structure of $Ag^+Bz \cdot AlCl_4^-$, determined by Amma and Turner [6], is somewhat similar to that of $Ag^+Bz \cdot ClO_4^-$ in that the $Ag^+$ ion seems bonded to the C—C bond of the ring at a distance of 2.57 Å. However, only one benzene is associated with each $Ag^+$, and the $Ag^+$—Cl interactions are important. Furthermore the asymmetry that is indicated in the $Ag^+$ to C—C bond indicated in Fig. 17-1 is more pronounced: one $Ag^+$—C distance is 2.47, the other 2.92 Å.

The $Ag^+$ ion (and also the $Cu^+$ ion) forms complexes also with a number of olefins, including ethylene, cyclooctatetraene, and cyclooctadiene. (See Section 4-1 and GR2 for further discussion.)

## 17-3 DETAILED STRUCTURE OF TWO-WAY $Ag^+$ COMPLEXES

Let us now consider, as an example, the detailed structure of the $Ag^+ \cdot C_2H_4$ complex, first recalling (cf. Section 4-1) that ethylene can act either as a $b\pi$-donor by donation of an electron from its bonding $\pi$ MO or as an $a\pi$-acceptor by acceptance of an electron into its antibonding excited $\pi^*$ MO. These MOs, depicted in Fig. 4-2, are of the LCAO forms

$$\pi, \pi^* = \frac{(2p\pi_a \pm 2p\pi_b)}{\sqrt{2}\sqrt{1 \pm S}}. \tag{17-1}$$

Now note the electron configuration $\cdots 4d^{10}5s$ of the Ag atom. Evidently in $Ag^+$ ($\cdots 4d^{10}$) as a $v$-acceptor the acceptor orbital is $5s$. However, $Ag^+$ can also act as an $n$-donor from its filled $4d$ shell, into ethylene $\pi^*$ as an acceptor MO in $Ag^+C_2H_4$ or in $Ag^+Bz$ into a benzene excited MO ($\phi_4^*$ or $\phi_5^*$ in Fig. 4-1). The possibilities of two-way action of this kind using metal ion $d$ AOs were first pointed out by Dewar in a discussion [7].

Using Ar and Un for any aromatic or unsaturated molecule, respectively, we have

$$\Psi = a\Psi_0 \begin{pmatrix} \text{Ar, Ag}^+ \\ \text{or} \\ \text{Un, Ag}^+ \end{pmatrix} + b\Psi_1 \begin{pmatrix} \text{Ar}^+\text{—Ag} \\ \text{or} \\ \text{Un}^+\text{—Ag} \end{pmatrix} + c\Psi_2 \begin{pmatrix} \text{Ag}^{2+}\text{—Ar}^- \\ \text{or} \\ \text{Ag}^{2+}\text{—Un}^- \end{pmatrix}. \tag{17-2}$$

Here the $\pi$-$v$ action with $Ag^+$ as the acceptor is given by $\Psi_1$, the reverse action by $\Psi_2$. In $Ag^{2+}$ the electron configuration is $\cdots 4d^9$, but we may ask which type of $d$-orbital is used in the bonding? Also, why does it not take too much energy to ionize $Ag^+$ to form $Ag^{2+}$? Further, we note that the action is sacrificial and thus unfavorable for Ar or Un either as donor or as acceptor. In spite of all this the complexes seem to have considerable stability. Why? Perhaps as more important than $\Psi_2$ we should consider $c\Psi_{12}(\text{Ar} \underset{\sigma}{\overset{\pi}{\leftrightarrows}} \text{Ag}^+)$, the double-bonded covalent structure. (See Section 17-11.)

A partial answer to the question about the energy to form $Ag^{2+}$ may be that the environmental cooperative action of the solvent and the double Coulomb energy involved in the $Ag^{2+} \cdot Bz^-$ interaction (twice as large as if the charge were only +1) are sufficient to recover most of the extra energy.

Let us now digress briefly to consider the classification of *d*-orbitals. There are three types, which we shall call $d\sigma$, $d\pi$, and $d\delta$. These are shown in Fig. 17-2, together with the more familiar labeling as $d_{z^2}$, etc., and the classification under $D_{6h}$ symmetry. Note carefully the axes, since different projections are shown in the several figures.

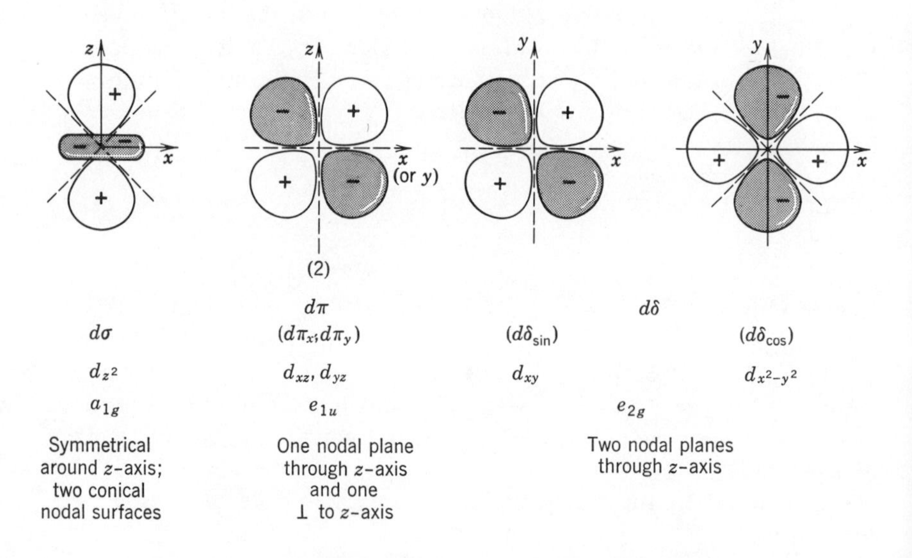


**Figure 17-2** Classification of *d*-orbitals. (Nodal surfaces indicated by - - - - - - - ).

In the $Ag^+Et$ (Et = ethylene) complex, the *d* AO that is involved in the $Ag^+$ donor action must be one that matches the $\pi^*$-acceptor MO of Et. The axes are defined in Fig. 17-3 (the $C_2H_4$ plane is perpendicular to the paper).

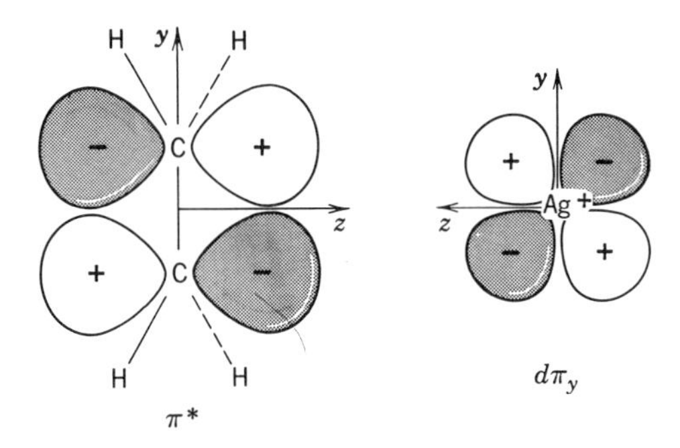


**Figure 17-3** Interacting orbitals for $\Psi_2(Ag^{2+}—Et^-)$.

We see that the $d\pi_y$ AO of $Ag^+$ overlaps very nicely with the $\pi^*$-orbital of Et so that the proper configuration of $Ag^{2+}$ in $\Psi_2$ in the complex is $\cdots 4d^{10}4d\pi_y{}^{-1}$; that is, $4d\sigma^2 4d\pi^4 4d\delta^4$ minus one $4d\pi_y$ electron.

Now for the $Ag^+Bz$ complex, for $Ar^-$ in (17-2), an appropriate benzene acceptor MO ($\phi_4^*$ of Fig. 4-1) is sketched below:

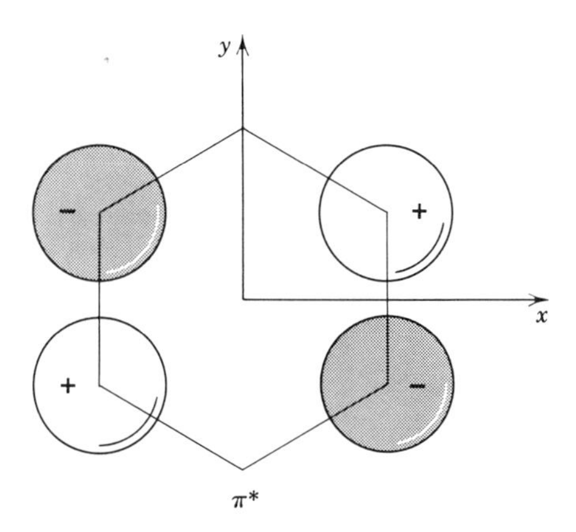

(Alternatively, $\phi_5^*$ of Fig. 4-1 could be used.) If the $d\delta_{\mathrm{sin}}$ orbital on $Ag^+$ were large enough, it would match this benzene MO ($\phi_4^*$) nicely. However, since $Ag^+$ is relatively small ($r_{Ag^+} = 1.26$ Å), a much better match of orbitals (between $\pi^*$ and the $4d\pi_y$ $Ag^+$ AO) is obtained if the $Ag^+$ is located over, for example, the right-hand C—C bond. This configuration is at the same time favorable for the $\pi$-donor, $v$-acceptor action, since the $5s\sigma$-acceptor AO of Ag then matches nicely the right-hand side of the $\pi$-donor MO, $\phi_2$:

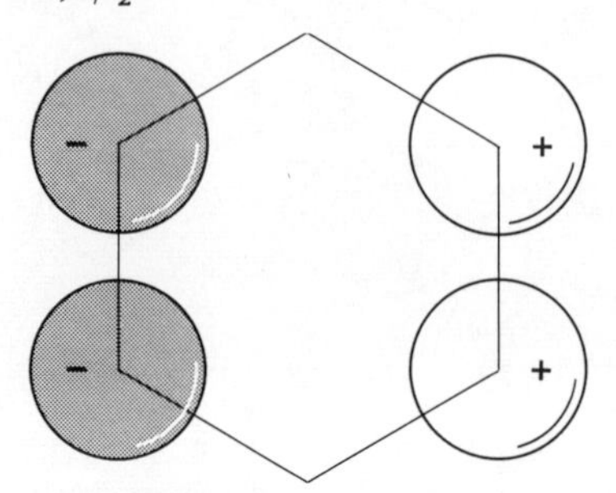


The small size of the $Ag^+$ $d$ AOs is presumably important here.†

## 17-4 OTHER TWO-WAY ACCEPTORS

Other good two-way metal-ion acceptors include $Cu^+$, $Au^+$, $Hg^{2+}$, Pt(II) [i.e., $Pt^{2+}$], Au(III), $Tl^+$, $Cd^{2+}$. An inspection of this list reveals that it contains the strongest members of those listed as class-*b* acceptors by Ahrland, Chatt, and Davies [9]. Chatt has done a great deal of work on metal-ion $\pi$-complexes. His *a, b* classification is rather closely related to Pearson's concept of "hard" and "soft" acids and bases [10].

Chatt, et al. [9], classified acceptors as follows:

Case a: an acceptor that prefers N, O, F to P, S, Cl, etc.
Case b: an acceptor that prefers P, S, Cl, etc., to N, O, F.

It appears that two-way acceptors belong to Chatt's case b because P, S, and Cl have low-lying *d*-orbitals so they can function not only as *n*-donors but also as *v*-acceptors in a two-way action, whereas N, O, and F have empty *d*-orbitals only at higher energies and so do not function

---

† For comparison we find that Slater [8] lists the radius of maximum charge density from SCF calculations for the $3d$-orbital of Cu as 0.32 Å, compared to a value of 1.03 for the $4s$-orbital. Comparable calculations for Ag are not available, but we might expect the relative sizes of the $4d$- to $5s$-orbitals to be similar to those for the $3d$- and $4s$-orbitals of Cu. Furthermore, it seems clear that the radius of the $4d$-orbitals in $Ag^+$ is much smaller than that of the $2p$-orbital in C ($\sim$0.7 Å) [8].

efficiently as *v*-acceptors. Thus, for example, although $Ag^+$ forms one-way complexes with $NH_3$, it prefers $PH_3$, with which a two-way complex is possible. It is important to note that the 3*d*-orbitals of P are lowered in energy when it acts as an *n*-donor, leaving it with a partial positive charge, making it much more effective as a simultaneous *d*-acceptor. This sort of cooperative, or "synergistic," effect between the donor and acceptor functioning is characteristic of two-way action.

## 17-5 AMPHODONORS AND AMPHOCEPTORS

In classical acid-base chemistry a compound, such as $Al(OH)_3$, that can react either with hydrogen ions from an acid to give a salt plus water or with $OH^-$ ions from a base to give $AlO_2^-$ plus water is called amphoteric. By analogy we may well call a two-way interactor that is primarily an acceptor, such as $Ag^+$, an *amphoceptor* and a two-way interactor that is primarily a donor, an *amphodonor*.

To summarize, two-way donor-acceptor action occurs readily for a given D,A pair if—

1. The acceptor is an amphoceptor and the donor is an amphodonor. We note that the amphoceptor A may also form a fairly strong complex with a one-way donor.
2. For two-way action, however, the amphoceptor requires an amphodonor.

We must therefore look into the classification of donors. We find that one-way *n*-donors include $R_3N$, $R_2O$, RF, $F^-$, etc. The *n*-amphodonors include $R_3P$, $R_2S$, $R_2Se$, etc., RCl, RBr, RI, etc. These amphodonors probably all use mainly their $d\pi$-orbitals for the acceptor action (cf. Fig. 17-4).

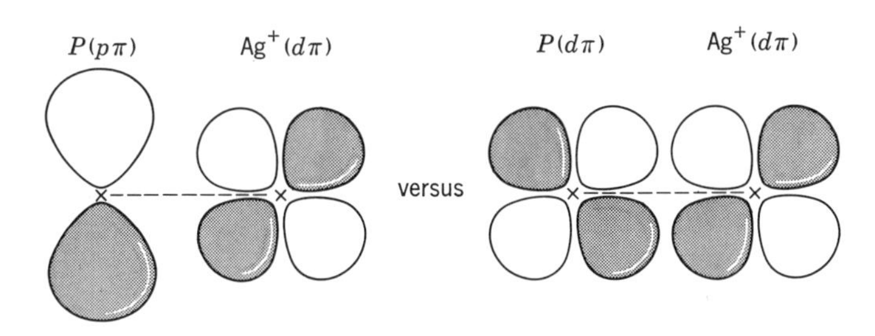


**Figure 17-4** Contrasting the $p\pi$- and $d\pi$-acceptor orbital overlap abilities for $R_3P$. Note that the overlap between the $P(d\pi)$ and $Ag^+(d\pi)$ orbitals appears to be more intimate than is the $P(p\pi)$, $Ag^+(d\pi)$ overlap.

The molecules CO, pyridine, and $PF_3$ are important and typical representatives of three varieties of amphodonor. Pyridine and CO are $n\sigma$-donors and $a\pi$-acceptors. The compound $PF_3$ is an $n\sigma$-donor (like $NH_3$) but only if at the same time it can act as a $d$-acceptor; the fact that it can be so good a $d$-acceptor is a consequence of the strong P → F polarity in the PF $\sigma$-bonds, making the P atom positive.

## 17-6 CARBON MONOXIDE AS AN AMPHODONOR

Let us first consider the complexes formed by CO, whose electronic configuration is

$$\underset{(1s_O)}{1\sigma^2} \quad \underset{(1s_C)}{2\sigma^2} \quad \underset{B}{3\sigma^2} \quad \underset{n_O}{4\sigma^2} \quad \underset{B}{1\pi^4} \quad \underset{n_C}{5\sigma^2} \quad \underset{A}{2\pi^0}, \quad {}^1\Sigma^+ \tag{17-3}$$

In (17-3) the bonding character of each MO is indicated by the symbol under it ($B$ for bonding, $A$ for antibonding, $n_O$ and $n_C$ for nonbonding orbitals mostly on O or C, respectively). Thus $3\sigma$ is *the* bonding $\sigma$ MO, $1\pi$ (twofold degenerate) is the bonding $\pi$ MO. The $3\sigma$ MO is strongly polar in the direction O → C, whereas $1\pi$ is strongly polar in the direction C → O so that it is about 25% on C and 75% on O. The $4\sigma$ and $5\sigma$ MOs are nearly nonbonding, the $4\sigma$ being concentrated largely on the O atom, whereas the $5\sigma$ MO is about 85% on the C atom and is described to a fairly good approximation as a $\sigma$ lone-pair MO. The lowest empty MO $2\pi$ is strongly antibonding and again strongly polar but now about 75% $2p\pi_C$ and 25% $2p\pi_O$, as required by orthogonality to $1\pi$. Perhaps the best single approximate valence representation of CO is as $:C^-{\equiv}O^+:$ (isoelectronic with :N≡N:), but this is not very good; for one thing, the actual dipole moment is nearly zero, which would not follow from this polar representation.

The $5\sigma$ MO qualifies for $n$-donor action. The oxygen lone-pair MO ($4\sigma$) is too deeply buried to be considered. Even so, the minimum ionization potential of CO (which corresponds to removal of a $5\sigma$-electron) is 14.01 eV, which is exceptionally high for effective donor action [cf. with 9.24 eV for benzene, 10.15 eV for $NH_3$ (but 12.59 eV for $H_2O$) from Section 9-6]. Thus CO is expected to be very weak as a one-way donor.

If CO functions as an amphodonor, however, the $5\sigma$-electron would be removed in the limit of complete donor action while at the same time an electron would be accepted from an amphoceptor into the $2\pi$-orbital, which, as we saw, is largely on the C atom. The result, characteristic of the ideal limiting case of maximum two-way action, is a good balance of charge—approximately zero net charge on the C atom and on the CO molecule. As a result of the first donor action the C atom would acquire a positive charge, making it and the molecule act as a very good acceptor for the reverse charge transfer.

## 17-7 BORINE CARBONYL

As a specific example consider borine carbonyl:

$$\underset{\left[\underset{BH_3}{\downarrow\uparrow}\right]}{OC{:}(g) + \frac{1}{2}B_2H_6(g)} \rightleftharpoons OC{:}BH_3(g) \qquad (17\text{-}4)$$

In view of the high value of $I_{CO}$, it would seem quite remarkable that this reaction could go if the CO acted as a one-way donor. Chatt has suggested [9] that $BH_3$ is a class b acceptor because of ability to act not only as a v-acceptor but also to some extent as a hyperconjugative quasi-$\pi$ donor in a two-way action. If so, the wavefunction might be approximated by

$$\Psi = a\Psi_0(OC, BH_3) + b\Psi_1(\overset{+}{O}C\text{–}B^-H_3) + c\Psi_2(OC^-\text{–}BH_3^+). \qquad (17/5)$$

Here the quasi-$\pi$ donor action of the $BH_3$ is represented in $\Psi_2$. To understand this action we consider the three B—H bonds in $BH_3$ as equivalent to one $\sigma$-bond between a hybrid boron $\sigma$ AO (mostly $s$, with some $p\sigma$) and a quasi-$\sigma$ $H_3$ group MO (denoted by $[\sigma]$) plus two quasi-$\pi$ bonds formed from the two $2p\pi$ AO's of B ($2p\pi_x$ and $2p\pi_y$) and the two matching quasi-$\pi$ group orbitals of $H_3$, denoted by $[\pi]$.† (See Fig. 17-5.) The above description is intended to refer to a pyramidal $BH_3$ which is strongly favored for dative bonding to the CO. For the pure dative structure $\Psi_1$ we would expect tetrahedral angles, but since no doubt the coefficient $a$ of $\Psi_0$ is fairly large (probably even $a > b$) in this loose complex, the geometry is probably that of a low pyramid with angles greater than tetrahedral. If one electron is transferred from a $[\pi]$ MO of $BH_3$ into a $2\pi$ MO of CO, the latter can form a bond with the remaining electron in the $[\pi]$ MO from which the electron was transferred; this could be either $[\pi_x]$ or $[\pi_y]$. (See Armstrong and Perkins [11] for some recent theoretical calculations.)

It is of interest to compare the CO force constant in $BH_3CO$ with that in free CO (19.0 millidynes/Å). Bethke and Wilson [11] found $k_{CO}$ = 18.97 in $BH_3CO$, indicating no appreciable weakening of the CO bond by the sacrificial $\pi$-acceptor action in the CO. This indicates that the coefficient $c$ of $\Psi_2$ is after all not large.

---

† The $\sigma$-$\pi$ classification is made with respect to the threefold-symmetry axis of $BH_3$.

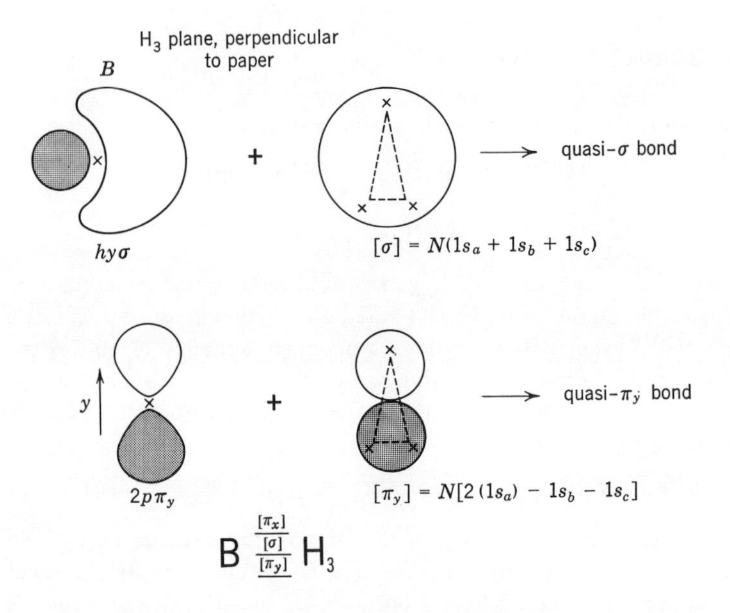


**Figure 17-5** Hyperconjugation description of bonding in $BH_3$.

We note that Chatt's classification of $BH_3$ as a case-b acceptor implies that $BH_3$ prefers P to N. As an example consider $R_3P:BH_3$. Here the $R_3P$ is like the CO, but the acceptor action by the $R_3P$ is into the $3d\pi$ AO of $R_3P$; thus it is not sacrificial and so should be favored over the action between $BH_3$ and CO.

## 17-8 THE METAL CARBONYLS

People have tried to understand compounds such as $Ni(CO)_4$ (tetrahedral symmetry) for a long time. The best attempt so far seems to be that by Nieuwpoort [13]. He has presented the results of LCAO-MO theoretical computations, and a very helpful discussion, for the following carbonyls: $Ni(CO)_4$, $[Co(CO)_4]^-$, and $[Fe(CO)_4]^{2-}$. These complexes apparently provide a very good illustration of two-way action by CO as a $5\sigma$-donor and $2\pi$-acceptor.

The electron configuration of the normal Ni atom is $\cdots 3d^8 4s^2$, with lowest state $^3F_4$, and both $Co^-$ and $Fe^{2-}$ should probably have the same normal state. Since in nickel and the neighboring transition elements the $4s$- and $3d$-orbitals have roughly the same binding energy, the $3d^{10}$, $^1S$ state is not very much higher in energy, and we shall take this state as our standard of reference in the subsequent discussion.

If the Ni atom had one special axis of symmetry ($z$) in the complex, we could write the $d^{10}$ configuration as $d\sigma^2 d\pi^4 d\delta^4$. Although this notation is not really applicable to the tetrahedral $Ni(CO)_4$, we shall nevertheless use it in a schematic or model discussion in which we assume interaction of only one CO with the Ni atom. This discussion should give a qualitatively valid idea of the true structure of the donor-acceptor action in $Ni(CO)_4$. The electron configuration of Ni is now

$$\cdots\ 3d\sigma^2 3d\pi^4 3d\delta^4 4s^0 4p\sigma^0 4p\pi^0 .$$

The donor action of Ni to the $2\pi$ CO MO must be from its $d\pi$ AOs whereas the electron accepted from the $5\sigma$ CO MO should go into a hybrid Ni AO made up predominately of $4s$ strongly mixed with $3d\sigma$ and somewhat with $4p\sigma$ (see next paragraph).

Mixing $4s$ with $3d\sigma$ yields two hybrid AOs, one of which is directed along $z$ strongly toward the CO and overlaps its $5\sigma$ MO strongly, while the other is larger around its middle so that electrons in it are completely nonbonding as far as the CO is concerned: see Fig. 17-6. Let us call these $b\sigma$ (i.e., bonding $\sigma$) and $n\sigma$ (nonbonding $\sigma$). It now appears that it would be best if we start by assuming the Ni atom to be in the configuration $n\sigma^2 3d\pi^4 3d\delta^4 b\sigma^0 4p\sigma^0 4p\pi^0$; then the CO lone pair can donate an electron into the empty $b\sigma$ AO to give a maximum $\sigma$-bonding effect. (Here we are neglecting for the sake of simplicity some further hybridization of $b\sigma$ with $4p\sigma$.) Nieuwpoort [13] has obtained some interesting results that can be stated in terms of our model. If we write the wavefunction as

$$\Psi = a\Psi_0(\mathrm{OC, Ni}) + b\Psi_1(\mathrm{OC}^+ \overset{\sigma}{\text{—}} \mathrm{Ni}^-) + c[\Psi_2(\mathrm{Ni}^+ \overset{\pi_x}{\text{—}} {}^-\mathrm{CO}) + \Psi_2(\mathrm{Ni}^+ \underset{\pi_y}{\text{—}} {}^-\mathrm{CO})], \tag{17-6}$$

he finds that $b^2$ corresponds to about 0.55 of complete transfer of one $\sigma$-electron into $\Psi_1$, and $c^2$ to about 0.21 of complete transfer of one $\pi_x$ and one $\pi_y$ electron or together 0.42 of one $\pi$-electron; that is, the Ni—C bond is about 55% of a full dative $\sigma$-bond, strengthened by about 42% of a $\pi$-bond dative in the reverse sense. In addition, Nieuwpoort finds about 12% (for $\pi_x$ plus $\pi_y$) contribution from a structure with a

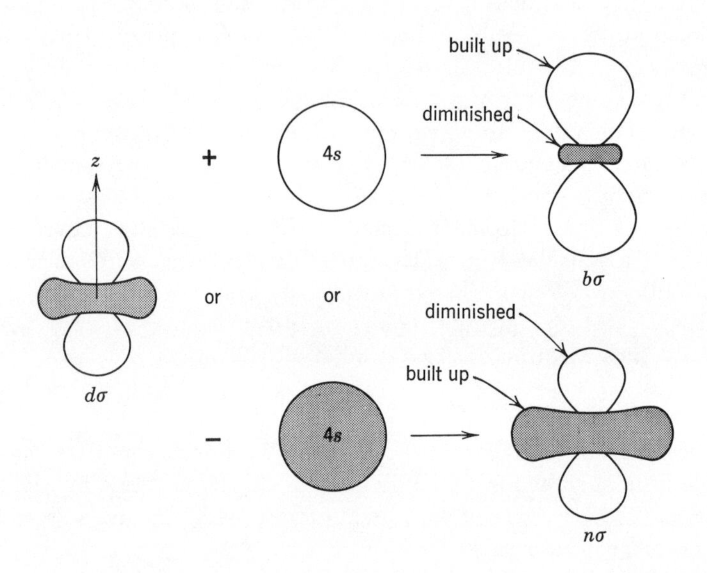


**Figure 17-6** The effect of mixing $s$ and $3d\sigma$ orbitals.

"forward" $\pi$-bond from CO acting as a sacrificial $\pi$-donor from its $1\pi$-shell to Ni as a vacant-orbital $p\pi$-acceptor.

For $[Fe(CO)_4]^{2-}$ he finds about the same coefficients, except that the amount of $\pi$ back bonding to CO increases to 132% of a single $\pi$-bond; that is, 66% of one $\pi_x$- *and* one $\pi_y$-bond, as is understandable when we note that the negative charge ($Fe^{2-}$) may be expected to make the Fe a better donor than the uncharged Ni atom.

These results can also be surveyed in terms of a population analysis. For $Ni(CO)_4$ the results are

| | |
|---|---|
| $3d^{8.3}(4s4p)^{2.7}$ | $3\sigma^{7.1}\ 4\sigma^{8.0}\ 5\sigma^{6.7}\ 1\pi^{15.5}\ 2\pi^{1.7}$ |
| Ni (net gain of one electron) | Four CO (net loss of one electron) |

compared to

| | |
|---|---|
| $3d^{10}$ (or, better, $3d^9 4s$) | $3\sigma^{8.0}\ 4\sigma^{8.0}\ 5\sigma^{8.0}\ 1\pi^{16.0}\ 2\pi^{0}$ |

for the starting Ni + 4CO. We see that the $4\sigma$ lone pair on the oxygen is not appreciably affected; the $3\sigma$ loses some population (this is sacrificial dative action), but the greatest $\sigma$-change is in the transfer from the $5\sigma$ CO MO to the $4s$ and $4p$ (or $b\sigma$ and $4p$) AOs of Ni. The $\pi$-electron transfer appears to be mostly from the $3d\pi_x$ and $3d\pi_y$ atomic orbitals on Ni to the $2\pi_x$ and $2\pi_y$ CO MOs.

The structure of $Ni(CO)_4$ is thus somewhat like

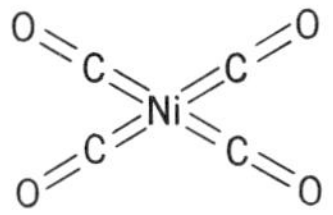

a formulation often suggested in the past, but the contributions from the $\sigma$ and $\pi$ charge-transfer bonds are unequal and are both less than for full bonds (there is much no-bond structure); thus the Ni—C linkage is not an ordinary double bond.

**TABLE 17-1**

**C—O and M—C Vibrational Frequencies in Some Carbonyls[a]**

| Infrared and Raman | $Ni(CO)_4$ | $[Co(CO)_4]^-$ | $[Fe(CO)_4]^{2-}$ | CO | $H_2CO$ |
|---|---|---|---|---|---|
| $\nu_{CO}$, $a_1$[b] | 2128 | 1883 | – | 2143 | 1743 |
| $f_2$[b] | 2057.7 | 1918 | 1786 | | |
| $\nu_{MC}$ $a_1$ | 380 | 439 | 464 | | |
| $f_2$ | 423 | 530 | 550 | | |
| $k_{CO}$ (millidynes/Å) | 17.6 | 14.4 | 12.6 | 19.0 | 12.1 |
| $k_{MC}$ (millidynes/Å) | 1.98 | 3.22 | 3.36 | | |

[a] Data from Nieuwpoort [13]. The units of $\nu_{CO}$, etc., are $cm^{-1}$.
[b] Symmetry classification under $T_d$ symmetry.

Some pertinent experimental data from the vibrational spectra are included in Table 17-1. To see the sacrificial effect of the back $\pi$-bonding on the CO we may compare the weighted average frequency $(\nu_a + 3\nu_{f_2})/4$ in the complex with that in the free molecule or, better, we may compare the CO stretching-force constants $k_{CO}$ in the next to last line. The comparison with $H_2CO$, where there is only one $\pi$-bond, is also relevant, since in the extreme case that the CO in a carbonyl accepted two electrons into the $2\pi$ MO (one into $2\pi_x$ and one into $2\pi_y$) these would about half-cancel the effects of the four bonding electrons in the $1\pi$ MO, leaving something like one $\pi$-bond as in $H_2CO$. (This discussion is somewhat oversimplified.) The most significant trends in the above table seem to be the increase in $k_{MC}$ and decrease in $k_{CO}$ as the amount of $\pi$ back bonding increases through the series.

## 17-9 OTHER AMPHODONORS

Some other amphodonors are tabulated below:

| | | |
|---|---|---|
| $\underline{R_3P}$ | $R_2S$ | RCl |
| $R_3As$ | $R_2Se$ | RBr |
| $R_3Sb$ | $R_2Te$ | $\underline{RI}$ |

The underlined donors are recognized as the strongest in each (vertical) group, but the reason for this seems not to be known.

Another interesting donor type is that of the isonitriles, RNC:. The structure of these molecules is similar to that of CO. If we approximate the latter as $\overset{+}{O}{\equiv}C^-$:, as discussed above, we may consider the analogous valence structure for the isonitriles to be $R{-}\overset{+}{N}{\equiv}C^-$:.

In the donor type $R_3P$ above R can be F. The $PF_3$ molecule is a surprisingly good amphodonor. Why? The high electronegativity of F causes it to remove charge from the P in the $\sigma$-bonding:

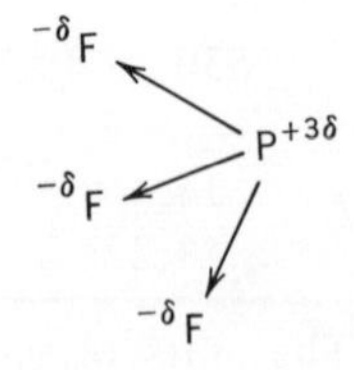

The resulting considerable positive charge on the P atom strongly enhances its $\pi$-acceptor capability in a two-way complex, apparently outweighing the unfavorable effect on its $\sigma$ lone-pair donor capability.

A somewhat similar discussion applies to the $\phi_3P$ molecule ($\phi$ = phenyl) which is a good two-way donor. In this case presumably the possibility of intramolecular dative conjugation between the phenyl rings and any $d\pi$ charge that is acquired by P enhances the $d\pi$-acceptor capacity of the phosphorus atom.

Pyridine is a well-known amphodonor in complex ions. All our discussion so far has emphasized its action as an $n$-donor in one-way $\sigma$-action. However, the empty $\pi^*$-orbital is able to accept charge in sacrificial back-bonding $\pi$-action. Since $\pi^*$ is less antibonding for Py than for CO, this $\pi$-acceptor action is less strongly sacrificial for the aromatic ring and may be more favorable than that in CO.† At any rate both Py and CO as amphodonors combine lone-pair (hence increvalent) $\sigma$-donor action with sacrificial $\pi$-action—a situation that is considerably more favorable than that for purely $\pi$-amphodonors like ethylene, benzene, or acetylene in which both donor and acceptor actions are sacrificial.

## 17-10 MORE EXAMPLES OF COMPLEXES WITH TWO-WAY ACTION

Many examples of two-way donor-acceptor action are found among more complicated coordination compounds. Here we shall list just a few of these. First there are the complexes between ethylene (or other Un) with platinum. A planar arrangement, except for the ethylene, whose long axis stands perpendicular to the plane, is characteristic; for example,

$$\left[\begin{array}{ccc} C_2H_4 & & Cl^- \\ & Pt^{2+} & \\ Cl^- & & Cl^- \end{array}\right]^- K^+.$$

The two-way action between the platinum (or other amphoceptor ion) and the amphodonor ethylene is, roughly (see top of next page); for an accurate description we should need an overall MO approach. Here the $C_2H_4$ plane is perpendicular to the line from the midpoint of the $C_2H_4$ to the Pt. The $Cl^-$ ions also function as amphodonors ($n$-donors and

† In this connection we may ask whether the form and polarity in Py of $\pi^*$ is as favorable as the polarity in CO of $2\pi$. On the other hand, Py as an $n$-donor is surely stronger than CO. Further: is Py an anomalously stronger base toward transition metal ions than toward $H^+$?

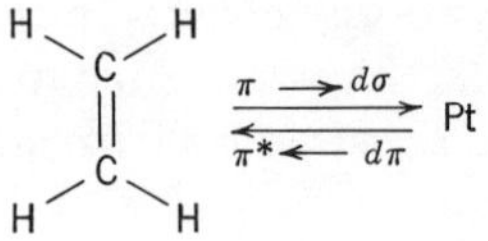

$H_2C{=}CH_2$ $\pi \rightarrow d\sigma$ Pt; $\pi^* \leftarrow d\pi$

$d\pi$-acceptors, but mainly $n$-donors) toward the $Pt^{2+}$; or with considerable justification we can describe the Pt—Cl bonding largely in terms of ordinary polar Pt—Cl $\sigma$-bonds like the $\sigma$-bond in HCl.

Another rather similar example is the tetrahedral Ni complex including the amphodonor $\phi_3P$ (cf. Section 17-9):

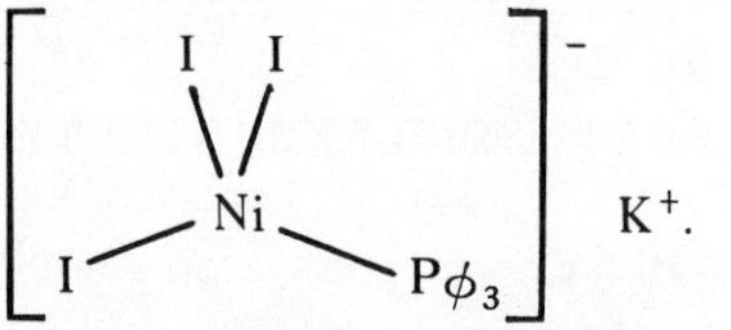

$$[NiI_3(P\phi_3)]^- \; K^+.$$

Neutral Pt complexes include the very stable

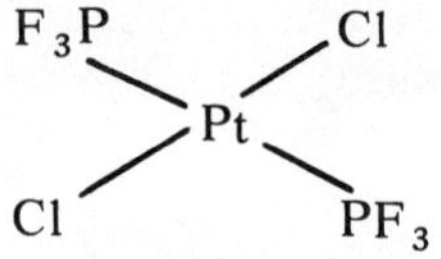

$$(F_3P)ClPtCl(PF_3)$$

(cf. Section 17-9) and

$$(H_4C_2)ClPt(\mu\text{-}Cl)_2PtCl(C_2H_4)$$

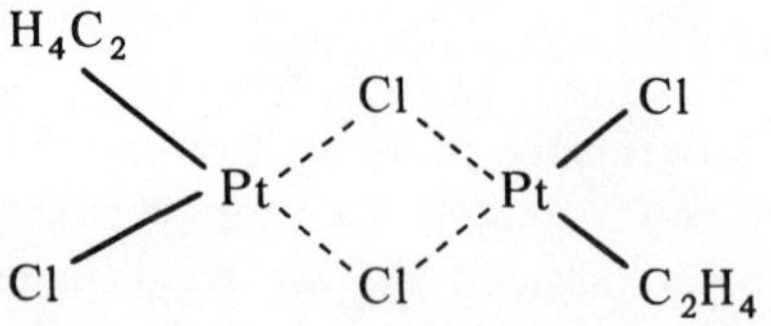

These are just a few examples selected out of very many known complexes.

## 17-11 TWO-WAY CHARGE TRANSFER AS PARTIAL DOUBLE BONDING

In describing two-way charge-transfer complexes a needed further contribution to $\Psi$ (namely, $\Psi_{12}$, a covalent double-bond structure) has not yet been discussed:

$$\Psi = a\Psi_0(\mathrm{D, A}) + b\Psi_1(\mathrm{D}^+ \overset{\sigma}{-} \mathrm{A}^-) + c\Psi_2(\mathrm{D}^- \overset{\pi}{-} \mathrm{A}^+) + d\Psi_{12}(\mathrm{D{=}A}). \qquad (17\text{-}7)$$

To illustrate the meaning of $\Psi_{12}$, let us consider the amine oxides (Section 18-2) as examples. For pyridine-*N*-oxide, intramolecular two-way action occurs since the $2p\pi$-orbital of O can donate back into the $\pi$-system of the ring:

(Py⟶O) ⟷ $\mathrm{C_5H_5}\overset{+}{\mathrm{N}}\overset{\sigma}{-}\mathrm{O}^-$ ⟷ $(-)\mathrm{N}\ \ \mathrm{O}^+$ ($\pi$)

$\psi_0$ $\qquad$ $\psi_1$ $\qquad$ $\psi_2$

Let us examine the orbitals involved in $\Psi_1$ and $\Psi_2$ in order to picture the electron distribution that results from this two-way action. For $\Psi_1$ we have clearly a $\sigma$-bond. As a result of the back donation in $\Psi_2$ the oxygen $2p\pi$-orbital contains only one electron; thus it can pair with the electron in a $\pi^*$-acceptor MO of the Py ring:

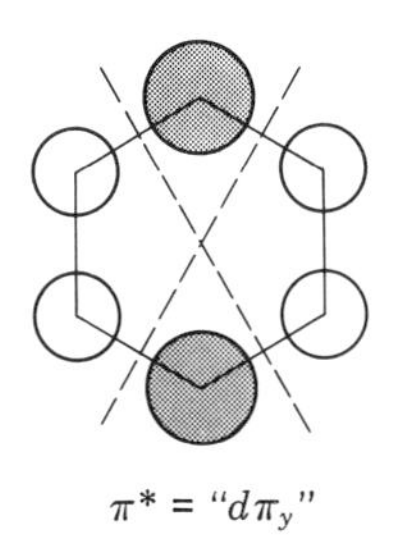

$\pi^* = $ "$d\pi_y$"

This $\pi^*$MO in its LCAO formulation is fairly strong on the N atom; hence in $\Psi_2$ we have a $\pi$-bond between the N and O, or, more accurately, between the pyridine ring and the O. Now suppose (as a purely hypothetical

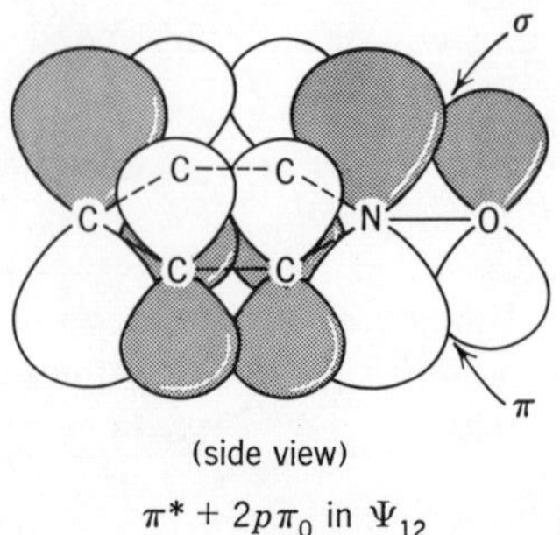


(side view)

$\pi^* + 2p\pi_0$ in $\Psi_{12}$

case) that $a = 0$ and $d = 0$ but $b = c$ in (17-7). This combination of equal amounts of $\Psi_1$ and $\Psi_2$ would have the characteristics of *one-half* of a double bond, since $b = c$ says that $\Psi$ is only half $\Psi_1$ and half $\Psi_2$. On the other hand, $\Psi_{12}$, which combines the characteristics of both $\Psi_1$ and $\Psi_2$, represents a *full* double bond; that is, for $a = b = c = 0$ but $d = 1$ we would have a full double bond, whereas the case $a = d = 0$ with $b = c$, though similar, corresponds to only a half double bond. Actually it seems likely that in pyridine *N*-oxide the largest coefficient is $b$ but that $a$, $d$ and perhaps also $c$ are all important. It is of interest to note that $\Psi_0$ and $\Psi_{12}$ are rather nearly orthogonal, as also are $\Psi_1$ and $\Psi_2$, whereas the nonorthogonality of $\Psi_1$ to $\Psi_0$ and to $\Psi_{12}$, also of $\Psi_2$ to $\Psi_0$ and $\Psi_{12}$, is fairly high.[†] Presumably the interaction matrix elements, as we found earlier in the one-way case, are approximately proportional to the nonorthogonalities just mentioned.

As another example, consider the two-way complex Py → $Ag^+$. Here the wavefunctions are

N, $Ag^+$ ⟷ $\overset{+}{N}$–Ag

$\Psi_0$ $\Psi_1$

N $\underset{\sigma}{\overset{\pi}{\rightleftarrows}}$ $Ag^+$ ⟷ (–) N $Ag^{2+}$ "$d\pi$"

$\Psi_{12}$ $\Psi_2$

[†] The nonorthogonalities in the latter cases are proportional to the overlap integral between a donor and an acceptor orbital, whereas in the former they are proportional to the *product* of two such overlap integrals.

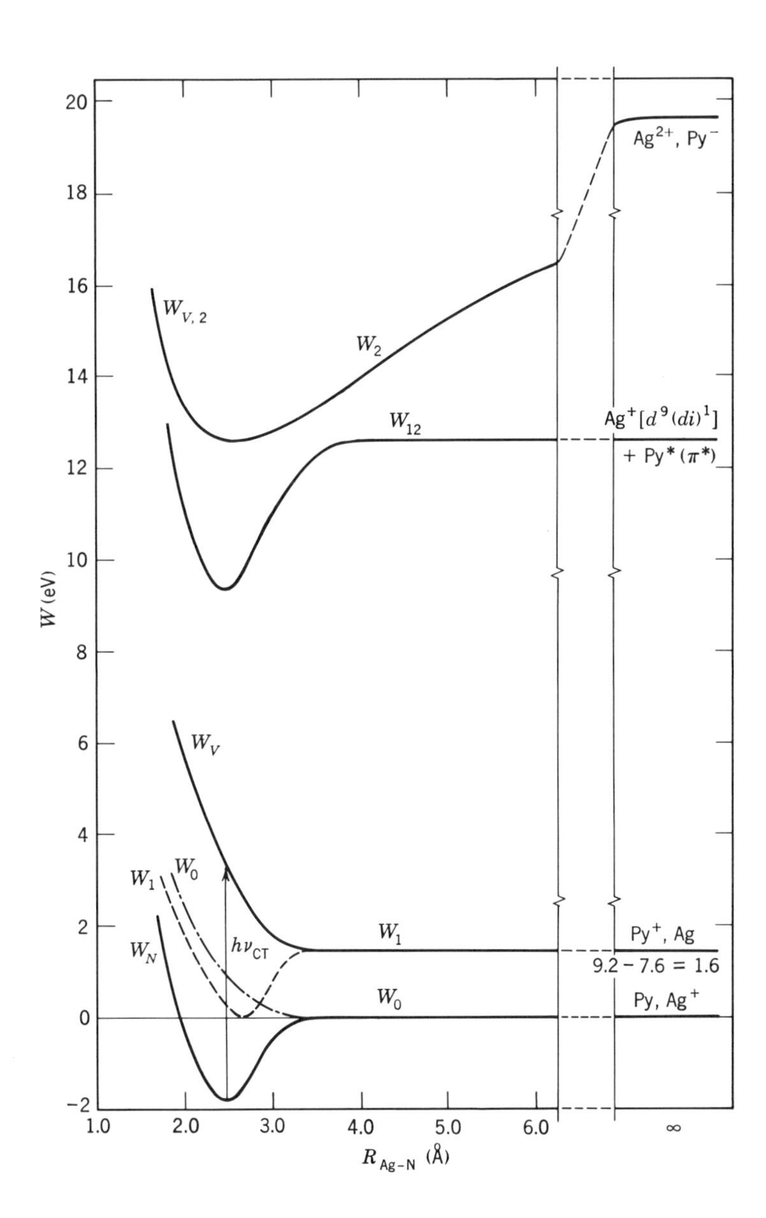


**Figure 17-7** Potential-energy curve for pyridine · $Ag^+$.

This example is similar to (Py⟶O), except that $Ag^+$ uses a $d\pi$-orbital. Fig. 17-7 shows an estimation of the potential curves for the Py⟶$Ag^+$ system. We estimate the energies at $R_{DA} = \infty$ using $I_{Py} = 9.2$ eV; $I_{Ag} = E_{Ag^+} = 7.6$ eV; $I_{Ag^+} = 21.5$ eV; $E_{Py} = 0$ eV. The location of the valence state for $\Psi_{12}$ is estimated as the average of the energy for $Ag^+$ to go from $d^{10}$ to $d^9 s$ and to $d^9 p$ plus the excitation energy corresponding to the $\pi^* \leftarrow n$ excitation of Py. This state is then strongly stabilized by the valence interaction to form the double bond, and also interacts strongly with the $W_2$ curve (see Fig. 17-7). However, we see that the bonding of the complex probably comes primarily from mixing of $\Psi_1$ and $\Psi_0$, with only secondary mixing of $\Psi_2$ and $\Psi_{12}$, with both of the latter probably being about equally important.

## 17-12 REFERENCES


[1] L. E. Orgel, *An Introduction to Transition Metal Chemistry,* Wiley, New York, 1960.

[2] C. J. Ballhausen, *Introduction to Ligand Field Theory,* McGraw-Hill, New York, 1962.

[3] C. K. Jørgensen, *Absorption Spectra and Chemical Bonding in Complexes,* Pergamon, Oxford, 1962.

[4] H. J. Taufen, M. J. Murray, and F. F. Cleveland, *J. Am. Chem. Soc.*, **63**, 3500 (1941).

[5] H. G. Smith and R. E. Rundle, *J. Am. Chem. Soc.*, **80**, 5075 (1958).

[6] R. W. Turner and E. L. Amma, *J. Am. Chem. Soc.*, **88**, 3243 (1966). See also ibid., **85**, 4046 (1963); ibid., **88**, 1877 (1966).

[7] M. Dewar, *Bull. soc. chim. de France,* **18**, C79 (1951).

[8] J. C. Slater, *Quantum Theory of Atomic Structure,* vol. 1, McGraw-Hill, New York, 1960, p. 210.

[9] S. Ahrland, J. Chatt, and N. R. Davies, *Quart. Rev.*, **12**, 265 (1958).

[10] R. G. Pearson, *J. Am. Chem. Soc.*, **85**, 3533 (1963).

[11] D. R. Armstrong and P. G. Perkins, *J. Chem. Soc.*, 1044 (1969).

[12] G. W. Bethke and M. K. Wilson, *J. Chem. Phys.*, **26**, 1118 (1957).

[13] W. C. Nieuwpoort, Ph.D. thesis, University of Amsterdam, 1965.

*CHAPTER* **18**

# *Intramolecular Donor-Acceptor Action*

## 18-1 INTRAMOLECULAR ONE-WAY AND TWO-WAY DONOR-ACCEPTOR ACTION

Intramolecular donor-acceptor action of two kinds has already been discussed briefly in Section 4-2. One kind can be formulated as R,Q action between an odd-electron donor R and acceptor Q. Here $\Psi_0$ is a covalent and $\Psi_1$ an ionic structure. All polar single bonds, as in HCl, $NH_3$, or the C—H bonds in benzene, are examples of $R\sigma,Q\sigma$ action.

A second kind is that of $n\pi,a\pi$ action as in chlorobenzene or anisole—or $b\pi,a\pi$ action as in nitrobenzene. Here $\Psi_0$ is a no-$\pi$-bond structure involving two $\pi$-islands, (one consisting of the ring $\pi$ electrons, the other consisting of the $\pi$ electrons on the substituent) whereas in the dative structure $\Psi_1$ an electron has been transferred from one $\pi$-island (from Cl as $n\pi$-donor in chlorobenzene but from $C_6H_5$ as $b\pi$-donor in nitrobenzene) to another ($C_6H_5$ as an $a\pi$-acceptor in chlorobenzene, $NO_2$ as an $a\pi$-acceptor in nitrobenzene).

Actually the second type does not occur alone but only *accompanied* by the first type. There are two cases: (a) both types of charge transfer may go in the same direction (for example, nitrobenzene); but usually (b) they go in opposite directions (as in chlorobenzene). The latter case exemplifies a type of two-way charge transfer that is different from that discussed in Chapter 17. However, the concepts of amphodonors and amphoceptors introduced in Section 17-5 can also be used here: in chlorobenzene or anisole the Cl or $OCH_3$ group is a $\pi$-island amphodonor, the $C_6H_5$ group is a $\pi$-island amphoceptor.†

## 18-2 INTRAMOLECULAR TWO-WAY ACTION IN PYRIDINE-*N*-OXIDE AND ALKYL PHOSPHINE OXIDES

In the molecule pyridine-*N*-oxide

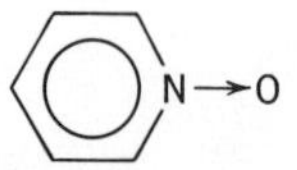

N→O

(see also Section 17-11) the pyridine is an amphodonor ($n\sigma$-donor and $a\pi$-acceptor) toward the O atom, whereas the O atom is an amphoceptor ($v\sigma$-acceptor and $n\pi$-donor). It is interesting to compare this compound with $(CH_3)_3N \rightarrow O$, where (aside from a weak hyperconjugation effect) only one-way $n\sigma$-action is possible within the molecule. Although both pyridine-*N*-oxide and the aliphatic amine oxides act as strong one-way lone-pair *intermolecular* donors at the O atom, the latter are much the stronger bases toward both $H^+$ and $I_2$—as is readily understandable since the O atom, which must be strongly negative as a result of the $N \rightarrow O$ intermolecular *n*-donor action, has in the pyridine-*N*-oxides lost part of this negative charge through the reverse $\pi$-action. The O atom is, however, a much stronger donor in the amine oxides than in the ethers or alcohols because of the enhanced negative charge that results from this intramolecular two-way action with the nitrogen atom.

Another similar case occurs in the phosphine oxides, where even in $(CH_3)_3P \rightarrow O$ there is two-way action because of the $d\pi$ vacant orbital of the P atom. These then because of the intramolecular two-way action are not as good one-way donors at the O atom as they would otherwise be.

## 18-3 COMPETITION BETWEEN INTRAMOLECULAR AND INTERMOLECULAR DONOR-ACCEPTOR ACTION: PYRIDINE-*N*-OXIDE AND THE ANILINES

† In nitrobenzene, $C_6H_5$ is a double (*R* and $b\pi$)-donor; the $NO_2$ group is a double ($\sigma$ and $\pi$) acceptor.

The weaker donor action of pyridine-*N*-oxide toward $I_2$ may also be discussed in terms of a competition between the $C_6H_5N$ ring as an intramolecular $a\pi$-acceptor and the $I_2$ molecule as an intermolecular $a\sigma$-acceptor, for the $n\pi$ lone-pair electrons of the O atom. The back charge transfer from the O atom to the ring in $n\pi$-donor action reduces the negative charge that the O atom has acquired in intramolecular $\sigma$-bond charge transfer from the nitrogen; this increases its ionization energy and makes it a poorer intermolecular donor in comparison with the aliphatic amine oxides. In agreement with this expectation Kubota [1] showed that $-\Delta H$ is much smaller (5.85 kcal/mole) for the formation of $I_2$ complexes of

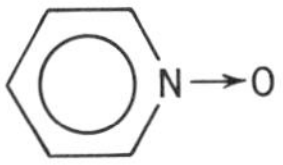

N→O

pyridine-*N*-oxide than for those of $R_3N \rightarrow O$ (10.0 kcal/mole for $R = CH_3$).

Still another example that illustrates the competition between intramolecular and intermolecular charge-transfer action is afforded by the dimethylaniline · $I_2$ complex. The intramolecular action in dimethylaniline is illustrated by

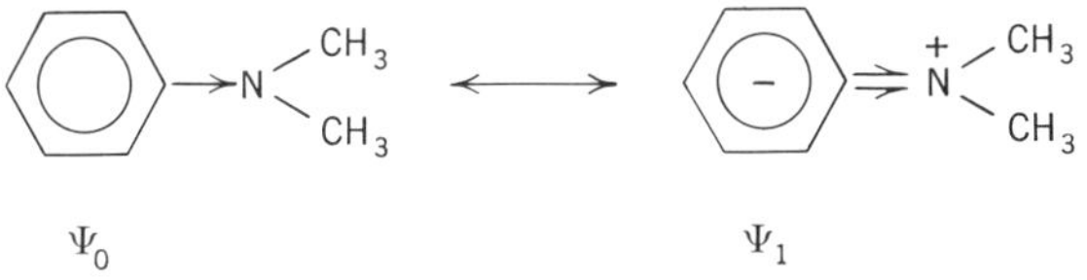

$\Psi_0$ $\longleftrightarrow$ $\Psi_1$

[The arrows pointing from the $C_6H_5$ toward the N atom indicate the polarity (from positive to negative) of the bonds.] Now bring in $I_2$ as an $a\sigma$ intermolecular acceptor. We may expect aniline or dimethylaniline to tend to act as an $n$-donor toward $I_2$, using the N-atom lone-pair electrons just as in $R_3N \cdot I_2$. However, it is just these electrons that are acting to make the $NR_2$ or $NH_2$ group an $n\pi$-donor toward the $C_6H_5$ group in intramolecular $n\pi,a\pi$ action. If these $\pi$-island electrons did not conjugate with the ring (that is, if $\Psi_1$ above did not mix with $\Psi_0$), we would expect a pyramidal arrangement of the three bonds from the N atom, just as in $R_3N$, so that aniline would be nonplanar. Actually it is believed to be distinctly nonplanar. However, even though it is nonplanar, the lone-pair electrons are still an "almost $\pi$-island," and although the intramolecular charge-transfer action is somewhat reduced by the nonplanarity, it still occurs, competing with any intermolecular charge-transfer action to $I_2$ and thus reducing the base strength of aniline toward $I_2$ as compared with that for a comparable aliphatic amine. (See Tsubomura [2].)

## 18-4 STRUCTURE OF $BX_3$; COMPETITION OF INTRAMOLECULAR DONOR-ACCEPTOR ACTION WITH INTERMOLECULAR ACCEPTOR ACTIONS

Now let us consider some details about the internal structures of *v*-acceptors of the type $BX_3$ and $AlX_3$, using a resonance-structure description for the insight it gives, even though an overall MO description would be less clumsy. These molecules are planar and symmetrical if X is a halogen; for example

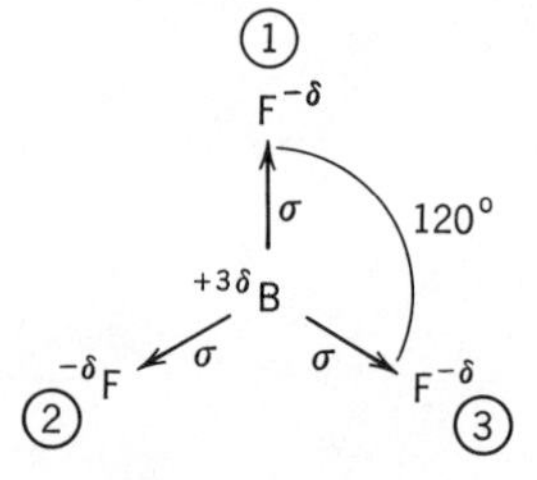

In $BF_3$ there is an important $R\sigma,Q\sigma$ charge transfer from B to the Fs through strong polarity in the $\sigma$-bonds. Using localized $\sigma$-bonds, these are formed from boron AOs, which are approximately trigonal $[s^{(1/3)}p^{(2/3)}]$ and $2p\sigma$ fluorine atom AOs. From the odd-odd charge-transfer point of view using resonance structures, the B—F $\sigma$-bonds are described by

$$\Psi = a\Psi_0(\text{B—F}) + b\Psi_1(\text{B}^+\text{F}^-) + \cdots.$$

Let us take local axes at each atom with the $x$-axis perpendicular to the plane of the molecule and the $z$-axes along the $\sigma$-bonds. Besides the electrons in the $\sigma$-bonds there are three lone pairs—one $2s$ and two $2p\pi$—on each F atom. One of the latter is in a $2p\pi_x$ AO. If we consider just one B—F bond, the F atom is in a field of approximately cylindrical symmetry (somewhat modified by the planar field of the $BF_3$ that distinguishes $2p\pi_x$ from $2p\pi_y$). The B atom has used all three of its valence electrons in $\sigma$-bonding, so *it* has a vacant $2p\pi_x$ AO.

As a result the $BF_3$ molecule is all set up for $\pi_x$ donor-acceptor action from F to B, thus transferring some of the negative F-atom charge back to B. Describing this action in terms of resonance structures we have

$$\Psi = a\Psi_0(\text{BF}_3) + b\Psi_1(\pi_{\text{F}1}^{-1}\pi_\text{B} + \pi_{\text{F}2}^{-1}\pi_\text{B} + \pi_{\text{F}3}^{-1}\pi_\text{B}).$$

Here $\Psi_0(BF_3)$ is the wavefunction for $BF_3$ with three strongly polar $\sigma$-bonds; $\pi_{\text{F}i}^{-1}$ represents the removal of one electron from the $2p\pi_x$ AO of the $i$th F atom; this electron is transferred to the $2p\pi_x$ orbital of $B(\pi_B)$. The above $\Psi$ can also be described by the shorthand method shown below.

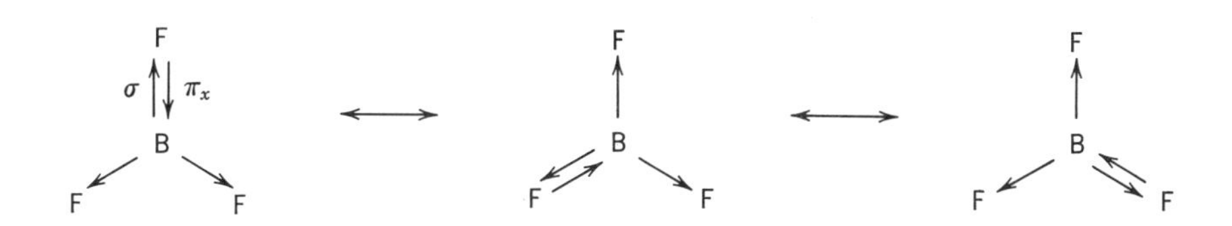


Here each arrow indicates charge transfer in the direction of the arrow. (See Pauling [3] for further discussion.)

Now let us ask about the extent of the $\pi$ charge-transfer action. Cotton and Leto [4] made a careful estimate of the energy that is gained in $BX_3$ and $AlX_3$ as a result of the partial $\pi$ back charge transfer, using the resonance-structure method. Their results are summarized in Table 18-1. We see that the $\pi$-bond stabilization that is due to intramolecular charge transfer into the "vacant" orbital on boron is 48 kcal/mole, or about 2 eV for $BF_3$ and significantly less for $BCl_3$.

It is clear that in correspondence to the energy effects just noted the "vacant" orbital on boron in $BF_3$ contains an appreciable population of electrons. Hence it does not accept from an intermolecular donor as readily as if it were completely empty. Empirically, the strengths of the $BX_3$ acceptors with $R_3N$ are in the order $BCl_3 > BBr_3 \gg BF_3$. We can

**TABLE 18-1**

**Stabilization Energies in $BX_3$ and $AlX_3$[a,b]**

| Energy Type | $BF_3$ | $BCl_3$ | $BBr_3$ | $AlCl_3$ | $AlBr_3$ |
|---|---|---|---|---|---|
| σ-Bond | 626 | 465 | 406 | 442 | 381 |
| π-Bond | 47.8 | 29.8 | 25.7 | 31.3 | 27.6 |

[a] Data from [4].
[b] Units are kilocalories per mole.

understand this order by realizing that the vacant orbital in a *v*-acceptor often must first be *made* vacant before it can accept electrons in an intermolecular action. From Table 18-1 we see that it will cost about 2 eV in energy to prepare $BF_3$ for this action, considerably more than for $BCl_3$.

Of course in a complex some compromise must be reached between the two competitive (intramolecular and intermolecular) actions. A good indication of the extent of this compromise is the geometrical configuration of the D—$BX_3$ complex. If the intermolecular donor action were complete, the XBX angles in the $BX_3$ group (now $B^-X_3$) would be expected to be tetrahedral (109°28') instead of 120° as in the isolated molecule. The experimental data (see Table 5-1) show that an intermediate compromise geometry is indeed found.

## 18-5 LONE PAIRS AND VACANT ORBITALS

In neutral *v*-acceptors we may expect to find as just illustrated for $BX_3$ that very often in the natural state of the molecule the *v*-orbital is not entirely vacant. Similarly in *n*-donors the lone pairs are usually not fully lone. In Section 17-6 we have seen that in carbon monoxide the lone pair is about 85% on carbon and about 15% on oxygen. It is, however, slightly antibonding, which makes it more readily donated.

In pyridine there are many pairs of electrons in valence-shell σ MOs, and these MOs extend over all the five carbon atoms as well as the nitrogen atom. However, according to theoretical calculations by Clementi [5], the most easily ionized of these σ MOs is concentrated about 75% on the nitrogen atom. Thus, rather surprisingly, there is an MO that is a passable approximation to a lone pair MO, though far from being a perfect one. Nevertheless when pyridine functions as an *n*σ-donor it may actually, by using other, deeper lying σ MOs to some extent, in effect use a hybrid donor MO that, though not 100% lone, is more nearly so than the most easily ionized $\sigma$ MO.

In $R_3N$ the lone pairs are more typically lone. Even here, however, the donor orbital extends very appreciably over the Rs. Since the interaction with the Rs may be antibonding, the fact that the orbital is not exactly lone may even improve its donor action.

## 18-6 BORINE AND TRIMETHYLBORON

In contrast with $BF_3$, we note that in $BH_3$ the vacant orbital is expected really to be vacant. This fact may help to account for the stability of $BH_3CO$. We have seen from Table 18-1 that for $BF_3$ or $BCl_3$ to act as $v$-acceptors we must first supply from 1 to 2 eV to make the acceptor orbital vacant. Borine does not require such preliminary preparation; this fact may largely explain why it is able to accept from a weak $n$-donor such as carbon monoxide when $BF_3$ cannot. However, as we have seen in Section 17-7, a hyperconjugative $[\pi]$ back donation from $BH_3$ to CO (but which would be much weaker for $BF_3$) may also help.

In $(CH_3)_3B$, on the other hand, the $CH_3$ groups (acting like F atoms) may interact with the vacant $\pi$-orbital by intramolecular dative hyperconjugation from the quasi-$\pi_x$ orbital of each $CH_3$ group. It seems likely that it is the resulting dative hyperconjugation energy that stabilizes the $B(CH_3)_3$ monomer over the dimer—in contrast to borine, which is not stable alone but is found as the dimer, $B_2H_6$.[†] Internal hyperconjugation in $(CH_3)_3B$ should result in a decrease in intermolecular acceptor strength as compared with $BH_3$, although in a hypothetical complex $OC \cdot B(CH_3)_3$ there should still be some intermolecular hyperconjugation like that in $OC \cdot BH_3$, though weaker. (The $B{<}^{C}_{C}{-}C$ bonds should function like the $B{<}^{H}_{H}{-}H$ bonds, but less strongly according to experience with C—C versus C—H hyperconjugation in purely carbon compounds.)

Added in proof: recent theoretical calculations indicate that hyperconjugation, while present, is not decisive in accounting for the monomeric character of $(CH_3)_3B$ [6].

---

[†] Similarly the fact that $BF_3$ and $BCl_3$ are monomeric may be ascribed to the stabilizing action of dative conjugation in the monomer as described in Section 8-4.

## 18-7 REFERENCES

[1] T. Kubota, *J. Am. Chem. Soc.*, **87**, 458 (1965).
[2] H. Tsubomura, *J. Am. Chem. Soc.*, **82**, 40 (1960).
[3] L. Pauling, *The Nature of the Chemical Bond,* 3rd ed., Cornell University Press, Ithaca, N.Y., 1960.
[4] F. A. Cotton and J. R. Leto, *J. Chem. Phys.*, **30**, 993 (1959).
[5] E. Clementi, *J. Chem. Phys.*, **46**, 4731 (1967).
[6] A. H. Cowley and W. D. White, *J. Am. Chem. Soc.*, **91**, 34 (1969).

# *Postscript*

While the manuscript for this book was in preparation a great many papers have been published in this field. One development, in particular, has been the realization that electrostatic attraction forces exist even between nonpolar molecules such as benzene and iodine. This realization and other related considerations have raised again the question of the relative importance of charge-transfer forces in determining the properties of the complex. After the completion of the main body of this book, we have prepared the following paper (submitted to the *Journal of the American Chemical Society* for publication) discussing some of these topics. Since this paper is not yet available as a reprint, we include it here in the text as a postscript.

# *Molecular Compounds and Their Spectra. Some General Considerations**


## ABSTRACT

The question of the meaning and relative importance of classical electrostatic (Coulomb and polarization) and charge-transfer (CT) forces for the stability and dipole moments of electron donor-acceptor (EDA) complexes is critically examined, and discussed in some detail for $\pi \cdot \pi$ and $\pi \cdot$halogen complexes. The importance of distinguishing between complexes with donor and acceptor both sacrificial, which are usually weak, and complexes with increvalent (lone pair) donors, which are often strong, is emphasized. It is concluded that while classical electrostatic forces make significant contributions to the stability of donor-acceptor complexes they perhaps are of predominant importance only for the weakest complexes. It is pointed out that although London dispersion forces necessarily contribute to the stability of vapor-state complexes, their effects are approximately cancelled out in the formation of complexes in solution.


## INTRODUCTION

A moot question in the understanding of the structure and spectra of


* This work was assisted by the Office of Naval Research, Physics Branch, under Contract No. N00014-67-A-0285-0001, and by a grant from the National Science Foundation, GP-9284.

EDA complexes is the extent to which classical electrostatic forces, as against CT forces, contribute to the energy of formation and dipole moments of EDA complexes [1–6]. This question which we examine in more detail below, is difficult to resolve quantitatively because both kinds of force are often qualitatively expected to contribute to these properties. Briegleb earlier accounted for the stability of certain $\pi \cdot \pi$ complexes, in particular nitro-compound complexes, by electrostatic forces, but later [7] adopted Mulliken's view that CT forces are of predominant importance except in hydrogen-bonded complexes. However, Mulliken now agrees that electrostatic forces may be responsible to an important and very likely sometimes predominant (although not exclusive) extent for $\pi \cdot \pi$ and other relatively weak complexes, as well as for H-bonded complexes; whereas for strong $n \cdot \sigma$ and $n \cdot v$ complexes, CT forces are clearly predominant [8]. For this reason he has adopted the general descriptive name "electron donor-acceptor complexes" used by Briegleb in his well-known book [7], instead of the name "charge-transfer complexes" which he at first introduced.

## COMPARISON OF THE RELATIVE IMPORTANCE OF CHARGE-TRANSFER AND OTHER FORCES

Classical electrostatic forces include both Coulomb (e.g. dipole-dipole) and polarization, or "induction," (e.g. dipole-induced-dipole) attractive forces. Hanna [1,2] has recently shown that quadrupole-induced-dipole forces may be comparable in importance to CT forces in explaining the stability of benzene $\cdot I_2$ and of similar $b\pi \cdot a\sigma$ [9] complexes. He has also shown that the observed dipole moment of the benzene $\cdot I_2$ complex must be caused, to a non-negligible extent, by the quadrupole-induced-dipole polarization forces, whereas earlier Mulliken had neglected this possibility and concluded that the observed dipole moment must result entirely from CT forces. (See also Mantione [5]).

Besides Hanna, other authors [3–6] have also recently argued the importance, if not preponderance, of Coulomb and polarization forces for the stability of EDA complexes, especially those of the $b\pi \cdot a\pi$ type ($\pi$ donor and $\pi$ acceptor) [9]. Mantione [5] has also pointed out that polarization forces can generate dipole moments in such complexes even if neither partner has a dipole moment; in particular, she presents calculations on hydrocarbon · TCNE complexes. For the napthalene · and pyrene · TCNE complexes, she obtains results in agreement with observed dipole moments. However, this agreement is obtained only by using charges of approximately +0.4 e and –0.4 e on the C and N atoms of each

CN group in TCNE. On the basis of SCF calculations on related molecules, it seems very improbable that the CN groups contain such large local dipoles. Thus, while polarization must contribute, it seems unlikely that it makes the major contribution in these cases.

Some authors [3,4] have sought to invoke the nonclassical London dispersion forces as making major contributions to the energies of formation of EDA complexes. In the usual case of complexes in *solution*, however, such contributions must ordinarily cancel out to a large extent, since when a complex is formed in solution the gain in intracomplex dispersion energy is approximately balanced by a loss of solvent-donor and solvent-acceptor dispersion energies [10]. In the vapor state, of course, the dispersion-force contributions must be important.

Some authors [5,6] point out that the wavefunctions of a complex, including those of its CT states, can be expressed in terms of expansions using only donor and/or acceptor (including *excited state*) wavefunctions, with no CT terms in the expansion. While this point is valid, it does not contradict the concept of charge transfer, nor is it very helpful in understanding the electronic distribution. It is somewhat analogous to saying, for example, that the electronic wavefunctions for a molecule can be expressed in terms of excited functions of just one of its atoms instead of by the usual LCAO expressions using AOs of more than one atom. (For example, $\Psi$ of HF can be represented using AOs of the F atom only, or $\Psi$ of $N_2$ can be represented using AOs of just one—either one—of the two N atoms.) The reason for using the LCAO procedure for molecules, or for the inclusion of dative $(D^+A^-)$ CT functions in representing the $\Psi$s of EDA complexes, is that such procedures make use of selected members of an overcomplete set of non-orthogonal functions and are more economical—more rapidly converging—than those using a complete orthogonal set of functions. When truncation is necessary, as it is for an approximate description, such a selection procedure represents a sensible and rational mode of description. It is only in this context that one properly speaks of CT forces.

Incidentally, the effect of classical polarization forces on the wavefunction of a complex can be expressed in terms of an expansion in terms of CT functions, so that classical polarization forces could well be included under the heading of CT forces. For this purpose the no-bond wavefunction $\Psi_0$ in the usual treatment [10] would be taken as unpolarized. Here, however, we wish to use the term CT forces only for the nonclassical or valence-theoretical CT forces, so that the no-bond wavefunction $\Psi_0$ is defined to include classical polarization.

Let us now examine some considerations bearing on the relative importance of charge transfer and electrostatic forces for some specific types of complexes, particularly the weak $b\pi \cdot a\pi$ and $b\pi \cdot a\sigma$ complexes [9] and then conclude with a few comments on the stronger $n \cdot a\sigma$ complexes (lone pair donor, $\sigma$ acceptor) [9].

## $b\pi \cdot a\pi$ COMPLEXES

In the $b\pi \cdot a\pi$ and $b\pi \cdot a\sigma$ complexes, both donor and acceptor are *sacrificial;* that is, electron transfer involves loss of a bonding electron from the $b\pi$ donor and gain of an antibonding electron by the $a\pi$ (e.g. TCNE) or $a\sigma$ (e.g. $I_2$) acceptor. Hence CT forces should be especially weak for complexes of these types. In fact most such complexes *are* weak: the values of properties such as the formation constant $K$, the enthalpy of formation $-\Delta H$, and the intensities of CT bands, are for the most part relatively small, especially when compared to values for strong $n \cdot a\sigma$ complexes. In this connection, see Tables 1 to 3, which give representative samples for $b\pi \cdot a\pi$ and $b\pi \cdot a\sigma$ complexes from Briegleb's extensive tables [7].

As discussed above, the stability of these weak complexes must be attributed in appreciable part to the action of classical Coulomb and/or polarization forces. On the other hand, the same considerations or causes which require CT forces to be predominantly responsible for the stability of strong $n \cdot v$ and $n \cdot \sigma$ complexes must still be at work in weaker complexes, hence CT forces must make at least *some* contribution to the stability of weak EDA complexes. Some of these considerations are now summarized.

The fact that all properties which are related to the strengths of $b\pi \cdot a\pi$ complexes are usually fairly well correlated [7] with familiar measures of electron donor and acceptor tendencies (especially with the magnitudes of ionization potential of donor and electron affinity of acceptor) seems to be a major argument in favor of the importance of CT forces for the strengths of the stronger of these complexes. In particular, the values of these properties increase in a fairly regular way through a series of complexes involving different donors with a common acceptor as the ionization potential of the donor decreases. Such trends, which explain changes in strength between molecular pairs for which there is no obvious reason to suppose much change in the electrostatic interaction, support the existence of nonnegligible CT forces in such complexes.

Consistent with this view is the usual absence of evidence of complex formation between like or nearly like molecules—even those which might be expected to exhibit some electrostatic forces of attraction; for example,

**TABLE 1**

**Some $K_x$ Values[a] in $CCl_4$ Solution at 20°C**

| Donor ($b\pi$) | Acceptor | | | | |
|---|---|---|---|---|---|
| | TCNE[b] ($a\pi$) | Chloranil ($a\pi$) | *p*-Benzoquinone ($a\pi$) | TNB[b] ($a\pi$) | $I_2$ ($a\sigma$) |
| Benzene | 10.7 | 3.4 | – | 24 | 1.55 |
| Hexamethylbenzene | 1530 | 96 | 6.0 | 74 | 15.7 |
| Phenanthrene | 71.5 | 49 | – | 123 | 4.5[c] |

[a] Values from Briegleb [7]. $K_x$ is the equilibrium constant calculated from equilibrium concentrations expressed in mole fraction units.
[b] Abbreviations: TCNE, tetracyanoethylene; TNB, trinitrobenzene.
[c] From J. Peters and W. B. Person, *J. Am. Chem. Soc.*, **86**, 10 (1964).

**TABLE 2**

**Some $-\Delta H_x$ Values[a] (kcal) for Complexes in $CCl_4$ Solution**

| Donor | Acceptor | | | | |
|---|---|---|---|---|---|
| | TCNE | Chloranil | *p*-Benzoquinone | TNB | $I_2$ |
| Benzene | 3.55 | 1.65 | 1.8 | 1.7 | 1.3 |
| Hexamethylbenzene | 7.75 | 5.35 | (1.8) | 4.7 | 3.73 |
| Phenanthrene | 4.30 | 3.65 | 1.8 | 4.3 | 1.6[c] |

[a] Values from Briegleb [7], except for phenanthrene $\cdot I_2$ [c] (See footnote c of Table 1).

between two molecules of trinitrobenzene or of benzene, or between benzene and the very weak acceptor nitrobenzene (in contrast to the definite formation of a benzene complex with the stronger acceptor trinitrobenzene). The fact that complete electron transfer occurs (usually with some assistance for example, in polar solvents or in crystal formation) in $b\pi \cdot a\pi$ inner complexes built from especially strong donors and acceptors forms a natural climax to the enhanced importance of CT forces in the corresponding outer complexes. (On this point see the further discussion near the end of this section.)

Some comments on Tables 1 to 3 are now in order. More complete

**TABLE 3**

**Some Absorption Intensity[a] Values for the CT Band in $CCl_4$ Solution[b]**

| | TCNE[b] | Chloranil | *p*-Benzoquinone | TNB | $I_2$ |
|---|---|---|---|---|---|
| Benzene | 3570 | 2350 | | 5855 | 15000 |
| Hexamethylbenzene | 4390 (6230) | 2650 (3950) | 2016 (2960) | 2150 (3000) | 6700 (8430) |
| Phenanthrene | – | 887 | | 1375 | 7100[c] |

[a] Values given are $\epsilon_{max}$, from [7], corrected to be proportional to D (in the case of those in parentheses) by multiplication by the ratio of $\nu_{max}$ in benzene to $\nu_{max}$ in hexamethylbenzene.
[b] TCNE in $CH_2Cl_2$ solution.
[c] From reference given in footnote c of Table 1.

tables would show a fairly steady progression in $K$ and $-\Delta H$ (and usually in CT band intensity) in each series of complexes between a given acceptor and a series of donors beginning with benzene and substituting methyl groups to end with hexamethylbenzene. As we expect if CT forces are important, increasing stability accompanies increasing donor ability as measured by decreasing donor ionization potential $I_D$ (which drops from 9.24 $e$V in benzene to 7.85 $e$V in hexamethylbenzene). However, it can also be argued that the increases in $-\Delta H$ and $K$ result in considerable part from increasing electrostatic effects caused as the methyl groups are substituted in the benzene ring. For example, the relatively large $K$ and $-\Delta H$ for the hexamethylbenzene · TCNE complex might perhaps be so explained.

More direct measures of the extent of CT should perhaps be the CT band intensities. According to CT theory, dipole strengths $D$ of the CT bands should increase with increasing extent of CT in the ground state of the complex. The oscillator strength $f$ of a band is proportional to $\int \epsilon_\nu d\nu$, taken over the band, and this in turn, for bands of equal half-width, is proportional to the peak absorptivity $\epsilon_{max}$ in the band. On the other hand, $f$ is proportional to $\nu_{max} D$; thus $D$ is approximately proportional to $\epsilon_{max}/\nu$. Thus, to compare $D$ values for benzene and hexamethylbenzene complexes, one may compare $\epsilon_{max}/\nu$ values; in Table 3 values of $\epsilon_{max}$ corrected by the ratio of $\nu_{max}$ values for benzene and hexamethylbenzene are therefore given (in parentheses) for hexamethylbenzene. It is seen that these values show for hexamethylbenzene a moderate increase in dipole strength for the CT band over that for benzene in the case of the TCNE and chloranil complexes, indicating a moderate increase in extent of CT [11]. Comparison of the data for the hexamethylbenzene complexes of chloranil with that of the benzene complexes shows much greater differences in $K$ and $-\Delta H$ than occur for $\epsilon_{max}$ values, suggesting that perhaps the stabilities of chloranil complexes are being influenced by electrostatic forces.

The fact that stabilities ($K$, $-\Delta H$) for complexes of a given acceptor with phenanthrene ($I_D = 8.1$ $e$V) sometimes increase and sometimes decrease from those for hexamethylbenzene ($I_D = 7.85$ $e$V) suggests that other than CT effects are also important. However, the variability in these results is not obviously accounted for by electrostatic forces. The intensities for the phenanthrene and hexamethylbenzene complexes with $I_2$ are approximately equal, as expected, but the intensity results from the other complexes are variable, suggesting again that an explanation involving both CT and electrostatic forces is needed.

In connection with the present discussion, the following point deserves mention. Clearly it would be preferable to compare the predictions of theory with vapor phase data on complexes rather than with the solution data of Tables 1 to 3. However, such comparisons with the limited vapor data now available are not very enlightening. Kroll [12], in his vapor-phase study of complexes of TCNE with methylated benzene donors, finds very little change in the intensity of the CT band—if anything there is some decrease in intensity—as the number of methyls increases, suggesting little change in CT through that series. Although these data contradict the conclusion reached from the variation in solution intensities of $b\pi \cdot$TCNE complexes, it is not clear that the vapor-based conclusion is to be preferred.

The intensities of the CT band as obtained in the vapor phase are considerably lower than those obtained in solution. The reason for this phenomenon is not clear. One suggestion is that the solvent cage around the complex in solution confines it so that it is under some pressure, which results in higher overlap and increased CT [12a].

The experimental observation that the intensities of the CT bands of typical $b\pi \cdot a\pi$ complexes increase greatly with increasing external pressure [13,14] is exactly what is predicted from a theory in which CT forces play a small but not negligible part in accounting for the stabilities of complexes. Namely, let the wavefunction of the complex be written in the form

$$\Psi(D \cdot A) = a\Psi_0(D, A) + b\Psi_1(D^+—A^-) + \cdots, \tag{1$a$}$$

or more generally

$$\Psi(\mathrm{D}\cdot\mathrm{A}) = a\Psi_0(\mathrm{D},\mathrm{A}) + \sum_i b_i\,\Psi_i(\mathrm{D}^+\text{—}\mathrm{A}^-) + \cdots, \tag{1b}$$

with $b^2/a^2$ (or $\Sigma\, b_i^2/a^2$) << 1 for a weak complex. The extent of CT action is measured for a normalized function (1*a*) by $F_{1N} = b^2 + abS_{01}$, the fraction of charge transferred from donor to acceptor in the complex; $S_{01}$ is the overlap of $\Psi_1$ with $\Psi_0$. The magnitude of CT, and with it the intensity of CT bands, is expected to depend on the extent to which donor and acceptor overlap. If the overlap is small in the absence of external pressure, as is expected for very weak complexes, it should be especially sensitive to external pressure, since for a small enough initial overlap of two molecules, the overlap must increase exponentially as they are squeezed together [15].

The relation observed, for a series of complexes formed from a series of donors and any one acceptor, between the energy change $h\nu_{\mathrm{CT}}$ corresponding to the CT band (in particular, the *first* CT band) of one of the complexes and the minimum ionization potential $I_{\mathrm{D}}$ of the donor, is another possible source of information concerning the amount of charge transfer. For weak complexes, to the extent that (1*a*) is accurate, this relation is expected to be of the form:

$$h\nu_{\mathrm{CT}} = I_{\mathrm{D}} - C_1 + \frac{C_2}{I_{\mathrm{D}} - C_1} \tag{2}$$

with $C_1$ and $C_2$ perhaps substantially constant for the given series of complexes [16]. Here $C_1$ is determined predominantly, for weak complexes, by the classical Coulomb energy of attraction, $-C$, between $\mathrm{D}^+$ and $\mathrm{A}^-$. The second term $C_2/(I_{\mathrm{D}} - C_1)$ is the CT resonance energy term, and is related to $b/a$ of (1); in the absence of charge transfer, both $C_2$ and $b$ would be zero. A detailed discussion of the application of (2) to an evaluation of the extent of CT in $b\pi\cdot a\pi$ and $b\pi\cdot a\sigma$ complexes is given in [10]. On the whole, no certain conclusions are reached in this way.

For most of the known $b\pi\cdot a\pi$ complexes, $h\nu_{\mathrm{CT}}$ is in the visible. From the theory [10] one sees that when $I_{\mathrm{D}} - C_1$ of (2), hence $\nu_{\mathrm{CT}}$, is especially small, the amount of CT must increase, roughly as $(\nu_{\mathrm{CT}})^{-2}$. Thus, for the tetramethylparaphenylenediamine (TMPPD) complex with chloranil (Chl), with $\nu_{\mathrm{CT}}$ in the infrared [17] near 11,500 $\mathrm{cm}^{-1}$, the amount of CT should be roughly five times as large as for benzene · TCNE (which [18] has $\nu_{\mathrm{CT}}$ about 26,000 $\mathrm{cm}^{-1}$), or about 10% CT if there is 2% for benzene · TCNE. That this estimate is if anything an underestimate is indicated by the ease of formation of the ions $\mathrm{TMPPD}^+$ and $\mathrm{Chl}^-$ in even moderately polar solvents. Taking this reasoning into account, it seems to be established that while the amount of CT is small, or even very small, for the weaker $b\pi\cdot a\pi$ complexes, it becomes large in those where $I_{\mathrm{D}}$ is sufficiently small.

## $b\pi \cdot$ HALOGEN COMPLEXES

In the 1:1 complexes of iodine or of other halogens (ICl, $Br_2$, $Cl_2$) acting as $a\sigma$-acceptors with $b\pi$-donors, CT action in both donor and acceptor is again sacrificial, and one expects weak complexes. In fact (see Tables 1 to 3), $K$ and $-\Delta H$ are again observed to be relatively small for these complexes. However, the CT bands are surprisingly strong, suggesting the existence of rather large charge transfer. Other measures of the fraction of charge transferred ($F_{1N}$) suggest that the extent of CT is not very large. For example, the shift of the halogen-halogen stretching frequency suggests $F_{1N} \simeq 0.03$ for benzene $\cdot I_2$, while the dipole moment for the complex, ignoring any quadrupole-induced dipole [1] suggests $F_{1N} \simeq 0.07$ to 0.11, depending on the assumptions made for the geometry of the complex [10]. Such estimates for $F_{1N}$ are consistent with a CT resonance energy of from 0.7 to 2.5 kcal/mole [10], which is comparable to the observed $-\Delta H_f$ of about 1.5 kcal/mole [7]. Hence we may conclude that the extent of CT is probably rather small, but that it may still contribute appreciably to the stability of the complex in this case.

On the other hand, pure quadrupole spectroscopy date [19,20], likewise data on Br-Br interatomic distance [21], for the benzene $\cdot Br_2$ complex in the solid state, indicate that the amount of charge transfer to the $Br_2$ here is very small. To be sure, the solid complex is not the 1:1 complex which we have been discussing, but is built of chains of alternating benzene and bromine molecules, with an axial orientation of each $Br_2$ along the six-fold axes of the two benzene molecules between which it lies [21]. Still, it seems likely that if there is very little charge transfer in the solid complex, the same is true in the 1:1 complex [22].

However, it is not clear just how much increase in the length of the Br—Br bond (or change in the quadrupole resonance frequency) should be expected for a nonnegligible amount of charge transfer. It seems reasonable to expect that the increase in Br—Br distance might be approximately linear with $F_{1N}$. For the strong aliphatic amine $\cdot X_2$ complexes ($F_{1N} \simeq 0.4$) the increase in the X—X bond length from that in the free halogen is about 0.25 Å. Hence we might expect an increase in Br—Br length for benzene $\cdot Br_2$ of about 1/8 to 1/4 of that value or from 0.03 to 0.06 Å ($F_{1N}$ from 0.05 to 0.10). This increase is only slightly greater than the experimental uncertainty in the X-ray work, and we suspect that this argument can be made conclusive only if a very careful X-ray study is made. If there is nonnegligible CT in the benzene $\cdot Br_2$ crystal, we also expect the benzene—Br distance to be less than the van der Waals distance. Experimentally, this distance is found to be 3.36 Å compared with 3.65 Å expected for the sum of van der Waals radii. Hence we believe the X-ray results are ambiguous but are also consistent with small but nonnegligible CT. A similar statement applies [20] to the quadruple resonance results.

## COMPLEXES OF $n \cdot a\sigma$ TYPE

There is no question that classical Coulomb and polarization forces can play only a minor part as compared with CT forces in accounting for the large observed $-\Delta H$ and $K$ values and dipole moments [10] of the strong $n \cdot a\sigma$ complexes of iodine with the aliphatic amines. As a check on this conclusion, Dr. M. Itoh has very kindly computed the classical dipole-induced-dipole contributions to the stabilization energies and the dipole moments of $NH_3 \cdot I_2$ and of $(CH_3)_3N \cdot I_2$, following the procedure used by Hanna [1], and has obtained the results given in Table 4. Hence there must indeed be considerable charge transfer in these stronger complexes; judging from the dipole moments, $F_{1N} \simeq 0.4$. Moreover, $K_x$, $-\Delta H$, and CT band intensity all increase together in these complexes with increasing donor strength, in agreement with the CT theory.

Because so many of the properties of the $n \cdot a\sigma$ complexes appear to be correlated as logical extensions of the CT theory of weak complexes, it is clear that some CT also occurs in the weaker complexes. We con-

**TABLE 4**

**Polarization Contributions to Stability and Dipole Moments of Amine · Iodine Complexes[a]**

| | Major Computed Polarization Contributions | | | | Observed | |
|---|---|---|---|---|---|---|
| Complex | Approx. 1 | | Approx. 2 | | | |
| | $\mu$(D) | $W$(kcal) | $\mu$(D) | $W$(kcal) | $\mu$(D) | $-\Delta H$(kcal) |
| $H_3N \cdot I_2$ | 0.91 | 0.34 | 1.96 | 1.58 | ~6.4 | 4.8 |
| $(H_3C)_3N \cdot I_2$ | 0.30 | 0.04 | 0.57 | 0.13 | ~6.0 | 10.2 |

[a] The approximations 1 and 2 are those of Hanna [1]. The data used are: $\mu$ of $NH_3$, 1.5 D, $\mu$ of $(CH_3)_3N$, 0.63 D; dimensions of $(CH_3)_3N \cdot I_2$, [21]; $I_2$ polarizability, [1]; observed $\mu$, [10], Table 6-2). $W$ = energy. In the computations, a point dipole located at the mid-point of the NH or CN bond was assumed.

cur with the conclusions of Hanna [1] that the extent of CT action may have been over-estimated for weak complexes in the past, but that it still involves forces whose magnitude is at least comparable to the electrostatic forces for most weak complexes. The predominance of CT forces in the $n \cdot a\sigma$ iodine complexes make them a useful limiting test case. The rather smooth variation of CT-dependent properties of the complexes from the weak $b\pi \cdot a\pi$ or $b\pi \cdot a\sigma$ complexes to the strong $n \cdot a\sigma$ complexes indicates that the extent of CT varies from very little ($F_{1N} \simeq 0.01$) to large ($F_{1N} \simeq 0.4$) in a similar way.

## REFERENCES AND NOTES


[1] M. W. Hanna, *J. Am. Chem. Soc.*, **90**, 285 (1968).

[2] M. W. Hanna and D. E. Williams, *J. Am. Chem. Soc.*, **90**, 5358 (1968); and also see J. L. Lippert, M. W. Hanna and P. J. Trotter, *J. Am. Chem. Soc.*, to be published, 1969.

[3] M. J. S. Dewar and C. C. Thompson, Jr., *Tetrahedron Suppl.*, **7**, 97 (1966).

[4] M. Mantione, in B. Pullman, *Molecular Associations in Biology*, Academic Press, 1948, and references given there.

[5] M. Mantione, *Theoretica Chim. Acta*, (Berl.), **11**, 119 (1968); *Int. J. Quantum Chem.*, to be published, 1969.

[6] J. P. Malrieu and Pl Claverie, *J. de chim. phys.*, **65**, 735 (1968).

[7] G. Briegleb, *Elektronen-Donator-Acceptor-Komplexe*, Springer-Verlag, Göttingen, 1961.

[8] R. S. Mulliken, *J. chim. phys.*, **61**, 26 (1964).

[9] Many authors used Dewar's term "$\pi$ complexes" for complexes with $\pi$ donors (or acceptors). However, a more precise terminology [10] (see also reference 8) which specifically indicates both the donor type and the acceptor type seems preferable.

[10] R. S. Mulliken and W. B. Person, *Molecular Complexes, A Lecture and Reprint Volume*, John Wiley and Sons, New York, 1969.

[11] However, the TNB complexes fail to fit into the picture, indicating that the theoretical explanation must be more complex. Actually, the CT intensity expression *is* more complex [10], but not in such a way as to offer a ready explanation. In the case of the $I_2$ complexes of the methylated benzenes, if the "axial model" is correct (see next Section), the usual theory is inapplicable [10], so that the observed decrease in intensity with methylation is not relevant evidence on the extent of CT. In general, contributions to the intensity from admixture of the CT state with locally excited states must be considered.

[12] M. Kroll, *J. Am. Chem. Soc.*, **90**, 1097 (1968).

[12a] J. Prochorow and A. Tramer, *J. Chem. Phys.*, **44**, 4545 (1966).

[13] J. R. Gott and W. G. Maisch, *J. Chem. Phys.*, **39**, 2229 (1963).

[14] H. W. Offen and collaborators, *J. Chem. Phys.*, **42**, 430 (1965); *ibid.*, **45**, 269 (1966); *ibid.*, **47**, 253, 4446 (1967) and later papers.

[15] The argument given here is probably conclusive, but *possibly* not, since $b^2/a^2$ depends on the overlap integral for a bonding donor MO $\phi_D$ not with an occupied acceptor MO, but with a normally unoccupied $A^-$ acceptor MO $\phi_{a^-}$ (see Ref. 10, Secs. 3-4 and 3-5). If the overlap of $\phi_d$ with the relatively larger orbital $\phi_{a^-}$ were appreciable when neutral A (with its smaller

orbital, $\phi_a$) is not yet overlapping neutral D, the argument would fail. This situation seems improbable for $b\pi \cdot a\pi$ complexes, where most likely $\phi_a$– is in fact scarcely bigger than the size of the acceptor. (The *size* of the acceptor is all that is relevant here; it is not necessary that any particular overlap *integral* of D and A MOs be non-zero.)

[16] See Ref. 10, (9-7) and (9-4), or see R. S. Mulliken and W. B. Person, *Ann. Rev. Phys. Chem.*, **13**, 107 (1962).
[17] R. Foster and T. J. Thomson, *Trans. Faraday Soc.*, **58**, 860 (1962).
[18] R. E. Merrifield and W. D. Phillips, *J. Am. Chem. Soc.*, **80**, 2778 (1958).
[19] H. O. Hooper, *J. Chem. Phys.*, **41**, 599 (1964). This paper also finds evidence of very little charge transfer in the complexes of *p*-xylene with $CBr_4$ and $CCl_4$ but these $b\pi \cdot a\sigma$ complexes are of an extremely weak type so that very little charge transfer would be expected.
[20] D. F. R. Gilson and C. T. O'Konski, *J. Chem. Phys.*, **48**, 2767 (1968).
[21] O. Hassel, *Mol. Phys.*, **1**, 241 (1958).
[22] The fact that the $Br_2$ is symmetrically located in the solid complex, so that, unlike the case of the 1:1 complex with its unsymmetrical location of the $Br_2$, no dipole moment can be created, should not inhibit (though perhaps it could modify) the occurrence of charge transfer).

[Reprinted from the Journal of the American Chemical Society, 72, 600 (1950).]


[CONTRIBUTION FROM THE PHYSICS DEPARTMENT, UNIVERSITY OF CHICAGO]

# Structures of Complexes Formed by Halogen Molecules with Aromatic and with Oxygenated Solvents[1]

BY ROBERT S. MULLIKEN

## I. Introduction

It is a familiar fact that iodine forms violet-colored solutions in certain solvents, brown solutions in others. In some solvents, for example benzene and methylated benzenes, it forms solutions of intermediate color. The most usual, although not universally accepted, explanation of the brown solutions has been that the altered color results from formation of molecular complexes.

Recently Benesi and Hildebrand,[1a] using spectroscopic methods, have shown definitely that benzene and mesitylene form 1:1 complexes of considerable stability with iodine. For solutions of iodine in these substances as solvents, they find that 60 or 85%, respectively, of the dissolved iodine is present in the complexes. This work lends support to earlier strong but less conclusive evidence by Hildebrand[1b] and others for the presence of 1:1 complexes in solutions of iodine in alcohol and other "brown" solvents.

On the other hand, the close agreement in form and intensity between the spectrum of iodine in the vapor state and in "violet" solvents (aliphatic hydrocarbons, carbon tetrachloride, etc.) indicates that complexes are not formed in these.[2,3,4] Other evidence[1a] supports this.

In the present paper, the existence of specific 1:1 complexes will be *assumed as established* in all cases except for the "violet" solvents.[5] An at-

(1) This work was in part assisted by the ONR under Task Order IX of Contract N6ori-20 with the University of Chicago.

(1a) H. A. Benesi and J. H. Hildebrand, THIS JOURNAL, **70**, 2382 (1948); **71**, 2703 (1949).

(1b) Hildebrand and Glascock, THIS JOURNAL, **31**, 26 (1909).

(2) See refs. 1a, 3, 4 and additional references given in refs. 1a, 3.

(3) G. Kortüm and G. Friedheim, *Z. Naturforschung*, **2a**, 20 (1947): comparison of absorption spectra of iodine in vapor, in cyclohexane solution, and in ether solution, from visible to λ2330.

(4) E. Rabinowitch and W. C. Wood, *Trans. Faraday Soc.*, **32**, 540 (1936): comparative graph of extinction coefficient ε in visible for vapor, vapor plus foreign gas at 50 cm. pressure, and carbon tetrachloride solution; and useful brief survey. By using foreign gas, the usual errors due to band structure below the convergence point of the bands are overcome, and the resulting ε curve is perhaps the most reliable measure available for the vapor. In bromine,[4] foreign gas causes a considerable increase in ε, but internal evidence indicates that this effect is absent in iodine.

(5) Further work will of course be desirable in order to establish the correctness of this assumption as directly as possible for examples of the second and third of the types of complexes considered here.

tempt will then be made to diagnose the observed absorption spectra so as to explain the nature of the chemical binding in these complexes. To anticipate briefly, iodine complexes falling into three distinct spectroscopic and chemical types will be distinguished, namely complexes with (1) simple benzene derivatives; (2) ethers, alcohols and water; (3) ketones. It will also be shown, from their analogous spectroscopic behavior in solution, that bromine, and iodine bromide and chloride, probably form analogous complexes. The present conclusions are in general agreement with the views of Fairbrother, and of Benesi and Hildebrand who suggest that the complexes result from "an acid–base interaction in the electron-donor sense" in which iodine functions as the acid, or electron-acceptor. However, the present proposals are more specific as to both electronic and geometrical structure.

## II. Spectra and Structure of Halogens in Vapor and Inert Solvents

Before considering solvent complexes, it will be useful to review critically, and clarify somewhat, existing knowledge of the structure of the halogens and their spectra in vapor and in inert solvents. The theory of the normal and lower excited electronic states has been extensively discussed, in particular by the writer.[6] It has been shown that each of the heavier halogens must possess a remarkably large number of excited electronic states in the energy range up to 6 ev., corresponding to numerous absorption transitions extending from infrared throughout the ordinary ultraviolet. However, most of these transitions are forbidden in the homopolar halogens by rigorous selection rules, while most of the remainder are predicted to be weak. In this way, the relative simplicity of the observed spectra can be understood. But it should be emphasized that the theory is completely definite and reliable[7] as to the existence and general nature of the numerous predicted excited levels. The importance of these levels in the present connection is the possibility that some of them may give rise to absorption transitions of appreciable intensity under suitable environmental conditions, even when the same transitions are absent or too weak to be noticed in pure vapor at low pressure.

A brief survey of the electronic structures involved will be useful here and in the later discussion of the natures of the various halogen complexes. In terms of MO's (molecular orbitals), the electron configuration for the outermost electrons in the ground state of any homopolar halogen, and the over-all state type, are

$$\ldots\sigma_g np)^2\ \pi_u np)^4\ \pi_g np)^4,\ {}^1\Sigma_g^+ \qquad (1)$$

Here $\sigma_g np$ is a bonding MO related to the $np\sigma$ bonding valence orbital of the halogen atom ($n$ = 3, 4, 5 for Cl, Br, I, respectively), while $\pi_u np$ and $\pi_g np$ are, respectively, bonding and antibonding $\pi$ MO's. The pair of bonding electrons in the $\sigma_g np$ MO forms the bond, while the eight $\pi$ electrons, which correspond in valence-bond theory to four unshared pairs of $\pi$ electrons on each atom, taken together produce little effect on the bond strength (probably a slight weakening).

The *lowest excited* group of electronic states or substates in all the homopolar halogens is

$$\ldots\sigma_g np)^2\ \pi_u np)^4\ \pi_g np)^3\ \sigma_u np,\ {}^3\Pi_{2u},\ {}^3\Pi_{1u},\ {}^3\Pi_{0^-u},\ {}^3\Pi_{0^+u},\ {}^1\Pi_u \qquad (2)$$

Because of the presence of the strongly antibonding $\sigma_u np$ MO here, these states are in part weakly bound, in part have repulsive potential energy curves. Transitions to three of these states are responsible for the well-known moderately intense visible (and infrared in bromine and iodine, or near-ultraviolet in chlorine and fluorine) absorption bands and continua.

In iodine, by far the strongest transition in the long wave length part of the spectrum is to the ${}^3\Pi_{0^+u}$ member of the group of states in (2). This is a highly characteristic but highly anomalous type of transition, since it runs counter to ordinary selection rules. It is nevertheless satisfactorily explainable by quantum theory, essentially in terms of strong intraatomic spin-orbit coupling which is present especially in iodine.[6]

The numerous additional predicted low-energy states mentioned previously are obtained by various distributions of the ten outer electrons among the four MO's which are present in (2). From the observed visible bands (transition (1) → (2)), the approximate energy difference between the MO's $\pi_g np$ and $\sigma_u np$ is known, and that between $\sigma_g np$ and $\sigma_u np$ can be obtained from ultraviolet spectra near λ2000; that between $\pi_u np$ and $\pi_g np$ can be estimated. With these data, the mean positions of the electronic states associated with each of the electron configurations given below in (3) can be estimated. The order in (3) is that of increasing estimated mean energy, for iodine. For iodine, the predicted states nearly all lie within 6 ev. of the ground level; for the other halogens, the predicted spread is somewhat greater.

$$\begin{array}{l} \sigma_g np)^2\ \pi_u np)^3\ \pi_g np)^4\ \sigma_u np;\ (g, 5) \\ \sigma_g np)^2\ \pi_u np)^4\ \pi_g np)^2\ \sigma_u np)^2;\ (g, 4) \\ \sigma_g np)^2\ \pi_u np)^3\ \pi_g np)^3\ \sigma_u np)^2;\ (u, 10) \\ \sigma_g np)^2\ \pi_u np)^2\ \pi_g np)^4\ \sigma_u np)^2;\ (g, 4) \\ \sigma_g np)\ \pi_u np)^4\ \pi_g np)^4\ \sigma_u np);\ (u, 3) \end{array} \qquad (3)$$

For each configuration, the even (g) or odd (u) character of the states is indicated in parentheses, likewise the number of different electronic states or substates associated with the configuration. In (1), the one state is $g$; in (2), the states are all $u$. In view of the rigorous selection rule $g \longleftrightarrow u$ for homopolar molecules, absorption from the ground state can occur only to $u$ states, except in the presence of external perturbations which destroy the symmetry. Transitions to many of the other states are also forbidden for isolated molecules; and in addition, many of the remaining *allowed* transitions are predicted to be weak. Most of the electronic states in (3) are unstable, that is, correspond to repulsion curves, but this in itself does not make them less active spectroscopically; it means merely that they would give continuous instead of discrete absorption spectra.

Recently Kortüm and Friedheim have re-examined the iodine vapor absorption down to about λ2300. The visible bands obey Beer's law, as does the beginning of an absorption in the farther ultra-

(6) R. S. Mulliken, especially *Phys. Rev.*, **46**, 549 (1934), *e. g.*, Table III; **57**, 500 (1940); also *J. Chem. Phys.*, **8**, 234, 382 (1940). For recent experimental work and interpretation and a valuable survey, see also P. Venkateswarlu and R. K. Asundi, *Indian J. Phys.*, **21**, 101 (1947), and forthcoming paper by P. Venkateswarlu.

(7) The theory is of course not strictly quantitative as yet, but it is probably safe to say that actual electronic levels lie within 1 or 2 ev. of the predicted positions in nearly all cases predicted to lie in the range 0–6 ev. The quantitative aspects of the predictions are based largely on empirical data—see paragraph containing Eq. (3).

violet whose intensity is still rising at the limit of absorption of these authors.

In addition, they report an extremely weak absorption, with maximum at λ2670, which does *not* obey Beer's law: its extinction coefficient $\epsilon$ increases with pressure; another group of investigators find the same absorption in saturated vapor. Kortüm and Friedheim attribute this to loosely-bound $I_4$ molecules (see following paragraph with respect to the same or a similar transition in solution). Bands with similar behavior are known in oxygen gas and liquid, and attributed to loosely-bound $O_4$. An equally satisfactory explanation would be that the bands are due to $I_2$–$I_2$ collisions. On either explanation, it is highly probable that the observed transition is one of those mentioned above which, while forbidden or extremely weak for isolated iodine molecules, becomes allowed under the perturbing action of intermolecular forces.

The same authors have investigated iodine absorption in cyclohexane, a "violet" solvent. Here the visible bands appear with a slight increase in intensity and a slight shift of wave length toward the red.[8] The Beer's law ultraviolet absorption whose intensity is still rising at the limit of observation (λ2300) appears as before, but with 100 times the extinction coefficient it shows in the vapor; there it is weak but not forbidden, here it is strong.[9] Finally there is a rather weak absorption with maximum near λ3100, whose $\epsilon$ increases with concentration; this is evidently similar to and perhaps identical with the pressure-sensitive vapor absorption with peak at λ2670. The same absorption was also observed earlier by Gróh and Papp[2] and attributed to $I_6$ molecules.

Similar effects occur for bromine, but sensitivity to pressure and to inert solvents is much increased. For the *visible* absorption (*cf.* (2)), foreign gases[10] cause increases in intensity up to 30% (as compared with practically none for iodine[4]), inert solvents (carbon tetrachloride, chloroform, cyclohexane) cause increases of 30–60%[4,11,12,13] and in liquid bromine an increase of about 135% is observed.[14] In the *ultraviolet*, two or three transitions which are very weak or absent in the pure vapor at low pressure are brought out with appreciable intensities by inert foreign gases; probable maxima occur at about λ3130, 2700 and 2270.[10] The last of these is also observed very weakly in bromine vapor ($\epsilon_{max.} \approx 5$),[15] and with high intensity ($\epsilon \approx 1000$) in liquid bromine.[14] In the inert solvents mentioned above, an absorption maximum two or three times as strong as that in the visible is observed in the range λ2500–2750.[12] This may be the same as the λ2700 maximum observed under the influence of foreign gases.[16] Most of the maxima brought out by foreign gases or inert solvents apparently obey Beer's law, but there are indications of additional weak bands[15,12] which, like the iodine maximum near λ2670,[3] do *not*, and so may be probably attributed to loosely-bound $Br_4$ molecules or to $Br_2$–$Br_2$ collisions.

Summarizing, the foregoing discussion shows that both allowed and forbidden or nearly forbidden vapor transitions in the halogens are enhanced in intensity or made allowed, sometimes rather strongly, by the action of inert gases and of inert liquid solvents, especially in the case of bromine. In addition, at least one normally forbidden transition is made weakly allowed by dimerization or halogen–halogen collisions. In the former type of transitions, Beer's law appears to be obeyed for any one particular environment; in the latter, not.

## III. Spectra and Structure of Halogen Complexes with Aromatic Solvents

In aromatic solvents, the visible iodine absorption is somewhat shifted toward higher frequencies. The absorption intensities are also slightly enhanced, but scarcely more than in inert solvents (see Table I).

TABLE I

IODINE VISIBLE ABSORPTION[a]

| Inert solvents | | | Aromatic solvents | | |
|---|---|---|---|---|---|
| | $\lambda_{max}$ | $\epsilon_{max}$ | | $\lambda_{max}$ | $\epsilon_{max}$ |
| Vapor plus foreign gas[4] | 5200 | 820 | Benzotrifluoride | 5120 | 870 |
| *n*-Heptane | 5200 | 910 | Nitrobenzene[b] | 5000 | |
| Cyclohexane[3] | 5270 | 1070 | Benzene | 5000 | 1040 |
| $CCl_4$[11] | 5200 | 980 | Toluene | 4970 | 1020 |
| $CCl_4$ | 5170 | 930 | *o*-Xylene | 4970 | 1060 |
| $CS_2$ | 5180 | 1120 | *p*-Xylene | 4950 | 1080 |
| $CH_3CHCl_2$ | 5030 | 870 | Mesitylene | 4900 | 1185 |

[a] The values of λ and $\epsilon$ are for the position of maximum absorption at room temperature. A better measure of absorption intensity would be the value of $\int \epsilon(\nu)d\nu$ integrated over the whole absorption band, or the corresponding *f* value. The data are from ref. 1a except as indicated by other reference numbers. [b] From a review by J. Kleinberg and A. W. Davidson, *Chem. Revs.*, **42**, 601(1948).

(8) Other authors[1,11] find similar results in *n*-heptane, carbon tetrachloride, carbon disulfide, dichloroethane, etc. The degree to which the intensity is increased varies somewhat, however. In most cases the wave-length shift is very small, or in one case toward shorter wave lengths (see Table I).

(9) Possibly this is due to a wave-length shift of the long wave length tail of a very strong vapor absorption which occurs at shorter wave lengths, rather than to a solvent intensification effect.

(10) N. S. Bayliss and A. L. G. Rees, *Trans. Faraday Soc.*, **35**, 792 (1939): bromine vapor plus foreign gases.

(11) A. E. Gillam and R. A. Morton, *Proc. Roy. Soc. (London)*, **124A**, 604 (1929): halogens in solution.

(12) R. G. Aickin, N. S. Bayliss and A. L. G. Rees, *ibid.*, **169A**, 234 (1938): bromine in solution.

(13) N. S. Bayliss, A. R. H. Cole and B. G. Green, *Australian J. Sci. Res.*, **1A**, 472 (1948): visible bromine bands in solution. See especially the oscillator strengths (*f* values) in Table I.

(14) D. Porret, *Proc. Roy. Soc. (London)*, **162A**, 414 (1937): liquid bromine. The spectra of solid bromine, iodine, and chlorine at very low temperatures have been studied by A. Nikitina and A. Prikhotko, *Acta Physicochimica U. R. S. S.*, **11**, 633 (1939).

(15) R. G. Aickin and N. S. Bayliss, *Trans. Faraday Soc.*, **34**, 1371 (1938): bromine vapor.

(16) It is of course possible that different transitions are favored by different environmental conditions; also, in view of the numerous theoretically demonstrated excited levels, that any given observed maximum may be composite.

Table II

Ultraviolet Absorption of Halogen–Aromatic Complexes

| Iodine complexes[a,b] | | | | Bromine complexes[c] | | | |
|---|---|---|---|---|---|---|---|
| | $\lambda_{max}$ | $\epsilon_{max}$ | $f$ | | $\lambda_{max}$ | $\epsilon_{max}$ | $f$ |
| Benzotrifluoride | <λ2800 | | | Chlorobenzene | 2960 | <Benzene | |
| Benzene | 2970 | 9770 | | Benzene | 2930 | 6800 | 0.18 |
| | 2980 | 9200 | 0.23[a] | Toluene | 2975 | >Benzene | |
| Toluene | 3060 | 8400 | | | | | |
| *o*-Xylene | 3190 | 8400 | | | | | |
| *p*-Xylene | 3150 | 7400 | | | | | |
| Mesitylene | 3330 | 8300 | | | | | |

[a] Ref. 1a, except ref. 17 for the second set of benzene figures. [b] It is to be noted that the $\epsilon$ data are per mole of iodine, not of complex.[1] When allowance is made for the partial dissociation of the complex, the $\epsilon_{max}$ values for the complex become 16,000 in benzene, 10,000 in mesitylene, and $f$ becomes 0.38 for the benzene–iodine complex. [c] Refs. 12, 17.

In view of the slightness of the changes, and in view of the peculiar and characteristic nature of the visible iodine absorption (see discussion following (2)), there can be little doubt that the iodine molecule is present in its aromatic complexes without more than a small change in its electronic structure.[16a] This statement holds, moreover, not only for the ground state (1), but also for the excited states (2) of the *visible* absorption.

On the other hand, the iodine–aromatic complexes show a very strong ultraviolet absorption near λ3000 (see Table II)[1a,17] which is not found for iodine in inert solvents, nor for the aromatic molecules by themselves. Since the visible bands show so little change, and since inert solvents do not cause vapor-forbidden iodine transitions to appear with at all nearly so high an intensity as the new bands here under discussion, it seems very improbable that this can be attributed to the iodine part of the complex.[17] It must then be due to the aromatic part of the complex, or else to a transition to an excited state belonging somehow to the complex as a whole.

It may be recalled that benzene itself shows a *weak* ultraviolet absorption ($\epsilon_{max} \approx 100$ for benzene in heptane) with maximum about λ2600. As is well known, this corresponds to an electronic transition *forbidden* by electronic selection rules alone, but weakly allowed by vibrational-electronic interaction. As is also well known, the analogous transition becomes *moderately strongly* allowed in certain benzene analogs (*e. g.*, pyridine) and derivatives (*e. g.*, aniline, where $\epsilon_{max} \approx 1000$) in which the hexagonal symmetry present in benzene has been lost. In unsymmetrical methyl derivatives such as toluene and the xylenes, it becomes *weakly* allowed. In most substituted benzenes, the absorption peak is shifted somewhat toward longer wave lengths.

The foregoing facts suggest that the observed absorptions of the iodine–aromatic complexes near λ3000 are essentially the same as the forbidden λ2600 transition in benzene, but now made strongly allowed by the close presence of the iodine molecule in some orientation which destroys the hexagonal symmetry. The data in Table II on the ultraviolet absorption spectrum of bromine in benzene and its derivatives now likewise find a probable interpretation in terms of the existence of analogous bromine complexes with analogous spectra.

As Benesi and Hildebrand have shown,[1a] the iodine–mesitylene complex is a much tighter one than the iodine–benzene complex, a fact which can be correlated with the much greater shift of the wave length (though not of the intensity) of the absorption peak as compared with that of the aromatic molecule by itself. Benesi and Hildebrand have concluded that the entire series of changes from benzotrifluoride to mesitylene (*cf.* Table II, also the smaller changes in Table I) are correlated with increasing ability of the variously substituted benzenes to act as electron donors. This idea is in entire agreement with what is known of the relative ionization energies of the various molecules: 9.24 ev. for benzene, 8.92 ev. for toluene, and probably about 8.3 for the xylenes and 8.1 for mesitylene.[18] It also receives support from work of Fairbrother,[19] who finds that iodine in benzene shows an apparent dipole moment of 0.6 *D*, in *p*-xylene one of 0.9 *D*; and one of 1.3 *D* in the "brown" solvent dioxane. These and other data can be understood if in the wave functions of the aromatic complexes there are appreciable amounts of resonance structures of the type $Ar^{+}\cdot I_2^{-}$, where Ar denotes the aromatic molecule.

It is of interest to inquire whether conclusions as to the geometrical configuration of the iodine–aromatic and probable bromine–aromatic complexes can be reached using electronic structure theory. Suggestive is the fact that van der Waals forces taken alone would probably favor the closest

(16a) The data for aromatic solvents in Table I should of course represent averages for associated and unassociated iodine (mostly the latter in benzotrifluoride, mostly the former in mesitylene[1a]). But if the unassociated molecules absorb nearly like iodine in inert solvents, the conclusion just stated is easily seen to be unaffected.

(17) N. S. Bayliss, *Nature*, **163**, 764 (1949). However, Bayliss attributes the strong absorption of both bromine and iodine near λ3000 in benzene to a greatly shifted strong halogen transition which in the free halogens occurs below λ2000.

(18) *Cf.* W. C. Price, *Chem. Revs.*, **41**, 257 (1947).

(19) F. Fairbrother, *Nature*, **160**, 87 (1947); *J. Chem. Soc.*, 1051 (1948), where also are references to earlier work by others. Note that if the benzene complex is 40% dissociated,[1a] this would make the dipole moment 0.77 *D* for molecules of the benzene–iodine complex itself.

possible packing of the two molecules.[20] This would place the iodine molecule with its center on the six-fold symmetry axis of the benzene molecule and its axis parallel to the plane of the benzene ring, at a distance of about 3.4 Å. above the latter.[21] The additional polar forces involved in a structure of the type $Ar^+ \cdot I_2^-$ would also probably give maximum attraction for this same model.

The interatomic distance in $I_2$ (2.67 Å.) closely matches the distance (2.78 Å.) between opposite carbon atoms in the ring. On constructing the $Ar \cdot I_2$ model just proposed, using an "atom kit," it appears that there would be little if any preference for the iodine axis to lie across two opposite sides of the ring as compared with the alternative of being parallel to a line joining two opposite carbon atoms. Moreover, there would apparently be little if any steric hindrance in methylated benzenes if the iodine axis should be directly above a methyl group, at least provided the latter were allowed to orient itself suitably by rotating about the bond joining it to the ring.

The indicated model has the right symmetry so that the iodine molecule can break down the selection rules for the λ2600 transition of benzene in qualitatively the same way that the methyl or amino groups do (mildly) in *p*-xylene or (more strongly) in *p*-phenylenediamine. Thus it might explain, at least qualitatively, the observed high intensities in Table II.

However, the observed intensities are so very high as to indicate a need for further consideration. A plausible quantum-mechanical explanation is the following. In iodine near λ1800 or λ2000 there occurs an extremely intense absorption, of the so-called N → V type, accompanying a transition from the ${}^1\Sigma_g^+$ ground state (see (1)) to a ${}^1\Sigma_u^+$ state belonging to the last of the electron configurations listed in (3).[6] The electric moment of this transition vibrates parallel to the line joining the iodine nuclei. Also in benzene near λ1800 there occurs an extremely strong transition ($f$ = 0.8)[22] of the N → V type, one of whose two electric moment components can vibrate parallel to that of the iodine N → V transition, if the geometrical model of the complex is that proposed above. Further, the upper electronic states of the λ1800 and λ2600 absorptions of benzene (also of a third transition near λ2000) all belong, in terms of MO theory, to a single electron configuration.[23] Now even though the benzene-iodine complex in its ground state is but loosely bound, leaving the electronic structures of the two partners but little altered, nevertheless strong interaction, connected with a resonance between the N → V transitions of the two partners, may well occur between their respective excited states, provided the geometrical model is that assumed above. Further, the influence of this interaction may reasonably be expected to extend also to the other benzene excited states of the same electron configuration, including the upper state of the λ2600 bands. In some such way the large wave-length shifts from λ2600, and especially the high intensities, of the absorptions in Table II might well be accounted for.

It is of interest to inquire further into the forces which bring about the halogen–aromatic complexes. Van der Waals forces can help, but are not sufficient, since equally large or larger van der Waals forces must be present between the halogens and some of the inert "violet" solvent molecules, as Professor Hildebrand has pointed out to the writer. *Small admixtures* of $Ar^+ \cdot I_2^-$ resonance structures, as mentioned above, give a reasonable explanation of the observed stabilities in view of, (1), the relatively low ionization energies of unsaturated as compared with saturated organic molecules[18,24]; (2), the considerable electron affinity which the iodine molecule must possess[25]; and, (3), the possibility of weak covalent binding between $Ar^+$ and I or $I_2^-$. What is here meant by $Ar^+ \cdot I_2^-$ is mainly a mixture of the two resonance structures[25a]

[structures: benzene ring with + charge, I⁻ and I (I bonded to ring) and benzene ring with + charge, I⁻ and I]

with two similar structures having the opposite iodine negative, plus lesser amounts of other pairs of similar structures with different locations of the + charge and of the unpaired Ar electron. The $I^- I \longleftrightarrow I I^-$ part of this resonance corresponds to a Pauling 3-electron bond in an $I_2^-$ ion, so that the above four structures can also be thought of in terms of resonance between *two* structures

(20) See especially J. H. de Boer, *Trans. Faraday Soc.*, **32**, 10 (1936); J. H. de Boer and G. Heller, *Physica*, **4**, 1045 (1937). The stronger polarizability of benzene *in* than perpendicular to the plane of the ring, and of iodine *along* than perpendicular to its axis would *per se* favor an end-on arrangement of the iodine against the edge of the benzene ring, with its axis in the plane of the latter. But the closer approach afforded by the arrangement described in the text would very probably give greater stability in view of the inverse sixth-power forces involved. See also F. London, *J. Phys. Chem.*, **46**, 305 (1942).

(21) This is based on a van der Waals radius of 1.8 Å. for the iodine atoms[11] and 1.70 Å. (half the interplanar distance in graphite) for the benzene carbon atoms. A slight shortening, due to the $Ar^+ \cdot I_2^-$ attractive forces is then assumed.

(22) J. R. Platt and H. B. Klevens, *Chem. Revs.*, **41**, 301 (1947).

(23) G. Nordheim, H. Sponer and E. Teller, *J. Chem. Phys.*, **8**, 455 (1940).

(24) The same reasoning suggests that olefins, especially methylated and conjugated olefins, which have relatively low ionization energies,[18] may form similar complexes. It is of interest that Fairbrother[19] reports that iodine dissolved in cyclohexene and diisobutylene respectively shows dipole moments of 1.1 and 1.9 *D*. *Cf.* also Ref. 25a.

(25) The electron affinity $E_m$ of the iodine molecule is given by $E_m = E_a - D_m + D_m^-$, where $E_a$ is the electron affinity of the iodine atom, and $D_m$ and $D_m^-$ are the energies of dissociation of iodine and of $I_2^-$, respectively. The value of $D_m^-$ (3-electron bond) can fairly safely be estimated as half of $D_m$. Using the known figures $E_a$ = 3.14 and $D_m$ = 1.54 electron-volts, one obtains $E_m$ = 2.37 ev. However, since the I–I distance in the complex is believed to be nearly that for unassociated iodine rather than that appropriate to $I_2^-$, the energy of the resonance structure $Ar^+ \cdot I_2^-$ in the actual complex must be somewhat higher. A rough estimate of the form of the potential curve for $I_2^-$, assuming $R$ = 3.5 Å. and $\omega$ = 10 cm.$^{-1}$ for equilibrium, gives, for $R$ = 2.67 Å. as in iodine, $E_m$ = 1.8 ev. See also H. S. W. Massey, "Negative Ions," Cambridge University Press, 1938.

(25a) The proposed electronic structures are of the same general type as those suggested by Pauling for the complexes of $Ag^+$ with olefins, etc. (S. Winstein and H. J. Lucas, This Journal, **60**, 836 (1938).) Electronic structures for $Bz \cdot I_2$ more or less similar to those below have been proposed by Dewar (*J. Chem. Soc.*, 406 (1946)) and by Fairbrother.[19] See also Williams, *Phys. Z.*, **29**, 174 (1929).

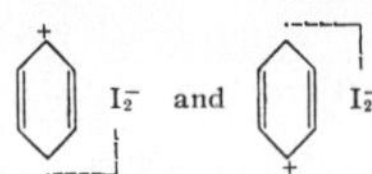

Thus far only one particularly plausible geometrical model, which may be called model I, has been considered. Other models which may be worth looking at briefly are: (II), with all atoms coplanar, the iodine molecule resting end-on against one side or corner of the benzene ring; (III), with the iodine resting waist-on against one side of the benzene plane, its axis perpendicular to the latter; (IV), with the iodine standing upright on the benzene plane, its axis coincident with the sixfold symmetry axis of benzene; (V), a filled-doughnut model similar to (IV) except that the two iodine atoms are on opposite sides of the benzene plane (suggested by Professor J. R. Platt). Other models involving oblique angles seem too improbable to need consideration.

Models II and III are of the same symmetry ($C_{2v}$) as I, so that both are effective in destroying the hexagonal symmetry of isolated benzene, thus making λ2600 of benzene into an allowed transition, as required. Models IV and V preserve the symmetry of benzene, IV partially (symmetry $C_{6v}$), and V completely (symmetry $D_{6h}$), so that in both cases the λ2600 transition of benzene remains forbidden. These models are then probably ruled out on that account. Model V also seems unlikely for other reasons.

Model II involves weaker electrostatic attractions than I in most of the $Ar^{+}\cdot I_2^{-}$ resonance structures. In addition, it can be shown[26] that for model II an excited state of $I_2^{-}$ would be necessary in order to meet symmetry requirements a type of consideration customarily, but not always safely, ignored in discussions of resonance structures. Thus it appears fairly sure that model II must be ruled out as compared with I, where electrostatic and probably van der Waals attractions are more favorable, and where the ground state of $I^{-}$ *does* meet the symmetry requirements.

For model III, the normal state of $I_2^{-}$ again meets the symmetry requirements, but the electrostatic situation, although more favorable than for model II, is less favorable than for model I. Further, even though the forbiddenness of the λ2600 transition is removed in this model, nevertheless resonance between the N → V transitions of benzene and iodine is not possible because their electric moments are now perpendicular to each other; hence our proposed explanation of the strikingly high intensity of the λ3000 absorption would not be valid for model III.

Thus the original model I appears to be definitely the most probable geometrical arrangement for halogen–benzene complexes. The same reasoning and conclusions hold for iodine–mesitylene complexes, and similar reasoning and conclusions for complexes with other substituted benzenes.[26a] The indicated model is obviously of possible interest in connection with the mechanism of halogen substitution in the benzene ring.

(26) The $Bz\cdot I_2$ wave function (Bz = benzene) is symmetrical with respect to all symmetry operations *of the model*, for any of the models. According to quantum mechanics, resonance is possible only between wave functions having identical symmetry properties. But for model II, the wave function of $Bz^{+}\cdot I_2^{-}$ formed using ground-state $I_2^{-}$ is antisymmetric to the plane of the molecule, because the orbital of the π electron removed from Bz is antisymmetric, while that of the σ electron added to iodine is symmetric, to this plane. Hence ground-state $Bz^{+}\cdot I_2^{-}$ in model II cannot resonate. The desired resonance can be obtained only if the $I_2^{-}$ is excited to a state having one less π electron and one more σ electron.

(26a) For a discussion of other benzene complexes, see G. Briegleb, "Zwischenmolekulare Kräfte und Molekülstruktur," F. Enke, Stuttgart, 1937. Of interest here are complexes like that formed by benzene with nitrobenzene, where according to Briegleb the two rings lie one above the other with their planes parallel,—a model which is similar to our benzene-iodine model I, and may very likely be explained at least partially in a similar way (*cf.* especially J. Weiss, *J. Chem. Soc.*, 245 (1942), who, however, suggests nearly 100% ionic structures, which surely cannot be correct). (According to Briegleb, the dipole moment of the nitro group polarizes the benzene in the plane of the ring, where the polarizability is especially high, and thus stabilizes the complex.) In complexes like those of benzene with $BX_3$ and $AlX_3$ (X = halogen), where the large dipole moments indicate a pyramidal structure for the $BX_3$ or $AlX_3$, a partial $Ar^{+}$-$MX_3^{-}$ structure with dative bond between the Ar and an M electron (*cf.* Dewar's π complexes) seems likely. Andrews and Keefer have very recently proposed a similar interpretation for Ar·Ag complexes (THIS JOURNAL, **71**, 3644 (1949); see also Ref. 25a).

*Note added in proof.*—An attractive alternative to the explanation above for the strong $Ar\cdot I_2$ absorption near λ3000 is that this is an *intermolecular charge-transfer spectrum*,[26b] essentially $Ar\cdot I_2 \rightarrow Ar^{+}I_2^{-}$. If the $Ar\cdot I_2$ wave function contains a small amount of $Ar^{+}I_2^{-}$ resonance structures, as discussed above, it follows that there must exist low excited st tes whose wave functions are *principally* $Ar^{+}I_2^{-}$ but contain small amounts of $Ar\cdot I_2$. Transition to the lowest of these would correspond essentially to the jump of an electron from the most easily ionized molecular orbital of the aromatic molecule to a previously unoccupied iodine molecular orbital. The oscillator strength of the transition can be shown[26b] to be given for Model I by

$$f = 1.085 \times 10^{11}\nu Q^2; \quad Q^2 \approx 2S^2z^2, \qquad (4)$$

where $S$ is the overlap integral between the two molecular orbitals, and $z$ is the distance at which their overlap is a maximum, measured from the center of Ar toward that of $I_2$. If we assume $z$ = 1.7 Å., then to match the observed $f$ of 0.23 for benzene (*cf.* Table II), $S$ = 0.33 is required. Unfortunately, this $S$ value appears improbably high (an upper limit for reasonable estimates appears to be about 0.1–0.2).

Nevertheless, the foregoing explanation may be worth considering, especially since it seems qualitatively well suited to explain the color phenomena which are characteristic of so many organic molecular complexes.[26c]

If the foregoing explanation should be correct, the spectroscopic arguments given above for and against the various models of $Ar\cdot I_2$ must of course be modified. For model I in the new explanation, the electric moment of the λ3000 absorption would vibrate along the line joining the Ar and $I_2$ centers.

## IV. Spectra and Structure of Halogen Complexes with Solvents of Type RR'O

Iodine forms brown solutions in ether, alcohols,

(26b) On *interatomic* charge-transfer spectra, see R. S. Mulliken, *J. Chem. Phys.*, **7**, 20 (1939); R. S. Mulliken and C. A. Rieke, Reports on Progress in Physics, The Physical Society, London, **8**, 231 (1941); E. Rabinowitch, *Rev. Mod. Phys.*, **14**, 112 (1942). Eq. (4) is obtained by treating the Ar and $I_2$ molecular orbitals as if they were *atomic* orbitals in the *atomic* orbital method for interatomic charge-transfer spectra.

(26c) See for example W. Brackmann, *Rec. trav. chim. pays-bas*, **68**, 147 (1949). Neither Brackmann nor Weiss[26a] gives, in the writer's opinion, a satisfactory explanation of colo phenomena in molecular complexes.

water, dioxane, pyridine and certain other similar solvents.[26d] For some of these the spectroscopic evidence is conflicting, probably because of chemical reactions, or in some cases, it may be surmised, because of benzene impurity whose iodine complex gives rise to strong absorption near λ3000. Allowing for these possibilities, the evidence[1a,3,4,27] seems to be consistent with the existence of a class of complexes between iodine and molecules of the type ROR′ (R,R′ = hydrogen or alkyl or other group), with a characteristic absorption having its maximum between λ4500 or 4600 and 4800.[27a] This is of about the same intensity as, but is at considerably shorter wave length than, the iodine λ5200 absorption in inert solvents.

The most reliable data seem to be those for iodine in ethyl ether.[1a,3] Here the visible absorption has its maximum at λ4620[1a] or λ4650,[3] with peak intensity $\epsilon_{max}$ = 880[1a] or 870.[3] The total absorption in the band is somewhat greater than for the visible band in violet solvents, since the absorption region is broader. The most probable explanation appears to be that (1) a 1:1 complex is formed[1b]; (2) the absorption near λ4600 corresponds to essentially the same electronic transition, occurring in the iodine part of the complex, as the visible absorption of iodine alone or of iodine in aromatic complexes; but (3) the complex is a tighter one, involving a greater change in the structure of the molecule, than in the aromatic complexes.

As for each other type of complex, we should carefully consider whether the change in electronic structure may be mainly in the ground state, or mainly in the excited electronic state, or more or less equally in both. Usually this question can be answered fairly well when a correct explanation of the spectrum is obtained, since this involves an identification of the nature of the electronic structure in both electronic states, for an atomic arrangement which is that of the ground state.[28]

In addition to the visible absorption, there is for the ether complex a stronger ultraviolet absorption ($\epsilon_{max}$ = 2450) with peak at λ2480.[3] Kortüm and Friedheim[8] also report indications of the non-Beer's-law absorption near λ3100 which they find in cyclohexane solution and attribute to $I_4$.

The considerable shift and broadening of the iodine visible absorption in the ROR′·$I_2$ complexes probably means that the valence structure of the iodine molecule in the complex has been very appreciably changed. If one asks what sort of a change this may be, the writer finds only one obvious answer, namely, a structure in which an electron of the "lone pair" of the oxygen atom is partially transferred to the iodine, in a manner to be described below. The ionization potential of this electron, which undoubtedly corresponds[29] to the observed minimum ionization potential of ROR′, at least for all ROR′ complexes in which R is a hydrogen atom or an alkyl group, varies considerably depending on R and R′. For example, it is 12.61 e. v. for water, 10.7 e. v. for ethyl alcohol, 10.2 e. v. for ether.[18] In analogy with the complexes formed by the substituted benzenes, one may then expect to find the stabilities of the complexes to increase markedly in this order. Fairbrother's finding[19] of an apparent dipole moment of 1.3 *D* for iodine dissolved in the double ether dioxane gives support to the type of structure just indicated.

The existence of stable RR′O·$I_2$ complexes can be understood if their wave functions, though mainly of van der Waals RR′O·$I_2$ structure, attain very appreciable percentages of two resonance forms of the type

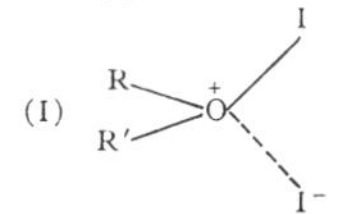


and of two other forms of the type

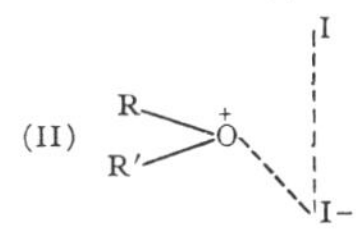


Type I resonance corresponds to one homopolar O–I bond plus one ionic $O^+I^-$ bond, Type II to a three-electron I–$I^-$ half-bond ($I_2^-$ structure) plus one $O^+I^-$ bond. The percentage of Type I structures in the wave function should depend in part on the strength of the $O^+$–I bond, and this in turn must depend on the O–I distance. For van der Waals contact, the O–I distance may be about 3.2 Å.[30] The actual distance may be estimated as somewhat less, say 3.0 Å. If the normal O–I bond length is 2.0 Å., and the O–I bond energy follows a Morse curve, the $O^+$–I bond energy at 3.0 Å. should be about one-third to one-fourth that for the normal bond length. Although this is not

(26d) Presumably pyridine forms a complex more or less similar to that of benzene, or else one of an onium type. The pyridine complex Py·$I_2$ is known also in crystalline form.

(27) J. Kleinberg and A. W. Davidson, *Chem. Revs.*, **42**, 601 (1948).

(27a) However, for iodine dissolved in ethyl alcohol glass at liquid hydrogen temperatures, the absorption maximum is reported to be shifted to about λ4000, with an increase in intensity (A. Prikhotko, *Acta Physicochimica U. R. S. S.*, **16**, 125 (1942)), perhaps due to a tightening of the complex, or possibly to a cage effect. Cage effects (N. S. Bayliss and A. L. G. Rees, *J. Chem. Phys.*, **8**, 377 (1940)) no doubt play some part in all solution spectra. However, such phenomena as the transition to a brown color when moderate amounts of a "brown" solvent are added to a "violet" iodine solution (*cf.*, *e. g.*, Ref. 1b) indicate that the color change here is primarily due to complex formation rather than cage effect. Further study will nevertheless be desirable.

(28) In interpreting any absorption spectrum it is important to keep always clearly in mind that the geometrical structure involved during the absorption process is that of the ground state. This is because according to the Franck–Condon principle, the peak of any molecular absorption region corresponds to the occurrence of an electronic transition with very little change in the configuration of the nuclei.

(29) R. S. Mulliken, *J. Chem. Phys.*, **3**, 506 (1935).

(30) *Cf.* A. L. G. Rees, *ibid.*, **16**, 995 (1948). However, Rees's values should probably not be accepted without reserve.

large, it may be enough to promote Type I resonance and help stabilize the complex to an appreciable extent.

The I–I axis probably lies *perpendicular* to the ROR′ plane in the manner shown, since the angles involved appear to give the best opportunity for O–I bonding by the available $O^+$ electron in the Type I resonance structures, and for $I–I^-$ bonding in the Type II structures. Coplanar arrangements of R′RO and iodine are unfavorable for reasons of symmetry, since, then (1), no $O^+$–I bond at all can be formed in the Type I structures; (2) in the Type II structures an excited state of $I_2^-$ would be necessary, just as in $Ar^+ \cdot I_2^-$.[26] Other geometrical models for the complex also appear to be unfavorable.

If, assuming the above model, the resonance hybrid contains a total of say 10% (reckoned in terms of *squares* of coefficients in the wave function) of the four resonance structures indicated, the I–I bond would be slightly "sprung." An examination, if feasible, of the Raman and infrared spectra might throw further light on this.[30a] When one writes out the electron configuration in terms of molecular orbitals, one finds that the two molecular orbitals involved in the visible absorption, and the oxygen non-bonding orbital, are particularly strongly affected. This happens in such a way as might very well account for the observed shift of the λ5200 iodine band toward shorter wave lengths, although quantitative computations would be needed to say positively that the observed shift is what one would expect. The above type of structure would mean that the interaction of the two partners of the ether complex is fairly strong *both* in the ground state and in the λ4620-excited electronic state.

An attempt to interpret the first ultraviolet peak does not seem particularly promising as an easy source of further clear-cut evidence on the structure of the complex. However, further study of the ultraviolet spectra of various complexes of the type $R'RO \cdot I_2$ might give interesting results. Perhaps the λ2480 absorption of the ether complex is of the charge-transfer type (*cf.* Note at end of Section III).

The above-proposed structure for $R'RO \cdot I_2$ complexes seems to correspond to an attractive mechanism, as follows, for such rapidly reversible equilibria as that between $I_2$, HOI and HI in aqueous solution

$$I_2 + H_2O \rightleftarrows H_2O \cdot I_2 \rightleftarrows H_2\overset{+}{O}I + I^- \rightleftarrows HOI + HI$$

with analogous reactions for the other halogens.

Spectroscopic data relevant to oxonium complexes of other halogens than iodine are also available. For example, iodine chloride (vapor absorption maximum at λ4700) forms brown solutions in inert solvents ($CCl_4$, $CHCl_3$) with visible absorption maximum at λ4600 ($\epsilon_{max}$ about 150–160). But in oxygenated solvents (ether, ethyl acetate, acetic acid) it forms yellow solutions in which the first absorption maximum occurs at λ3500–3600, although without change of peak intensity.[31] Presumably we have here very stable complexes of the same structure as the iodine complexes, but in which the structure has gone rather far toward that of a normal valence compound of oxonium type. Iodine bromide shows a similar behavior (red solutions, $\lambda_{max} \approx 4900$, $\epsilon_{max} \approx 350$–400 in inert solvents, yellow solutions, $\lambda_{max} \approx 4000$, $\epsilon_{max} \approx 350$–400 in ethyl alcohol and the like).

The shift of the visible band in iodine chloride and iodine bromide solutions is so large that one might hesitate to identify it with confidence as belonging to the halogen molecule in the complex. However, in view of the fact that it is in both cases, just as in iodine, the first absorption of any considerable intensity, and that in all three cases its peak intensity is nearly independent of the type of solvent, the identification becomes rather convincing. Further, the steady shift of this band, in the case of iodine, in going from inert solvents to aromatic solvents to RR′O solvents, re-inforces this identification.

Brief mention should be made here of a different well-known kind of complex formed by the halogens, namely, the $A^+X_3^-$ type. In their work on ICl, Gillam and Morton[31] concluded that ICl in HCl and $Na^+Cl^-$ solutions respectively forms $HICl_2$ and $Na^+ICl_2^-$. The $AICl_2$ or $ICl_2^-$ structure shows a characteristic absorption ($\lambda_{max} = 3420$, $\epsilon_{max} = 275$) which, probably by chance, is similar though not identical in position and intensity to that of the R′OR.ICl complexes. (The familiar $I_3^-$ complex absorbs near λ3500.)

Presumably similar to the $RR'O \cdot I_2$ complexes are the very stable deeply colored $RR'S \cdot I_2$ and $RR'S \cdot Br_2$ complexes, several of which are known in crystalline form.[31a] The halogen is recoverable from these on heating.

### V. Iodine–Acetone Complex

Although an absorption maximum at λ4600 has been reported for iodine in acetone,[31b] Benesi and Hildebrand did not find this.[1a] According to them, the first maximum occurs at λ3630, with an intensity very much higher ($\epsilon_{max} = 6100$) than for the first maximum of iodine in any of the solvents discussed hitherto.[32] [*Added in proof.—*

(30a) A rather similar case is that of the $Ag^+$·olefine complexes,[24a] where Raman studies (Taufen, Murray, and Cleveland, This Journal, **63**, 3500 (1941)) have confirmed the predicted loosening of the C=C bond.

(31) ICl: A. E. Gillam and R. A. Morton, *Proc. Roy. Soc. (London)*, **132A**, 152 (1931); IBr: A. E. Gillam, *Trans. Faraday Soc.*, **29**, 1132 (1933).

(31a) Fromm and Raizics, *Ann.*, **374**, 90 (1910).

(31b) F. H. Getman, This Journal, **50**, 2883 (1928). According to Getman, acetone and acetophenone are brown solvents. Getman reports that acetone solutions of iodine lose their color on standing.

(32) Maxima at λ3630 (and at λ2820) have been reported by Walls and Ludlam for iodine in ethyl alcohol,[27] but Batley[3] reports no maximum except at λ4470 if reactions are avoided. Cennano[3] reports maxima at λ3600 and at λ2970 for iodine in $CHCl_3$, $CH_3OH$, and in acetone. The most likely explanation of the λ3600 and λ2970 maxima here is that they are due to small amounts of the *strongly-absorbing* $I_3^-$ and benzene complexes, resp., present as impurities.

However, in further studies (private communication) they find that this is probably due to $I_3^-$ formed in a rapid reaction. By using mixed acetone–$CCl_4$ solutions, they obtain evidence for a moderately shifted visible absorption, which perhaps after all may be in harmony with German's observations.[31b]]

It is of interest to consider briefly the probable structure of a 1:1 complex of iodine with acetone. An examination of possible resonance structures suggests that those which could cause complex formation are of a type $RR'CO^{+}\cdot I_2^-$ very similar to those in the type $RR'O\cdot I_2$, except that the non-bonding oxygen orbital involved in the complex formation has its axis *in the RR'CO plane*, whereas in $RR'O^{+}\cdot I_2^-$ the non-bonding oxygen orbital had its axis *perpendicular* to the RR'O plane

$$\begin{matrix} R \\ R' \end{matrix} \!\!> C = \overset{+}{O} \begin{matrix} \diagup I \\ \diagdown I^- \end{matrix}$$

As compared with complexes of the RR'O type,[33] an additional stabilizing factor is present here. Namely, because of its planar structure, there must be a conjugation or hyperconjugation effect between some of the iodine non-bonding $\pi$ orbitals (see (1)) and the orbitals occupied by the $\pi$ bonding electrons in the C=O double bond. This effect could be represented by resonance structures, but in a rather cumbersome way; it is more easily seen using MO's (molecular orbitals). This might give a very appreciable further stabilization if the O–I bond distance is sufficiently short.

[*Added in proof*.—The reported complexes formed[1b] with iodine by the brown solvents ethyl acetate and acetic acid might be of either or both of the RR'O and RR'C=O types.]

**Acknowledgments.**—The writer is indebted to Professor J. H. Hildebrand and to Chicago colleagues for helpful discussion and criticism.

## VI. Summary

The absorption spectra of solutions of iodine in aromatic and oxygenated solvents are interpreted on the assumption that they are spectra of 1:1 complexes of iodine with solvent molecules. (The correctness of this assumption at least for certain aromatic complexes has been shown by Benesi and Hildebrand.) The extent to which the $\lambda5200$ absorption region, highly characteristic of the iodine molecule in vapor and in inert solvents, is shifted toward the ultraviolet and altered in shape or intensity, is used as the principal basis for a division of these complexes into three classes. By valence-theoretical considerations a unique probable geometrical and electronic structure is obtained for each class.

The probable structures are as follows. In general agreement with considerations advanced by Benesi and Hildebrand, and others, the electronic wave functions of all three classes are believed to contain resonance components of the general type $A^{+}I_2^-$ or $A^{+}I^{-}I$. In the $Ar\cdot I_2$ complexes (Ar = benzene or methylated benzene), the iodine molecule lies above the plane of the benzene ring with its axis parallel to the latter. In the $R'RO\cdot I_2$ complexes the iodine molecule stands against the oxygen atom, with its axis perpendicular to the R'RO plane. The probable acetone–iodine complex is similar except that the iodine axis is coplanar with the $\begin{matrix} R \\ R' \end{matrix}\!\!>C=O$ skeleton. The polar forces, which are present in all three cases, are aided in the first class by partial C–I bonding and in the second and third classes by partial $O^{+}$–I bonding, and further in $\begin{matrix} R' \\ R \end{matrix}\!\!>C=O\begin{matrix} \diagup I \\ \diagdown I \end{matrix}$ by conjugation between the C=O and the I–I $\pi$ electrons. The indicated structures are suggestive as to reaction mechanisms for the halogens.

The ultraviolet spectra, especially of the aromatic complexes, are discussed. The very intense absorption of the aromatic complexes near $\lambda3000$ is attributed to a transition in the aromatic part of the complex. It is suggested that this transition, although nearly forbidden in the aromatic molecule, is made strongly allowed by strong interaction between excited states of the two partners in the complex. As an alternative, it is suggested that this absorption, as also the color of other organic molecular complexes, may be due to an *intermolecular charge transfer process* during light absorption.

Bromine in benzene and toluene solutions shows spectra which indicate that it forms complexes of the same kind as described above for iodine. Iodine chloride and bromide solution spectra in R'RO solvents also indicate the formation of complexes similar to those formed by iodine, but tighter.

Brief comments are made on some other organic molecular complexes.

The nature of the spectra of iodine and bromine in vapor and inert solvents is reviewed and it is hoped somewhat clarified.

CHICAGO, ILLINOIS RECEIVED NOVEMBER 11, 1949

(33) The minimum ionization energies of ethyl ether (10.2 e. v.) and acetone (10.1 e. v.) are apparently about equal,[18] and since these probably correspond in both cases to removal of a non-bonding oxygen electron (see R. S. Mulliken, *J. Chem. Phys.*, **3**, 506 (1935), on RR'O; **3**, 564 (1935) on RR'CO), this indicates that this process is about equally easy in both cases.

# 25. THE INTERACTION OF ELECTRON DONORS AND ACCEPTORS*

Robert S. Mulliken
The University of Chicago
Chicago, Illinois

## INTRODUCTION

In chemical theory, the usefulness of G. N. Lewis's classification[1] of molecules as bases and acids has become increasingly evident. In turn, the same molecules which are good Lewis bases are in general also good Brönsted bases. But good Lewis acids are in general an entirely different (and larger) class of molecules than Brönsted or ordinary acids, which may be briefly described as molecules which when dissolved in suitable solvents act as proton donors.[2] Sidgwick's "acceptor," "donor" terminology[3] is probably preferable for reasons of clarity[2] to the (Lewis) "acid," "base" terminology, and will be used from here on.

The concepts of electron donors and acceptors may be generalized to include, more or less, reducing and oxidizing agents respectively; with this generalization, they correspond closely to Ingold's categories of nucleophilic and electrophilic reagents.[4] Although Lewis' original concept [1,3] of giving or accepting of electron pairs is generalized in the present point of view, and odd-electron donors and acceptors are also admitted to the fold, the etymology of the words donor, acceptor remains equally as appropriate as before.

Until recently, in spite of various approaches, no precise general statement seems to have been given in quantum-mechanical terms of the functioning of donors and acceptors in their chemical interactions. This gap can apparently be filled in a surprisingly simple way, by first writing an approximate expression for the wave function of any interacting pair of molecules in which one is a donor and one an acceptor.[5,6]

## RESONANCE THEORY OF DONOR-ACCEPTOR INTERACTION

As a simple long-familiar example to establish the pattern, one may consider the interaction of two univalent unlike atoms, say H and Br. Here the partners unite to form a stable joint system or molecule, with a wave function approximately of the form

$$\psi = a\psi(H - Br) + b\psi(H^+, Br^-) + c\psi(H^-, Br^+) , \qquad (1)$$

with $a^2 > b^2 > c^2$. In $\psi(H^+, Br^-)$, the H atom serves as donor, the Br atom as acceptor; in $\psi(H^-, Br^+)$ the Br atom is donor, the H atom acceptor. This illustrates that any atom may act either as donor or as acceptor, but the fact that $b^2 > c^2$ means that H is better as a donor, Br as an acceptor. However, the fact that $\psi(H - Br)$ is the predominant term in (1) shows that the donor and acceptor properties of H and Br are not as strong in HBr as their covalent bonding properties. For the pair Na + Br, the analogue of (1) is

$$\psi = a\psi(Na^+, Br^-) + b\psi(Na - Br) + c\psi(Na^-, Br^+), \qquad (2)$$

*This work was assisted by ONR under Task Order IX of Contract N6ori with The University of Chicago.

[1] Lewis, G. N., "Valence and The Structure of Atoms and Molecules," The Chemical Catalog Company, New York, 1923; Lewis, G. N., J. Franklin Inst., 226: 293, 1938; Luder, W. F., and Zuffanti, S., "The Electronic Theory of Acids and Bases, New York, 1946

[2] For a review on what is meant by acids, bases, donors, and acceptors, see Bell, R. P. Qu. Rev. Chem., 1: 113, 1947

[3] Sidgwick, N. V., "The Electronic Theory of Valency," Oxford University Press 1929

[4] Cf. Ingold, C. K., Chem. Rev., 15: 225, 1934

[5] See Mulliken, R. S., J. Am. Chem. Soc., 72: 600, 1950, where the present work began in an effort to explain the ultraviolet spectra of iodine-benzene and related molecular complexes.

[6] A detailed paper by the writer covering part of the material presented here, together with a theoretical discussion of the spectra of molecular complexes, appears in J. Am. Chem. Soc., 74: 811, 1952

SOURCE: Paper 25 of ONR Report on September 1951 Conference on "Quantum-Mechanical Methods in Valence Theory."

with $a^2 \gg b^2 \gg c^2$, corresponding to the fact that Na is a very strong donor (and Br a fairly strong acceptor). The last term is so unimportant in Equation (2) that it may as well be dropped; even the second term is probably of little importance.

Generalizing from Equations (1) and (2), one may write for any pair of entities of which either may be an atom, atom-ion, molecule, molecule-ion, or even a larger aggregate, for example, a metal,

$$\psi = a\psi_0(AB) + b\psi_1(A^-B^+) + \ldots\ldots, \tag{3}$$

in which A denotes the acceptor B, the donor. As a simple familiar example, A may be $BMe_3$ (trimethylboron) and B may be $NH_3$ (ammonia), which unite to form a stable "molecular compound" $Me_3B{\cdot}NH_3$. Here probably $b^2 > a^2$, with the predominant structure $\psi_1$ a "dative" structure $A^-{-}B^+$, and the subordinate structure $\psi_0$ a no-bond structure A,B. But in a multitude of loose molecular complexes to which Equation (3) is also applicable, it is clear that $b^2 \ll a^2$. An example is the complex between Bz (benzene) as donor and iodine ($I_2$) as acceptor.

In $Me_3B{\cdot}NH_3$, $Bz{\cdot}I_2$, and other similar cases where A and B are even-electron closed-shell[7] entities, Equation (3) may be written in the more specific form

$$\psi = a\psi_0(A,B) + b\psi_1(A^- {-} B^+) + \ldots\ldots \tag{4}$$

Such molecular compounds and complexes between closed-shell entities, owing their stability to a partial transfer of negative charge from B to A, may be called charge-transfer compounds or complexes.

In the above discussion, both odd-electron (H, Br, Na) and even-electron ($BMe_3$, $NH_3$, Bz, $I_2$) donors and acceptors have been encountered. The general theory is the same for odd-odd, odd-even, and even-even donor-acceptor pairs, with or without ionic charges, with differences only in the ways in which covalent and ionic binding appear in the terms $\psi_0$ and $\psi_1$ of Equation (3). Thus in Equation (1), $\psi_0$ is a covalent function, $\psi_1$ a function with an ionic but no covalent bond, while in Equation (4), usually $\psi_0$ is a no-bond, $\psi_1$ a dative function. In all cases, the donor-acceptor system $A{\cdot}B$ is stabilized by resonance between $\psi_0$ and $\psi_1$, with maximum stabilization if a and b are about equal.

In case A and B are similar or identical in donor and acceptor powers, resonance with a further term $c\psi_2$ becomes important in Equation (3) (cf. Equations (1), (2)). Additional types of resonance, in particular with $\psi$'s involving excited states of A and/or B, must in general also be included.

## THE STRENGTHS OF DONORS AND ACCEPTORS

The preceding discussion leads to the question as to what factors determine the strengths of A and B as donors or acceptors. In general, a strong donor B or acceptor A is one which tends to make b/a large in Equation (3). Since $\psi$ of Equation (3) refers to the ground state of AB, b/a tends to be larger the lower the energy of the structure $\psi_1(A^-B^+)$ relative to $\psi_0(AB)$. The energy of $A^{-\bullet} + B^+$ is obviously lower the smaller the ionization energy I of B, and the larger the electron affinity E of A. Hence small I makes for a good donor and large E for a good acceptor. Additional factors are often important; in particular, the closeness to which the electrical centers of gravity of oppositely charged ions can approach each other or, briefly, the mutual "approachability". Thus a good donor should have small I and/or good approachability, and a good acceptor, large E and/or good approachability. For the case that A and B are atoms or atom-ions, good approachability goes with small size.[8]

When two molecular entities A and B unite to form a stable compound $A{\cdot}B$ with electronic structure describable by Equation (3), the resulting configuration of the nuclei, or

---

[7] This term will be used for any system with all electrons paired.

[8] A more detailed discussion is given in the reference mentioned in footnote 6.

skeleton, is in general a severe compromise between the "natural" skeletons corresponding to pure $\psi_0$(AB) and pure $\psi_1(A^-B^+)$. This is because, in general, A and $A^-$ differ strongly in their natural skeletons, likewise B and $B^+$. Hence in a compromise structure A·B, either A and B or else $A^-$ and $B^+$, or both, are in general necessarily strongly distorted.

The molecular compound $H_3N \cdot BMe_3$ may be considered as an example. Here, the "natural" skeleton of structure $\psi_0$ in the compound involves a low pyramid for $NH_3$ and also for $BMe_3$, the planar form of free $BMe_3$ having been distorted (even in $\psi_0$) by mutual repulsion between $NH_3$ and $BMe_3$ because they are close together; while structure $\psi_1$ corresponds to a skeleton with tetrahedral angles about the N and B atoms, similar to that in $H_3C-CMe_3$. The actual skeleton here, although a compromise, should be closer to that of $\psi_1$ in view of $b^2 > a^2$.

In a loose molecular complex, for example $Bz \cdot I_2$, only small distortions of the natural skeletons of the partners Bz and $I_2$ are expected, since the compromise structure is nearly pure $\psi_0(Bz, I_2)$. But as a corollary, $I_2^-$ and $Bz^+$ in $\psi_1(I_2^- - Bz^+)$ in $Bz \cdot I_2$ may be strongly distorted from their natural skeletons. Actually the distortion is small for $Bz^+$, whose natural skeleton differs only slightly from that of Bz. But for $I_2^-$ it is large, since the equilibrium distance is much larger for $I_2^-$ than for $I_2$.

These considerations lead to the important point that in using I and E as partial measures of donor and acceptor strengths, one must not take values of these quantities corresponding to passage from B with its natural skeleton to $B^+$ with its natural skeleton, or from A in its natural shape to $A^-$ in its natural shape; instead, one must take so-called vertical values of I and E corresponding to no change in skeleton ($I^{vert}$ and $E^{vert}$). Moreover, one must take $I^{vert}$ and $E^{vert}$ corresponding to such deformed skeletons for both A and B as exist in the final compromise skeleton of A·B. In the special case of loose complexes, this specification reduces very nearly to taking $I^{vert}$ and $E^{vert}$ for the natural skeletons of A and B.

It is evident that values of $I^{vert}$ and $E^{vert}$ will be valuable in interpreting and/or predicting donor and acceptor strengths. Values of I for molecules are known in many cases from electron impact experiments and especially from the study of Rydberg series in vacuum ultraviolet absorption spectra[9]; corrections to $I^{vert}$ may usually be estimated satisfactorily. Values of E and especially $E^{vert}$ are less available.

The foregoing considerations must be modified in the case of donor-acceptor reactions where the result is not an addition compound but a displacement reaction (see next two sections).

## MAJOR CLASSES OF DONORS AND ACCEPTORS

It is convenient to group donors and acceptors into several classes. The most familiar donors are the onium bases like $NH_3$ and $OH_2$, which contain unshared or nonbonding electrons (lone pairs) of reasonably low I (onium or n donors). This class may be taken to include also negative ions like $Cl^-$, $OH^-$, $CN^-$ (n' donors).

These closed-shell n donors unite particularly strongly with closed-shell acceptors of the vacant-orbital or v type, such as $BMe_3$, $BF_3$, $AlCl_3$, $HgCl_2$, O atom. For the neutral n donors, the loss of an electron by the donor B and its gain by the acceptor A involves an increase of both covalent and ionic valence for both partners, and the resulting dative bond helps strongly to stabilize the compound A·B.

Two other groups of strong acceptors are (1) positive ions like $Ag^+$ (but not $H_3O^+$), which may be classed as v* acceptors, and (2) radicals with a vacancy in their outer shell, e.g. Cl, OH, CN. The v* acceptors differ from the neutral v acceptors in that acceptance of an electron increases only covalent but not ionic valence.

---

[9] Cf., e.g., Price, W. C., Chem. Rev., 41: 257, 1947; Walsh, A. D., Qu. Rev. Chem., 2: 73, 1948

Important are $\pi$ donors and acceptors, where the electron ionized or accepted is one of the unsaturation ($\pi$) electrons of an unsaturated or aromatic organic compound. Here the relatively low I values of $\pi$ donors[9] are a favorable factor, but the low degree of localization of the $\pi$ electrons is unfavorable for approachability, and the fact that $\pi$ ionization weakens the $\pi$ bonding more or less is also unfavorable. The occurrence of good $\pi$ acceptors is apparently limited to cases where attached strongly electrophilic atoms or groups ($NO_2$, C = O, $SO_3H$, etc.) have sucked negative charge (via $\pi$-bond resonance) out of other $\pi$-electron-bearing atoms or groups, thus increasing the electron affinity of the latter for acceptance of an additional $\pi$ electron (examples: trinitrobenzene, maleic anhydride). Since the $\pi$ molecular orbital into which the accepted electron goes is one which is more or less antibonding, and little localized, $\pi$ acceptors are ordinarily weak.

An important class is that of the d or dissociative acceptors like HCl, MeCl, MeOH, also positive ions like $H_3O^+$, $NH_4^+$, and other protonated bases. A somewhat related class is that of the x acceptors, typified by the halogen molecules $X_2$ and XY. The x and d classes are mostly weak acceptors. They owe their acceptor character to the electronegative atoms they contain.

For the case of x acceptors, this is seen as follows. If the acceptor A is a compound XY of two atoms or radicals X and Y, then from Figure 1, if $E_X > E_Y$,

$$E_{XY}^{vert} = E_X + D^- - D + \rho - \rho^- = E_{XY} + \rho - \rho^- , \tag{5}$$

the values to be used for $\rho$ and $\rho^-$ in any given actual compound A·B being those corresponding to that compromise X-Y distance R which exists in that compound. From Equation (5) and Figure 1, it is seen that $E^{vert}$ increases sharply with increase in R, that is, with loosening of the X-Y bond, reaching $E_X$ as asymptotic value. From Equation (5), it is seen that the acceptor behavior of x acceptors depends primarily on large $E_X$, such as occurs particularly if X is a halogen atom. Small D, as in $Cl_2$ or $I_2$, also helps.

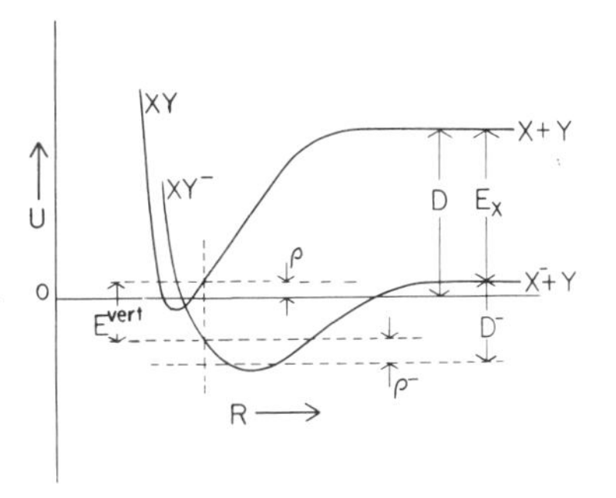


Figure 1 - Potential energy curves for an x acceptor XY and its negative ion, as a function of distance R between X and Y, to illustrate Equation (5). The curves are schematic; actually $D^-$ is expected to be roughly D/2 for halogen molecules.

As judged by the dissociation constants of their complexes with Bz,[10] the following order of strength is observed for the halogens as x acceptors: $ICl > I_2 > Br_2 > Cl_2$. Except for ICl (where perhaps dipole-polarization of the Bz gives extra stability to Bz·ICl), this order is intelligible in terms of Figure 1; we need only suppose that $\rho^-$ for R equal to its equilibrium value in the halogen molecule increases from $I_2$ to $Cl_2$, as is reasonable.

## CHARGE-TRANSFER COMPLEXES AND CHARGE-TRANSFER REACTIONS, ASSISTED AND UNASSISTED

Frequently when a given A and B alone form only a loose charge-transfer complex, if any at all, the intervention of an additional substance or agency can produce a strong complex or a chemical reaction (charge-transfer reaction). These may be called assisted charge-transfer complexes and assisted charge-transfer reactions.

---

[10] Keefer, R. M., and Andrews, L. J., J. Am. Chem. Soc., 72: 4677, 5170, 1950; 73: 462, 1951

As an example, consider the interaction of the strong n donor $NH_3$ and the d acceptor HCl. Since the vapor of $NH_4Cl$ apparently consists of separate molecules of $NH_3$ and HCl, it is empirically evident that the complex $H_3N{\cdot}HCl$ must be so loose that it is broken up by thermal energy even at fairly low temperatures. This is understandable in terms of Figure 1, since for a loose complex, $E^{vert}$ for HCl would probably be strongly negative, so that HCl per se should be an extremely weak acceptor. (However, a very loose complex due to hydrogen bonding should occur.) But if $NH_3$ and HCl are brought together in the presence of a solvent which contains dipole molecules (for example, ammonia), solvation greatly assists charge-transfer, leading to solvated $NH_4^+Cl^-$ as end product. Letting S denote solvent, or solvent molecules, the overall reaction is

$$NH_3 + HCl + S \rightarrow (H_3N{\cdot}HCl)S \rightarrow \begin{cases} NH_4^+Cl^-S & (6a) \\ \text{or} & \\ NH_4^+S + Cl^-S, & (6b) \end{cases}$$

the second possibility (complete dissociation) being realized only with strong high-dielectric-constant solvents. Processes such as (6a) and (6b) can be called solvent-assisted ionic-dissociative charge-transfer reactions.

Noteworthy special cases of solvent-assisted donor-acceptor interaction pairs are these practically-important cases where either the donor or the acceptor molecules are also solvent molecules; examples, $H_3N$ - HCl in $NH_3$, and $H_2O$ - HCl in $H_2O$.

The cases $H_2O$ - HR in $H_2O$, R being any suitable radical, include all the acids, in the ordinary sense of aqueous H-acids.

In general, the behavior of H-acids in water is to be described from the present viewpoint as water-assisted d-acceptor behavior toward $H_2O$ as an n donor. This implies ionic-dissociative charge-transfer. Relative d-acceptor strengths now evidently no longer depend on $E^{vert}$ of the free acceptors, but largely on the relative net energies of the reactions

$$H_2O + HR + aq \rightarrow H_3O^+aq + R^-aq. \tag{7}$$

A phenomenon which may be called symmetrization is of frequent occurrence in the last stages of (usually assisted) charge-transfer reactions. Examples are:

$$NH_3 + H_3O^+aq \rightarrow NH_4{}^+aq + H_2O;$$

$$Cl^-aq + ICl \rightarrow ICl_2{}^-aq;$$

$$F^-S + BF_3 \rightarrow BF_4^-S.$$

In these examples, B and A are respectively $NH_3$ and $H_3O^+aq$, $Cl^-aq$ and ICl, and $F^-S$ and $BF_3$ (S means solvent or, perhaps, a positive ion such as $Na^+$ which is an extremely reluctant acceptor). Near-symmetrization and partial symmetrization also are frequent: example, $AlCl_3$ + $Br^- \rightarrow AlCl_3Br^-$.

In the category of ionic-dissociative charge-transfer reactions, an important type in addition to the solvent-assisted reactions is that of crystallization-assisted ionic-dissociative charge-transfer reactions: example, the condensation of $NH_3$ + HCl vapor to form crystalline $NH_4^+Cl^-$. Such reactions might also be called polymerization-assisted, since the crystal may be thought of as a stable polymer of the per se unstable salt molecule $NH_4^+Cl^-$.

## INNER, OUTER, AND MIDDLE COMPLEXES IN CHARGE-TRANSFER REACTIONS

For both assisted and unassisted charge-transfer reactions, further insight can be obtained by plotting curves of energy against a reaction coordinate, say C. By this is meant, roughly, the distance between the donor B and acceptor A, but with the understanding that at each value of C both the electronic structures and the skeletons of A and B have adjusted themselves adiabatically (i.e., smoothly and continuously) to give minimum total energy.

Two major patterns for this adiabatic reorganization may be distinguished: ionic-dissociative, where covalent bonds present in the original partners are dissolved and ions are formed, as for example when $NH_3 + HCl \rightarrow NH_4^+Cl^-$; and associative, as for example when $H_3N + BMe_3 \rightarrow H_3N^+ - B^-Me_3$ or $Cl^- + AlCl_3 \rightarrow AlCl_4^-$.

Figures 2 and 3 contain several curves of energy U, for charge-transfer processes involving a donor B and an acceptor A at least one of which is a closed-shell neutral molecule, plotted against the reaction coordinate C. The curves are schematic but illustrate probably the main types of behavior.

The forms of the curves in Figure 2 are based on the following considerations. At large separations, a = 1, b = 0 in Equation (3) or (4). With decreasing separations, if b remained 0, exchange repulsion associated with $\psi_0$ would cause U to rise more and more steeply, beginning as soon as A and B overlap appreciably. Actually, as soon as they begin to overlap, b in general begins to increase and resonance between $\psi_0$ and $\psi_1$ increases. This and various polarization effects between A and B tend to cut down more and more the rapid rise of U which would occur for pure unpolarized $\psi_0$.[11]

The net outcome should be curves like those in Figure 2, the typical case being one with an "outer complex" and an "inner complex", with a maximum between. These maxima, or barriers, may be identified with "activated complexes" of chemical kinetics. The inner complexes when not inaccessibly high, yet not stable enough to be end products, may be identifiable with "reaction intermediates".

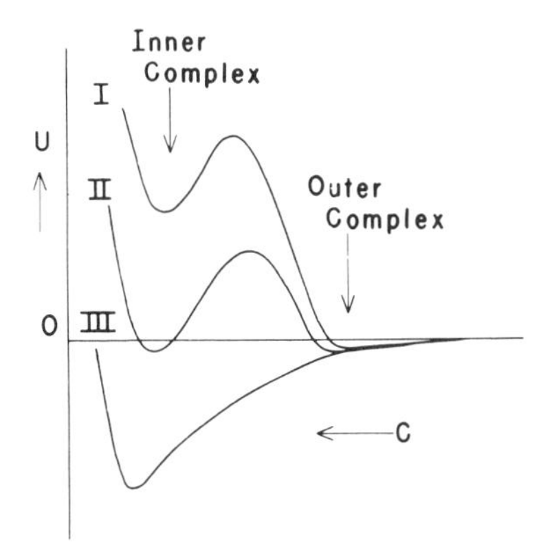


Figure 2 - Energy curves U(C) for unassisted interaction (inert solvent or none) between a donor B and an acceptor A (qualitative only). C = reaction coordinate, increasing from 0 toward 1 as $b^2/a^2$ in Equation (3) increases. The "outer complex" (C small), when present, corresponds to $b^2 \ll a^2$, the "inner complex" (C = 1) to $b^2 \approx a^2$ or $b^2 > a^2$.

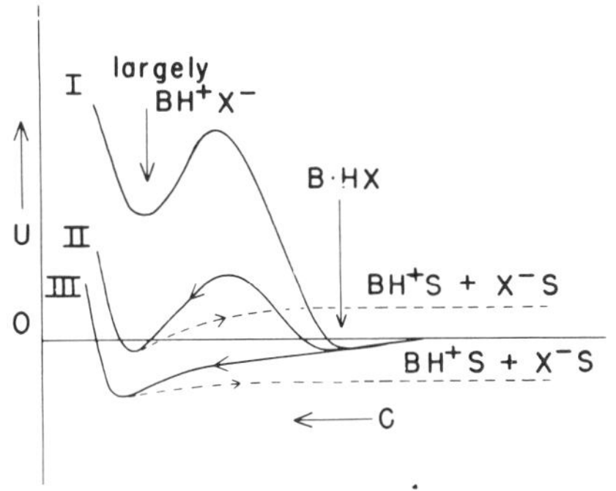


Figure 3 - Energy curves U(C) for ionic-dissociative interaction between a donor B (e.g. $NH_3$) and a d acceptor HX. (The curves would be similar for d acceptors in general.) (Curve I is unassisted.) Curve II is the same as Curve I but is assisted by a weak and rather low-dielectric solvent (lower branch of curve is for dissociation of $BH^+X^-S$ into solvated ions (S = solvent)). Curve III is like Curve II but for strong and high dielectric solvent. With the curves shown, little dissociation is expected with Curve II, nearly 100% with Curve III.

---

[11] In some cases, valence bond rearrangements such as can occur without charge-transfer (as e.g., in $H_2 + I_2 \rightarrow 2HI$) may also contribute to cutting down U by resonance.

It seems probable that among donor-acceptor pairs of class $BX_3 + NR_3$ and related classes, reacting without solvent assistance, examples of reaction curves of all the types I, II, and III of Figure 2, and/or variations of these, occur.[12] In Figure 2, additional variations would include curves with no inner minimum. Another possibility is a curve like I but with a small barrier outside the outer minimum. A possible example is the loose $I_2 \cdot Pr$ complex (Pr = propylene), concerning which Freed reports[13] that iodine in solution at 77°K in propane (an inert solvent) upon addition of Pr, develops the color of the $I_2 \cdot Pr$ complex only slowly, indicating a small outer barrier.

The preceding discussion needs amendment in one respect, namely that besides charge-transfer forces, additional attractive forces—associated with the $\psi_0$ component of the wave function—may participate in creating loose complexes. For every pair of neutral entities A,B, London dispersion forces give of course at least a weak van der Waals attraction. In addition, classical electrostatic forces must tend to give loose complexes whenever A and/or B is an ion or has a dipole moment. The so-called hydrogen bond[14] falls in this category.

It is generally believed that hydrogen bonding is usually primarily electrostatic, but that there is some additional stabilization by resonance forces of the kind here called charge-transfer forces. In the extreme case of $HF_2^-$ formed from $HF + F^-$, assisted symmetrized charge-transfer binding apparently predominates over pure electrostatic binding. In a similar way, charge-transfer binding apparently predominates over electrostatic polarization-binding in the formation of $I_3^-$ from $I_2 + I^-$.[15]

Instead of Figures 2 and 3, another kind of diagram in which both positive and negative values are assigned to the reaction coordinate C is useful for charge-transfer reactions such as

$$I^-aq + I_2 \rightarrow (III)^-aq \leftarrow I_2 + I^-aq. \quad (8)$$

or

$$HO^-aq + MeI \rightarrow (HOMeI)^-aq \rightarrow HOMe + I^-aq. \quad (9)$$

Figure 4 shows typical curves for reactions like these. Curves I and II correspond to the case where the reaction proceeds continuously to the right over an activation barrier or barriers, as in the typical displacement reaction (9). Reaction (8) corresponds to a special case of Curve III of Figure 4 with equal values of U for large positive and negative values of C. The stable complex at the middle of the figure in Curve III may be called a "middle complex."

The charge-transfer-process idea, with the aid of Figures 2 to 4, appears capable of giving plausible indications about reaction-paths for a great variety of chemical reactions.

## SURFACE INTERACTIONS

It seems probable that the present approach will be found helpful in understanding adsorption of molecules by solids. A number of instances have been reported where spectroscopic phenomena similar to those observed with $Ar \cdot X_2$ complexes are found for apparent adsorption-complexes.[16] In general, the existence of donor-acceptor interactions at surfaces, and their importance for heterogeneous catalysis, are recognized possibilities.[17]

---

[12] See especially the papers of H. C. Brown and collaborators, mainly in the J. Am. Chem. Soc.

[13] Freed, S., and Sancier, K. M., J. Am. Chem. Soc., 74: 1273, 1952.

[14] See Pauling, L., "The Nature of the Chemical Bond," Ch. IX, Cornell University Press, 1940

[15] Cf. Pimentel, G. C., J. Chem. Phys., 19: 446, 1951 for a discussion of the linear ions $(FHF)^-$ and $I_3^-$ essentially as symmetrized charge-transfer complexes. In $I_3^-$, partial trivalency of the central I atom probably also assists.

[16] Jura, G., Grotz, L., and Hildebrand, J. H., Abstract No. 128 of Division of Physical and Inorganic Chemistry, Am. Chem. Soc. Meeting of September 1950.

[17] Cf. for example C. Walling, Abstract No. 130, loc. cit.

Adsorption would correspond to formation of an outer complex, chemical reaction to passage to an inner complex. An important special case is that where the solid is a metal.

VAN DER WAALS ATTRACTIONS

Although charge-transfer resonance forces should be particularly strong between unlike atomic or molecular entities of which one is strongly a donor and the other strongly an acceptor, they must operate also between the members of each pair of like entities, since every such entity has, at least very weakly, both donor and acceptor properties. Further, it is easily seen[6] that, for loose complexes, a molecular entity can act as donor or acceptor or both toward any number of neighboring entities simultaneously. Moreover, charge-transfer forces are forces of attraction only. In all these respects, charge-transfer forces resemble the dispersion forces which F. London used so successfully in explaining van der Waals attractions. Hence it appears that they must supplement dispersion and other attraction forces (e.g. hydrogen bonding, other dipole-dipole, and dipole-polarization forces) in helping to explain cohesion between molecules.

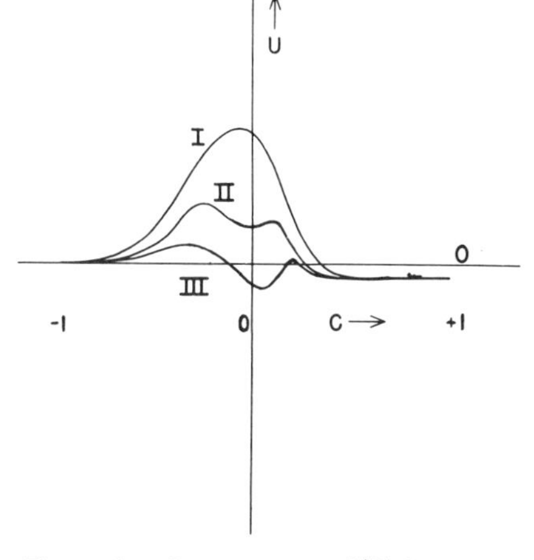


Figure 4 - Energy curves U(C) for charge-transfer reactions of the type of Equations (8) and (9). "Middl complex" at C = O of Curve III. C = -1, 0, and +1 here correspond more or less to C = 0, 0.5, and 1 of Figures 1 and 2. In each curve, shallow depressions (outer complexes—cf. Figures 1 and 2) near C = ±1 are an additional possibility.

They differ from the other forces mentioned in that they fall off faster (exponentially, like valence forces) with increasing distance; they operate only if there is appreciable overlapping of the wave functions of the interaction partners. (Conversely, they should increase rapidly under compression.)

But the most interesting feature of charge-transfer forces for molecular cohesion may be their rather remarkable orientational properties, which are a result of quantum-mechanical symmetry requirements[18] on the wave functions $\psi_1 (A^- - B^+)$. The more conventional electrostatic and dispersion forces also, to be sure, have orientational properties, at least for nonspherical molecules, but of a relatively orthodox character. Charge-transfer forces often favor different and from the orthodox viewpoint surprising orientations. Thus, they may often be influential or decisive in molecular crystals in establishing modes of packing of molecules which otherwise would not be intelligible. Possible effects of charge-transfer forces for cohesion energy or crystal-structure in a few examples will now be considered.

For some time the fact that rare gas crystals have the face-centered-cubic in preference to the hexagonal-close-packed structure has been a minor mystery. It has recently been shown that this cannot be understood on the basis of dispersion forces, even with inclusion of higher-order effects.[19] Since with rare gases, no other types of polarization forces have been

[18] Quantum-mechanical symmetry and other requirements, and their application to $Bz \cdot I_2$, $Bz \cdot Ag^+$, and other complexes, are discussed in detail in papers cited in footnotes 5 and 6.

[19] Axilrod, B. M., J. Chem. Phys., 19: 719, 724, 1951

recognized, it would seem that the orientational properties of charge-transfer forces may perhaps offer a new possibility of explanation.[20]

For the hydrogen halides, which form molecular crystals, London's dispersion forces account well for the observed sublimation energies. Even dipole-dipole forces have been shown to be a minor factor. In this case, it can be shown that charge-transfer forces also should be of very minor or negligible importance, since these compounds when unassisted are very weak d acceptors (see above) and also very weak donors.

For the crystals of aromatic hydrocarbons, say benzene or naphthalene, one may reasonably expect greater relative importance for charge-transfer forces, since these molecules are good $\pi$ donors and should also be at least weak $\pi$ acceptors. This is especially true of the larger molecules, as can be seen from the fact that with increasing molecule size, I and E approach equality; in the limiting case of graphite, I equals E (but, of course, approachability diminishes). It can be shown that charge-transfer forces disfavor stacking of benzene molecules one directly above another. If there is a similar effect in graphite, this may explain the observed fact that neighboring sheets of atoms in graphite crystals are displaced from vertical stacking by half a ring diameter.[6] The fact that the van der Waals distance between graphite layers is smaller than the distance between parallel aromatic ring planes in other crystals[21] might well also be a charge-transfer effect. The odd ways in which various aromatic molecules are packed in their crystals might well also be attributable to orientational charge-transfer forces.

[20] For any pair of adjacent atoms of, say, argon in a crystal, there should be a small resonance component of the type

$$X \equiv \psi_1 (A^+ - A^-) + \psi_1' (A^- - A^+),$$

in which $A^+$ is $^2P$ and $A^-$ is $^2S$. The cylindrically symmetrical ($^1\Sigma_g^+$-form, or at least $O_g^+$-form) charge distribution for $X$ departs appreciably from that of two superimposed spherically symmetrical distributions about the two atoms. This should apparently give rare gas atoms new possibilities of being lined up in a polarized fashion. However, there is serious question whether $A^-$ (even in the stabilizing presence of $A^+$) is an admissible structure. But with the exclusion of other possibilities, the present charge-transfer mechanism seems worth further investigation.

[21] *Cf.* Pauling, L., "Nature of the Chemical Bond," last page of Chap. V, Cornell University Press, 1940

[Reprinted from the Journal of the American Chemical Society, 64, 811 (1952).]


[CONTRIBUTION FROM THE DEPARTMENT OF PHYSICS, UNIVERSITY OF CHICAGO]

# Molecular Compounds and their Spectra. II[1]

BY ROBERT S. MULLIKEN

(1) A simple general quantum-mechanical theory is presented for the interaction of electron acceptors and donors (Lewis acids and bases) to form 1:1 or $n$:1 molecular compounds ranging from loose complexes to stable compounds. This puts into more accurate or more general form ideas which have been in frequent use for some time. The theory involves resonance between no-bond structures (A,B) and dative structures ($A^-$–$B^+$), where A is an acceptor atom, molecule, or ion and B is a donor atom, molecule, or ion. Two classes of donors ($\pi$, and $n$ or onium, bases) and three classes of acceptors ($\pi$, $v$ or vacant-orbital, and $d$ or dissociative) are particularly considered; $i$ (ionic) donors and acceptors are also mentioned. General and specific factors governing the strengths of interaction between acceptors and donors of various classes are deduced from the theory. (2) A special class of intense electronic absorption spectra characteristic of molecular compounds A·B, and non-existent for either partner A or B alone, is predicted. These are called charge-transfer spectra. (3) The forces which lead to complex-formation may be called charge-transfer forces. They may be of comparable importance to London's dispersion forces in accounting for van der Waals attractions. They have characteristic specific orientational properties of possible importance for the manner of packing of molecules in liquids, in molecular crystals, in heterogeneous systems, and in biological systems. They may also be important in adsorption. They should increase under compression and thus contribute to compressibilities. The effect of charge-transfer forces in lowering activation barriers for chemical reactions is briefly discussed. (4) The benzene–iodine and the $BX_3 \cdot NR_3$ types of molecular compound and the $Ag^+$ complexes are considered in detail. The characteristic absorption peak of the benzene–iodine and related complexes near λ3000, discovered by Benesi and Hildebrand, is identified with the predicted charge-transfer absorption. Its position and intensity are in good agreement with the theory. Theoretical considerations based on symmetries of quantum-mechanical wave functions often favor unsymmetrical geometrical configurations of molecular complexes. For example, they point to an off-axis position for the $Ag^+$ in the $Ag^+$–benzene complex, a result which is supported also by empirical evidence.

## I. Introduction

The appearance in many cases of strong color on bringing together two colorless or nearly colorless organic compounds is well known. The effect is generally attributed to a loose reversible association ("molecular compound" or "molecular complex") of the original molecules in a definite ratio, most often 1:1. Equilibrium constants, heats of formation and other thermodynamic data for many such complexes have been established.

There is similar and other evidence for complexes between inorganic and organic molecules; for example, between halogen molecules and organic compounds,[2] and between $Ag^+$ and aromatic or unsaturated compounds.[3,4] Finally, such compounds as $R_3N \cdot BX_3$ may be viewed[5] as colorless molecular complexes of unusually high stability.

Many authors have discussed the structure of molecular complexes. For stable compounds like $R_3N \cdot BF_3$, the dative bond structure $R_3N^+$–$B^-F_3$ is generally accepted. Likewise, the view that a molecular complex is the result of the combination of an electron donor or base, say B, with an electron acceptor or Lewis acid,[4a] say A, is widely held.

Notable contributions to the theory of molecular complexes have been made by Weiss[6] and by Brackmann.[5] Some of Dewar's ideas,[7] particularly on $\pi$ complexes, are relevant. So also is Pauling's 1938 suggestion[3] that the $C_2H_4 \cdot Ag^+$ complex is made stable by resonance between a predominant no-bond structure and structures of the type $>C^+$—$C<$ (with Ag attached to the second C).

Finally, a 1942 note by Woodward,[8] in which he advanced the idea of an "*intermolecular semi-polar bond*" in molecular complexes, comes close to the theme of the present paper; Woodward also refers to similar work by Bateman.

Neither Weiss's nor Brackmann's views appear acceptable *in toto*, but a selection and combination of their ideas with some extensions appears to lead to a satisfactory theory. This will be given below, including an explanation of the colors of molecular complexes and related phenomena, matters not treated satisfactorily by previous writers.

Weiss proposed that all molecular complexes have an essentially ionic structure $B^+A^-$, and pointed out that a *low ionization potential* for the base B, and a *high electron affinity* for the Lewis acid A, should then favor a stable complex. He attributed the color of molecular complexes to intense charge-resonance spectra arising within the ions in the complex.

Brackmann attributed molecular complex formation to "complex resonance," meaning quantum-mechanical resonance between a no-bond structure and a structure with a bond between the two partners A and B, but made no clear statement about ionic character in the latter structure. Brackmann insisted that (assuming colorless partners) the complex *as a whole* determines the color, that is, that the light absorption causing color is not localized in one of the partners. This is an important and, according to the present analysis, an essentially correct idea. Brackmann also emphasized that reversible formation of a resonance complex, by reducing the activation barrier, may often be a preliminary step in an irreversible chemical reaction.

The writer recently[2d] discussed the structure of

(1) This work was assisted in part by the ONR under Task Order IX on Contract N6ori-20 with the University of Chicago. Presented, in part, as Paper No. 122 at the Symposium on Generalized Acids and Bases of the meeting of the Division of Physical and Inorganic Chemistry of the American Chemical Society at Chicago, Ill., on Sept. 5, 1950. A summary has been published in *J. Chem. Phys.*, **19**, 514 (1951). Ref. 2d is to be regarded as paper I of the present series.

(2) (a) H. A. Benesi and J. H. Hildebrand, THIS JOURNAL, **70**, 2382 (1948); **71**, 2703 (1949); (b) T. M. Cromwell and R. L. Scott, *ibid.*, **72**, 3825 (1950); (c) R. M. Keefer and L. J. Andrews, *ibid.*, **72**, 4677, 5170 (1950); **73**, 462 (1951); (d) R. S. Mulliken, *ibid.*, **72**, 600 (1950).

(3) S. Winstein and H. J. Lucas, *ibid.*, **60**, 836 (1938); 1:1, 2:1, and 1:2 $Ag^+$ complexes with ethylene, benzene, etc.

(4) L. J. Andrews and R. M. Keefer, *ibid.*, **71**, 3644 (1949); **72**, 3113, 5034 (1950).

(5) W. Brackmann, *Rec. trav. chim.*, **68**, 147 (1949).

(6) J. Weiss, *J. Chem. Soc.*, 245 (1942), and to some extent later papers.

(7) M. J. S. Dewar, *Nature*, **156**, 784 (1945); *J. Chem. Soc.*, 406, 777 (1946); "The Electronic Theory of Organic Chemistry," Oxford Clarendon Press, London, 1949.

(8) R. B. Woodward, THIS JOURNAL, **64**, 3058 (1942).

the complexes formed by halogen molecules ($X_2$ or XY) with aromatic (Ar) and other solvents, in terms of resonance between a predominant no-bond structure (Ar, $X_2$) and small admixtures of structures of the type ($Ar^+–X_2^-$). As was shown by Benesi and Hildebrand and others, the Ar·$I_2$ complexes, in addition to visible absorption attributable to the $I_2$ in the complex, have an intense characteristic ultraviolet peak near λ 3000.[2] This the writer at first attributed to a modified Ar absorption, but later, in a "Note added in Proof," to an Ar → $X_2$ electron-transfer process,[2d] in harmony with Brackmann's idea that any complex should have a characteristic absorption of its own.

The present paper develops this idea further as part of a general discussion of the structure of complexes and compounds between Lewis acids and bases, and of related matters. Because of the great variety and complexity of actual molecular situations, the discussion should be regarded at many points as approximate and tentative.

## II. Structure and Charge-transfer Spectra of Molecular Complexes

The viewpoint indicated in Section I can be put into quantum-mechanical form by writing the wave function of the ground state N of any molecular compound A·B as

$$\psi_N = a\psi_0 + b\psi_1 + \cdots \qquad (1)$$

The acceptor A and donor B may in general be any suitable pair chosen from atoms, atom-ions, molecule-ions, molecules, or perhaps even solids, but in the present section with the limitation that both are in totally symmetrical singlet electronic ground states.[9]

In Eq. (1), $\psi_0$ is (with respect to covalent bonding at least) a "*no-bond*" wave function $\psi$(A,B). It has the form

$$\psi_0 = \psi(A,B) = \mathcal{A}\psi_A\psi_B + \cdots \qquad (2)$$

where $\mathcal{A}$ denotes that the product $\psi_A\psi_B$ of the wave functions of A and B is to be made antisymmetric in all the electrons, and the terms indicated by + · · · · represents such small modifications[10] as might be expected from hitherto recognized types of of polarization effects. In Eq. (1), $\psi_1$ is a "dative" wave function corresponding to *transfer of an electron* from B to A accompanied by the establishment of a (usually weak, because of the distance between A and B) covalent bond between the odd electrons in $A^-$ and $B^+$. That is

$$\psi_1 = \psi(A^- - B^+) + \cdots \qquad (3)$$

where the + · · · again indicates small modifying terms.[10]

(9) If A or B is an atom or atom-ion, this means a *closed-shell* electronic structure in ordinary terminology. If A or B is a molecule or molecule-ion, this also means a closed-shell structure if the description is given in terms of *molecular orbitals*. If, however, the description is given in terms of *atomic orbitals*, it means that all valence electrons are paired in electron-pair bonds.

(10) It will be convenient to consider these well-recognized effects as included in $\psi_0$ of Eq. (2) or $\psi_1$ of Eq. (3), in order to distinguish them from the effect of the interaction between $\psi_0$ and $\psi_1$. Strictly speaking, these two kinds of effects are not always entirely independent; thus if A and B in $\psi_0$ are oppositely charged ions, some part of their mutual polarization is expressed by the terms in $\psi_1$ in Eq. (1).

In Eq. (1), the + · · · indicates additional terms $c\psi_2$ + · · ·.[11] However, in the present Section $\psi_N$ will be approximated by the sum of the first two terms alone. If $\psi_N$ is then normalized so that $\int \psi_N^2\, dv = 1$, the coefficients $a$ and $b$ are related by

$$\left.\begin{aligned} a^2 + 2abS + b^2 = 1 \\ \text{where}\quad S \equiv \int \psi_0\psi_1 dv \end{aligned}\right\} \qquad (4)$$

We next consider force and energy relations. For loose complexes, second-order perturbation theory will give an adequate approximation. Then[12]

$$\left.\begin{aligned} W_N \equiv \int \psi_N H \psi_N\, dv \approx W_0 - \frac{(H_{01} - SW_0)^2}{(W_1 - W_0)} + \cdots \\ \text{where}\quad W_0 \equiv \int \psi_0 H \psi_0\, dv;\ W_1 \equiv \int \psi_1 H \psi_1\, dv \\ H_{01} \equiv \int \psi_0 H \psi_1\, dv \end{aligned}\right\} \qquad (5)$$

$H$ is the *exact* Hamiltonian operator for the entire set of nuclei and electrons. $W_0$ is equal to the sum of the separate energies of A and B, modified by any energy of attraction arising from ionic, ion–dipole, dipole–dipole, hydrogen bridge, London dispersion, or classical-type polarization forces, also by any energy of repulsion arising from exchange repulsion forces. $W_1$ has a similar meaning, but includes also attraction energy of ionic and covalent bonding.

The *resonance energy* in the ground state due to interaction of $\psi_1$ with $\psi_0$ is now given by $W_0 - W_N$ of Eq. (5). This should be large if $(H_{01} - SW_0)^2$ is large, which in general is true only if $\psi_0$ and $\psi_1$ overlap strongly (and are of the same symmetry—see below) and if $W_1 - W_0$ is reasonably small. The *energy of formation* of the complex is

$$Q = (W_A + W_B) - W_N = (W_A + W_B - W_0) + (W_0 - W_N) \qquad (6)$$

The *charge-transfer forces* corresponding to the resonance energy may either be *assisted* or *opposed* by the forces of familiar type mentioned above, according as $W_A + W_B \gtrless W_0$.

In using Eq. (6) for an actual molecular complex, it is necessary to take values of $H_{01}$, $W_1$, $W_0$ appropriate to the actual geometrical configuration of the complex in equilibrium (or with only its zero-point vibration energy) in its ground state. However, Eq. (5) and (6) can of course *also* be used for other configurations, for example, in constructing a curve or surface showing how $W_N$ changes as two separated molecules A and B approach each other.

Second-order perturbation theory yields the following approximate relation for the coefficients in Eq. (1)[12]

$$\rho \equiv b/a \approx -(H_{01} - SW_0)/(W_1 - W_0) \qquad (7)$$

Eq. (4) can then be used to get $a$ and $b$ individually.

(11) If A and B are neutral molecules, a term $\psi_2$ of structure $A^+–B^-$ (B acting as acid and A as base) may be of appreciable importance. A second excited state $\psi_F$ usually somewhat higher than $\psi_E$ of Eq. (2) may then be important for the spectrum of the complex; $\psi_E$ and $\psi_F$ as well as $\psi_N$ will then be mixtures of $\psi_0$, $\psi_1$ and $\psi_2$. Frequently also, further terms of structure $A^-–B^+$ derived from low excited states of $A^-$ or sometimes of $B^+$ may be of appreciable importance (see *e.g.*, ref. 34).

(12) Equations (5), (7), (10) and (11) are generalizations of familiar expressions. They may be derived by, for example, following the procedure used by H. Margenau and G. M. Murphy, "The Mathematics of Physics and Chemistry," Van Nostrand Co., Inc., New York, N. Y., 1943, in obtaining Eq. (11-106) and (11-108); modified, however, by starting from the secular equation (11-97) instead of (11-104).

An essential requirement in Eq. (1) is that $\psi_1$ shall be of the same group-theory species as $\psi_0$; otherwise $H_{01}$ and $S$ are zero and there is no resonance. This means usually that $\psi_1$ must be (1), of the same spin type as $\psi_0$; (2) of the same orbital species under the group-theoretical classification corresponding to the over-all symmetry of the complex as a whole. With our supposition that $\psi_A$ and $\psi_B$ in Eq. (2) are both totally symmetrical singlet states,[9] $\psi_0$ is necessarily also a totally-symmetrical singlet state of the symmetry of the complex, and $\psi_1$ must then be of this same type. Requirement (1) may be somewhat relaxed in the case of heavy atoms (*e.g.*, iodine) with very strong spin-orbit coupling; but in this event, $\psi_1$ must definitely be of the same *spin-orbit* species as $\psi_0$.

If the complex has no over-all symmetry, requirement (2) vanishes. However, unless $\psi_1$ is of the same group-theory species as $\psi_0$ under the species-classification of whatever *approximate* symmetry exists in the neighborhood of the A–B interaction zone, $H_{01} - SW_0$ will still be too small to yield a stable complex.

The symmetry requirements just stated should often be important in determining the geometrical arrangement of the partners in a complex. Thus if for a particular geometrical configuration the lowest-energy state of $A^-$–$B^+$ is of a different group-theory species than $\psi_0$, one must go to an *excited* state of $A^-$–$B^+$ to find an acceptable $\psi_1$. Because of the inverse proportionality of the resonance energy to $W_1 - W_0$ (see Eq. (5)), the situation just described is unfavorable for a stable complex. However, there may then exist a different geometrical configuration whose symmetry permits ground-state $A^-$—$B^+$ to interact with $\psi_0$ and thus to serve as $\psi_1$. If so, then, other things being equal, this geometrical configuration will be favored for the actual complex. Several examples of the application of these considerations are described in a previous paper[2d] and in the later sections of the present paper.

Aside from the symmetry considerations just presented, no theoretical analysis of the magnitudes to be expected in various cases for $H_{01} - SW_0$ will be attempted here. Suffice it to say that it is theoretically reasonable in general to expect $H_{01} - SW_0$ to be of adequate size to account for the observed phenomena in terms of the present theory. Conversely, estimates[2d] of other forces[10] to which complex-formation has often been attributed indicate these to be of inadequate size, particularly in such cases as $Bz{\cdot}I_2$ where the component molecules do not possess even dipole fields.

If A·B is a loose complex between closed-shell systems, with little overlapping between A and B, the covalent bond in the $\psi_1$ structure $A^-$–$B^+$ is necessarily weak. It must then be formulated in accordance with Heitler–London theory. However, the bond need not necessarily be interatomic, that is, between electrons of two specific atoms. It may instead be intermolecular; or between an atomic and a molecular electron. Thus, for example, if B is Bz (benzene), the odd electron in $B^+$ which is to form a bond may most conveniently be described as occupying an MO (molecular orbital) of the benzene ring as a whole—one of the $\pi$MO's. (This is the viewpoint used by Dewar[7] in speaking of $\pi$ complexes.) Then if, for example, A is $I_2$, the odd electron in $A^-$ may similarly be described as occupying an MO of $I_2^-$. In this case we have an intermolecular electron-pair bond.[13] In the $Bz{\cdot}Ag^+$ complex the bond in $\psi_1$ (structure $Bz^+$–Ag) is between a $Bz^+$ electron in an MO and an Ag electron in an AO.

In more stable complexes or compounds such as $BF_3{\cdot}NR_3$ or $R_3NO$, the electron-pair bond in $\psi_1$ (in these cases $\psi_1$ is probably lower in energy than $\psi_0$) is to a large extent localized between the N and B atoms or the N and O atoms, respectively.

For tight complexes or stable compounds, Eq. (5), (7) are no longer more than qualitatively correct.[14] A fairly good approximation can then be obtained if Eq. (1) is replaced entirely by an MO description for the complex or molecule as a whole, including a rather strongly polar N–B or N–O bonding MO occupied by two electrons. For *loose* complexes, such an MO description in terms of MO's of the complex as a whole would not be a good approximation.

In loose molecular complexes we expect $a^2 >> b^2$ in Eq. (1). In compounds such as $BX_3{\cdot}NR_3$ or $R_3NO$, $a$ and $b$ should be more nearly equal, probably with $b^2 > a^2$.

If the ground state electronic wave function is given by Eq. (1), it necessarily follows that there exists an *excited state* function $\psi_E$ of the form

$$\psi_E = a^*\psi_1 - b^*\psi_0 + \cdots \quad (8)$$

with $a^* \approx a$, $b^* \approx b$. In the present Section, $\psi_E$ will be approximated by the first two terms on the right of Eq. (8).[11] Then, corresponding to Eq. (4),

$$a^{*2} - 2a^*b^*S + b^{*2} = 1 \quad (9)$$

In the approximation of second-order perturbation theory, the following relations hold[12]

$$W_E = W_1 + \frac{(H_{01} - SW_1)^2}{(W_1 - W_0)} + \cdots \quad (10)$$

$$-\rho^* = b^*/a^* = -(H_{01} - SW_1)/(W_1 - W_0) \quad (11)$$

The existence of an *intense absorption spectrum* corresponding to the transition $\psi_N \rightarrow \psi_E$ can be predicted, and its total absolute intensity[15] approximately computed. Since if $a^2 >> b^2$, $\psi_N$ has nearly pure no-bond character and $\psi_E$ nearly pure ionic character, the spectrum associated with the transition may then be called an *intermolecular charge-*

(13) Alternatively, one may speak of a bond between a particular carbon atom $\pi$ electron and a particular iodine atom electron, but if so, it is necessary to use a cumbersome description in terms of resonance among numerous bond-structures of this type.

(14) The predictions made about state E and the N → E transition in this Section should therefore for $F_3B.NMe_3$ and similar stable complexes be regarded with some reserve. The non-localized MO viewpoint suggests that E may (for the equilibrium $R$ of $F_3B.NMe_3$) be a rather high-energy state, in which case it may exist only in mixture with several other excited states.

(15) The spectrum $\psi_N \rightarrow \psi_E$ should in general appear as a broad band or group of bands. The present computation refers to the total integrated absorption intensity, as discussed by R. S. Mulliken, *J. Chem. Phys.*, **7**, 14 (1939); see Mulliken and Rieke, "Reports on Progress in Physics, of the Physical Society, London," Vol. VIII, p. 231, 1941, for some corrections.

*transfer spectrum:* light absorption causes an electron to jump from B to A.[16]

It will be noted that the predicted charge-transfer spectrum is characteristic of the molecular complex A.B as such, and cannot be attributed to either of the partners A or B, being in this respect in agreement with one of the ideas advanced by Brackmann. Additional, *intramolecular*, spectra of A and B, more or less modified by their association, are of course also to be expected.

Frequently, intramolecular and charge-transfer spectra may overlap, or sometimes also interfere quantum-mechanically (that is, their excited states may partially mix). In such cases, it may not be possible to identify charge-transfer spectra unambiguously or uniquely.

To obtain the predicted intensity[15] of the charge-transfer absorption, the quantum-mechanical dipole moment $\mu_{EN}$ of the transition may first be computed. This is given by

$$\mu_{EN} = -e\int\psi_E\Sigma \mathbf{r}_i\psi_N\, dv \qquad (12)$$

where $\mathbf{r}_i$ is the vector distance of the $i^{th}$ electron from any convenient origin.[17] Using Eqs. (1) and (8), Eq. (12) gives

$$\mu_{EN} = a^*b\mu_1 - ab^*\mu_0 + (aa^* - bb^*)\mu_{01} \qquad (13)$$

where

$$\mu_1 \equiv -e\int\psi_1\Sigma \mathbf{r}_i\psi_1\, dv \qquad (14)$$

$$\mu_0 \equiv -e\int\psi_0\Sigma \mathbf{r}_i\psi_0\, dv \qquad (15)$$

$$\mu_{01} \equiv -e\int\psi_1\Sigma \mathbf{r}_i\psi_0\, dv \qquad (16)$$

From the orthogonality condition $\int\psi_N\psi_E dv = 0$ and Eqs. (1) and (8), the following relation is obtained

$$\left.\begin{array}{l}(a^*b - ab^*) = -(aa^* - bb^*)S \\ \text{where } S \equiv \int\psi_0\psi_1\, dv\end{array}\right\} \qquad (17)$$

Making use of Eq. (17), Eq. (13) can be rewritten in the convenient form

$$\mu_{EN} = a^*b(\mu_1 - \mu_0) + (aa^* - bb^*)(\mu_{01} - S\mu_0) \qquad (18)$$

As we shall see below, the term in $\mu_1 - \mu_0$ is the main one.[18] The magnitude of $\mu_1 - \mu_0$ is easily estimated; it is essentially the change in the ordinary permanent dipole moment which would be produced by displacing one electron from a particular orbital in B to a particular orbital in A, with the nuclei held fixed.[17] Letting $\mathbf{r}_B$ and $\mathbf{r}_A$ denote the average positions of the electron in the B or A orbital, respectively, then $\mu_1 - \mu_0$ is, at least very nearly, $-e(\mathbf{r}_A - \mathbf{r}_B) = e(\mathbf{r}_B - \mathbf{r}_A)$. Thus $(\mu_1 - \mu_0)$ should be of the order of magnitude of 10 debye units. The magnitude of the factor $a^*b$ should be between 0.1 or 0.2 and 0.7 in all molecular complexes sufficiently stable to be detected (see discussion of the benzene–iodine complex in Section III below).

In evaluating $\mu_{01}$, it will be sufficient for present purposes to use approximations for $\psi_0$ and $\psi_1$. If we describe the structure of each partner in terms of MO's of that partner, the process $\psi_0 \rightarrow \psi_1$ involves the jump of one of a pair of outer electrons, initially occupying an MO $\phi_B$, in B, into a previously unoccupied MO $\phi_A$ in A. This occurs in such a way that the second electron, left in $\phi_B$, remains paired to the electron now in $\phi_A$, but now by a covalent bond. (If one or both partners are atoms, read AO instead of MO.) On substituting the expressions for $\psi_0$ and $\psi_1$ conforming to these specifications into Eq. (16), and integrating, one finds, to a close approximation[19]

$$\left.\begin{array}{l}\mu_{01} - S\mu_0 \approx eS(\bar{\mathbf{r}}_B - \bar{\mathbf{r}}_{AB}) \\ \text{where } \mathbf{r}_B \equiv \int\phi_B\mathbf{r}\phi_B\, dv;\ S_{AB}\bar{\mathbf{r}}_{AB} \equiv \int\phi_A\mathbf{r}\phi_B d\tau \\ S \equiv \int\psi_0\psi_1\, dv \approx 2^{1/2}(1 + S^2_{AB})^{-1/2}S_{AB};\ S_{AB} \equiv \int\phi_A\phi_B d\tau\end{array}\right\} \qquad (19)$$

In Eq. (19), $\bar{\mathbf{r}}_{AB}$ is the average position of an electron having a charge distribution of the form of the overlap of the MO's $\phi_A$ and $\phi_B$, and is therefore located between $\mathbf{r}_B$ and $\mathbf{r}_A$. In Eq. (19), $S$ of Eq. (4) or (17) has been approximately evaluated in terms of the overlap integral $S_{AB}$ of the MO's $\phi_A$ and $\phi_B$.

Putting the first of Eqs. (19) into Eq. (18), with $\mu_1 - \mu_0 \approx e(\bar{\mathbf{r}}_B - \bar{\mathbf{r}}_A)$ as discussed above, one obtains

$$\mu_{EN} = a^*be(\bar{\mathbf{r}}_B - \bar{\mathbf{r}}_A) + (aa^* - bb^*)eS(\bar{\mathbf{r}}_B - \bar{\mathbf{r}}_{AB}) \qquad (20)$$

In case the molecular complex has an axis of symmetry running through the centers of $\phi_A$ and $\phi_B$, then both terms in Eq. (20), and so $\mu_{EN}$, are directed along that axis. The first term is directed from A toward B, and if $a > b$ the second is directed likewise and the two terms add. For loose complexes ($a^2 >> b^2$) the first term is much the larger, mainly because of the smallness of $S$ if $\phi_A$ and $\phi_B$ do not overlap strongly.[18]

To obtain the total intensity of the $N \rightarrow E$ absorption in terms of the so-called oscillator strength $f$ of the transition, one may use

$$f = (4.704 \times 10^{-7})\nu(\mu_x^2 + \mu_y^2 + \mu_z^2) \qquad (21)$$

where $\mu_x$, $\mu_y$, $\mu_z$ here refer to the $x$, $y$ and $z$ components of the vector $\mu_{EN}$ *in debye units*, and $\bar{\nu}$, *in cm.*$^{-1}$, is a suitably weighted average wave number over the $N \rightarrow E$ band or bands (roughly the value of $\nu$ at the peak of intensity).[15]

(16) The intermolecular charge-transfer spectra discussed here are related to the *interatomic* charge-transfer spectra discussed by R. S. Mulliken (*J. Chem. Phys.*, **7**, 20 (1939) and later papers; see especially Mulliken and Rieke, Ref. 15). For a review on electron-transfer spectra, see E. Rabinowitch, *Rev. Modern Phys.*, **14**, 112 (1942). One of the earliest recognized types is that involved in the photographic process: $[Ag^+Br^-] \rightarrow [AgBr]$ (Lenard, 1909; Fajans, 1922).

(17) Equation (12) is based on the usual assumption that the positions of the nuclei do not change during the electronic transition N → E. According to the Franck–Condon principle, this introduces but minor errors into the computed total absolute intensity (see Mulliken, and Mulliken and Rieke, ref. 15, for detailed analysis). Note that because of the rigorous orthogonality of the (true exact) wave functions $\psi_E$ and $\psi_N$, $\mu_{EN}$ is independent of the origin of coördinates for the $\mathbf{r}_i$, and (when the assumption mentioned above is used) does not involve the nuclear coördinates. Note that throughout Eqs. (12)–(20), all the $\mu$'s, like the $\mathbf{r}$'s, are *vector quantities*.

(18) In the "Note added in Proof," on p. 605 in ref. 2d, it was erroneously assumed that the first term was less important than the second.

(19) The evaluation proceeds by taking $\psi_0$ and $\psi_1$ as antisymmetrized MO-product functions as follows (*cf.* R. S. Mulliken, *J. Chem. Phys.*, **8**, 234 (1940), Section III, for notation and a similar discussion; the MO's of the two *molecules* A and B play the same role in the present case as the AO's of the two *atoms* in the discussion cited):

$$\psi_0 \approx \mathfrak{N}_0(n!)^{-1/2}\sum_P(-1)^P P\phi_B\alpha(1)\phi_B\beta(2)\phi_3\alpha(3)\cdots\phi_n\beta(n)$$

$$\psi_1 = 2^{-1/2}(1 + S^2_{AB})^{-1/2}(\psi_I + \psi_{II})$$

$$\psi_I \approx \mathfrak{N}_1(n!)^{-1/2}\sum_P(-1)^P\phi_B\alpha(1)\phi_A\beta(2)\phi_3\alpha(3)\cdots\phi_n\beta(n)$$

$$\psi_{II} \approx \mathfrak{N}(n!)^{-1/2}\sum_P(-1)^P\phi_A\alpha(1)\phi_B\beta(2)\phi_3\alpha(3)\cdots\phi_n\beta(n)$$

The normalizing factor in the expression for $\psi_1$ is obtained in the usual manner (see Eq. (19) for definition of $S_{AB}$); $n$ is the total number of electrons in the complex; $\phi_3\cdots\phi_n$ are A and B MO's occupied by electrons which remain undisturbed during the process $\psi_0 \rightarrow \psi_1$. (Actually $\phi_3$ and $\phi_4$, and so on, are identical in pairs, but the electrons occupying them differ in spin —$\alpha$ or $\beta$.)

Substituting the above expressions for $\psi_0$ and $\psi_1$ into Eq. (16), one obtains (aside from certain factors including $\mathfrak{N}_0\mathfrak{N}_1$ which on multiplying together yield a factor of 1, either exactly or very nearly: compare the reference cited above) the following

$$\mu_{01} = -Se(\bar{\mathbf{r}}_B + \bar{\mathbf{r}}_{AB} + \bar{\mathbf{r}}_3 + \cdots \mathbf{r}_n)$$

In this expression, $S$ means $2^{1/2}(1 + S^2_{AB})^{-1/2}S_{AB}$ (*cf.* Eq. (19)); the near or exact equality of $S$ of Eq. (17) to this is readily obtained for the case that $\psi_0$ and $\psi_1$ are as given above. Now it is readily seen that if $\psi_0$ is as given above, $\mu_0$ is given by

$$\mu_0 = -e(\bar{\mathbf{r}}_B + \bar{\mathbf{r}}_B + \bar{\mathbf{r}}_3 + \cdots \bar{\mathbf{r}}_n)$$

from which and the expression just given for $\mu_{01}$, the first of Eqs. (19) follows at once.

Here it is important to point out that, since Eqs. (1) and (8) give in general only somewhat rough approximations for $\psi_N$ and $\psi_E$,[11] Eq. (20)–(21) are correspondingly rough. However, it would seem that they should in general be reliable as to order of magnitude. The most essential fact is that they predict high intensities for N → E transitions even in loose complexes.

* * * *

The preceding discussion can be extended to $n$:1 molecular complexes, provided they are loose so that one resonance structure is predominant. Eq. (1) may then be generalized to

$$\psi_N = a\psi_0 + \Sigma b_i\psi_i + \cdots \qquad (22)$$

provided $a^2 > \Sigma b_i^2$. Thus for a complex A·B·A or B·A·B (for example, B = benzene, A = $Ag^+$ or $I_2$), provided the two A's are in equivalent locations with respect to the B or the two B's with respect to the A, we have

$$\psi_N = a\psi_0 + b(\psi_1 + \psi_1') + \cdots \qquad (23)$$

In Eq. (23), $\psi_0$ is a no-bond structure and $\psi_1$ and $\psi_1'$ are structures $A{\cdot}B^+{-}A^-$ and $A^-{-}B^+{\cdot}A$ or $B^+{-}A^-{\cdot}B$ and $B{\cdot}A^-{-}B^+$.

The resonance energy for a loose $n$:1 complex should be given approximately by a sum of terms of the form of $W_0 - W_N$ of Eq. (5), one for each $\psi_i$ in Eq. (22). For a 1:2 or 2:1 complex this means merely multiplying the result of Eq. (5) by a factor 2. More accurately, there will be a saturation effect diminishing somewhat the resonance energy of an $n$:1 complex, but this should not become important as long as $\Sigma b_i^2$ is sufficiently small compared with $a$.[2] Repulsions between the different A's attached to a single B, or the B's attached to an A, may cause a further diminution.

From the empirical fact that molecular compounds of the same 1:1 composition for which there may be evidence in solution often appear also as crystalline solids, it seems probable that charge-transfer forces operate in more or less localized fashion in much the same way in such solids (which may also be regarded as $n$:$n$ complexes, $n$ exceedingly large) as in 1:1 or $n$:1 complexes in vapor or solution. However, no attempt at a theoretical analysis will be made here.

Another situation to which it seems probable that the present theory can in many cases be extended is that of an electron acceptor or donor adsorbed on or reacting with a metal or solid, the latter then acting as an electron donor or acceptor, respectively.

## III. The Benzene–Iodine and Related Loose Complexes

The preceding discussion can be clarified by discussion of the benzene–iodine complex. As Benesi and Hildebrand have shown,[2] this has an intense characteristic absorption near λ 2900. It will be convenient to begin by computing the $f$ value for this absorption on the hypothesis that it is an intermolecular charge-transfer spectrum.

If we assume the most compact and most probable model,[2d] with the iodine molecule resting on the benzene molecule with its axis parallel to the plane of the benzene and its center on the sixfold axis of the benzene, then $\mu_{EN}$ lies along the latter axis. Calling this the $z$ axis, Eq. (20) becomes

$$\mu_{EN} = a^*be(\bar{z}_B - \bar{z}_A) + (aa^* - bb^*)eS(\bar{z}_B - \bar{z}_{AB}) \quad (24)$$

Here benzene is the donor (B) and iodine the acceptor (A),[2a–d] and $\bar{z}_A$ and $\bar{z}_B$ are evidently the $z$ values of points on the $z$ axis near the centers of the two molecules. Allowing for some compression by the attractive forces forming the complex, $\bar{z}_B - \bar{z}_A$ may be estimated[2d] as 3.4 Å. and $\bar{z}_B - \bar{z}_{AB}$ as 1.7 Å. The value of $S$ is very uncertain, but 0.1 seems reasonable.[20]

We next need values of $a$, $a^*$, $b$ and $b^*$. That of $b$ can be estimated to be roughly 0.17 from the dipole moment $\mu_N$ of the benzene–iodine complex, as determined from Fairbrother's data.[21] From this and the assumed value of $S$ one obtains

$$a = 0.97,\ a^* = 0.99,\ b^* = 0.27$$

substituting in Eq. (22), one obtains

$$\mu_{EN} = (2.69 - 0.75) = 3.45\ D$$

The two terms 2.69 and 0.75 correspond, respectively, to the first and second terms in Eq. (24). From this value of $\mu_{EN}$, together with $\nu = 3.36 \times 10^4$,[2a] and using Eq. (21), one obtains $f = 0.19$. The observed value[22,23] is 0.30. The agreement is satisfactory in view of uncertainties in both the theoretical and observed values.

The above $b$ value of 0.17, giving $b^2 = 0.028$, corresponds to 2.8% of ionic character in state N.

For further understanding, Fig. 1 has been constructed to indicate how the energies $W_N(R)$ and $W_E(R)$ of states N and E may vary with the distance $R$ between the centers of the benzene and iodine molecules, and how $W_N(R)$ and $W_E(R)$ may have arisen as the result of a resonance interaction between states $\psi_0$ and $\psi_1$ of Eq. (2) with respective energies $W_0(R)$ and $W_1(R)$. The energy $W_N(\infty) = W_0(\infty)$ is taken as zero. $W_0(R)$ corresponds to constant energy with decreasing $R$ down to 3.7 Å., then a slight dip (dispersion force attraction), then a pronounced rise (exchange

(20) Note that $S = 2^{1/2}(1 + S^2_{AB})^{-1/2}S_{AB}$ by Eq. (19). In ref. 2d, "an upper limit for reasonable estimates of $S_{AB}$ [there called $S$] appears to be about 0.1–0.2."

(21) First the value $\mu_N = 0.72\ D$ is obtained by the method indicated in footnote 19 of ref. 2d, revised in accordance with ref. 2b to the basis that iodine in pure benzene is 70% complexed. Now theoretically, if $\psi_N$ is given by Eq. (1)—see also Eq. (15), (16)—we have

$$\mu_N = \int\psi_N M\psi_N\, dv = a^2\mu_0 + b^2\mu_1 + 2ab\mu_{01}$$

But $\mu_0 = 0$, and using Eq. (19) for $\mu_{01}$, with $\bar{z}_B - \bar{z}_{AB} = \frac{1}{2}(\bar{z}_B - \bar{z}_A)$ in accordance with the values of $\bar{z}_B - \bar{z}_{AB}$ and $\bar{z}_B - \bar{z}_A$ assumed above, and putting $\mu_1 = e(\bar{z}_B - \bar{z}_A)$, we have $2\mu_{01} = S\mu_1$. Hence

$$0.72 = \mu_N = (b^2 + abS)\mu_1 = (b^2 - 0.1b)\mu_1 = 16.33(b^2 - 0.1b)$$

if we put $a = 1$ (nearly enough correct), $S = 0.1$, and $\mu_1 = e(\bar{z}_B - \bar{z}_A) = 16.33\ D$, using the values of $S$ and $(\bar{z}_B - \bar{z}_A)$ assumed in the text above. From this, $b = 0.17$. However, mainly because of its strong dependence on the uncertain quantity $S$, this value of $b$ cannot be considered reliable.

(22) An unpublished $f$ value of 0.21 per mole of *total* $I_2$ in benzene by Green and Rees, quoted by N. S. Bayliss (*J. Chem. Phys.*, **18**, 292 (1950)) is probably the most reliable (see ref. 2, Table II, for summary of earlier data). Assuming that this $f$ value comes only from the 70% of the $I_2$ which is associated (*cf.* ref. 21), the $f$ value for the complexed molecules is 0.30.

(23) The observed value[22] is measured in benzene solution, whereas the computed value is applicable to isolated molecules, as in a vapor. However, experience indicates that actually there may be little difference between vapor and solution $f$ values (*cf.* L. E. Jacobs and J. R. Platt, *J. Chem. Phys.*, **16**, 1137 (1948), including references to earlier work).

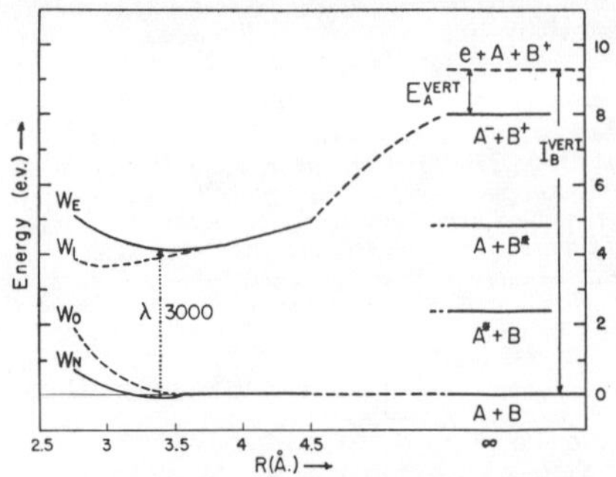


Fig. 1.—Some potential energy curves $W(R)$ for the benzene-iodine complex as a function of the distance $R$ between the centers of the two molecules, assuming Model $R$ described in Section III of the text. Close to the curve $W_E$ there should be a second curve coming also from $A^- + B^+$ (see text). Additional curves (not drawn in) come from $A^* + B$, $A + B^*$, etc.

repulsions). But at the smaller $R$ values resonance between $\psi_0$ and $\psi_1$ depresses $W_0$ to become $W_N$ and raises $W_1$ to become $W_E$. This interaction is expected now to increase rapidly with decreasing $R$ because of increase of $H_{01}^2$ and, to a lesser extent, decrease of $W_1 - W_0$, in Eq. (5). Figure 1 was drawn by first sketching in $W_0$ on the basis of qualitative considerations, then $W_1$ using roughly quantitative considerations now to be detailed, then drawing $W_N$ and $W_E$ on the basis of the expectations just outlined.[24]

The $W_1$ curve was based on the following: (1) For $R = \infty$, the energy is higher for $A^- + B^+$ than for $A + B$ (see Fig. 1) by the amount

$$I_B^{\text{vert}} - E_A^{\text{vert}}$$

$I_B^{\text{vert}}$ and $E_A^{\text{vert}}$ are the ionization energy of the benzene molecule and the electron affinity of the iodine molecule, for *vertical processes*. For benzene, $I_B$ is 9.24 e.v., and $I_B^{\text{vert}}$ must be practically the same.[25] For iodine, $E_A^{\text{vert}}$, which should be considerably less than $E_A$, has been estimated as 1.8 e.v.[26] (2) As $A^-$ and $B^+$ approach, $W_1(R)$ drops because of the Coulomb attraction energy $e^2/R$ until perhaps $R = 3.4$ Å. (3) At sufficiently small $R$ values, covalent binding between the odd electron on $I_2^-$ and that on $Bz^+$ should lower the energy somewhat further[25] (perhaps 0.3–0.5 e.v.), but also, (4), exchange repulsions set in between the closed shells of electrons in the two molecules, and finally predominate, causing $W_1(R)$ to rise again. $W_1(R)$ in Fig. 1 for $R < 3.3$ Å. has been drawn in qualitative agreement with considerations (3) and (4).[26]

One expects some modification in the magnitudes of effects (1) and (2) if the molecules approach not in the vapor state, but in solution. A quantitative treatment would require an elaborate analysis, but for a non-ionizing medium of low dielectric constant, a rough consideration indicates that the modifications required may not be large. They have been ignored in constructing Fig. 1, which is intended only to be illustrative. A further correction, also ignored, would allow for the fact that the positive and negative charges on $Bz^+$ and $I_2^-$, respectively, are not concentrated at the centers of the two molecule-ions.

(24) The heat of association of $Bz \cdot I_2$ is about 1.4 kcal. as determined by Cromwell and Scott[2b] from equilibrium data in solution.

(25) See discussion of models beginning four paragraphs below for some further details.

(26) See Ref. 2d, footnote 25. Actually, the value 1.2 e.v. was used so as to raise curve $W_1$ in Fig. 1 to fit the observed $\nu_{NE}$. Without changing $E_A^{\text{vert}}$ from 1.8 e.v., a similar adjustment could have been made by assuming a greater exchange repulsion (larger effect 4) for $W_1$ than for $W_0$, attributable to the larger size of $I_2^-$ in $\psi_1$ than of $I_2$ in $\psi_0$.

By the Franck–Condon principle, the peak frequency $\nu_{NE}$ of the N → E absorption should correspond closely to $W_E - W_N$ measured *vertically up* at the $R$ value (assumed in Fig. 1 to be 3.4 Å.) of the minimum of curve $W_N$. Curve $W_E$ in Fig. 1 was adjusted to make $\nu_{NE}$ agree with the observed value of 33,600 cm.$^{-1}$ (4.17 e.v.). This was done using items (1)–(4) stated above for curve $W_1$, except that $I_A^{\text{vert}}$ was taken as 1.2 e.v.[26]; $W_E$ at 3.4 Å. was taken 0.15 e.v. above $W_1$.[27] The minor numerical adjustments made here are not unreasonable. This and the satisfactory agreement noted above between observed and computed $f$ values give considerable support to the essential validity of the present theory in explaining the λ 2900 absorption.

A brief consideration of the expected complete absorption spectrum of $Bz \cdot I_2$ is now in order. In general for a complex A·B this should include: (a) absorption characteristic of A and B; (b) several charge-transfer spectra, corresponding to various excited states of $B^+$ and $A^-$. At the right of Fig. 1 ($R = \infty$), energy levels are shown for the two lowest spectroscopically important excited states of $I_2 + Bz$. In the free molecules, these give rise to absorption near λ 5200 ($I_2$) and near λ 2600 (Bz).[2] In the complex, absorption occurs[2] with a maximum near λ 5000. The slight shift from λ 5200 for free $I_2$ to λ 5000 for $I_2$ in the complex tells us that the distance between the $W(R)$ curves drawn from A + B, *i.e.*, $W_N$, and from $A^* + B$ (not shown in Fig. 1) increases by about 0.1 e.v. in going from $R = \infty$ to $R = 3.4$ Å., and means that $A^* + B$ has slightly less tendency to form a complex than A + B. Location of an additional absorption by $Bz \cdot I_2$ near λ 2600 corresponding to λ 2600 of benzene would be of considerable interest.[28]

Thus far, the *resting* model, R (see second paragraph of this Section) has been assumed. Arguments for this will now be given, also some necessary details about the electronic structures of $Bz^+$ and $I_2^-$. First, axes $x$ and $y$ may be taken in the benzene plane as indicated in Fig. 2, with $z$ up from this plane. Next, it is necessary to notice certain sub-types of model R: namely $R_x$ and $R_y$, with iodine axis parallel to $x$ or $y$ axis, respectively, both with symmetry $C_{2v}$; and intermediate models with symmetry $C_2$. These sub-types probably differ very little in energy and spectroscopic properties, even in methylated benzenes (*cf.* p. 604 of ref. 2d).

Fig. 2.—Benzene $\pi$ MO's and their species classifications under perturbing field of symmetry $C_{2v}$ with $xz$ and $yz$ as symmetry planes. The black and white circles indicate $2p_z$ carbon AO's which are, respectively, positive or negative on the positive-$z$ side of the Bz plane, and of opposite sign on the opposite side. The sizes of the circles indicate magnitudes of the coefficients of the AO's.

The arguments for model R are as follows.[29] (1) It is the *most compact* model, thus permitting maximum Coulomb

(27) By second-order perturbation theory, if $\nu_{NE}$, and $b$ of Eq. (1), are known, then the vertical frequency $\nu_{01}$ can be computed; further, the energy intervals $W_E - W_1$ and $W_0 - W_N$ should be nearly equal [*cf.* Eq. (4), (5), (7), (10)]. The result given in the text was obtained in this way.

(28) This transition appears to have been found in recent work in this Laboratory; see J. S. Ham, J. R. Platt and H. M. McConnell, *J. Chem. Phys.*, **19**, 1301 (1951). In ref. 2d, the writer at first attributed the λ 2900 absorption to modified λ 2600 absorption by the Bz in the complex, but then in a "Note added in Proof" suggested a charge-transfer process as a possible alternative.[18] Bayliss (*cf.* *J. Chem. Phys.*, **18**, 292 (1950)) attributed it to a modification of the intense $I_2$ absorption near λ 2000 by action of the solvent.

(29) See also ref. 2d, noting, however, that some of the arguments for model $R$ (there called Model I) under the assumption that λ 2900 is a Bz absorption are now no longer valid.

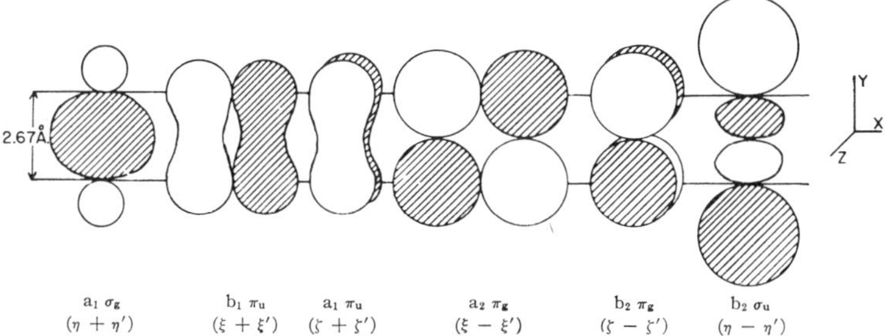


Fig. 3.—Semi-schematic drawing of forms of outer shell iodine ($I_2$ or $I_2^-$) MO's under perturbing field of symmetry $C_{2v}$ for $R_y$ model of $Bz \cdot I_2$, with $xz$ and $yz$ as symmetry planes. Shaded and unshaded volumes in the orbitals indicate positive and negative regions, respectively. Inner nodes are omitted. The characterizations in parentheses correspond to Eq. (25).

attraction energy in $W_1$. (2) $\psi_0$ is the totally symmetrical singlet ($^1A_1$) type, and (see Section II, sixth and eighth paragraphs) $\psi_1$ has to be of the same type. For model R, this condition can be satisfied using ground-state $Bz^+$ and $I_2^-$.

The lowest ionization potential of Bz involves removal of an electron from a $\pi$ MO. There are three of these in Bz, each occupied by two electrons. Their forms may be approximated by suitable linear combinations (see Fig. 2) of carbon atom $2p_z$ AO's. The MO's $b_1$ and $b_2$ in Fig. 2 are of equal ionization energy (9.24 volts, the minimum ionization potential of Bz) in free benzene, while $a_1$ is less easily ionized. (The notation assumes model $R_x$ or $R_y$.) Removal of one electron from the $b_1$ or $b_2$ MO gives, respectively, a $^2B_1$ or a $^2B_2$ state of $Bz^+$.

The six outer shell MO's of $I_2$ or $I_2^-$, in the forms they should assume in model $R_y$, may be approximated (normalization factors are omitted)[30] as

$$\left.\begin{array}{ll} a_1\sigma_g \approx \eta + \eta'; & a_2\pi_g \approx \xi - \xi' \\ b_1\pi_u \approx \xi + \xi'; & b_2\pi_g \approx \zeta - \zeta' \\ a_1\pi_u \approx \zeta + \zeta'; & b_2\sigma_u \approx \eta - \eta' \end{array}\right\} \quad (25)$$

Here $\xi$, $\eta$, $\zeta$ refer to 5p iodine atoms AO's with their axes, respectively, parallel to $x$, $y$ and $z$ and their positive ends suitably chosen (see Fig. 3). For each MO in (25), the first symbol gives the classification under the $C_{2v}$ symmetry of model $R_y$ of $Bz \cdot I_2$, the second that for the isolated iodine molecule ($D_{\infty h}$ symmetry).[31] In the latter, the two $\pi_u$ forms are equal in energy, as are also the two $\pi_g$ forms.

The iodine molecule in its ground state has two electrons in each of the MO's of Fig. 3 except the strongly antibonding $b_2\sigma_u$, which is empty.[31] In the ground state of $I_2^-$, the odd electron goes into the $b_2\sigma_u$ MO, giving a $^2B_2$ state.[32] The required $^1A_1$ state of $Bz^+$–$I_2^-$ to serve as $\psi_1$ for model $R_y$ can be obtained if and only if the $Bz^+$ is also in its $^2B_2$ state.[30,33,34] It will be noted further that this combination permits favorable overlapping between the odd electrons in the $Bz^+$ $b_2$ MO and the $I_2^-$ $b_2$ MO, to form the indicated weak $Bz^+$–$I_2^-$ bond.[30,32]

Several additional models for $Bz \cdot I_2$ will now be examined to see why they are less probable than model R. They are also of interest because analogous models may be set up for $Bz \cdot Ag^+$ (see Section VI) and other Bz complexes, with relative stabilities depending on the electronic structure and the geometry of the acceptor involved. They or their analogs are of further interest as possible activated states or reaction intermediates for substitution reactions.[35]

Model E, the *edgewise model*, has the $I_2$ axis parallel to the Bz sixfold axis but located alongside one edge (model $E_x$) or corner ($E_y$) of the Bz ring. The symmetry is $C_{2v}$ in either case For models E just as for Models R, a $\psi_1$ satisfying the symmetry requirements can be obtained from ground-state $Bz^+$ and $I_2^-$. However, the $Bz^+$ and $I_2^-$ charge centers are a little farther apart now (especially for $E_y$), making $W_1 - W_0$ somewhat larger, so that Models E should probably be somewhat less stable than Model R (*cf.* Eq. 5).

In both of models $R_x$ and $E_x$, the $I_2$ axis lies in the $xz$ plane of Fig. 2, but parallel or perpendicular to the $xy$ plane in $R$ and $E$, respectively. A third model (or range of models) with the $I_2$ axis still in the $xz$ plane, but inclined to the $xy$ plane, is intermediate between models $E_x$ and $R_x$.[30] This *oblique* model $O_x$ has symmetry $C_s$. It seems likely that models $R_x$ and $E_x$ both correspond to distinct energy minima (with that of $R_x$ the lower) relative to moderate deformations (except rotations of $R_x$ to $R_y$). If so, model $O_x$ would correspond to a mild energy maximum between these two minima. In contrast to the case of $Bz \cdot I_2$, the analogs of models R and E for $Bz \cdot Ag^+$ (Section VI) probably correspond to energy maxima, with minimum energy for the analog of Model $O_x$. Model $O_y$, intermediate between $R_y$ and $E_y$ may be important for substitution reactions (see later paper[35a]). Two further

(30) For model $R_x$, the labels $b_1$ and $b_2$ in Eq. (25) and Fig. 3 and the axes $x$ and $y$ in Fig. 3 would be exchanged, and the ground state of $I_2^-$ would be of species $^2B_1$. This combined with $^2B_1$ of $Bz^+$ would then give $\psi_1$. The weak $Bz^+$–$I_2^-$ bond in $\psi_1$ for the $R_x$ model would be between a $b_1$ electron of $Bz^+$ and a $b_1\sigma_u$ of $I_2^-$.[32]

(31) See ref. 2d, Section II.

(32) In addition, $I_2^-$ must have five low-energy unstable excited states, some of which are of possible interest for $Bz^+$–$I_2^-$ resonance structures. Each is obtained by putting two electrons into $b_2\sigma_u$ and taking one electron out of one of the other five MO's of Eq. (25). For each of models $R_x$ and $R_y$,[30] the $Bz^+$–$I_2^-$ bond in $\psi_1$ should be appreciably stabilized by some admixture[34] of structures involving two of these excited $I_2^-$ states.

(33) This combination also gives a $^3A_1$ state. Further, the $^2B_1$ state of $Bz^+$ combined with the $^2B_2$ ground state of $I_2^-$ gives a $^1A_2$ and a $^3A_2$ state. All these states must give $W(R)$ curves with the same asymptote as $W_E$ of Fig. 1, but lying somewhat above the latter (not shown in Fig. 1).

(34) For model Ry, $I_2^-$ in its lowest excited state, which is of type $^2B_2$ with the odd electron in $b_2\pi_g$, can also interact with $^2B_2$ of $Bz^+$ o give a second acceptable $\psi_1$. Further, excited $I_2^-$ of the type $^2B_1$ with the odd electron in $b_1\pi_u$ can interact with $^2B_1$ of $Bz^+$ to give a third acceptable $\psi_1$. No doubt those higher-energy $\psi_1$'s contribute appreciably to the complete resonance structure of $\psi_N$. For model Rx, a similar pair of excited $\psi_1$'s exists.

(35) The models here called R, L, E and A were called I, II, III and IV in ref. 2d.

(35a) Probably in *Proc. Nat. Acad. Sci.*

groups of models, with the iodine axis *lying* in the benzene plane, either pointing toward the center of the ring ($L_x$, $L_y$) or *tangential* to an edge or corner of the ring ($T_x$, $T_y$), are easily seen to be unfavorable for steric and/or symmetry reasons.

Finally, Model A, the *axial model*, with the $I_2$ axis coincident with the Bz sixfold axis and of symmetry $C_{6v}$, deserves mention. Although at first sight attractive,[2c] this model does not satisfy the symmetry requirements for $\psi_1$ unless *excited* $I_2^-$ (excitation energy perhaps 2.5 e.v.)[32] or excited $Bz^+$ (excitation energy perhaps 4 or 5 e.v.) is used. Further, the mean separation between the charge centers in $Bz^+$ and $I_2^-$ is greater here than for any of Models R, O or E. Thus Model A must very probably be ruled out.

Besides $Bz{\cdot}I_2$, numerous other related $Ar{\cdot}X_2$ and Ar·XY complexes have been studied, beginning with $Ms{\cdot}I_2$ (Ms = mesitylene).[2a,2c,2d] For $Ms{\cdot}I_2$ the visible $I_2$ absorption peak is at λ 4900 and the charge-transfer peak at λ 3330 as compared with λ 5000 and λ 2900, respectively, for $Bz{\cdot}I_2$; and the equilibrium constant of the complex is increased. All these facts are qualitatively in agreement with the theory in this paper. For Ms, $I_B$ is about 1.1 e.v. less than for Bz;[2] hence, other things being equal, curve $W_1$ in Fig. 1 should be lower for $Ms{\cdot}I_2$ than for $Bz{\cdot}I_2$ by this amount. However, there are other changes: $R$ should be smaller at equilibrium, curve $W_N$ should have a deeper minimum, and the distance of curve $W_E$ above $W_1$ should be increased. The last mentioned changes may well be the main reason why the shift in wave length of the $\nu_{NE}$ absorption peak is less than the predicted shift (to λ 4100) that one would obtain on the basis of the change in $I_B$ alone. Less easily understandable is the fact that the $\nu_{NE}$ *intensity* for $Ms{\cdot}I_2$ is *less* than for $Bz{\cdot}I_2$, whereas Eq. (22) would tend[36] to indicate the contrary.

The spectroscopic and equilibrium constant data on other $Ar{\cdot}X_2$ and Ar·XY complexes,[2] while apparently all in general agreement with the present theory, show a number of interesting features worthy of further systematic experimental study and theoretical analysis. It seems probable that the geometrical arrangement corresponds to model $R$ in all or most cases. The order of stability $ICl > I_2 > Br_2 > Cl_2$ for Bz complexes[2c] is understandable by charge-transfer theory except for Bz·ICl, which would be expected to resemble $Bz{\cdot}Br_2$ in stability. This anomaly may perhaps be explained by additional, dipole polarization, forces exerted by the ICl, lying in an $O$ rather than an $R$ position.

The present conclusions differ from those tacitly adopted by other authors who, for complexes like $Bz{\cdot}I_2$, have indicated structures of partially $BzI^+I^-$ character[37]; for example, Dewar[7] wrote $BzI^{\delta+}I^{\delta-}$. Polarization of this kind would be immediately excluded by symmetry except for a model with the two iodine atoms in non-equivalent positions, and as we have seen above, these models for $Bz{\cdot}I_2$ seem improbable. Moreover, even in such a model, for example Model A, the present analysis indicates that the polarization effect would consist only in a somewhat unequal distribution of negative charge over the two iodine atoms in the minor $Bz^+{-}I_2^-$ resonance component ($\psi_1$), with only extremely slight (quadrupole-quadrupole) polarization in the main ($Bz{\cdot}I_2$) component.[38]

If $Bz{\cdot}I_2$ could be dissolved in a sufficiently strongly ionizing medium, one might expect to obtain $(BzI)^+$ plus $I^-$, with predominantly $Bz^+{-}I$ structure in the positive ion.[39] A related case is that of iodine dissolved in pyridine (Py), an ionizing medium, where $(PyI)^+$ ions are present,[40] presumably in equilibrium with undissociated $Py{\cdot}I_2$.[41] The structure of the $(PyI)^+$ ions is doubtless predominately $Py^+{-}I$ with the iodine attached to the nitrogen atom of the pyridine.[42,43] With respect to predominant bond structure, this conclusion differs from that of previous authors, who speak of unipositive iodine in this and related compounds.[40]

## IV. Factors Determining Lewis Acid and Base Strengths. Molecular Compounds as Reaction Intermediates

At this point it will be useful to draw some general conclusions from the discussion of $Bz{\cdot}I_2$ in Section III, particularly as to the factors determining Lewis acid and base,[44] or electron donor and acceptor,[45] strengths. For this purpose, Fig. 1 may be taken as typical for any Lewis acid–base pair A, B in which A and B are weakly-interacting neutral molecules in totally-symmetrical singlet states. The dissociation constants of A·B for a given B with different A's, or for a given A with different B's, are the natural measures of Lewis acid or base strengths. However, it must be kept in mind[44] that the strength of a (Lewis) acid is not quite a

(36) Mainly by reason of increased $b$ in Eq. (1), corresponding to the greater stability of $Ms.I_2$.

(37) In particular, see F. Fairbrother, *Nature*, **160**, 87 (1947), *J. Chem. Soc.*, 1051 (1948).

(38) In Model A, $\psi_1$ would be $Bz^+{-}I_2^-$ just as in Model R. However, this can be written as a combination of two structures $Bz^+{-}II^-$ and $Bz^+I^-I$ (with a bracket linking Bz and the second I), of which the latter would probably predominate in view of the weakness of the $Bz^+{-}I$ bond in the actual complex.

(39) The alternative structure $Bz,I^+$ is disfavored not only by its no-bond character but also because for good resonance with $Bz^+{-}I$, the $I^+$ would have to be in an excited singlet state.

(40) R. A. Zingaro, C. A. VanderWerf and J. Kleinberg, THIS JOURNAL, **73**, 88 (1951), and other references there cited.

(41) *Cf.* ref. 2d, footnote 26d.

(42) Ionization of Py may be effected either by removal of a π electron as for Bz, or of a lone-pair N atom electron, probably with nearly the same ionization energy. This gives a choice of two quite different $\psi_1$'s for $Py{\cdot}I_2$, the one leading to a π-type complex as in benzene, the other to an onium-type complex. The latter is favored by the fact that it permits more strongly localized binding in $\psi_1(Py^+{-}I_2^-)$, and becomes *strongly* favored in an ionizing medium.

(43) The structure $Py^+{-}I$ would be favored over $Py,I^+$ for the same reasons[39] as $Bz^+{-}I$ over $Bz,I^+$ in $(BzI)^+$, but even more strongly because the $N^+$ in the $Py^+$ should be able to form a fairly strong localized bond to iodine without disturbing the aromatic π structure.

(44) G. N. Lewis, *J. Franklin Inst.*, **226**, 293 (1938). Lewis defined a basic molecule as "one that has an electron-pair that may enter the valence shell of another atom to consummate the electron-pair bond," an acid molecule as "one which is capable of receiving such an electron-pair into the shell of one of its atoms." The present viewpoint as embodied in Eq. (1), emphasizing partial transfer of one electron from base to Lewis acid with accompanying (weak or strong) interatomic or intermolecular dative bond formation, essentially translates Lewis's definition into quantum-mechanical terms but also (almost automatically, because of the form of Eq. (1)) broadens its scope considerably, although in a way consistent with the spirit of Lewis's viewpoint.

(45) For a valuable critical review on acids and bases, with considerable emphasis on terminology, see R. P. Bell, *Quart. Rev. Chem.*, **1**, 113 (1949).

unique absolute quantity, but depends appreciably on specific features of its interaction with the base with which it is paired.

Since the dissociation constant of any complex A·B is determined to a large extent by its heat of formation, attention may be turned to this quantity, which in turn should usually depend mainly on the resonance energy as given by Eq. (5).

If curve $W_0$ in Fig. 1 is fairly flat, $W_1 - W_0$ depends mainly on the location of curve $W_1$. Clearly $W_1$ is the lower, and so B the better base, the smaller is $I_B^{vert}$; and $W_1$ is the lower and so A the better acid, the larger is $E_A^{vert}$. $I_B$ and $E_A$ are, respectively, properties of B and A alone, and are of major importance[6,2d] for base or acid strength. The somewhat different quantities $I_B^{vert}$ and $E_A^{vert}$, however, depend at least slightly on the equilibrium configuration of the complex.

For a given distance $R$ between centers of A and B, the Coulomb energy in $\psi_1$ may depend considerably on the detailed charge distribution within $B^+$ and $A^-$, and on the orientation of A to B. Thus if the charges are largely localized on two neighboring atoms, as on the N and B atoms in $R_3B{\cdot}NR_3$ (*cf.* Section V), the Coulomb energy is greater than $e^2/R$, whereas if the charges are widely dispersed as in Ar·Nt (Ar = aromatic, Nt = nitrated aromatic), it is less than $e^2/R$. To a large extent these Coulomb attractions, as well as the covalent bonding and the exchange repulsions between $A^-$ and $B^+$, may be broken down into additive contributions to acid and base strengths; but, in part, specific factors remain. Finally, *all* the energy effects associated with curve $W_1$ are affected by the properties of any medium in which the complex may be dissolved.

The forms of the $W_0$ and $W_N$ curves may now be reconsidered more carefully. If A and/or B are "soft" molecules, curve $W_0$ rises less steeply than if both are "hard" molecules, and curve $W_N$ then tends to have a deeper minimum. Hence, "softness" in A or B, respectively, should tend to make it a better acid or base. Two special cases may be mentioned. If there is a general softness in curve $W_0$ down to distances close to those of equilibrium in curve $W_N$, we may have an exothermic complex with no activation energy. If there is a hardness in curve $W_0$ at relatively large distances (for example as a result of steric hindrance) but this levels off or even relaxes at shorter distances, there may then be in curve $W_N$ an activation barrier followed by a more or less deep energy minimum. The complex may then be either exothermic or endothermic. Even if moderately endothermic, it may then exist in appreciable concentration in equilibrium with its dissociation products. If strongly endothermic, it may be important as an "activated" complex.

In general, the many-dimensional surface which is the generalization of curve $W_0$ of Fig. 1 does not rise indefinitely in all directions when various atoms or groups in A and B are brought closer together. Instead, there exist activation energies for various chemical reactions. However, complex-formation depresses part and probably most of the many-dimensional surface $W_N$ below $W_0$, and probably increasingly so as A and B are brought closer (increase of $H_{01}^2/(W_1 - W_0)$ in Eq. (5)). This should lower the activation barriers for some if not all possible chemical reactions between A and B. Thus complex-formation (especially if the complex is fairly tightly bound) may often be the precursor of a chemical reaction—an idea earlier set forth clearly by Brackmann[5] and doubtless others. Such a reaction may in many cases proceed so rapidly at ordinary temperatures that the reversible complex is never isolated.

As an example, it occurs to one that the observed benzene–halogen complexes may be intermediates in the halogenation of benzene.[2] This could be true without implying that halogenation necessarily proceeds over an activation barrier in the direction of the particular coördinate $R$ in Fig. 1. Rather, it could proceed over a quite different barrier or "activated complex." The point is that much or all of the $W_N$ surface, including this other barrier, may be lower than it would be without existence of a $Bz{\cdot}X_2$ complex.

## V. Compounds of the $BX_3{\cdot}NR_3$ Type

In Section III and previously,[2d] several examples of loose molecular compounds between neutral closed-shell molecules have been discussed. As has already been indicated, the stable compounds $BR_3{\cdot}NR_3$ and the like can be understood in terms of the same general theoretical framework. However, there are important quantitative differences. Figure 4, for $BF_3{\cdot}NMe_3$ (Me = methyl), has been constructed to aid in understanding these. It must be emphasized that Fig. 4 is largely schematic.

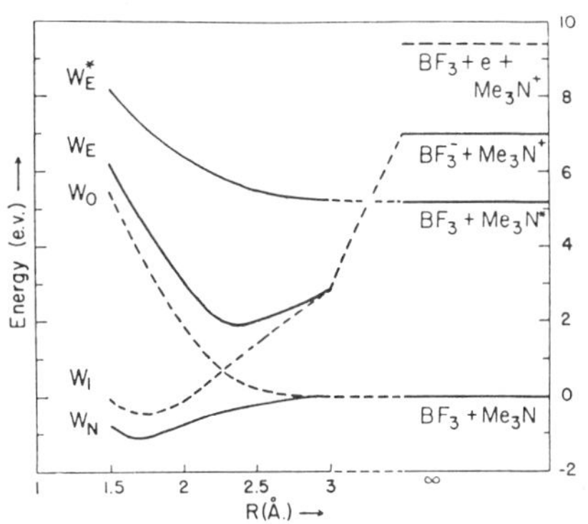


Fig. 4.—Schematic diagram of potential curves for the compound $BF_3{\cdot}NMe_3$ as a function of the distance $R$ between the centers of the B and N atoms, with the remaining atoms so located at all $R$ values as to minimize the energy *for curve* $W_N$. The diagram is largely qualitative. However, the position of the level $BF_3 + Me_3N^*$ is approximately correct, where $Me_3N^*$ represents the first excited singlet state of $Me_3N$.

The energy curves in Fig. 4 are all intended to correspond to an adjustment of the shapes and dimensions of the $BF_3$ and $NR_3$ groups at each value of the B–N distance $R$ to be such as to minimize the energy of $W_N$, but *not* of the other $W$ curves. This preserves the relationship of *verticality* between the other curves and $W_N$, as is needed for an analysis of the interactions of $\psi_0$ and $\psi_1$ to give $\psi_N$ and $\psi_E$, and for a discussion of the absorption spectra of $BF_3{\cdot}NMe_3$.

An important point is that if $BF_3$ and $NMe_3$ approach each other with their symmetry axes coincident, then the no-bond wave function $\psi_0$, and the dative structure $F_3B^--N^+Me_3$ formed from $F_3B^-$ and $N^+Me_3$ in their respective ground states, are both totally symmetrical singlet states of the whole molecule. Thus ground-state $F_3B^--N^+Me_3$ can serve as $\psi_1$. It appears altogether probable that as $R$ decreases from $\infty$, the two molecules approach with their axes always in coincidence. Further, as $R$ decreases, the B atom must move out of the $F_3$ plane toward the N atom and the N atom also move toward the B atom, the tendency being to set up tetrahedral angles around both. This tendency would be fully realized in the equilibrium state of pure $\psi_1$, in which the charges and the covalent bond are largely localized on the B and F atoms. For pure $\psi_0$, the equilibrium geometry would be quite different, with some electrostatic attraction between the boron and nitrogen atoms at larger $R$ values, but strong repulsion between the two molecules if $R$ gets at all small. The actual ground state is expected to have a wave function and an equilibrium geometrical arrangement which are compromises between those of $\psi_1$ and $\psi_0$.

According to chemical experience, $\psi_1$ predominates in $\psi_N$ ($b^2 > a^2$ in Eq. (1)). This implies a crossing of the curves $W_0$ and $W_1$ as shown in Fig. 4, and a resultant excited state curve $W_E$ with a minimum at a large $R$ value and a steep rise at smaller $R$. This should lead to a broad charge-transfer absorption spectrum in the ultraviolet.[14]

Besides $W_N$ and $W_E$, Fig. 4 shows a $W(R)$ curve (marked $W^*_E$) for an excited singlet state of $BF_3 \cdot NMe_3$ related to the first excited singlet state of $NMe_3$. The latter, which gives rise to well-known absorption of $NMe_3$ near λ 2400, almost certainly corresponds to excitation of an electron from the lone pair of the nitrogen atom.[46] This lone pair becomes a $B^--N^+$ bonding pair in the molecule $BF_3.NMe_3$, with a resultant increase in the binding energy of its electrons, so that the corresponding λ 2400 absorption should shift strongly toward shorter wave lengths. The $W_E{}^*$ curve in Fig. 4 has been drawn to be in qualitative agreement with this expectation.

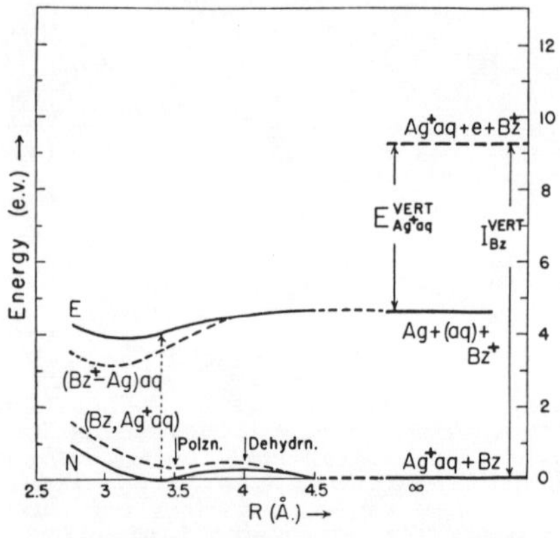


Fig. 5.—Predicted potential curves for the interaction of a benzene molecule with an $Ag^+$ ion in aqueous silver nitrate solution, as a function of the distance $R$ in the model described in the text (qualitative only). At the right, $Ag^+$ aq. designates a hydrated $Ag^+$ ion, while Bz indicates a benzene molecule surrounded by solution. $Bz^+$ designates a benzene molecule ionized vertically, which in this case implies that the environing molecules are in the same positions and orientations as for curve N of the complex; and that the medium is *only electronically* polarized by the $Bz^+$.

(46) *Cf.* R. S. Mulliken, *J. Chem. Phys.*, **3**, 506 (1935).

## VI. The $Ag^+$ Complexes

Examples of a type of complex in which one partner is an atom–ion are those formed[3,4] by $Ag^+$, acting as a Lewis acid, with ethylene, benzene, or related compounds as bases. Here the partners are totally symmetrical singlet structures[9] as before, and the theory of Section II is applicable. State $\psi_0$ has the no-bond structure (B, $Ag^+$) and $\psi_1$ the bonded structure $B^+$–Ag.

The $Ag^+$ complexes occur both in aqueous solutions, as for example when benzene is dissolved at low concentrations in aqueous $AgNO_3$, and in non-aqueous solutions, as when $AgClO_4$ is dissolved in benzene. In the former case, the $NO_3^-$ ions may be assumed to be completely dissociated from the ions of $Ag^+$ or of $Ag^+ \cdot Bz$. In the latter, the $ClO_4^-$ ions presumably remain attached to the $Ag^+$ or $Ag^+ \cdot Bz$ ions, but in a relatively passive role.

Let us consider the aqueous case, for which a *schematic* $W(R)$ diagram is given in Fig. 5. This is similar to Fig. 1, except for the following points. (1) The electron affinity of the Lewis acid here is much larger because the acid is a positive ion, the electron affinity being now the ionization potential of the silver atom. (2) But there is no $e^2/R$ attraction in state $\psi_1$ or state $E$. (3) Allowance must be made for sizeable solvation and polarization effects.

In state $\psi_0$, as $R$ decreases, we may imagine a Bz molecule to approach an $Ag^+$aq ion. As the Bz penetrates the hydration sphere surrounding the $Ag^+$, the $\psi_1$ curve may rise at first as water molecules are displaced, then fall somewhat as polarization of the Bz by the $Ag^+$ becomes appreciable (*cf.* Fig. 5). The resulting complex, probably endothermic if it were pure $\psi_0$, perhaps becomes thermoneutral or exothermic by resonance with $\psi_1$ giving $\psi_N$.

To permit correct adjudication of the resonance of $\psi_1$ with $\psi_0$, curve $W_1$ in Fig. 5 (labeled $(Bz^+-Ag)$aq), is drawn for the situation that all the nuclei, including those of immediately neighboring water molecules, are for each $R$ value located at the positions they assume for curve $W_N$ of the actual complex. The position of curve $W$ at $R = \infty$ is determined using $I_{Bz}{}^{vert}$ and $E_{Ag^+aq}{}^{vert}$. $I_{Bz}{}^{vert}$ is the ordinary $I$ of Bz minus slight corrections; $E_{Ag^+aq}{}^{vert}$ is the ordinary $I$ of Ag (7.5 e.v.) minus a considerable correction (estimated as 2.9 e.v. in Fig. 5) corresponding to loss of hydration energy by, (a), displacement of some $H_2O$ by Bz; (b), diminished attraction or change to repulsion for the remaining $H_2O$ when $Ag^+$ becomes Ag without change in positions of these $H_2O$. At sufficiently small $R$, covalent bonding tends to lower $W_1$ somewhat. Finally, resonance interaction with $\psi_0$ leads to the $N$ and $E$ curves of Fig. 5.

Although Fig. 5 is only schematic, it is perhaps only fair to state that no intense ultraviolet charge-transfer spectrum, such as Fig. 5 would suggest, has yet been identified for $Bz \cdot Ag^+$.

For Bz in aqueous $AgNO_3$ solution, it is known that 2:1 complexes $Ag^+BzAg^+$ are present in appreciable amounts in addition to the predominant 1:1 complexes.[3,4]

A somewhat similar discussion could be given for the $Bz \cdot Ag^+ClO_4^-$ complex in benzene solution, with $ClO_4^-$ more or less taking the place of the water in $Ag^+$aq.

A question of considerable interest for $Bz \cdot Ag^+$aq. and $Bz \cdot Ag^+ClO_4^-$ is the location of the $Ag^+$ with respect to the Bz ring. Previous writers[3,4] have made the plausible suggestion that the $Ag^+$ lies above the middle of the ring on its symmetry axis (Model A of Section III). This, however, is probably excluded by symmetry considerations very similar to those for Model A of $Bz \cdot I_2$. Since $\psi_0$ is of type $^1A_1$, $\psi_1$ will give no resonance unless it is also $^1A_1$. But since $Bz^+$ has a nodal plane through the symmetry axis (*cf.* Fig. 2 and Section III), an electronic state of Ag must be used which has a like plane. The lowest suitable state is a $^2P$ or $^2D$ state requiring about 4 e.v. excitation energy. This makes the energy of $\psi_1$ so high that resonance with $\psi_0$ is weakened (*cf.* Eq. (5)).

Similar reasoning makes it improbable that the $Ag^+$ ion lies alongside the Bz in the latter's plane (*cf.* Model E of $Bz{\cdot}I_2$). For in this case a *second* nodal plane of the $Bz^+$, in the ring plane ($xy$ plane of Fig. 2) gives $\psi_1$ a nodal plane in this same location, making it incapable of resonance with $\psi_0$ unless a $^2$P or $^2$D excited state of Ag is used in $\psi_1$.

However, if the $Ag^+$ is pulled to one side from the Bz symmetry axis, toward the region between but somewhat above two carbon atoms of the ring (symmetry $C_s$—*cf.* Model $O_x$ of $Bz{\cdot}I_2$), fairly strong resonance of $\psi_1$ with $\psi_0$ may reasonably be expected.[47] In this orientation, one of the two states of $Bz^+$ (*cf.* Fig. 2) can give with Ag a totally symmetrical singlet as required for $\psi_1$.[48]

When the present work was presented at a recent meeting,[49] this model was proposed on the basis of the preceding arguments. In the discussion following, Professor R. E. Rundle stated that in unpublished work[50] on the crystal structure of $Bz{\cdot}AgClO_4$ (white crystals out of benzene solution) he had located the $Ag^+$ ions in exactly the position mentioned. In the crystal, each $Ag^+$ is located in this way with respect to *two* Bz, and each Bz is similarly located with respect to *two* $Ag^+$. Such a crystal is of course not the same thing as a 1:1 complex, but (*cf.* also the last paragraphs of Section III), a preliminary consideration of the symmetry requirements indicates that these may still operate mainly in a localized fashion to give approximately the same favored orientations (Model O) as for an isolated $Bz{\cdot}AgClO_4$.

Recent papers of Andrews and Keefer[4] show the following equilibrium constants $K$ for 1:1 $Ag^+$ complexes in aqueous $AgNO_3$ solution: benzene, 2.4; toluene, 2.95; xylenes, 2.6–3.0; mesitylene, 1.8. The attainment of a maximum $K$ for toluene and the xylenes and the low value for mesitylene would be difficult to understand if the $Ag^+$ were located as in Model A on the symmetry axis of the aromatic ring. But if it is located as in Model O, the observed $K$ values are at once intelligible as the result of a balance between ($a$), a tendency toward increasing basicity with increasing number of methyl groups, and, ($b$), steric hindrance by methyl groups to attainment by the $Ag^+$ of one of its favored locations between two carbon atoms. Space models indicate that steric hindrance may well somewhat destabilize the two locations next adjacent to a methyl group. This leaves six "good" locations for benzene, four for toluene, three for *o*-xylene, two for *m*-xylene and *p*-xylene, none for mesitylene.

(47) The distance $R$ between centers of $Ag^+$ and Bz is larger in Model O than in Model A, but this is not of itself important here, in contrast to the case of $Bz{\cdot}I_2$, where minimum $R$ makes $W_1$ lower because of the Coulomb attraction between $Bz^+$ and $I_2^-$.

(48) By similar reasoning, the most likely position for the $Ag^+$ in the $Ag^+$-ethylene complex[1] is directly above the center of the ethylene, *on* the twofold axis perpendicular to the plane of the latter.

(49) Reference 1.

(50) Now published: R. E. Rundle and J. H. Goring, THIS JOURNAL, **72**, 5337 (1950).

The reasoning just given is reinforced by the contrasting behavior of the halogen–aromatic complexes: their stability increases steadily from benzene to toluene to the xylenes to mesitylene. This is readily understood in terms of Model R for the $Ar{\cdot}I_2$ complex, where space models indicate no appreciable steric hindrance for any of the methyl-substituted benzenes.[2d]

## VII. Further Examples and Further Discussion of the Interaction Strengths of Lewis Acids and Bases

The theory given in Section II is applicable to a wide variety of chemical entities A·B. In Section II, it was restricted to cases in which A and B are even-electron, closed-shell molecules, ions, or atoms.[9] A preliminary discussion of the factors governing the strengths of A and B as Lewis acids and bases has been given in Section IV. In Sections III, V and VI, examples of particular types of molecular complexes or compounds have been considered. The present Section is devoted to a generalization of the discussion, including the application of the theory of Section II to the case that A and B are odd electron entities, and to the case that they are identical. First, the closed-shell A and B types will be reviewed.

Two important classes of *neutral-molecule* bases are the weak $\pi$ bases (unsaturated or aromatic hydrocarbons or amines, etc.), and the often strong $n$ (or *onium*) bases ($NR_3$, $OR_2$, etc.).[51] For $\pi$ bases, the donated electron in the dative structure $\psi_1(B^+\text{–}A^-)$ of Eq. (1) comes from a more or less bonding MO; for $n$ bases, from a non-bonding MO occupied in the original base by a lone pair of electrons. Some molecules, for example pyridine[42] and acetone, have the possibility of functioning either as $\pi$ or as onium bases.

Neutral-molecule Lewis acids or acceptors include $\pi$ acceptors (nitroaromatic compounds, maleic anhydride, etc.), which combine especially with $\pi$ bases or donors to form loose complexes,[52] and $v$ (vacant-orbital) acceptors ($BX_3$, $BR_3$, $AlX_3$, $SnX_4$, etc.) which combine especially with $n$ donors to form fairly stable compounds.

They also include the halogens, which combine with either $\pi$ or $n$ bases (*cf.* Section III and ref. 2d) to form loose complexes. Further, the hydrogen halides, although they are strong acids in the Brönsted sense,[45] act as weak Lewis acids rather like the halogens. The Lewis-acid character of the halogens and hydrogen halides is attributable to the incompletely satisfied electronegativity of their halogen atoms; and since the acquisition of a negative charge by halogen can take place, characteristically, only concurrently with a pronounced loosening of the covalent binding, these and other similar Lewis acids, *e.g.*, the alkyl halides, may be called $d$ (*dissociative*) acids. Their weak acidity can become strong only by dissociation (into either

(51) This nomenclature is related to that of Dewar ("$\pi$ complexes") and to that currently used by J. R. Platt and others in describing ultraviolet absorption spectra of unsaturated molecules ("$\pi$–$\pi$ transitions" and "$n-\pi$ transitions"); M. Kasha, Faraday Society Discussions, 1950, No. 9, p. 14.

(52) See *e.g.*, J. Landauer and H. M. McConnell, THIS JOURNAL, **74**, in press (1952).

atoms or ions; this will be analyzed further in a later paper[35a]).[53]

The acid strengths of $\pi$ acids also are often attributable mainly to the presence of unsatisfied electronegative atoms. In $v$ acids, the presence of incompletely satisfied electronegative atoms, while not essential (consider *e.g.*, $BMe_3$,—Me-methyl), greatly increases acid strength (as *e.g.*, in $BCl_3$). This is understandable from the fact that the chlorine atoms withdraw negative charge from the boron atom, and so must increase the electron affinity $E$ of the latter's vacant 2p orbital, thus increasing acid strength (*cf.* Section IV). However, halogen and similar atoms in $BX_3$ also at the same time feed some negative charge into the vacant orbital by a familiar resonance effect, thus tending to decrease its $E$; *but* apparently the *net* effect is usually to increase acid strength.[54]

*Ionic* ($i$) bases include $X^-$, $CN^-$, $OH^-$, etc. Atomic or ionic ($i$) Lewis acids include O, $Ag^+$, $Li^+$, $H^+$. (In the case of the O atom, an excited singlet state must be used.) Ionic acids or bases are necessarily accompanied by partner ions of opposite sign, which being themselves bases or acids, tend to neutralize the acids or bases which they accompany. This tendency is largely avoided only in ionizing media where after solvation of both ions the neutralizing partner is ionized away, or in the case that the partner ion (*e.g.*, $K^+$, a very weak Lewis acid, or $ClO_4^-$, a very weak base[44]) is of large size and low charge.

As already suggested, it is possible to regard *odd-electron* systems $A'$ (*e.g.*, Cl, CN, OH, $NH_2$, $C_2H_5$) and $B'$ (*e.g.*, H, Na, NO, $NO_2$) as Lewis acids and bases for which the theory of Section II is applicable. (Primes are used here for odd-electron acids and bases to distinguish them from the more usual even-electron ones denoted by A and B.) The "complexes" $A'{\cdot}B'$ are now usually stable compounds. The theory of the structure and charge-transfer spectra of such compounds is the same as for the complexes A·B, *provided* $\psi_0$ in Eq. (1) is taken to refer to a pure ionic structure $(A')^-(B')^+$ and $\psi_1$ to a pure covalent structure $A'-B'$.

However, even this modification of the previous treatment can be avoided if the basis of reference is shifted from the odd-electron entities $A'$ and $B'$ to a pair of even-electron entities A and B such that A is $(B')^+$ and B is $(A')^-$. For example, if $B'$ is Na or H and $A'$ is Cl, then A is $Na^+$ or $H^+$ and B is $Cl^-$. In terms of such a choice of A and B, the entire discussion in Section II is applicable, at least in its qualitative aspects. Just as in previous examples,[55] either $\psi_0$ or $\psi_1$ may predominate in the ground-state wave function $\psi_N$ of Eq. (1). For example, $\psi_0(Na^+,Cl^-)$ predominates in diatomic NaCl, $\psi_1$(H–Cl) in HCl. For the excited-state function $\psi_E$, the relations are reversed. These simple diatomic examples illustrate how the (in general intermolecular) charge-transfer spectra N → E of Section II become identical in special cases with the more familiar interatomic charge-transfer spectra (N → V spectra).[56]

Returning to complexes and compounds between neutral molecules, it may be recalled from Section IV that the strength of any Lewis acid–base interaction is governed largely by the characteristics of the dative structure $\psi_1(A^-–B^+)$. Especially favorable are low ionization energy $I$ of B and close approach between the centers of gravity of the charges on $B^+$ and $A^-$. High vertical electron affinity $E$ for A is also helpful. Close approach between $B^+$ and $A^-$ lowers the energy of $\psi_1$ both through increased Coulomb interaction and through strong covalent binding. Low energy of $\psi_1$, as well as close approach *per se*, favor strong resonance of $\psi_1$ with $\psi_0$.

In molecular compounds between $n$ bases and $v$ acids, conditions for close approach are especially favorable (see Sections IV, V). The relative binding strengths of different compounds of this type should depend largely on the $I$ and the approachability of the base, and on the $E$ and the approachability of the acid. According to H. C. Brown,[57] base strength, for $BX_3$, $AlX_3$ and the like as reference acids, decreases in each of the series $NR_3 > PR_3 > AsR_3 > SbR_3$; $OR_2 > SeR_2 > TeR_2$. In each series $I$ decreases in the order given, tending to *increase* base strength, but at the same time the size of the onium atom increases, decreasing its approachability and tending to decrease base strength. If the latter effect predominates, Brown's result can be understood. Brown's observation that base strength varies in the order $NMe_3 > NH_3 > NF_3$ is readily understandable because $I$ is considerably less for $NMe_3$ than for $NH_3$,[58] and would be expected to be less for $NH_3$ than for $NF_3$. (The observed diminished binding strength of compounds containing two bulky alkyl groups is of course also understandable, in terms of diminished approachability.) Likewise Brown's order of base strength $NR_3 > OR_2 > ClR$ is understandable because $I$ is smaller for $NR_3$ than for $OR_2$.[58] Further, Brown's order of acid strength $BF_3 > BH_3 > BMe_3$ is understandable because $E$ would be expected to decrease in this order[54] (see discussion in an earlier paragraph). It is not clear, however, why $AlMe_3$ should be a stronger Lewis acid than $BMe_3$ toward onium bases, as Brown reports.

An interesting case is that of borine carbonyl $OC{\cdot}BH_3$, where a somewhat stable compound is formed in spite of the high $I$ of CO (14.5 volts).

(53) Actually, the functioning of the hydrogen halides and similar $d$ acids as Lewis acids may be so weak as to be negligible, since $\psi_1$ must be high in energy ($E^{vert}$ is probably *negative* for such molecules in loose complexes). It nevertheless makes sense to classify them as $d$ acids in view of their behavior in strong solvents. See also R. Ferreira, *J. Chem. Phys.*, **19**, 794 (1951).

(54) H. C. Brown (see ref. 57) attributes the decreasing order of acid strength $BBr_3 > BCl_3 > BF_3$ to increasing resonance in this order; the anomalous weakness of $B(OMe)_3$—much weaker than $BMe_3$—he attributes likewise to this type of resonance.

(55) It should be noted that ionic binding now occurs in $\psi_0$ instead of in $\psi_1$. The case differs in this respect from that where A and B are neutral molecules.

(56) The case of charge-transfer spectra in molecules with homopolar bonds ($H_2$, $C_2H_4$, etc.)[16] corresponds to the special situation where A and B are identical (*cf.* Section VIII). For specific discussion of the states and spectra of diatomic molecules of the type here considered, see R. S. Mulliken, *Phys. Rev.*, **50**, 1017, 1028 (1936); **51**, 310 (1937).

(57) H. C. Brown, lecture at ONR-AEC-sponsored symposium at University of Chicago, Feb. 21–23, 1951; and numerous published papers, especially in THIS JOURNAL.

(58) See for example W. C. Price, *Chem. Revs.*, **41**, 257 (1947); A. D. Walsh, *Quart. Rev. Chem.*, **2**, 73 (1948). For example, $H_2O$, 12.7, $(C_2H_5)_2O$, 10.2; $NH_3$, 10.8, $(CH_3)_3N$, 9.4; HCl, 12.84, $C_2H_5Cl$, 10.89. For $\pi$ bases, ethylene 10.50, propylene 9.70; butadiene 9.07; benzene 9.24, toluene 8.92; naphthalene 8.3.

Here apparently a very favorable approachability situation balances the high $I$ in the $\psi_1$ resonance structure $(OC)^+$–$B^-H_3$.[59]

The present theory also gives a tempting explanation of the ease with which gaseous carbon monoxide attacks solid ferrous metals. One may assume an initial attack involving strong development of a $\psi_1$ resonance structure $M^-$–$(CO)^+$, where M indicates solid metal; the CO functions as an M base, the metal as a $v$ acid. On closer approach, the initial complex changes smoothly into a predominant M=C=O structure, while additional CO molecules attack until the gaseous carbonyl is released. But this is only speculative, and probably wrong.

As compared with the $n$ bases, the $\pi$ bases are in general aided by low I values (10.50 for $C_2H_4$, 9.24 for benzene, with considerably lower values[58] for methylated or conjugated compounds or larger-ring aromatics), but are much more limited as to approachability. Maximum approachability is to be expected between $\pi$ bases and $\pi$ acids, arranged (in the case of aromatics) with their planes parallel, so that the diffusely distributed positive charge in $B^+$ in $\psi_1(B^+$–$A^-)$ is as close as may be to the also diffusely distributed negative charge in $A^-$. For the association of aromatic $\pi$ bases Ar with $v$ acids like $BR_3$, the localized negative charge on the boron in $B^-R_3$ in $\psi_1(Ar^+$–$B^-R_3)$ is at a disadvantage for approach to the diffusely distributed positive charge in $Ar^+$. Taking Bz·$BX_3$ as typical, there is an additional handicap: a symmetrical location of the $BX_3$ with its axis coincident with the Bz axis gives no resonance at all, for quantum-mechanical symmetry reasons similar to those which disfavor a symmetrical structure for Bz·$Ag^+$(*cf.* Section VI); while the unsymmetrical location required (as in Bz·$Ag^+$) for resonance may be hampered (more than for Bz·$Ag^+$) by steric interference between the $X_3$ and Bz planes. The two unfavorable specific factors indicated may be adequate to explain why $BX_3$ and $AlX_3$ do not form complexes with aromatic bases.[60] On the other hand, pyridine,[42] undoubtedly acting as an $n$ rather than as a $\pi$ base, does form compounds with $BX_3$.

The iodine and other halogen molecules because of their simpler shape and the special nature of their electronic configuration should be more adaptable than $BX_3$, and indeed they apparently act as Lewis acids both toward $\pi$ bases (Section III) and toward $n$ bases such as $R_2O$ and probably $R_2CO$, with in each case a different but always a compact geometrical configuration satisfying the symmetry requirements of that case.[2] The absence of known complexes of the halogens with certain bases is probably explainable by too great reactivity: interaction may pass rapidly through complex formation to irreversible chemical change.

(59) W. Gordy, H. Ring and A. B. Burg, *Phys. Rev.*, **78**, 517 (1950), conclude that the no-bond structure contributes 40–50% to $\psi_N$, the structure $(OC)^+$–$(BH_3)^-$ about 30% (this includes the two structures $O^+$≡C-$B^-H_3$ and O=$C^+$-$B^-H_3$).

(60) In the literature, it is sometimes indicated that molecular compounds Ar·$BX_3$ or Ar·$AlX_3$ *do* exist (see *e.g.* F. Briegleb, cited in footnote 26a of Ref. 2d), but according to H. C. Brown (see Brown, Pearsall and Eddy, This Journal, **72**, 5347 (1950)), this is not true.

## VIII. Self-complexes, Intermolecular Forces and Compressibility

The limiting case where A and B are identical is of some interest. In this case, Eq. (1), (8) and (5) take[11] the special forms

$$\psi_N = a\psi_0 + b(\psi_1 + \psi_1') + \cdots \qquad (1a)$$

$$\left.\begin{aligned}\psi_E &= c(\psi_1 - \psi_1') + \cdots \\ \psi_F &= a^*(\psi_1 + \psi_1') - b^*\psi_0 + \cdots\end{aligned}\right\} \qquad (8a)$$

$$W_N = W_0 - \frac{2(H_{01} - SW_0)^2}{(W_1 - W_0)} + \cdots \qquad (5a)$$

Here if $\psi_1$ is $A^-$–$B^+$, $\psi_1'$ is is $A^+$–$B^-$. Comparing Eq. (5a) with Eq. (5), and referring also to Eq. (6), it is seen that very appreciable charge-transfer forces tending to form complexes must exist between like just as between unlike molecules. Empirically, however, it seems clear that these forces are generally weaker for like molecules. This may be understood by the fact that ordinarily if A is a strong Lewis acid, it is a weak base, and conversely. However, in some cases a strong base is at the same time a strong Lewis acid (*e.g.*, metals, where $I = E$, and polycyclic aromatic hydrocarbons), while in other cases (*e.g.*, rare gases, $N_2$, $CH_4$) a weak base is at the same time a weak Lewis acid.

The interaction of two Bz molecules may be considered as an example. One can approach this as the limiting case in a series $Me_n$Bz·Bz$(NO_2)_{n'}$ with $n$ and $n'$ decreasing to zero ($Me_n$Bz = $n$-methylbenzene, Bz$(NO_2)_{n'}$ = $n'$-nitrobenzene). There is evidence[62] that very weak complexes of the indicated type exist, *e.g.*, for $n = 0$, $n' = 3$. One may surmise that in Bz·Bz itself, the forces, while not negligible, are too weak to give appreciable concentrations of Bz·Bz in the liquid at room temperature. To the extent that heats of formation are decisive, one might then suppose that these considerably exceed $kT$ for larger values of $n + n'$, but fall below $kT$ for Bz·Bz itself.

It may still be that in crystalline Bz, at least at low temperature, the charge-transfer forces are important in determining how the molecules are stacked. Considering a single pair Bz·Bz, if the two molecules were stacked with their axes coincident and their planes parallel, the symmetry would be $D_{6h}$. In $\psi_1$ and $\psi_1'$ of Eq. (1a), ground-state $Bz^+$ and $Bz^-$ would be of types $^2E_1$ and $^2E_2$, respectively. Now although the proper combination of these $(\psi_1 + \psi_1')$ gives rise to a very considerable number of states of Bz·Bz, no one of these is of the species $^1A_{1g}$ required for resonance with $\psi_0$. Hence a different stacking of the two Bz molecules is indicated. It seems likely that an arrangement obtained by sliding one Bz about half a ring diameter over the other would give maximum resonance.[61]

One thus sees that in pure liquids and molecular crystals, for example liquid benzene and crystalline benzene and naphthalene, charge-transfer forces should favor definite types of packing and definite orientations. They may thus afford an explanation of the way in which for example certain aromatic molecules are tilted at odd angles in their crystals.

The possibility emerges that charge-transfer forces may often share with London's well-known dispersion forces in accounting for the familiar van der Waals cohesive forces between molecules, especially in systems containing more than one component. Charge-transfer forces share with dispersion forces the property of approximate additivity: the fact that one molecule is bound to another does not prevent a third from being attracted,

(61) It may be relevant that in crystalline graphite, exactly this sort of a displacement between successive planes is observed. Since each graphite plane is essentially a giant aromatic molecule (here with $I = E$), it seems possible that charge-transfer forces may be at least partially responsible.

steric factors permitting. (See the discussion of $n$:1 complexes at the end of Section II.) Dispersion-force attractions tend to be largest in orientations bringing maximum polarizabilities into play, while the orientational properties of charge-transfer forces, as shown in Sections II and III, are governed by considerations of quantum-mechanical symmetry of molecular wave functions. The two kinds of forces are thus different in this respect, and probably the sharpness of the orientational effect is stronger for the charge-transfer forces.

From what has preceded, it appears likely that charge-transfer forces may often be of the same order of magnitude as dispersion forces. Perhaps dispersion forces usually predominate for interactions between like molecules, but charge-transfer forces often in solutions and other systems in which molecules of different kinds are present together. Charge-transfer forces may well also be important in heterogeneous systems and in adsorption phenomena, and may afford new possibilities for understanding intermolecular interactions in biological systems. Definite conclusions about these points must, however, wait upon further investigations.

Even if it should turn out that charge-transfer forces are relatively unimportant for one-component liquids and crystals not under strong internal or external pressure, the theory indicates that if in any way a substance is sufficiently compressed, the charge-transfer forces should increase rather rapidly. (See here also the discussion of "hard" and "soft" molecules in Section IV.) Such compression might perhaps be effected by internal ionic forces in the case of a partially ionic crystal, by unusually strong dispersion or dipole forces, or by strong external pressure. The existence of charge-transfer forces should in general contribute considerably—in an anisotropic manner because of their strong orientational properties—to compressibilities. Here it is relevant to recall the studies of Gibson and Loeffler[61] who found marked shifts in the locations of the absorption spectra of aniline-polynitrobenzene and similar solutions under pressures of 1000 atmospheres; these are spectra which are attributable to the presence of very loose charge-transfer complexes.[62]

(62) R. E. Gibson and O. H. Loeffler, This Journal, **61**, 2877 (1939); **62**, 1324 (1940).

Chicago 37, Illinois Received June 11, 1951

[Reprinted from the Journal of Physical Chemistry, 56, 801 (1952).]


# MOLECULAR COMPOUNDS AND THEIR SPECTRA. III. THE INTERACTION OF ELECTRON DONORS AND ACCEPTORS[1]

ROBERT S. MULLIKEN

*Department of Physics, The University of Chicago, Chicago 37, Illinois*

*Received March 19, 1952*

Extending earlier work,[2] a classification of electron acceptors and donors each into a number of types is given in Section II. In an extension of Sidgwick's nomenclature, donors D and acceptors A are here defined (see Section I) as all those entities during whose interaction transfer of negative charge from D to A takes place, with the formation as end-product either of an additive combination $A_m \cdot D_n$ or of new entities. In all cases of 1:1 interaction, the wave function $\psi$ of A·D (and, formally at least, of the end-products also in the dissociative case) is of the approximate form given in equation (1) with appropriate ionic or covalent bonding (or no bonding) between D and A, and between $D^+$ and $A^-$, depending on whether A and/or D are closed-shell molecules or ions, or radicals.[3] Donors and acceptors as here defined correspond closely to nucleophilic and electrophilic reagents as defined by Ingold or, except for the inclusion here of donor and acceptor radicals, correspond rather well to bases and acids as defined by G. N. Lewis. In Section III, the applicability of an extension of eq. (1) to crystalline molecular compounds is considered briefly.[3] A brief discussion and listing of possible or probable known charge-transfer spectra[2,4] of donor-acceptor molecular complexes are given in Section IV and Table VI. Sections V–VIII contain further elucidation of matters discussed in Sections I–II and in ref. 2.

The energy $U$ of interaction between a donor and acceptor as a function of a charge-transfer coördinate $C$ (a kind of reaction coördinate, so defined as to increase from 0 to 1 with increasing transfer of electronic charge from D to A) is studied in Section IX for interactions between donor–acceptor pairs of the various classes defined here. In many cases, there should be two important minima in the $U(C)$ curve, namely, one for a loose "*outer complex*" for small $C$, and one for a tighter "*inner complex*," often of ion-pair character, for large $C$ (see Figs. 1–2). In any particular case, one of these is the stable form, while the other is an excited or activated state (lower in energy, however, than the "activated complex" which usually intervenes between them). However, in many cases where the donor and acceptor form only a loose outer complex or none at all in the vapor state or in an inert solvent, the inner complex may become the stable form under the coöperative action of a suitable active solvent. The latter functions by solvation of the inner complex or its ions, either acting mainly electrostatically, or in some cases acting as (or, with the assistance of) an auxiliary acceptor or donor (double complex formation). The formation of ion-pair clusters or ionic crystals (*e.g.*, $NH_3 + HCl \rightarrow NH_4^+Cl^-$) can play the same role as that of an electrostatically functioning solvent in stabilizing the inner complex of a donor–acceptor pair. In a few interaction types, a "*middle complex*" is important (*cf.* Fig. 3), corresponding either to an activated complex or intermediate in a reaction such as those involving a Walden inversion, or to a stable association product as in $I_3^-$ or $HF_2^-$. Section X contains improvements and errata for the previous papers of this series. The Appendix, consisting of Tables III–VI, contains detailed descriptions of the various donor and acceptor types and of their modes of functioning.

## I. The Interaction of Donors and Acceptors

In chemical theory, the usefulness of G. N. Lewis's broad conception[5,6] of what should be meant by the words *acid* and *base* has become increasingly evident. However, since these words (especially *acid*) are commonly used with narrower meanings, it may be wisest to follow Sidgwick[7] in referring to Lewis acids and bases as (electron) acceptors and donors, respectively.[8]

(1) This work was assisted by the ONR under Task Order IX of Contract N6ori with The University of Chicago.

(2) II of this series: R. S. Mulliken, *J. Am. Chem. Soc.*, **74**, 811 (1952).

(3) See also Paper No. 25 in ONR Report on September, 1951, Conference on "Quantum-Mechanical Methods in Valence Theory." This Report is obtainable from L. M. McKenzie, Head, Physics Branch, Office of Naval Research, Department of Navy, Washington 25, D. C.

(4) I of this series: R. S. Mulliken, *J. Am. Chem. Soc.*, **72**, 600 (1950).

(5) G. N. Lewis, "Valence and The Structure of Atoms and Molecules," The Chemical Catalog Company (Reinhold Publ. Corp.), New York, N. Y., 1923, pp. 142, 133, 113, 107; *J. Franklin Institute*, **226**, 293 (1938).

(6) W. F. Luder and S. Zuffanti, "The Electronic Theory of Acids and Bases," John Wiley and Sons, Inc., New York, N. Y., 1946.

(7) N. V. Sidgwick, "The Electronic Theory of Valency," Oxford University Press, 1929, in particular, p. 116, for the definition of donors and acceptors.

(8) For a valuable review, see R. P. Bell, *Quart. Rev. Chem.*, **1**, 113 (1947).

While Lewis's ideas grew largely out of inorganic chemistry, similar ideas were developed more or less independently in the field of organic chemistry, culminating in Lapworth's categories of anionoid and cationoid reagents, or Ingold's[9] of nucleophilic and electrophilic reagents. The latter correspond closely to Lewis bases and acids, respectively, except that they add reducing agents rather generally to the former and oxidizing agents to the latter. Usanovich[10] also proposed a similar classification. Luder and Zuffanti (ref. 6, Chap. 4) elaborated an approach similar to Ingold's, but substituted the term "electrodotic" for Ingold's "nucleophilic." They stated that "both acids (primarily in Lewis's sense) and oxidizing agents are electron acceptors," and "are electrophilic (reagents)"; and that "both bases and reducing agents are electron donors" and "are electrodotic (reagents)."

In defining basic and acidic molecules, Lewis (although his essential idea seems to have been distinctly broader) emphasized as characteristic the *sharing of an electron pair*, furnished by the base and accepted by *an atom* in the acid. Sidgwick adopted the same definition for donors and acceptors.

Ingold[9] said: "reagents which donate their electrons to, or share them with, a foreign atomic nucleus may be termed nucleophilic"; those "which acquire electrons, or a share in electrons, previously belonging to a foreign molecule or ion, may be termed electrophilic." Luder and Zuffanti (ref. 6) stated that "an acid accepts a share in an electron pair held by a base; an oxidizing agent takes over completely the electrons donated by a reducing agent," and made a corresponding statement regarding bases and reducing agents. It is proposed here to use Sidgwick's simple and almost self-explanatory terms "donor" and "acceptor" to mean essentially the same things as Ingold's "nucleophilic reagent" and "electrophilic reagent," or Luder and Zuffanti's "electrodotic reagent" and "electrophilic reagent."

More precisely, (electron) *donors* D and *acceptors* A are here defined as *all those entities such that, during the interaction between a particular species of D and a particular species of A entities, transfer of negative charge from D to A takes place, with the formation as end-products either of additive combinations or of new entities. The additive combinations may be 1:1, m:1, 1:n, or in general m:n combinations.*

This definition is one which becomes extremely natural when one attempts to express the familiar ideas of donor–acceptor interaction in quantum mechanical symbols.[2] The wave function $\psi$ of the (stable or transitory) 1:1 complex A·D then takes in general the approximate form

$$\psi \approx a\psi_0(\mathrm{AD}) + b\psi_1(\mathrm{A^-D^+}) \qquad (1)$$

with appropriate ionic or covalent bonding (or no bonding) between A and D, and between $A^-$ and $D^+$, depending on whether A and/or D are closed-shell molecules or ions, or radicals.[3]

Equation (1) describes the partial transfer of an electron from D to A; the ratio $b^2/a^2$ varies between 0 for no transfer and $\infty$ for complete transfer. Equation (1) does not necessarily demand either (1) sharing of an electron pair, or (2) that the transferred electron shall come strictly from one particular atom in D and go strictly to one particular atom in A. However, it does not exclude these as important special cases. Lewis's and Sidgwick's use of restrictions (1) and (2) in setting up their formal definitions of acids and bases, or acceptors and donors, though seeming natural at the time, has had some tendency to inhibit others from making the fullest or freest use of the inherent possibilities of the donor–acceptor concept. Also, without an explicit quantum mechanical formulation, the nature of *partial* electron transfer has tended to appear rather obscure; in particular, the validity of the donor–acceptor interaction concept for explaining the many loose organic molecular complexes. has not always been seen in a clear light.[2]

Ingold's definitions dropped the first of the two limitations in Lewis's, but did not clearly dispose of the second. The present definition, as given above in words and in quantum mechanical form, definitely drops both. In dropping the second limitation, it permits one, when this is appropriate, to think in terms of intermolecular donor–acceptor action between molecules as wholes—an idea which may be considered as a slight generalization of one used by Dewar in connection with his concept of $\pi$-complexes.

The simple quantum mechanical viewpoint expressed in eq. (1) makes clearer the justifiability of the inclusion by Ingold and others of bases and reducing agents in a single class (donors) and of Lewis acids and oxidizing agents in another (acceptors). It goes further in indicating that there is perhaps even no fundamental need in terms of theory to distinguish bases and reducing agents as subclasses of the class "donor" or Lewis acids and oxidizing agents as subclasses of the class "acceptor." These distinctions now appear as perhaps matters of practical convenience rather than of basic theory. However, the question of course deserves much more thorough consideration. A point to be kept in mind is perhaps the fact that the terms "oxidation" and "reduction" are generally used for over-all processes, and often in more or less formalistic ways with respect to assignments of charges to atoms; whereas very often (not always) the concepts of Lewis acids and bases (or electrophilic and nucleophilic reagents) are used in connection with (real or supposed) actual mechanisms of reactions.

In terms of Lewis's definition calling for the sharing of an electron pair, acceptors such as $BF_3$ and $H^+$ (if it existed free) were clear cases of typical acids. Oddly enough, Lewis appears to have found it a little awkward to include the H-acids, for example HCl, as acids at all *per se* (that is, in the absence of ionizing solvents)—as can be inferred

(9) C. K. Ingold, *Chem. Revs.*, **15**, 225 (1934), especially pp. 265–273; *J. Chem. Soc.*, 1120 (1933), especially the footnote on p. 1121.

(10) Usanovich in 1939 (see ref. 6, p. 14, for a summary in English) defines an acid as any substance capable of giving up cations or of combining with anions, and a base as any substance capable of giving up anions or of combining with cations.

from Lewis's 1938 paper,[5] and from Luder and Zuffanti's book.[6]

Luder and Zuffanti describe H-acids HQ as secondary acids, in the sense that, while they are not really Lewis acids themselves, they are capable of supplying a Lewis acid (namely $H^+$) to a base. [That is, they are donors of the Lewis acid $H^+$, or proton donors.] These secondary Lewis acids are thus thought of as composites of a primary Lewis acid ($H^+$) and a base ($Q^-$); and in a similar way various other composites of real or conceivable primary Lewis acids and bases can be described as secondary Lewis acids. (The same or similar composites could equally be described as secondary bases.) For example, various covalent compounds RQ, likewise such typical compounds of primary Lewis acids and bases as $H_3N{\rightarrow}BMe_3$, may be considered as secondary Lewis acids (or as secondary bases). (See also Table VI, footnote *d.*) The matter of the classification of H-acids and various other molecules RQ as acceptors will be discussed rather thoroughly below.

Meanwhile, it may be noted that Ingold had no hesitation in classifying molecules such as HQ and RQ as electrophilic reagents. (From a similar point of view, Usanovich classified HQ and RQ as acids.[10]) In so doing, Ingold pointed out briefly (ref. 9, p. 269) that electrophilic reagents can be subdivided into classes in various ways, one such subdivision being into two types which may be called associative acceptors (*e.g.*, $BF_3$) and dissociative acceptors (*e.g.*, HQ or RQ).

In a preceding paper,[2] a classification of donors and acceptors into a number of distinct types was outlined. This has been somewhat revised, extended, and clarified in Tables I–VI below. Using this classification, the present formulation (like Ingold's) seems among other things to make it easier to treat under a unified scheme and viewpoint the action of one and the same donor molecule in such different activities as are expressed by, for instance

$$H_2O + BF_3 \longrightarrow H_2O{\rightarrow}BF_3 \qquad (2)$$

and

$$H_2O + HCl \xrightarrow{aq} H_3O^+aq + Cl^-aq \qquad (3)$$

The first action (with a good L-acid) is purely associative, the second (with a good H-acid) is an ionogenic displacement reaction; from the viewpoint of the H-acid acceptor, it is dissociative. What the two reactions have in common is (1) partial transfer of an electron from the donor $H_2O$ to the acceptor, and (2) formation of a (somewhat incomplete, or partial) dative bond between the donor and the acceptor. In the second reaction, of course, an additional thing happens: namely, splitting of the acceptor, during the acceptation process, into an ion $Cl^-$ and a (partially positive) H atom, it being the latter, rather than the acceptor as a whole, which unites covalently with the (strongly positively charged) $H_2O$ to form the dative bond.[11]

(11) How best to describe what happens in reactions like (3) has of course long been a moot question. See for example ref. 7, pp. 68, 114, on the similar case of $NH_3$ + HCl.

While the commonly used description of the second reaction as a proton transfer reaction is very convenient, it is open to two criticisms: (1) it is only formal or schematic, in the sense that at no time during the process of transfer is the proton really more than partially free from an electron (in other words, the proton carries a large fraction of an electron with it during its transfer); (2) it ignores the important fact of concurrent partial electron transfer from the $H_2O$. If (2) is ignored, the fruitful possibility of a unified common classification of the two reactions as electron donor–acceptor reactions is thrown away. A more detailed analysis of reaction (3) is given in Section VII.

Here it may further be noted that Luder and Zuffanti in describing H-acids as secondary Lewis acids have adopted the same proton transfer viewpoint as was used by Brønsted and Lowry. In the present viewpoint (as also apparently in Ingold's concept of electrophilic reagents), the H-acids HQ, together with other molecules RQ, are classified directly as acceptors in their own right along with primary Lewis acids, such as $BF_3$. According to this viewpoint, while it is true that an acceptor like HCl can be regarded formally as a compound of another acceptor $H^+$ and a donor $Cl^-$, physically according to quantum mechanics the structure is believed to be much more nearly covalent than ionic.

Since in ordinary chemistry the free ion $H^+$ does not occur (though to be sure it can exist in gas discharges), it appears to be more realistic to classify it as a virtual than as an actual acceptor under normal conditions.

## II. The Classification of Donors and Acceptors and Their Interactions

A general classification of donors and of acceptors each into several fairly well-marked types is given in Tables III and IV in the Appendix; each type is there characterized in detail. An orienting survey is given in Table I in this Section. A general scheme displaying the chief modes of interaction of some of the most important donor–acceptor pair types is presented in brief condensed form in Table II, and in detail in Table V in the Appendix. Table V is supplemented in Table VII in the Appendix by a listing of numerous individual examples, together with remarks in the case of certain pair-types about special features of their behavior. The examples include cases of molecular complex formation (D·A), molecular compound formation (D→A), and bimolecular displacement reactions (A + D → B + C) between donors D and acceptors A. (The use of the words "complex" and "compound" here is not intended as a definitive proposal.)

The notation used in the Tables is rather fully explained in their footnotes, but a few general remarks about it here may be worth while. First of all, the symbols have been chosen with considerable care to be as simple and brief as is consistent with making them reasonably self-explanatory, convenient for speaking, writing or printing, and free from possibilities of confusion with other symbols likely to be used in the same context.

Table I

Types of Donors and Acceptors[a]

1. Donors ($D$)

| | | | | |
|---|---|---|---|---|
| Major Types: | $n$ (onium) | $b\pi$ ($\pi$ donor) | $b\sigma$ ($\sigma$ donor) | R (radical) |
| Some sub-types: | $n$ (neutral onium)<br>$n'$ (onium anion) | | | |

2. Acceptors ($A$)

| | | | | |
|---|---|---|---|---|
| Major Types: | $v$ (vacant-orbital) | $\pi$ | $\sigma$ | Q (radical) |
| Some sub-types: | $v$ (neutral $v$)<br>$v^*$ ($v$ cation) | $x\pi$ ($\pi$)<br>$k\pi$ (ketoid $\pi$) | $x\sigma$ (halogenoid $\sigma$)<br>$h\sigma$ (H-acid)<br>$k\sigma$ ($\sigma$)<br>$h\sigma^*$ (cationic H-acid)<br>$k\sigma^*$ ($\sigma$ cation) | |

[a] For further details, including subclassification according to associative and dissociative modes of functioning (denoted by subscripts *a* and *d*, respectively), see Tables III and IV in the Appendix (see also Table II). The $n$ donors and $v$ acceptors always function associatively.

While the broad categories of "donor" and "acceptor" denote modes of functioning, it is convenient to divide donors and acceptors each into classes based on their structure before interaction. The first broad division of donors is into even lone-pair or onium ($n$ and $n'$), even bonding-electron ($\pi$ and $\sigma$), and odd-electron (R) radical donors. Similarly, acceptors are classified as even vacant-orbital ($v$ and $v^*$), even bonding-electron ($\pi$, $\sigma$ and $\sigma^*$), and odd-electron (Q) radical acceptors. The individual classes mentioned (there are also others, but those mentioned are the most important) can conveniently be further divided in some cases into subclasses (for example, the $\sigma$ acceptor subclass $h\sigma$ consists of all neutral molecule H-acids). The boundaries between classes (or subclasses) are not always sharp, because the structures of actual donors and acceptors are often more or less intermediate between those of two or more classes (see, for example, Table IV, Remarks column). It should also be noted that the same molecule, especially if it is a large molecule with various parts, may function under different circumstances sometimes as one kind of a donor, sometimes as another; or again as one or another type of acceptor. For example, even so small a molecule as $H_2O$ functions on occasion either as an $n$ donor or as an $h\sigma_d$ acceptor; and probably sometimes as a $b\sigma_d$ donor, or even perhaps in other ways.

For each of the structure-based classes or subclasses of donors and acceptors, there is at least one, and there are often two, characteristic modes of functioning or behavior. When there are two modes of functioning, one of these is associative, the other is dissociative (indicated by a and d in Table II). In Tables III–VI in the Appendix, these are symbolized by adding subscripts (a or d) to the class or subclass symbol. The combined symbols are then regarded as denoting subclasses or sub-subclasses. For some classes or subclasses, only the associative, or else only the dissociative, mode is usual.

For a given donor or acceptor functioning in the dissociative mode (this most often occurs only in the presence of "environmental coöperation," that is, assisting electrostatic or other forces or agencies, for example, those due to an ionizing solvent: see Section VI), one of its covalent bonds is broken, commonly in a displacement reaction in which ions are formed or exchanged. Actual separation of the atoms or ions formerly joined by the broken covalent bond does not necessarily occur, however. It does *not* occur when the $\pi$ bond of a double bond is broken ($b\pi_d$ donors, and $k\pi_d$ acceptors), since a $\sigma$ bond then always remains. Even in the case where a $\sigma$ bond (single bond) is broken, if two ions are formed, they may still cohere if in the presence of a non-ionizing solvent or in the vapor state (but they are seldom formed under these circumstances), or if in an ionic crystal; commonly, however, they become separated by the process of electrolytic dissociation in an ionizing solvent.

For a donor or acceptor functioning in the associative mode, a new covalent bond, either incipient (in loose complexes) or more or less fully developed, and either interatomic or intermolecular, is formed. Only lone-pair donors and vacant-orbital acceptors are capable of functioning in the associative mode to form strong fairly fully developed new covalent bonds. When bonding-electron ($\sigma$ and $\pi$) donors or acceptors function

Table II

Major Donor-Acceptor Reaction Types

| Donor Types | Acceptor Types: $v$ | $v^*$ | $x\pi$ | $k\pi$ | $x\sigma$ | $h\sigma$ | $k\sigma$ | $h\sigma^*$ and $k\sigma^*$ |
|---|---|---|---|---|---|---|---|---|
| $n$ | aa | aa | aa | ad | aa<br>$ad_s$ | $ad_s$ | $ad_s$ | $ad_r$ |
| $n'$ | aa | aa | aa<br>ad | aa<br>ad | aa | aa<br>$ad_r$ | $ad_r$ | $ad_r$ |
| $b\pi$ | aa | aa<br>da | aa | aa | aa<br>$dd_s$ | aa<br>$dd_s$ | | |
| $b\sigma$ | $da_s$ | $da_r$ | | | aa | | | |

*Explanation.* For each entry in the table, the first symbol refers to the donor, the second to the acceptor. The symbol a denotes associative, d dissociative, behavior. In associative behavior of a donor or acceptor, all the original bonds remain unbroken (and in some cases a new bond is formed). In dissociative behavior of $\sigma$ donors or acceptors, a single bond ($\sigma$ bond) is broken during the reaction; for $\pi$ donors or acceptors, the $\pi$ component of a double bond is broken, but since a single bond then remains, the link is not split. Processes involving dissociative behavior usually require environmental assistance (*cf.* Section VI). The subscript s denotes that a salt (undissociated or dissociated) is formed. The subscript r means that a displacement reaction occurs. In all cases where there is no subscript, the donor and acceptor cohere upon interaction. (However, often the product may react further.) For further details, see Tables V and VI.

associatively, it is only with a loosening (partial breaking) of their bonds; if the interaction becomes really strong (usually under environmental coöperation), it passes over into the dissociative mode, with more or less complete rupture of one bond.

The class and subclass symbols have been so chosen that they can be simplified, adapted, or extended in various ways according to convenience. For example, one may speak of $b\pi$ reagents, or of $b\pi$ or simply $\pi$ donors; one may speak of $x\sigma$ acceptors, or collectively of $\sigma$ acceptors (including $x\sigma$, $h\sigma$, $k\sigma$, $l\sigma$, $h\sigma^*$, and $k\sigma^*$, acceptors), or collectively of $\sigma_d$ acceptors. The notations $v^*$ and $\sigma^*$ for unipositive cation acceptors and $n'$ for anion acceptors can readily be extended to multiply charged ions (*e.g.*, $v^{**}$, $v^{***}$, $n''$, $x\pi^*$, etc.). The superscripts * and ′ have been used, rather than + and −, because in some discussions one may wish to refer to donors and acceptors of these types to or from which electrons have been added or subtracted (for example, $n'^{+}$ or $v^{*-}$).

Finally, it should be pointed out that several changes, believed to be considerable improvements, have been made in the notation used in previous papers.[2,4] Among other changes, D (for "donor") has been substituted for B (which suggests "base"), and the acceptors formerly called $d$ have now become the subclasses $x\sigma$, $h\sigma$, and $k\sigma$ of the $\sigma$ class of acceptors.

It should be emphasized that the classification scheme as here proposed is still more or less tentative and incomplete. In particular, the classes R and Q and their interactions have been treated only sketchily. It should also be emphasized that the primary purpose of the detailed Tables in the Appendix is not to attempt an authoritative classification of actual molecules and reactions, but rather to give probable or plausible examples to illustrate the functioning of the various donor and acceptor types. The writer hopes he may be forgiven if some of the examples appear unrealistic to experts who are much more familiar than he with the relevant experimental facts.

Time alone can show just how useful a detailed classification such as that given here may be, or to what extent, or in what areas or what connections, this or a similar classification may have value.

## III. Loose Molecular Complexes in the Solid State

The structure of any loose 1:1 molecular complex or compound between *neutral closed-shell entities* can be described in terms of wave functions essentially as[2]

$$\psi_N \approx a\psi_0(D, A) + b\psi_1(D^+\text{–}A^-) \qquad (1a)$$

where $\psi_0(D, A)$ represents a no-bond structure and $\psi_1(D^+\text{–}A^-)$ a dative structure for the donor–acceptor pair D, A. A more general expression not limited to neutral closed-shell entities has already been given at the beginning of Section I as eq. (1).

Equation (1a) is easily generalized to cover $n$:1 and other cases.[2] The fact that solid crystals are abundantly known[12] with the same 1:1 (or in general $m$:$n$) composition as for individual molecules of complexes in solution or in vapor can also be understood. It is only necessary to assume that even in such a crystal (the same applies also to an individual $n$:1 or $m$:$n$ complex) the predominant intermolecular forces are local, pair-wise, donor–acceptor interactions between each donor molecule and its nearest sufficiently near acceptor neighbors, and between each acceptor molecule and its nearest sufficiently near donor neighbors. This assumption is quantum mechanically entirely reasonable, and it appears safe to take the fact of the very frequent occurrence of complexes in solution and as crystalline solids with the same stoichiometric composition, as very strong empirical evidence of its correctness. Further evidence from the spectra of solid complexes will be reviewed in Section IV. Cooperative effects involving more remote neighbors (and finally the entire crystal) must also of course exist, but apparently these are of secondary importance in typical cases.

As has been pointed out previously,[2,3] donor–acceptor interactions, even though relatively weak, should exist also in molecular crystals built from a single molecular species, and often these forces should have orientational properties. In particular they should tend to cause aromatic molecules to be stacked in such a way that the planes of adjacent molecules, although often parallel, are displaced from being directly superposed.[13] This is what is observed in the crystals both of aromatic molecules of a single species and of aromatic molecular complexes. Such molecules are often stacked like a pack of cards along an axis, but with the planes of the molecules all inclined to the stacking axis.[14] Examples of the two cases are the crystals of hexamethylbenzene (stacking angle 44°27′) and quinhydrone (1:1 quinone–hydroquinone; stacking angle 34°). On the other hand, for the crystals of $(CH_2Br)_6C_6$ (hexabromomethylbenzene) the molecules are strung directly above one another along the axis (stacking angle 0°). This is understandable on the basis of steric effects, including the fact that the bromine atoms are so big as to keep the benzene planes much farther apart than in hexamethylbenzene and thus greatly to reduce the charge-transfer orientational forces.

(12) P. Pfeiffer, "Organische Molekülverbindungen," 2nd edition, F. Enke, Stuttgart, 1927

(13) Reference 2, Secs. VII, VIII; and J. Landauer and H. McConnell, *J. Am. Chem. Soc.*, **74**, 1221 (1952).

(14) K. Nakamoto, *ibid.*, **74**, 390, 392, 1739 (1952). These papers include a convenient brief survey of several examples illustrating the modes of packing of aromatic molecules in crystals.

### IV. Charge Transfer Spectra

As has been pointed out previously,[2] if eq. (1a) represents the ground state N of a 1:1 molecular complex, there must also exist an excited state E with

$$\psi_E \approx a^*\psi(D^+\text{–}A^-) - b^*\psi(D, A) \qquad (4)$$

Then if, for example, the ground state has predominantly no-bond structure ($a^2 \gg b^2$), this excited state must have predominantly dative structure. It was also pointed out that on the basis of quantum mechanics the absorption spectrum of the complex must include (in addition to the individual spectra of D and A, somewhat modified by their interaction) a band (normally in the visible or ordinary ultraviolet) corresponding to an absorption jump from state N to state E, and characteristic of the complex as a whole. This was called an intermolecular charge transfer spectrum. Especially notable is the fact that the theory predicts the possibility of highly intense charge transfer absorption even for very loose complexes.[2]

As a prime example, the intense absorption near λ2900 in the spectra of the loose complexes of benzenoid hydrocarbons with molecules of the halogens was discussed in detail in ref. 2. Tentative identification of an intense ultraviolet absorption of solutions of iodine in ethyl ether as being a charge transfer spectrum of an ether–iodine complex was also made.[4] Work just completed by Mr. J. S. Ham in this Laboratory, showing that this spectrum definitely belongs to a 1:1 ether–iodine complex, supports this view. Mr. Ham has also measured a new ultraviolet absorption in the spectrum of solutions of iodine in *t*-butyl alcohol, and has identified it as a charge transfer spectrum of a 1:1 complex.

During the past year, other investigators (especially Andrews and Keefer) have published several papers on molecular complexes and their spectra, in which equilibrium constants were determined and which revealed the possible or probable presence of charge transfer spectra. The wave lengths of a number of such spectra are listed under the appropriate complexes in Table VI, together with literature references. In many cases, the charge transfer identification was mentioned tentatively by the authors of the papers cited.

In addition, there are numerous less recent papers on organic complexes in which color changes have been noted, and a few for which spectra have been mapped. Further, attention should be called to the absorption spectra of inorganic complexes and ions in solution and in crystals, a subject briefly reviewed by Rabinowitch in 1942.[16] Most of the spectra of stable complex anions such as $NO_3^-$, $CH_3COO^-$, $MnO_4^-$, and of complex cations of similar stability, may best be considered as normal molecular spectra, but it may be noted that normal molecular spectra include *interatomic* charge transfer spectra.[17] In addition, anions in solutions, in particular simple ions like $Cl^-$, contain spectra which have long been attributed to electron transfer from the anion to environing molecules, but to which Platzmann and Franck have recently given a new and rather different interpretation.[18] Earlier (1926), Franck and collaborators had shown that the absorption spectra of alkali halide vapors correspond to an interatomic electron transfer process. Still earlier, the concept of electron transfer was used in interpreting the spectra of ionic crystals.[16] The high intensity short wave length absorption spectra of many of the less stable cations and anions in solution have been described by Rabinowitch as electron transfer spectra.[16]

In the earlier work, it was supposed that electron transfer spectra occur only corresponding to transfer of an electron from a negative ion (or perhaps an electron donor like $H_2O$—*cf.* Table V of ref. 16) to a positive ion. The more recent work[17,2] corresponds to a broader concept in which electron transfer spectra may occur corresponding to electron transfer between any two atomic or molecular entities, even if these are uncharged. It appears probable that a re-examination of the structure and spectra of complex ions from the present point of view will be very fruitful.

Returning to the organic complexes, a recent paper by Nakamoto[14] is of particular interest. Nakamoto examined long wave length spectra which appear to be the charge-transfer spectra of certain 1:1 crystalline molecular complexes (among others, quinone:hydroquinone), in polarized light. For 1:1 complexes between aromatic or unsaturated $\pi$ donors and $\pi$ acceptors, the theory[2] predicts that the electric vector should be polarized with a large component perpendicular to the planes of two interacting adjacent molecules which are parallel to each other, and this is what Nakamoto found. As was pointed out in Section III, it is reasonable to suppose that the charge transfer forces in such crystalline complexes act essentially pairwise between neighboring molecules. The spectra should then be similar to those for individual 1:1 complexes in solution, although appreciably modified by co-operative effects involving non-neighbor molecules. Thus Nakamoto's work appears to support both the charge transfer interpretation of the characteristic spectra of molecular complexes, and the idea of primarily pairwise-acting charge transfer forces in crystalline complexes.

A further point of interest is that in aromatic crystals composed of a single molecular species (*e.g.*, hexamethylbenzene), although the stacking of the molecules at an angle to the stacking axis

(16) E. Rabinowitch, *Rev. Modern Phys.*, **14**, 112 (1942). Review on electron transfer and related spectra.

(17) R. S. Mulliken, *J. Chem. Phys.*, **7**, 201 (1939), and later papers, especially R. S. Mulliken and C. A. Rieke, *Reports on Progress in Physics* (London Physical Society), **VIII**, 231 (1941).

(18) R. Platzmann and J. Franck, *L. Farkas Memorial Volume.*

(*cf.* Section III above) is explainable by charge transfer forces (and no other explanation is evident), there is in the longer wave length part of the spectrum no charge transfer absorption band like that for quinone–hydroquinone. This is theoretically not unreasonable, since for self-complexes[13] the charge transfer forces should be much weaker, and the charge-transfer absorption in general much weaker and at shorter wave lengths, than for the much stronger complexes between unlike molecules. Whether, however, the existence of sufficiently strong charge transfer forces to account for the stacking of the molecules of a self-complex at a considerable angle is quantitatively compatible, in terms of theory, with the absence of any indication of a charge transfer spectrum, is a matter which should be investigated further. Meantime, it may be useful to assume this compatibility at least as a guiding hypothesis in further experimental studies.

## V. The Strengths of Donors and Acceptors

The factors determining the strengths of donors and acceptors in the formation of relatively loose addition complexes have been discussed previously.[2,3] The importance of low vertical ionization potentials $I^{vert}$ for strong donors, and of high vertical electron affinities $E^{vert}$ for strong acceptors, was stressed. It was pointed out that $I^{vert}$ and $E^{vert}$ should be taken corresponding to a nuclear skeleton which is that of the actual complex. This is a compromise (often a severe one) between the often very different skeletal structures that would occur for molecules with electronic structures corresponding to the separate resonance components $\psi_0$ and $\psi_1$ of eq. (1a). The importance of other factors, in particular mutual approachability, in determining interaction strengths, was also emphasized. For example, approachability is especially good between $n$ (or $n'$) donors and $v$ (or $v^*$) acceptors, or between $\pi$ donors and $\pi$ acceptors, but not between $\pi$ donors and $v$ acceptors.

When donors and acceptors interact not in the associative but in the dissociative mode, the factors determining donor and acceptor strengths are altered considerably. Leaving aside entropy factors for the moment, the important factors can be seen by writing equations for the various terms involved in the net heat of reaction. For ionogenic displacement reactions like (3), one immediately finds that low $I$ (here not $I^{vert}$) values are favorable for good donors, and high $E$ (not $E^{vert}$) for good acceptors.

Specifically, for the reaction of an associative donor D with a dissociative $\sigma$ acceptor RQ

$$D + RQ + sl \longrightarrow DR^{+}sl + Q^{-}sl \qquad (3a)$$

the heat of reaction is evidently

$$\left.\begin{array}{l} E_Q + H_{Q^-} + H_{DR^+} + (B_{DR^+} - B_{RQ}) - I_D \\ = (H_{DR^+} + B_{DR^+} - I_D) + (H_{Q^-} - B_{RQ} + E_Q) \end{array}\right\} \qquad (3b)$$

where the $B$'s denote bond dissociation energies and the $H$'s solvation energies. Similarly, for the reaction of a dissociative $\sigma$ donor RQ with an associative acceptor A (*e.g.*, a $v$ acceptor)

$$RQ + A + sl \longrightarrow R^{+}sl + QA^{-}sl \qquad (3c)$$

the heat of reaction is

$$\left.\begin{array}{l} E_A + H_{R^+} + H_{QA^-} + (B_{QA^-} - B_{RQ}) - I_R \\ = (H_{R^+} - B_{RQ} - I_R) + (H_{QA^-} + B_{QA^-} + E_A) \end{array}\right\} \qquad (3d)$$

In eqs. (3b) and (3d), it is not $I_D^{vert}$ and $E_A^{vert}$ which matter, but $I_D$ and $E_Q$ in the former, $I_R$ and $E_A$ in the latter. The difference between the energy of the bond which is formed and that which is broken also has some influence. It is easily seen why solvation of the ions is so often the decisive factor in making ionogenic reactions possible.

As stressed by Lewis,[5] it is not feasible to arrange bases and Lewis acids into unique orders of strength valid for all Lewis acid–base reactions. It is usually concluded as a corollary that it is futile to try to arrange donors or acceptors in any universally valid quantitative orders of strength, except for acid and base strengths in the familiar case of H-acids interacting with bases in solutions. However, with the present classification of donors and acceptors into a number of fairly well-marked types, perhaps it will be worth while to see whether roughly quantitative scales of donor and acceptor strength can be set up for the interactions of donors of a particular type with acceptors of a particular type; and then to look for different scales for other pairs of types.

## VI. Conditional and Unconditional Donor and Acceptor Behavior[19]

Beyond the classifications already mentioned, donor–acceptor reactions may be characterized as either unconditional or absolute (occurring between the members of the donor–acceptor pair without assistance, for example, in vapor or in inert solvents); or as conditional or contingent (requiring the presence of environmental coöperation).

Environmental coöperative action[20] may take various forms, among which it will be convenient to distinguish two principal cases. One of these, designated "es" in Tables I–VI (see footnote *c* of Table III), is that of essentially electrostatic environmental action. In the other, two molecules coöperate in reacting as acceptors with a third molecule as donor, or *vice versa* (for examples, see Tables V, VI).

Usually "es" involves either attachment of solvent molecules to the donor–acceptor reaction product (that is, solvation; usually with dissociation into solvated ions), or else polymerization (formation of ion-clusters in solution, or of an ionic crystalline solid). The role of es in conditional reactions is to stabilize the reaction products sufficiently to make the reaction possible.

Dissociatively functioning donors ($b\pi_d$ and $b\sigma_d$ donors) and acceptors ($\pi_d$, $\sigma_d$ and $\sigma^*_d$ acceptors) seldom behave dissociatively without environmental coöperation; in other words, they are *usually* conditional.

(19) For a general survey of reaction rates and mechanisms, especially in solution, reference may be made to L. P. Hammett, "Physical Organic Chemistry," McGraw-Hill Book Co., Inc., New York, N. Y., 1940, and to Glasstone, Laidler and Eyring, "The Theory of Rate Processes," McGraw-Hill Book Co., Inc., New York, N. Y., 1941.

(20) See also E. D. Hughes' discussion (*Trans. Faraday Soc.*, **34**, 185 (1938)) of "constitutional effects . . ." and "environmental effects in nucleophilic substitution."

It will be noted that the terms "conditional" and "unconditional" describe alternative modes of functioning of donors and acceptors, just as do the terms "dissociative" and "associative," introduced in Section II and indicated in Tables V–VI by subscripts d and a. It would therefore be appropriate to add further subscripts, say c and u, for conditional and unconditional modes, respectively, leading to symbols such as $x\sigma_{ua}$, $x\sigma_{cd}$, $h\sigma_{cd}$, etc. In most cases, however, only one of the alternatives u or c occurs for any one donor or acceptor class and a or d subclass, and so these subscripts have been omitted in the symbols in the Tables.

Nevertheless under some circumstances, to indicate special types of conditional functioning, it may at times prove useful to employ subscript symbols. Symbols such as $x\sigma_{cd}$, $x\sigma_{es.d}$, $x\sigma_{vd}$, $x\sigma_{sl.d}$, $x\sigma_{xt.d}$, are therefore suggested to denote, respectively, conditional functioning of unspecified character, of electrostatic character, of auxiliary $v$-acceptor character, of solvent character (which in turn may be of pure es, donor or acceptor, or mixed, character), and ionic-crystal-es character; with at the same time dissociative functioning in all.

The $n'$ donors and the $v^*$ and $d^*$ acceptors, being ionic, are seldom encountered except under es conditions. However, in Tables III and IV, this fact has been embodied in the definitions of these types. For example, the anionic chlorine donor is defined as $Cl^-$es, which may mean $Cl^-$ surrounded by $Na^+$ (and so on) in a crystal, or $Cl^-$ attached to a single $Na^+$ in vapor or in an inert solvent, or hydrated $Cl^-$ ($Cl^-$aq) in water solution. With these definitions, $n'$ donors and $v^*$ and $d^*$ acceptors are here regarded as usually unconditional.

A further point is that the "es" in an $n'$ donor or $v^*$ acceptor is intended in general to have qualitative rather than quantitative meaning, so that, for example, it is not necessary to account for a fixed number of water molecules when a $Cl^-$aq donor reacts.

Another related point is the fact that when an ion is solvated, the solvating solvent molecules may function either in a purely es role as discussed in the two preceding paragraphs, or in a more or less strongly donor or acceptor role as, for example, the two $NH_3$ in $Ag^+(NH_3)_2$aq, and the $SO_2$ in $Cl^-SO_2$. If the solvent molecules function definitely as donors or acceptors, they are conceived of here as forming integrated structural parts of larger entities, such as, for example, the $\sigma^*_d$ acceptor $Ag^+(NH_3)_2$aq, or the $\sigma'_d$ donor $Cl^-SO_2$. In practice, of course, border-line cases are frequent. In such cases, it may be best to be guided by convenience, and to some extent custom, even at the risk of arbitrariness. For example (*cf.* Table IV), $Ag^+$aq is most conveniently regarded as a $v^*$ acceptor, rather than as a $\sigma^*_d$ acceptor like $Ag^+(NH_3)_2$aq, even though it may be that the $\sigma^*_d$ classification would be closer to the truth. Similarly, $Cl^-SO_2$ may often conveniently be regarded like $Cl^-$aq as an $n'$ rather than as a $\sigma'_d$ donor. On the other hand, it is probably unwise to class the typical stable $\sigma^*_d$ acceptor $H_3O^+$aq as a $v^*$ acceptor ($H^+$aq).

Conditional $\sigma_d$ acceptors ($\sigma_{cd}$ acceptors) like HQ or RQ are, *by themselves*, strictly speaking, not acceptors or Lewis acids at all; it is really not they, but HQ plus es or RQ plus es, which function as $\sigma_{cd}$ acceptors. The usual omission of reference to es may be considered as a convention adopted in the interest of simplicity. It is particularly important to remind oneself of this when using the ordinary concept of H-acids, a concept of whose very existence this tacit convention is an intrinsic part.

Some types of donors and acceptors typically function only unconditionally, and others only conditionally. Acceptors of the $x\sigma$ subclass, however, although unconditional when functioning associatively ($x\sigma_{ua}$ acceptors), often act in the presence of ionizing solvents as conditional acceptors ($x\sigma_{cd}$ acceptors). Thus if iodine is dissolved in water, it seems probable that unconditional formation of the loose reversible $n{\cdot}x\sigma_a$ complex $H_2O{\cdot}I_2$ first occurs, very rapidly; and that then, with the es assistance of water, reaction occurs over an activation barrier to a structure $H_2OI^+I^-$ (of $n{\to}x\sigma_d$ type), thence to $H_2OI^+$es + $I^-$es; whereupon the $\sigma^*_d$ acceptor $[H_2OI]^+$es is attacked by the $n$ donor $H_2O$ to form HOI + $H_3O^+$aq; and so on. There seems to be evidence that in this complex and reversible series of donor–acceptor reactions, the iodine is present predominantly as $H_2O{\cdot}I_2$. In the example just discussed, it will be noted that the water acts sometimes in an es role, sometimes as an $n$-donor.

## VII. Detailed Comparison of Dissociative and Associative Donor–Acceptor Reactions

To clarify further the brief comparison in Section I between associative and dissociative charge transfer reactions, it is instructive to follow each of the reactions eq. (2) and (3) through from beginning to end in terms of a varying linear combination of two resonance structures $\psi_0$ and $\psi_1$ of the kind specified in eq. (1a). In both reactions, $a$ of eq. (1a) decreases and $b$ increases as the reaction proceeds. To make matters fully clear, additional details are needed to describe the metamorphoses, during reaction, of the internal structures of the original donor and acceptor in $\psi_0$ and of their ions in $\psi_1$.[11,19]

For reaction (3), eq. (1a) becomes

$$\psi_N \approx a\psi_0(H_2O,HCl) + b\psi_1(\underset{aq}{H_2O^+\text{—}HCl^-}) \quad (4)$$

Here the formulation in terms of $HCl^-$ is admittedly artificial, especially since $E^{vert} \ll 0$ in the first stages of the reaction (see Table VI, discussion under reaction-type 14). Actually, with detailed resonance structures given as in (5) below, no mention of $HCl^-$, or of $\psi_0$ or $\psi_1$, is really essential. However, the introduction of the concepts which these respresent is valuable in showing the parallelism between reactions (2) and (3).

Neglecting some minor resonance structures, and leaving to be tacitly understood the presence and es action of the solvent, reaction (3) must go somewhat as follows

$$\psi(H^{+0.4}H^{+0.4}O^{-0.8}, H^{+0.2}Cl^{-0.8}) = \psi_0(H_2O, HCl) =$$

$$\psi_0{:}\begin{Bmatrix} 0.8\ H_2O & & H\text{———}Cl \\ 0.2\ H_2O\cdots\cdots & & H^+\cdots Cl^- \end{Bmatrix} \longrightarrow$$

$$\left\{\begin{matrix} \psi_0{:}\begin{Bmatrix} 0.39\ H_2O \leftrightarrow H\text{———}Cl \\ 0.35\ H_2O\cdots\cdots H^+\cdots\cdots Cl^- \end{Bmatrix} \\ \psi_1{:}\begin{Bmatrix} 0.25\ H_2O^+\text{——}H \leftrightarrow Cl^- \\ 0.01\ H_2O^+\cdots\cdots H^- \leftrightarrow Cl^- \end{Bmatrix} \end{matrix}\right\} \longrightarrow \quad (5)$$

$$\begin{Bmatrix} 0.5\ \psi_0{:}\ H_2O\cdot\cdot H^+\cdots\cdots\cdots Cl^- \\ 0.5\ \psi_1{:}\ H_2O^+\text{—}H\ \cdots\cdots\ Cl^- \end{Bmatrix} =$$

$$\psi(H^{+0.5}H^{+0.5}O^{-0.5}H^{+0.5}\cdots\cdots Cl^{-1}) = \psi(H_3O^+\cdots\cdots Cl^-) = \psi[(H_2O)^{+0.5}(HCl)^{-0.5}].$$

The numbers preceding the individual resonance structures are rough estimates or guesses of their relative weights. The symbols —, ···, and ↔ indicate covalent bonding, electrostatic attraction, and non-bonded repulsion, respectively. The first and last lines contain various types of summary of the estimated detailed charge distributions in the initial and final stages.[21]

A comparison between various formulations of the initial and final stages in (5) permits various descriptions of the over-all effects of the reaction. One point which is interesting is that although the $H_2O$ molecule as a whole donates an estimated $0.5e$ to (the two parts of) the HCl, the O atom in $H_2O$ donates only $0.3e$ of this, and moreover still remains negatively charged in the final ion $H_3O^+$. Also, the Cl atom, in attaining its final charge of $-1.0e$, starts with an estimated $-0.2e$ taken from its original partner, picks up a further $-0.3e$ from this partner during the action, and gains $-0.5e$ more from the original $H_2O$, of which the O supplies $-0.3e$ and the two hydrogens each $-0.1e$.

In scheme (5), the intermediate stage shown probably represents an activated state. In an analogous formulation of some of the other reactions in Table VI (see in particular reaction-types 13, 14 and 17 there), the intermediate stage certainly corresponds to an activation barrier over which the reaction proceeds slowly (rate measurements have been made in many examples). The fact that reactions such as $H_2O + HCl \rightarrow H_3O^+Cl^-$ do not occur in the vapor phase probably means that without solvent or crystallization assistance the final fully ionic stage of the reaction, although presumably lower in energy than the intermediate activated state, would be higher in energy than the initial stage before reaction. The way in which the energy varies with degree of reaction would then correspond to curve I in Fig. 1, below (see Section IX for a further analysis).

For comparison with (5), a similarly formulated description of the initial and final stages of reaction (2) is given in (6). Again the reader is warned that the estimated numbers are uncertain (though enlightened) guesses.

$$\psi(H^{+0.4})_2O^{-0.8},\ B^{+1.8}(F^{-0.6})_3 = \underset{\text{planar}}{\psi_0(H_2O,\ BF_3)} \longrightarrow$$

$$\left\{\begin{matrix} 0.2\psi_0\colon\ H_2O \cdots BF_3 \\ 0.8\psi_1\colon\ H_2O^+ \overline{\cdots} BF_3^- \end{matrix}\right\} = \qquad (6)$$

$$\psi(H^{+0.6})_2O^{-0.4},\ B^{+1.3}(F^{-0.7})_3 = \underset{\text{pyramid}}{\psi[(H_2O)^{+0.8}(BF_3)^{-0.8}]}$$

The fact that $H_2O{\cdot}BF_3$ is an extremely powerful H-acid, *i.e.*, $h\sigma_d$ acceptor, yielding, with a base D, $DH^+$es + $(HOBF_3)^-$es, is readily understandable if the charge distribution in the final product is somewhat as shown.

## VIII. Symmetrization

Attention should be called briefly to a familar phenomenon which is characteristic of the final stages of many donor–acceptor interactions. When for instance a donor of structure $R_nZ$ interacts with a $k\sigma_d$ acceptor of structure RQ to give $R_{n+1}Z^+ + Q^-$, the original donor and acceptor at first approach without special cognizance of the common possession of R atoms; but during the last stages of the reaction, more or less internal readjustment takes place in the $R_nZ$ structure in such a way that all $n + 1$ R atoms become equivalent, with some extra gain in stability thereby. Processes of this kind and of the type $RQ + YQ_n \rightarrow R^+ + YQ_{n+1}^-$ may be called *symmetrization* processes. Some typical examples (sl = solvent) are

$$\left.\begin{matrix} RX + BX_3 \longrightarrow R^+sl + BX_4^-sl \\ n\,XR + n\,NR_3 \longrightarrow (NR_4^+X^-)_n\ \text{solid} \\ Cl^-aq + ICl \longrightarrow ICl_2^-aq \\ Br^-sl + AlCl_3 \longrightarrow AlCl_3Br^-sl \end{matrix}\right\} \qquad (7)$$

The last example is typical of the frequently occurring phenomenon of partial, or near, symmetrization.

## IX. Inner, Outer and Middle Complexes

**Energy of Donor–Acceptor Pair as Function of Charge Transfer Coördinate.**—In a general consideration of the possible modes of interaction of a D,A (donor–acceptor) pair, it is instructive to plot the energy of interaction $U$ against a reaction coördinate $C$. Without defining it precisely, let $C$ be a quantity which increases continuously with charge transfer (that is, with $b/a$ in eq. (1a)), subject to the added specification that the nuclear skeleton be so adjusted for each value of $C$ as to make $U$ as small as possible. Thus $C$ may be called a charge transfer reaction coördinate, or simply a charge transfer coördinate.

A convenient scale for $C$ runs from $C = 0$ for $b/a = 0$ (D and A not yet in contact) to $C = 1.0$ for some state of maximum charge transfer. Figures 1 and 2 show several plausible forms for $U(C)$, in which the left-hand minimum of $U$ may be taken as defining $C = 1$, while in the region at the right, $C \rightarrow 0$ as D and A separate and $U \rightarrow 0$. For those curves in which a maximum of $U$ occurs between $C = 0$ and $C = 1$, the position of this maximum may conveniently be taken as defining $C = 0.5$.

As will be shown below, Figs. 1 and 2, or variations on them, are probably applicable to most D–A interaction types. For a few types, however, the $U(C)$ curves must look like those in Fig. 3, and it is then convenient to define $C$ as ranging from $-1$ through 0 to $+1$ as $b/a$ goes from 0 to 1 to $\infty$.

As compared with the geometrically defined reaction coördinates used by Eyring, Polanyi and others, the charge transfer coördinate $C$ is less concrete in that it is based on theoretical quantities which are not accurately known. Further, one cannot be sure that the normal path of every charge transfer reaction has to be one in which $C$ steadily increases; however, geometrically defined reaction coördinates are not necessarily free from a similar difficulty. In any event, $C$ is extremely convenient for present purposes, in that by its use the degree of completion of a donor–acceptor reaction is described in terms of a single coördinate.

**Inner and Outer Complexes.**—The broad division of the modes of functioning of donors and

(21) The structure $H_2O^+ \cdots H^- \leftrightarrow Cl$, with estimated coefficient only 0.01 in the intermediate stage in (5), is mentioned here since, although unimportant in this example, analogous structures should be of considerable importance in general (for example, in $H_2O + I_2$ and other $n + x\sigma_d$ reactions (*cf.* Table VI)). The importance of this structure relative to $H_2O^+$–$H \leftrightarrow Cl^-$ is probably at a maximum at the beginning of the reaction ($c$ nearly zero).

acceptors into associative and dissociative was discussed in Sections II and VI. It is useful at this point to introduce a further, derived, concept. Donor–acceptor *pairs* may be said to be functioning associatively if both partners are functioning associatively, or dissociatively if (at least) one partner is functioning dissociatively.

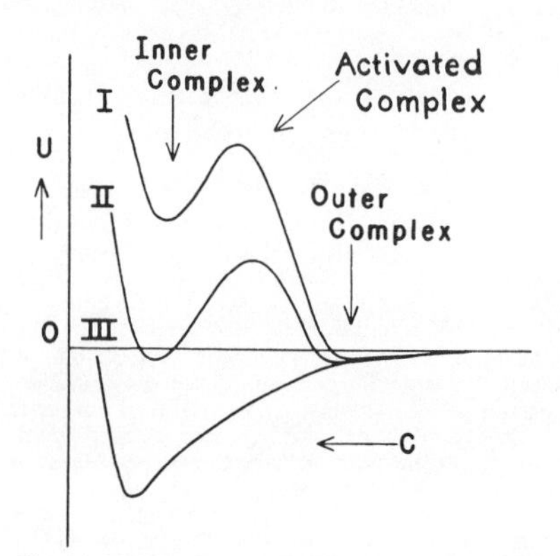


Fig. 1.—Energy curves $U(C)$ for unassisted interaction (inert solvent or none) between a donor D and an acceptor A. (Qualitative only; $C$ = reaction coördinate, increasing from 0 toward 1 as $b^2/a^2$ in eq. (3) increases.) The "outer complex" ($C$ small) corresponds to $b^2 << a^2$, the "inner complex" ($C = 1$) to $b^2 \approx a^2$ or $b^2 > a^2$, and the "activated complex" ($C = 0.5$) is intermediate.

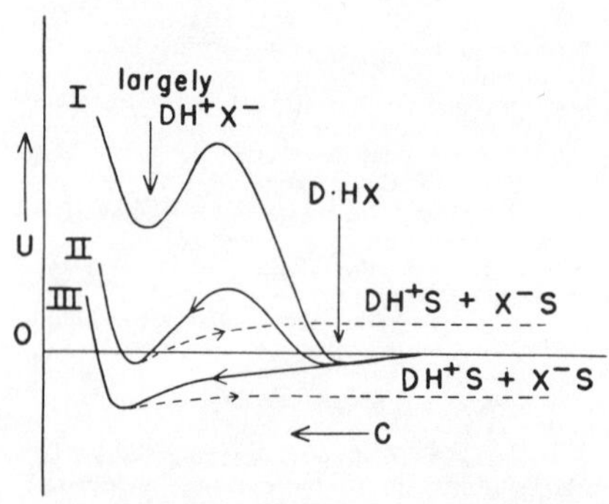


Fig. 2.—Energy curves $U(C)$ for assisted interaction between a donor D and a $\sigma_d$ acceptor HX (the curves would be similar for Group 3, 4 or 5 D,A pairs in general): curve I, unassisted; curve II, same assisted by a weak and rather low-dielectric solvent (dotted branch of curve is for dissociation of $DH^+X^-S$ into solvated ions (S = solvent)); curve III, like curve II but for strong and high-dielectric solvent.

Table II gives a survey of the usual observed behaviors of important types of D,A pairs. Behavior aa (see Table II) is associative. All other behaviors (ad, da and dd) are dissociative; it may be recalled here that dissociative functioning usually occurs only with environmental assistance. Pairs in which the donor is of class $n$ or $n'$ and the

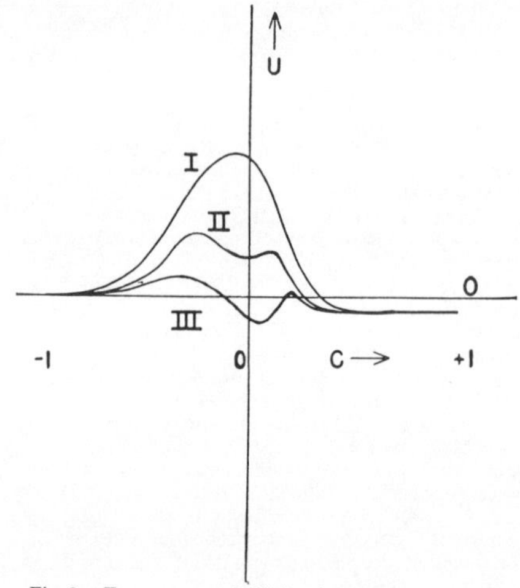


Fig. 3.—Energy curves $U(C)$ for charge transfer reactions of the type of eq. (8) and (9). "Middle complex" near $C = 0$ of curve III. $C = -1$, 0 and +1 here correspond more or less to $C = 0$, 0.5 and 1 of Figs. 1 and 2. In each curve, shallow depressions (outer complexes—*cf.* Figs. 1 and 2) near $C = \pm 1$ are an additional possibility.

acceptor of class $v$ or $v^*$ function only associatively; these will be called Group 1. For all other types of D,A pairs, either associative or dissociative functioning is in principle possible, and is frequently realized (see Table II). Among these latter types, dissociative functioning involves in some cases typical displacement reactions other than salt formation (subscript r in Table II); these will be called Group 2. In others it involves ion-pair (salt) formation (subscript s in Table II); these will be called Group 3. In still others (Groups 4 and 5) it involves formation of an addition product in which the dissociative action is the breaking of a $\pi$ bond of a double bond, leaving a $\sigma$ bond, however, so that the atoms formerly linked by the double bond are still held together. Group 3 includes the important types $n + \sigma$ and $b\pi + \sigma$ (see Sections 13, 14, 18, 19 of Table VI); some examples are

$$NH_3 + HCl \longrightarrow NH_4^+Cl^- \text{ (or } NH_4 + Cl^- \text{ in solution)}$$

$$Ar + HF + BF_3 \longrightarrow ArH^+BF_4^- \text{ in HF solution}$$

In Group 4 (types $n$ or $n' + \pi$, *cf.* Section 9 of Table VI), internal ionization occurs, while in Group 5 (type $b\pi + v^*$, *cf.* Section 7 of Table VI), an ionic complex or intermediate of considerable stability is formed.

A review of the behavior patterns just outlined and of other experimental facts suggests that those for Groups 3, 4 and 5 can be understood in terms of $U(C)$ curves similar to those shown in Figs. 1 and 2, with associative action corresponding to the shallow "*outer complex*," dissociative action to the deeper "*inner complex*," and with in general an "*activated complex*" corresponding to the maximum between these minima. Later in this Section, qualitative quantum theoretical reasoning will be presented in support of curves of the general forms indicated, but the discussion immediately following will be more in terms of empirical evidence.

The curves of Fig. 1 are intended to apply to D,A interactions *without* environmental assistance. Among the curves in Fig. 1, curve I may be fairly typical for Groups 3, 4 and 5, while curve III (with no distinction between inner and outer complex) is probably typical for stable Group 1 pairs (Groups 1 and 2 will be discussed further at the end of this Section). Figure 2 is intended to apply to D,A interactions of Groups 3, 4 and 5 with varying degrees of environmental assistance. Variations on Figs. 1 and 2 are of course also to be expected, including cases with almost no outer complex, or with no inner complex, or with an additional small activation barrier outside the outer complex.[22]

To make it more definite, Fig. 2 has been labeled to correspond to a Group 3 D,A pair. Curve I in Fig. 2 (which is essentially the same as curve I in Fig. 1) may be taken as typical for the case of little or no environmental coöperation. Curves II and III may then represent the same D,A pair in the presence of two degrees of strong environmental influence, for example of the electrostatic influence of a polar solvent. The shift from a curve like I to one like II or III is just what is to be expected under the influence of a polar environment, since such an environment must lower the energy *more and more as $C$ increases*, since increasing $C$ corresponds to increasing ionic character in the wave function. If the environmental influence is sufficiently strong, the inner complex may become the stable form of the D,A pair when with less or no environmental influence the outer complex (or the separate molecules) may represent the stable form.

If the inner complex becomes the energetically more stable form, more or less complete separation into solvated ions may thereupon occur as a secondary process (dashed curves in Fig. 2) if the assisting agent is an ionizing solvent. This electrolytic dissociation may be regarded as taking place with $C$ essentially constant at the value 1 if the inner complex ($C = 1$) already corresponds to maximum charge transfer and 100% ionic character. (The $C$ scale thus is not applicable to the dashed curves.)

If the assisting process is ionic-crystal formation (*e.g.*, $NH_3 + HCl \rightarrow NH_4^+Cl^-$ crystal), it is a sort of polymerization rather than a separation of the ions which takes place. Curves II and III, but not the dashed curves, are now still more or less applicable. The process is still "dissociative" in the basic sense used here that a covalent bond originally present is broken.

The fact that for a given D,A pair under a particular set of environmental conditions one usually observes *either* an inner complex *or* an outer complex is understandable in terms of Figs. 1 and 2. The further fact that, if environmental conditions are suitably varied, many D,A pairs can be found to exhibit either associative (outer complex) or dissociative (inner complex) behavior is also readily understandable in terms of Fig. 2 and the discussion given above. To make matters more explicit, examples of each of the main D,A pair types which can function dissociatively will now be considered (for further examples, see Table VI).

The known experimental facts on the Group 3 D,A pair $NH_3 + HCl$ (type $n + h\sigma$) are consistent with Fig. 2 using a type I curve for the vapor state and a type III curve for this pair either in solution in a polar solvent (say, liquid ammonia), or in a crystal (solid $NH_4^+Cl^-$). As is well known, solid $NH_4^+Cl^-$ on vaporization gives a vapor consisting of unassociated $NH_3$ and HCl molecules[23]; it has also been shown that if thoroughly dry $NH_3$ and HCl gases are brought together at room temperature, they do not react to form $NH_4^+Cl^-$ crystals.[24] These facts are understandable if the energy of the unassisted $NH_4^+Cl^-$ inner complex of curve I is so high that the number of such ion-pairs in equilibrium with $NH_3$ + HCl vapor at room temperature is too small to lead to the formation of crystal nuclei.

It may, however, be assumed on theoretical grounds that some small fraction of the molecules at any moment are associated in the form of an $H_3N \cdot HCl$ outer complex which should be held together by weak $N \cdots H$ hydrogen bonding plus dispersion forces. Theoretical considerations indicate, however, that such outer complexes as may exist for $NH_3$ + HCl and other D,A pairs of the types $n + h\sigma$, $n + k\sigma$, $\pi + h\sigma$, and $\pi + k\sigma$, are *not* stabilized to any appreciable extent by *charge transfer forces;* in other words, that $C = 0$ for such outer complexes. A further discussion of these points is given in the introductions to Sections 14 and 19 of Table VI.

The system $H_2O$ + HCl should be similar to $NH_3$

(22) An example of the last-mentioned case is the loose $I_2 \cdot$propylene associative complex concerning which Freed and Sancier (*J. Am. Chem. Soc.*, **74**, 1273 (1952)) report that iodine in solution at 77°K. in propane, an inert solvent, upon addition of propylene, develops the color of the iodine·propylene complex only slowly, indicating a small outer barrier.

(23) See. W. H. Rodebush and J. C. Michalek, *J. Am. Chem. Soc.*, **51**, 748 (1943).

(24) According to Spotz and Hirschfelder (*J. Chem. Phys.*, **19**, 1215 (1951)), $NH_3$ and HCl gases on mixing react to form Liesegang rings of solid $NH_4^+Cl^-$, but the authors' theoretical analysis indicates that there must be a clustering of very large numbers of D,A pairs before crystals begin to form. More recently, W. H. Johnston and P. J. Manno have reported (*Ind. Eng. Chem.*, **44**, 1304 (1952)) that no reaction at all occurs, even after hours of waiting, between thoroughly dried $NH_3$ and HCl gases.

These results may be understood as follows. In terms of Fig. 2, a single $NH_4^+Cl^-$ ion-pair would correspond to the high-energy inner complex of curve I. If enough such ion-pairs could be clustered together, their mutual electrostatic attractions would lead to curves of lower and lower energy (per pair) until, at some critical cluster size, some curve (perhaps one similar to II) would be reached, for which the cluster would be big enough to grow spontaneously. Actually, however, any extensive clustering (and this would be rare) would have to be in the first instance of *molecule-pairs* $H_3N \cdot HCl$ (outer complexes); and it is clear that the activation energy barring the way to a simultaneous transformation of enough of these to ion-pairs would probably be so high that it could not ordinarily occur at an appreciable rate. On the other hand, it is understandable that $H_2O$ molecules should assist clustering and reduce the activation energy for transformation to an $NH_4^+Cl^-$ cluster, although it is not at all clear why water vapor should be quantitatively adequate to produce precipitation of solid $NH_4^+Cl^-$.

Consideration of curve I for $NH_3$ + HCl leads to the interesting possibility that at sufficiently high temperatures and pressures, gaseous $NH_3$ + HCl may contain a large proportion of $NH_4^+Cl^-$ ion-pairs.

+ HCl, but since $H_2O$ is a weaker base than $NH_3$, the energy of the ion-pair inner complex (structure $H_3O^+Cl^-$) in the vapor system might be expected to be still higher than for $NH_3$ + HCl. In a polar solvent (say, $H_2O$), however, the inner complex evidently is sufficiently stabilized so that all the HCl + $H_2O$ goes over to this form (and at the same time dissociates into ions). But in contrast to the D,A pair $NH_3$ + HCl, the stability of the inner complex is here evidently not sufficient to permit formation of an ionic crystal. In other words, even an infinite clustering of $H_3O^+Cl^-$ monomers does not bring curve I of Fig. 2 down quite low enough—although presumably it comes rather close to doing so.

The D,A pair ethyl ether plus iodine ($I_2$) is an example of the Group 3 type $n + x\sigma$ for which the existence (in solutions in inert solvents) of an outer complex stabilized by charge transfer forces seems to be well established by spectroscopic evidence.

In the case of another $n + x\sigma$ D,A pair, namely, pyridine (Py) plus iodine in the rather polar solvent pyridine,[2] there is satisfactory evidence for ionization into $(PyI)^+$ and $I^-$, indicating inner complex formation. Here the electrical conductivity increases slowly with time, indicating a slow approach toward equilibrium by passage from outer to inner complex *over a barrier of considerable height*, in harmony with a $U(C)$ curve similar to II of Fig. 2. Further, the limiting conductivity varies with dilution, again suggesting a definite equilibrium between Py,$I_2$ pairs in an outer complex (here it seems reasonable to suppose a *rapid* equilibrium $Py + I_2 \leftrightarrows Py{\cdot}I_2$) and in an inner complex ($PyI^+I^-sl \rightleftarrows PyI^+sl + I^-sl$).

It seems probable that the system $H_2O + I_2$ in water solution involves a similar but much more rapid set of equilibria (plus side equilibria). References on the D,A pairs just discussed are given in Table VI, Section 13.

Continuing with Group 3 examples, there is good evidence for the existence, in non-polar solvents, of loose outer complexes between $b\pi$ donors (*e.g.*, benzene and the alkylbenzenes) and $\sigma$ acceptors of all subclasses ($h\sigma$ acceptors, *e.g.*, HCl; $k\sigma$ acceptors, *e.g.*, methyl alcohol; and $x\sigma$ acceptors, *e.g.*, $I_2$ and other halogens). Just as for $n + \sigma$ D,A pairs, however, it is very probable from theoretical considerations that the $b\pi$ outer complexes observed with $h\sigma$ and $k\sigma$ acceptors are hydrogen bonded and not charge transfer complexes; whereas for the $b\pi{\cdot}x\sigma$ outer complexes, there is abundant spectroscopic evidence that these are true charge transfer complexes. (See Table VI, Sections 18 and 19, for further discussion and references on outer complexes of the $b\pi + \sigma$ types.)

Further, it is rather well established (*cf.* Table VI, Section 19) that with suitable environmental coöperation, $b\pi$ donors with $\sigma$ acceptors form salts which conform completely to the idea of inner complexes as outlined in this paper. These salts (either as ion-pairs, or dissociated) probably exist as reaction intermediates momentarily present in low concentration, or in some cases as stable entities reversibly producible. Illustrative of the $b\pi + h\sigma$ inner complexes are those formed from the alkylbenzenes and the hydrogen halides with an auxiliary acceptor (this action may be regarded as one type of environmental coöperation—see Section VI)

$$Xy + HF + BF_3 \rightleftarrows (XyH)^+BF_4^- \text{ in HF solution}$$

$$Tol + HCl + AlCl_3 \rightleftarrows (TolH^+)AlCl_4^- \text{ in toluene solution}$$

where Xy means xylene; Tol, toluene. In the first of these reactions, the $v$ acceptor $BF_3$, plus the polar solvent HF, furnish the environmental coöperation which stabilizes the inner complex $(XyH^+)F^-$ of Xy and HF.

[**Digression on Double Charge Transfer Reactions.**—Reactions in which there is an auxiliary acceptor may be thought of either as assisted dissociative charge transfer reactions, as has been done above, or as dissociative "double charge transfer reactions." The latter description is appropriate because there is partial charge transfer from a donor molecule to both of the two acceptor molecules involved.

It is worth noting explicitly here that in such reactions each molecule is behaving typically in its donor or acceptor role. Thus in the first reaction, Xy functions as a dissociative donor, in the sense that one of the original $\pi$ bonds is broken. (More accurately stated, there is a breaking up of the original aromatic resonance, such that the main resonance structures in the carbonium ion $XyH^+$ contain two $\pi$ bonds instead of three.[25]) Further, the acceptor HF functions dissociatively in the same way as if it had reacted with an $n$ donor, such as $NH_3$; but with the additional circumstance that the resulting $F^-$, instead of remaining free, becomes simultaneously an $n'$ donor toward the $v$ acceptor $BF_3$.]

There seems to be every reason to believe that $k\sigma$ and $x\sigma$ acceptors can under suitable conditions function in the same way as $h\sigma$ acceptors toward aromatic hydrocarbons; for example, that an ionic inner complex may exist as an intermediate in the dry chlorination of, say, benzene, as shown

$$Bz + Cl_2 + FeCl_3 \longrightarrow [(BzCl^+)(FeCl_4^-)] \longrightarrow PhCl + HCl + FeCl_3$$

The foregoing examples have all been from Group 3. Turning to Group 5, the interaction of $b\pi$ donors with $v^*$ acceptors apparently can likewise give rise to either outer or inner complexes (see Table VI, Section 7). Thus Bz (benzene) forms what appears to be an outer complex with $Ag^+$ and similar $v^*$ acceptors, under suitable conditions of environmental assistance; for example, when $Ag^+ClO_4^-$ is dissolved in Bz, the complex is formed in the presence of the $ClO_4^-$ which presumably remains in the ion-pair $(Bz{\cdot}Ag^+)ClO_4^-$. Here the $Ag^+$ is believed not to be on the symmetry axis of the Bz but to be held, relatively loosely, above the middle of one C–C bond of the ring.[2] On the other hand, in certain solutions in polar solvents, Bz is believed to react with $v^*$ acceptors present in small concentrations, *e.g.*, $NO_2^+sl$ in sulfuric acid solution, or $R^+sl$ in certain other solutions ($R^+$ denoting

(25) However, there is partial restabilization, probably strong, of the carbonium ion by hyperconjugation of the $\pi$ system with the $CH_3$ or similar group in $XyH^+$ (see forthcoming LCAO MO calculations on $BzH^+$ by L. W. Pickett in coöperation with the author).

an alkyl carbonium ion), to form reaction intermediates such as $(BzNO_2)^+sl$ and $(BzR)^+sl$, in which undoubtedly the added radical becomes attached by a fairly normal bond to one carbon of the Bz. The resulting carbonium ion presumably has the same kind of structure as $BzH^+$ discussed above, in which in effect one of the original $\pi$ bonds has been broken, so that the carbonium ion is an inner complex in the same sense as for D,A pairs of types $b\pi + h\sigma$ or $b\pi + k\sigma$. In fact, if we disregard the accompanying anions, the same inner complex is in principle obtainable from a $b\pi + k\sigma$ reaction as from a $b\pi + v^*$ reaction, in case the $k\sigma$ acceptor is of a structure RQ which can ionize to form $R^+$ and $Q^-$ freeing the $R^+$ to function as a $v^*$ acceptor. If $H^+$ were capable of actual existence as a $v^*$ acceptor, a similar comparison could be made between $b\pi + h\sigma$ and $b\pi + H^+$. Comparing the reactions $Bz + Ag^+$ and the hypothetical $Bz + H^+$, one notes that the former gives an outer complex, the latter would give an inner complex, the difference doubtless being explainable in terms of the difference in relative strengths of Ag and H bonds and in the sizes of the $Ag^+$ and $H^+$ ions. [ADDED IN PROOF.—Possibly another model for $BzH^+$, with the $H^+$ located above and between two ring carbons (*cf.* $Bz{\cdot}Ag^+$) should also be considered. In a recent conversation, M. Dewar advocates this model.]

The plausibility of the postulated forms of the $U(C)$ curves in Fig. 2 on the basis of the experimental evidence has now been shown. A qualitative proof that they are also in agreement with theoretical expectations from quantum mechanics will now be undertaken.

The theory of the formation of outer complexes by charge transfer forces has been given in ref. 2 (for a brief review, plus some extensions, see Sections I, IV and V above). In the important case of closed-shell donors and acceptors, the theory is based on the idea that as a donor and an acceptor come (or are forced) more and more closely together, the initial no-bond function $\psi_0(D,A)$ becomes increasingly admixed with a dative function $\psi_1(D^+—A^-)$. Taken by itself, $\psi_0$ would give a steadily increasing exchange repulsion for increasingly close approach.[26] If, however, the corresponding repulsion energy at first rises sufficiently slowly, it may be temporarily overcome by resonance between $\psi_0$ and $\psi_1$ so that a charge transfer outer complex is formed.

But if the energy of $\psi_1$ is too high relative to that of $\psi_0$, or the matrix element of energy for the resonance of $\psi_0$ and $\psi_1$ is too small, the charge transfer forces may be too weak to give rise to an outer complex. Theoretical considerations indicate that this situation occurs for certain Group 3 D,A pairs ($n$ and $b\pi$ donors with $h\sigma$ and $k\sigma$ acceptors) where the acceptors have no (or a negative) electron affinity. However, outer complexes nevertheless often occur in such cases as a result of other forces, in particular hydrogen bonding (see specific discussion of examples of Group 3 D,A pairs in preceding paragraphs). In other Group 3 cases where the acceptor has an appreciable electron affinity ($n$ and $b\pi$ donors with $x\sigma$ acceptors), outer complexes definitely attributable to charge transfer forces occur.

(26) In complex formation in general, valence bond rearrangements such as can occur without charge transfer (as, *e.g.*, in $H_2 + I_2 \rightarrow 2HI$) should also, at close approach, contribute to cutting down $U$, by resonance.

At distances of approach closer than those for the outer complex, $U(C)$ should at first rise with decreasing distance because of the predominance of $\psi_0$. However, even in cases where the charge transfer forces are slow in getting started, they must eventually come into play; that is, $C$ increases above zero, $\psi_1$ mixes more and more into $\psi_0$, and resonance between $\psi_0$ and $\psi_1$ begins to cut down the rate of energy rise. Finally there must come a point where $\psi_1$ begins to predominate over $\psi_0$. The fact then becomes important that the energy of $\psi_1$ decreases with decreasing separation of D and A—up to the point where relevant atoms come within normal bond distances of each other, after which, of course, it increases. In view of these facts, it is readily seen that the energy may be expected to go over a maximum (activated complex) and then fall to a minimum corresponding to the inner complex. In this way, $U(C)$ curves like curve I in Fig. 2 are seen to be in agreement with theoretical expectations for D,A pairs interacting without environmental coöperation. Simple theoretical arguments given earlier then show that, with suitable environmental coöperation, curve I should be replaced by curves like II and III.

Although the reasoning just given is generally applicable for donors and acceptors capable of functioning dissociatively, and having closed-shell structures, certain additional details regarding the dissociative process are also relevant. These can be explained in terms of an example. For this, the case of $H_2O + HCl$, examined in Section VII (see especially eqs. (5) there) is useful. The qualitative discussion which follows is equally applicable to the unassisted D,A pair or to the same pair under environmental assistance. (In Section VII, solution in water was assumed.)

It is convenient to begin by supposing the HCl to be placed near the $H_2O$ with the H directly between the Cl and the O, and with the Cl at a distance from the O approximately equal to that in the (ionic but undissociated) inner complex or reaction product $H_3O^+Cl^-$. For this state of affairs, $C$ would still be only a little larger than zero. If now $C$ is allowed to increase, the H atom moves toward the O atom. The O–Cl distance, though regulated by the specification that it so adjust itself as to keep $U$ as low as possible for any value of $C$, very likely remains nearly constant. As $C$ increases, $U$ tends to increase because (1) the non-bonded repulsions within $\psi_0$ (see eqs. (5)) increase; (2) the amount $(b^2/a^2)$ of the higher-energy wave function $\psi_1$ increases. On the other hand, the following factors work increasingly to *diminish* $U$ as $C$ increases: (1) *electrostatic polarization* within $\psi_0$ (that is, increase of the $H_2O \cdots H^+ \cdots Cl^-$ resonance component within $\psi_0$); (2) decrease in the energy of $\psi_1$ because of increasing attractions and decreasing repulsions as the H moves toward the O atom; (3) decrease in the energy of $\psi_1$ because of

TABLE III

ELECTRON DONOR (D) TYPES[a]

| Symbol | Name | Essential Structure[b, c] | Nature of Donor Action[c]: General Character | Nature of Donor Action[c]: VB Description[d] | Nature of Donor Action[c]: MO Description[d] | Examples |
|---|---|---|---|---|---|---|
| $n$ | Onium donor | Neutral even system containing relatively easily ionized atomic lone pair | Association for valency increase | Partial dative transfer (*cf.* eq. (1a)) from lone pair, as follows: $n + A \rightarrow \text{Res.}\begin{Bmatrix} n,A \\ n^{+}\!-\!^{-}A \end{Bmatrix}$ | $n\!: + A \rightarrow n\!: A$ | Amines, alcohols, ethers, ketones, nitriles, CO, sometimes halides, sometimes $SO_2$ |
| $n'$ | Onium anion donor | Even ion $Q^-$ or usually $Q^-$es containing easily ionized atomic lone pair | Same as $n$ | Same as $n$ | Same as $n$ | Es-solvated anions[c] of H-acids, *e.g.*, $I^-$, $CH_3COO^-$, $OH^-$, $NH_2^-$ |
| $b\pi$ | $\pi$ donor | Neutral even system containing easily ionized bonding $\pi$ electrons | $b\pi_a$: loose dative association; $b\pi_d$: Strong dative association essentially with formation of one $\sigma$ bond and loss of one $\pi$ bond | $b\pi_a$: Partial dative transfer of electron from any $\pi$ bonding pair, with resonance of donor action among all $\pi$ electrons | $b\pi_a$: Partial dative transfer of electron from pair occupying most easily ionized bonding $\pi$ MO | Aromatic (Ar) and unsaturated (Un) hydrocarbons, and their substitution products with electron-releasing substituents |
| $b\sigma$ | $\sigma$ donor | Neutral even system RQ with rather weak and polar R–Q bond (polarity $R^+Q^-$) | $b\sigma_a$: Loose dative association (rare); $b\sigma_d$: Ionogenic displacement reaction, usually es-assisted | $b\sigma_d$: Partial dative transfer of electron from R–Q bonding pair to acceptor with accompanying partial transfer of electron R → Q, and liberation of $R^+$: $^{+\delta}R$ / $^{-\delta}Q$ (α, β arrows to A) $\xrightarrow{es}$ $R^{+1}es$ plus $(Q^{\alpha-1}\!-\!A^{-\alpha})^{-1}es$; $(\beta = 1 - \alpha - \delta)$ | | Alkyl and aralkyl halides, esters, etc., especially if $R^+$ is resonance-stabilized; also solid metals (*e.g.*, Na, Mg, Cu) |
| R | Radical donor (reducing radical) | System with relatively easily ionized odd electron | | | | Univalent metal atoms, H atom, alkyl, aralkyl, and other easily ionized radicals |

[a] See Section II for general explanation. [b] "Essential Structure" refers of course, in complicated cases, only to the region where the donor or acceptor action takes place. [c] The symbol es (*cf.* Section VI) denotes the presence of a source of electrostatic forces which stabilizes the system in question: usually an attached polar solvent molecule or molecules, or sometimes a companion ion or ions, as in ion-pairs, ion-clusters, or solid salt. [d] VB means valence bond; MO means molecular orbital. Res. denotes resonance between the structures indicated.

"dative polarization" (that is, readjustment of the relative proportions of the resonance components—probably increase of $H_2O^+\!-\!H \leftrightarrow Cl^-$ relative to $H_2O^+\cdots H^- \leftrightarrow Cl$, the latter being probably at maximum importance near $C = 0$)[21] within $\psi_1$; (4) resonance between $\psi_1$ and $\psi_0$.[26] The net outcome of the factors outlined should probably be $U(C)$ curves like those in Fig. 2, perhaps like I for no environmental assistance and like II or III in the presence of such assistance.

**The *n,v* Molecular Compounds.**—The Group 1 D,A pairs (see Sections 1, 2 and 5 in Table VI) fall into a rather different behavior class from the Group 3–5 pairs so far considered. They comprise *the* typical donor–acceptor pairs of Lewis and Sidgwick, for which direct association yields a stable datively-bonded addition product. Here curves like II or III of Fig. 1, or variations on them, are evidently applicable. For a case like $BCl_3 + NM_3$, curve III seems probable.[27] For $BR_3 + NR'_3$ using bulky R and R′ groups to cause steric hindrance, as in some of the work of H. C. Brown and his collaborators, curves more like II seem likely, or possibly even like I in very extreme cases.

For Group 1 D,A pairs, the terminology "outer complex" and "inner complex" can still be used, if desired, in case two minima are present in the $U(C)$ curve; but if so, it must not be assumed to have the same structural implications as for Group 3–5 pairs.

**Middle Complex.**—For D,A pairs in which one partner is an ion and the other a neutral closed-shell system, particularly if the donor is of the $n'$ type, one finds $U(C)$ curves like those in Fig. 3, or variations on them. Figure 3 is applicable, for example, to D,A interactions such as

Type $n' + k\sigma_d$ (of Group 2):

$$HO^-aq + MeI \longrightarrow (HOMeI^-)aq \longrightarrow HOMe + I^-aq \quad (8)$$

Types $n' + h\sigma_a$ and $n' + x\sigma_a$:

$$\begin{aligned} F^-sl + HF &\longrightarrow (FHF)^-sl \longleftarrow FH + F^-sl \\ I^-sl + I_2 &\longrightarrow (III)^-sl \longleftarrow I_2 + I^-sl \end{aligned} \quad (9)$$

Curves I and II in Fig. 3 correspond to Group 2 cases where a reaction proceeds continuously to the right over an activation barrier or barriers near $C = 0$ (activated complex or complexes), as in the typical displacement reaction (8). Curve III is applicable in cases where association occurs giving a stable complex ion, as in (9). Here the minimum near $C = 0$ may be called a "middle complex." Reactions (9) correspond to special cases of curve III having equal values of $U$ for numerically equal positive and negative $C$, and not necessarily having any activation barriers at the left and right of $C = 0$. Actually, it seems probable that there may be no such barrier for $HF_2^-$, at least in some solvents.

(27) This is supported by recent work of D. Garvin and G. B. Kistiakowsky, *J. Chem. Phys.*, **20**, 105 (1952).

TABLE IV

ELECTRON ACCEPTOR (A) TYPES[a]

| Symbol | Name | Essential Structure | Nature of Acceptor Action[b] | | | Examples | Remarks[c] |
|---|---|---|---|---|---|---|---|
| | | | General Character | VB Description | MO Description | | |
| $v$ | Vacant-orbital acceptor | Neutral even system in which an orbital or orbitals of relatively high $E$ (electron affinity) are vacant | Association for valency increase | Partial dative acceptance (*cf.* Eq. (1a)) of electron into vacant orbital, as follows: $D + v \longrightarrow$ Res. $\left\{\begin{matrix} D,A \\ D^{+}\text{—}{}^{-}v \end{matrix}\right\}$ | $D{:} + v \longrightarrow D{:}\ v$ | $BMe_3$, $AlMe_3$, $BX_3$, $AlX_3$, $FeX_3$, $ZnCl_2$, $HgCl_2$, O atom, $SnCl_4$ (?), $SiF_4$(?) | A1, A2 |
| $v^*$ | Vacant-orbital cation acceptor | Even ion $R^+$ or usually $R^+$es with High-$E$ localized orbital vacant | Same as $v$ | Same as $v$ | Same as $v$ | $Ag^+$es, $NO_2^+$es, $HSO_3^+$es, carbonium cations $Ak^+$ or $Ak^+$es | A1, B1, B2, B3, B4 |
| $h\sigma^*_d$ and $k\sigma^*_d$ | Cationic dissociative $\sigma$ acceptor | Even cation of structure $DH^+$es (type $h\sigma^*$), $DR^+$-es (type $k\sigma^*$), or in general $(D\rightarrow)_m R^{n+}$es (*generalized* $k\sigma^*$ type) | Always dissociative ($\sigma^*_d$). Displacement reaction: acceptance of stronger base in place of weaker | In general, $p\overline{D} + (D\rightarrow)_m R^{n+}es \rightarrow (D\rightarrow)_q(\overline{D}\rightarrow)_p R^{n+}es + pD$. In particular, if acceptor is $DR^+$es or $DH^+$es, 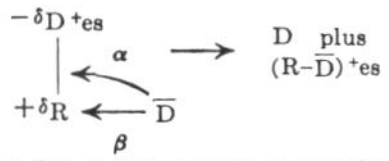 $-^{\delta}D^+$es, $+^{\delta}R \xleftarrow{\beta} \overline{D}$ ($\alpha$) $\longrightarrow$ D plus $(R\text{–}\overline{D})^+$es. Here $D^+$ and D play the same roles that Q and $Q^-$ play in $\sigma_d$ acceptor action. | | $h\sigma^*_d$: $H_3O^+$es, $NH_4^+$es; $k\sigma^*_d$: $Ag(NH_3)_2^+$es. $Ag^+$aq may also belong here, but is more conveniently regarded as $v^*$ | B1, B2, B3, B4 |
| $x\pi$ | $\pi$ acceptor | Neutral even system containing bonding $\pi$ electrons relatively strongly held | $x\pi_a$ or $k\pi_a$: Loose dative association | $x\pi_a$ or $k\pi_a$: Partial acceptance of $\pi$ electron by any $\pi$ bonding pair, converting the latter partially into an odd $\pi$ electron (which gives dative bonding to donor) plus $\pi$ lone pair; with resonance of this action among all $\pi$ electrons | $x\pi_a$ or $k\pi_a$: Partial acceptance of electron into lowest-energy (highest-$E$) antibonding $\pi$ MO, with dative bonding to donor | $x\pi$: Aromatic or unsaturated hydrocarbons with electronegative or electrophilic substituents. For example, trinitrobenzene, maleic anhydride | |
| $k\pi$ | Ketoid $\pi$ acceptor | Neutral even system usually of structure Z=O (or resonating Z=O) containing strongly polar $\pi$ bond(s) | $x\pi_d$ or $k\pi_d$: Association (for better saturation and electronegativity satisfaction) with the formation of one $\sigma$ and the breaking of one $\pi$ bond | $k\pi_d$: Partial electron acceptance by both R and Q with loss of R–Q $\pi$ bond and formation of R–D bond: 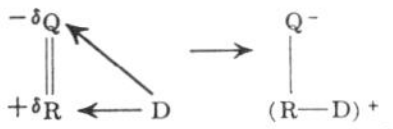 $-^{\delta}Q = +^{\delta}R \leftarrow D \longrightarrow Q^-\text{—}(R\text{—}D)^+$ usually followed by rearrangements or further reaction | | $k\pi$: RHCO, RR′CO, RCN, $CO_2$, $SO_2$(res. 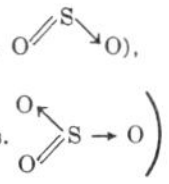 O=S→O), $SO_3$ (res. O←S(=O)→O) | C1, C2 |

This last statement is based on the fact that at larger $C$ values, the attraction between $F^-$ and HF must be a matter of (exceptionally strong) primarily electrostatic hydrogen bonding; at smaller $C$ values, strong charge transfer binding (plus symmetrization—see Section VIII) must also set in.[28]

A viewpoint similar to that of Fig. 3 can also be used (as an alternative to Fig. 1) in certain Group 1 cases of type $n' + v$, $n + v^*$, and the like. Consider for example

$$F^-sl + BF_3 \longrightarrow BF_4^-sl \longleftarrow BF_3 + F^-sl$$

with a different F atom as $F^-$ on the right and on the left. These reactions may be mapped as in Fig. 3 III, with $C = 0$ corresponding to $BF_4^-sl$. (More appropriately, there should be *four* instead of just two directions leading out from $C = 0$.) Another example is

$$NH_3 + AgNH_3^+sl \longrightarrow Ag(NH_3)_2^+sl \longleftarrow NH_3Ag^+sl + NH_3$$

Here again Fig. 3 III might be used.

**Reaction Paths and Mechanisms.**—The charge transfer process idea, with the aid of Figs. 1–3, appears capable of giving plausible indications about reaction paths for a great variety of chemical reactions. In any such consideration, careful attention must be paid to the quantum-mechanical symmetry requirements of refs. 2 and 4; these requirements impose important restrictions which may prove to be valuable in deciding between otherwise plausible possibilities. In general, great CAUTION will be needed in attempting to apply the foregoing ideas to the determination of what actually may happen in specific reactions, since the possible paths, even for a single over-all reaction, are often numerous, diverse in type, and compli-

(28) *Cf.* G. C. Pimentel, *J. Chem. Phys.*, **19**, 446 (1951), for a discussion (using non-localized MO methods) of the linear ions $(FHF)^-$ and $I_3^-$ essentially as symmetrized charge-transfer complexes. (In $I_3^-$, partial trivalency of the central I atom probably also assists.) It is not certain that the free ion $I_3^-$ is exactly linear and symmetrical; if it is not, corresponding minor modifications of obvious character are needed in curve III of Fig. 3.

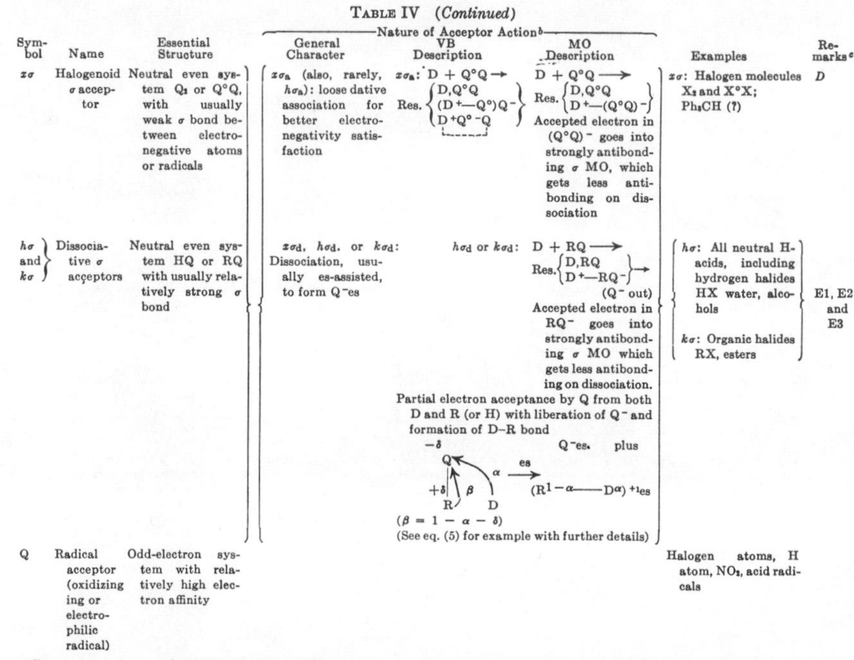

TABLE IV (*Continued*)

| Symbol | Name | Essential Structure | Nature of Acceptor Action[b]: General Character | Nature of Acceptor Action[b]: VB Description | Nature of Acceptor Action[b]: MO Description | Examples | Remarks[c] |
|---|---|---|---|---|---|---|---|
| $x\sigma$ | Halogenoid $\sigma$ acceptor | Neutral even system $Q_2$ or $Q^\circ Q$, with usually weak $\sigma$ bond between electronegative atoms or radicals | $x\sigma_a$ (also, rarely, $h\sigma_a$): loose dative association for better electronegativity satisfaction | $x\sigma_a$: $D + Q^\circ Q \rightarrow$ Res. $\{D,Q^\circ Q;\ (D^+—Q^\circ)Q^-;\ D^+Q^{\circ -}Q\}$ | $D + Q^\circ Q \longrightarrow$ Res. $\{D,Q^\circ Q;\ D^+—(Q^\circ Q)^-\}$ Accepted electron in $(Q^\circ Q)^-$ goes into strongly antibonding $\sigma$ MO, which gets less antibonding on dissociation | $x\sigma$: Halogen molecules $X_2$ and $X^\circ X$; $Ph_3CH$ (?) | *D* |
| $h\sigma$ and $k\sigma$ | Dissociative $\sigma$ acceptors | Neutral even system HQ or RQ with usually relatively strong $\sigma$ bond | $x\sigma_d$, $h\sigma_d$, or $k\sigma_d$: Dissociation, usually es-assisted, to form $Q^-$es | $h\sigma_d$ or $k\sigma_d$: $D + RQ \longrightarrow$ Res. $\{D,RQ;\ D^+—RQ^-\} \rightarrow$ ($Q^-$ out) | Accepted electron in $RQ^-$ goes into strongly antibonding $\sigma$ MO which gets less antibonding on dissociation. Partial electron acceptance by Q from both D and R (or H) with liberation of $Q^-$ and formation of D–R bond: $Q^{-\delta}$, $R^{+\delta}$, $\beta$, $\alpha$, D $\xrightarrow{es}$ $Q^-$es plus $(R^{1-\alpha}—D^{\alpha})^{+1}$es ($\beta = 1 - \alpha - \delta$) (See eq. (5) for example with further details) | $h\sigma$: All neutral H-acids, including hydrogen halides HX water, alcohols; $k\sigma$: Organic halides RX, esters | E1, E2 and E3 |
| Q | Radical acceptor (oxidizing or electrophilic radical) | Odd-electron system with relatively high electron affinity | | | | Halogen atoms, H atom, $NO_2$, acid radicals | |

[a] See notes *a*, *b*, *c*, and *d* of Table III. [b] There appear to be two main driving forces behind acceptor action: (1) the tendency of atoms with vacant orbitals to pick up electrons to form additional bonds (valency increase); (2) the tendency of electronegative atoms in a molecule to become negatively charged (electronegativity satisfaction). The first of these is largely confined to *v* acceptors, while the second is present for most acceptors. In $\pi$-electron systems resonance effects are also important. In *all* types of acceptors acceptor strength increases on increased loading with electronegative atoms. [c] The following remarks pertain to cited cases: A1, These and $v^*$ are *the* typical Lewis acids; A2, really the vacant orbital is not quite vacant, because of resonance structures like $X_2 \rangle B^- = X^+$ or $Cl^+ = Hg^- = Cl^+$, and to this extent *v* acceptors are $k\pi$ acceptors like RR′C=O and O=C=O; B1, The $v^*$ acceptors are typical Lewis acids. Luder calls the $\sigma^*$ acceptors secondary Lewis acids; B2, Ions of structure $R^+$sl are classified as $v^*$ or as $k\sigma^*$ according as the solvent molecules (sl) are held primarily by es or by donor action and so regarded as accessory or as constitutive. In doubtful cases, the $v^*$ classification is the more convenient. Ions of structure $H^+$sl are almost always most properly called $h\sigma^*$ rather than $v^*$; B3, However even definitely $k\sigma^*$ or $h\sigma^*$ acceptors are often conveniently regarded *by courtesy* as $v^*$, by ignoring sl in $R^+$sl or $H^+$sl; for example, $H_3O^+$aq may be regarded as $H^+$; B4, The type $h\sigma^*$ comprises precisely the conjugate acids $DH^+$ of all neutral bases D. Examples include among others $ArH^+$ and $UnH^+$. However, since $UnH^+$ is at the same time $R^+$ (*e.g.*, if Un is $C_2H_4$, $UnH^+$ is $C_2H_5^+$), acceptors of this type function more often associatively as $v^*$ acceptors in their own right than dissociatively as $h\sigma^*_d$ acceptors; C1, $k\pi$: The $k\pi$ and *v* acceptors are not separated by any sharp boundary, but a continuous range of intermediate cases is possible (see remark A2 on *v* acceptors); C2, Some readers may question the structural formulations given here for $SO_2$ and $SO_3$, but it is the writer's considered opinion that they correspond to the major resonance components, and that, for instance, structures using *d* orbitals to give hexavalent sulfur with more double bonds are present only to a lesser extent; D, $x\sigma$: The halogen molecules $X_2$ and $X^\circ X$ are characterized by (a) weak $\sigma$ bonds, (b) $E^{vert} > 0$ though small, but increasing strongly during dissociation to limiting value $E_X$; E1, The $h\sigma$ class consists precisely of all the neutral H-acids; E2, The halides HX and RX are characterized by (a) strong polar $\sigma$ bonds, (b) $E^{vert} \ll 0$, but becoming $> 0$ during dissociation, with limiting value $E_X$; E3, HQ and RQ are sometimes described as secondary Lewis acids, conceived as formed from the primary Lewis acid $H^+$ or $R^+$ and the base $Q^-$.

cated, and dependent on temperature, pressure, the nature of the medium, and other factors.

## X. Corrections and Improvements on Previous Papers

Some improvements in the donor and acceptor classes and classification symbols of paper II[2] are described in Section II above. An error in quoting spectroscopic data on $Et_2O \cdot I_2$ in paper I[4] is corrected in Table VI (reaction-type 13, footnote *v*). A revision of Fig. 5 of paper II is described in Table VI (reaction-type 7, footnote *m*). An overlooked reference of some importance relevant to paper I is: Childs and Walker, *Trans. Faraday Soc.*, **34**, 1506 (1938), on the spectra of bromine in benzene, acetic acid, water and ethyl alcohol.

In paper I, on page 606, two pairs of resonance structures given for $R'RO \cdot I_2$, and called (I) and (II), are not independent; pair (II) should be

TABLE V: NATURE OF PRODUCTS IN SOME DONOR-ACCEPTOR REACTIONS[a]

| Acceptor Type | Donor Type: $n$ | $n'$ (usually $Q^-$es) | $b\pi$: $b\pi_a$ | $b\pi_d$ | $b\sigma$ (commonly RQ): $b\sigma_a$ | $b\sigma_d$ | R $\{R_n, R_\pi, R_\sigma\}$ |
|---|---|---|---|---|---|---|---|
| $v$ | DATIVE (1) | <Sy>-DATIVE (2) | Dative (3) | | | es: $R^+[Qv]^-$es (4) | |
| $v^*$ (Usually $R^+$es) | <Sy>-DATIVE (5) | DATIVE forming RQ; or Sy-DATIVE (6) | Dative (7) | ($b\pi$)–R $\sigma$-bond formed datively, with loss of one $\pi$ bond (7) | | $R^+{}_D$es + $QR_A$ (8) | (25) |
| $\pi$: $x\pi$ $\{x\pi_a, x\pi_d\}$ | Dative (9); Q–($x\pi$) $\sigma$-bond formed datively, with loss of one $\pi$-bond (10) | | Dative (11) | | | | |
| $\pi$: $k\pi$ (usually Z=O) $\{k\pi_a, k\pi_d\}$ | Dative (9); <es>: $n^+$-Z-O$^-$ → (9) | es-dative (10); Q-Z-O$^-$es <→> (10) | Dative (11) | | | (12) | |
| $\sigma$: $x\sigma$ (Q°Q) $\{x\sigma_a, x\sigma_d\}$ | Dative (13); es: $[nQ°]^+Q^-$es <→> (13); es,$v$: $[nQ°]^+[vQ]^-$ es <→> | Sy-dative (16) | Dative (18) | es: $[\pi Q°]^+Q^-$es <→>; es,$v$: $[\pi Q°]^+[vQ]^-$es <→> (18) | Dative (20) | | |
| $\sigma$: $h\sigma$ (HQ) $\{h\sigma_a, h\sigma_d\}$ | Unstable or H-bonded (14); es: $[nH]^+Q^-$es (14) | Sy-dative (16); $Q_DH + Q^-{}_A$es (17) | H-Bonded (19) | es: $[\pi H]^+Q^-$es (19) es,$v$: $[\pi H]^+[vQ]^-$es (19) | | $(Q_DHQ_A)^- + R^+$ (21); $h\sigma_d$: Perhaps es: $Q_DH + R^+Q^-{}_A$es (21) | $R^+Q^- + H$ |
| $\sigma$: $k\sigma$ (RQ) $\{k\sigma_a, k\sigma_d\}$ | Unstable (14); es: $[nR]^+Q^-$es (14) | $Q_DR + Q^-{}_A$es (17) | $k\sigma_a$: Unstable or H-bonded (19) | | | | $R^+{}_DQ^- + R_A$ (25) |
| $\sigma$: $l\sigma_d$($n°v$) | $nv + n°$ (15) | | | | | | |
| $\sigma^*$: $x\sigma^*{}_d$ ($[Q°D]^+$es) | | | | | | | |
| $\sigma^*$: $h\sigma^*{}_d$ ($[HD]^+$es) | $[nH]^+$es + D (22) | QH + D (23) | | | | Perhaps QH + $R^+$es + D (24) | |
| $\sigma^*$: $k\sigma^*{}_d$ ($[RD]^+$es) | $[nR]^+$es + D | QR + D (23) | | | | | |
| Q $\{Q_v, Q_\pi, Q_\sigma\}$ | | (25) | Dative (25) | | | | RQ or $R^+Q^-$ (25) |

[a] For detailed descriptions of donor (D) and acceptor (A) types and notation, see Tables III and IV. Subscripts a and d appended to donor and acceptor symbols of the $\sigma$ and $\pi$ classes indicate associative (a) or dissociative (d) functioning. For any donor–acceptor type, Table V describes briefly the nature of the initial reaction-product or products for a direct bimolecular reaction between donor and acceptor. The word "dative" indicates a *loose* dative complex, the word "DATIVE" a dative complex or compound which in typical examples is strongly bound. The *numbers in parentheses* in Table V refer to Sections of Table VI where specific examples of the reaction-types indicated are listed.

In Table V, certain symbols and punctuation marks are used only with definite explicit meanings, as follows: parentheses are used with a brief structural description of a donor or acceptor type; <. . .> means that the indicated behavior *may* or *may not* occur in a given example depending on circumstances, and/or that it occurs in some examples but not in all; → means that an indicated reaction product is an intermediate (stable or unstable) which normally reacts further (either intramolecularly and/or, usually, with other molecules). *Subscripts* D and A serve, when necessary, to identify different R and/or Q radicals in the reaction-products in terms of their origin in the donor or acceptor reactant. The abbreviation es or es,$v$ followed by a colon indicates that the reaction occurs conditionally (in most cases, only conditionally)—see Section VI—through coöperative action of agents acting electrostatically (es), or through the coöperation of an auxiliary $v$ acceptor plus es agents (es,$v$). The symbol Sy, standing for symmetrization, denotes a process in which, when an incoming donor or acceptor joins a partner containing identical (or sufficiently similar) atoms, all the like atoms become mutually equivalent (or nearly so) in position and internal electronic distribution (see Section VIII).

Tables V and VI are not exhaustive: (a) additional D and A types exist, *e.g.*, $n''$ and $\sigma'$ donors, $v^{**}$, $v^{***}$ and $\pi^*$ acceptors, etc.; (b) additional reaction-types corresponding to some of the blanks in Table V exist, but only those types are listed in Table V of which examples are given in Table VI; (c) reaction-types involving R donors and Q acceptors are indicated only very sketchily. Tables V and VI are *not intended to imply that the actual reaction mechanism* for a given donor and acceptor is necessarily always (or perhaps ever) the direct bimolecular mechanism indicated in the tables. However, the reaction-types listed are believed to be correct at least as to over-all driving force and over-all result, although whether for any given individual example they go at all, or reach an equilibrium, or go to completion, depends of course on the acceptor and donor strengths of the particular reactants. The reaction-types listed are believed also to correspond in most cases to reaction mechanisms which are likely to be realized in practice at least under some circumstances. As is well known, over-all bimolecular reactions often proceed actually in steps, with the monomolecular or quasi-monomolecular ionization process RQ → $R^+$sl + $Q^-$sl as an initial step, the solvent (sl) action involved being sometimes purely electrostatic (es) and sometimes partly or wholly a donor or acceptor action. The $v^*$ or $\sigma^*{}_d$ acceptor $R^+$sl and/or the $n'$ or $\sigma'_d$ donor ($\sigma'_d$ is of structure $n'v$) then react further. (The ionization process is the reversal of a donor–acceptor reaction of type $v^*$ or $\sigma^*{}_d$ plus $n'$ or $\sigma'_d$ → RQ.) In a few examples in Tables V and VI, processes involving the assistance of an auxiliary acceptor or donor (usually a $v$ acceptor) have been indicated. No opinion is implied here as to whether ever, or how often, the actual reaction mechanism in such processes is termolecular.

dropped. Resonance between the two equivalent forms (I) gives $I_2^-$ 3-electron bonding *together* with $O^+$–I electron-pair bonding, the two effects being nearly but not quite additive so long as the $O^+$–I bonding is weak.

**Acknowledgment.**—The writer is indebted to various people for helpful suggestions and criticisms, and particularly to Dr. Harden McConnell.

## Appendix

### Detailed Characterization of Donor and Acceptor Classes and Their Interaction Types, with Examples

TABLE VI

SOME EXAMPLES OF DONOR–ACCEPTOR COMPLEXES (D·A), COMPOUNDS (D→A), AND PROBABLE REACTIONS (D + A → PRODUCTS), MOSTLY IN SOLUTION, TOGETHER WITH APPROXIMATE WAVE LENGTHS λ OF PROBABLE CHARGE TRANSFER ABSORPTION SPECTRA PEAKS OF SOME OF THE COMPLEXES D·A[a,b,c]

**1, Type $n + v^d$**

$Et_2O + BF_3 \longrightarrow Et_2O{\to}BF_3$
$Me_3N + BMe_3 \longrightarrow Me_3N{\to}BMe_3$
$Me_3N + {}^1/_2Al_2Cl_6 \longrightarrow Me_3N{\to}AlCl_3$
$MeCN + BF_3 \longrightarrow MeCN{\to}BF_3$
$MeBr + {}^1/_2Ga_2Cl_6 \longrightarrow MeBr{\to}GaCl_3$[e]
But *not* $HCl + AlCl_3 \longrightarrow HCl{\to}AlCl_3$[f]
$Me_3P{\to}O + BF_3 \longrightarrow Me_3P{\to}O{\to}BF_3$[g]
$Me_3N{\to}O$; Res. $OS{\to}O$; Res. $O_2S{\to}O$

**2, Type $n' + v$ (usually $Q^-$es $+ v$)**

$F^-aq + BF_3 \longrightarrow BF_4^-aq$
$HO^-aq + BF_3 \longrightarrow HOBF_3^-aq$
$Na^+Cl^-(solid) + AlCl_3 \longrightarrow Na^+AlCl_4^-(solid)$

**3, Type $b\pi + v$**

$b\pi_a$: Complexes of the type $b\pi_a{\cdot}v$ are not very stable[h]

**4, Type $b\sigma + v$**

$b\sigma_d$: $RF + BF_3 \longrightarrow R^+es + BF_4^-es$ or $R^+BF_4^-$(R = alkyl) in RF as es solvent

$NOCl + AlCl_3 \longrightarrow NO^+AlCl_4^-(solid) \rightleftarrows NO^+sl + AlCl_4^-sl$, *e.g.*, in liquid NOCl[i]

$COCl_2 + AlCl_3 \longrightarrow COCl^+sl + AlCl_4^-sl$, *e.g.*, in liquid $COCl_2$

For reactions like the above, direct reaction and initial ionization of the solvent are alternative mechanisms, either of which may be predominant depending on the particular case and circumstances.

**5, Type $n + v^*$**

$CH_2I_2 + Ag^+aq \rightleftarrows CH_2I_2{\cdot}Ag^+aq$[j]
(here perhaps $Ag^+aq$ is more nearly of $k\sigma^*_d$ than $v^*$ type)

$Me_3N + Me^+es \longrightarrow Me_4N^+es$
(perhaps in MeF solution, after $MeF + BF_3 \longrightarrow Me^+es + BF_4^-es$)

$2NH_3 + Ag^+NO_3^-(crystal) \xrightarrow{es} (H_3N{\to})_2Ag^+es + NO_3^-es$
(in aq or aqueous ammonia, with aq as es)

**6, Type $n' + v^{*d}$**

$Cl^-sl + Ph_3C^+sl \rightleftarrows Ph_3CCl$[k]
in liquid $SO_2$, nitromethane, or acetone (sl action in $Cl^-$sl partly dative: see under $n' + k\pi$ (Part 10 of this Table))

$I^-aq + Ag^+aq \longrightarrow AgI(solid) + aq$
(see remarks on $Ag^+aq$ under $n + v^*$ (Part 5 of this Table))

$2CN^-aq + Ag^+aq \longrightarrow [Ag(CN)_2]^-aq$

**7, Type $b\pi + v^*$**

$b\pi_a$: $Bz + Ag^+ClO_4^- \rightleftarrows Bz{\cdot}Ag^+ClO_4^-$, and $(Bz{\cdot}Ag^+{\cdot}Bz)ClO_4^-$?
(in benzene solution)

$MePh + Ag^+aq \rightleftarrows MePh{\cdot}Ag^+aq$ (λ2300?)[m] and $aqAg^+{\cdot}MePh{\cdot}Ag^+aq$
(in aqueous solution; here see remark on $Ag^+aq$ under $n + v^*$ (Part 5 of this Table))

$b\pi_d$: $Bz + NO_2^+es + HSO_4^-es \longrightarrow [BzNO_2]^+es + HSO_4^-es \longrightarrow PhNO_2 + H_2SO_4$
(in sulfuric acid solution, with sulfuric acid as es)

$Bz + R^+es + AlCl_4^-es \longrightarrow [BzR]^+es + AlCl_4^-es \longrightarrow PhR + HCl + AlCl_3$ (?)
(*e.g.*, perhaps EtCl solution of $AlCl_3$ ($\longrightarrow Et^+es + AlCl_4^-es$) + Bz)[b]

**8, Type $b\sigma + v^*$**

$b\sigma_d$: $MeEt_2CH + Me_2CH^+ \longrightarrow MeEt_2C^+ + Me_2CH_2$[n]
(Here in the $b\sigma_d$ donor RQ, R is $MeEt_2C$, Q is H)

Zn metal + $Ag^+sl \longrightarrow Zn^{++}sl$ + Ag metal

Na metal + $NH_4^+sl \longrightarrow Na^+sl$ + Na(metal)—$NH_4$ which decomposes to give $NH_3$ and $H_2$

**9, Types $n + x\pi$ and $n + k\pi$**

$x\pi_a$: $NH_3$ + dinitrobenzene $\rightleftarrows$ $NH_3{\cdot}$dinitrobenzene

$k\pi_a$: $Me_3N{\to}O + SO_2 \longrightarrow Me_3N{\to}O{\to}SO_2$[g]

$k\pi_d$: $H_2O + CO_2 \overset{aq}{\rightleftarrows} \left(\text{Res. } O{=}C\begin{smallmatrix}O^-\\OH_2^+\end{smallmatrix}\ aq\right) \rightleftarrows \text{Res. } O{=}C\begin{smallmatrix}O^-\\OH\end{smallmatrix}\ aq + H_3O^+aq$

(or perhaps $CO(OH)_2$ is formed in the first step)

$NH_3 + RCHO \longrightarrow \left(RHC\begin{smallmatrix}O^-\\NH_3^+\end{smallmatrix}\right) \longrightarrow RHC\begin{smallmatrix}OH\\NH_2\end{smallmatrix} \longrightarrow$

(In this and the first following, or probably both the following, examples, the over-all process is of the type $b\sigma_d + k\pi_d$, with an *intramolecular* $b\sigma_d$ action. If the action proceeds in a *single step*, it should be classified as $b\sigma_d + k\pi_d$.)

$H_2O + SO_3 \xrightarrow{aq} \left(\text{Res. } \begin{smallmatrix}O\\O\end{smallmatrix}{>}S\begin{smallmatrix}O^-\\OH_2^+\end{smallmatrix}\ aq\right) \longrightarrow \begin{smallmatrix}O\\O\end{smallmatrix}{>}S\begin{smallmatrix}OH\\OH\end{smallmatrix}\ aq$

$HCl + SO_3 \longrightarrow HCl{\to}SO_3$ or $ClSO_2OH$ (?) (chlorsulfonic acid)[o]

TABLE VI (*Continued*)

**10, Types $n' + x\pi$ and $n' + k\pi$**

$x\pi_d$: $MeO^-sl + O_2N$–$C_6H_2(NO_2)_2$–$OEt \xrightarrow{sl}$

Res. $\left[ (O^-)_2N^+{=}C\langle^{CH=C(NO_2)}_{CH=C(NO_2)}\rangle C\langle^{OEt}_{OMe} \right] sl^p$

(in solution, with sl = es)

$k\pi_a$: $(Cl^-{\cdot}SO_2)sl$ or $(Cl^-{\rightarrow}SO_2)sl$
(sl action on $Cl^-$ partly es, partly dative)
$(Cl^-{\cdot}CH_3NO_2)sl$ and $(Cl^-{\cdot}Me_2CO)sl$ in nitromethane and acetone

$k\pi_d$: $CN^-aq + RCHO \longrightarrow \left(RHC\langle^{O^-}_{CN}\ aq\right) \xrightarrow{H_3O^+aq} RHC\langle^{OH}_{CN}$

$OH^-aq + CO_2 \longrightarrow HCO_3^-aq$
$Na^+Cl^-(solid) + SO_3 \longrightarrow Na^+[ClSO_3]^-(solid)$

**11, Types $b\pi + x\pi$ and $b\pi + k\pi$[a]**

$b\pi_a{\cdot}x\pi_a$: $PhNH_2$ + *s*-trinitrobenzene $\rightleftarrows$ $PhNH_2{\cdot}$*s*-trinitrobenzene (λ 4000)[q]
Bz + *s*-trinitrobenzene $\rightleftarrows$ Bz·*s*-trinitrobenzene (λ 2800)[r]

$b\pi_a{\cdot}k\pi_a$: $MBz + SO_2 \rightleftarrows MBz{\cdot}SO_2$ (λ 2840)[s]
(MBz = various methylated benzenes)
Bz + oxalyl chloride $\rightleftarrows$ Bz·oxalyl chloride (λ 2700?)[t]
Hydroquinone + quinone $\rightleftarrows$ hydroquinone·quinone (quinhydrone) (about λ 5600 in crystal)[u]

**12, Type $b\sigma_d + k\pi_d$**

(see under $n + k\pi$, Part 9)

**13, Type $n + x\sigma$**

$x\sigma_a$: *t*-butyl alcohol + $I_2 \rightleftarrows$ *t*-Bu alcohol·$I_2$ (λ 2330)[v]
$Et_2O + I_2 \rightleftarrows Et_2O{\cdot}I_2$ (λ 2480)[v]
$RX + X_2 \longrightarrow RX{\cdot}X_2$ (λ 3000–3500)[va] (X = Br or I)

$x\sigma_a$ and $x\sigma_d$: $Py + I_2 \rightleftarrows Py{\cdot}I_2 \overset{slow}{\rightleftarrows} [PyI]^+py + I^-py$; etc.[w]
(In pyridine solution. In $Py{\cdot}I_2$, Py may be acting as a *mixed* $n$ and $b\pi_a$ donor.)

$H_2O + I_2 \rightleftarrows H_2O{\cdot}I_2 \overset{aq}{\rightleftarrows} [H_2OI]^+aq + I^-aq \overset{H_2O}{\rightleftarrows} HOI + H_3O^+aq + I^-aq$ (and $I^-aq + I_2 \rightleftarrows I_3^-aq$, etc.)

$NH_3 + I_2 \rightleftarrows H_3N{\cdot}I_2 \overset{aq\ am}{\rightleftarrows} [H_3NI]^+es + I^-es \rightleftarrows NH_2I + NH_4^+es + I^-es$[x]

**14, Types $n + h\sigma$ and $n + k\sigma$[d,y]**

$n{\cdot}h\sigma_a$ and $n{\cdot}k\sigma_a$: Theoretically, since almost certainly $E^{vert} << 0$ for HX (X = halogen) and probably for most HQ and RQ, it is probable that any loose complexes of the type $n{\cdot}h\sigma_a$ and $n{\cdot}k\sigma_a$ which may exist are in most cases due to electrostatic rather than charge transfer forces (*cf.* Sections VI, IX). Formation of onium acids or salts ($h\sigma_d$ or $k\sigma_d$ behavior) usually (if not always) occurs only with es assistance ($h\sigma_{ed}$ or $k\sigma_{ed}$ behavior), either by formation of an ionic crystal or by solvation. For example, $NH_4^+Cl^-$ solid, and $NH_4^+sl + Cl^-sl$, are stable, but individual $NH_4^+Cl^-$ molecules in vapor (*cf.* refs. 23 and 24 and Section IX) or in inert solvents are apparently not stable (nor apparently is a loose complex $NH_3{\cdot}HCl$ very stable). It may be, however, that individual salt molecules are in *some* cases stable without es assistance; for example, perhaps $Me_4N^+Cl^-$ and the like in benzene solution.

$h\sigma_d$: $H_2O + HCl \xrightarrow{aq} H_3O^+es + Cl^-es$ (es = aq)
$H_2O + HBr \xrightarrow{SO_2} H_3O^+es + Cl^-{\cdot}SO_2$ (es = $SO_2$)
$H_3N + HOH \overset{aq}{\rightleftarrows} NH_4^+es + OH^-es$ (es = aq)

$k\sigma_d$: $Me_3N + MeCl \xrightarrow{slow} [Me_4N]^+Cl^-$ (slightly soluble solid)
(in Bz or $PhNO_2$ solution)

**15, Type $n + l\sigma$[d]**

$l\sigma_d$: $Me_2CO{\rightarrow}BCl_3 + Py \longrightarrow Py{\rightarrow}BCl_3 + Me_2CO$

**16, Types $n' + x\sigma_a$ and $n' + h\sigma_a$**

$x\sigma_a$: $I^-aq + I_2 \rightleftarrows I_3^-aq$
$h\sigma_a$: $F^-aq + HF \longrightarrow HF_2^-aq$ (*cf.* also $b\sigma + h\sigma$ below (Part 21) concerning $HCl_2^-$)

**17, Types $n' + h\sigma_d$ and $n' + k\sigma_d$[b]**

$h\sigma_d$: $NH_2^-es + HCPh_3 \longrightarrow NH_3 + CPh_3^-es$
(in liquid ammonia solution)
$EtO^-es + HCH_2COOEt \rightleftarrows EtOH + [CH_2COOEt]^-es$
(in EtOH solution)

$k\sigma_d$: $aqI^- + MeI \longrightarrow IMe + I^-aq$
$aqOH^- + MeI \xrightarrow{slow} H\overset{*}{O}Me + I^-aq$
$aqCl^- + CH_2{-}CH_2$ (bridged by O) $\longrightarrow$
$Cl{-}CH_2{-}CH_2{-}O^-aq + HOH \longrightarrow Cl{-}CH_2{-}CH_2{-}OH + OH^-aq$

This example is of the type
$(n') + (k\sigma_d) \longrightarrow (n') + (h\sigma_d) \longrightarrow$ etc.

**18, Type $b\pi + x\sigma$**

$b\pi_a{\cdot}x\sigma_a$: $MBz + X^\circ X \rightleftarrows MBz{\cdot}X^\circ X$ (λ 2900)[ya]
(MBz = various methylated benzenes, X°X = $Cl_2$, $Br_2$, $I_2$, ICl)
Naphthalene + $X_2 \rightleftarrows$ naphthalene·$X_2$ (λ 3500)[yb]
$MC_2H_4 + I_2 \rightleftarrows MC_2H_4{\cdot}I_2$ (λ 2700–3000)[z]
($MC_2H_4$ = various methylated ethylenes)
$Bz + Ph_3CH \longrightarrow Bz{\cdot}Ph_3CH$[a]

TABLE VI (*Continued*)

$b\pi_d + x\sigma_d$: Bz + $Cl_2$ + $FeCl_3$ ⟶ ([BzCl]$^+$[$FeCl_4$]$^-$) ⟶ PhCl + HCl + $FeCl_3$ ⟶

$C_2R_4$ + $Cl_2$ + $AlCl_3$ ⟶ ([$C_2R_4Cl$]$^+$[$AlCl_4$]$^-$)

(double charge transfer reactions of type $b\pi_d + x\sigma_d + v$)

**19, Types** $b\pi + h\sigma$ and $b\pi + k\sigma$

$b\pi_a \cdot h\sigma_a$: There is evidence for loose complexes of the types $b\pi_a \cdot h\sigma_a$ (*e. g.*, methylated benzenes with HCl)[aa] and $b\pi_a \cdot k\sigma_a$ (*e.g.*, Bz·MeOH).[ab] These, however, are almost certainly members of a large class of weakly H-bonded es complexes,[ac] and not dative complexes. Presumably the H of MeOH or of HCl is attracted by the somewhat negatively charged carbons of the benzene ring (the C–H bonds are believed to have polarity $C^-H^+$). This would explain the increasing solubility with increasing methylation observed by Brown and Brady[aa] for HCl in MBz, since increasing methylation sends negative charge increasingly into the ring. Brown and Brady attribute the effect to increasing basicity, which would indeed give the same result, and which (interpreting basicity as meaning charge transfer donor strength) is indeed here believed responsible for the increasing stability with methylation of the superficially closely analogous MBz + $X_2$ complexes of type $b\pi_a \cdot x\sigma_a$; but it appears improbable that charge transfer donor–acceptor interaction is appreciable for typical $h\sigma$ and $k\sigma$ acceptors without es solvent assistance (see remarks above under Types $n + h\sigma$ and $n + k\sigma$, Part 14).

$b\pi_d, h\sigma_d$: $C_2R_4$ + $HHSO_4$ $\overset{es}{\rightleftarrows}$ [$C_2R_4H$]$^+$es + $HSO_4^-$es

(in sulfuric acid as es)

Ar + HF $\overset{HF}{\rightleftarrows}$ [ArH]$^+$es + $F^-$es [or $HF_2^-$es)[ad]

Ar + HF + $BF_3$ $\rightleftarrows$ [ArH]$^+$es + $BF_4^-$es

(in liquid HF as es)[ae]

PhMe + HCl + $^1/_2Al_2Cl_6$ $\rightleftarrows$ [MePhH]$^+$[$AlCl_4$]$^-$ etc

(in toluene solution at low temperatures)[af]

**20, Type** $b\sigma + x\sigma$

$b\sigma_a \cdot x\sigma_a$: Cyclopropane + $I_2$ $\rightleftarrows$ cyclopropane·$I_2$ (λ 2400)[i]

**21, Types** $b\sigma + h\sigma$ and $b\sigma + k\sigma$

$b\sigma_d, h\sigma_a$: anisyl chloride + HCl ⟶ (anisyl)$^+$ + $HCl_2^-$ (?) (in liquid HCl)[b]

$b\sigma_d, h\sigma_d$: possibly (?) $NO_2OH$ + $HHSO_4$ $\overset{es}{\longrightarrow}$ $NO_2^+$es + $HSO_4^-$es + $H_2O$

(in sulfuric acid as es)

**22, Type** $n + h\sigma^*$

$h\sigma_d^*$: $R_3N$ + $H_3O^+$aq ⟶ [$R_3NH$]$^+$ + $H_2O$

$H_2O$ + [$H_2OI$]$^+$aq ⟶ $H_3O^+$aq + HOI (*cf.* Type $n + x\sigma$, Part 13)

**23, Types** $n' + h\sigma^*$ and $n' + k\sigma^*$

$h\sigma_d^*$: $MeCOO^-$aq + $H_3O^+$aq ⟶ MeCOOH + $H_2O$

$k\sigma_d^*$: $OH^-$aq + $Me_3S^+$aq ⟶ MeOH + $Me_2S$

**24, Type** $b\sigma_d + k\sigma^*_d$

Ca metal + $COCl^+$sl ⟶ CO (from metal—CO?) + $Ca^{++}$sl + $Cl^-$ (in $COCl_2$ solution)

**25, Some Types** Involving R or Q

$R_n + v^*$: H + $H^+$ $\overset{sy}{\longrightarrow}$ $H_2^+$ (in gas)

$R_n + k\sigma_d$: Na + MeCl ⟶ $Na^+Cl^-$ + Me (in gas)

$n' + Q_v$: $Cl^-$ + Cl $\overset{sy}{\longrightarrow}$ $Cl_2^-$ (in gas)

$b\pi_a + Q_\sigma$: MBz + $Ph_3C$ ⟶ MBz·$Ph_3C$

(MBz = methylated benzenes)

$R_n + Q_v$: Na + Cl ⟶ $Na^+Cl^-$ (near λ 3000)[ag]

$R_n + R_n$: H + H ⟶ $H_2$ (λ 1010)[ah]

$Q_v + Q_v$: I + I ⟶ $I_2$ (λ 1800)[ah]

$R_n + Q_v$: H + I ⟶ HI (est. λ 1280)[ah]

(These last four examples illustrate *interatomic* charge transfer spectra.)

[a] For numerous examples of organic and organic–inorganic molecular complexes and compounds, mostly in the solid state, see P. Pfeiffer, "Organische Molekülverbindungen," 2nd edition, F. Enke, Stuttgart, 1927. [b] For references and discussion of a great many examples of donor–acceptor reactions and their mechanisms see L. P. Hammett, "Physical Organic Chemistry," McGraw-Hill Book Co., Inc., New York, N.Y., 1940. For additional examples, see Luder and Zuffanti, ref. 6. [c] The expression "Res." means a mixture of equivalent structures of the type given. The dative bond arrow symbol, as in D→A, represents not a *pure* dative structure but a resonance mixture of pure dative structure with some no-bond structure (*cf.* Eq. (1a)). [d] W. F. Luder (*J. Chem. Phys.*, **20**, 525 (1952)) describes on the one hand the 1:1 reaction types $n' + v^*$ and $n + v$ as neutralization processes (*e.g.*, $CN^-$ + $H^+$ → HCN and $Me_2CO$ + $BCl_3$ → $Me_2CO$→$BCl_3$) and on the other hand the types $n + h\sigma_d$ and $n + l\sigma_d$ (*e.g.*, HOH + HCN → $H_3O^+$ + $CN^-$ and Py + $Me_2COBCl_3$ → $PyBCl_3$ + $Me_2CO$) as displacement processes, with $H^+$ and $BCl_3$ in the first pair regarded as primary Lewis acids, HCN and $Me_2COBCl_3$ in the second pair as secondary Lewis acids (*cf.* Section I for related comments). [e] H. C. Brown, H. Pearsall and H. P. Eddy, *J. Am. Chem. Soc.*, **72**, 5347 (1950). [f] *Ibid.*, and H. C. Brown and H. Pearsall, *ibid.*, **73**, 4681 (1951); R. L. Richardson and S. W. Benson, *ibid.*, **73**, 5096 (1951). [g] A. B. Burg and W. E. McKee, *ibid.*, **73**, 4590 (1951). [h] Regarding Ar·$AlX_3$, see R. E. Van Dyke, *ibid.*, **72**, 3619 (1950); D. D. Eley and P. J. King, *Trans. Faraday Soc.*, **47**, 1287 (1951). H. C. Brown and W. J. Wallace have found Bz·$Al_2Br_6$ (private communication). [i] A. B. Burg and D. E. McKenzie, *J. Am. Chem. Soc.*, **74**, 3143 (1952). [j] L. J. Andrews and R. M. Keefer, *ibid.*, **73**, 5733 (1951). [k] *Cf.* N. N. Lichtin and P. D. Bartlett, *ibid.*, **73**, 5530 (1951). [m] R. M. Keefer and L. J. Andrews, *ibid.*, **74**, 640 (1952). For the toluene·$Ag^+$aq complex, Keefer, and Andrews' analysis shows (1) an absorption peak near λ 2650, of about double the intensity and at slightly longer wave lengths than a corresponding peak of toluene by itself; (2) rapidly rising intensity at shorter wave lengths toward a much more intense peak (not reached) which might lie at about λ 2300. This second absorption is here tentatively identified as the charge-transfer (toluene→$Ag^+$) absorption. ($Ag^+$aq also shows a strong peak at somewhat shorter wave lengths, which may tentatively be identified with $H_2O$→$Ag^+$ charge transfer.)

In ref. 2, Fig. 5 suggests that the (toluene→$Ag^+$) charge transfer peak might be expected near λ 3100. However, in constructing Fig. 5, previously existing estimates of the solvation energy of $Ag^+$ (see for example O. K. Rice, "Electronic Structure and Chemical Binding," McGraw-Hill Book Co., Inc., New York, N. Y., 1940: on p. 402 the solvation energy of $Ag^+$ in water is given as 106 kcal./mole) were overlooked. Making use of these, curve E of Fig. 5 would be raised by 1 or 2 ev. (perhaps by 2 ev. at

TABLE VI (*Continued*)

large $R$ and 1 ev. at small $R$), and could very well be consistent with a charge transfer peak near λ 2300.

For an important recent survey on $Ag^+$ complexes, see G. Salomon, in the book, "Proceedings of Symposium on Cationic Polymerization and Related Complexes," edited by P. H. Plesch (Cambridge, 1952).

[n] P. D. Bartlett, F. E. Condon and A. Schneider, *J. Am. Chem. Soc.*, **66**, 1531 (1944). [o] *Cf.* H. H. Sisler and L. F. Audrieth, *ibid.*, **61**, 3392 (1939), for references and studies on this and related examples. [p] Branch and Calvin, "The Theory of Organic Chemistry," Prentice-Hall Inc., New York, N. Y., 1941, p. 481. [q] J. Landauer and H. McConnell, *J. Am. Chem. Soc.*, **74**, 1221 (1952). And see ref. 15 in text of Section III, above. [r] H. McConnell and D. M. G. Lawrey, submitted to *J. Am. Chem. Soc.* [s] L. J. Andrews and R. M. Keefer, *ibid.*, **73**, 4169 (1951). [t] B. D. Saksena and R. E. Kagarise, *J. Chem. Phys.*, **19**, 994 (1951). [u] See K. Nakamoto, *J. Am. Chem. Soc.*, **74**, 1739 (1952), on polarized light spectra of this and other $b\pi_a \cdot k\pi_a$ complexes in solid state. (See text, Section IV, for further details.) [v] Regarding $Et_2O \cdot I_2$, *cf.* refs. 4 and 2 in text. In the first reference (p. 606), $\epsilon_{max}$ for the λ2480 peak was in error; it should be about 7760. Confirmed by recent work by J. S. Ham, extended also to ($t$-butyl alcohol)$\cdot I_2$ (see text, Section IV). [va] R. M. Keefer and L. J. Andrews, *J. Am. Chem. Soc.*, **74**, 1891 (1952). [w] L. F. Audrieth and E. J. Birr, *ibid.*, **55**, 668 (1933); R. A. Zingaro, C. A. VanderWerf and J. Kleinberg, *ibid.*, **73**, 88 (1951). See also ref. 2 in text. [x] R. K. McAlpine, *J. Am. Chem. Soc.*, **74**, 725 (1952). [y] See also Section IX of text. [ya] *Cf.* refs. 4, 2, in text. [yb] Blake, Winston and Patterson, *J. Am. Chem. Soc.*, **73**, 4437 (1951). [z] L. J. Andrews and R. M. Keefer, *ibid.*, **74**, 458 (1952); S. Freed and K. M. Sancier, *ibid.*, **74**, 1273 (1952). Various substituted ethylenes and butadienes, also (F and S) cyclopropane$\cdot I_2$. [aa] H. C. Brown and J. D. Brady, *J. Am. Chem. Soc.*, **71**, 3573 (1949); **74**, 3570 (1952); solubilities of HCl in methylated benzenes. However, E. K. Plyler and D. Williams (*Phys. Rev.*, **49**, 215 (1936)) found but little shift in the HCl infrared fundamental for benzene solutions as compared with HCl vapor, whereas for other solvents (*e.g.*, nitrobenzene, ethers) large shifts were found: W. Gordy and P. C. Martin, *J. Chem. Phys.*, **7**, 99 (1939). [ab] L. H. Jones and R. M. Badger, *J. Am. Chem. Soc.*, **73**, 3132 (1951). [ac] Work by many authors. Some recent papers including key references to earlier papers are: S. Searles and M. Tamres, *ibid.*, **73**, 3704 (1951), and references on infrared spectra in footnotes *aa* and *ab*. [ad] Klatt, *Z. anorg. allegem. Chem.*, **234**, 189 (1937); M. Kilpatrick, unpublished work. [ae] D. A. McCaulay and A. P. Lien, *J. Am. Chem. Soc.*, **73**, 2013 (1951). M. Kilpatrick, unpublished work. [af] H. C. Brown and H. W. Pearsall, *J. Am. Chem. Soc.*, **74**, 191 (1952). [ag] *Cf.* R. S. Mulliken, *Phys. Rev.*, **51**, 327 (1927). [ah] *Cf.* R. S. Mulliken and C. A. Rieke, *Reports on Progress in Physics* (London Physical Society), **8**, 249 (1941).

## DISCUSSION

H. C. BROWN.—From a chemical viewpoint the interactions of iodine and hydrogen chloride with various types of donor molecules exhibit close similarity. Thus, both substances interact with aromatic nuclei to form relatively unstable 1:1 complexes.

$$ArH + I_2 \rightleftharpoons ArH \cdots I\text{—}I$$

$$ArH + HCl \rightleftharpoons ArH \cdots H\text{—}Cl$$

With somewhat stronger bases, such as ethyl ether, they interact to form stabler 1:1 complexes.

$$R_2O + I_2 \rightleftharpoons R_2O\text{—}I\text{—}I$$

$$R_2O + HCl \rightleftharpoons R_2O\text{—}H\text{—}Cl$$

Although there is stronger bonding between the two components, and the bonds between the iodine atoms and between hydrogen and chlorine must be correspondingly weakened, the situation has evidently not reached the point where the halide ion is ionized. In the case of pyridine, a still stronger base, ionization occurs.

$$C_5H_5N + I_2 \rightleftharpoons C_5H_5NI^+ \, I^-$$

$$C_5H_5N + HCl \rightleftharpoons C_5H_5NH^+ \, Cl^-$$

Finally both "acids" unite with halide ions.

$$C_5H_5NI^+ \, I^- + I_2 \rightleftharpoons C_5H_5NI^+ \, I_3^-$$

$$C_5H_5NH^+ \, Cl^- + HCl \rightleftharpoons C_5H_5NH^+ \, HCl_2^-$$

The phenomena are so similar that I question the desirability of assigning the acceptor iodine ($x\sigma$) to an entirely different class ($h\sigma$) solely on the basis of a postulated difference in the nature of the bonding between hydrogen or iodine and the donor molecule. If it is desirable to organize these phenomena into distinct classes, I believe that the classification should be based primarily upon experimental similarities and differences, rather than upon hypothetical differences in the nature of bond involved. (Frankly, I doubt whether our understanding of and agreement in the field of chemical bonding has yet reached the point where we can safely use it as a basis for classification of donor-acceptor interaction.)

To illustrate this point let us consider the complexes formed by aromatic hydrocarbons with hydrogen chloride on the one hand and by hydrogen chloride–aluminum chloride on the other. In the case of the hydrogen chloride–aromatic complexes, we observe that they are both formed and dissociated rapidly at −80°, they are colorless, they do not exchange deuterium with the aromatic, and they do not conduct the electric current. On the other hand, the hydrogen chloride–aluminum chloride aromatic complexes form and dissociate slowly at −80°, are intensely colored, exchange deuterium rapidly, and are excellent conductors of the electric current.

We have applied the term $\pi$-complexes to the hydrogen chloride-aromatic species and other related complexes, and have proposed the term $\sigma$-complexes for derivatives of the $HCl\text{-}AlCl_3$ type.

···HCl   [ $CH_3$ ... + ... H H  ↔  $H^+$ H—C—H ... H H ] $AlCl_4^-$

$\pi$-complexes  $\sigma$-complex

Apart from the preference one may have for his own terminology and classification, it appears to me far safer to base any classification upon similarities and differences in chemical behavior which are directly observable in the laboratory, rather than upon assumed similarities and differences in chemical bonding.

Finally I should like to call attention to the dangers involved in adopting any elaborate system of notation such as that proposed by Professor Mulliken. Experts working in the field would have little difficulty in mastering and utilizing a large group of specialized symbols. There is little doubt that communication among the specialists would be somewhat facilitated by such a shorthand once it is mastered. However, there is also little doubt that this increased ease of communication among the specialists will be accompanied by a serious loss in communication with workers in other fields of chemistry who would doubtless find too burdensome the task of mastering and using so many specialized symbols. Since the phenomena associated with acids and bases (or donors and acceptors) are of the utmost importance to practically all fields of chemistry, it is clear that any new phenomena and ideas which are uncovered in this area should be readily communicable to all chemists who may wish to coördinate these findings with their own investigations. I question whether the relatively minor advantage of an increased ease of communication among specialists would be a reasonable price to pay for a serious decrease in the ease of communication with workers in related areas.

I

R. S. MULLIKEN.—Professor Brown's feeling that my classification of donors and acceptors may be excessively elaborate has caused me to revise the original mimeographed version of the manuscript prepared for the meeting. I have added two brief introductory tables (I and II in the printed version), which I think show that the basic classification ($\pi$, $\sigma$, $n$ and R donors, and $\pi$, $\sigma$, $v$ and Q acceptors) is relatively simple. The four elaborate and detailed tables of the original version have now been somewhat clarified and placed at the end as an Appendix.

However, perhaps Professor Brown also had in mind the fact that intermediate cases in the classification are frequent (see, for example, some of the Remarks in Table IV). He asked, in conversation, why it is necessary ever to classify such molecules as the alkyl halides as $\sigma$ donors; why may one not simply suppose that RX always functions (through its X atom) as an $n$ (onium) donor? Thus (here *cf.* Section 4 in Table VI) why not

$$RX + BX_3 \longrightarrow (RX{\rightarrow}BX_3) \longrightarrow R^+BX_4^-$$

presumably with an unstable intermediate analogous in structure to the stable product known from the work of Brown and his collaborators (*cf.* Table VI, Section 1 and footnote *e*) to be formed in such a reaction as

$$RBr + GaBr_3 \longrightarrow RBr{\rightarrow}GaBr_3?$$

Actually, both the suggested intermediate in the first and the final product in the second of these reactions must contain (among others) important resonance components of the *two types* $R{-}X^+{-}{}^-AX_3$ and $R^+X - {}^-AX_3$, of which the first corresponds to $n$ functioning and the second to $\sigma$ functioning of the donor.

Since it is not always easy to decide in such cases which (if either) resonance component strongly predominates, it is then often safest to make the classification by saying that the given donor or acceptor is functioning in a manner intermediate between those of two specified pure classes. I do not feel that even a somewhat frequent occurrence of intermediate cases destroys the value of a significant classification scheme, since much of this value lies in the insight one obtains just in the process of attempting a classification.

Referring again to the first of the two reactions mentioned above, it is clear that RX is functioning in the *over-all* reaction as a pure dissociative $\sigma$ (*i.e.*, $\sigma_d$) donor, even though in an intermediate stage it may be acting partly like an $n$ donor. One may, however, ask whether the reaction actually proceeds bimolecularly straight through in one sweep to the final ion-pair product, or in distinct stages of which the first might conceivably be a bimolecular reaction to form an intermediate of the type indicated above. Or does it go quasi-unimolecularly in a familiar way involving as a first step the ionization of RX in solution to $R^+sl + X^-sl$? These alternatives illustrate the fact that the functioning of a given donor (or acceptor) in a particular reaction which proceeds in distinct stages may be classified either in terms of its behavior in the initial stage, or in terms of the over-all reaction. As is well known, reactions of a given over-all type may in general proceed by any of a variety of mechanisms, often with two or more competing mechanisms operative at substantial rates in one and the same reaction vessel. As is also well known, our present knowledge as to precisely what mechanisms are actually most important in individual reactions under specified conditions is still very limited.

In the list of examples of donor–acceptor reactions in Table VI of my paper, my intention was to base each classification if possible on what seemed the most likely *first distinct stage* in the reaction under fairly usual conditions; but I may not have been wholly consistent in this. (My intention and point of view are explained in detail in the last paragraph of the notes to Table VI.) In any event, I believe that the proposed classification scheme, if it proves worth while at all, may be useful in the analysis and classification *both* of reaction mechanisms and of over-all reactions.

## II

With Professor Brown's comment that the acceptor types of which $I_2$ and HCl are representatives show closely parallel empirical behavior in the formation of loose complexes, I cannot seriously disagree. In fact, I originally (see ref. 2 in my paper) had classed these types together as *d* acceptors, but later (*cf.* footnote 53 of ref. 2, and ref. 3) after a good deal of thought concluded that they should be rated as distinct subclasses ($x\sigma$ and $h\sigma$) of the $\sigma$ acceptor class. This conclusion was based on the three following considerations.

(A) Much weaker acceptor properties are to be *expected theoretically* for the $h\sigma$ than for the $x\sigma$ group, in view of the much larger bond strengths and smaller electron affinities for the former; the relevance of these factors can be seen by reference to Fig. 1 and eq. (5) of ref. 3 taken in connection with the general theory of donor-acceptor interaction energies in ref. 2. (B) The loose HCl complexes do not appear to differ empirically in any essential way from numerous other loose complexes which are generally agreed to be hydrogen-bonded complexes of essentially electrostatic character: see Section 19 and related footnotes in Table VI. (C) The $x\sigma$ subclass is set apart by the fact that its loose complexes with both $n$ and $\pi$ donors show characteristic ultraviolet spectra of their own, of which the $\lambda$ 2900 absorption found by Benesi and Hildebrand is the classical example; and the best explanation (*cf.* ref. 2) appears to be that these are charge transfer spectra of definitely donor–acceptor complexes. No corresponding spectra of hydrogen halide complexes have been reported, and it is my prediction, derived from the theoretical considerations mentioned under (A), that no such spectra are likely to be found.

However, Professor Brown's comment has caused me to revise somewhat, for the printed version, the discussion as given in the original mimeographed copy, by making use of the terminology "outer complexes" for loose addition compounds of both the $x\sigma$ and $h\sigma$ types, regardless of whether charge transfer or other forces may be principally responsible for the formation of these complexes. This considerably simplifies the presentation and enlarges the applicability of Fig. 2. At the same time, I continue to maintain (subject always to correction if new evidence requires) that the $x\sigma$ and $h\sigma$ acceptors are very different in respect to the major forces which make them associate with donors, as well as in certain empirical, particularly spectroscopic, properties.* On the other hand, I believe that the fact that the $x\sigma$ and $h\sigma$ acceptors are in many respects similar in their complexing properties is appropriately recognized in classifying them both as subclasses of a single major class, the $\sigma$ acceptors.

## III

Professor Brown has proposed to use the term "$\pi$-complexes" for what I would call $b\pi$, $h\sigma$ *outer complexes*, and the term "$\sigma$-complexes" for what I would call $v$-assisted $b\pi$, $h\sigma$ *inner complexes* (with, in particular, HCl representing $h\sigma$). While I completely agree with Professor Brown's opinions† that his "$\pi$-complexes" are loose complexes and that his "$\sigma$-complexes" are intermediates of carbonium-salt type, with these two states of interaction separated by an activation barrier, I feel that his terminology is open to several objections. I would rather first classify the donor and acceptor each according to its initial structure, and then use terminology such as "outer complex" or "inner complex," as proposed in ref. 3 and further developed in the present paper, to distinguish between the two characteristic ("associative" and "dissociative") modes of interaction which I think occur for many types of donor, acceptor pairs.

* Another reason for setting apart the subclass $h\sigma$ is that it is identical with the important class of neutral molecule H-acids. Except for the distinctive chemical interest of the H-acids, the empirically and theoretically similar subclasses $h\sigma$ and $k\sigma$ (alkyl halides, *etc.*,) might well be lumped together.

† As expressed in more detail by Brown and Brady *J. Am. Chem. Soc.*, **74**, 3570 (1952). I am indebted to Professor Brown for letting me see a copy of this paper at the time of the Notre Dame meeting. In the present and in a previous paper (ref. 2), I have drawn heavily for examples, evidence, and suggested explanations on the very instructive and important work reported in numerous papers by Professor Brown and his students.

# Intermolecular Charge-Transfer Forces*)

Robert S. MULLIKEN

*Laboratory of Molecular Structure and Spectra, Department of Physics, University of Chicago, Chicago, Illinois, U. S. A.*

I want to introduce a discordant note by talking about doing some things in a very easy way. They are things which perhaps concern ordinary chemistry rather more than the solid state. I think they illustrate how a little quantum mechanics can go a long way, and how by a very simple formulation one may reach a great many conclusions. What I want to talk about may be described as a new type of cohesive force between molecules. In some ways it is not exactly new, since many isolated groups of examples have been familiar for a long time in both physics and chemistry. What is new is the use of a quantum mechanical description which is so simple as to cover an enormous variety of cases, including familiar cases, but more, to make some new predictions and to explain spectroscopic and other phenomena which previously have not been explained. Many of the applications are to problems of the mechanism and path of chemical reactions, and so rather outside the scope of this conference. However, there are applications also to intermolecular forces in liquids and molecular crystals and to electrical and spectroscopic properties of solutions and crystals.

The basic idea is the qualitatively familiar one that any two bits of matter, whether they are atoms or positive or negative ions, or molecules or even solids have a tendency to exchange electrons, that is, to act as electron donors (D, symbolically) or electron acceptors (A). These terms are used in the theory of semiconductors but they also apply in orthodox chemical phenomena. Quantum mechanically, the interaction of an electron donor and acceptor may be formulated by saying that when any two such entities get together their joint or combined wave function may be expressed in the form

$$\psi(\mathrm{D, A}) = a\psi_0(\mathrm{DA}) + b\psi_1(\mathrm{D^+A^-}) \,. \quad (1)$$

In many cases, the first term is the main term and corresponds to just the ordinarily recognized forces between molecules, including London dispersion forces and any electrostatic interactions between dipole moments, and so on, if these are present. The special feature of our wave function (1) is the inclusion of the second term in which an electron has gone over from the donor to the acceptor. Some-

*) For further details, one may refer to R.S. Mulliken: J. Am. Chem. Soc. **72** (1950), 600; **74** (1952), 811; J. Phys. Chem. **56** (1952), 801. See also Paper No. 25 in ONR Report on September, 1951, Conference on "*Quantum-Mechanical Methods in Valence Theory,*" issued by the Office of Naval Research, Department of Navy, Washington 25, D. C.

SOURCE: Proceedings of the International Conference on Theoretical Physics, Kyoto and Tokyo, September 1953.

times if the donor and acceptor are nearly equal in strength, there may be an additional important term:

$$\psi(D, A) = a\psi_0(DA) + b\psi_1(D^+A^-) + c\psi_2(D^-A^+) + \cdots\cdots. \qquad (2)$$

The coefficient $b$ is equal to or greater than $c$, and $a$ in most cases is greater than the others. If D is a good donor and A a good acceptor, the last term in Eq. (2) can be dropped as unimportant. If D and A are identical atoms or molecules, then of course $b = c$. Some examples may make this clear.

A very simple familiar case is that of the diatomic molecule. (Or instead of one atom, we may substitute a radical, such as a methyl radical). For instance, in the case of HCl, we have

$$\psi(HCl) = a\psi_0(H-Cl) + b\psi_1(H^+Cl^-) + \cdots\cdots. \qquad (3)$$

Here the main term $\psi_0$ corresponds to a covalent bond, which may be expressed by a Heitler-London wave function or by a localized molecular orbital type wave function. The second term is ionic. The hydrogen molecule might be written in the same way, except that in this case we would have the third term as in Eq. (2), with the coefficient $c = b$. In the case of HCl, however, the hydrogen atom is an electron donor to the chlorine and the third term may be neglected without much harm. Eq. (3) illustrates the case where the donor and the acceptor are atoms or radicals having odd numbers of electrons.

The cases of special interest which were not so well understood before are cases where two complete molecules come together: two molecules which each have a closed-shell electronic state. An example is the benzene-iodine system (benzene is $C_6H_6$); here it is believed that a 1:1 molecular complex is formed. The wave function for this may be written as

$$\psi(Bz \cdot I_2) = a\psi_0(Bz \cdot I_2) + b\psi_1(Bz^+ - I_2^-) + \cdots\cdots. \qquad (4)$$

In the first term we have nothing but London dispersion force and quadrupole interactions because each of the molecules separately has no dipole moment and no ionic charge. The second term represents a state in which an electron has passed from a benzene molecular orbital to an iodine molecular orbital. Now empirically, iodine when dissolved in benzene acts as if its molecules have a dipole moment. This and other evidence can be understood if some of the benzene and iodine molecules have united to form molecular complexes with a structure just as given by our formulation in Eq. (4); the dipole moment may be attributed to the $Bz^+$—$I_2^-$ term in Eq. (4). In fact, with this formulation and using the observed dipole moment,— in this case $\mu = 0.70$D for the molecules of the 1:1 complex,— we can get the coefficient $b^2 = 0.03$, then $a^2 = 0.97$.

There are various other examples, for instance, benzene with trinitrobenzene. Trinitrobenzene is benzene with three $NO_2$ groups substituted. The symmetrical trinitrobenzene by itself has no dipole moment, but here again, when this is dissolved in benzene (which also has no dipole moment by itself), a dipole moment is observed. This may be explained in the same way as in the case of the benzene-iodine system.

Another example is pyridine-iodine where analogous phenomena are observed which may be explained in a similar way. In this case, a nitrogen atom is present replacing one carbon and one hydrogen of the benzene. Here there is already a dipole moment for the pyridine, but when the iodine is dissolved in pyridine a much larger dipole moment is observed, indicating that $b^2$ may be as large as perhaps 0.2~0.25 for the $Py \cdot I_2$ complex.

It appears that in these and numerous other cases the molecular complex is stabilized by a resonance between the two primitive wave functions like $\psi_0$ and $\psi_1$ of Eq. (1) or (4). Corresponding to this energy stabilization, we may speak of a *charge-transfer force* between the molecules. This type of force should be quite universal between any pair of molecules, although it may be very small in many cases. Undoubtedly it is very appreciable when the molecules are rather unlike and one is a good donor and the other a good acceptor, but it should still exist between like molecules to some extent. For that reason it may be of some importance in determining the structure of molecular crystals, in addition to other known types of force. In order to estimate the stabilization corresponding to the wave function of Eq. (1) or (4), we may use second order perturbation theory, by which the stabilization energy would be something like

$$\Delta E = -\frac{W_{01}{}^2}{W_1 - W_0}. \tag{5}$$

At present the idea of stabilization of molecular association by charge-transfer forces is very much in an empirical stage. We would hardly have predicted quantum mechanically in advance that such a resonance term as that approximated by Eq. (5) would be very important in many cases. The main evidence for its importance is the discovery that this sort of interaction is frequently needed to explain the association of molecules. However, we can begin after the fact to see perhaps how it might come about theoretically. The matrix element $W_{01}$ should be roughly proportional to an overlap integral, namely the integral over all space of the product of the molecular orbital of the donor molecule from which the electron is transferred, times the molecular orbital of the negative ion of the acceptor molecule into which the electron has gone:

$$W_{01} = C\int \phi_D \phi_{A^-} d\tau. \tag{6}$$

The analysis shows that in Eq. (6) it should be the molecular orbital of the *negative ion*. Now any negative ion is notoriously larger than the corresponding neutral molecule, e.g., $H^-$ is large compared with H. Furthermore, in a case like this, the electron must go into an *antibonding* molecular orbital of the acceptor molecule, which tends to cause a repulsion between the atoms, and to make a very big orbital; for instance, the lowest-energy *antibonding* orbital for the two hydrogen atoms in a hydrogen molecule is much larger than the corresponding bonding orbital. So perhaps we can understand how the overlap we are considering may be rather large even for two molecules which are scarcely closer than in ordinary collision contact. Otherwise, it should be quite surprising that the matrix element can be so large.

Now as I have mentioned earlier, charge-transfer forces should operate also between molecules in a molecular crystal. As an example, we may consider a crystal of benzene or some other aromatic hydrocarbon. For these molecules the ionization potential is quite low; that is an important condition for a good donor. But also the electron affinity seems to be very considerable, meaning that such molecules may be fairly good acceptors too. Now if one goes from benzene to bigger and bigger aromatic molecules, one gets finally to graphite. In graphite the electron affinity and ionization potential are equal and the ionization potential is very low, making it a good donor, while also the electron affinity is very high making it a good acceptor. So these forces may be of some importance in determining how the molecules in such crystals stick together, and are oriented in the crystals (see what follows).

There are two other aspects of the simple theory which we get from the wave function of Eq. (1) or (4). I should like to mention these briefly. One is the matter of orientation properties. The type of formulation of Eq. (1) or (4) offers the possibility that the size of the coefficient $b$ may be dependent on the orientation of the two molecules in forming the complex. The coefficient $b$ might be very large for one orientation and zero for some other orientation. As applied to crystals, this has interesting possibilities. Of course, the London dispersion forces and the other electrostatic forces have orientational properties too. But now we get a new kind of orientational property, so we have a force which may tend to bring about a different sort of orientation. One can see that the coefficient $b$ should be proportional to the matrix element $W_{01}$, while this in turn is approximately proportional to the overlap according to Eq. (6). You can see that this overlap integral may easily be zero by symmetry. I could easily draw some pictures of molecular orbitals and show some examples. One such example is that of the complex between the benzene molecule and the silver ion. (In solution, complexes are formed between hydrocarbons having $\pi$-electrons and various positive ions.) In this case the $Ag^+$, as acceptor, after gaining an electron, is a neutral silver atom, and so the orbital $\phi_{A^-}$ in this case is $5s$, which is totally symmetrical. The benzene orbital from which the electron comes is necessarily antisymmetric to a plane perpendicular to the plane of the molecule. The overlap integral is then zero if the silver ion is put above the middle of the benzene ring. So we conclude that it should *not* be on the axis of the ring, where chemists previously supposed it was, since $b$ would then be zero, but we predict it must be off axis; and by looking to see where overlap is at a maximum, we see that a reasonable location should be above one of the C—C bonds of the ring. This is exactly where it was found by Rundle by X-ray measurements in crystals of the benzene-silver perchlorate complex. Of course the crystal is not a 1 : 1 complex, but more nearly an $\infty : \infty$ complex. It may best be considered as an interlocking system of 2 : 1 and 1 : 2 complexes, each of which can be described by a rather obvious generalization of an equation like (4). Each benzene according to Rundle has an $Ag^+$ above one C—C bond on one face of the ring, and another below an opposite C—C bond on the other face (1 : 2 relation), while at the same time each $Ag^+$ lies between two benzene rings (2 : 1 relation).

Finally, the theory predicts that there should be some very interesting characteristic spectra for donor-acceptor molecular complexes. If the wave function (4) describes the ground state of the benzene-iodine molecular complex, there must be according to quantum mechanics a complementary excited state with a wave function approximately of the form

$$\psi_E = -b\psi_0(\text{Bz}\cdot\text{I}_2) + a\psi_1(\text{Bz}^+ - \text{I}_2^-) + \cdots\cdots. \quad (7)$$

Since $a$ is large, $b$ small, the excited state should be nearly all ionic (97% ionic if the ground state is 3% ionic). Further, one can predict a spectroscopic transition from $\psi$ to $\psi_E$, and can compute approximately its frequency and intensity, and predict its polarization. This sort of transition is an intermolecular one, which cannot appear in the spectra of the separate molecules. The total absorption spectrum of the complex is then predicted to be a sum of the spectra of the separate molecules, benzene and iodine, with some shift due to their interaction, plus this new additional "intermolecular charge-transfer" spectrum.

Now, the theoretical computation predicts that this transition should be very intense even for a very loose molecular complex. We have some information about the energy of formation of the benzene-iodine complex indicating that it is about 0.06 eV, while for the pyridine-iodine complex this energy comes to about 0.3 eV*). However, even with the very small interaction for the benzene-iodine complex, there is observed a very intense spectrum in the ultraviolet which agrees in its observed position and intensity with that calculated for this kind of transition. The polarization has not been observed in this case, but for some other molecular complexes in the solid state for which both donor and acceptor are aromatic molecules, Dr. Nakamoto has made an interesting study of the polarization of the absorbed light. According to his results, the polarization is more or less perpendicular to the molecule planes in the long wave length part of the spectrum, which can be identified with the predicted intermolecular charge-transfer transition. This is in contrast to the usual spectra of aromatic molecules in the ultraviolet where the polarization is characteristically *in* the molecule plane. These results correspond exactly to what is expected from our theory.

It is interesting to ask where the high intensity of charge-transfer transitions comes from. The oscillator strength or $f$ value is of the order of 0.3 for this

*) This is of course the *excess* energy of formation over that corresponding to the usual van der Waals (London dispersion force) attraction. The latter might amount to roughly 0.1 eV between an $I_2$ and a benzene or pyridine molecule.

transition in the benzene-iodine system, in spite of the very weak binding. Where does this intensity come from? I would say that it is stolen from the spectroscopic absorption continuum. If the two molecules of the complex were separate, this intensity would belong in the continuum, in the ionization continuum for the benzene. But in the complex, the electron instead of escaping completely from the benzene, has stopped on the iodine molecule, and the corresponding transition intensity is taken away from that of the continuum.

---

## DISCUSSION

**M. Kubo:** You said that a complex between trinitrobenzene and benzene is polar. I believe that trinitrobenzene has an *apparent* finite dipole moment in benzene solution. Is that right?

**Mulliken:** Yes.

**Kubo:** But I think there is another possible method of explanation that trinitrobenzene is nonpolar and has a high atomic polarization.

**Mulliken:** I have assumed that some allowance was made for this, for example possibly by a comparison between the polarization of trinitrobenzene in benzene as compared with that in another inert solvent such as heptane, but I am not certain. The reference is to J. Weiss (J. Am. Chem. Soc. **245** (1942), 52).

**Kubo:** In measuring the dipole moment by solution method, we subtract the atomic and electronic polarization from the total polarization of a solute to obtain orientation polarization, from which dipole moment is calculated. Usually we assume the atomic polarization to be equal to say 10 % of the electronic polarization. Hence it is very difficult to decide whether a molecule is perfectly nonpolar or has a small finite moment. The temperature dependence of polarization does not give good results for solution method, owing to the solvent effect and narrow temperature range available.

**Mulliken:** The possible uncertainty here is one which occurs in most cases where dipole moments are determined from observed polarization in solution, without temperature variation. Fortunately in the case of the evidence for dipole moments for the *iodine* complexes, atomic polarization for the iodine molecule does not exist. The determination of dipole moments of molecular complexes forms an interesting field which has as yet been very little studied, but to which more attention should be given. While I am not in a position to give a critical judgment on the existing experimental evidence for the presence of a dipole moment for trinitrobenzene in benzene, I have personally very little doubt that there is such a dipole moment, in view of other evidence.

**Kubo:** Are there any definite experimental evidence in favor of the existence of a complex between trinitrobenzene and benzene other than the dipole moment measurement ?

**Mulliken :** Yes, between various aromatic hydrocarbons, or amines and various other aromatic molecules containing nitro-groups, there is a new spectrum having the characteristics of an intermolecular charge-transfer spectrum; this is discussed in two or three recent publications from our laboratory. The presence of this new spectrum is shown usually by the fact that one gets a new color on mixing such substances.

**H. Fröhlich :** A critical test would be the temperature dependence.

**Kubo :** But in this case (dipole moment in solution) the temperature range is very narrow. If you measure the dielectric constant of gas, you have a very wide range of temperature. But in this case, (the trinitrobenzene dissolved in the benzene), the range of temperature is narrow.

**Fröhlich :** Nevertheless there should be temperature dependence, if you assume a dipole moment.

**K. Nukasawa :** (1) The electron affinity of acceptors is an important quantity but difficult to estimate. Is there any way to estimate this ?

(2) When the donor and acceptor are situated in convenient position for electrons to migrate, there will occur necessarily the exchange repulsion between them. Is there any criterion which would be more effective, exchange repulsion or charge-transfer attraction ?

**Mulliken :** (1) All electron affinities are of course relatively small (perhaps mostly 1—3 ev), so that their variation from one electron acceptor to another is much less important than the variations in ionization potential for donors. From chemical behavior and stability of molecular complexes we may reach some qualitative conclusions about relative electron affinities of acceptor molecules; in general the presence of oxygen or other electronegative atoms in a molecule (as e. g. in trinitrobenzene or maleic anhydride) makes molecules better acceptors and so presumably means increased electron affinity. Quantitative methods for determining molecular electron affinities are not very well developed, but there seem to be possibilities for the future.

(2) The equilibrium configuration of any molecular complex must depend on a balance between the charge-transfer forces and the exchange repulsions. The relative magnitudes of these forces differ greatly in different cases. In many cases (but usually only with the assistance of a polar medium or some other assisting agency) the charge-transfer forces overcome completely the exchange repulsions corresponding to formation of an essentially ionic "inner complex"; this change may also be regarded as a chemical reaction.

# The Interaction of Electron Donor and Acceptor Molecules*

R. S. MULLIKEN

*Laboratory of Molecular Structure and Spectra, Department of Physics*
*The Uuiversity of Chicago, Chicago 37, Illinois, U.S.A.*

Interactions between chemical molecules occur with varying degrees of intensity. The weakest interactions lead to pure vapors, liquids, solids, or solutions in which the identities of the original molecules are essentially preserved. These "physical" interactions can be understood in terms of the classical electrostatic forces and quantum-mechanical London dispersion forces.

"Chemical" interactions lead to the *association* of molecules in definite proportions to form "molecular compounds" or "molecular complexes"; or to chemical *reactions* in which new chemical molecules are formed. A large part of the "chemical" interactions can be understood as interactions of electron donor with electron acceptor molecules. If this type of interaction goes forward only to a limited extent, molecular complexes or compounds are formed. If it proceeds further, it may go over into a chemical reaction.

There are, of course, many intermediate and mixed cases. In the formation of loose molecular complexes, "physical" as well as "chemical" forces are involved. Stronger molecular complexes may, especially when assisted by a favorable environment, act as stepping-stones or "intermediates" in chemical reactions.

Wave functions for a familiar type of molecular compound, and for a typical molecular complex, can be written as follows:

$$\psi(\mathrm{Me_3N \cdot BF_3}) \approx a\underset{\text{no-bond}}{\psi(\mathrm{Me_3N,\ BF_3})} + b\underset{\text{dative}}{\psi(\mathrm{Me_3N^+ - BF_3})}\,,$$

$$\psi(\mathrm{Bz \cdot I_2}) \approx a\psi(\mathrm{Bz,\ I_2}) + b\psi(\mathrm{Bz^+ - I_2^-})\,.$$

In these expressions, the first molecule (trimethylamine, $Me_3N$; or benzene, Bz or $C_6H_6$) is the electron donor, the second is the electron acceptor. The wave function of the compound or complex is written as a linear combination of a "no-bond" function involving only "physical" forces between the molecules, and a "dative" function in which an electron has been transferred from the donor to the acceptor molecule, leaving each molecule with an odd electron so that a covalent bond is formed. In the first of the above two examples, there is little doubt that $b^2 > a^2$, perhaps even $b^2 \gg a^2$. In the second example, $b^2 \ll a^2$; from the dipole moment of $Bz \cdot I_2$, a value $b^2 \approx 0.03$ has been estimated, but possibly $b^2$ may be somewhat larger.**

* The first part of this paper was similar to the paper on Intermolecular Charge-Transfer Forces given at the Kyoto Conference, and only an outline of this part is given here. For further details, one may refer to the Kyoto paper and to R. S. Mulliken: J. Am. Chem. Soc. **72** (1950) 600; **74** (1952) 811; J. Phys. Chem. **56** (1952) 801. See also Paper No. 25 in ONR Report on September, 1951, Conference on "*Quantum-Mechanical Methods in Valence Theory*", issued by the Office of Naval Research, Department of Navy, Washington 25, D. C.

** The magnitude of $b^2$, rather than of $b$, is the best measure of the amount of dative character in the total wave function. Roughly, $a^2 + b^2 = 1$ if the total $\psi$ is normalized to unity, as is assumed here.

SOURCE: Symposium on Molecular Physics at Nikko, 1953.

Still other examples of loose complexes are

$$\psi(C_2H_4 \cdot Ag^+) \approx a\psi(C_2H_4\,,\ Ag^+) + b\psi(C_2H_4{}^+ - Ag)\,;$$

$$\psi(Bz \cdot Ag^+) \approx a\psi(Bz,\ Ag^+) + b\psi(Bz^+ - Ag)\,.$$

Evidence showing the existence of the indicated complexes* is found for solutions in aqueous media, *e.g.*, in aqueous $Ag^+NO_3{}^-$ solution. A complex of this type also occurs when $Ag^+ClO_4{}^-$ is dissoved in benzene, in which it is very soluble; from such a solution, stable white crystals of $Bz \cdot Ag^+ClO_4{}^-$ complex can also be obtained.

An important principle governing the donor and acceptor activities of molecules is that, other things being equal, a low minimum ionization potential makes a molecule a good donor, and a high electron affinity makes a molecule a good acceptor. Negative ions should then tend to be good donors, positive ions to be good acceptors. Other factors, such as size and shape, and degree of localization of the donor or acceptor function, are of course also important. The two principles stated above may conveniently be called the ***energy principle*** and the ***accessibility principle.***

In every case of the formation of a molecular complex with a wave function of the form

$$\psi(D \cdot A) \approx a\psi(D,\ A) + b\psi(D^+ - A^-)\,,$$

where D is the donor and A the acceptor molecule, an essential condition is that the two molecules be so oriented that $\psi(D,\ A)$ and $\psi(D^+ - A^-)$ belong to the same symmetry species with respect to a group-theoretical classification based on the geometrical symmetry of the complex for the particular orientation in question. If the two $\psi$'s belong to different symmetry species, $b$ is necessarily zero. On the other hand, for certain orientations $b$ should be a maximum. As was pointed out in the Kyoto conference report (see Eq. 6 of that report), $b$ should tend to be a maximum when the *overlap integral* between the donor molecular orbital (highest filled orbital of the donor) from which the electron has come in $\psi(D^+ - A^-)$, and the $A^-$ orbital (lowest unfilled orbital of the acceptor) into which it has gone, is a maximum. It should of course be noted that maximum overlap depends on the distance between D and A as well as on the orientation, and that certain orientations do not permit as close approach as others.

The applicatian of this *orientation principle* to the $Bz \cdot Ag^+$ complex was outlined in the Kyoto conference report, and it was pointed out that the most probable predicted location for the $Ag^+$ is above any one of the three C—C bonds of the Bz, on either side of the Bz plane. It was also mentioned that these are exactly the locations actually found by Rundle in an X-ray structural analysis of $Bz \cdot Ag^+ClO_4{}^-$ crystals. Similar considerations applied to the $C_2H_4 \cdot Ag^+$ complex indicate that here the $Ag^+$ should be located centrally above the middle of the C—C bond, on either side of the $C_2H_4$ plane.

Further types of molecular complexes include those formed by nitro-compounds as acceptors, for example

* In the case of $\psi(C_2H_4 \cdot Ag^+)$ and especially in similar complexes with $Cu^+$ or $Pt^{++}$ instead of $Ag^+$, there is good evidence that $\psi(C_2H_4 \cdot A^+)$, where $A^+$ is the acceptor ion, contains a third important term $c\psi(A^{++} - C_2H_4{}^-)$ in which $C_2H_4$ now acts as the acceptor and $A^+$ as the donor. (See M. Dewar; Bull. Soc. chim. de France **18**, C. 71 (1951): see Dewar's remarks in the discussion on p. C. 79.)

$$\psi(\text{Bz}\cdot\text{TNB}) \approx a\psi(\text{Bz, TNB}) + b\psi(\text{Bz}^+ - \text{TNB}^-)\ ,$$

where TNB means symmetrical trinitrobenzene. In the cases of $\psi(\text{Bz}\cdot\text{TNB})$, $\psi(\text{Bz}\cdot\text{I}_2)$, $\psi(\text{Bz}\cdot\text{Ag}^+)$, and various others, it appears probable that there is not much localization, in the dative part of the wave function, of the positive charge within the donor molecule ion, of the negative charge within the acceptor ion (or atom), or of the covalent bond.

However, the *p*-nitroaniline crystal seems to furnish an interesting example of complex-formation in which the donor, acceptor, and covalent bond activities are all rather strongly localized, each in a different location. In crystalline molecular complexes, usually *each* molecule may interact with *two* neighbors, one on each side.* For example, in crystalline $\text{Bz}\cdot\text{AgClO}_4$ each Bz interacts with an $\text{Ag}^+$ on each side of the Bz plane, and each $\text{Ag}^+$ with a Bz on each side. However, each individual interaction can still be represented by a wave function in the same way as before. These "local" wave functions can then be combined to form a wave function for the entire crystal. In the *p*-nitroaniline crystal, each molecule is linked with neighbors in such a way that it acts as a donor toward one, with the donor action localized on the $\text{NH}_2$ group, and as an acceptor toward another, with the acceptor action localized on the $\text{NO}_2$ group. The total wave function of the crystal would then be made up from a no-bond function for each individual molecule, together with dative wave functions for pairs of properly oriented neighboring molecules, with the bond structure shown in the left figure for each such dative interaction.

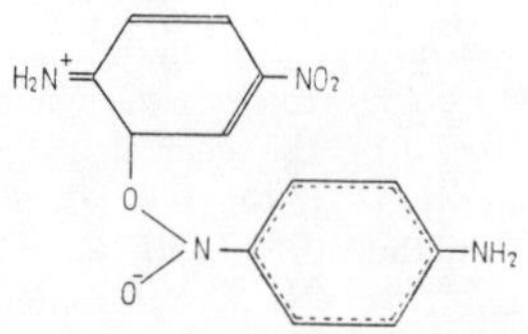

The donor action is largely on the N atom of the $\text{NH}_2$ group of one molecule, the acceptor action on the N atom (which in the normal $\text{NO}_2$ structure would be $\text{N}^+$) of the $\text{NO}_2$ group of the other molecule. The covalent action involves a change in the aromatic bond structure in the ring of the first molecule, but its most striking feature is a cross linkage between one carbon of this ring and one oxygen of the $\text{NO}_2$ group of the other ring.

In the complexes of aniline ($\text{C}_6\text{H}_5\text{NH}_2$) and substituted anilines with the nitrobenzenes, *e.g.*, $\text{C}_6\text{H}_5\text{NH}_2\cdot\text{TNB}$, which are known in solution, it seems possible that there is more or less localization of the donor, acceptor, and covalent functions, similar to the localizations in the p-nitroaniline crystal. However, there is no evidence to show whethere or not this is important. In complexes like $\text{Bz}\cdot\text{TNB}$, there seems to be no reason to expect localization, except such as is inherent in the forms of the $\pi$ MO's of $\text{Bz}^+$.

Nitro compounds form loose molecular complexes not only with electron donors, such as aromatic hydrocarbons or amines, but also with other nitro compounds. Briegleb attributed the stability of *all* nitro complexes to electrostatic forces resulting from the strong local dipole moment of each $\text{NO}_2$ group, plus dispersion forces. It now appears

* A different pattern is shown by the phenoquinone crystal, which is built up of *separate* units each consisting of a 2:1 molecular complex, one molecule of quinone (the acceptor) being sandwiched between two p-cresol, *i.e.*, $\text{p-C}_6\text{H}_4(\text{OH})_2$, donor molecules (see T. T. Harding and S. C. Wallwork: Acta Crystallographica **6** (1953) 791).

probable that Briegleb's original explanation is largely valid for complexes between different nitro compounds, but that in complexes between nitro compounds and donor molecules, donor-acceptor forces largely predominate over electrostatic forces.

As was mentioned at the beginning of this account, donor-acceptor interactions may lead either to the formation of molecular complexes, or to chemical reactions. There is reason to believe that in many reactions, the first step may be the formation of a loose complex ("*outer complex*") in which a small amount of charge transfer has occurred. Then a second step may lead to an "inner complex", and this step may be followed by completion of a chemical reaction, finally involving transfer of a complete ionic charge. It seems probable that the principles outlined above for the formation of loose complexes can be used to obtain insight into *reaction paths* in such reactions.

Before discussing reaction paths, it will be desirable to illustrate the meaning of the term "inner complex". In liquid HF the following equilibrium occurs:*

$$C_6H_6 + HF \rightleftarrows C_6H_7^+ F^-$$

The action is similar to

$$NH_3 + HCl \rightarrow NH_4^+ Cl^-$$

in water solution. In each case a donor molecule ($C_6H_6$ or $NH_3$) interacts with a weak acceptor (HF or HCl), *with the assistance of a polar medium* which favors formation of ions, to form an "inner complex". In principle, the same thing could happen for benzene plus $I_2$ or $Cl_2$ in a polar medium;

$$C_6H_6 + Cl_2 \rightleftarrows C_6H_6 \cdot Cl_2 \qquad \text{"outer complex"}$$

$$C_6H_6 \cdot Cl_2 \underset{\text{polar medium}}{\rightleftarrows} (C_6H_6Cl)^+ Cl^- \qquad \text{"inner complex".}$$

In this case, the outer complex is known, but the inner complex is apparently not stable. With the acid molecules HF and HCl there is evidence that a very loose outer complex can exist (for example, when HCl gas is dissolved in $C_6H_6$, or better, toluene), but that also, in a suitable very polar medium, the inner complex (or its ions) can be formed. What is involved can be shown qualitatively by the following potential curves:

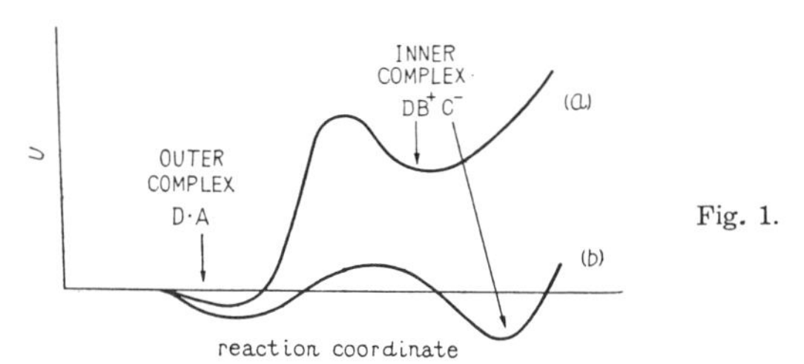


Fig. 1.

* Cf. M. Kilpatrick and F. E. Luborsky, J. Am. Chem. Soc. **75** (1953) 577, and references given there. Actually this type of reaction does not go appreciably to the right for benzene itself in pure HF. It goes strongly for hexamethylbenzene in HF. For benzene it goes if assisted by $BF_3$ acting as an "auxiliary acceptor":

$$C_6H_6 + HF + BF_3 \rightleftarrows C_6H_7^+ BF_4^-.$$

In the absence of a polar medium (curve a) the inner complex, while it may have a potential minimum, is too high in energy to be formed (examples, $NH_3+HCl$ in the pure dry vapor state), but the outer complex may be more or less stable as a loose complex. In the presence of a sufficiently strong polar medium (curve b) the inner complex, in which often the acceptor anion $A^-$ is broken up so as to yield an ion-pair $DB^+C^-$, may be more stable than the outer complex. (Further, the inner complex may largely dissociate into ions in the polar medium, but this proecss may be considered as a secondary effect for our purposes.)

An important class of reactions to which many of the foregoing considerations can be applied is that of electrophilic substitution reactions on aromatic molecules. (Similar considerations can be applied to addition reactions on unsaturated molecules.) Ingold's "electrophilic" and "nucleophilic" reagents are respectively electron acceptor and donor molecules. An example of an electronphilic substitution reaction is the nitration of benzene in strong acid solution. According to the work of Hughes, Ingold and co-workers,* the essential steps are

$$C_6H_6+NO_2^+ \rightarrow (C_6H_6NO_2)^+;$$

$$(C_6H_6NO_2)^+ + A^- \rightarrow C_6H_5NO_2+HA.$$

In this reaction, $NO_2^+$ is an "electrophilic reagent" or electron-acceptor molecule. As we have seen previously, benzene is a donor molecule. HA may be $HNO_3$ or perhaps $H_2SO_4$.

Guided by the considerations presented above. we can form a plausible idea of what may happen during the course of this reaction. First, we should keep in mind that the reaction occurs in a polar medium containing various ions (perhaps $NO_3^-$, $H_3O^+$, $HSO_4^-$, $H_3SO_4^+$, and others) and neutral molecules ($H_2O$, $H_2SO_4$, $C_6H_6$ etc.). As the $NO_2^+$ approaches the $C_6H_6$, at first a loose outer complex $Bz \cdot NO_2^+$ may be formed, with

$$\psi(Bz \cdot NO_2^+) \approx a\psi(Bz,\ NO_2^+) + b\psi(Bz^+ - NO_2),$$

somewhat like the $Bz^+ \cdot Ag^+$ complex described above. For this loose complex, there may be no strong tendency for the $NO_2^+$ to localize near any particular atom of the Bz molecule. If the $U$ curve is somewhat like curve (b) in the Fig. 1 given above, passage to the inner complex would be over a low activation barrier. During this process, $b/a$ would increase strongly, the $NO_2^+$ (which by itself should be linear like $CO_2$) would become bent, and solvent molecules surrounding the $NO_2^+$ would be pushed aside permitting the $NO_2^+$ to come closer to the Bz. Finally in the inner complex the charge of the $NO_2^+$ would be largely transferred to the Bz and the $NO_2$ would become strongly attached to the $Bz^+ (b^2 \gg a^2)$. During the process of attachment, the $NO_2$ would become strongly bent, and the attachment would be largely localizd into a $\sigma$ bond from the N atom to one of the carbon atoms, giving an inner complex of approximately the structure shown in the left.

$$\left[ C_6H_6 <^{NO_2}_{H} \right]^+$$

A similar structure is accepted by most people for the

* C. K. Ingold: "*Structure and Mechanism in Organic Chemistry*", (Cornell University Press, Ithaca, New York, 1953).

$C_6H_7^+$ion within the inner complex $C_6H_7^+F^-$ formed by $C_6H_6+HF$ as described above.* The union of the donor $C_6H_6$ and the acceptor $NO_2^+$ to form the strongly bound molecule-ion $(C_6H_6NO_2)^+$ is in some respects analogous to the union of $Me_3N$ and $BF_3$ to form the stable molecular compound $Me_3N \cdot BF_3$, mentioned earlier.

However, one more step is now needed to complete the substitution reaction. This must apparently consist in an attack by a donor molecule (most likely an anion $A^-$) on $(C_6H_6NO_2)^+$ now acting as an acceptor, as follows:

$$(C_6H_6NO_2)^+ + A^- \rightarrow C_6H_5NO_2 + HA\,.$$

For the complete reaction, say

$$C_6H_6 + NO_2^+ + A^- \rightarrow C_6H_5NO_2 + HA\,,$$

the potential curve may be somewhat as follows:

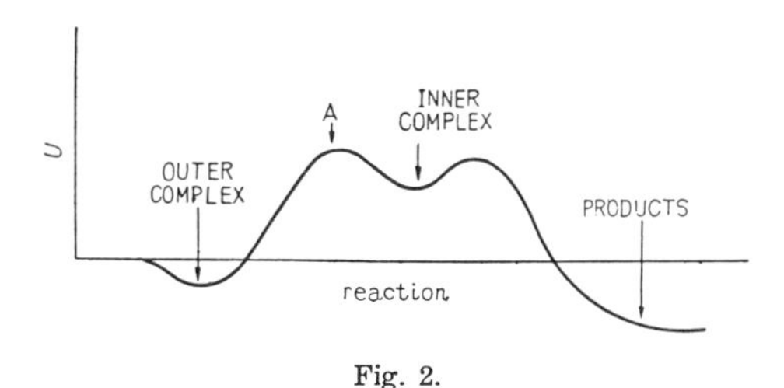


Fig. 2.

Here the inner complex or "reaction intermediate", not being an ion-pair, is not stabilized by the polar medium, relative to $C_6H_6+NO_2^+$ (unlike the situation to which the Fig. 1 refers).

Supposing that the Figure just given is more or less correct, it would be interesting to know how the $C_6H_6$ and $NO_2^+$ are oriented relative to each other at each stage in the reaction. No attempt will be made here to give an explicit answer, but it seems worth while to point out that the *orientation principle* mentioned earlier should be valuable in predicting approximately what relative orientations the molecules should assume at each stage of approach; or at least, in *elminating from consideration* many reaction paths which otherwise would seem plausible. Especially important would be to determine what orientation, at the highest point A of the $U$ curve,** would make the latter, that is, the predicted activation energy, as low as possible.

Considerations similar to the foregoing are applicable also to electrophilic substitution reactions in which the reagent is a neutral molecule, *e.g.*, $Cl_2$ or one of the other halogen molecules. In such reations, the inner complex is an ion-pair, and so is stabilized, and the reaction assisted, by a polar medium, *e.g.*, water (see Fig. 1). Water can also assist here in the second stage of the reaction:

* See Pickett, Muller and Mulliken: J. Chem. Phys. **21** (1953) 1400, and subsequent longer paper, for further discussion.

** The organic chemical evidence (see Ingold's book) seems to indicate that an activation barrier exists for this reaction, and that the barrier is higher before than after the inner complex.

$$\begin{cases} C_6H_6+Cl_2 \rightarrow C_6H_6\cdot Cl_2 \quad \text{(outer complex)} \\ C_6N_6\cdot Cl_2 \rightarrow \left[ \bigcirc <^{Cl}_{H} \right]^+ Cl^- \quad \text{(inner complex)} \end{cases}$$

$$(C_6H_6Cl)^+Cl^- + H_2O \rightarrow C_6H_5Cl + H_3O^+ + Cl^-.$$

Instead of a polar medium, an auxiliary acceptor can also act as an assisting agent, for example perhaps*

$$C_6H_6\cdot Cl_2 + AlCl_3 \rightarrow \left[ \bigcirc <^{Cl}_{H} \right]^+ AlCl_4^- \quad \text{(inner complex).}$$

The examples so far considered involve *inter*molecular donor-acceptor reactions. The idea of partial charge transfer from a donor to an acceptor *within* a molecule also has many possibilities of application. Recently, Nagakura and Tanaka at Tokyo and, independently, Stuart at Oxford, have used the idea in interpreting various aspects of the structure and spectra of substituted aromatic molecules. The general idea is that for electron-donor substituents (OH, $NH_2$, etc.) there is a partial electron transfer to the aromatic ring (or rings) acting as acceptor, while for acceptor substituents ($NO_2$, CHO, etc.) there is a transfer trom the ring acting as donor to the substituent. Recent work of Fukui and coworkers** on "frontier electrons" seems also to be related to the same point of view.

* The example illustrates the principle, but may not be a realistic one.
** See paper by Fukui and review by G. Araki in the Proceedings of the Conference.

[Reprinted from the Journal of the American Chemical Society, 76, 3869 (1954).]


[CONTRIBUTION FROM THE LABORATORY OF MOLECULAR STRUCTURE AND SPECTRA, DEPARTMENT OF PHYSICS, THE UNIVERSITY OF CHICAGO]

# Molecular Compounds and Their Spectra. IV. The Pyridine–Iodine System[1]

BY C. REID[2a] AND R. S. MULLIKEN[2b]

RECEIVED JANUARY 27, 1954

The visible and ultraviolet absorption spectra of dilute solutions of iodine plus pyridine in heptane have been studied, and the existence of an equilibrium with a 1:1 molecular complex $Py \cdot I_2$ ("outer complex") was demonstrated [$K = 290$ at 16.7°, where $K$ means $(Py \cdot I_2)/(Py)(I_2)$]. The corresponding changes in heat content, entropy, and free energy (at 17°) in formation of the complex were determined to be −7.8 kcal./mole, −15.5 cal./deg. mole, and −3.3 kcal./mole, respectively. The location and intensities of the $I_2$ band ($\lambda_{max}$ 4220 Å., $\epsilon_{max}$ 1320) and of the charge-transfer band ($\lambda_{max}$ 2350, $\epsilon_{max}$ 50,000) of $Py \cdot I_2$ were determined. The λ 4220 band shifts gradually, and increases in intensity, on adding more pyridine to the aforementioned heptane solutions, until for pure pyridine solutions it has reached about λ 3890, with $\epsilon_{max}$ 2120, provided the solutions are not too dilute in iodine. These changes can most probably be attributed to a somewhat increased polarity and stability of the $Py \cdot I_2$ "outer complex" in the polar solvent pyridine than in the non-polar solvent heptane. There is no evidence of the presence of the "inner complex" $(PyI)^+I^-$ in more than small concentrations, but conductivity studies by Kortüm and Wilski indicate that appreciable small concentrations of its ions $(PyI)^+$ and $I^-$ are present in pure pyridine solutions of iodine. Additional studies in *very dilute* solutions of iodine in pyridine show further interesting spectroscopic changes, which are discussed, but we feel that further experimental study will be needed using extreme precautions toward exclusion of side-reactions, moisture or impurities.

## Introduction and Survey

Recent studies have confirmed older ideas that in its violet solutions, iodine exists essentially free, but that in its brown solutions it forms 1:1 molecular complexes with the solvent.[3] The strong visible absorption of $I_2$ vapor with maximum at λ 5,200 is essentially unchanged in "violet" solvents, but in solutions where it forms complexes this peak is shifted toward shorter wave lengths; this accounts for the altered color. In addition, a new very intense peak characteristic of the complex, first noted by Benesi and Hildebrand for aromatic solvents, appears at shorter wave lengths, usually in the ultraviolet. The interpretation of this new peak as a *charge-transfer* spectrum has proved important for a clearer understanding of the electronic structure of these complexes.[3]

There is evidence[4–7] that iodine forms especially stable complexes with pyridine and related compounds. Waentig[4] reported golden crystals, which he attributed to $Py \cdot I_2$, crystallizing from a saturated solution of iodine in pyridine. From heats of solution Hartley and Skinner[5] estimated the heat of formation of $Py \cdot I_2$ in solution to be about 7.95 kcal./mole, much larger than for other types of iodine complexes. Similarly, the enhancement of the dipole moment in the formation of $Py \cdot I_2$ is exceptionally large.[6] Further, the change in the infrared spectrum of Py when it goes into $Py \cdot I_2$ is much greater[5] than the corresponding effect in the case of complex-forming solvents of other types.

Audrieth and Birr[8] reported that solutions of iodine in pyridine show high electrical conductivities, which slowly increase with time to asymptotic

(1) This work was assisted by the Office of Ordnance Research under Project TB2-0001(505) of Contract DA-11-022-ORD-1002 with The University of Chicago.

(2) (a) On leave of absence from The University of British Columbia, 1952–1953. Present address: Department of Chemistry, The University of British Columbia, Vancouver, Canada. (b) On leave of absence from The University of Chicago, 1952–1953; Fulbright Research Scholar at Oxford University, 1952–1953.

(3) See R. S. Mulliken, (a) THIS JOURNAL, **72**, 600 (1950); **74**, 811 (1952); (b) *J. Phys. Chem.*, **56**, 801 (1952), for quantum-theoretical interpretation of molecular complexes and their spectra, and a comprehensive review. These are I, II and III of the present series.

(4) (a) P. Waentig, *Z. physik. Chem.*, **68**, 513 (1909); Chatelet, *Ann. chim.*, [11] **2**, 12 (1934); H. Carlsohn, *Z. angew. Chem.*, **45**, 580 (1932); **46**, 747 (1933).

(5) K. Hartley and H. A. Skinner, *Trans. Faraday Soc.*, **26**, 621 (1950).

(6) Y. K. Syrkin and K. M. Anisimowa, *Doklady Akad. Nauk. SSSR*, **59**, 1457 (1948); G. Kortüm, *J. chim. phys.*, **49**, C127 (1952); G. Kortüm and H. Walz, *Z. Elektrochem.*, **57**, 73 (1953).

(7) D. L. Glusker, H. W. Thompson and R. S. Mulliken, *J. Chem. Phys.*, **21**, 1407 (1953), and references given there; also further unpublished results of Mr. Glusker. Also W. Luck, *Z. Elektrochem.*, **59**, 870 (1952), especially table IV.

(8) L. F. Audrieth and E. J. Birr, THIS JOURNAL, **55**, 668 (1933).

values. According to them the molar conductivity based on $I_2$ is so high in dilute solutions that it cannot be explained by simple dissociation into $I^+$ (or $PyI^+$) and $I^-$. They suggested instead the formation of a ternary electrolyte

$$Py \cdot I_2 \rightleftharpoons Py^{++} + 2I^-$$

However, recent work of Kortüm and Wilski,[9] using very great precautions to keep moisture excluded, indicates that iodine in freshly prepared solutions in pure pyridine at concentrations in the neighborhood of $10^{-4}$ molar gives only a small conductivity, though larger than for most iodine complexes.[10] They find, however, that this increases with time, and attributes the effect to a slow iodination in the ring; the effect is strongly catalyzed by platinum sponge.

Kleinberg, VanderWerf and associates[11] have made a spectrophotometric investigation of solutions of iodine in pyridine (also in quinoline). They too conclude that a very slow iodination in the ring occurs; this should liberate $I^-$ ions, which may form $I_3^-$ ions with $I_2$.

Mulliken[12] in 1952 suggested that when $I_2$ is dissolved in pyridine the following should be considered as the primary reactions

$$Py + I_2 \underset{\text{fast}}{\rightleftharpoons} Py \cdot I_2 \text{ "outer complex"} \quad (1)$$

$$Py \cdot I_2 \rightleftharpoons (PyI)^+I^- \text{ "inner complex"} \quad (2)$$

$$(PyI)^+I^- \underset{\text{fast}}{\rightleftharpoons} PyI^+ + I^- \quad (3)$$

The "outer complex" $Py \cdot I_2$ in (1) would be a molecular complex of the usual type. The "inner complex" in (2) would be an essentially ionic structure (N-iodopyridinium iodide). It was suggested that, in iodine solutions in pyridine, the pyridine has a double role, acting as an electron donor toward $I_2$ in reaction (1), and as a polar medium in assisting reactions (2) and (3).

The present research was undertaken in the hope of studying these two roles of pyridine separately by a spectrophotometric investigation, first, of equilibrium (1) at varying low concentrations of Py and $I_2$ in a non-polar solvent medium; second, of the combined equilibria (1), (2) and (3) in a polar medium (perhaps pyridine itself, or preferably a different polar solvent). These two phases of the present work are reported in sections I and II below.

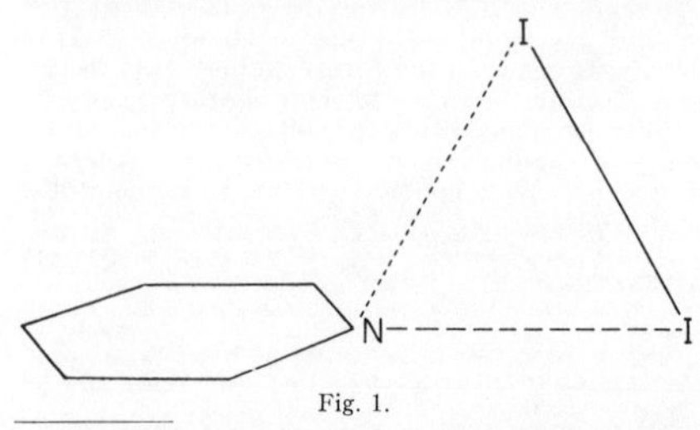

Fig. 1.

In section I equilibrium (1) was successfully studied in heptane solution. The visible iodine band of the outer complex $Py \cdot I_2$ was located at the exceptionally strongly shifted position of λ 4,220 (for free iodine it is at λ 5,200), and the expected charge-transfer band at λ 2,350. The equilibrium constant for (1), and the heat of formation of $Py \cdot I_2$, were determined. This work confirms other indications[4,7] that $Py \cdot I_2$ is an exceptionally tightly bound outer complex. Taking into consideration the observed dipole moments[4] of Py (2.28 *D*) and of $Py \cdot I_2$ (4.5 *D*), and assuming a geometrical structure[13] somewhat as shown in Fig. 1, one can estimate that the outer complex $Py \cdot I_2$ may easily have as much as perhaps 25% dative character. That is, in the type of formulation given by Mulliken[3]

$$\psi(Py \cdot I_2) \approx a\psi_0(Py, I_2) + b\psi_1(Py^+ - I_2^-) \quad (4)$$
$$\qquad\qquad \text{no-bond} \qquad\qquad \text{dative}$$

with $a^2 \approx 0.75$, $b^2 \approx 0.25$. In eq. 4, because of the asymmetry (Fig. 1) and unusual strength of the complex, the dative function $\psi_1$ may be already approximately of the structure $C_5H_5\overset{+}{N} - I \; I^-$ with the $N^+$ bonded to one I atom nearly in the Py plane (N-iodopyridinium ion) leaving the other I atom as an $I^-$ above the plane.[14] An outer complex with an exceptionally large amount of dative character may well account for the fact[7] that complex formation causes greater changes in the infrared spectrum in the case of Py than for any other known cases (except the related picolines).

When the work reported in section II was undertaken, it was with the thought,[12] suggested by the conductivity studies of Audrieth and Birr,[8] that in pure Py, acting as a polar medium, (a) equilibrium for reaction (2) lies almost completely to the right; but (b) the reaction proceeds only very slowly, over a high potential barrier; and that as fast as $(PyI)^+$-$I^-$ is formed, reaction 3 proceeds largely to the right. However, the recent work of Kortüm and Wilski[7] indicates that ions $PyI^+$ and $I^-$ are formed at once in $I_2$ solutions in Py, in definite relatively small equilibrium concentrations, and that a later slow increase in ionic concentration is due to slow side-reactions. Taken in connection with our spectrophotometric results in sections I and IIA and the discussion presented in IIA, the work of Kortüm and Wilski indicates that in the absence of side-reactions most of the iodine would remain as $Py \cdot I_2$, but that a small portion of it has at once undergone reaction (2) followed by (3), or else perhaps the direct ionization

(9) G. Kortüm and H. Wilski, *Z. physik. Chem.*, **202**, 35 (1953). See also Kortüm, ref. 6.

(10) They find an ionic dissociation constant $(PyI^+)(I^-)/(Py \cdot I_2)$ of about $4.6 \times 10^{-8}$, which corresponds to about 2% ionization at $10^{-4}$ molar iodine. This may be compared with $1.2 \times 10^{-11}$ for $(H_2OI)^+ \cdot (I^-)/(H_2O \cdot I_2)$ as determined by R. P. Bell and E. Gelles [*J. Chem. Soc.*, 2734 (1951)] and smaller values (see ref. 9) for the benzene and dioxane complexes. However, it seems not impossible that some of the alcohols may have larger values [*cf.* L. I. Katzin, *J. Chem. Phys.*, **21**, 490 (1953)].

(11) (a) R. Zingaro, C. A. VanderWerf and J. Kleinberg, This Journal, **73**, 88 (1951); (b) J. Kleinberg, E. Colton, J. Sattizahn and C. A. VanderWerf, *ibid.*, **75**, 447 (1953).

(12) Reference 3a, p. 818; ref. 3b, pp. 812, 819.

(13) This is based on general considerations previously advanced by Mulliken (ref. 3).

(14) The Py would then be acting as an *n* donor in the terminology of ref. 3b. However, to a slight extent, it probably acts simultaneously as a π-donor (like benzene in its iodine complex; *cf.* footnote 42 on page 818 of ref. 3a. $\psi_1$ in eq. 4 would then involve a mixture of mainly *n* with a little π-donor action by the Py

$$Py \cdot I_2 \rightleftarrows PyI^+ + I^- \quad (5)$$

Further discussion will be given in section IIA.

## Experimental

C.P. pyridine was refluxed with chromium trioxide for several hours to remove traces of picolines, dried by NaOH, and distilled from magnesium perchlorate. C.P. iodine was sublimed and kept in a desiccator. Solvents were purified by the methods described by Potts.[15] Absorption measurements were made in a Beckman spectrophotometer, using 10-, 1-, 0.0296- and 0.0109-cm. cell thicknesses. Apart from the use of cells with fairly well fitting lids, no precautions were taken to avoid moisture uptake during a run. No lids at all were possible in experiments using spacers to decrease cell thickness.

## I. The $Py \cdot I_2$ Complex in a Non-polar Solvent

The equilibrium (1) was studied in very dilute solutions (<0.1% Py + $I_2$) in heptane (>99.9% by weight). As pyridine in increasing but small amounts is added to a (violet) dilute solution of iodine in heptane, the solution goes through a reddish color to golden brown. The uncomplexed $I_2$ peak at 5,200 Å. diminishes and is replaced by a new and somewhat higher but otherwise very similar peak at 4,220 Å. (*cf.* Fig. 2). The peaks are well

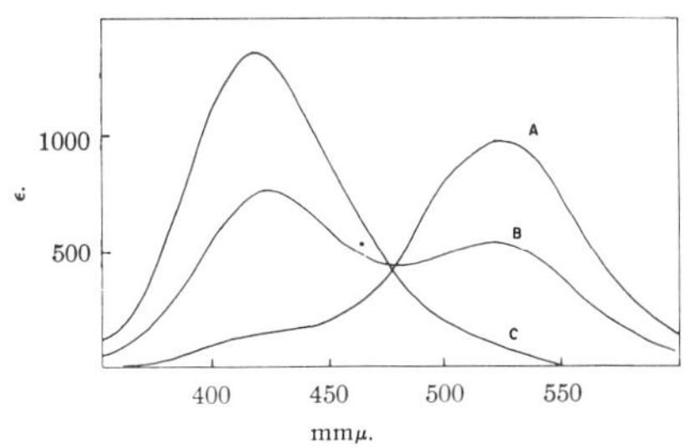


Fig. 2.—Plot of extinction coefficient ($\epsilon = (1/lc) \log_{10} (I_0/I)$, where $l$ = cell thickness, and $c$ = formal molarity based on total iodine added) against wave length for 0.0005 $M$ iodine solutions in heptane, with increasing amounts of pyridine. Room temperature, cell thickness = 1.00 cm.; A, pyridine 0.0005 $M$; B, pyridine 0.005 $M$; C, pyridine 0.25 $M$.

enough separated for a fairly accurate determination of the equilibrium constant $K$

$$K = \frac{(Py \cdot I_2)}{(Py)(I_2)} \text{ liters/mole}$$

From the $K$ values at 2° (649), 16.7° (290), and 41° (101)—*cf.* Fig. 3—a graph was made (Fig. 4) from which in the usual way the heat of dissociation of $Py \cdot I_2$ was calculated to be 7.8 ± 0.2 kcal./mole. It is of some interest that this result agrees closely with the value 7.95 kcal./mole estimated from the heat of solution of $I_2$ in pure Py by Hartley and Skinner.[3] From the available data the free energy and entropy changes for reaction 1 were also computed, the results, in conventional units, at 17°, being $\Delta F = -3.3$, $\Delta H = -7.8$, $\Delta S = -15.5$.

McConnell, Ham and Platt[16] have predicted that

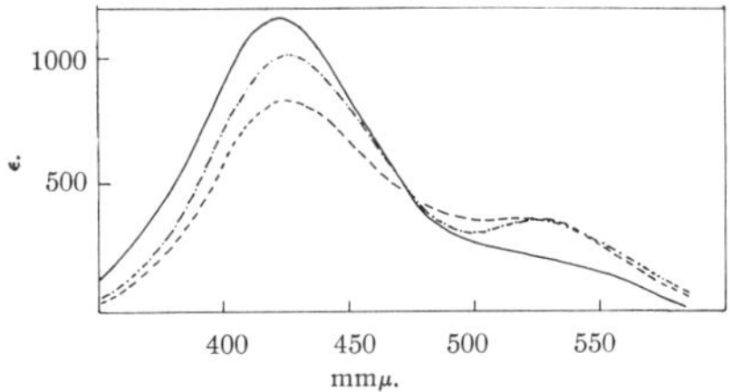


Fig. 3.—Plot of formal extinction coefficient (see Fig. 2) against wave length of $I_2$ + Py in heptane for a series of temperatures; cell thickness = 1.00 cm.

| | | |
|---|---|---|
| - · - · - 2° | Py = 0.005 $M$ | $I_2$ = 0.0005 $M$ |
| ——— 16.7° | Py = 0.025 $M$ | $I_2$ = 0.000625 $M$ |
| - - - - - 41° | Py = 0.025 $M$ | $I_2$ = 0.000625 $M$ |

The equilibrium shifts strongly toward $Py \cdot I_2$ as the temperature is lowered, but the pyridine concentration has been lowered in the 2° experiment so that both peaks are measurable.

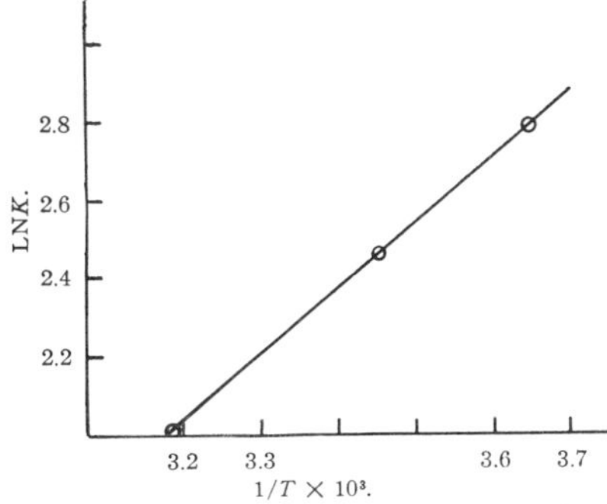


Fig. 4.—Plot of $K = (Py \cdot I_2)/(Py)(I_2)$ (liters/mole) against $1/T$ for the equilibrium between iodine and pyridine in heptane.

the charge-transfer peak of $Py \cdot I_2$ should occur at 38,000 kayser (2,635 Å.), on the basis of an electron impact value of 9.8 volts[17] for the ionization potential of pyridine. Use of the same ionization potential in an equation given by Hastings, Franklin, Schiller and Matsen,[18] which fits a great number of iodine complexes closely, gives a similar prediction (38,300 kayser or 2,610 Å.). A search of this region using $Py \cdot I_2$ in heptane at concentrations of 0.01 $M$ in Py and 0.0005 $M$ in $I_2$, with thin cells to avoid excessive pyridine absorption, showed such a band with peak at 42,600 kayser (2,350 Å.). The extinction coefficient is sufficiently large ($\epsilon$ = 50,000) that no difficulty was experienced in locating this band in spite of the considerable pyridine absorption in this region.

(15) W. J. Potts, *J. Chem. Phys.*, **20**, 809 (1952).

(16) H. McConnell, J. S. Ham and J. R. Platt, *J. Chem. Phys.*, **21**, 66 (1953).

(17) Hustrulid, Kusch and Tate, *Phys. Rev.*, **54**, 1037 (1938). Stevenson and Schissler in unpublished work have recently obtained 9.85 volts (private communication from D. P. Stevenson).

(18) S. H. Hastings, J. L. Franklin, J. C. Schiller and F. A. Matsen, THIS JOURNAL, **75**, 2900 (1953). The form of their equation is based on Mulliken's theoretical discussion in ref. 3a.

The charge-transfer band is shown in Fig. 5, in which the pyridine absorption was automatically cancelled out by using as a blank part of the heptane + pyridine solution which had been used to dissolve iodine. No correction for free $I_2$ was needed, in view of its very low concentration and small absorption near λ 2,350.

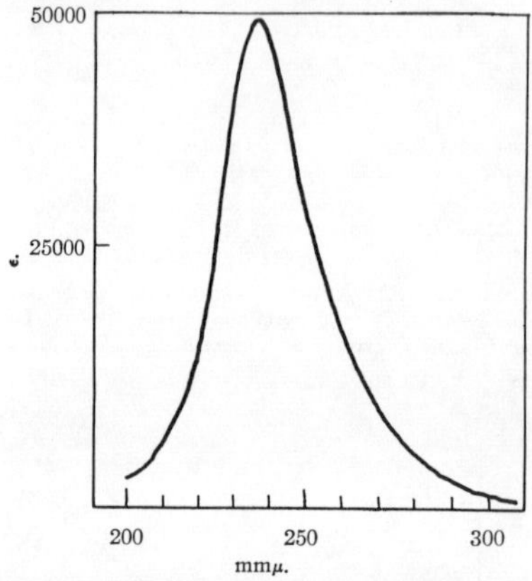


Fig. 5.—The Py·$I_2$ charge-transfer spectrum at room temperature. The absorption of pyridine which lies in this region was cancelled out exactly by dividing a 0.05 *M* pyridine solution in heptane into two parts, adding iodine (0.0005 *M*) to one half, and using the other half as a blank; cell thickness = 0.0296 cm. Free $I_2$ is negligible in its effect.

The fact that the observed charge-transfer band is at somewhat shorter wave lengths than predicted may perhaps be connected with the exceptionally high stability of the Py·$I_2$ complex. The validity of the predictions mentioned above is dependent on an approximate constancy of certain parameters in the equations used. Although this constancy is apparently surprisingly well fulfilled for most iodine complexes,[18] it has no obvious theoretical basis. On the other hand, a value of 10.3 volts for the ionization potential would give a prediction corresponding to the observed position of the charge-transfer band.

Perhaps the observed 9.8 volts is the first π-ionization potential, but the relevant potential, which should correspond not to a π but to a non-bonding (*i.e.*, "onium" or *n*) ionization potential essentially of the N atom,[14] is higher. However, the absorption spectrum of pyridine suggests that the π and *n* ionization potentials are actually almost equal. This statement is based on the fact that, taking the means of the frequencies of transitions to corresponding singlet and triplet states,[19] the frequencies of the first "*n*–π" and "π–π" transitions are nearly equal. But even if the two ionization potentials are equal one should bear in mind that it has never been proved that, in the so-called *n*–π transitions in the aza-substituted aromatics, the transition is really from a true localized non-bonding (*n*) orbital of the N atom. It would be safer to call such transitions σ–π transitions, where the σ orbital may be only partly localized on the N atom. The appropriate localized N atom true *n* ionization potential required in predicting the location of the charge-transfer band would then correspond to a weighted mean of several σ-ionization potentials and might be appreciably greater than the minimum σ ionization potential.

(19) *Cf. e.g.*, J. H. Rush and H. Sponer, *J. Chem. Phys.*, **20**, 1847 (1952), Table VII.

## II. The System Pyridine Plus Iodine in Polar Solvents

**A. The Transition to Pure Pyridine as Solvent.**—When, in a dilute solution of iodine plus pyridine in heptane, the pyridine concentration is gradually increased, the λ 4,220 Py·$I_2$ iodine band begins to shift toward shorter wave lengths, and its extinction coefficient increases. The relations between position and ε of the band maximum, and pyridine concentration, are shown in Fig. 6, for a fixed concentration (0.0005 molar) of iodine. (At these concentrations, practically all the iodine should be complexed.) The position of the band approaches a limiting value of 3,890 Å., and $\epsilon_{max}$ a limiting value of 2,120 in pure pyridine.[20] These changes, as distinct from some of the phenomena to be described in Part C of this section, are reversible: dilution of the solution with heptane results in a return of the position of the band to λ 4,220 with corresponding diminution in intensity.

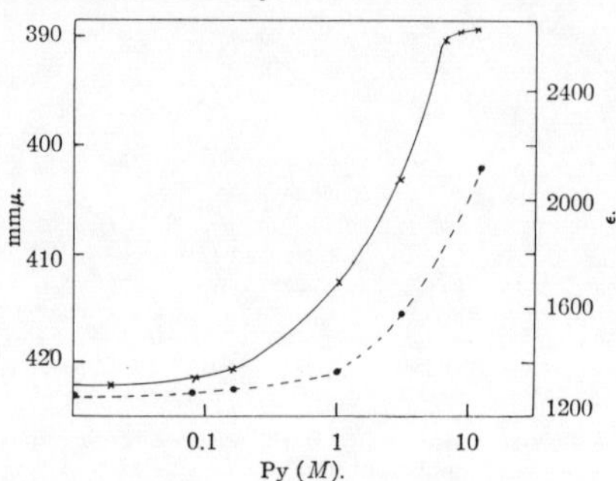


Fig. 6.—Variation in position (×——×) and extinction coefficient (●- - -●) of the Py·$I_2$ absorption band with increasing pyridine concentration. $I_2$ = 0.0005 *M*; solvent, heptane; room temperature.

Attempts were made also to see what happens to the "charge-transfer" band at 2,350 Å. as the pyridine concentration is increased. Unfortunately, even using special thin cells (0.001-cm.) constructed by putting a rolled lead spacer between quartz plates the experiments could be carried out only up to 1.5 *M* Py (see Fig. 7). At this Py concentration, with 0.06 *M* iodine, the position of the charge-transfer band appears to be shifted to about 2,450 Å. No appreciable change in the ratio of the pyridine

(20) Kleinberg and collaborators (ref. 11b), for iodine at 2 × $10^{-4}$ molar in pyridine, find $\lambda_{max}$ = 383–380 mμ and $\epsilon_{max}$ = 2600–2700. The moderate difference between their results and ours at 5 × $10^{-4}$ molar can be understood in terms of our findings at high dilution, as reported in section IIC and Fig. 8.

molar absorption to that of the charge-transfer band could be detected.

In connection with the interpretation of the foregoing observations, some unpublished infrared work of Glusker[7] on solutions containing Py and $I_2$ is highly relevant. He finds no appreciable difference between the modified Py infrared bands in $CS_2$ solutions very dilute in Py and in those much more concentrated, up to $I_2$ solutions in pure Py. This strongly indicates that these modified bands are due to essentially the same Py·$I_2$ entity whether the solvent is an inert one ($CS_2$) or pure Py. The gradual shift of the λ 4,220 Py·$I_2$ band in heptane to λ 3,890 in Py solvent may now probably be attributed to a gradual clustering of polar Py molecules around the strongly polar Py·$I_2$ molecules, causing these to become more polar [increased $b$ in eq. 4] and more stable; but the infrared evidence indicates that these changes cannot be very large.[21]

It was suggested earlier[12] that the "inner complex" of structure $(PyI)^+I^-$ may be so much stabilized by the polar solvent pyridine as to be present in predominant amount in that solvent. But according to the preceding paragraph, it appears that Py·$I_2$ remains predominant even in pure Py, and this suggests that $(PyI)^+I^-$ if present is only in small amounts. The definite presence[9] of the *ions* of $(PyI)^+I^-$ in small concentrations does, however, presumably indicate that a correspondingly small amount of the inner complex itself is present in accordance with eq. 3.[22]

**B. Pyridine Plus Iodine in Other Polar Solvents.**—In order to differentiate between specific effects due to excess pyridine and effects due to increasing polarity of the solvent as pyridine is added to heptane solutions of iodine, attempts were made to study the pyridine–iodine complexes in other polar solvents.

Experiments in which pyridine was added to iodine dissolved in methanol were inconclusive because the very strong MeOH·$I_2$ charge-transfer band showed that most of the iodine was complexed with methanol rather than with pyridine. This was true up to concentrations of 4–5% of pyridine, beyond which it was impracticable to go.

When Py·$I_2$ solutions in Py were diluted with water, precipitation of golden-yellow "Py·$I_2$" crystals resulted.[23] Examination of the resulting solutions after filtration showed no trace of the characteristic Py·$I_2$ bands, but only $I_3^-$ bands and visible and charge-transfer bands attributable to small amounts of complexes of $I_2$ with the solvent.[24] Apparently the solid Py·$I_2$, or perhaps[23] $(PyI)^+I^-$, phase is but little soluble in these solvents.

**C. Very Dilute Solutions of Iodine in Pure Pyridine.**—In pure pyridine at concentrations below about 0.001 molar in iodine, the position and particularly the intensity of the λ 3,890 band become increasingly concentration-dependent (*cf.* Fig. 8), a fact which was not observed by Kleinberg and

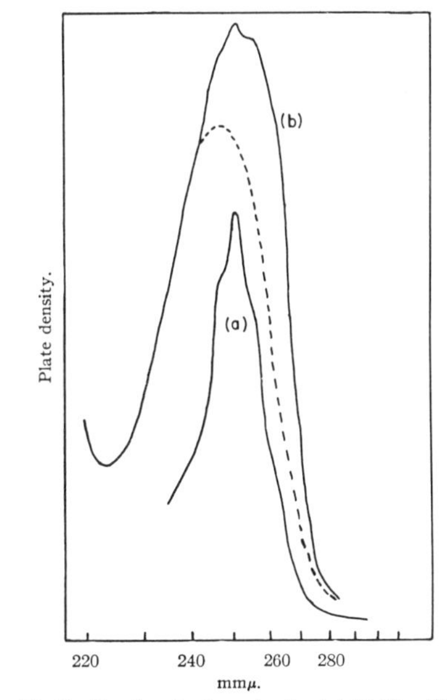


Fig. 7.—Densitometer trace showing (a) 1.5 $M$ pyridine, (b) 1.5 $M$ pyridine + 0.06 $M$ iodine showing charge-transfer peak shifted to about 2,450 Å. and superposed on the pyridine absorption. The dotted curve gives the estimated shape of the charge-transfer band.

(21) Or perhaps the observed continuously shifting peak is the result of a *superposition* in changing proportions of two distinct bands; if so, these may most probably be attributed to unsolvated and fully solvated Py·$I_2$ molecules. The limiting positions λ 4220 and λ 3890 are so close together (unlike those of the Py·$I_2$ and free $I_2$ iodine peaks in Fig. 1) that the superposition of two such bands would give a single peak. Another conceivable explanation of the λ 3890 peak, namely, that it might correspond to a superposition of the λ 4220 Py·$I_2$ peak and the λ 3600 $I_3^-$ peak can almost certainly be ruled out because these peaks are too far apart. [A very small, probably negligible, amount of the very strongly absorbing ion $I_3^-$ should be present in equilibrium in accordance with eqs. 6 and 7 of section IIC. In addition, the possible presence of a trace of water or other impurities should give rise to additional $I_3^-$, but probably not enough to affect the observed absorption appreciably except for the very low $I_2$ concentrations discussed in section IIC.]

(22) The present work does not throw doubts on the concept of an "inner complex" as discussed in reference 1b, but indicates that the inner complex of Py·$I_2$ is *not so low in energy* as was at first surmised. (Consideration of the system $I_2$ + $H_2O$ similarly indicates that there, too, the inner complex $(H_2OI)^+I^-$ in water solution is a structure of higher energy than the outer complex $H_2O·I_2$.) It is conceivable, however, that the solvated inner complex or ion-pair $(PyI)^+I^-$, while separated by a considerable activation barrier from the lower-energy outer complex, may be somewhat unstable with respect to the interposition of a Py molecule between the $(PyI)^+$ and $I^-$ ions, so that instead of the equilibria (2) and (3) one has something like

$$Py·I_2 + Py \rightleftarrows (PyI)^+(Py)I^- \quad (2')$$

$$(PyI)^+(Py)I^- \rightleftarrows PyI^+ + I^- + Py \quad (3')$$

all participants in (2′) and (3′) being of course solvated. If more than one Py is interposed between $(PyI)^+$ and $I^-$, the Py may be regarded simply as a dielectric medium separating the ions.

(23) An X-ray study of these crystals would be of interest. It seems possible that they may be built from $(PyI)^+$ and $I^-$ ions (*cf.* ref. 3b, section VI, and the discussion of $NH_4^+$ + $Cl^-$ crystals on p. 811 of section VIII), although their insolubility in water seems to indicate the contrary.

(24) L. I. Katzin [*J. Chem. Phys.*, **21**, 490 (1953)] has studied the spectra of solutions of iodine in water and the alcohols and has demonstrated the presence of $I_3^-$, probably due largely, however, to the presence or formation of $I^-$ from impurities.

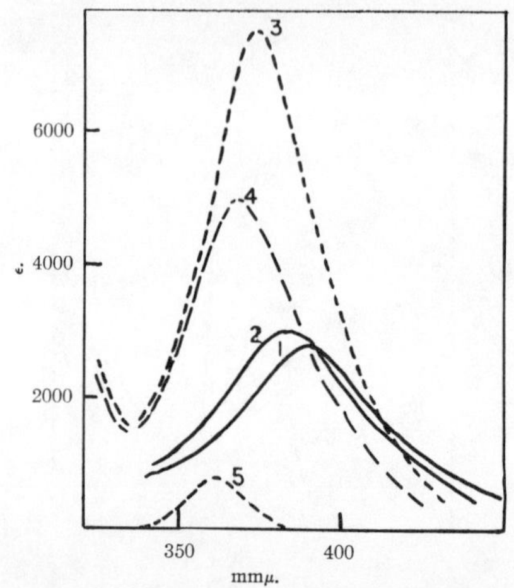


Fig. 8.—Increasing dilution (1–5) of $I_2$ solution in pure Py. Solid curves 1 and 2 are essentially due to Py·$I_2$. Curve 3 shows the $I_3^-$ curve nearing its maximum value. In curves 4 and 5 the $I_3^-$ intensity falls again presumably because of dissociation into $I^-$ ions. All ε values are based on formal $I_2$ concentration: 1, 0.025 *M;* 2, 0.0005 *M;* 3, 0.00005 *M;* 4, 0.0000125 *M;* 5, 0.00000612 *M;* 6, 0.00000306 *M.*

his associates[11b,20] because their cell thickness could be changed only by a factor of two, whereas it was varied by a factor of >1,000 in the experiments here described.

Strong dilution of more concentrated pyridine solutions (>0.07 molar), or the preparation of more dilute ones from pyridine and solid iodine, results in an *instantaneous* shift in the band maximum, accompanied by an increase in extinction coefficient (*cf.* Fig. 8, curves 1–3). If the dilution is to between $10^{-4}$ and $10^{-5}$ molar, the maximum shifts to 3,680 Å. and the apparent extinction coefficient based on $I_2$ rises to a maximum value of 9,000. The simultaneous appearance of a characteristic band of nearly double the extinction coefficient at 2,875 Å. makes it fairly certain that the maximum at 3,680 Å. is due to $I_3^-$ ions. (The usual absorption spectrum of $I_3^-$ consists of two peaks, one at λ 3,650 and one of nearly double as great peak intensity at λ 2,950.[25]) The observed maximum extinction coefficient suggests that under optimum conditions about four $I_2$ molecules yield one $I_3^-$ group. This would indicate that about half the iodine remains as Py·$I_2$, but that about half has reacted instantaneously in some way involving the formation of $I^-$ followed by

$$\text{Py}\cdot I_2 + I^- \rightleftarrows I_3^- + \text{Py} \qquad (6)$$

If dilution is continued below $10^{-5}$ molar, a new phenomenon is observed. The $I_3^-$ peak near λ 3,650 diminishes rapidly in intensity (*cf.* Fig. 8, curves 4 and 5), and no new band appears in the visible or ultraviolet to take its place. This change occurs just below the concentration range where Audrieth and Birr[8] reported the onset of anomalously high conductivity,[26] and may be attributed to redissociation of $I_3^-$ ions, by a reversal of eq. 6 accompanied by a passage of eq. 5 to the right, as is to be expected at sufficiently high dilutions; the net result would be

$$I_3^- + \text{Py} \longrightarrow (\text{PyI})^+ + 2I^- \qquad (7)$$

Conceivably also

$$\text{Py} + \text{PyI}^+ \rightleftarrows 2\text{Py}^+ + I^- \qquad (8)$$
$$\text{Py}^+ + \text{Py} \longrightarrow \text{Py}^+\text{Py} \qquad (9)$$

In all the reactions (5–9), the ions should of course be solvated. It is of interest that the ion $\text{Py}^+$, containing an odd electron, should be paramagnetic. Such an ion in Py solution should at once acquire extra stability by the formation with Py according to eq. 9 of an interesting ion of biphenyl-like structure with a three-electron bond between the two nitrogen atoms and further stabilized by various kinds of conjugation or resonance; this ion would still be paramagnetic.

Since our results were obtained under conditions of moisture-exclusion less rigorous than those of Kortüm and Wilski, it seems possible that the almost instantaneous $I_3^-$ ion production which we report in the $10^{-4}$ to $10^{-5}$ molar concentration range may be moisture-dependent. Or conceivably it may have been due somehow to impurities or to side-reactions which had occurred in spite of all precautions. It is known that in water or the lower alcohols $I_3^-$ ions in erratic amounts are instantaneously formed from dissolved iodine, probably largely as a result of the presence of impurities and/or side-reactions which form $I^-$.[24]

If we suppose, however, that our solutions were free from side-reactions or other foreign sources of $I^-$, and that the only important equilibria involved were (1), (5) and (6), it can be shown that the relative concentration of $I_3^-$, that is, the ratio $(I_3^-)/(\text{Py}\cdot I_2)$, should be small and approximately constant in the higher ranges of total iodine concentration, but should slowly *diminish,* not strongly increase as we observed, at high dilutions. Hence it may be that our results at high dilutions were due to impurities or side-reactions which somehow gave rise to $I^-$ in relative concentrations which became large enough to form spectroscopically noticeable amounts of $I_3^-$ near $10^{-4}$ molar iodine concentration[27]; the observed subsequent redissociation of this $I_3^-$ at higher dilutions according to eq. 7 would be exactly what one should expect. Conceivably (although this seems very unlikely) the rise in $I^-$, hence in $I_3^-$, below $10^{-4}$ molar, might be due to reactions (8)–(9). To obtain a more reliable understanding of what actually happens at high dilution, further investigations will be required.

CHICAGO, ILLINOIS

(25) (a) A. D. Awtrey and R. E. Connick, THIS JOURNAL, **73**, 1842 (1951); (b) R. E. Buckles, J. P. Yuk and A. I. Popov, *ibid.*, **74**, 4379 (1952).

(26) The occurrence of eqs. 7–9 would account for the anomalously high conductivity observed by Audrieth and Birr without assuming the presence of $\text{Py}^{++}$ ions as they did. However, since these results of Audrieth and Birr were obtained from aged solutions, after occurrence of what other investigators (ref. 9 and 11) consider to be a slow ring iodination liberating $I^-$ ions, it would seem that their results may not be relevant to what occurs in pure Py without side-reactions.

(27) Our results and conclusions at higher concentrations in pyridine, and our results in heptane solution, are not called in question by this possibility.

EXTRAIT DU:
RECUEIL DES TRAVAUX CHIMIQUES DES PAYS-BAS
publ. par la Koninklijke Nederlandse Chemische Vereniging [La Haye, Hollande]
**T. 75** — No. 6 — juin 1956.
Soc. anon. d'éditions scientifiques D. B. CENTEN, Amsterdam.

535.33 : 54-386 : 530.145

# MOLECULAR COMPLEXES AND THEIR SPECTRA. VI. SOME PROBLEMS AND NEW DEVELOPMENTS.

BY

ROBERT S. MULLIKEN
(Physics Department, University of Chicago).

Although the charge-transfer theory probably affords a correct general account of major aspects of the structure and spectra of molecular complexes, its application to the detailed structures of particular types of complex presents a variety of problems whose definitive solution will require much new experimental and theoretical research, but should then yield many interesting conclusions. The effects of London dispersion forces and classical electrostatic forces as well as charge-transfer forces must always be taken into account. Orientation isomerism, acceptation cross-sections of acceptor molecules, paraffin-halogen interactions and their contact charge-transfer spectra, intensity problems, the blue shift of the iodine visible band in the spectra of iodine complexes, and complexes with two charge-transfer spectra, are discussed below.

## 1. Introduction.

I shall consider here some aspects of the application of quantum-mechanical charge-transfer theory [1]), with emphasis on unsolved problems, on simple approximate relations, and on new developments. I believe that charge-transfer theory gives an essentially correct general account of the structure and spectra of molecular complexes. However, its application to the detailed geometry of particular types of complex presents a great variety of unsolved problems. The elucidation of these will require much new experimental and theoretical research which, however, promises to be very rewarding for the understanding not only of molecular complexes but also of intermolecular forces and of the intimate mechanisms of chemical reactions.

---

[1]) See *R. S. Mulliken,* (a) J. Am. Chem. Soc. **72,** 600 (1950); **74,** 811 (1952); (b) J. Phys. Chem. **56,** 801 (1952); (c) *C. Reid* and *R. S. Mulliken,* J. Am. Chem. Soc. **76,** 3869 (1954); (d) J. de chim. phys. **51,** 341 (1954): (e) Symposium on Molecular Physics at Nikko, Japan, 1953, p. 45; (f) J. Chem. Phys. **23,** 397 (1955).

For loose 1 : 1 complexes between even-electron donor (D) and acceptor (A) molecules, the ground state wave function $\psi_N$ may be written approximately as:

$$\Psi_N = \Psi_0(D,A) + b\,\Psi_1(D^+ - A^-) \quad . \quad . \quad . \quad . \quad (1)$$

The coefficient $b$ (whose square is a measure of the amount of charge transfer) should be approximately [2]:

$$b = cS/(W_1 - W_0) \quad . \quad . \quad . \quad . \quad . \quad . \quad (2)$$

where c is roughly a constant and $W_1$ and $W_0$ are the energies of hypothetical pure dative and pure no-bond wave functions $\Psi_1$ and $\Psi_0$ ($W_1 > W_0$). $S$ is the overlap integral between, (*a*), the donor orbital $\Phi_D$ out of which (in $\Psi_1$) an electron has been donated and, (*b*), the acceptor orbital $\Phi_{A^-}$ of the *negative ion* of the acceptor molecule into which the same electron has been accepted in $\Psi_1$:

$$S = \int \Phi_D \Phi_A \, dv \quad . \quad . \quad . \quad . \quad . \quad . \quad . \quad . \quad . \quad (3)$$

In general, $\Phi_D$ and $\Phi_A$ are MOs (molecular orbitals).

## 2. Overlap and Orientation Principle: Orientation Isomerism.

A consideration of Eqs. (1)–(3) leads to the *orientation and overlap principle* [1], according to which the partners in a donor–acceptor complex tend to assume a relative orientation such as to make $S$ of Eq. (3) a maximum. For an orientation such that $S$ is zero, $b$ also is zero by Eq. (2) and, by Eq. (1), charge-transfer interaction disappears. For the $Bz \cdot I_2$ complex (Bz = benzene) the most plausible orientation [1a] is one with the $I_2$ lying symmetrically above the Bz with its axis parallel to the Bz plane; either lengthwise, or crosswise, or perhaps rotating freely. However, recent X-ray evidence by *Dallinga* [3] on $Bz \cdot I_2$ and $Ms \cdot I_2$ in solution (Ms = mesitylene), and infrared work by *Collin* and *d'Or* [4], if on $Bz \cdot Cl_2$, point to a less symmetrical (perhaps highly mobile) location for the halogen molecule in the halogen-benzene and related complexes. The orientation of the partners in such complexes is thus open to question. Recent work of *Ham* [5])

[2]) Eqs. (2) and (5) follow from the equations of Ref. 1a if it is recognized that $H_{01}$ of Eqs. (5) there should be roughly proportional to $S$ ($S$ *here* corresponds to $S_{AB}$ of Eqs. (19) *there*).

[3]) *G. Dallinga*, unpublished work on $Bz \cdot I_2$.

[4]) *J. Collin* and *L. d'Or*, J. Chem. Phys. **23**, 397 (1955); and report at this Conference.

[5]) *J. S. Ham*, J. Am. Chem.Soc. **76**, 3875 (1954).

suggests the existence of two different favored orientations, a more compact one at liquid nitrogen temperatures, a looser one at room temperature; but much more work is needed. Such orientation isomerism may well be of frequent occurrence.

It is important to ask whether Eq. (1) should not be replaced by the more general equation

$$\Psi_N = \Psi_0 + \Sigma\, b_{ij} \Psi_1 (D_i^+ - A_j^-) \quad . \quad . \quad . \quad . \quad . \quad (4)$$

where $D_i^+$ and $A_j^-$ include excited states of $D^+$ and $A^-$; Eq. (2) would then hold for each $b_{ij}$. Although the *lowest energy* $\Psi_1$, assumed in Eq. (1), is favored by the factor $1/(W_1 - W_0)$ in Eq. (2), it is conceivable that $S$ (and possibly $c$) might increase so as to outweigh this factor for at least one or two of the excited $\Psi_1(D^+ - A^-)$'s. If so, the simple orientation and overlap principle based on Eq. (1) would become untenable, and in any event it seems likely that more than one $\Psi_1$ will frequently be of some importance [1f]. However, a consideration of some examples indicates that actually there may be a fairly general tendency for $S$ to decrease as $(W_1 - W_0)$ increases [6], so that Eq. (1) may after all be usually fairly adequate and the simple orientation principle valid. The whole matter deserves further study.

## 3. Charge-transfer Spectra; Acceptation Cross-sections.

For a complex whose ground state is represented by Eq. (1) there should be an excited state [1a] with wave function given approximately by:

$$\Psi_E = \Psi_1 - b^* \Psi_0; \; b^* = c^* S/(W_1 - W_0) \quad . \quad . \quad . \quad (5)$$

Corresponding to a spectroscopic absorption process from $\Psi_N$ to $\Psi_E$ there should be an *intense* "charge-transfer band" with frequency given [7] by

$$h\nu = W_E - W_N = (W_1 - W_0)\,(1 + b^2 + b^{*2}) \quad . \quad . \quad (6)$$

Using Eqs. (2) and (5) for $b$ and $b^*$, and writing

$$(c^2 + c^{*2}) S^2 = 2\,\beta^2; \; W_1 - W_0 = I_D - A \quad . \quad . \quad . \quad . \quad (7)$$

one obtains

$$h\nu = (I_D - A) + 2\beta^2/(I_D - A) \quad . \quad . \quad . \quad . \quad . \quad (8)$$

Here $I_D$ is the ionization energy of the donor, and $A$ represents,

---

[6]) In general, one must consider $S$ (or $H_{01}$) of Ref. 1a (cf. Ref. 2).

[7]) The right hand expression of Eq. (6) follows from Eqs. (5), (7), (10) and (11) of Ref. 1a if the approximate expressions for $b$ and $b^*$ given here in Eqs. (2) and (5) are used.

theoretically, the electron affinity $E_A$ of the acceptor plus the net energy of attraction between $D^+$ and $A^-$ in $\Psi_1$ (with $D^+$ and $A^-$ at the distance and orientation existing in the ground state of the complex). It may be noted that $c^{*2}$ may be several times as large as $c^2$, tending to make $2\beta^2$ fairly large.

Eq. (8) with constant $A$ and $\beta^2$, first derived by *Hastings et al.*, has been found empirically [8] to fit surprisingly well the observed frequencies of the charge-transfer bands of the complexes of a large number and variety of donor molecules with any one acceptor ($A$ is somewhat different for different acceptors). From the theoretical derivation of Eq. (8) one would not have predicted a constant $A$, nor, especially, a constant $\beta$, and it is certainly not surprising that for some donors the observed charge-transfer frequency deviates appreciably from Eq. (8) with the $A$ and $\beta$ values which fit most of the other donors.

Similar remarks apply to the still simpler empirical relation [9]

$$h\nu = I_D - B \qquad (9)$$

where $B$ is a characteristic constant for any acceptor. Although Eq. (8) represents a segment of a parabola, the curvature is not great in the rather limited range of $I_D$ values for observed complexes, and when one plots the points for various donors, it is readily seen how a straight line relation as in Eq. (9) can fit them about equally well.

## 4. Contact Charge-Transfer.

As is well known, the van der Waals radius of an atom is determined by the sudden onset of a repulsive force when two atoms come closer than the sum of their van der Waals radii. This "exchange repulsion" sets in when the orbitals of the outer electrons begin to overlap appreciably, with a repulsion energy proportional to the square of the overlap integral between the overlapping orbitals [10].

Because of the non-spherical shape of molecules, it will be more convenient to speak here in terms of van der Waals volume, rather than radius. For an acceptor molecule in contact with a donor molecule, the *van der Waals volume* is determined by the overlap properties of the outermost MOs (for example, $\pi_u$, $\sigma_g$, and especially the mildly antibonding $\pi_g$ in the case of the iodine molecule). However, in charge-transfer interaction with a donor molecule, what counts is not the van

---

[8]) *S. H. Hastings, J. L. Franklin, J. C. Schiller,* and *F. A. Matsen,* J. Am. Chem. Soc. **75,** 2900 (1953).

[9]) *H. McConnell, J. S. Ham* and *J. R. Platt,* J. Chem. Phys. **21,** 66 (1953); and *J. R. Platt* (private communication). See also *C. van de Stolpe,* Thesis Amsterdam 1953; *A. Bier,* Rec. trav. chim. **75,** 866 (1956).

[10]) Cf. e.g. *R. S. Mulliken,* J. Am. Chem. Soc. **72,** 4493 (1950), Section V.

der Waals volume, but what may be called the *electron-acceptation volume*. This volume corresponds to overlap of a donor MO, not with the MOs normally present in the acceptor, but with a further, "acceptation". MO ($\Phi_A^-$ in Eq. 3) which is occupied only in the negative ion of the acceptor. For iodine, the acceptation MO is a $\sigma_u$ MO which is *strongly antibonding*, a characteristic which should make it much larger than the outer MOs in neutral iodine. In general, the acceptation volume should greatly exceed the van der Waals volume for all acceptors which use an antibonding acceptation MO (namely all $\sigma$ and $\pi$ acceptors in the classification of Ref. [1b]). In other words, $S$ of Eq. (3) may be large even for *loose contact* of a donor with such an acceptor.

The foregoing discussion gives a basis for understanding the fact [11]) [8]) that bromine and iodine dissolved in saturated hydrocarbons absorb strongly in the ultraviolet to wave lengths as long as 2600 Å, whereas iodine vapor shows very little absorption at longer wave lengths than 2000 Å. *Evans* [11]) has shown that when n-heptane is added to solutions of iodine in perfluoroheptane (the spectral distribution for iodine absorption in pure perfluoroheptane solution is practically the same as in iodine vapor), the absorption extending out to 2600 Å at once appears, and merely increases in intensity as more heptane is added. As Evans has pointed out, this behavior indicates clearly that the extended absorption is not a "medium effect" but is due to specific interactions between n-heptane and iodine molecules. However, recent work of *Kortüm* and *Vogel* [12]) strongly indicates that there is no appreciable complex formation between iodine and such molecules as n-heptane and cyclohexane, and *Evans*' spectroscopic studies are most easily interpreted in the same way. Dr. *L. E. Orgel* and the writer [13]) have therefore concluded (making more definite a suggestion of *Evans*) that halogen molecules which are merely *near to* or in *contact with* saturated hydrocarbons are responsible for most of the observed intense absorption between $\lambda$2100 and $\lambda$2600, and that this absorption belongs to a hydrocarbon-halogen charge-transfer spectrum [14]).

A review of the theory indeed shows that Eqs. (1)-(9) are all valid, and that fairly intense charge-transfer absorption should be possible, for donor-acceptor pairs which are merely close or in contact, even if

---

[11]) *D. F. Evans*, J. Chem. Phys. **23**, 1424, 1426 (1955). Further, *S. Freed* and *K. H. Sancier*, J. Am. Chem. Soc. **74**, 1273 (1952) have shown the existence of a very loose complex of cyclopropane with iodine at low temperatures, with a definite charge-transfer peak at 2400 Å.

[12]) *G. Kortüm* and *W. M. Vogel*, Z. Electrochem. **59**, 16 (1955).

[13]) See forthcoming paper.

no actual molecules of complex are present; that is, even if the equilibrium constant for complex formation is zero. The only requirement is that the overlap integral $S$ defined by Eq. (3) shall differ sufficiently from zero at van der Waals contact. A corollary of the observation of contact charge-transfer spectra is that there must also be a small amount of charge-transfer (since $b > 0$ in Eq. 1 if $S > 0$) from donor to acceptor in the ground state of a pair of molecules such as heptane and iodine when in contact[14].

### 5. Intensity Problems.

Although the observed intensities of the charge-transfer spectra of molecular complexes are high, as predicted by theory, there are some unexplained disagreements in detail. For example, in the iodine complexes of the methylated benzenes, the intensities *decrease* in intensity with increasing stability, whereas theory predicts exactly the reverse if the complexes are all of like structure. Perhaps an explanation can be found when full account is taken of the possibility that the observed spectra: (1) are due to varying mixtures of orientation isomers, perhaps often with high amplitude relative vibrations or rotations between the partners in a complex; (2) include contributions from contact charge-transfer spectra[15]. It also seems possible that the presently accepted molar absorption coefficients evaluated by the method of *Benesi* and *Hildebrand* are unreliable[16].

### 6. Blue Shift of the Iodine Visible Band in Iodine Complexes.

The outer-electron-MO configurations of $I_2^-$, $I_2$, and the upper state of the visible absorption bands of $I_2$ are, in brief, respectively as follows:[1a]

$$\ldots \sigma_g^2 \pi_u^4 \pi_g^4 \sigma_u; \quad \ldots \sigma_g^2 \pi_u^4 \pi_g^4; \quad \ldots \sigma_g^2 \pi_u^4 \pi_g^3 \sigma_u \quad . \quad . \quad . \quad . \quad (10)$$

In $I_2$ complexes, as compared with free $I_2$, the visible $I_2$ band is shifted toward shorter wave lengths ("blue shift") by an amount which is somewhat correlated with the stability of the complex. Although other factors must also enter, the following seems reasonable as the major cause of these shifts. As was pointed out in Section 4, the $\sigma_u$ MO is strongly antibonding and therefore large, and its presence in excited $I_2$

---

[14]) Because of the high observed intensities, we believe that these spectra are not restricted to the, at any moment relatively few, molecule-pairs which in collisions have been forced together closer than to van der Waals' distances.

[15]) For some other proposals see *H. Murakami*, Bull. Chem. Soc. Japan **26**, 441 (1953) and later papers.

[16]) See *R. L. Scott*, in the report on this Conference, and the thereto appended discussion remarks by *Mulliken* on some recent conclusions by *Orgel*.

should very considerably increase the effective size of the molecule. When the iodine molecule, paired off with a close partner in a complex, is excited by visible light absorption ($\sigma_u \leftarrow \pi_g$), its suddenly swollen size introduces an exchange repulsion between it and the partner molecule. (Or, one may say, the $\sigma_u$ electron collides with the partner molecule.) This repulsion energy is added to the usual energy of the excited iodine molecule, giving a blue shift in the (vertical, or Franck-Condon-peak) absorption frequency.

If the foregoing idea is correct, the "blue shift" should be greater the more *intimate* the contact, or overlap, of the partners in the normal state of a complex. (It need not necessarily be quite so closely correlated with the equilibrium constant or even the heat of formation). If so, the following conclusions can be stated: (1) in the methylated benzene complexes (small blue shifts [1])) there is only relatively very loose contact at room temperature, but (moderate blue shifts [5])) closer contact at low temperatures; (2) in the onium, especially in the pyridine and the amine, complexes (large blue shifts [1])), contact is relatively intimate even at room temperature [17]). It is worth noting here that there is a rough parallelism between iodine blue shifts and the extent to which the infrared spectra of the donor molecules in the classes of iodine complexes mentioned are altered by complex formation.

It seems likely that a thorough study, not only of the blue shifts, but also of the intensities, of the iodine intramolecular visible band in iodine complexes may be made fruitful for an increased understanding of orientation and overlap in these complexes.

## 7. Complexes with Two charge-Transfer Spectra.

According to Eq. (4), the ground state of a complex may in general be stabilized by resonance with more than one dative function $\psi_1$. It follows that, in general, Eq. (5) is applicable to any one of a number of excited states involving charge transfer. A favorable situation for the occurrence of two observable charge-transfer spectra exists in some of the complexes formed by the substituted benzenes as donors. The lowest ionization potential $I$ of benzene is twofold, that is, there are two independent donor MOs $Q_D$ of equal $I$. In the substituted benzenes, these two MOs must differ in $I$, with the largest difference when the substituent is an amino group, the next largest perhaps when it is a hydroxyl or alkoxyl group.

---

[17]) Normally, solution of non-polar molecules in inert solvents produces a *red* shift. *Relative to this expectation,* the nearly zero shifts for iodine in such solvents as heptane and $CCl_4$ must be considered as small blue shifts. These would be expected according to the considerations of this section combined with those of section 4.

Experimentally, Dr. *Norman Smith* working with Prof. *W. G. Brown* in Chemistry at Chicago has examined the absorption spectra of numerous complexes of this type with chloranil as the acceptor [18]. As an example, he finds, for dimethylaniline as donor, two bands, one at 6450 Å, the other at 3450 Å, which he identifies as charge-transfer spectra. Further, Mr. *P. A. de Maine* at Chicago has made a careful study of the complex of anisole with iodine, and again finds two charge-transfer spectra, one at 3450 Å, the other at 2950 Å [18]; and gets other similar results. Dr. *Orgel* has also developed some general theoretical relations about the splitting between the two low *I*'s in substituted benzenes, of which he has made interesting application to various observed charge-transfer spectra [19].

A point of considerable interest is that the dative structures $\psi_1$ corresponding to the two *I*'s of a substituted benzene should favor different orientations of the acceptor molecule. Taking into account also the theory of the intensities of charge-transfer spectra [1]) it seems likely that each of the two spectra in the examples just discussed belongs (mainly, at least) to a different isomeric form of the complex.

---

[18]) Work in course of publication.
[19]) *L. E. Orgel*, J. Chem. Phys. **23**, 1352 (1955).

[Reprinted from the Journal of the American Chemical Society, **79**, 4839 (1957).]


[CONTRIBUTION FROM THE LABORATORY OF MOLECULAR STRUCTURE AND SPECTRA, DEPARTMENT OF PHYSICS, THE UNIVERSITY OF CHICAGO]

# Molecular Complexes and Their Spectra. VII. The Spectrophotometric Study of Molecular Complexes in Solution; Contact Charge-transfer Spectra[1]

BY L. E. ORGEL AND R. S. MULLIKEN

RECEIVED APRIL 18, 1957

The application of spectrophotometric techniques to the determination of the equilibrium constants and extinction coefficients of molecular complexes in solution is discussed and complications due to the presence of several 1:1 complexes with different orientations and to "contact" charge-transfer absorption by pairs of molecules contiguous to each other are emphasized. It is pointed out that values of equilibrium constants and extinction coefficients for loose complexes as determined by the method of Benesi and Hildebrand need reinterpretation or revision. Absorption by pairs of molecules in the complete range of cases from statistical contacts to 1:1 complexes is discussed in terms of a simple model. The resulting equations are used to show that the apparent anomaly of decreasing extinction coefficient with increasing methylation in the charge-transfer spectra of the iodine complexes of the methylated benzenes can be removed when it is recognized that a considerable part of the absorption in the more loosely associated cases is probably due to contact pairs rather than complexes. Nitro compound complexes are also discussed qualitatively. We believe that the simple model used here helps to clarify some long-standing difficulties, emphasized especially by Bayliss, in the attribution of spectral changes in solution to the formation of loose complexes.

## Isomerism in 1–1 Complex Formation[2]

In this paper we shall discuss the interpretation of measurements of the absorption spectra of weak complexes, particularly charge-transfer complexes, in solution. We shall be concerned with two factors[2] which may complicate the analysis of the experimental data, namely, the existence of several geometrically and/or electronically different 1:1 complexes in equilibrium, and the occurrence of contact charge-transfer absorption. (What we mean by "several different 1:1 complexes in equilibrium" is explained near the end of this Section.)

The usual method of determining the equilibrium constant and extinction coefficient for a 1:1 complex in solution is the one first proposed by Benesi and Hildebrand.[3] They use the mass action relation

$$K = x_C/(x_D - x_C)(x_A - x_C) \quad (1)$$

where $x_C$ is the mole fraction of complex; $x_D$ and $x_A$, respectively, are the total (*i.e.*, complexed plus uncomplexed) mole fractions of the donor and ac-

(1) This work was assisted by the Office of Ordnance Research under Project TB2-0001(505) of Contract DA-11-022-ORD-1002 with The University of Chicago.

(2) Attention already has been called briefly to the effects of the presence of more than one 1:1 complex by E. Grunwald and J. E. Leffler (see Ross, Labes and Schwarz, THIS JOURNAL, **78**, 343 (1956), footnote 2).

(3) H. A. Benesi and J. H. Hildebrand, *ibid.*, **71**, 2703 (1949). For an improved procedure, and other comments, see R. L. Scott, ref. 9 below.

ceptor which form the complex; and $K$ is the equilibrium constant. They also use the usual expression for the optical density due to the complex

$$d(\lambda) \equiv \log_{10} I_0(\lambda)/I(\lambda) = (C)l\epsilon_C(\lambda) \qquad (2)$$

where $\epsilon_c(\lambda)$ is the molar extinction coefficient of the complex at any wave length $\lambda$ where it absorbs, (C) is its concentration in moles/liter, and $l$ is the path length in cm. From these they derive the relation, valid provided $x_D >> x_A$ and neither D nor A absorbs at $\lambda$

$$1/\epsilon_A(\lambda) \equiv (A)l/d(\lambda) = \{1/\epsilon_c(\lambda)\} + \{1/[K\epsilon_c(\lambda)]\}\{1/x_D\} \qquad (3)$$

which permits $K$ and $1/\epsilon_c(\lambda)$ to be determined from the slope and the intercept of the line obtained by plotting $(A)l/d(\lambda)$ against $1/x_D$ using the experimental data. The quantity $\epsilon_A(\lambda)$ is the *apparent* molar extinction coefficient of A at $\lambda$ based on its *total* concentration (A), in moles/liter.

If there are several different 1:1 complexes, each with a different equilibrium constant and spectrum, relations analogous to (1) and (2) may be derived, namely

$$K' = x'_C/(x_D - x'_C)(x_A - x'_C) \qquad (1a)$$

$$d(\lambda) = (C')l\epsilon'_c(\lambda) \qquad (2a)$$

where

$$K' = \Sigma_i K_i,\ x'_C = \Sigma_i x_{Ci},\ \text{and}\ (C') = \Sigma_i (C_i) \qquad (4)$$

and

$$\epsilon'_c(\lambda) = \Sigma_i K_i \epsilon_{ci}(\lambda)/K' \qquad (5)$$

$K_i$, $x_{Ci}$ or $(C_i)$, and $\epsilon_{ci}(\lambda)$ are the equilibrium constant, mole fraction or concentration, and molar extinction coefficient of the i'th complex.

Relation (1a) follows directly from (4) and the definition

$$K_i = x_{Ci}/(x_D - \Sigma_i x_{Ci})(x_A - \Sigma_i x_{Ci})$$

Relation (2a) follows from the expression for the optical density of a mixture of absorbing molecules

$$d(\lambda) = \Sigma_i (C_i) l \epsilon_{ci}(\lambda)$$

after substituting for $(C_i)$ using $(C_i)/(C') = x_{Ci}/x_C' = K_i/K'$ and then using eq. 5 which defines $\epsilon_c'(\lambda)$.

It follows from (1a) and (2a) that for a system in which several different 1:1 complexes are present, the application of Benesi and Hildebrand's method will yield only a single $K$ and a single $\epsilon_c(\lambda)$ just as if only a single complex were present having

$$K = \Sigma K_i \qquad (4)$$

$$\epsilon_c(\lambda) = \Sigma K_i \epsilon_{ci}(\lambda)/K \qquad (5)$$

Hence the apparent constants which are determined are a total equilibrium constant and a weighted average extinction coefficient. Moreover, these results should be *independent of the wave length* employed in the analysis, no matter whether it belongs to the absorption spectrum of one complex or another, or to the superposed absorptions of more than one complex; but corrections must be made at wave lengths where there is absorption by free D or A molecules, for example by using the Ketelaar modification[4] of the Benesi and Hildebrand equation.

The foregoing analysis shows that the molar extinction coefficients as determined by the Benesi–Hildebrand or similar procedure are directly valid in the sense required by Mulliken's theory[5] only in the case that a single complex alone is present. For example, suppose there are two 1:1 complexes of different geometrical structure and each with its own distinct charge-transfer band (as is to be expected theoretically in certain cases), and suppose further that these two complexes have equal equilibrium constants $K_1$ and $K_2$, and equal peak extinction coefficients $(\epsilon_{c1})_{max}$ and $(\epsilon_{c2})_{max}$. The application of the Benesi–Hildebrand method will then yield a $K$ equal to $2K_1$ or $2K_2$ and peak extinction coefficients equal to just half the true values for the individual complexes if the bands of the two complexes do not overlap. Probable examples (except that there is no reason to believe that $K_1 = K_2$) are the aniline–chloranil complexes studied by N. Smith.[6]

Furthermore, an observation that equilibrium constants, determined using bands in different regions of the spectrum (*e.g.*, visible and ultraviolet or infrared and visible) and attributable to complex formation, are identical does not necessarily mean that these absorption bands are due to a single complex with a well-defined orientation.

From the fact that, if a single complex is present, $\Delta H/R$ is equal to the slope of the line obtained by plotting $\log K$ against $1/T$, it can be shown using eq. 4 that, if several complexes are present

$$\Delta H'(T) = \Sigma K_i \Delta H_i / K'$$

Under these conditions the plot of $\log K'$ against $1/T$ is not linear (unless the $\Delta H_i$ are all identical and temperature-independent) and in order to determine $\Delta H'(T_0)$, the heat of formation at $T_0$, the slope of the *tangent* at $T = T_0$ must be taken in place of the slope of the straight line which would be obtained if only one complex (with $\Delta H$ independent of $T$) were present.

These results show that, if measurements are carried out at only one temperature, the method of Benesi and Hildebrand does not enable one to distinguish between a system in which one well-defined complex exists and a system in which several or an infinite range of complexes are present. However (unless the $\Delta H_i$ are all identical), the $K_i$ will vary in different ways with temperature so that (a) the effective extinction coefficient $\epsilon_c'(\lambda)$ will be temperature dependent; (b) $\Delta H'$ will be temperature dependent.

Explicit formulas for the temperature dependence of these quantities if several complexes are present are

$$\epsilon'_c(\lambda,T) = \epsilon'_c(\lambda,T_0)\left\{1 + \left[\frac{1}{RT} - \frac{1}{RT_0}\right]\Sigma_i H_i\left[\frac{\epsilon_i(\lambda)}{\epsilon'_c(\lambda,T_0)} - \frac{1}{K'(T_0)}\right]\right\} \qquad (7)$$

$$\Delta H'(T) = \Delta H'(T_0)\left\{1 + \left[\frac{1}{RT} - \frac{1}{RT_0}\right]\Sigma_i\left[\frac{\Delta H_i^2}{\Delta H'(T_0)} - \frac{\Delta H_i}{K'(T_0)}\right]\right\} \qquad (8)$$

(4) J. A. A. Ketelaar, C. van der Stolpe, A. Goudsmit and W. Dzcubas, *Rec. trav. chim.*, **71**, 1104 (1952).

(5) R. S. Mulliken, This Journal, **74**, 811 (1952), and references therein. Paper II of present series.

(6) *Cf.* L. E. Orgel, *J. Chem. Phys.*, **23**, 1352 (1955).

if $\Delta H'(T)$ is derived from a study of the variation of $K'$ with $T$.

A good criterion for the presence of a single complex is the temperature-independence of the total oscillator strength integrated over a charge-transfer band.[7] The constancy of $\Delta H'$ with $T$ is another test for the presence of a single complex.

Some ambiguity may arise in the use of the term 1:1 complex. From the point of view of thermodynamics it is convenient to consider the total concentration of complexed molecules and to ignore differences in configuration which may occur. If this is done there is no need to reinterpret the measured extinction coefficients, equilibrium constants and heats of formation. However, the immediate consequences of the electronic theory of complex formation are always deduced for fixed geometrical configurations, so that an analysis of the kind given above is essential before the predictions of the theory can be tested empirically.

There is a closely related point raised by the foregoing discussion, namely, that, in weak complexes, thermal oscillations of large amplitude and sometimes rotations are likely to occur. Whether the variations in molecular structure arising from these are attributed to the vibrations and rotations of a single complex or to the statistical distribution of the molecules in a continuum of configurations is a matter of taste.

The important point, which is independent of any convention, is that the observed properties of complexes are statistical averages over all attainable configurations in thermal equilibrium. The conclusion applies especially to loose complexes and with even greater force to those properties of pairs of contiguous molecules which are discussed in the next section. It should be noted that the extinction coefficients are likely to be particularly sensitive to orientation, and the wave length or wave lengths of maximum absorption much less so.[5]

### Contact Charge-transfer

If we say that D and A form a 1:1 complex in an inert solvent S we mean that the number of adjacent DA pairs is in excess of the number to be expected as a result of random encounters under the influence of van der Waals forces only. (Similarly for m:n complexes.) However, it can be shown that "contact" charge-transfer absorption may occur during random encounters whenever a donor and an acceptor are sufficiently close to one another.[8] Summarizing our conclusions concerning the possibility of "contact" absorptions, which have already been reported in some detail,[8] we may say that, provided the overlap integral between appropriate donor and acceptor orbitals is appreciable even for pairs of molecules in loose contact or close to one another, then charge-transfer absorption can occur even if no stable complex is formed. The theoretical basis of this result and some of its consequences for the interpretation of the charge-transfer absorption of iodine dissolved in hydrocarbons will be discussed below.

We must now analyze the effect of contact absorption on the validity of Benesi and Hildebrand's method. We shall consider first a very much oversimplified model for a three-component system of non-complexing D, A and S molecules. We represent the molecules by equal spheres forming a close-packed liquid and suppose that the optical absorption in some wave length region is proportional to the concentration of DA contacts and define a "molar extinction coefficient" $\epsilon_{DA}(\lambda)$ and a molar concentration (DA) for these. As "contacts" we count every pair of adjacent D and A molecules, whether or not one or both of its members also belong to other DA pairs. Then, if $V$ is the mean molar volume

$$d(\lambda) = (DA)l\epsilon_{DA}(\lambda) = 12\epsilon_{DA}(\lambda)x_A x_D l/V = 12\,\epsilon_{DA}(\lambda)\,(A)lx_D$$

Or

$$1/\epsilon_A(\lambda) \equiv (A)l/d(\lambda) = [1/12\epsilon_{DA}(\lambda)][1/x_D]$$

The factor twelve has been introduced to allow for the fact that in this model each molecule has twelve nearest neighbors. If we applied Benesi and Hildebrand's method (*cf.* eq. 3) to such a system we would deduce that $K = 0$ and, incorrectly, that $\epsilon_c = \infty$.

The restrictive conditions of this simple model may be relaxed greatly without altering the main conclusion. Provided that the classes of contacts, not necessarily identical, which contribute to the intensity at the wave lengths studied are all such that their number depends linearly on $x_D$, then a Benesi and Hildebrand analysis gives a straight line passing through the origin when $(A)l/d(\lambda)$ of eq. 3 is plotted against $1/x_D$.

In fact this would still be true even if D and A interacted strongly to form specific complexes, provided only that (DA) were proportional to $x_D$. The latter contingency would be realized if (leaving aside purely steric factors which would equally be present in the case of random contacts) the complexing ability of the donor molecule in a 1:1 complex remained undiminished for the formation of 1:2 or higher complexes, and if $x_D >> x_A$. Actually, the *charge-transfer* complexing ability of a donor (or acceptor) molecule for an additional *favorably-oriented* partner should not be much diminished in very loose 1:1 complexes, but *should* be more and more sharply reduced in more and more stable 1:1 complexes.[6]

In stable 1:1 complexes, besides a considerable degree of saturation of charge-transfer forces, more or less site-saturation must occur, varying with the types of donor and acceptor. For example, if the donor and acceptor are both of the $\pi$ type (*e.g.*, aniline and chloranil), there are probably only two really favorable sites (above and below the plane of the molecular skeleton) for the exertion of appreciable charge-transfer forces between the donor and the acceptor; thus only 2:1 and 1:2 complexes (or perhaps also 1:1:1:1. . . . . .1:1 columns) might be expected to occur.

From the foregoing, it would seem that if one considers a series of D,A pairs with decreasing charge-transfer forces, there should be a gradual

(7) $\epsilon_c(\lambda)$ itself should not be quite temperature-independent even when only a single complex is present, since the increase in the amplitude of thermal vibrations caused by an increase in the temperature broadens any absorption band.

(8) R. S. Mulliken, *Rec. trav. chim. Pays-Bas*, **75**, 845 (1956). Paper VI of present series.

relaxation of both force-saturation and (because of thermal disorientation) site-saturation, leading to a gradual and continuous transition from the limiting case of saturated 1:1 complexes where eq. 1 and so the Benesi–Hildebrand eq. 3 should hold (provided $x_D >> x_A$, as is required for the validity of that equation), to the limiting case where the statistical relation

$$x_{DA} = \alpha x_A x_D, \text{ or } (DA) = \alpha(A)x_D \qquad (9)$$

is obeyed. In eq. 9, $\alpha$ is the average number of possible contact (*i.e.*, next-neighbor) sites for a D molecule around any A molecule under conditions of loose thermal contact.

Corresponding to Benesi and Hildebrand's eq. 3, one now has

$$1/\epsilon_A(\lambda) \equiv (A)l/d(\lambda) = [1/\alpha\bar{\epsilon}_{DA}(\lambda)][1/x_D] \qquad (10)$$

In eq. 10, we have written $\bar{\epsilon}_{DA}(\lambda)$, rather than $\epsilon_{DA}(\lambda)$, in view of the fact that, in general,[5,8] $\epsilon_{DA}(\lambda)$ should vary strongly with the relative orientation of D and A, as well as with the mean distance between them. Thus a suitable average value of $\epsilon_{DA}(\lambda)$, denoted by $\bar{\epsilon}_{DA}(\lambda)$, is needed in eq. 10. If $\alpha$ can be estimated, the *slope* of the graph of $1/\epsilon_A(\lambda)$ against $1/x_D$ gives $\bar{\epsilon}_{DA}(\lambda)$.

It is clear that the relation between $x_D$ and the concentration of DA pairs [$x_C$ in (1), $x_{DA}$ in (9)] may be intermediate between (1) and (9) if the presence of the one D molecule in a particular orientation in the neighborhood of an A molecule reduces the probability of a second D molecule attaching itself to A but does not entirely prohibit this from happening. In this case the Benesi and Hildebrand method does not necessarily give a linear plot of $1/\epsilon_A(\lambda)$ against $1/x_D$. However, if the absorption can be considered as a sum of complex and contact absorption, obeying relations (1) and (9), respectively, it can be shown (see below, under "Generalized Benesi–Hildebrand equation") that, so long as $K$ is independent of concentration, a straight line must be obtained. In Fig. 1 the upper and lower dashed lines represent the results to be expected for the plot of $1/\epsilon_A(\lambda)$ against $1/x_D$ for pure complex and pure contact absorption, respectively. The solid line represents the result to be expected for the actual analysis under the assumption of additivity between complex and contact absorption. The smaller $K$ is for complex formation, the closer does the solid line approach the lower dashed line. This means that as complex formation becomes weaker the extinction coefficients are more and more overestimated. As $K$ tends to zero the apparent extinction coefficient tends to infinity.

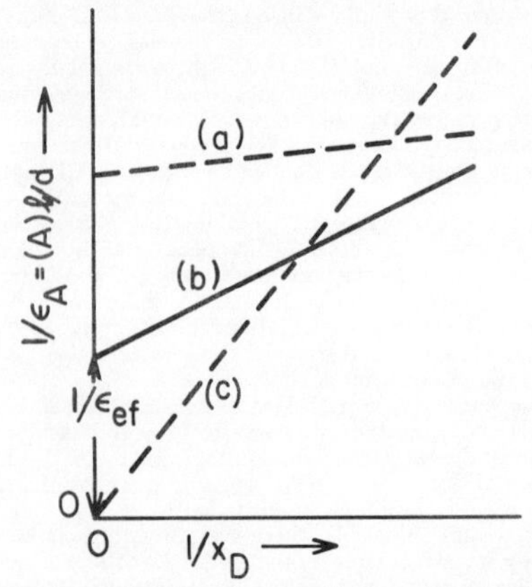


Fig. 1.—Generalized Benesi–Hildebrand graphs. The solid line (b), with intercept $1/\epsilon_{ef}$ and slope $1/K\epsilon_{ef}$ illustrates the general case of eq. 15, for $K \approx \alpha\bar{\epsilon}_{DA}$. The dashed lines correspond to close approaches to the two limiting cases where the charge-transfer spectrum is due to: (a) complexes only, eq. 3; (c) contacts only, eq. 10. For (a), $K >> \alpha\bar{\epsilon}_{DA}$, the intercept is $\approx 1/\epsilon_c$, and the slope is $\approx 1/K\epsilon_c$. For (c), $0 \approx K << \alpha\bar{\epsilon}_{DA}$, the intercept is $\approx 0$, and the slope is $\approx 1/\alpha\bar{\epsilon}_{DA}$.

In practice it is likely that the situation is often very complicated since both of the difficulties, namely, the presence of several 1:1 complexes and the occurrence of contact absorption, which we have discussed, are likely to be important in the same system. If complex formation is weak, two or more geometrical configurations of the DA pair, or even a continuous range of them, may have almost equal probabilities of occurring; in this event $\epsilon_c'(\lambda)$ becomes a weighted mean over these various configurations. In some of these configurations further association is prohibited so that (1) is valid; to others (9) or an intermediate relation must apply. Yet in all cases the charge-transfer absorption is likely to be in the same spectral region since, provided the mutual electrostatic attraction of the donor and acceptor ions in the dative excited state does not vary too much, the energy of the transition depends mainly on $I_D - E_A$, the difference between the ionization potential of the donor and the electron affinity of the acceptor.

According to the theory, DA pairs in those configurations whose concentrations are given by eq. 1 will often have larger extinction coefficients than those in other configurations, since the partners in the former case will usually be closer together and overlap more. However, there is no reason to believe that the extinction coefficients for pairs in contact, that is pairs in configurations whose concentrations are determined by eq. 9, are necessarily small. It is therefore necessary to reconsider many of the published values of extinction coefficients and perhaps equilibrium constants in the light of this analysis.

At this point we may remark that, as has been emphasized by Scott,[9] Benesi and Hildebrand's method is probably not in general the most suitable one for analyzing the optical data, since other linear relations can be derived on the basis of (1) and (2), which under some circumstances allow more accurate determinations of $K$ and $\epsilon$ to be made.[9] These methods, however, are also subject to error for the reasons already suggested unless contact absorption is allowed for.

**Generalized Benesi–Hildebrand Equation.**—For a solution of a donor D and an acceptor A in an inert solvent, let us assume that CT (charge-

(9) R. L. Scott, *Rec. trav. chim., Pays-Bas*, **75**, 787 (1956). Also unpublished work of J. Petruska.

transfer) absorption is of just two kinds: (1) by 1:1 complexes C, with CT band molar extinction coefficient $\epsilon_c(\lambda)$; (2) by pairs of "free" D and A molecules in contact, with extinction coefficient $\bar{\epsilon}_{DA}(\lambda)$ per mole of pairs. The concentration of complexes is governed by eq. 1; let us suppose that the spectroscopically effective concentration of contact CT pairs is governed by the following modification of eq. 9.

$$x_{DA} = \alpha(x_D - x_C)(x_A - x_C) \quad (11)$$

Equation 11 assumes that every *free* DA pair is effective (and equally so) in contact CT interaction regardless of whether more than one D is in contact with an A, or *vice versa*, but that D and A molecules bound in a 1:1 complex are completely ineffective both for the formation of higher-order CT complexes and in contact CT interaction with further D or A molecules. The second assumption may be questioned; however, it is apparently equivalent to the setting up of a simple *model* which *arbitrarily* divides all pairs into two classes, namely, CT-saturated "1:1 complexes" and CT-unsaturated "contact-pairs,"—whereas actually there may in the case of weak CT forces be a fairly continuous shading off from 1:1 and higher-order "complexes" to "contacts." In effect, the model artificially abstracts all the saturatedness from all DA pairs and concentrates it in a limited number of pairs called 1:1 complexes in which saturation is complete. In any event, the model makes it possible to obtain some simple results which form the basis for Fig. 1.

Comparing eq. 1 and 11, we have

$$x_{DA} = (\alpha/K)x_C; \text{ hence } (DA) = (\alpha/K)(C) \quad (12)$$

For the CT-spectral optical density $d$ (*cf.* eq. 2) one now has

$$d(\lambda) = [(C)\epsilon_c(\lambda) + (DA)\bar{\epsilon}_{DA}(\lambda)]l = (C)\epsilon_{ef}(\lambda)l \quad (13)$$

with

$$\epsilon_{ef}(\lambda) = \epsilon_c(\lambda) + \alpha\bar{\epsilon}_{DA}(\lambda)/K \quad (14)$$

For later purposes, it will be convenient to define

$$\rho \equiv \alpha\bar{\epsilon}_{DA}/\epsilon_c \quad (14a)$$

Then

$$\epsilon_{ef} = \epsilon_c(1 + \rho/K); \quad \epsilon_c = K\epsilon_{ef}/(K + \rho) \quad (14b)$$

Since $d = (C)\epsilon_{ef}l$ as given by eq. 13 is formally the same as eq. 2, a mixed solution of D and A conforming to the model assumed above must obey exactly the same formal equations, including the Benesi–Hildebrand eq. 3, as if contact CT did not exist, the only difference being that $\epsilon_c(\lambda)$ is replaced by $\epsilon_{ef}(\lambda)$. Hence, instead of eq. 3, one has

$$1/\epsilon_A(\lambda) \equiv (A)l/d(\lambda) = [1/\epsilon_{ef}(\lambda)] + [1/K\epsilon_{ef}(\lambda)][1/x_D] \quad (15)$$

From eqs. 14 it is now seen that for $\bar{\epsilon}_{DA} = 0$ (no contact CT spectrum), $\epsilon_{ef} = \epsilon_c$ as is tacitly assumed in using the Benesi–Hildebrand (or any other) type of analysis which attributes the CT spectrum solely to complexes. On the other hand, for $K = 0$ in eqs. (14), $\epsilon_{ef} = \infty$ but $K\epsilon_{ef} = \bar{\epsilon}_{DA}\alpha$, so that eq. 15 goes into eq. 10, the equation for pure contact CT absorption. For intermediate cases, the generalized Benesi–Hildebrand eq. 15, with $(A)l/d(\lambda)$ varying linearly with $1/x_D$ for (D) >> (A), always holds.

A point of considerable interest is that, according to the foregoing analysis, the slope $(1/K\epsilon_{ef})$ combined with the intercept $(1/\epsilon_{ef})$ of the graph of eq. 15 yields a *correct* value of the equilibrium constant $K$ for complex formation even if the value of $\epsilon_{ef}$ is larger than $\epsilon_c$ because of contact CT absorption. Thus (subject, however, to the arbitrariness inherent in the simple model assumed above) the $K$ values reported in the literature based on the Benesi–Hildebrand method may still have much real significance even for loose complexes.

The foregoing analysis, and in particular eq. 15, are equally valid for any temperature. It follows as an important corollary that $\Delta H$ values obtained in the usual way, from the variation of Benesi–Hildebrand $K$ values with temperature, should have the same degree of significance as the $K$ values themselves, even for loose complexes.

Various other interesting deductions can be made. For example, if $K$ is large, the theory[5] leads one to expect the peak value of $\bar{\epsilon}_{DA}$ to be considerably smaller than that of $\epsilon_c$ and to occur at a somewhat different wave length, probably at a shorter wave length in most cases; but if $K$ is small enough, $\bar{\epsilon}_{DA}(\lambda)$ and $\epsilon_c(\lambda)$ should become more nearly the same.

When D and A are in solution in a third substance S as solvent, the A molecules may be considered as falling into three classes, those in complexes (fraction $F_C$, where $F_C = (C)/(A) = x_C/x_A$), those in DA contact pairs (fraction $F_{ct}$), and those which are "free," that is, in contact only with S molecules (fraction $F_f$). We have been considering only the D,A charge-transfer spectrum, consisting of complex and contact contributions. There should also be an SA and a DS charge-transfer spectrum, but in the case of inert solvents these should usually be only contact CT spectra at shorter wave lengths.

Now referring to eqs. 13 and 12, and recalling that the apparent extinction coefficient $\epsilon_A$ of A in a solution containing a donor is $d/(A)l$, one readily obtains

$$\epsilon_A = \epsilon_c F_C + \bar{\epsilon}_{DA}\alpha F_C/K = \epsilon_c F_C + \epsilon_c F_C \rho/K \quad (16)$$

where the terms $\epsilon_c F_C$ and $\bar{\epsilon}_{DA}\alpha F_C/K$ are the respective contributions of complex and contact absorption to $\epsilon_A$. From eq. 16, it is seen that the *fractional contributions* to $\epsilon_A$ by complexes ($\Phi_C$) and contact pairs ($\Phi_{ct}$) are

$$\Phi_C = K/(K + \rho); \quad \Phi_{ct} = \rho/(K + \rho) \quad (17)$$

For *tight* complexes, *e.g.* Py·$I_2$ with $K = 290$ at 17°, $\Phi_{ct}$ becomes unimportant. Equations 16–17 are true *independent of concentration*, *e.g.*, even in pure donor as solvent [$x_D = 1$ in eq. 15 and $F_f = 0$], provided $K$ remains constant and (D) >> (A). (Actually, $K$ is in general rather different for different solvents.)

Approaching $\epsilon_A$ from a different viewpoint, it is obvious that

$$\epsilon_A = \epsilon_c F_C + \bar{\epsilon}_{DA}\alpha F_{ct}$$

Unfortunately, because of groupings $AD_n$ with $n > 1$ [also $A_nD$ unless (D) >> (A)], whose importance varies with concentration, it is not possible to give simple general expressions for $F_{ct}$ and

$F_f$. However, in the special case of pure donor solvent ($F_f = 0$), $F_{ct} = 1 - F_C$, so that

$$\epsilon_A = \epsilon_c F_C + \bar{\epsilon}_{DA}\alpha(1 - F_C) \quad (18)$$

Hence, comparing eq. 16 and 18, $F_C/K = (1 - F_C)$, and

$$F_C = K/(1 + K) \quad (19)$$

in *pure donor as solvent* with A in small concentration. It is noteworthy that $F_C$ in eq. 19 is independent of $\bar{\epsilon}_{DA}$, $\alpha$ or $\rho$. Combining eq. 16 and 19, one has

$$\epsilon_A = \epsilon_c K/(1 + K) + \bar{\epsilon}_{DA}\alpha/(1 + K) = \epsilon_c[K/(1 + K) + \rho/(1 + K)] \quad (20)$$

**Theoretical Basis for Contact Charge-transfer Absorption.**—We must now justify our statement that charge-transfer absorption can occur whenever a donor molecule and an acceptor molecule are in contact. This is easily done, for although Mulliken in his theoretical treatment of charge-transfer spectra has couched his discussion in terms of stable complexes, his methods are quite general. His demonstration that absorption bands corresponding to intermolecular charge-transfer transitions should occur when a donor-acceptor complex is formed applies equally well to pairs of molecules in contact or even merely sufficiently near to each other. It does not depend on the ability of the charge-transfer forces to overcome the exchange repulsions between the components, but on the existence of a non-zero overlap integral between donor and acceptor orbitals. This point has been discussed recently by Mulliken (see particularly eq. 2 and 3 in ref. 8).

Mulliken[5] has shown that an approximate expression for the transition moment, the square of which determines the intensity of the charge-transfer transition, is

$$\mu_{EN} = a^*be(\bar{r}_D - \bar{r}_A) + 2^{1/2}(aa^* - b^*b)eS(\bar{r}_D - \bar{r}_{DA}) \quad (21)$$

when $S$ is the overlap integral of the D and A orbitals involved, and the vectors $\bar{r}_D$, $\bar{r}_A$ and $r_{DA}$ locate the charge centers of these orbitals and of the overlap charge, respectively, and where, $a$, $b$, $a^*$, $b^*$ are the coefficients in the wave functions $a\psi_0 + b\psi_1$ and $a^*\psi_1 - b^*\psi_0$ for the ground and excited states, respectively.

It is not clear whether the first or second term in (21) is the more important, since $a$ and $a^*$ are approximately unity and $b$ should probably be roughly proportional to $S$. (For stable complexes, Mulliken supposes that the first term is the more important.) This point is of considerable interest not only for the theory of the spectra but also for that of the ground state, since the quantity $b^2$ is a measure of contact charge-transfer between D and A in the ground state. If the second term in (21) predominated over the first for loose contact it would be possible for the charge-transfer band to appear with moderate intensity even if charge-transfer from a donor in loose contact with an acceptor were small. We do not know whether this is to be expected or not.

J. N. Murrell and one of the writers (RSM) are investigating this question further; they are also examining another possibility suggested by Dr. Murrell, namely, that contact charge-transfer spectra may owe much of their intensity to mixing of the CT excited state of the DA pair with a nearby excited state (say E) of the donor, state E being of such a nature that an intense spectroscopic transition occurs between it and the ground state of the donor. Details will be presented in a later paper.

We have assumed in eq. 9 and 11 that, if a number of acceptor molecules are clustered around a donor molecule, the absorption spectrum is the superposition of those to be expected for each of the pairs separately, excluding only those which are "complexed." Mulliken's theory shows[5] that this approximation should be quite good provided that the charge-transfer band is not too sensitive to changes of solvent. This can be seen most easily for a group $DA_1A_2$ of two identical acceptor and one donor molecules. There are two ionic structures $D^+A_1^-A_2$ and $D^+A_1A_2^-$ corresponding to the dative structure $D^+A^-$ of a simple DA pair, and these can be mixed together. The energy required to produce either of them is equal to that required to produce the simple $D^+A^-$ structure in the same orientation except for a small "solvent effect" due to the replacement of a solvent molecule by an acceptor molecule. Usually there is little interaction between the two ionic structures and so the resonance energies are approximately additive and the excitation energies correspond closely to those for simple DA pairs in appropriate orientations. A similar argument shows that the intensities are approximately additive. In our simple model these approximate relations become exact for the DA "contact pairs," but do not hold at all for the DA 1:1 "complexes."

We believe that the present discussion does much to clarify a difficulty which has been felt by various people for a long time. Bayliss in particular has argued against complex formation as the explanation of the spectroscopic changes which occur when iodine is dissolved in various solvents. Recently,[10] in a valuable discussion rather closely related to that given here, he has analyzed very clearly the difficulties in distinguishing spectroscopically between complexes and "physical perturbations."

The present analysis shows that one important type of spectra, namely, intermolecular charge-transfer spectra, can occur as a result of specific interactions between donor and acceptor molecules even if the equilibrium constant for complex formation approaches or reaches zero, and indicates how a continuous range of cases from mere contact through loose complexes to tight complexes can show similar spectroscopic behavior.

**Halogen Complexes.**—Bromine and iodine dissolved in saturated hydrocarbons absorb strongly in the ultraviolet to wave lengths as long as 2600 Å., whereas iodine vapor shows hardly any absorption at longer wave lengths than 2000 Å.[11] Evans has shown that when $n$-heptane is added to solutions of

(10) N. S. Bayliss and C. J. Brackenridge, This Journal, **77**, 3959 (1955). Another valuable and relevant paper on "solvent effects in organic spectra" is that by N. S. Bayliss and E. G. McRae, *J. Phys. Chem.*, **58**, 1002 (1954).

(11) D. F. Evans, *J. Chem. Phys.*, **23**, 1426, 1429 (1954). Further, Freed and Sancier, This Journal, **74**, 1273 (1952), have shown the existence of a very loose complex of cyclopropane with iodine at low temperatures, with a definite charge-transfer peak at 2400 Å.

iodine in perfluoroheptane (the spectral distribution in this range for iodine absorption in pure perfluoroheptane solution is practically the same as in iodine vapor), the absorption extending out to 2600 Å. at once appears, and merely increases in intensity as more heptane is added. As Evans has pointed out, this behavior indicates clearly that the extended absorption is not a medium effect but is due to specific interactions between *n*-heptane and iodine molecules. On the other hand, recent work of Kortüm and Vogel[12] indicates that there is no appreciable complex formation between iodine and such molecules as *n*-heptane and cyclohexane, and Evans' spectroscopic studies are most easily interpreted in the same way. We have therefore concluded (making more definite a proposal of Evans) that halogen molecules which are merely *near to* or in *contact with* saturated hydrocarbons are responsible for most of the observed intense absorption beyond 2100 Å., and that this absorption belongs to a hydrocarbon–halogen contact charge-transfer spectrum.

Although the observed intensities of the charge-transfer spectra of actual halogen complexes as determined by the Benesi–Hildebrand method are high, as predicted by theory, there are some unexplained disagreements in detail. For example, in the series of iodine complexes of methylated benzenes the intensities so determined decrease as the stability increases,[13] whereas theory predicts the reverse[5] if the complexes are all of like structure. One explanation might be that the spectra are due to varying mixtures of orientation isomers, a possibility that receives some support from low temperature studies.[14] A more general explanation must include also an allowance for contact absorption, whose effect, in a series of geometrically similar complexes, should be very roughly inversely proportional to the equilibrium constant for complex formation. It is therefore encouraging that the decrease of apparent extinction coefficient with increasing equilibrium constant is much smaller for iodine chloride complexes than for the weaker iodine complexes.[15] This is shown in Fig. 2. It seems probable therefore that the apparent disagreement between Mulliken's predictions and the observed behavior of the intensities in alkylbenzene–iodine complexes is due to the method of interpreting the data rather than to any inadequacy of the theoretical treatment.

This conclusion receives strong support when the experimental data are examined in terms of eqs. 14b and 15. Using $K$ and $\epsilon_{ef}$ values based mainly on work of Benesi and Hildebrand, $\epsilon_c$ has been calculated by eq. 14b for each of several assumed values of $\rho$. The following Table I compares the calculated $\epsilon_c$ values for solutions of iodine as acceptor with benzene, mesitylene or hexamethylbenzene as donor in carbon tetrachloride.

It is seen that for $\rho = 4$ or 5, the $\epsilon_c$ values become concordant with theoretical expectations, in that they increase with methylation. It should be noted especially that these $\rho$ values are very reasonable for aromatic molecules around an iodine molecule. Thus if $\bar{\epsilon}_{DA}$ were equal to $\epsilon_c$, $\rho$ would become equal to $\alpha$, and $\alpha$ values of 4 or 5 are reasonable. However, $\bar{\epsilon}_{DA}$ should definitely be less than $\epsilon_c$, hence $\alpha > \rho$. Further, $\bar{\epsilon}_{DA}/\epsilon_c$ should decrease with increasing strength of complexing. Also, $\alpha$ should apparently decrease somewhat with increasing methylation of benzene. Hence it seems likely that the true values of $\rho$ should decrease with increasing strength of complexing in the methylated benzenes. While it is not possible from the data to determine exact values for $\alpha$, $\bar{\epsilon}_{DA}$, and $\epsilon_c$, a plausible guess might correspond to $\rho = 4$ or 5 for benzene, $\rho = 3$ or 4 for hexamethylbenzene.

Referring to eq. 17, one sees that if $\rho = 5$ for benzene–iodine "contacts" in solution, then, independently of benzene concentration, approximately one-fourth of the charge-transfer band intensity is attributable to "complexes" and three-fourth to "contacts." However, it must be kept in mind that this result has been obtained by using a decidedly artificial model.

It is interesting to examine the spectrophotometric data on the CT spectra of solutions of iodine in

(12) G. Kortüm and W. M. Vogel, *Z. Electrochem.*, **58**, 15 (1955).

(13) (a) R. M. Keefer and L. J. Andrews, This Journal, **74**, 4500 (1952); **77**, 2164 (1955). Also (b) Tamres, Virzi and Searles, *ibid.*, **75**, 4358 (1953).

(14) J. S. Ham, *ibid.*, **76**, 3875 (1954).

(15) N. Ogimachi, L. J. Andrews and R. M. Keefer, *ibid.*, **77**, 4202 (1955).

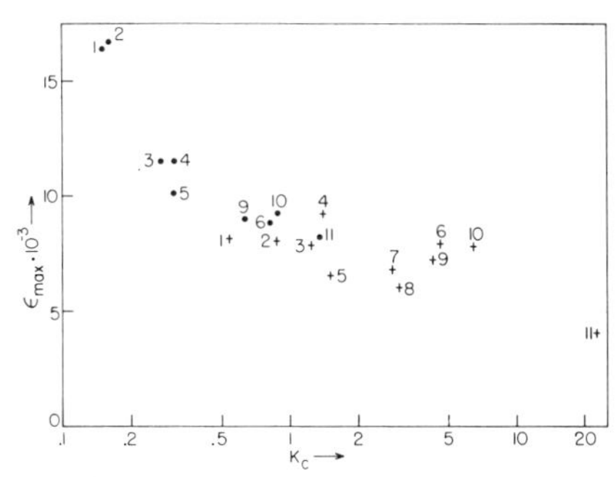


Fig. 2.—The variation of $\epsilon_{max}$ with $K_c$ for iodine[13] and iodine monochloride[13, 15] complexes of 1, benzene; 2, toluene; 3, *o*-xylene; 4, *m*-xylene; 5, *p*-xylene; 6, mesitylene; 7, 1,2,2-trimethylbenzene; 8, 1,2,3-trimethylbenzene; 9, durene; 10, pentamethylbenzene; 11, hexamethylbenzene. $I_2$ complexes ●, ICl complexes +. Note that $K_c$ differs from $K$ of eq. 1 in that it is based on concentrations (C), (D), (A) instead of mole fractions; however, the values of $K_c$ are usually approximately proportional to those of $K$, and Fig. 2 would not be qualitatively changed if $\epsilon$ were plotted against $K$. $\epsilon$ values are now to be interpreted as $\epsilon_{ef}$ values.

Table I

Peak Values of $\epsilon_c$ Computed from Eq. 14b[a]

| | $K$ | Peak $\epsilon_{ef}$ | Peak $\epsilon_c$ for $\rho = 0$ | $\rho = 2$ | $\rho = 3$ | $\rho = 4$ | $\rho = 5$ |
|---|---|---|---|---|---|---|---|
| Benzene | 1.72 | 15400 | 15400 | 7120 | 5610 | 4630 | 3940 |
| Mesitylene | 7 2 | 9300 | 9300 | 7280 | 6570 | 5980 | 5490 |
| Hexamethyl-benzene | (15.5) | (7700) | (7700) | 6820 | 6460 | 6110 | 5820 |

[a] The data for benzene and mesitylene are from ref. 3, those for hexamethylbenzene are estimated from ref. 13.

pure aromatic hydrocarbon solvents. While it is not true in general that $K$ should be independent of solvent, it is apparently true by chance that $K$ is nearly the same for the equilibrium of $I_2$ with benzene in $CCl_4$ solution and in pure benzene.[16] This is illustrated by the fact that the CT peak value of $\epsilon_A$ for $I_2$ in pure benzene (9770) agrees closely with the value computed using Benesi and Hildebrand's $K$ and $\epsilon_{ef}$ as given in Table I. There is a rather similar agreement for iodine in mesitylene ($\epsilon_A$ = 8300 for iodine in pure mesitylene).[3]

While these agreements are interesting, they have no immediate bearing on the question under discussion here; they are equally consistent with any value of $\rho$ in Table I. However, it is worth noting some further implications for the case of iodine in pure benzene of our tentative conclusion that $\rho = 5$ may be roughly correct.[17] From eq. 19, $F_C = 0.63$ (independent of $\rho$). But from eq. 18 these 63% of $I_2$ molecules which are "complexed" contribute only (3940)(0.63) = 2500, while the remaining 37% of $I_2$ molecules forming "contact" pairs contribute (3940)(5)(0.37) = 7270 to the total observed $\epsilon_A$ of 9770. The more tightly bound complexed iodine molecules are then relatively much less effective spectroscopically, because of their assumed saturation, than those involved in loose contact pairs.

Although one might be inclined to say that this conclusion is unreasonable, it is nevertheless what our model requires in view of the observed $K$ value, and there seems to be no really good reason to reject it. While actually there must be a continuous gradation between looser and tighter contacts, of various orientations, for the several benzene molecules surrounding each iodine molecule, the model arbitrarily replaces these by a definite fraction of close, relatively low-energy, relatively favorably oriented, saturated 1:1 contacts called complexes, each loosely surrounded by non-interacting benzene molecules, plus a remaining fraction of iodine molecules having loose and random contacts and CT interactions each with $\alpha(=\rho/\bar{\epsilon}_{DA})$ benzene molecules. In an X-ray investigation of solutions of $I_2$ in benzene and mesitylene, Dallinga has obtained some information indicating that the average interactions are quite loose.[18]

(16) However, see T. M. Cromwell and R. L. Scott, who have reviewed Benesi and Hildebrand's data and conclude that $K$ = 1.9 in $CCl_4$ but $K$ = 2.3 in pure benzene (also, $K$ = 1.4 in *n*-heptane), with $\epsilon_{ef}$ = 14,000. These changes do not affect our conclusions appreciably.

(17) Perhaps $\rho$ = 4 would be better, but use of this value would not change our essential conclusions.

(18) G. Dallinga, *Acta Cryst.*, **7**, 665 (1954). Abstract only.

**Nitro Compound Complexes.**—Just as for $I_2$ complexes, Foster and Hammick have found that the apparent molecular extinction coefficients for *sym*-trinitrobenzene complexes decrease as the complexes become more stable,[19] contrary to the predictions of the simple charge-transfer theory. Ross and his co-workers at first concluded that equilibrium constants for the formation of complexes between nitro compounds and aromatic donors determined by optical methods are smaller than those given by more direct methods.[20] However, their further investigations[21] make this conclusion uncertain. Studies of the temperature dependence of the extinction coefficient both for the trinitrobenzene–aniline and trinitrobenzene–naphthalene complexes show that when the temperature is raised the apparent extinction coefficient increases considerably as the equilibrium constant decreases.[21]

This is exactly what would be expected according to eq. 14 if the observed spectra are partly due to the contact charge-transfer, provided (as is very probable) that $\epsilon_c$ does not change much with temperature and that $\alpha\bar{\epsilon}_{DA}$ decreases more slowly with temperature than $K$ (or else, less likely, increases). However, further experimental work over a wide range of temperatures obviously is needed. (Among other things, one should consider whether the possible presence of more than one isomeric complex could also play a role here.)

According to an analysis by Mulliken,[8] the possibility of contact charge-transfer is based on the fact that the acceptation orbital in the negative ion of an acceptor may be considerably larger than the van der Waals size of the neutral acceptor molecule. Since charge-transfer spectra become possible as soon as the donor orbital begins to overlap this orbital of the acceptor negative ion, this interaction can occur even at greater than van der Waals distances. However, different acceptors vary in the size of the acceptation orbital, which is particularly large for halogen molecules. For nitro compounds it may not be as much larger than the outer orbitals of the neutral acceptor molecule as for the halogens, which suggests that contact charge-transfer spectra may be less important for complexes of nitro-compounds than for those of the halogens.

CHICAGO, ILLINOIS

(19) R. Foster and D. H. Hammick, *J. Chem. Soc.*, 2685 (1954).

(20) S. D. Ross, M. Bassing and I. Kuntz, THIS JOURNAL, **76**, 4176 (1954).

(21) S. D. Ross, private communication.

[CONTRIBUTION FROM THE DEPARTMENT OF THEORETICAL CHEMISTRY, UNIVERSITY CHEMICAL LABORATORY]

# Molecular Complexes and their Spectra. IX. The Relationship between the Stability of a Complex and the Intensity of its Charge-transfer Bands[1]

BY J. N. MURRELL

RECEIVED MARCH 2, 1959

The matrix elements of the Hamiltonian and of the transition moment operator which occur in the theory of charge-transfer complexes are examined in more detail than has previously been attempted. The contribution to the intensity of the charge-transfer band arising from the interaction of the charge-transfer state with the ground state is compared with the contributions expected from the excited states of the donor and acceptor. It is shown that the donor-excited states will contribute the greatest intensity except when, for reasons of symmetry, they do not interact with the charge-transfer state. Contact charge-transfer absorption will be due almost entirely to the interaction of the charge-transfer and donor-excited states. It is proposed that in a series of related complexes, the relative behavior of the intensity of the charge-transfer band and the stability of the complex depends on the variation of the difference in energy between the most stable configuration and the configuration giving the greatest contribution to the intensity: these configurations are not usually identical. Examples are given to show that the intensity of the charge-transfer band may increase or decrease as the complex becomes more stable.

## 1. Introduction

Charge-transfer (c-t.) absorption bands have been observed for pairs of molecules which form crystalline molecular complexes (*e.g.*, the quinhydrones),[2] for molecules which appear to form a stable complex in solution but which give no crystalline complex (*e.g.*, iodine and benzene)[3] and for molecules

(1) This work was initiated at the Department of Physics, The University of Chicago, whilst the author held a Fellowship from the Commonwealth Fund of New York.

(2) L. Michaelis and S. Granick, THIS JOURNAL, **66**, 1023 (1944).

(3) H. A. Benesi and J. H. Hildebrand, *ibid.*, **70**, 2832 (1948).

SOURCE: *J. Am. Chem. Soc.*, 81, 5037 (1959).

which do not form a stable complex in solution (*e.g.*, iodine and *n*-heptane).[4] To cover these these three cases Orgel and Mulliken[5] have proposed the existence of both "complex" and "contact" c-t. spectra, the former being associated with a stable complex and the latter being due to absorption of light when the two molecules come together during a chance encounter. Thus the quinhydrones possess complex c-t. spectra, iodine and *n*-heptane give rise to contact spectra and the spectrum of iodine and benzene probably possesses a mixture of the two.

Mulliken's original theory of charge transfer complexes[6] brought out a general parallel between the binding energy of the complex and the intensity of the charge-transfer band. As Orgel and Mulliken[5] have pointed out, this simple theory does allow for the possibility of having charge-transfer absorption even though no stable complex is formed. However, it has still to be established that this is more than a theoretical possibility. In other words, can there be sufficient overlap of the donor and acceptor orbitals during a chance collision to give an observable charge-transfer band when this overlap is not sufficient to give a stable complex? Further, the intensity of the charge-transfer band until now has been attributed to the "mixing" of the charge-transfer state and the ground state which occurs when the two molecules are close enough for the donating and accepting orbitals to overlap. There are, however, other sources of this intensity, namely, the excited states of the donor or acceptor. There may be more mixing of the charge-transfer states with these excited states than there is with the ground state.

By examining the matrix elements of the Hamiltonian and transition moment operator between the wave functions of the separate donor–acceptor pair in more detail than has been attempted hitherto, we hope to clear up some of the questions which have been raised above.

Since, as we shall see, the matrix elements in their exact form are complicated, we shall, in the spirit of perturbation theory, evaluate them to the first order in small quantities. Unlike the more conventional perturbation theory, however, the terms contributing to the matrix elements are not simply expressible in powers of some perturbation parameter. Instead they are made up of integrals involving orbitals of the donor (D) and the acceptor (A), and although these individually go to zero as D and A are separated, they do so at different rates. To introduce some rigor into the theory, we shall define integrals which contain the overlap electron density $\phi_d\ \phi_a$ between a D orbital $\phi_d$ and an A orbital $\phi_a$ to be of first order in small quantities for the purposes of the pertubation theory. This definition will include the overlap integrals[7]

$$S_{ad} = <\phi_a(1)\phi_d(1)> \quad (1.1)$$

nuclear attraction integrals of the type

$$H_{ad}(D) = <\phi_a(1)\phi_d(1)|Z_d/r_{1d}> \quad (1.2)$$

where $Z/r_1$ is the potential field acting on electron 1 due to the nuclei of the donor (or the acceptor); the two electron integrals

$$G_{ad,a'a''} = <\phi_a(1)\phi_d(1)\,|1/r_{12}|\phi_{a'}(2)\phi_{a''}(2)> \quad (1.3)$$

(and $G_{ad,d'd''}$), where $\phi_{a'}$ and $\phi_{a''}$ may, or may not, be the same as $\phi_a$; and the transition moment between $\phi_a$ and $\phi_d$

$$M_{ad} = <\phi_a(1)\phi_d(1)|\mathbf{r}_1> \quad (1.4)$$

However, the two electron integral

$$<\phi_a(1)\phi_d(1)|1/r_{12}|\phi_{a'}(2)\phi_{d'}(2)> \quad (1.5)$$

will be regarded as second order since it depends on the interaction of two overlap densities.

The integrals representing the interaction of two charge densities, one on D the other on A, do not fall into the above classification; they include both one and the two electron integrals

$$H_{aa'}(D) = <\phi_a(1)_a\phi'(1)|Z_d/r_{1d}> \quad (1.6)$$

and

$$G_{aa',dd'} = <\phi_a(1)\phi_{a'}(1)|1/r_{12}|\phi_d(2)\phi_{d'}(2)> \quad (1.7)$$

Although these also go to zero as D and A are separated, they do so as some inverse power of the separation, whereas integrals 1.1 to 1.5 vary inverse exponentially with the separation of the donor and acceptor. It is therefore convenient to regard integrals 1.6 as zeroth order quantities.

**2. Evaluation of the Matrix Elements of the Hamiltonian.**—It will be assumed that we know the eigenfunctions for the donor–acceptor pair when the two molecules are far apart. They can be written as a product of eigenfunctions for the separate donor and acceptor providing that the exchange of electrons between the two molecules is allowed for. Thus we write

$$\Psi_r = \mathcal{A}\theta_{dr}(i)\theta_{ar}(j) \quad (2.1)$$

where

$$\mathcal{H}(i,d)\theta_{dr}(i) = E_{dr}\theta_{dr}(i) \quad (2.2)$$

and

$$\mathcal{H}(j,a)\,\theta_{ar}(j) = E_{ar}\theta_{ar}(j) \quad (2.3)$$

$\mathcal{H}$ (i, d) contains all the terms in the Hamiltonian which depend only on the coördinates of electrons i and the nuclei of D: $\mathcal{H}$(j,a) contains the terms which depend only on the coördinates of electrons j and the nuclei of A. $\mathcal{A}$ is the antisymmetrizing operator.

When D and A come close together the functions $\psi_r$ will cease to be eigenfunctions of the complete Hamiltonian and, in addition, they no longer form an orthogonal set. However, we can still expand the perturbed wave functions in terms of this non-orthogonal set as

$$\Psi_r' = \Psi_r + \sum_{s \neq r} a_{sr}\ \Psi_s \quad (2.4)$$

the coefficient $a_{sr}$ being given by perturbation theory as

$$a_{sr} = \frac{\mathcal{H}_{sr} - S_{sr}\mathcal{H}_{rr}}{\mathcal{H}_{rr} - \mathcal{H}_{ss}} \equiv \frac{K_{sr}}{\Delta E_{sr}} \quad (2.5)$$

If the complete Hamiltonian is now expanded as

$$\mathcal{H} = \mathcal{H}(i,d) + \mathcal{H}(j,a) - \sum_i Z_a/r_{ia} - \sum_j Z_d/r_{jd} + \sum_{ij} 1/r_{ij} + Z_aZ_d/r_{ad} \quad (2.6)$$

(4) D. F. Evans, *J. Chem. Phys.*, **23**, 1436 (1954).
(5) L. E. Orgel and R. S. Mulliken, THIS JOURNAL, **79**, 4839 (1957).
(6) R. S. Mulliken, *ibid.*, **74**, 811 (1952).
(7) The bracket (< >) notation is used in this paper to represent integration over the coördinates of all the electrons involved.

then making use of the relationships (2.1) to (2.3) we find that $K_{sr}$ is given by

$$K_{sr} = N^{-1}\langle\Psi_s - S_{si}\Psi_r| - \sum_i Z_a/r_{ia} - \sum_j Z_d/r_{jd} + \sum_{ij} 1/r_{ij}|\theta_{dr}(i)\theta_{ar}(j)\rangle \quad (2.7)$$

$N$ is a normalizing constant such that $N^{-2}$ is equal to the number of ways of dividing the electrons into two groups of numbers $n_i$ and $n_j$.

To continue further it is necessary to be explicit about the function $\psi_s$. If $\psi_s$ differs from $\psi_r$ in having different numbers of electrons occupying the orbitals of D and A, then an expansion of the type (2.1) will have the form

$$\Psi_s = \mathcal{A}\theta_{ds}(i')\,\theta_{as}(j') \quad (2.8)$$

where the number of electrons i′ is not equal to the number of electrons i (and $n_j \neq n_{j'}$). Alternatively, one can use an expansion of the type

$$\Psi_s = \mathcal{A}\chi_{s\mu}(i)\,\chi_{s\nu}(j) \quad (2.9)$$

in which the electrons i occupy the orbitals which make up the function $\chi_{s\mu}$, but these orbitals belong both to D and to A. Under these circumstances neither $\chi_{s\mu}$ nor $\chi_{s\nu}$ will in general be an eigenfunction of either $\mathcal{H}$ (i,d) or $\mathcal{H}$ (j,a). Whichever expansion is adopted the expression for $K_{sr}$ remains complicated, involving a summation either over i′j′ or over $\mu\nu$. However, the general structure of these matrix elements can be seen by selecting a few simple examples. We shall begin by studying the matrix elements in the case when $\Psi_r$ and $\Psi_s$ are functions of the coördinates of just one electron. In this way we hope to obtain the leading term in the expression for $K_{sr}$. We shall then consider a more sophisticated example, choosing four-electron wave functions, to bring out the nature of the additional terms in $K_{sr}$.

**The One-electron Approximation.**—Suppose that in the ground state (D,A) the electron occupies an orbital $\phi_\delta$ of D and in the charge transfer state it occupies $\phi_\alpha$ of A. To determine the coefficient $a_{\delta\alpha}$ in the perturbed orbital (or state, in the one-electron approximation) $\phi_{\alpha'}$, where

$$\phi_{\alpha'} = \phi_{\delta\alpha} + a_{\delta\alpha}\phi_\delta \quad (2.10)$$

we assume that $\phi_\alpha$ is an eigenfunction for an electron moving in the electrostatic field $V$(A) of A and that the perturbing field is $V(D^+)$ of $D^+$. The total Hamiltonian for the system will then be

$$H = V(A) + V(D^+) \quad (2.10a)$$

Since we are using only one-electron wave functions, the matrix element of $K$, given by 2–7, has the form

$$K_{\delta\alpha} = \langle\phi_\delta - S_{\delta\alpha}\phi_\alpha|V(D^+)|\phi_\alpha\rangle \equiv V_{\delta\alpha}(D^+) - S_{\delta\alpha}V_{\alpha\alpha}(D^+) \quad (2.11)$$

If $\phi_\delta$ and $\phi_\alpha$ are not orthogonal, we can introduce the normalized overlap density $\phi_\delta\phi_\alpha/S_{\delta\alpha}$ and write

$$K_{\delta\alpha} = S_{\delta\alpha}\langle(\phi_\delta\phi_\alpha/S_{\delta\alpha}) - \phi_\alpha^2\,|V(D^+)\rangle \equiv S_{\delta\alpha}W_{\delta\alpha}(D^+) \quad (2.12)$$

$K_{\delta\alpha}$ will now depend not only on the magnitude of the overlap integral $S_{\delta\alpha}$, but also on how nearly the normalized overlap density is to being the same as $\phi_\alpha^2$. For example, if $\phi_\delta$ and $\phi_\alpha$ are orbitals of the same size, then the center of the overlap density lies midway between D and A. However, if one of the orbitals is much smaller than the other the overlap density will be concentrated near the smaller of the two. Figure 1, which shows the electron density $[(\phi_\delta\phi_\alpha)/S_{\delta\alpha} - \phi_\alpha^2]$, for two is orbitals of the type $\phi = (\zeta^3/\pi)^{1/2}e^{-\zeta r}$, plotted along the D–A axis, illustrates this point. It follows that $W_{\delta\alpha}(D^+)$ will be favored by having $\phi_\delta$ small by comparison with $\phi_\alpha$. One other point of interest arising from the figure is that when $\phi_\alpha$ is much smaller than $\phi_\delta$ the electron density is localized on A and is almost independent of the D–A separation.

The integral $W_{\delta\alpha}(D^+)$ will go to zero as the D–A separation goes to infinity, although this limiting value is approached more rapidly the smaller $\phi_\alpha$ is in comparison with $\phi_\delta$. It follows that $K_{\delta\alpha}$ tends to zero more rapidly than $S_{\delta\alpha}$.

In the case that $S_{\delta\alpha} = 0$, then $K_{\delta\alpha} = V_{\delta\alpha}[D^+)$, and again $K_{\delta\alpha}$ will be favored by having the donor orbital small in comparison with the acceptor.

To find the coefficient $a_{\alpha\delta}$ in the perturbed donor orbital

$$\phi_{\delta'} = \phi_\delta + a_{\alpha\delta}\phi_\alpha \quad (2.13)$$

we proceed in a similar manner to that described above, except that we emphasize $\phi_\delta$ to be an eigenfunction of the donor and the perturbing field to be that of the acceptor, $V$(A). As in (2.11) and (2.12) we have

$$K_{\alpha\delta} = V_{\alpha\delta}(A) - S_{\alpha\delta}V_{\delta\delta}(A) \equiv S_{\alpha\delta}W_{\alpha\delta}(A) \quad (2.14)$$

It is now seen that the very condition that favored $K_{\delta\alpha}$, namely, having $\phi_\delta$ small in comparison with $\phi_\alpha$, is the one which militates against $K_{\alpha\delta}$. Moreover, if D and A are both neutral molecules, then whereas $V(D^+)$ falls off as the inverse power of the distance from D, $V$(A) falls off as the inverse exponential of the distance from A.

Outside a sphere which contains essentially all the electrons of A we can take $V(A) = 0$. It follows that the general condition that there should be no stabilization of the ground state is that the overlap density between the donating and accepting orbital should not penetrate the electron density of A in its ground state.

Instead of evaluating $K_{\alpha\delta}$ and $K_{\delta\alpha}$ separately we could have derived one from the other by making use of the orthogonality condition $\langle\phi_{\alpha'}\phi_{\delta'}\rangle = 0$. If $\Delta E_{\delta\alpha}$ is the energy required to transfer an electron from $\phi_\delta$ to $\phi_\alpha$, then to the first order in small quantities we have

$$\langle\phi_{\alpha'}\phi_{\delta'}\rangle = \langle\phi_\alpha + (S_{\delta\alpha}W_{\delta\alpha}(D^+)/\Delta E_{\delta\alpha})\phi_\delta|\phi_\delta - (S_{\delta\alpha}W_{\alpha\delta}(A)/\Delta E_{\alpha\delta})\phi_\alpha\rangle = \frac{S_{\delta\alpha}}{\Delta E_{\delta\alpha}}(\Delta E_{\delta\alpha} - W_{\alpha\delta}(A) + W_{\delta\alpha}(D^+)) \quad (2.15)$$

It follows that if these states are orthogonal

$$W_{\alpha\delta}(A) - W_{\delta\alpha}(D^+) = \Delta E_{\delta\alpha} \quad (2.16)$$

If $W_{\alpha\delta}$(A) is small then $-W_{\delta\alpha}(D^+) \simeq \Delta E_{\delta\alpha}$: the perturbed ground state is then $\phi_{\delta'} = \phi_\delta$, and the perturbed c-t. state is $\phi_{\alpha'} = \phi_\alpha - S_{\alpha\delta}\phi_\delta$.

**The Four-electron Approximation.**—We wish now to see how the matrix elements of $K$ develop as we introduce more freedom into our wave functions. The particular freedom that is required is that of allowing the donated electron to exchange with the other electrons of the acceptor and for the hole left in the donor orbitals to exchange with any other

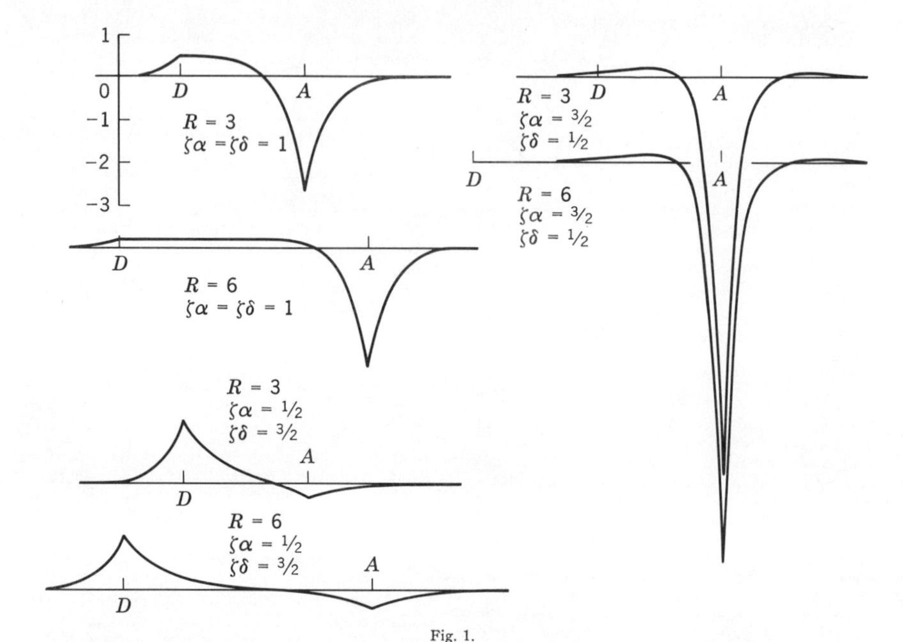


Fig. 1.

holes in the donor orbitals. The simplest wave function which will allow us to note the consequences of such exchanges and which in addition represents the interaction of closed shell molecules (the most common case of interest), will be one which is a function of the position of four electrons.

Let us write the ground state wave function as

$$\Psi_0(\mathrm{D,A}) = |\phi_d(1)\bar{\phi}_d(2)\phi_a(3)\bar{\phi}_a(4)| \quad (2.17)$$

this being a normalized Slater determinant represented for brevity by its principal diagonal. An unbarred orbital will be taken to contain an electron of $\alpha$-spin, a barred orbital one of $\beta$-spin. If $\phi_{d'}$ is a vacant oribtal of D and $\phi_{a'}$ a vacant orbital of A, then we can represent typical singlet excited states of the D–A pair by the wave functions

$$\Psi_1(\mathrm{D^*,A}) = \sqrt{\frac{1}{2}}\{|\phi_d(1)\bar{\phi}_{d'}(2)\phi_a(3)\bar{\phi}_a(4)| - |\bar{\phi}_d(1)\phi_{d'}(2)\phi_a(3)\bar{\phi}_a(4)|\} \quad (2.18)$$

$$\Psi_2(\mathrm{D,A^*}) = \sqrt{\frac{1}{2}}\{|\phi_d(1)\bar{\phi}_d(2)\phi_a(3)\bar{\phi}_{a'}(4)| - |\phi_d(1)\bar{\phi}_d(2)\bar{\phi}_a(3)\phi_{a'}(4)|\} \quad (2.19)$$

A typical c-t. state will then be

$$\Psi_3(\mathrm{D^+A^-}) = \sqrt{\frac{1}{2}}\{|\phi_d(1)\bar{\phi}_{a'}(2)\phi_a(3)\bar{\phi}_a(4)| - |\bar{\phi}_d(1)\phi_{a'}(2)\phi_a(3)\bar{\phi}_a(4)|\} \quad (2.20)$$

We shall be interested in comparing the amounts of ground state $\Psi_0$, donor excited state $\Psi_1$ and acceptor excited state $\Psi_2$, which are introduced into the c-t. state $\Psi_3$ when the two molecules come close together

If $\Psi_3$ is expanded in the form of (2.1) in which we write

$$\theta_{d3}(i)\theta_{a3}(j) = \sqrt{\frac{1}{2}}\{\phi_d(1)|\bar{\phi}_{a'}(2)\phi_a(3)\bar{\phi}_a(4)| - \bar{\phi}_d(1)|\phi_a(2)\phi_a(3)\bar{\phi}_a(4)|\} \quad (2.21)$$

(the requirement that this function be an eigenfunction of the total spin operator $S^2$ necessitates the use of two terms), then the perturbation term in the Hamiltonian will be

$$-Z_a/r_{1a} - \sum_{j=2,3,4} (Z_d/r_{jd} - 1/r_{1j}) + Z_aZ_d/r_{ad} \quad (2.22)$$

$Z_a$ represents the effective nuclear charge of A stripped of its two electrons in $\phi_a$, and $Z_d$ is the effective nuclear charge of D stripped of its two $\phi_d$ electrons. Evaluating the matrix elements of $K$ from (2.7), we have

$$K_{03} = \sqrt{2}\,\{{}_dH_{a'}(\mathrm{D}) + G_{da',dd} - S_{da'}[H_{a'a'}(\mathrm{D}) + G_{a'a',dd}] - S_{da}[H_{aa'}(\mathrm{D}) + G_{aa',dd}]\} \quad (2.23)$$

$$K_{13} = H_{d'a'}(\mathrm{D}) + G_{d'a',dd} - S_{d'a'}(H_{a'a'}(\mathrm{D}) + G_{a'a',dd}) - S_{d'a}(H_{aa'}(\mathrm{D}) + G_{aa',dd}) + S_{da'}(H_{dd'}(\mathrm{A}) + 2G_{dd',aa}) - S_{da}G_{dd'aa'} + G_{da',dd'} \quad (2.24)$$

$$-K_{23} = H_{da}(\mathrm{D}) + G_{da,dd} - S_{da}(H_{aa}(\mathrm{D}) + G_{dd,aa}) - 2S_{da'}(H_{aa'}(\mathrm{D}) + G_{dd,aa'}) \quad (2.25)$$

The field acting on electron i due to the nuclei and electrons of D is

$$-Z_d/r_{ia} + 2\phi_{da}(1)/r_{i1} \quad (2.26)$$

It follows that we can replace $H_{da'}(\mathrm{D}) + G_{da',dd}$ by $V_{da'}(\mathrm{D^+})$. Similarly $H_{dd'}(\mathrm{A}) + 2G_{dd',aa}$ can be replaced by $V_{dd'}(\mathrm{A})$. The matrix elements then take the form

$$K_{03} = \sqrt{2}\{V_{da'}(\mathrm{D^+}) - S_{da'}V_{a'a'}(\mathrm{D^+}) - S_{da}V_{aa'}(\mathrm{D^+})\} \quad (2.27)$$

$$K_{13} = V_{d'a'}(\mathrm{D^+}) - S_{d'a'}V_{a'a'}(\mathrm{D^+}) - S_{d'a}V_{aa'}(\mathrm{D^+}) + S_{da'}V_{dd'}(\mathrm{A}) + G_{da',dd'} - S_{da}G_{dd'aa'} \quad (2.28)$$

$$-K_{23} = V_{da}(\mathrm{D^+}) - S_{da}V_{aa}(\mathrm{D^+}) - 2S_{da'}V_{aa'}(\mathrm{D^+}) \quad (2.29)$$

If we now look at the c-t. state from the point of view of $\Psi_0$, $\phi_d$ is the donating and $\phi_{a'}$ the accepting orbital: from the point of view of $\Psi_1$, however, it is $\phi_{d'}$ that is the donating orbital, whilst from the point of view of $\Psi_2$ it is $\phi_a$ which is the accepting orbital. We notice the appearance of the terms $V_{\delta\alpha}(D^+) - S_{\delta\alpha}V_{\alpha\alpha}(D^+)$ in all three matrix elements, where $\phi_\delta$ is the donating and $\phi_\alpha$ is the accepting orbital. These are just the terms that we obtained in the one-electron approximation.

Other terms in the matrix elements arise from our allowing the donated electron to exchange with the other electrons of the donor or acceptor. For example, we can construct $\Psi_3$ from $\Psi_0$ by the electron transfer $\phi_d \rightarrow \phi_{a'}$ followed by the exchange $\phi_a \leftrightarrow \phi_{a'}$. This gives rise to the contribution $-S_{da}V_{aa'}(D^+)$ in $K_{03}$. Similarly we construct $\Psi_3$ from $\Psi_1$ by $\phi_d \leftrightarrow \phi_{d'}$ followed by $\phi_{d'} \rightarrow \phi_{a'}$. This gives rise to contribution $S_{da'}V_{dd'}(A) + G_{da'dd'}$ in $K_{13}$. The other terms can be associated with similar combinations of transfer and exchange. It is noticed that all the terms over and above those of the one-electron approximation involve the product of an overlap integral, or overlap density, and the transition density between two orthogonal orbitals. Thus whereas the integral $V_{a'a'}(D^+)$ will vary as the inverse first power of the D–A separation (if D is a neutral molecule), $V_{aa'}(D^+)$ will vary as the inverse second power or higher power of this separation, depending on whether the transition density $\phi_a\phi_{a'}$ can be approximated by a dipole or a higher multipole. For this reason it is to be hoped that the matrix elements evaluated for the one-electron approximation (with suitable spin factors: $\sqrt{2}$ for $K_{03}$, 1 for $K_{13}$ and $-1$ for $K_{23}$) will give the most important contribution to the matrix elements.

**3. The Intensity of the C-T. Band and the Binding Energy of the Complex.**—We are now in a position to examine the questions posed in the introduction, namely, what is the most likely source of the intensity of the c-t. band, and can one observe c-t. absorption if no stable complex is formed?

Returning to the example of section 2 in which four-electron wave functions were used, we can write the perturbed ground state function

$$\Psi_0' = \Psi_0(DA) + a_{30}\Psi_3(D^+A^-) \qquad (3.1)$$

and the perturbed c-t. state

$$\Psi_3' = \Psi_3(D^+A^-) + a_{03}\Psi(DA) + a_{03}\Psi(D^*A) + a_{23}\Psi_2(DA^*) \qquad (3.2)$$

Since $\Psi_0'$ and $\Psi_3'$ must be orthogonal, we have, to the first order

$$a_{30} + a_{03} = -S_{03} \qquad (3.3)$$

and using this relationship the transition moment between these two states is then found to be

$$\mathbf{M}'_{03} = (\mathbf{M}_{03} - S_{03}\mathbf{M}_{00}) + a_{30}(\mathbf{M}_{33} - \mathbf{M}_{00}) + a_{13}\mathbf{M}_{01} + a_{23}\mathbf{M}_{02} \qquad (3.4)$$

The first two terms correspond to the expression for the c-t. transition moment given by Mulliken,[6] and discussed again in reference 5. Since the coefficient $a_{30}$ is a measure of the c-t. stabilization of the ground state, the second term in (3.4) has been taken to represent the "complex" c-t. transition moment whilst the first term represents the contact c-t. transition moment.

Using the wave functions (2.17) and (2.20) it easily can be shown that

$$\mathbf{M}_{03} - S_{03}\mathbf{M}_{33} = \sqrt{2}\{\mathbf{M}_{da'} - S_{da'}\mathbf{M}_{dd} - S_{da}\mathbf{M}_{aa'}\} \qquad (3.5)$$

The first two terms in (3.5) are those given by the one-electron approximation; after normalizing the electron density $\phi_d\phi_{a'}$ it can be seen that they are equal to the overlap integral $S_{da'}$, multiplied by the dipole moment of the electron density $[(\phi_d\phi_{a'}/S_{da'}) - \phi_d^2]$. Following the same reasoning as in section 2, the largest values of this moment will be obtained when $\phi_{a'}$ is small compared with $\phi_d$. However, since the reverse situation is more likely to be realized in practice, we can conclude that this dipole moment usually will be rather small.

The relative magnitude of the terms in (3.4) can best be seen by inserting the expressions for the coefficients $a_{30}$, $a_{13}$ and $a_{23}$ which are derived in section 2. Using the one-electron approximation with the correct spin factors, we have

$$M_{03}' = \sqrt{2}S_{da'}\{<(\phi_d\phi_{a'}/S_{da'}) - \phi_d^2|r> - <\phi_{a'}^2 - \phi_d^2|r> (\Delta E_{03})^{-1}W_{a'd}(A)\} + M_{01}S_{d'a'}(\Delta E_{13})^{-1}W_{d'a'}(D^+) - M_{02}S_{da}(\Delta E_{23})^{-1}W_{da}(D^+) \qquad (3.6)$$

The condition for $a_{30}$ to be zero and hence no stabilization of the ground state, is that the overlap density $\phi_d\phi_{a'}$ shall not penetrate the electron cloud of A. This condition is most likely to be satisfied if $\phi_{a'}$ is a diffuse orbital and $\phi_d$ is a compact orbital. However these are just the conditions which lead to small values for the "contact" c-t. transition moment as defined by Mulliken and Orgel. We conclude therefore that both contributions to the c-t. intensity which have been considered in previous papers in this series will be appreciably smaller for contact DA pairs than they will be for true complexes.

The third and fourth terms in (3.4) can also be considered as contributing to the contact c-t. transition moment. The magnitudes of $\mathbf{M}_{01}$ and $\mathbf{M}_{02}$ depend on the intensity of the transitions $D \rightarrow D^*$ and $A \rightarrow A^*$ of the donor and acceptor, respectively. For an intense absorption band ($f \sim 1$) occurring with an energy of 50,000 cm.$^{-1}$, the transition moment is about equal to that associated with the displacement of a unit charge through 6 Å. We see that for intense absorption bands $\mathbf{M}_{01}$ and $\mathbf{M}_{02}$ can be as large as the expected values for $\mathbf{M}_{33}$ - $\mathbf{M}_{00}$. The coefficients $a_{13}$ and $a_{23}$ depend primarily on the magnitude of the overlap integrals $S_{d'a'}$ and $S_{da}$, respectively. As a result of the more diffuse nature of excited state orbitals we can assume

$$S_{d'a'} > S_{da'} > S_{da}$$

For example, with Slater 1s orbitals having exponents $\zeta_d = \zeta_a = {}^3/_2$ and $\zeta_{d'} = \zeta_{a'} = {}^1/_2$, then at a D–A separation of 6 a.u. we have[8]

$$S_{d'a'} = 0.349; \; S_{da'} = 0.081; \; S_{da} = 0.005$$

It is clear that there may be appreciable interaction of the c-t. state with the donor-excited states even when there is no stabilization of the ground state. The interaction of the c-t. state with the acceptor-excited state is likely to be unimportant except when a very stable complex is formed, since the two molecules have to approach close enough for the ground state orbitals to overlap, and this will generally be opposed by the exchange repulsive forces.

(8) R. S. Mulliken, C. A. Rieke, D. Orloff and H. Orloff, *J. Chem. Phys.*, **17**, 1248 (1949).

There is some experimental evidence for the small part played by the acceptor-excited states. In the strong complexes between iodine and pyridine[9] or trimethylamine,[10] the visible band of iodine lies on the long wave length side of the c-t. band and has been shifted to the blue by an amount which is a little greater than the stabilization of the ground state. Any appreciable interaction between the iodine excited state and the c-t. state would have tended to give a red shift to this band.

In addition to the above arguments based on the magnitude of the overlap integrals, it is seen that $a_{13}$ (and $a_{23}$) depend on the energy of an overlap density in the electrostatic field of the donor positive ion, whereas $a_{30}$ depends on the energy of an overlap density in the field of a neutral molecule A. Moroever, the c-t. states and donor-excited states are likely to be closer in energy than are the c-t. and ground state. Everything points to the fact that it is the interaction of the c-t. state with the donor-excited states which gives the most important contribution to the c-t. intensity, and this can certainly operate under conditions in which no stable complex is formed.

In the discussion so far, no mention has been made of any restrictions to the matrix elements imposed by the symmetry of the complex. If the direction of polarization of the c-t. state is perpendicular to the direction of polarization of some donor-excited state, then there will be no mixing of the two. To illustrate the importance of symmetry in determining the intensity of a c-t. band, we will consider the example of the benzene–iodine complex.

The X-ray studies of Hassel[11] on crystals of the $Br_2$:Benzene complex suggest that the most stable configuration for such molecules is one in which the halogen molecule points down toward the center of the benzene ring, the whole molecule having axial symmetry. For such a geometry the c-t. states are polarized at right angles to the direction of polarization of any $\pi \rightarrow \pi^*$ transition of the benzene molecule, hence the excited states of the donor will not contribute to the intensity of the c-t. band. However, in solution there will no doubt exist other configurations for the complex with slightly higher energy which no longer have such a high symmetry. The halogen molecule may for example lie over one of the C–C bonds, it may be tilted away from the perpendicular to the benzene ring, or it may even lie in the same plane as the benzene ring. For these configurations of lower symmetry there is now the possibility that the direction of charge transfer is no longer orthogonal to the direction of the $\pi \rightarrow \pi^*$ transitions of the donor, hence the donor excited states can now contribute to the intensity of the c-t. band. We therefore have good reason to believe that in the benzene–iodine complex it is those configurations which have rather high energy which contribute the greatest intensity to the c-t. band: some of the higher energy configurations may only exist during an accidental contact of the donor and acceptor.

(9) C. Reid and R. S. Mulliken, This Journal, **76**, 3869 (1954).
(10) S. Nagakura, *ibid.*, **80**, 520 (1958).
(11) O. Hassel, *J. Mol. Phys.*, **1**, 241 (1958).

In the series of complexes between iodine and the methylated benzenes the intensity of the c-t. band decreases as the complex becomes more stable (see Table I). Orgel and Mulliken[5] ascribed this behavior to the existence of contact c-t. absorption: as the complex becomes more stable the number of contact configurations will decrease. In this paper I have put forward an explanation, based on the important influence of donor-excited states, of why the contact absorption may well be more intense than the absorption of the stable complexes.

Table I

The Charge-transfer Bands of Iodine and Chloranil with the Methylated Benzenes

The iodine complexes were examined in $CCl_4$: those of chloranil were examined in butyl ether, except that the first five hydrocarbons in the table were studied in the presence of the competing complex of chloranil and N,N-dimethylaniline.

| | Complexes with iodine[12] | | | Complexes with chloranil[13] | | |
|---|---|---|---|---|---|---|
| | $\lambda_{max}$ | $\epsilon_{max}$ | $K_{eq}$ | $\lambda_{max}$ | $\epsilon_{max}$ | $K_{eq}$ |
| Benzene | 292 | 16400 | 0.15 | 340 | 2180 | 0.30 |
| Toluene | 302 | 16700 | .16 | 365 | 1920 | 0.50 |
| *o*-Xylene | 316 | 12500 | .27 | 385 | 2090 | 1.05 |
| *m*-Xylene | 318 | 12500 | .31 | 390 | 2000 | 0.84 |
| *p*-Xylene | 304 | 10100 | .31 | 410 | 1960 | 0.89 |
| 1,2,4-Trimethylbenzene | .. | ... | .. | 420 | 1985 | 1.02 |
| 1,3,5-Trimethylbenzene | 332 | 8850 | 0.82 | 410 | 2250 | 1.17 |
| 1,2,3,4-Tetramethylbenzene | .. | ... | .. | 445 | 2585 | 2.65 |
| 1,2,3,5-Tetramethylbenzene | .. | ... | .. | 450 | 2495 | 2.47 |
| 1,2,4,5-Tetramethylbenzene | 332 | 9000 | 0.63 | 470 | 2320 | 3.02 |
| Pentamethylbenzene | 357 | 9260 | 0.88 | 480 | 2680 | 5.32 |
| Hexamethylbenzene | 375 | 8200 | 1.35 | 505 | 2880 | 9.08 |

In a series of related complexes, the relative behavior of the c-t. intensity and the stability of the complexes will depend roughly on the variation of the energy difference between the most stable configuration and the configuration giving the greatest c-t. intensity. To show that the behavior of the $I_2$–Benzene complexes is not always followed, although there are many examples which behave in the same way (for references see the review by McGlynn),[14] Table I also records the pertinent data for the chloranil–benzene complexes. For this series the c-t. intensity increases as the complex becomes more stable. It may well be that for these complexes, the donating and accepting orbitals, both being $\pi$ molecular orbitals, can only overlap to an appreciable extent when the two molecules lie one on the other in parallel planes. In this configuration the direction of charge transfer is perpendicular to the benzene ring. It is to be noted that the intensity of the c-t. band is much less than for the iodine complexes.

(12) L. J. Andrews and R. M. Keefer, This Journal, **74**, 4500 (1952).
(13) N. Smith, Ph.D. Thesis, University of Chicago, Chicago, Illinois.
(14) S. P. McGlynn, *Chem. Revs.*, **58**, 1113 (1958).

The series of complexes between *s*-trinitrobenzene and substituted anilines[15] occupy a position intermediate between the iodine–benzene and chloranil–benzene complexes. The intensity of the c-t. band is almost constant throughout the series, and what little variation there is, does not appear to be related to the variation in the equilibrium constant.

If we have a series of complexes in which the most stable configuration is one which allows the donor-excited states to contribute to the c-t. intensity, then we should also expect the intensity to increase as the complexes become more stable. Unfortunately there appear to be few examples in the literature which could be expected to satisfy these conditions (there is some evidence that the $I_2$–alcohol complexes may provide one example).[16] Most complexes which have been studied involve an aromatic molecule as a donor, and in these cases the direction of charge transfer in the most stable configuration is always roughly perpendicular to the plane of the aromatic molecule.

In the discussion so far we have considered the form of the interaction between the ground state and a typical c-t. state. However, it must be remembered that all c-t. states, and not only the one of lowest energy, can contribute to the stabilization of the ground state. In the complexes between iodine and a condensed aromatic hydrocarbon, we can expect that as the size of the hydrocarbon is increased, the overlap of the donating and accepting orbital will generally decrease, since the donor orbital is presumably spread over the whole molecule. The intensity of the c-t. band will therefore decrease and the stabilization of the ground state due to the lowest c-t. state probably will decrease also. However, as the hydrocarbon increases in size the number of c-t. states which can interact with the ground state is increased, and although their accompanying c-t. bands may be hidden beneath the absorption bands of the two components, they can all contribute to the stabilization of the complex. The net result of increasing the size of the aromatic molecule may therefore be to give more stable complexes which have weaker c-t. bands. This behavior is observed both for iodine and chloranil complexes (see Table II).

(15) A. Bier, *Rec. trav. chim.*, **75**, 866 (1956).

(16) P. A. D. deMaine, *J. Chem. Phys.*, **26**, 1192 (1957).

TABLE II

THE COMPLEXES OF IODINE AND CHLORANIL WITH SOME CONDENSED AROMATIC HYDROCARBONS

| | Complexes with iodine[17] | | | Complexes with chloranil[13] | | |
|---|---|---|---|---|---|---|
| | $\lambda_{max}$ | $\epsilon_{max}$ | $K_{eq}$ | $\lambda_{max}$ | $\epsilon_{max}$ | $K_{eq}$ |
| Benzene | 292 | 16400 | 0.15 | 340 | 2180 | 0.30 |
| Naphthalene | 360 | 2395 | 0.62 | 460 | 820 | 1.17 |
| Phenanthrene | 378 | 1492 | 1.06 | .. | .. | .. |
| Anthracene | 430 | 112 | 52.35 | 610 | 325 | 7.60 |

In conclusion therefore we can say that c-t. absorption can be observed even if no stable complex is formed but that the intensity of this absorption comes not from the ground state but from the donor-excited states. The relative behavior of the intensity of a c-t. band and the stability of the complex for a series of similar donors depends on the variation of the difference in energy between the most stable configuration and the configuration which gives the most intense c-t. band.

**Acknowledgment.**—I wish to thank Dr. R. S. Mulliken for the stimulating discussions and advice he has given me on this topic.

(17) R. Bhattacharya and S. Basu, *Trans. Faraday Soc.*, **54**, 1286 (1958).

CAMBRIDGE, ENGLAND

[Reprinted from the Journal of the American Chemical Society, 82, 5966 (1960).]


CONTRIBUTION FROM THE LABORATORY OF MOLECULAR STRUCTURE AND SPECTRA, DEPARTMENT OF PHYSICS, UNIVERSITY OF CHICAGO, CHICAGO 37, ILLINOIS]

# Molecular Complexes and their Spectra. XII. Ultraviolet Absorption Spectra Caused by the Interaction of Oxygen with Organic Molecules[1]

BY H. TSUBOMURA[2] AND R. S. MULLIKEN

RECEIVED APRIL 8, 1960

D. F. Evans and others have found that extra absorption spectra appear when oxygen is dissolved in some organic solvents. In the present research, extra absorption spectra caused by oxygen have been measured by bubbling oxygen into ethyl alcohol, dioxane, *n*-butylamine, benzene, mesitylene, pyrrole, triethylamine, aniline, N,N-dimethylaniline, etc. It has been found that the smaller the ionization potential of the organic solvent molecule, the longer the wave length at which the oxygen-induced band lies. This and other experimental results seem to indicate that the extra absorption bands are caused by charge-transfer interaction between oxygen as an electron acceptor and the organic solvents as electron donors, although no stable complexes are formed between them. Wave functions for various excited states of the $O_2$-donor pair including the charge-transfer states have been set up and the matrix elements of the Hamiltonian between these states estimated. It is concluded that the charge-transfer interaction between oxygen and an organic donor molecule can indeed give rise to charge-transfer absorption, with an intensity which is enhanced by interaction between the charge-transfer state and a singlet excited state or states of the donor. It is found that the charge-transfer state can also interact with triplet excited states of the donor and thereby cause the observed enhancements by oxygen of the singlet-triplet absorption bands.

## Introduction

Evans reported that oxygen dissolved in aromatic substances gives rise to absorption at wave lengths longer than for the aromatic compounds alone.[3a] Also Munck and Scott[4] found that saturated hydrocarbons and aliphatic alcohols and ethers show absorption in the near ultraviolet when oxygen is dissolved in them. They also reported that the absorption intensity in the case of cyclohexane is proportional to the partial pressure of oxygen in equilibrium with the solvent. Later, Evans[3b] remarked that the *maxima* which he obtained for the absorption of oxygen–aromatic compound systems might be due to instrumental error, although the existence of extra absorption at longer wave lengths was not in doubt. More recently, Evans[3c] measured the absorption spectra of aromatic compounds both in solution and in vapor with oxygen under high pressure dissolved in the solution or mixed with the vapor. He found additional small peaks at the long wave length tail of the diffuse and stronger absorption which he had observed before. He concluded that these weak bands are singlet-triplet absorption bands enhanced by oxygen. With regard to the stronger bands, he considered it more probable that they are due to charge-transfer absorption[5] caused by the interaction between oxygen and the aromatic compounds, with oxygen as an electron acceptor, rather than to transition from the ground to higher triplet states of the aromatic molecules.

Heidt and others[6] observed that oxygen dissolved in water showed absorption below 220 mμ, which they explained as due to hydrogen bonding between oxygen and water. However, from the close similarity of the absorption of the oxygen–water system to the oxygen-donor systems just discussed, the reason for the appearance of absorption would seem to be the same in both cases.

In this paper, further experimental and theoretical work has been done on the effect of oxygen upon the electronic absorption spectra of various organic compounds.

## Experimental

A Warren Spectracord was used for the measurements of the ultraviolet and visible absorption intensities. The substance to be studied (mostly organic pure liquid, sometimes solution in an organic solvent) was filled into a 1 cm. quartz cell with ground glass stopper, nitrogen was bubbled through for at least 1 minute, the stopper was immediately put in, a wax seal was applied and the absorption spectrum was recorded. Then oxygen was bubbled into the substance for at least 1 minute, the cell was sealed and the spectrum was again recorded. After that, nitrogen was bubbled again into the sample, and the absorption spectrum was once more recorded. For all substances except *n*-butylamine, the absorption spectra of the samples through which nitrogen was passed for the first and the second time agreed well with each other. This is evidence that the extra absorption bands found are not due to oxidation products. In the case of *n*-butylamine, a slight change was observed between the two spectra, indicating a slow irreversible reaction during the measurements.

**Materials.**—The oxygen used was the Extra Dry Grade of the Matheson Co., with minimum purity of 99.6%. The purification of iodine, *n*-heptane, N,N-dimethylaniline and its *o*- and *p*-methyl derivatives was the same as described elsewhere.[7] Ethyl alcohol was the pure absolute alcohol of the U. S. Industrial Chemicals Co., which seemed to contain about $2 \times 10^{-4}M$ of benzene as concluded from its ultraviolet absorption spectrum. This was used without further purification. Anilinium hydrochloride was prepared from purified aniline and hydrochloric acid. All other samples were Eastman organic chemicals. Aniline was dried with sodium hydroxide and distilled under reduced pressure with a column about one foot high. Mesitylene was distilled under reduced pressure with the same column. Dioxane, *n*-butylamine, pyridine and triethyl-

(1) This work was assisted by a grant from the National Science Foundation, for which the authors express their gratitude.

(2) On leave of absence from the Institute for Solid State Physics, the University of Tokyo, Azabu, Minatoku, Tokyo, Japan.

(3) (a) D. F. Evans, *J. Chem. Soc.*, 345 (1953); (b) D. F. Evans, *Chem. and Ind.* (*London*), 1061 (1953); (c) D. F. Evans, *J. Chem. Soc.*, 1351, 3885 (1957); *ibid.*, 2753 (1959).

(4) A. U. Munck and J. R. Scott, *Nature*, **177**, 587 (1956).

(5) R. S. Mulliken, THIS JOURNAL, **64**, 811 (1952); *J. Phys. Chem.*, **56**, 801 (1952).

(6) L. J. Heidt and L. Ekstrom, THIS JOURNAL, **79**, 1260 (1957); L. J. Heidt and A. M. Johnson, *ibid.*, **79**, 5587 (1957).

(7) H. Tsubomura, *ibid.*, **82**, 40 (1960).

amine were dried with suitable drying agents, such as calcium sulfate and sodium hydroxide, and distilled with a Podbielniak column, while benzene and dioxane were used without further purification. Eastman "Spectrograde" benzene also was used and the same result was obtained. All compounds that were distilled had constant boiling points. The purity of these compounds also was checked by examining their ultraviolet absorption spectra in the liquid state.

## Results

The absorption spectra of the samples into which nitrogen had been passed are subtracted from those of the same samples through which oxygen was passed. The resulting spectra for the absorption caused by oxygen are shown in Fig. 1. For all substances measured except ethyl alcohol and acetonitrile, the wave length region for the absorption measurements was limited by the strong absorption bands of the organic molecules lying in the near ultraviolet. The spectra shown in Fig. 1 have been measured down to the shortest measurable wave lengths. In all cases the absorption increased toward shorter wave lengths without reaching a maximum. Although the spectra for N,N-dimethylaniline and N,N-dimethyl-*p*-toluidine show an inflection, seemingly indicating that the wave lengths of the maxima are not far beyond the experimental limits, it seems probable that this is from instrumental error (most likely to stray light of longer wave length), because the spectra down to 340 m$\mu$ of the same materials diluted by *n*-heptane (see below) do not show any tendency which indicates that the curves are near their maxima. This was also confirmed by a measurement for N,N-dimethylaniline using a Cary Spectrophotometer, Model 14, which showed no maximum in that wave length region. This is an unfortunate situation, for we cannot obtain more than a rough idea about the integrated intensities and the vertical excitation energies of the absorption bands here concerned.

**Dilution of N,N-Dimethylaniline and Triethylamine with *n*-Heptane.**—Measurements of spectra have been made for N,N-dimethylaniline diluted with *n*-heptane in the volume ratios of 1:2, 1:5 and 1:10. It was found that the absorption intensity was approximately proportional to the volume fraction of N,N-dimethylaniline in the region of wave length above 425 m$\mu$. The discrepancy from proportionality then increases as the wave length becomes shorter (the apparent absorption intensity with higher concentration becomes relatively smaller). This again can be attributed to stray light. When the same measurements were made with a Cary spectrophotometer Model 14, an approximate proportionality was found down to 370 m$\mu$, although the spectrum with pure N,N-dimethylaniline showed a false peak (not shown by the dilute solutions) at shorter wave lengths.

The Ostwald coefficient of oxygen for *n*-heptane, that is, the equilibrium volume of oxygen at 1 atmosphere pressure at the measured temperature dissolved in a unit volume of heptane is $\lambda = 0.304$ (25°),[8] almost three times as large as that for aniline. Although the solubility of oxygen in N,N-dimethylaniline is not known, we may assume it to be somewhere between those in heptane and aniline. Then we may expect that dilution of aniline with *n*-heptane will somewhat increase the solubility of oxygen. Therefore, the experimental result that the absorption intensity is approximately proportional to the volume fraction of N,N-dimethylaniline seems to indicate that there is no specific formation of a complex between oxygen and the aniline, in other words, that the association constant $K$ is very nearly zero, since, if $K$ were appreciable, the absorption intensity would not continue to increase proportionally with the volume fraction of the aniline after the latter is already at greater concentration than the oxygen.

Similarly, it was found that triethylamine diluted with *n*-heptane in the volume ratios of 1:2, 1:4 and 1:8 showed absorption caused by oxygen approximately proportional to the volume fraction of triethylamine. This also indicates that even such a strong donor as triethylamine does not form a complex of appreciable strength with oxygen. All these results are quite similar to those of Evans[9] on the optical densities of the contact charge transfer absorption of iodine in mixtures of *n*-heptane and perfluoroheptane.

**Effect of Temperature.**—The absorption spectrum of N,N-dimethylaniline saturated with air was measured at 25° using the same substance saturated with nitrogen as the reference; the temperature then was raised to 40° and then cooled to 25°, the absorption spectra being measured each time. No change in the absorption spectrum was observed within experimental error. Similarly, the absorption in the 300 m$\mu$ region of the *n*-heptane solution of triethylamine ($^1/_4$ volume fraction) saturated with oxygen was found to be essentially temperature independent between zero and 20°.

To discuss the results obtained above, one should know the magnitude of change of the concentration of oxygen in the liquid with temperature. In the case of carbon tetrachloride, ether and acetone, for which the solubility of oxygen has been measured at various temperatures, the changes of the solubilities between 20 and 40° are only a few per cent.[8] Therefore, it is likely that the change of solubilities for N,N-dimethylaniline and triethylamine with similar temperature difference is also only a few per cent. Moreover, the change of oxygen concentration in the above measurements is believed to have been even smaller, because the cell was sealed with a small air gap left above the liquid. Therefore, the results obtained above may be taken approximately to mean that the absorbance at constant oxygen concentration essentially does not change with temperature. This means that the heats of formation of the oxygen complexes with N,N-dimethylaniline and triethylamine are practically zero. In other words, there is no formation of a definite complex in these cases.

For comparison, the contact charge transfer absorption of a 0.460 m$M$ *n*-heptane solution of iodine in the ultraviolet was measured at 20 and 63°. The optical density at 240 m$\mu$ at 20° was 0.72 and decreased at the higher temperature only by about 20%, a decrease which may be explained by the

(8) Gmelin "Handbuch der anorganischen Chemie," 8 Auf., Verlag Chemie, G. m. b. H., 1958.

(9) D. F. Evans, *J. Chem. Phys.*, **23**, 1424 (1955).

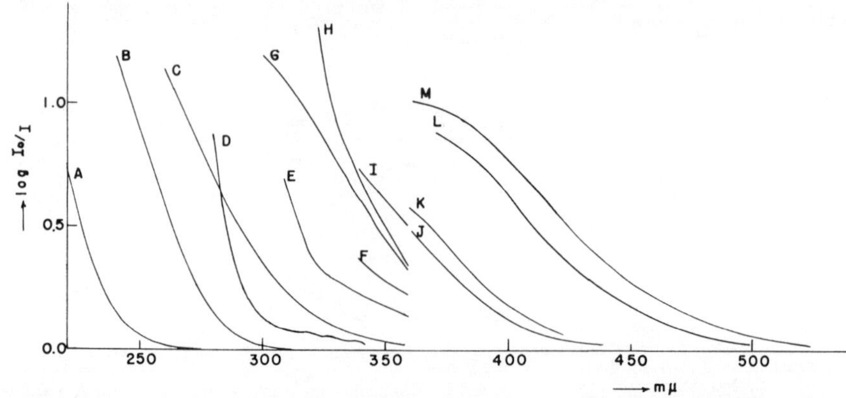


Fig. 1.—Absorption bands induced by saturating organic liquids with oxygen: A, ethyl alcohol; B, dioxane; C, *n*-butylamine; D, benzene; E, mesitylene; F, 1 vol. aniline and 2 vol. ethyl alcohol; G, pyrroel; H, triethylamine; I, aniline; J, N,N-dimethyl-*o*-toluidine; K, N,N-dimethyl-2,6-xylidine (the spectrum for this case may be less accurate than the others); L, N,N-dimethylaniline; M, N,N-dimethyl-*p*-toluidine.

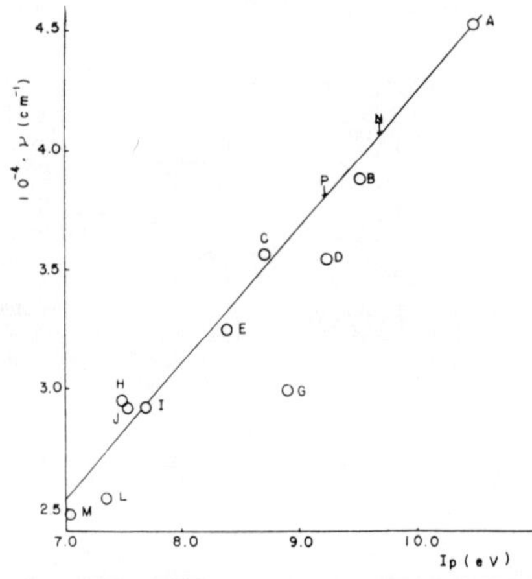


Fig. 2.—Relationship between the wave numbers of oxygen-induced absorption bands and the ionization potentials of organic molecules. The signs for the donors are the same as given in Fig. 1. The references for the ionization potentials are as follows: *n*-butylamine, triethylamine and aniline, K. Watanabe and J. R. Mottle, *J. Chem. Phys.*, **26,** 1773 (1957); dioxane, J. D. Morrison and A. J. C. Nicholson, *ibid.*, **20,** 1021 (1952); benzene, W. C. Price, *ibid.*, **3,** 256, 365, 439 (1935); pyrrole, W. C. Price and A. D. Walsh, *Proc. Roy. Soc.*, **A179,** 201 (1941); N,N-dimethylaniline, N,N-dimethyl-*p*-toluidine and N,N-dimethyl-*o*-toluidine, Professor J. R. Platt's estimation based on the relation between ionization potentials and the frequencies of the band lying in the shortest wave length region; all others, reference 15. The arrows in the figure marked with N and P show the ionization potentials of acetone and pyridine, respectively.

thermal expansion of the solution. This ultraviolet band has been explained as arising from charge-transfer interaction between the saturated hydrocarbon and iodine and called contact charge-transfer absorption,[10] because no evidence has been found for the formation of a true complex between iodine and saturated hydrocarbons. The present observation of no large temperature change of the absorption intensity also supports the conclusion that no complex is formed between iodine and heptane.

**Effect of Solvent.**—It has been found that 1 volume of aniline mixed with 2 volumes of ethyl alcohol shows extra absorption by oxygen in the same wave length region as does pure aniline (Fig. 1). Thus the hydrogen bonding OH . . . $NH_2C_6H_5$ does not seem to affect the absorption spectra of the aniline–oxygen system very much. However, the protonation of aniline affects it greatly. It was found that 1 g. of anilinium chloride dissolved in 2 cc. of ethyl alcohol containing 10% of water has no absorption due to oxygen in the region above 320 mμ, where aniline saturated with oxygen absorbs strongly.[10a]

## Discussion

As was mentioned in the "Experimental" part, the absorption of various organic donor compounds caused by oxygen disappears completely on passing nitrogen into the liquids. This indicates that the observed bands are due to weak interactions between oxygen and the organic compounds and not to formation of oxidation products.

We can see from Fig. 1, and also from Evans' and Munck and Scott's results, that the absorption

(10) L. E. Orgel and R. S. Mulliken, This Journal, **79,** 4839 (1957).

(10a) The solubility of oxygen in water does not decrease very much even in extremely concentrated electrolyte solutions,[9] and hydrochloric acid is known to have a relatively small effect on the solubility of oxygen in water. Therefore it can be assumed that the concentration of oxygen in the solution of anilinium hydrochloride is not much different from that in the aniline–ethyl alcohol mixture.

caused by the interaction of oxygen with organic molecules starts at longer wave lengths the smaller the ionization potential of the organic molecule. In order to see whether there is a more quantitative correlation, the wave numbers at which the extinction coefficients reach the value of 0.7 are plotted against the ionization potentials in Fig. 2. Of course, we must be careful in using such wave numbers as a substitute for the wave numbers of absorption maxima, because they depend not only on the transition energies but also on the unknown integrated absorption intensities and on the solubilities of oxygen in the liquids. Nevertheless, as can be seen in Fig. 2, an approximate linearity is found between ionization potentials and wave numbers. This is similar to the relationship found by McConnell, Ham and Platt[11] and by Hastings, Franklin, Schiller and Matsen[12] for iodine charge-transfer complexes. Theoretically, since the term arising from the second order perturbation energy is small for weak complexes, the frequency of the charge-transfer band is given approximately by $h\nu = I - E - W$, where $I$ is the (vertical) lowest ionization potential of the donor, $E$ is the (vertical) electron affinity of the acceptor and $W$ the electrostatic and other interaction energy between the molecules in the charge-transfer state. If we assume $W$ to be relatively constant for various donors, we expect a linear relationship between $\nu$ and $I$. It seems therefore very reasonable that the bands caused by oxygen are due to transitions to charge-transfer states.

Also, the fact that no absorption due to oxygen has been found in the case of anilinium chloride can be explained by the charge-transfer theory, because due to the effect of the proton bonded with the nitrogen lone pair, the ionization potential of anilinium hydrochloride must be much larger than that of aniline, and hence the charge-transfer band, if it exists at all, should be at shorter wave lengths.

Three other compounds have been found in the present work which show no extra absorption when they are saturated with oxygen. They are acetone, acetonitrile and pyridine. It is quite understandable that acetonitrile does not show a charge-transfer band with oxygen, because of its high ionization potential (12.23 ev.)[13]; any charge-transfer band must be in the vacuum ultraviolet region. The fact that pyridine, which forms a strong complex with iodine,[14] apparently does not show absorption due to oxygen can be understood if the charge-transfer absorption occurs in the region of strong absorption of pyridine. From the ionization potential of pyridine (9.12 ev.)[15] and the straight line in Fig. 2, the absorption from the oxygen–pyridine interaction is predicted to have an appreciable intensity at about 37,600 cm.$^{-1}$, which is indeed in the region of the strong absorption band of liquid pyridine. In agreement with this view, Evans[3c] found absorption which we can attribute to the charge-transfer band by dissolving oxygen in pyridine under high pressure. Similarly, from the ionization potential of acetone (9.69 ev.),[15] and the straight line in Fig. 2, the wave length where oxygen begins to absorb appreciably in acetone is predicted to be about 250 m$\mu$, which is entirely in the region of the absorption by the $n$-$\pi$ band of acetone. Further, Evans[3a] reported that no extra absorption was found in carbon tetrachloride, ethyl benzoate and benzaldehyde. This seems to be explainable in a similar way, although, in the case of carbon tetrachloride, the existence of any charge-transfer interaction with oxygen may be rather doubtful.

There is evidence that the interactions between oxygen and all the donor compounds studied here or by others are very weak. As already mentioned, the equilibrium constants and heats of formation for the oxygen–N,N-dimethylaniline and oxygen–triethylamine systems are negligible. It is also found that dissolved oxygen in benzene and in $n$-heptane has the same paramagnetic susceptibility as in its vapor state.[16] The weakness of the interaction is also indicated by the relative solubilities of the oxygen in various solvents. If there were a considerable stabilization by the donor–acceptor interaction between oxygen and organic molecules, the solubility of oxygen should increase with the basicity of the solvent. On the contrary, the Ostwald coefficient for aniline, for example (0.107 at 20°), is smaller than that for benzene (0.219 at 20°) and even smaller than that for carbon tetrachloride (0.2996 at 20°).[8] Such solubilities are similar to those for hydrogen and nitrogen in the same solvents and indicate that the internal pressures of the solvents are the main factors in determining the solubilities.[17] These facts certainly indicate that there is a negligibly small stabilization by the donor–acceptor interaction between oxygen and organic molecules. In this respect, the absorption spectra caused by oxygen in organic solvents are very similar to that of iodine in $n$-heptane and may be regarded as due to contact charge-transfer absorption. In both cases, the spectra are considered to arise from contact pairs in random orientations.[10]

It is a remarkable fact that oxygen does not seem to form complexes with such strong bases as triethylamine or aniline, whereas iodine forms very strong complexes with them.[7,18] Let us consider some possible reasons for this difference in behavior between iodine and oxygen. First, while the electron affinity of iodine is about 2.5 e.v.,[19] that of oxygen is possibly only 0.15 ev. or not much more.[20] However, it is rather difficult to estimate the vertical electron affinities at the equilibrium separations of the molecules. That of iodine is considered to be much less than 2 ev., probably about zero[21] and

(11) H. McConnell, J. S. Ham and J. R. Platt, *J. Chem. Phys.*, **21**, 66 (1953).

(12) S. H. Hastings, J. L. Franklin, J. C. Schiller and F. A. Matsen, This Journal, **75**, 2900 (1953).

(13) K. Watanabe and T. Nakayama, Technical Report, No. 1-a, University of Hawaii, 1958.

(14) C. Reid and R. S. Mulliken, This Journal, **76**, 3869 (1954).

(15) K. Watanabe, *J. Chem. Phys.*, **26**, 542 (1957).

(16) B. C. Eggleston, D. F. Evans and R. E. Richards, *J. Chem. Soc.*, 941 (1954).

(17) J. H. Hildebrand, "Solubility," Chem. Catalog Co., New York, N. Y., 1924; "Solubility of Non-Electrolytes," Reinhold Publishing Co., New York, N. Y., 1936.

(18) S. Nagakura, This Journal, **80**, 520 (1958).

(19) This can be calculated as (electron affinity of I atom) − (dissociation energy of $I_2$ molecule) + (dissociation energy of $I_2^-$ into I + $I^-$), where the last term may be assumed to be about one half of the $I_2$ dissociation energy.

(20) R. S. Mulliken, *Phys. Rev.*, **115**, 1225 (1959).

(21) M. A. Biondi and R. E. Fox, *ibid.*, **109**, 2012 (1958).

that of oxygen will be also smaller than 0.15 or possibly negative. Hence, we do not know how much difference there is between the vertical electron affinities of the two molecules, but it is probably small. Secondly, the lowest vacant molecular orbitals of oxygen are of the type $\pi_g 2p$, so that (since two such electrons are already present in $O_2$) the increase in size of $O_2$ in going to $O_2^-$ should be much less than that of $I_2$ in going to $I_2^-$, where the $\sigma_u 5p$ of the added electron should be very considerably larger than the orbitals present in $I_2$. Therefore, the orbital of the added electron in $O_2^-$ cannot overlap much with the orbitals of other molecules in contact with $O_2$. Thirdly, these $\pi_g$ orbitals of $O_2$ (when real orbitals are used) have two nodal planes passing through the center of the molecule and perpendicular to each other, so that the overlap integral between these orbitals and the lone pair orbitals of the amines and other $n$ donors which have cylindrical symmetry must at best be small.

If the oxygen-induced extra absorption is due to contact charge transfer, the absorption intensity must be proportional to the concentration of oxygen.[10] The oxygen molecules are considered to be surrounded by solvent molecules with random orientation. There would then doubtless be one or two special mutual orientations between oxygen and solvent molecules which permit the charge-transfer absorption to occur. We do not know how many solvent molecules on the average gave rise to absorption per oxygen molecule. However, it will not be meaningless to calculate the molar extinction coefficients $\epsilon$ in terms of the concentration of oxygen in the solutions. Following are some values taken at wave lengths near the limit of experimental observation; these give lower limits for $\epsilon_{max}$ values.

| | |
|---|---|
| For benzene | 96 at 280 mμ |
| For aniline | 164 at 340 mμ |
| For ethyl alcohol | 76 at 220 mμ |

It should be noted that the $\epsilon$ values just given are large compared with corresponding values for T−N bands (*i.e.* bands caused by transitions from the ground state to the lowest triplet state) enhanced by oxygen at atmosphere pressure as used here. The $\epsilon$ value for the T−N λ3400 band for benzene enhanced by dissolved oxygen can be estimated from present data to be about one. Also, according to high pressure measurements by Evans, the oxygen-induced longest wave length T−N bands for fluorene, naphthalene, phenanthrene, α-bromonaphthalene, 1,2,3,4-di-benzanthracene, etc., have absorbances of the order of, or less than, 0.2. These measurements have been made for chloroform solutions, with a path length of 7.2 cm., concentrations of donors ranging from 0.5 to 1.0 molar and with oxygen pressure from 50 to 70 atm. If the solubility of oxygen is assumed to be proportional to the pressure of oxygen and also if the absorbance is assumed to be proportional to oxygen concentration (both assumptions being partly verified by Evans' measurement of the absorption of naphthalene), then the $\epsilon$ value for the T−N bands of the aromatic molecules mentioned above is calculated to be of the order of or less than $0.2/7.2 \times 50 \times 0.0092 = 0.06$, where 0.0092 is the concentration of oxygen in chloroform saturated with one atm. of oxygen. For comparing the $\epsilon$ value, measured in about 1 *M* solutions of donors, with the others measured in pure donor liquids, a factor of about 6 to 8 should be multiplied in, and this gives about 0.4 to 0.5, which is of the same order of magnitude as the $\epsilon$ values for benzene. These values are about one hundredth of the $\epsilon$ values of the bands which we have identified as charge-transfer bands, even though the $\epsilon$ values for the latter, unlike those for the T−N bands, are not $\epsilon_{max}$ values. This large difference seems to make it very unlikely that the latter bands are also caused by oxygen-enhanced T−N transitions (that is, transitions to higher triplet states, in the case of aromatic compounds where enhanced known T−N bands already have been found). For it seems unlikely that there are enhanced T−N bands which are one hundred times stronger than those already observed; and the existence of additional near-by T levels would in some cases be difficult to understand theoretically.

### Further Theoretical Investigation of the Absorption Spectra Caused by Oxygen

We will further discuss the wave functions of the ground and various excited states of the oxygen–donor pair for the purpose of explaining the effect of oxygen upon the absorption spectra of organic molecules.

For simplicity, we will hereafter assume that the oxygen and D (donor) molecules form 1:1 pairs only. In actual solutions, it seems probable that an oxygen molecule interacts with more than one D molecule simultaneously. However, the discussion for the 1:1 interaction easily can be extended to the 1:*n* case and no substantial change seems to be needed in the conclusions which will be reached.

In Fig. 3 are shown schematically the low-energy states of oxygen, a D molecule and their 1:1 pair. The ground state of oxygen is $^3\Sigma^-_g$, where two unpaired electrons reside in the degenerate $\pi_g$ molecular orbitals. The ground state of D is usually a totally symmetric singlet and is denoted in Fig. 3 by $^1$A. The lowest *excited* singlet and triplet states of D are designated by $^1$E and $^3$E, which are considered to arise by promoting an electron from the highest filled molecular orbital $\phi_1$ of D to the lowest unfilled one $\phi_2$. In molecules such as benzene, there are degenerate orbitals which make the situation more complicated. In the present discussion we assume for simplicity that no such degeneracy exists. For $\pi_g$ we shall use the two real m.o.'s $\pi$ and $\bar{\pi}$ of the respective forms $F\cos\phi$ and $F\sin\phi$, where $F$ is of the form $f_a - f_b$, $a$ and $b$ referring to the two atoms ($f = cre^{-\alpha r}\sin\theta$). Omitting core electrons, the wave functions of the above mentioned states can then be written as shown: For oxygen, for $M_S = 1$

$$\Psi_{+1}(^3\Sigma^-_g) = \mathcal{A}\{(\pi\alpha)^{(3)}(\bar{\pi}\alpha)^{(4)}\} \quad (1)$$

For D

$$\Psi(^1\mathrm{A}) = \mathcal{A}\{(\phi_1\alpha)^{(1)}(\phi_1\beta)^{(2)}\} \quad (2)$$

$$\Psi(^1\mathrm{E}) = 2^{-1/2}\,\mathcal{A}\{(\phi_1\alpha)^{(1)}(\phi_2\beta)^{(2)} - (\phi_1\beta)^{(1)}(\phi_2\alpha^{(2)}\} \quad (3)$$

$$\Psi_{+1}(^3\mathrm{E}) = \mathcal{A}\{(\phi_1\alpha)^{(1)}(\phi_2\alpha^{(2)}\} \quad (4)$$

where $\mathcal{A}$ is the antisymmetrizer, and the numerical superscripts in parentheses refer to electrons number 1 to 4. In the triplet states, there are of course two other wave functions which correspond to $M_S$

= 0 and −1. In using eq. (1)–(4) we treat the complex as a four-electron system. Omission of the other electrons will cause no error in the essential conclusions to be reached.

**The Intensity of the Charge-transfer Band.**—We will first discuss the intensity of the charge-transfer (CT) band. As the oxygen-D pair is a very weak complex or a contact pair, the wave function of the ground state of this pair (denoted by $^3$A) can be represented by the fully antisymmetrized product of those for the individual molecules, $\Psi(^3\Sigma^-_g)$ and $\Psi(^1A)$. The over-all state is a triplet state. Similarly the state of the pair which corresponds to the $^1$E of D (denoted here by $^3$G) is a triplet and is represented by the antisymmetrized product of $\Psi(^3\Sigma^-_g)$ and $\Psi(^1E)$. The CT state is considered to arise by the transfer of an electron from $\phi_1$ to $\pi$ and/or $\bar{\pi}$. The relative probability of the transfer to either of the two orbitals depends on the structure of the pair and the axis of quantization of oxygen, and, by suitably choosing it, we can make in some cases the transfer to one of these m.o.'s nil. Hereafter, we assume for simplicity that the transfer occurs only to $\pi$. The CT state may be either a triplet or a singlet ($O_2^-$ and $D^+$ are both in doublet states). The triplet CT state with $M_S = 1$ is given by the function

$$\Psi_{+1}(^3CT) = \mathcal{A}\{(\phi_1\alpha)^{(1)}(\pi\beta)^{(2)}(\pi\alpha)^{(3)}(\bar{\pi}\alpha)^{(4)}\} \quad (5)$$

As both the ground state and this state are triplet, the transition between these states is multiplicity-allowed in the same way as in other molecular complexes.[5]

Murrell[22] recently pointed out that the observed intensities of contact CT bands, such as are caused *e.g.* by iodine dissolved in saturated hydrocarbons, are much stronger than probably expected from their quite weak CT interactions in the ground state and, therefore, the major part of the intensities of these bands probably are caused by interaction of CT states and donor excited states, so that the contact CT band borrows more or less its intensity from the strongly allowed transition between the donor ground state and the donor singlet excited state.

This explanation seems to be applicable also to the oxygen–donor CT absorption spectra. In general, the CT state may interact with several of the excited singlet states of the donor in the presence of $O_2$. We will discuss, as an example, the interaction with the lowest singlet excited state $^1$E of the donor. This latter state corresponds to the $^3$G state of the $O_2$–D pair and is represented by the wave function

$$\Psi_{+1}(^3G) = \mathcal{A}(2)^{-1/2}\{(\phi_1\alpha)^{(1)}(\phi_2\beta)^{(2)} - (\phi_1\beta)^{(1)}(\phi_2\alpha)^{(2)}\}(\pi\alpha)^{(3)}(\bar{\pi}\alpha)^{(4)} \quad (6)$$

According to second order perturbation theory, $\Psi(^3CT)$ should receive an admixture of $\Psi(^3G)$ with coefficient $-\beta/\Delta W$ where $\Delta W$ is the energy difference between the two states and $\beta$ is given as

$$\beta = (\Psi(^3G)|H|\psi(^3CT)) - (\Psi(^3G)|\Psi(^3CT))\ (\Psi(^3CT)|H|\Psi(^3CT))$$

(22) J. N. Murrell, This Journal, **81**, 5073 (1959). According to a private communication from Dr. Murrell, he has recently reached conclusions which are in part the same as ours on the enhancement of the T−N bands of organic molecules by oxygen (to be published in *Molecular Physics*).

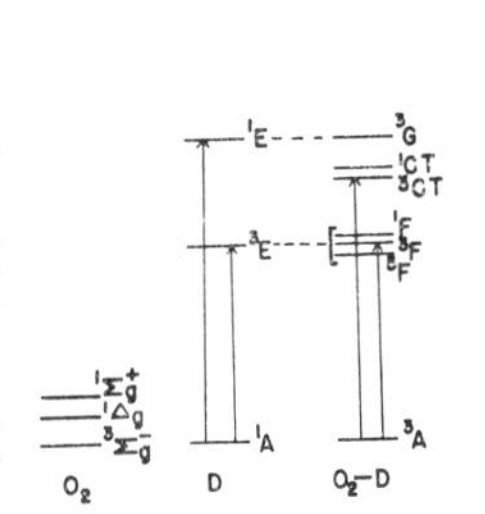

Fig. 3.—The energy levels of oxygen, a D molecule and their 1:1 pair.

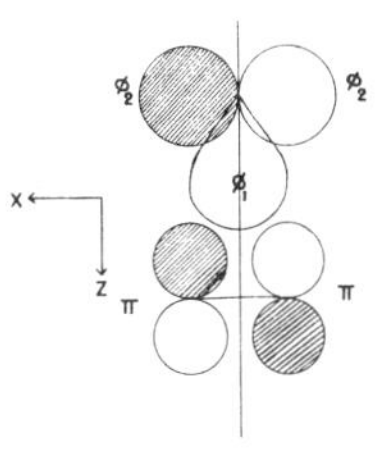

Fig. 4.—Schematic diagram indicating a particular arrangement for the donor orbitals $\phi_1$ and $\phi_2$ and the oxygen acceptor orbital $\pi$. The shadowed circles indicate the regions where the wave functions are negative.

in which $H$ is the Hamiltonian operator for the four electrons. As Murrell has done[22] it can be shown that $\beta$ is approximately given by the one-electron integrals

$$\beta = (2)^{-1/2}(V\pi\phi_2(D^+) - S\pi\phi_2V\pi\pi(D^+)) \quad (7)$$

where $V(D^+)$ is the one-electron integral for the potential energy of the charge distribution $\pi\phi_2$ or $\pi\pi$ in the field of the $D^+$ core and $S$ the overlap integral $\int \pi\phi_2 dv$. This is essentially the same as the corresponding matrix element for a contact pair of molecules which are both singlet in the ground states, except for the factor $1/\sqrt{2}$.

In certain cases where the pair has a structural symmetry, this integral may become zero, because of the antisymmetric nature of $\pi$ with respect to the reflection in the plane which bisects the O–O bond, depending on the symmetry of $\phi_2$. However, since we assume that $O_2$ and D are just in contact with each other, with random orientations, it is expected that in a certain number of the pairs the molecules are situated in such a way that $\beta$ does not vanish. Consequently, the $^3$CT − $^3$A band borrows $(\beta/\Delta W)^2$ times the intensity of the $^3$G − $^3$A band, which is the same as the $^1$E − $^1$A intensity of the donor. Let us assume that the latter band is a strongly allowed one and has $\epsilon$ of about 10,000. Then, as $\Delta W$ and $\beta$ may be taken to be of the order of 1 and 0.1 ev.,[23] respectively, the intensity borrowed is estimated as being of the order of $10{,}000 \times (0.1/1)^2 = 100$ which seems to be a reasonable value.

In cases where the transition to the lowest singlet excited state is an optically forbidden one (as in the case of benzene), the CT state must borrow spectroscopic combining power with the ground state from a higher singlet excited state or states.

**The Enhancement of T-N Bands of Donors by CT Interaction with an Oxygen Molecule.**—The state of the $O_2$–D pair in which D is in the $^3$E state

(23) $\beta$ may be estimated for instance by the proportionality of this kind of integral to the overlap integrals, and from the resonance integral (~4 ev.) between a C—C bond and the overlap integral between two Slater 2$\sigma$ orbitals of carbon and oxygen atoms separated by 3A. Approximately one half of the value estimated in this way is tentatively assumed here, in view of the presence of two nodal planes in the $\pi$ orbital.

and $O_2$ is in the ground state can make quintet ($^5F$), triplet ($^3F$) and singlet ($^1F$) states, as shown in Fig. 3. The $^3CT$ state can interact with $^3F$, whose wave function with $M_S = 1$ is given by

$$\Psi(^3F) = -(\zeta_1 + \zeta_2 - \zeta_3 - \zeta_4)/2 \quad (8)$$

where

$$\zeta_1 = \mathcal{A}\{(\phi_1\alpha)^{(1)}(\phi_2\alpha)^{(2)}(\pi\alpha)^{(3)}(\bar{\pi}\beta)^{(4)}\}$$
$$\zeta_2 = \mathcal{A}\{(\phi_1\alpha)^{(1)}(\phi_2\alpha)^{(2)}(\pi\beta)^{(3)}(\bar{\pi}\alpha)^{(4)}\}$$
$$\zeta_3 = \mathcal{A}\{(\phi_1\alpha)^{(1)}(\phi_2\beta)^{(2)}(\pi\alpha)^{(3)}(\bar{\pi}\alpha)^{(4)}\}$$
$$\zeta_4 = \mathcal{A}\}(\phi_1\beta)^{(1)}(\phi_2\alpha)^{(2)}(\pi\alpha)^{(3)}(\bar{\pi}\alpha)^{(4)}\}$$

The matrix element of the perturbation of $^3F$ by $^3CT$ is calculated as

$$\beta' = (\Psi(^3CT)|H|\Psi(^3F)) - (\Psi(^3CT)|\Psi(^3F))(\Psi(^3F)|H|\Psi(^3F)) \simeq V\pi\phi_2(O_2) - S\pi\phi_1 V\phi_2\phi_2(O_2)$$

Here $V(O_2)$ is the energy of the overlap distribution in the field of the neutral $O_2$ molecule. In general, $\beta'/\sqrt{2}$ is considered to be smaller than $\beta$ mentioned before, because the integrals contained in $\beta'$ are for the potential field of neutral $O_2$ while those in $\beta$ are for the potential field of $D^+$.[22] It will not be unreasonable to assume $\beta'/\sqrt{2}$ to be of the order of one third of $\beta$. Then, if the energy difference between $^3F$ and $^3CT$ is of the same order as that between $^3G$ and $^3CT$, the $^3F$–$^3A$ band can borrow about 0.002 of the intensity of the CT band, that is, approximately of the same order as the observed $\epsilon$ values of the T–N bands for various compounds in the presence of oxygen. Thus it seems probable that the main reason for the enhancement of the T–N bands by oxygen is the interaction between $^3CT$ and $^3F$, giving rise to the borrowing of intensity by the T–N band from the CT band, the intensity of the latter, in turn, coming at least partly from the strong singlet–singlet band of D in the presence of $O_2$. Although a quantitative evaluation of $\beta$ and $\beta'$ would be desirable for obtaining a clear conclusion, it seems likely that the values given above for them are of the correct order of magnitude. At least it is clear that the above described mechanism for the enhancement of the T–N bands by oxygen gives a much larger effect than possible alternatives, as is shown in the following discussions, and therefore seems to be by far the most plausible one at the present time.

It may be worth while to discuss whether there is an appreciable effect from the direct interaction between $^3G$ and $^3F$. The matrix element for the perturbation between these states contains integrals of the type $(\phi_2^i\pi^i|1/r_{ij}|\pi^j\phi_2^j)$ and smaller ones. In view of the fact that this is an exchange integral and hence a factor of the overlap integral $(\phi_2/\pi) \sim 0.01$ smaller than the resonance integrals $\beta$ and $\beta'$ mentioned before, the matrix element between $^3F$ and $^3G$ seems to be much smaller than $\beta\beta'$. Therefore, it can be concluded that the direct interaction between $^3F$ and $^3G$ has a much smaller effect on the intensity of the T–N band than the interaction between these states *through* $^3CT$.

**Other Possible Causes of Oxygen-induced Enhancement of T-N Bands of Donor Molecules.**—There are two other causes which might be considered to enhance the T-N bands of D molecules in the presence of $O_2$.

**(1) Transition between $^3F$ and $^3A$ without Interaction with Any Other States.**—Since these two states are of the same multiplicity, there might be an appreciable transition probability between them even when no perturbations by other states are taken into account. The transition moment is calculated as

$$\mu = (\Psi(^3F)|\sum_i er_i|\Psi(^3A))$$
$$= -\frac{e}{2}[(\phi_2|r|\pi)(\phi_2|\pi) + (\phi_2|r|\pi)(\phi_1|\pi) - 2(\phi_1|r|\pi)(\phi_1|\pi)] \quad (10)$$

where integrals containing $\bar{\pi}$ vanish for the reason mentioned before. The origin of $r$ must be taken at the middle of the molecules in order to cancel the moment which arises from the core charges, and this makes the integrals quite small. Moreover, the terms in (10) vanish in many cases where the contact pair has structural symmetry, and in other cases they cancel mostly with each other because of the opposite signs they have. The most favorable case may be shown schematically in Fig. 4, where the pair has $C_{2v}$ symmetry and $\phi_1$ and $\phi_2$ are respectively symmetric and antisymmetric with respect to reflection in the plane of symmetry. We can see that the $z$ component of the moment is generally very small, since the origin is taken at the center of the pair. Only the $x$ component of the first term remains, and it is of the order of $e(\pi|\phi_2)^2R$, where $R$ may be a little larger than the O–O distance and taken as 2Å. This gives a dipole moment of $3.8 \times 10^{-3}$ debye, and an $\epsilon$ value of 0.01. As this is smaller than the observed values for most cases, we can conclude that this effect is also too small to explain the enhancement.

**(2) Effect of the Inhomogeneous Magnetic Field of the Oxygen Molecule.**—This was suggested by Evans[3c] as one possible cause. Also, a similar effect has been supposed by some authors to be the cause of certain catalytic effects of paramagnetic substances. They consider that the inhomogeneous magnetic field of paramagnetic molecules or ions acts at the spins of two electrons in the absorbing (or reacting) molecules in different ways and may cause some mixing of singlet into triplet states. A similar mechanism was used by Wigner[24] to account for the conversion of para- into ortho-hydrogen by the magnetic field of ions, with the magnetic field acting on nuclear spins instead of electron spins.

However, such a direct action of the inhomogeneous magnetic field seems to be too small to account for the effect of $O_2$ in T–N intensities.

For instance, the matrix element between $^1E$ and $^3E$ for the perturbation of a magnetic field due to an oxygen molecule which is in contact with D is given by

$$\beta'' = \int \Psi(^1E)^*|\mathbf{H}\cdot\mathbf{S}|\Psi(^3E)dv \quad (11)$$

where $\mathbf{H}$ is the magnetic field due to the oxygen molecule, $\mathbf{S} = \sum_i \mathbf{s}_i$, and $\mathbf{s}_i$ is the spin operator for one of the two donor electrons (in $\phi_1$ and $\phi_2$). If we assumed the direction of the field to be that of the axis of the quantization $z$, then

$$\beta'' = \int \Psi(^1E)^*|H_zS_z|\Psi(^3E)_{M_S=0}\,dv$$

(24) E. Wigner, *Z. physik. Chem.*, **B23**, 28 (1933).

The calculation leads to

$$\beta'' = (\phi_1^* | H_z s_z | \phi_1) - (\phi_2^* | H_z s_z | \phi_2) \qquad (12)$$

In the integrations it is to be noted that $H_z$ is nonuniform. The order of magnitude of the integrals can be obtained by approximating them as the interaction energy between two magnetic dipoles, one at the center of oxygen and another at the center of $\phi_1$ or $\phi_2$. This gives $4\mu/R^3$ as a maximum value, where $\mu$ is the Bohr magneton and $R$ is the distance between the dipoles. From (12), it can be seen that the two integrals may largely cancel each other, so that $\beta''$ is smaller than the larger of the two integrals, hence less than at most $4\mu/R^3$. Taking $R$ as 2A, which undoubtedly is smaller than in any actual case, we obtain $\beta'' < 2.6 \times 10^{-5}$ ev. Then the perturbed wave function for the $^3$E state is mixed with that of $^1$E with a coefficient $\alpha = -\beta''/[W(^3E) - W(^1E)]$. Since the denominator is of the order of 1 ev., we obtain $\alpha < 2.6 \times 10^{-5}$. The intensity of the perturbed T–N band for a D.$O_2$ pair is now shown to be $\alpha^2$ times that of the $^1$E –$^1$A band. If the $\epsilon_{max}$ value for the latter band is of the order of $10^4$, that of the former is predicted to be $7 \times 10^{-6}$ at most. Hence it is clear that this effect of the inhomogeneous magnetic field of oxygen in enhancing the T–N band is wholly negligible.[24a]

**j-j Coupling Scheme for Charge Transfer Complexes.**—So far, all the discussions have been made under the tacit assumption that in the CT states the spins of the odd electrons of the two molecules are coupled together, so that the singlet and triplet states are completely distinct, except for the usually quite small spin-orbit coupling.

As emphasized by Prof. J. R. Platt in conversation, in some CT complexes the spin coupling in the CT states may be weak and these may be better described by a j-j-like coupling scheme, where the various states are mixed singlet and triplet. In that case, the CT states may interact with both the triplet and singlet excited states of D, even in complexes without paramagnetic acceptors.

In the cases discussed in this paper, the $O_2^-$ in the $O_2^-$-O$^+$ CT states must be in a $^2\pi$I state with two well-separated sub-states $^2\pi_3$J$_2$ and $^2\pi_1$J$_2$, so that the coupling should actually be j-j-like (for vertical transitions from the loose-contact ground state). However, it is easily shown that the net perturbing effects on the $^3$F and $^3$G wave functions should be practically the same as if the CT states were pure singlet and triplet, so that the results of the discussion given above are unaffected.

(24a) Note Added in Proof.–Papers by G. Porter and M. R. Wright, *J. Chem. Phys.*, **55**, 705 (1958), and H. Linschitz and L. Pekkarinen, *J. Am. Chem. Soc.*, **82**, 2411 (1960), contain material relevant to the above discussion. Porter and Wright have measured the decay rate of the triplet states of anthracene and naphthalene directly with flash photolysis and absorbance measurements. They found that either oxygen or nitric oxide dissolved in the solution increases the decay rate strongly. They also found that most paramagnetic metal ions increase the decay rate, but that the effect has no parallelism with their magnetic moments. They concluded that the decay is caused mainly by *intra*molecular conversion from the triplet to the singlet ground state, enhanced by the paramagnetic substances. Linschitz and Pekkarinen have gone further, pointing out that these quenchers of the triplet states may possibly cause charge-transfer interaction with the molecules in the T state, and the S–T interconversion could be caused mainly by the charge-transfer interaction. Related evidence is found in a recent paper by D. F. Evans (*Proc. Roy. Soc.* (*London*), **A255**, 55 (1960)) who reports that paramagnetic rare earth ions do not enhance appreciably the T–N absorption bands of anthracene.

**Quenching of Fluorescence of Organic Compounds by Oxygen.**—It is well known that oxygen quenches the fluorescence of certain molecules very strongly. For instance, the fluorescence of aromatic hydrocarbons in hexane solutions is quenched by dissolved oxygen at concentrations of the order of $10^{-3}$ $M$.[25] The fluorescence of benzene and anthracene in the vapor phase is also quenched very effectively by oxygen.[26] Further experimental results on oxygen quenching can be found, for instance, in Förster's book (p. 185–186)[27] and in Pringsheim's book (p. 332 *et seq.*).[28]

As a possible mechanism for this oxygen quenching, C. Reid[29] pointed out that oxygen might enhance the conversion of the fluorescent molecule from its singlet excited state to the triplet state. Triplet states have lifetimes so long that at room temperature the molecule usually loses its energy, without phosphorescent emission, by collision or by radiationless transitions. The conversion from singlet to triplet states usually is considered to occur mostly at isoenergetic vibrational levels of the two states, where a small matrix element between these states can give rise to a large amount of conversion. Kasha[30] was the first to discuss this S-T conversion enhancement. He found that the ratio of fluorescence intensity to phosphorescence intensity of halogenated aromatics decreases as the atomic number of the halogen atom increases and explained this fact by the increase in S-T matrix element caused by the enhanced spin-orbit coupling due to the field of the halogen atom.

Like some other authors, Reid considered the enhancement of the S–T intercombination in the presence of the oxygen molecule to be caused by the latter's magnetic field. As had been discussed earlier in this paper, this is very unlikely, and it appears that the mechanism should be revised in the following way. It has been concluded that the charge-transfer state of the oxygen-donor pair interacts with both the excited singlet and excited triplet states of the donor with much greater intensity than for the S–T interaction ordinarily present in molecules without heavy atoms. Therefore, the radiationless transition from the singlet excited state of the donor ($^3$G) to the $^3$CT state and that from the $^3$CT state to the triplet excited state of the donor ($^3$F) may be expected to take place with much larger probability than that of the S–T intersystem crossing in the absence of oxygen. Consequently, the over-all probability of the S–T conversion is enhanced by the charge-transfer interaction.

The lifetimes of strongly radiating singly excited states of molecules are of the order of $10^{-8}$ sec. which is $10^4$ times larger than the inverse of the the mean collisional frequency of molecules in solu-

(25) E. J. Bowen and A. Norton, *Trans. Faraday Soc.*, **35**, 44 (1939); E. J. Bowen and A. H. Williams, *ibid.*, 765 (1935).

(26) (a) E. J. Bowen and W. S. Metcalf, *Proc. Roy. Soc.* (*London*), **A206**, 437 (1951); (b) B. Stevens, *Trans. Faraday Soc.*, **51**, 610 (1955).

(27) T. Förster, "Fluorescenz organischer Verbindungen," Vandenhoeck und Ruprecht, Göttingen, 1951.

(28) P. Pringsheim, "Fluorescence and Phosphorescence," Interscience Publishers, Inc., New York, N. Y., 1949.

(29) C. Reid, *Quart. Rev. Chem.*, **205** (1958).

(30) M. Kasha, *Disc. Faraday Soc.*, **9**, 14 (1950).

tions at room temperature. Hence the excited molecules have ample probability of encountering oxygen molecules before they fluoresce even at a concentration of oxygen of $10^{-3}$ *M*.

It is interesting to point out that the above described theory has certain points in common with previous theories proposed by Kautsky and Weiss. Kautsky[31] explained the quenching action of oxygen by assuming an energy transfer from the excited molecule to the oxygen molecule, resulting in an excitation of oxygen from its ground state to its low $^1\Sigma_g^+$ or $^1\Delta_g$ state, accompanied by intersystem crossing in the organic molecule from a singlet to a triplet state. In this way the multiplicity of the system is held constant. This theory is in agreement with the present one in emphasizing the role of the triplet ground state of oxygen. But, in the present theory, the energy difference between the singlet and triplet states of the donor is dissipated as thermal energy (vibrational at first), and no excitation of the oxygen molecule to its metastable states is necessarily involved.

Weiss,[32] on the other hand, explained the quenching action by electron transfer of the type

$$D^* + O_2 \longrightarrow D^+ + O_2^-$$

with dissociation of the ions, mostly followed by subsequent chemical reaction. This theory has one feature in common with the present one in relating the quenching action of oxygen to charge transfer. However, although in aqueous solutions and with certain dye molecules the process proposed by Weiss seems to occur, such a complete electron transfer as he proposes is very unlikely in non-polar solutions, where the present theory of the mechanism of quenching seems to be much more adequate.[33]

**Acknowledgments.**—Technical assistance by Mr. J. M. Kliegman and Mr. R. P. Lang for the experimental part of the present work is gratefully acknowledged.

(31) H. Kautsky, *Trans. Faraday Soc.*, **35**, 216 (1939).

(32) J. Weiss, *ibid.*, **35**, 48 (1939); *ibid.*, **152**, 133 (1946).

(33) After the manuscript of this paper had been sent to the *Journal*, we learned of a paper by G. J. Hoijtink, *Mol. Physics*, **3**, 67 (1960), in which he discussed the effect of oxygen on the T–N bands of organic molecules. He took into account only the matrix element between the states we call $^3G$ and $^3F$, which we find to have smaller effect than the interaction of these states through $^3CT$. See also the reference to recent work by J. N. Murrell in ref. 22.

Reprinted from
ANNUAL REVIEW OF PHYSICAL CHEMISTRY
Vol. 13, 1962

# DONOR-ACCEPTOR COMPLEXES[1]

BY R. S. MULLIKEN AND WILLIS B. PERSON[2]
*Laboratory of Molecular Structure and Spectra, Department of Physics, University of Chicago, Chicago, Illinois*

The appearance late in 1961 of an excellent book by Briegleb (1) on the intended subject of the present review has made a comprehensive review at this time somewhat superfluous. We shall then refer the reader primarily to Briegleb's book, entitled *Elektronen-Donator-Acceptor-Komplexe*, as well as to some other recent reviews. Our remaining space will be devoted to several topics of current interest to us. Here we shall in part review very recent material not so fully covered in Briegleb's book, and in part we shall present our current thinking, which in some respects goes beyond or differs from that in the literature and in Briegleb's book.

## REVIEWS

Although the existence of molecular complexes has long been recognized [see Pfeiffer (2)], and the basis for understanding them in terms of acid-base theory has long existed [Lewis (3)], interest in the subject received a strong stimulus in 1949 with the discovery of a new absorption band in the ultraviolet spectrum of solutions of $I_2$ in benzene, which was characteristic of a complex [Benesi & Hildebrand (4)]. Not only did this observation provide a means to study this complex and the many others which show an analogous and characteristic new absorption band, but also its interpretation led to an extension of the Lewis acid-base theory in a quantum-theoretical form which provides the basis for the interpretation of a wide variety of phenomena associated with molecular complexes [Mulliken, first in a "Note Added in Proof" in (5), followed by (6, 7, 8)]. Since then the field has been extremely active. Briegleb (1) lists over 600 names in his author index, many of whom are authors of more than one paper, most of them published since 1950.

Several reviews of the subject have been published since 1950, many of them still useful as a supplement to Briegleb's book because of their differences in emphasis and general approach. The first review, by Andrews (10), emphasized experimental aspects, particularly for one-to-one complexes involving an aromatic hydrocarbon as one partner. His review is especially valuable for its historical coverage, particularly of the early ideas explaining

[1] Although in the advance publicity this article was listed as "Charge-transfer Complexes," we prefer the present more inclusive title, which allows us to include complexes in which charge-transfer stabilization is of minor importance compared to, for example, stabilization by electrostatic forces (see footnotes 3 and 5).

The survey of the literature extends through November 1961.

[2] Guggenheim Fellow, 1960–61. On leave from the Department of Chemistry, University of Iowa, Iowa City, Iowa.

why the complexes are stable, and also for its discussion of experimental techniques and its survey of types of interactions. The spectroscopic aspect of the subject was discussed by Orgel (11) in a general review of charge-transfer spectra with special emphasis on molecular complexes. Shortly after these two reviews appeared, Terenin (12) summarized the subject in a well-balanced review which was considered by scientists who read Russian to be one of the most useful summaries available. In 1958, McGlynn (13) reviewed the subject from the point of view of the theorist, with special emphasis on spectroscopic aspects, and pointed out some apparent difficulties in reconciling theory with experiment.

During the past two years, the number of reviews has more than doubled. One of the most interesting was a brief review by Briegleb & Czekalla (14) emphasizing spectroscopic studies. This review is especially valuable because it presents extensive new data which the authors make a determined effort to correlate quantitatively with theory. Much of this material, of course, appears in the book (1). McGlynn (15) has re-emphasized some of the aspects covered in his earlier review, with attention to biological applications. Interest in the application of the ideas of charge-transfer complexes to biological problems has been greatly stimulated by Szent-Györgyi (16). Booth (17) has published a qualitative review emphasizing some of the experimental aspects of the subject. Some of the theoretical aspects have been reviewed by Murrell (18), who gives special emphasis to the problem of the intensity of the charge-transfer band, and to intramolecular charge-transfer spectra. Tsubomura & Kuboyama (19) have written a short review for scientists who read Japanese. In one short section (10 pp) of a long review on molecular electronic absorption spectra, Mason (20) has given an excellent summary of the spectroscopic phases of the subject, with proper balance and emphasis, correct critical evaluation, and clear presentation of the main points. The yearly spectroscopic developments on complexes have usually been included as short sections of articles in the *Annual Review of Physical Chemistry*. [See Platt (21), Price (22), and Ramsay (23) for those in the past three years]. A very recent review (23a) by Andrews & Keefer deals with molecular complexes of the halogens.

## Review of Briegleb's Book

Briegleb's book (1) in relatively short space (some 280 pages) covers comprehensively and thoroughly a large segment of the rather rapidly growing field of 1:1 intermolecular complexes and their spectra. Complexes in which one or both partners are ions, and neutral-molecule complexes in which the acceptor is a boron or metal halide are, however, excluded, as are hydrogen-bonded complexes.[3] Complexes in which both donor and acceptor

[3] Hydrogen-bonded complexes are surely properly classified as donor-acceptor complexes (see footnote 5). In the last section of this review, it is also pointed out that, in their infrared spectra, hydrogen-bonded complexes show close parallelisms to charge-transfer complexes. For details on hydrogen bonding, see Pimentel & McClellan (9).

are $\pi$-electron systems, and complexes in which the acceptor is a halogen molecule are especially completely treated.

The relatively recent work of Briegleb and Czekalla (14), which first demonstrated the occurrence of charge-transfer spectra of 1:1 complexes in fluorescence, and also their clarifying work on phosphorescence spectra of 1:1 complexes, are fully reviewed. The recent premature death of Czekalla, which has cut short a brilliant career, is a matter for deep regret.

The following rather free translation of some passages from the preface of Briegleb's book contains several points of interest.

> The methods and the experimental and theoretical results of the numerous physico-chemical investigations carried out, especially in the last ten years, on electron-donor-acceptor complexes should be significant for a broadened understanding of chemical-catalytic processes, and for the numerous investigations on electron exchange between adsorbed molecules and a solid surface, especially in the case of semiconductors; also in connection with the interpretation of catalytic processes and likewise for bio-chemical processes; and for consideration of redox processes and of energy transfer and energy conduction mechanisms. Therefore, it seemed useful to give a survey. . . .
>
> We have restricted as far as possible the treatment of the quantum mechanical theory and attached more value to treating the existing *experimental* material from unified viewpoints, in order above all to make accessible the general foundations and relationships to a wider circle of interested people not only from chemistry and physics, but also from biochemistry, physiological chemistry, and general biology.

Extensive tables of data and interpretations of these, numerous useful figures, ample bibliographic references, and a good index are included. Professor Briegleb has succeeded admirably in attaining his stated objectives. His book covers the relevant literature fully through 1960 and contains also a number of 1961 references. It deserves mention that the book is the second[4] of a series of monographs on molecular compounds and coordination compounds, edited by Professor Briegleb among others. In the preface, Professor Briegleb makes special mention of the part played by Dr. W. Liptay in writing the last portion of the book.

Although of course we do not agree with every detail, Briegleb has presented the charge-transfer resonance theory very clearly and has applied it quantitatively in creating a unified picture of the subject treated. However, much remains to be done in the elaboration and refinement of the theory and its application to the vast variety of situations which occur in the realm of complexes. Nevertheless, one can go a long way with the simple theory without going seriously astray, and Briegleb's book should provide a very valuable guide.

### Relationship Between Charge-Transfer Band Frequency and Ionization Potential of the Donor

We wish now to discuss in some detail a few topics, most of which have been discussed very well by Briegleb (1). We shall, however, supplement his

[4] The first volume of the series is: H. L. Schläfer, *Komplexbildung in Lösung* (Springer-Verlag, Berlin-Göttingen-Heidelberg, 1961).

discussion of these topics, add emphasis, or sometimes present a different viewpoint.

As the first such topic, let us consider the relation between the frequency $\nu_{CT}$ of the charge -transfer band and the ionization potential $I_D$ of the donor. Empirically it has been found that a plot of $\nu_{CT}$ against $I_D$ gives a straight line [see all the reviews, also McConnell, Ham & Platt (24) and Foster (25)]. As McGlynn (15) comments, "This is surprising, and the almost religious belief which exists in this linearity must be cautioned against since deviations do exist. . . . " The discussion by Briegleb (1) [see also (14)] should do much to dispel the belief in a linear relation; however some further analysis seems necessary.

To this end, let us first briefly define some terms. The structure ($\psi_N$) of the normal state of a complex between two neutral molecules can be described in terms of resonance between a "no-bond" structure $\psi_0$(D, A) and a "dative" structure $\psi_1(D^+-A^-)$, of which the former is predominant in typical loose complexes. The $\psi$'s are wave functions. The energy in the ground state is given by:

$$W_N = W_\infty - \Delta H_f = W_\infty + G_0 - X_0 = W_0 - X_0 \qquad 1.$$

Here $W_\infty$ is the energy of the two separated molecules (usually in solution); $-\Delta H_f$ is the enthalpy of formation of the complex; $G_0$ is the sum of several terms including electrostatic energy (dipole–dipole, etc.) and van der Waals energy;[5] and $X_0$ is the resonance energy of interaction between the "no-bond" and "dative" states.

Similarly the energy of the excited state in which we are most interested, namely the charge-transfer state $\psi_{CT}$, is given by:

$$W_{CT} = W_\infty + I_D{}^v - E_A{}^v - G_1 + X_1 = W_1 + X_1 \qquad 2.$$

Here $I_D{}^v$ is the vertical ionization potential of the isolated donor; $E_A{}^v$ is the vertical electron affinity of the isolated acceptor; $G_1$ is the sum of several terms including the large electrostatic energy of interaction between the charges on $D^+$ and $A^-$, the much smaller valence energy of the dative $D^+–A^-$ bond, and a correction due to the fact that $D^+$ and $A^-$ are in configurations different from those of isolated $D^+$ and $A^-$ ions; $X_1$ is the resonance energy due

[5] $G_0$ is expected to be small and sometimes negative for weak complexes. For strong complexes the D–A distance is often much shorter than the van der Waals distance, so that $G_0$ is probably large and positive because of exchange (steric) repulsion. Usually $G_0$ is large enough so that $-\Delta H_f$ is by no means equal to $X_0$, thus complicating what would otherwise be a simple relationship between $\Delta H_f$ and the coefficients, $b$ and $a$, of $\psi_1$ and $\psi_0$ in $\psi_N$ [see Briegleb (1), p. 21 ff.].

If $X_0$ is larger in magnitude than $G_0$, the resonance with the dative structure is a major force stabilizing the complex, which is then properly called a charge-transfer complex. If $G_0$ is large in size relative to $X_0$, and negative, as for example in most hydrogen-bonded complexes, the forces holding the complex together are primarily electrostatic. The latter complexes can still be donor-acceptor complexes, but they have often been considered in a different class from charge-transfer complexes.

to interaction with the "no-bond" state. The energy change corresponding to the charge-transfer band is then equal to $W_{CT} - W_N$, or

$$h\nu_{CT} = (W_1 - W_0) + (X_1 - X_0) = I_D^v - (E_A^v + G_1 + G_0) + X_1 - X_0 \qquad 3.$$

For weak complexes, the resonance energy contribution to $h\nu_{CT}$ can be computed by perturbation theory:

$$X_1 - X_0 \approx \frac{\beta_0^2 + \beta_1^2}{W_1 - W_0} \qquad 4.$$

Here $\beta_0$ and $\beta_1$ are matrix elements:

$$\beta_0 = H_{01} - W_0 S_{01} = \int \psi_0 \mathcal{H} \psi_1 dv - W_0 \int \psi_0 \psi_1 dv. \qquad 5.$$

and

$$\beta_1 = H_{01} - W_1 S_{01}; \quad \text{hence} \quad \beta_0 - \beta_1 = S_{01}(W_1 - W_0) \qquad 6.$$

If we suppose $(E_A + G_1 + G_0)$ to be approximately constant for a given acceptor, and if $\beta_0^2 + \beta_1^2$ also is supposed constant, Equation 3 takes the following form:

$$h\nu_{CT} = I_D^v - C_1 + \frac{C_2}{I_D^v - C_1} \qquad 7.$$

This equation is that chosen by Briegleb (1), following earlier authors (1), to represent the data for each acceptor. When $h\nu_{CT}$ is plotted against $I_D$, he finds that the data can be fitted by curves of the form of Equation 7; he then evaluates $C_1$ and $C_2$. For example, for $I_2$ he finds $C_1 = 5.2$ ev and $C_2 = 1.5$ $(\text{ev})^2$.

However, it should not surprise anyone to find donors for which $h\nu_{CT}$ does not fit on such a curve. A few examples (pyridine·$I_2$, triethylamine·$I_2$) are apparent in Briegleb's figures [cf. (1, p. 78); or (14, Fig. 7)]. A partial explanation was offered by Collin (26), who noted that the correlation is made using from the literature ionization energies which are probably closer to adiabatic energies ($I_D^{ad}$) than to vertical energies ($I_D^v$); ($I_D^v > I_D^{ad}$) [see also Briegleb (1)]. But further, there is no theoretical reason to expect $G_1 + G_0$ and $\beta_0^2 + \beta_1^2$ to remain constant for all donors, even with the same acceptor.

Recently Yada, Tanaka & Nagakura (27) have made an extensive study of amine complexes with $I_2$. When the $h\nu_{CT}$ values for these complexes are plotted against $I_D$ (see Fig. 1), we see that there is a distinct difference in their behavior as contrasted with that for almost all other $I_2$ complexes. Not only is the curve for the amine data displaced from that for the other donors, but also its slope is different. A constant correction to $I_D^{ad}$ for verticality for all amines (which seems reasonable) does not cause the two curves to coincide.

The explanation for the behavior of the amine complexes is apparently that the values of the constants in Equation 7 are drastically different for amine complexes than for $\pi$ complexes. Furthermore, it seems apparent that $W_1$ is approximately equal to $W_0$ for amine complexes, so that the perturbation method no longer gives a good approximation for the resonance energy. Instead, use of the variation method gives

$$(h\nu_{CT})^2 = \left[\frac{W_1 - W_0}{1 - S_{01}^2}\right]^2 \left[1 + \frac{4\beta_1\beta_0}{(W_1 - W_0)^2}\right] \qquad 8.$$

[cf. Yada, Tanaka & Nagakura (27), and Briegleb (1)]. Here the terms are as

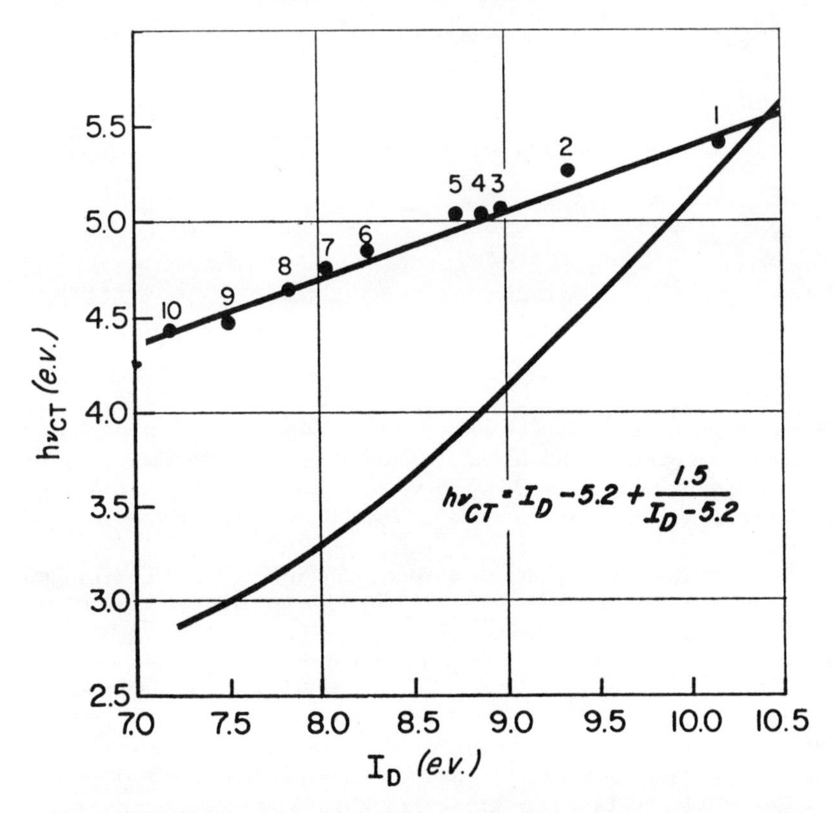


FIG. 1. Dependence of the transition energy, $h\nu_{CT}$, upon the ionization potential of the donor, $I_D$, for complexes of $I_2$. Values for the ionization potential are from photoionization studies listed by K. Watanabe, T. Nakayama, and J. Mottl in their *Final Report on Ionization Potential of Molecules by a Photoionization Method.* The lower curve is that found by Briegleb to fit the data for most $I_2$ complexes [see (1, Figure 33, p. 78)]. The data for the amine complexes from (27) are shown in the upper curve: 1. ammonia, 2. pyridine, 3. methylamine, 4. ethylamine, 5. *n*-butylamine, 6. dimethylamine, 7. diethylamine, 8. trimethylamine, 9. triethylamine, 10. tri-(*n*-propyl)amine.

defined earlier, and $W_1 - W_0 = I_D^v - C_1$ if the definition of $C_1$ used by Briegleb, and given above, is adopted.

When one tries to fit the data for the amines by an equation of the form of Equation 8, one finds this to be possible with the following parameters: $C_1 = E_A^v + G_1 + G_0 \approx 6.9$ ev, $S_{01} = 0.3$, and $\beta_0 = -2.5$ ev. These values are to be contrasted with the values for the other $I_2$ complexes ($C_1 = 5.2$ ev, $S_{01} = 0.1$, $\beta_0 = -0.6$ ev).

The reason these parameters are so different for amine complexes probably lies in the much closer approach of the donor and acceptor molecules in these $n$ donor, $\sigma$ acceptor complexes than in looser complexes. Close approach should cause $G_0$, $G_1$, $S_{01}$, and $\beta_0$ to become relatively large, in agreement with what is found.[6]

To summarize, the analysis of data for $h\nu_{CT}$ in terms of the approximate Equation 7 has been ably demonstrated by Briegleb (1, 14). Deviations can be expected for donor classes which are sufficiently different from each other. Usually these differences would result in slight shifts of the curve of Equation 7, but if the changes in parameters are such that $W_1$ is approximately equal to $W_0$, the slope of the curve fitting the data can be expected to decrease as the resonance interaction becomes more important with respect to the other terms determining $h\nu_{CT}$. In that case, Equation 8 must be used.

Although Equations 7 and 8 call for nonlinear relations of $h\nu_{CT}$ to $I_D$, one finds that their graphs are only slightly curved over the observed range of $I_D$. Thus, within experimental error, most of the data can be fitted by linear relations. For example, the data for the weak $I_2$ complexes are fitted by:

$$h\nu_{CT} = 0.87\, I_D^v - 3.6 \text{ (ev)} \qquad 9.$$

It has been noted with some puzzlement [see (20), for example] that the linear relationships which best fit the data for $I_2$ complexes have slopes somewhat less than unity [McConnell, Ham & Platt (24), and Foster (25)]. As must be clear by now, the values of the constants in a linear relation have no immediate theoretical significance.

Plotted in Figure 2 are Equation 7 and Equation 8, using corresponding values of the parameters for both ($S_{01} = 0.1$ in Equation 8, a plausible value for weak complexes). We see that Equation 9 gives a good empirical representation of either of the other two equations over the practical range for $I_D$. The fact that the slope is less than unity is a result of the nonzero resonance interaction.

## Lone Pair Donors

A recent significant trend has been an increase in the number of studies of complexes involving donor molecules with lone pairs of electrons ($n$ donors). These studies are important because the $n$-donor, $\sigma$-acceptor complexes are relatively stable ($-\Delta H_f$ is as high as 12 kcal/mole), and the geometrical

[6] It may seem surprising that $\beta_0$ is so high and $S_{01}$ so low for the amine complexes. However, this value of $S_{01}$ is approximately what is estimated from calculations using Slater orbitals (27), and it does not seem possible to juggle parameters to get a better fit.

structures seem fairly certainly established, at least for amine complexes. It is thus possible to observe more accurately than in other cases the effects on various experimental quantities of an increase in the strength of the donor-acceptor interaction. Furthermore, the $n$-donor action is largely localized at one atom, in contrast to the situation for $\pi$ donors, for example benzene. Many of our empirical ideas about the behaviour of complexes are, however, based on the extensive early studies with $\pi$ donors. For these complexes the strength of interaction varies over a relatively small range ($K_f$ ratios vary only by a factor of about 100). It is thus gratifying to have so much new data from studies of the strong $n$-donor complexes.

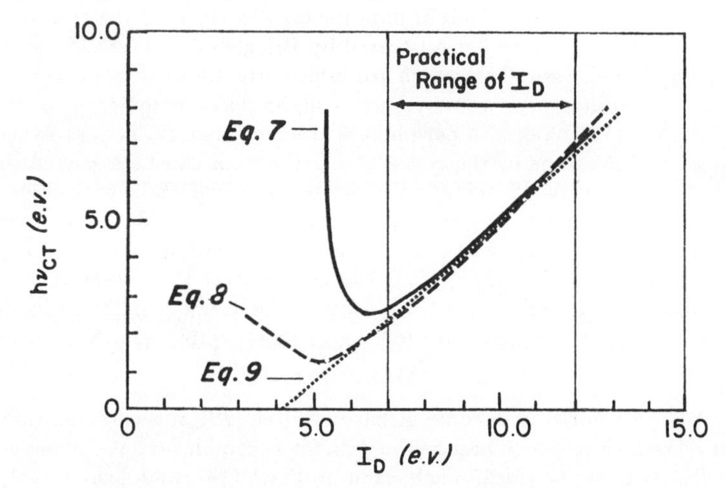


FIG. 2. Comparison of the three equations which give the relationship between $h\nu_{CT}$ and the ionization potential of the donor, $I_D$. —— Eq. 7, with $C_1=5.2$, $C_2=1.5$; – – – –, Eq. 8, with $W_1-W_0=I_D-5.2$, $\beta_0=-0.6$, $\beta_1$ from Eq. 6, $S=0.1$; · · · · ·, the empirical linear relation of Eq. 9. Note the close agreement between the three relations in the region of $I_D$ values for all donors studied so far (7 to 12 ev). This agreement is exact within the usual experimental uncertainties in $h\nu_{CT}$ and in $I_D$. Note also how the more exact Eq. 8 agrees with the empirical linear relation Eq. 9 to lower values of $I_D$.

Studies of new $n$ donors with $I_2$ as acceptor reported during the past year include: complexes with the sulfur and selenium analogues of 1,4-dioxane [McCullough & Zimmermann (28)], complexes with amides [Drago, Carlson, Rose & Wenz (29)], complexes with sulfides [Good, Major, Nag-Chadhuri & McGlynn (30); also Tamres & Searles (31)], and with sulfides and amides [Tsubomura & Lang (32)]. De Maine & Carapellucci (33) have repeated and extended some earlier studies of $I_2$ complexes with diethylether, with puzzling results.

As one example of the modification of ideas which may be necessary as more data from the strong $n$-donor, halogen complexes become available, we have already discussed the relation between $h\nu_{CT}$ and $I_D$ for the amine complexes. Examination of the same relation for some of the other $n$-donor complexes listed above suggests that modification of the Briegleb parameters of Equation 7 is necessary for these cases also. The data do not extend over as large a range of $I_D$ values as for the amines, however, so that the necessity for modification of parameter values is not quite so obvious.

As another example of an idea which may need to be modified in view of the new data from these strong complexes, consider the problem of the intensity of the charge-transfer absorption band. For complexes of $n$ donors with $I_2$, Tsubomura & Lang (32) show quite clearly that the intensity of the charge-transfer absorption band increases with increasing strength of interaction (as measured either by $\Delta H_f$ or by $K_f$). This relation does not always hold for the peak molar absorptivity, $\epsilon_m$, but it is found to be true for the integrated molar absorptivity or oscillator strength, $f$, and for the transition dipole moment $D$; $f$ and $D$ are the measures of intensity which have theoretical significance.[7] The intensity data for the $n$-donor complexes with $I_2$ are therefore in agreement with the original predictions by Mulliken (6).

Much has been written apout the failure of the complexes between aromatic donors and $I_2$ to follow the predicted intensity relation [see Mulliken (8), Orgel & Mulliken (34), Murrell (35), as well as most of the reviews since 1958]. For the strong complexes with $n$ donors, however, for which the geometry of the complex should be definite, the theory and the experimental results agree.

For the weaker aromatic iodine complexes, the geometry is much less certain (see below). If the "axial model" for the benzene-$I_2$ and similar complexes is correct, the application of the intensity theory leads to different predictions than would follow from the "resting model" originally assumed as most plausible by Mulliken.

Further, it should be noted that considerable uncertainty exists as to the

[7] The transition dipole moment $D$ is given in terms of experimental quantities by:

$$2.303 \int (\epsilon/\nu)d\nu = 8\pi^3 D^2/3hc,$$

$$\epsilon = (1/nl) \log_{10} (I_0/I).$$

Here $n$ = concentration in moles/liter and $l$ = path length in cm. If $\nu$ is measured in $cm^{-1}$, Tsubomura & Lang (32) give the approximate formulae:

$$f = 4.319 \times 10^{-9}[\epsilon_m \cdot \Delta\nu_{1/2}]; \qquad D = 0.0958 \left[\frac{\epsilon_m \Delta\nu_{1/2}}{\nu_m}\right]^{1/2}.$$

Here $\epsilon_m$ is the maximum absorptivity, $\Delta\nu_{1/2}$ is the width in $cm^{-1}$ of the band at half-intensity, and $\nu_m$ is the wave number of maximum absorption. The units of $D$ are Debyes. Theoretically, the transition dipole moment $D$ is given by $\int\psi_p(\mu_{op})\psi_q dv$, where $\psi_p$ and $\psi_q$ are the wave functions of the two electronic states involved, and $\mu_{op}$ is the dipole moment operator. The function $D$ usually has a behaviour similar to $\epsilon_m$, but the frequency dependence and the breadth of the band may result in a difference between the behaviour of $D$ and $\epsilon_m$ for a series of complexes.

exact magnitude of the supposed discrepancy between theory and experiment even for the aromatic donors. As noted previously, $I_2$ complexes with these donors do not vary over a wide range of strengths; thus, it is necessary to have accurate data if one is to observe trends. Errors in the determination of $\epsilon_m$ for weak complexes are notoriously large. Practically no data on integrated intensities (or transition dipole moments) have been reported. One must, therefore, conclude that the basis of our empirical knowledge of the variation of transition dipole moment with strength of interaction is as yet sketchy. Certainly the behaviour of the intensities of the charge transfer bands of the stronger complexes over a wide range of interaction strengths is in agreement with what is predicted by the simple theory.

## New and Interesting Acceptors

Much of our knowledge about the behaviour of donor-acceptor complexes comes from the study of complexes between various donor molecules and a standard acceptor molecule (say $I_2$). However, it is of considerable interest to examine the behaviour of complexes with a variety of acceptors. (See Briegleb (1) and Andrews (10) for lists of acceptors.) Each year brings several additions to the list of acceptors whose complexes have been studied. Ferstandig, Toland & Heaton (36) report a study of the complexes of pyromellitic dianhydride,

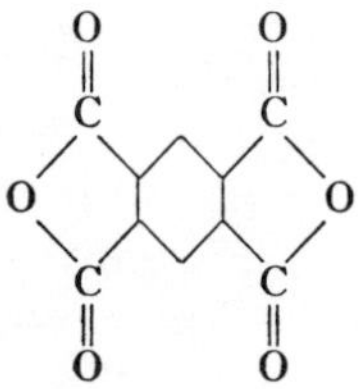

with aromatic donors. This compound acts as a $\pi$ acceptor and forms complexes with aromatic donors which have somewhat larger formation constants even than the relatively strong tetracyanoethylene complexes [see Merrifield & Phillips (37)]. It is thus one of the stronger $\pi$ acceptors.

Chowdhury has continued the studies of the $\pi$ acceptors tetrachlorophthalic anhydride (38) [cf. Chowdhury & Basu (39)] and quinone (40).

Czekalla & Meyer (41) have reported a direct measurement of the dipole moment in the charge-transfer state of the complex between hexamethylbenzene and tetrachlorophthalic anhydride. As expected from the theory, this moment is very large ($14 \pm 3D$). This interesting work is outlined in Briegleb's book (1, pp. 19 ff.).

Feldman & Winstein have reported in a communication (42) some studies on colored complexes of the tropylium ion, which is a very interesting $\pi$ acceptor. Further reports on these complexes are awaited with interest.

There has been considerable interest in the contact charge-transfer spectra observed when $O_2$ is dissolved in donor solvents. These studies follow work begun by Evans (43). The recent papers include studies by Tsubomura

& Mulliken (44), Evans (45), and Jortner & Sokolov (46). The latter also studied similar contact charge-transfer spectra of solutions of NO.

Among the most interesting studies of new acceptors are those reported recently on complexes between aromatic donors and the iodine atom [Rand & Strong (47); Strong, Rand & Britt (48); Strong & Perano (49); Porter & Smith (50); Gover & Porter (51); and Bridge (52)]. In these experiments, I atoms are produced by flash photolysis of $I_2$ solutions. The absorption spectrum, taken shortly after the flash, shows the expected decrease in absorption by the visible $I_2$ band, but there is an accompanying increase in absorption in other regions of the visible spectrum. This increase has been interpreted as the charge-transfer band of a complex between the I atom and the solvent. The donor solvents have been mostly aromatic hydrocarbons, although Gover & Porter (51) found similar absorption for solutions in alcohols and aliphatic halides, and some weak absorption in aliphatic hydrocarbons. However, they failed to observe such spectra in the *n*-donor solvents 1,4-dioxane, tetrahydrofuran, pyridine, and triethylamine, a fact which is somewhat puzzling but may perhaps be attributed to experimental complications.

These studies are especially significant because the I atom is the only acceptor whose electron affinity is reliably known. This fact provides the possibility of testing the theory for the frequency of the charge-transfer band, outlined above. Jortner & Sokolov (53) have attempted to evaluate electron affinities for other acceptors by comparing the frequency of the charge-transfer band of the complex between hexamethylbenzene and the I atom with those of complexes of hexamethylbenzene with other acceptors, e.g., the $I_2$ molecule. Here they write:

$$(h\nu_{CT})_{I_2} - (h\nu_{CT})_{I} = E_{I} - E_{I_2}^{v}$$

Unfortunately, this equation ignores several terms which must contribute to $h\nu_{CT}$ (see Eq. 3). These other terms are probably not the same for different complexes. In particular, $G_1$ may be expected to change as the size of the acceptor changes. Furthermore, the plot shown by Gover & Porter (51) of $h\nu_{CT}$ versus $I_D$ for the I atom complexes has a slope which is less than that for the $I_2$ complexes (Fig. 1). In fact, the curve is roughly parallel to that for the amine-$I_2$ complexes, suggesting that the resonance energy is quite large for I atom complexes. However, it is possible that some modification of Jortner & Sokolov's scheme (53) could be adopted to give vertical electron affinities for other acceptors.

Porter & Smith (50) discuss the significance for reaction mechanisms of the results on I atom complexes. The I atom complexes were discovered in studies of the mechanism of the recombination of I atoms, and suggest once more the importance of donor-acceptor complexes in reaction mechanisms.

Hassel and co-workers (54 to 58) have recently reported a series of x-ray studies on crystalline compounds between *n*-donor molecules (dioxane, dithiane, and $S_8$) and organic halides (iodoform and oxalyl bromide). They have interpreted their results as evidence for donor-acceptor complexes between the molecules mentioned. This premise is supported by the observation that the distances between the halogen atoms and the *n*-donor atoms are significantly shorter than van der Waals distances. There is some doubt, however, as to the importance of charge-transfer forces in these complexes.

Finally, we should like to mention the exciting work on the new cyanocarbon acceptors TCNQ (I), TCS-1 (II), and TCS-2 (III). Most of the work reported so far has been connected with the electrical properties of solid complexes (59 to 61), a topic which is just barely outside the (arbitrary) boundaries of this review. However, the subject is interesting both for itself and its potential practical applications, and for the light which studies of these semi-conducting or photo-conducting solid complexes should throw on our general understanding of complexes [Cairns (62)].

## Analysis of Data

It is often necessary to modify the Benesi-Hildebrand (4) procedure in order to analyze the data for strong complexes to obtain $K$ and $\epsilon$. This fact, plus increasing awareness of the importance of experimental errors and the necessity for an analysis of their effect on reported results, has helped keep up interest in methods of analysis. The most complete coverage of this subject is that in Briegleb's book (1). A portion of the material there was published by Liptay (63). Tamres (64) has recently reviewed the subject, with particular emphasis on the usual neglect of solvent competition and of the effect of changes of donor concentration. It appears that each author ought to carry out a complete error analysis, thus determining the sensitivity of his procedure to various factors involved, and if necessary modify his procedure in the light of his findings, so as to reduce his errors.

Here we wish also to enter a plea that more complete spectroscopic information be reported in studies on complexes. The present practice of giving only $\lambda_{max}$ and $\epsilon_{max}$ for bands is obviously insufficient. As indicated earlier, it is the value of $D$ or $f$ which is of most theoretical significance. Further, the shapes of absorption curves are often significant. Thus, it seems to us important that complete absorption curves be published, so that the information they contain be not lost.

## Geometry of Complexes

X-ray study of the structures of solid complexes continues to inform us about the mysteries of their geometry. We wish especially to discuss this topic because of the implication sometimes given that the failure of the experimental results to agree with early tentative predictions [Mulliken (6)] represents a serious failure of the theory.

For complexes between *n* donors and halogens the solid-state geometry seems well established [see Hassel (65)]. For amine-halogen complexes such as $R_3N \cdot XY$ or $C_6H_5N \cdot XY$, the atoms N–X–Y lie on a straight line which coincides with the symmetry axis of the donor molecule (XY may, e.g., be ICl or $I_2$). This geometry affords maximum overlap between the lone-pair donor orbital from which in the dative ($D^+$–$A^-$) structure the electron comes, and the $\sigma_u$ acceptor molecular orbital into which it goes. In view of the existence of 1:1 complexes of amines with halogens as well-defined units in the crystal, and in view also of the absence of contrary experimental evidence or clearly contradictory theoretical reasons, it seems fairly probable that the same geometry is characteristic of the corresponding 1:1 complexes in solution. We have assumed this to be true in our discussion of the amine-halogen complexes as given above.

Although the indicated linear arrangements N–X–Y were not anticipated theoretically—in fact a different arrangement was proposed as plausible by Reid & Mulliken (65a) in the case of $C_6H_5N \cdot I_2$—they involve no contradiction of the basic theory. Rather, they suggest that one must for the present rely largely on experimental evidence to determine the geometrical configurations of complexes.

In order to make theoretical predictions, it is usually necessary to make simplifying assumptions (*a*), by first deciding what particular dative wave function $\psi(D^+–A^-)$ is needed (or what dative functions, since more than one may be involved, according to the most general form of the theory); and (*b*), by assuming or estimating the magnitudes of various quantities such as $W_1 - W_0$, $G_0$, $X_0$, $\beta_0$, in Equations 1 to 8 or in their generalizations to include more than one dative wave function; (*c*) or both.

The early predictions (6) on the geometrical configurations of complexes of various types were based (*a*) on the simplifying assumption that only the one dative function of lowest energy is important, and (*b*) in some cases on plausible but not certain assumptions about the magnitudes of various terms in Equations 1 to 8. Hassel's work and some of the other current evidence tend to indicate that assumption (*a*) is in many cases too simple, and that regarding (*b*), we still have much to learn about the magnitudes of various parameters which appear in the theory. However, in the special case of amine-halogen complexes, the axially-symmetric model found in Hassel's work on the solid complexes is in accord with the simple assumption (*a*); and, even though this model was not anticipated on the basis of arguments concerning the energy parameters involved (65*a*), it in no way contradicts the basic theory, but merely shows the need of a more thorough understanding of the details of the electronic structures involved. The question at issue is not that of the basic theory, but of its detailed application, which involves a number of interesting factors that must differ from one type of complex to another.

For $\pi$ complexes the geometrical structures are those expected from a naive application of the theory. Wallwork (66) has reported studies of the structures of solid $\pi$ complexes. The donor and acceptor planes apparently lie parallel in the crystal, at approximately the van der Waals distance apart, or somewhat less. This configuration is what would be expected from the principle of maximum overlap for these relatively weak complexes.

The geometry of solid complexes between halogens and various donors has been studied extensively by Hassel and co-workers, and is reviewed by

Briegleb (1). Recently Hassel & Hope (67) have studied the structure of the $Py \cdot 2I_2$ complex. They found it to be composed of $[Py–I–Py]^+$ units and $I_3^-$ and $I_2$ units associated in the same way as in the $I_7^-$ ion. The N–I distance (2.16 A) is a little shorter than that found in the molecular complexes (from 2.26 to 2.31 A). Hassel, Rømming & Tufte (68) have investigated the 1:1 complex between picoline and $I_2$. They found it to be a true molecular complex, quite similar to the other amine complexes which they studied.

In addition to these studies the structures of solid complexes between $I_2$ and the sulfur and selenium analogue of dioxane have been reported by McCullough, Chao & Zuccare (69); Chao & McCullough (70, 71).

In all these complexes between *n* donors and halogens, two features are particularly striking. First, the D–X distance between the donor atom and the first halogen atom is significantly shorter than the van der Waals distance (from 0.5 to 1.3 A shorter). In fact, for the complexes between amines and iodine (and ICl) the N–I distance (from 2.26 to 2.31 A) is not very much longer than the sum of the covalent radii (2.03 A). Secondly, in most of these strong complexes the X–Y distance is longer than in the free halogen. For amine complexes, the lengthening is about 0.1 to 0.2 A. These results are in agreement with predictions of the theory.

The geometrical structures of the complexes between $\pi$ donors and halogens are still a puzzle. The structures found in the solid complexes between benzene and $Cl_2$ and $Br_2$ show the halogen molecule perpendicular to the ring, lying on a sixfold symmetry axis, and equidistant between two benzene rings. Chains of alternating benzene and halogen molecules run through the crystal. Alternate chains are staggered so that benzene rings from the adjacent chains surround the halogen of a particular chain, edge-on. This structure is what might be expected from considerations of closest packing of a crystal composed of positively charged disks and negatively charged rods in a 1:1 ratio.

The above-described structure of the solid complex suggests that also for the 1:1 complex in solution the halogen molecule may be oriented with its axis perpendicular to the ring and coinciding with the sixfold axis of the benzene ("axial model"). However, the suggestion is not convincing, since in the crystal each halogen molecule interacts with two or more benzene molecules, but with only one in the case of a 1:1 complex in solution. Experimentally, there seems to be as yet no convincing proof as to whether the axial model, or one of the models proposed early as most plausible (6), namely the "resting model" (halogen axis parallel to benzene plane), or the "oblique model," is correct.

The evidence for the axial model from infrared spectra (appearance of bands forbidden for the free molecules) at one time appeared to be quite convincing [see Ferguson (72, 73, 74) and the related evidence from the appearance of the X–X stretching vibration of a symmetrical halogen molecule reported in Collin & D'Or (75)]. However, the same brilliant argument which was suggested to explain the infrared spectrum appears to destroy the arguments about geometry of the complexes [Ferguson & Matsen (76)]. At present, the question as to the actual geometry of the 1:1 complexes between halogens and aromatic hydrocarbons in solution must probably be considered as an open one.

If the axial model is correct for the complex in solution, a different theoretical interpretation of the nature of the observed charge-transfer band and of the nature of the dative state or states which stabilize the complex in its normal state is required than would be necessary if the resting model or the oblique model is correct. If the axial model is correct, the dative structure which stabilizes the ground state of the complex by resonance [see Aono (77)] cannot be the same as the dative structure which is the principal contributor of the wave function of the excited state of the observed charge-transfer absorption band. The observed intensity of this band would then require a different theoretical explanation (which can, however, be given satisfactorily) than is needed in the more typical case (exemplified by the amine-halogen complexes) in which only one dative wave function is involved.

## Infrared Spectra

Finally, we would like to close our review of selected topics by some brief comments on the infrared spectra of complexes. This topic has been reviewed by Briegleb (1); again his coverage is good. Unfortunately, his use of the symbol, $\epsilon_m$, for the integrated molar absorption coefficient reported in his Table 38 may lead to some confusion. The symbol $\epsilon_m$ usually refers to the molar absorptivity, defined above. Infrared intensities are measured as the integrated quantity:

$$A = \frac{1}{nl}\int \ln_e (I_0/I)_\nu d\nu$$

It is the integrated intensity, $A$, which is reported in Briegleb's Table 38 under the column labeled $\epsilon_m$. The units are cm$^{-1}$ cm$^2$/millimole (or "darks"). Another typographical error appears in Equation VIII, 4, where the reduced mass $m_r$, should multiply the integrated molar absorption coefficient instead of divide it. Further confusion may arise in the use of the minus sign in that equation in front of the "added effective charge," $e_{c\,eff}$, which should be added to the existing effective charge for the isolated molecule, $e_{0\,eff}$.

Recent work not reported by Briegleb includes studies of the enhancement of vibrations of methylated benzenes [Chang & Ferguson (78)]; studies of complexes with pyridine and bipyridines [Popov, Marshall, Stute & Person (79)]; and studies of the infrared and Raman spectra of some trihalide ions [Person *et al.* (80)]. In addition a really complete review should include the many papers on the infrared spectra of molecules or ions which are complexed with metal ions. Again, we exclude these by an arbitrary definition of the boundaries of our subject.

There are several points which we want to discuss. First of all, there is the point of view from which one examines the experimental results. The experimental observation gives changes in frequency and in intensity; in order to relate these changes to changes in bonding in the donor or acceptor molecule, it is necessary to analyze them to get the changes in force constants $k$ and in the effective charges ($=d\mu/dr$) associated with the vibration. When relatively simple-minded procedures were applied to do this for the X–Y stretching vibration in D···X–Y complexes, it was found that the fundamental behaviour of $\Delta k$ and $\Delta(d\mu/dr)$ was very similar for a large number of complexes [see Person, Humphrey, Deskin, & Popov (81); Person, Humphrey & Popov (82); and Person, Erickson & Buckles (83)]. Indeed, when the mass effects are divided out, it is found that the relatively small absolute changes in the spectra of the halogens correspond to quite large changes in $k$ and $(d\mu/dr)$. These are quite similar to those in the X–H stretching vibration in hydrogen-bonded complexes.

On the other hand, one can see that many of the changes observed in the spectra of donors are relatively minor in terms of $\Delta k$ and $\Delta(d\mu/dr)$, even when they alter profoundly the appearance of the spectrum of the complexed donor; e.g., it has been suggested that the large change observed in the appearance of the spectrum of pyridine on complexing [see Glusker & Thompson (84) and Zingaro & Tolberg (85)] is the result of only small changes in the $k$'s and relatively small changes in $d\mu/dr$ [see (79)].

From this point of view, also, we can see that the changes in donor spectra may be very hard to interpret, since there are so many things which can cause frequency shifts without any very large change in force constants. For example, the change in symmetry on complexing can cause vibrations which are isolated in the free molecule to mix in the complexed molecule, resulting in large changes in the appearance of the spectrum.

The second point which we would like to discuss concerns the changes in $k$ and $(d\mu/dr)$ for the X–Y stretching vibration in the case where donor-acceptor complexes (D···X–Y) are formed. Analogous changes have concerned students of the hydrogen bond for many years [see Pimentel & McClellan (9)]. The similarities between these changes in halogen and hydrogen-bonded complexes suggest that we attempt to discuss them all with the same theory. We shall consider here only the intensity changes.

The first attempt (81) to explain the intensification of the X–Y stretching vibration was to suggest that the donor, because of the negative charge on its active atom (O or N, etc.), polarizes the X–Y bond: $D\cdots X^{(+\delta)}-Y^{(-\delta)}$. The added effective charge ($+\delta$) then contributes to an increase in intensity. Although this effect is undoubtedly present, Ferguson & Matsen (86) showed that it could not be quantitatively adequate, since the largest possible choice for a reasonable value of $\delta$ does not account for more than 10 per cent of the observed increase in intensity. The explanation suffers from the same difficulty that confronts the analogous explanation for the change in intensity of the O–H stretching vibration in hydrogen bonding.

For the halogen complexes, Ferguson & Matsen (76, 86) have given what seems to be a good explanation for the intensification. They argue that the change in intensity results from a varying electron affinity of the acceptor (when considering the X–Y stretching vibration). As a result, the contribution of the dative structure changes as the X–Y bond stretches, and there is an oscillating flow of electrons from donor to acceptor with the frequency of the X–Y stretching vibration. We do not agree with some of the details of Ferguson & Matsen's papers, but the general idea appears quite reasonable.

Although the similarity does not seem to have been recognized, the arguments of Ferguson & Matsen (76, 86) are much like those used by McKinney & Barrow (87) to explain the intensification of the O–H stretching vibration in hydrogen-bonded complexes. Their treatment, in turn, is similar to the earlier explanation given by Tsubomura (88). The recent paper by Jones & Simpson (89) discusses the general problem of the intensity of a vibration which changes the mixing of resonance structures contributing to the ground-state wave function for a molecule. Application of their results to the specific case of charge-transfer complexes leads to the same results as for the other papers.

Starting from any one of the papers, it is easy to derive the basic equation of Ferguson & Matsen (86). As mentioned above, it is possible to argue about the details of evaluating parameters in their equation, but the best choice gives calculated intensity values which are in good agreement with the experimental values for the stretching vibration in halogen complexes [Friedrich (90)].

The most interesting result of these treatments is that the intensification, both for the halogen complexes and for the hydrogen-bonded complexes, is the result of the varying contribution of the dative structure ($\psi_1$) as the X–Y bond length oscillates. The two types of complex may differ in the magnitude of the contribution from the dative structure, but the infrared intensity changes apparently result from the same phenomenon.

In conclusion, the above comments on the theory of infrared spectra seem quite applicable to the entire field of donor-acceptor complexes. With respect, also, to the general theory of such complexes and their spectra, one can still quarrel with its detailed application; however, after ten years, the general theory appears to be correct. It is hoped that attempts to apply the theory quantitatively will continue, and that its quantitative application will become routine. We believe that Briegleb's book (1) should provide a stimulus for such attempts.

## ADDENDUM

Our discussion of typical 1:1 complexes, and Briegleb's (1), have been in terms of resonance of a no-bond structure with a dative structure (or in general, with several dative as well as perhaps also locally excited structures). A valid alternative to this "resonance-structure method" is a method in which the structure is described in terms of molecular orbitals of the complex as a whole. The wave functions involved in the two methods can be shown to become identical in the limit of very loose complexes, but deviate increasingly for stabler complexes. In the "whole-complex MO method," MO's (molecular orbitals) of the entire complex are in general constructed as linear combinations of donor and acceptor MO's.

The whole-complex-MO method is clearly preferable in electron donor-acceptor compounds where a central acceptor cation or atom is surrounded by several donor molecules or anions as ligands, as e.g., in $Pt^{++}(C_2H_4)(NH_3)(Cl^-)_2$. However, we feel that for a unified treatment of 1:1 donor-acceptor complexes, the resonance-structure method is probably preferable.

Although the whole-complex-MO method deserves discussion in a complete survey, we had not intended to discuss it here. However, we feel that the present few remarks added in proof are necessary in view of the publication after the completion of this review of two interesting papers on a number of complexes between $\pi$ donors and $\pi$ acceptors, in which Dewar & Lepley (91) and Dewar & Rodgers (92) discuss their results in terms of the whole-complex-MO point of view. This point of view was propounded for $\pi$ complexes some time ago by Dewar (93), to whom is also due the important idea of two-way charge-transfer stabilization (e.g. $C_2H_4 \cdot Ag^+$ with resonance structures $C_2H_4^+ - Ag$ and $C_2H_4^- - Ag^{++}$).

In the two papers cited, Dewar *et al.* show linear plots, against $\nu_{CT}$, of the coefficient $x$ in the Hückel MO formula $I = \alpha - x\beta$ for the minimum ionization energy $I$ of their aromatic hydrocarbon $\pi$ donors. Except that $I$ is here represented by $x$, and ordinate and abscissa are exchanged, these graphs are of the same type as the straight line in Figure 2 representing Equation 9. The latter straight line, it may be recalled, is understandable as an approximation to the slightly curved line in Figure 2 representing Equation 7. In the discussion of Figures 1 and 2 and Equation 9 above, it was pointed out that the slope of the Equation 9 line, given by the coefficient of $I_D$ there, falls increasingly below 1 for increasingly strong complexes (cf. Fig. 1). Conversely, this slope should be near 1 for weak complexes, and $C_2$ in the resonance-structure theory (see Eq. 4) should be relatively small. Some of Briegleb's graphs (1, p. 76, 77) and his listed $C_2$ values (1, p. 77, Table 31) show that these conditions are fulfilled for the $\pi$, $\pi$ complexes he has surveyed. The graphs of Dewar *et al.*, when similarly plotted, also show slopes near 1.

A resulting important conclusion, which receives further support from the characteristically low intensities of their charge-transfer bands (1), is that the resonance interaction which stabilizes $\pi$, $\pi$ complexes is always relatively weak (weaker than for $\pi$, $\sigma$ and much weaker than for $n$, $\sigma$ complexes). The conclusion is based here on use of the resonance-structure method, but is equally valid in the whole-complex-MO method. As already noted, the two methods differ very little except in outward appearance in the case of weak complexes.

LITERATURE CITED


1. Briegleb, G., *Elektronen-Donator-Acceptor-Komplexe* (Springer-Verlag, Berlin-Göttingen-Heidelberg, 1961)
2. Pfeiffer, P., *Organische Molekülverbindungen*, 2nd Ed. (F. Enke, Stuttgart, Germany, 1927)
3. Lewis, G. N., *J. Franklin Inst.*, **226**, 293 (1938)
4. Benesi, H. A., and Hildebrand, J. H., *J. Am. Chem. Soc.*, **71**, 2703 (1949); *J. Am. Chem. Soc.*, **70**, 2832 (1948)
5. Mulliken, R. S., *J. Am. Chem. Soc.*, **72**, 600 (1950)
6. Mulliken, R. S., *J. Am. Chem. Soc.*, **74**, 811 (1952)
7. Mulliken, R. S., *J. Phys. Chem.*, **56**, 801 (1952)
8. Mulliken, R. S., *Rec. trav. chim.*, **75**, 845 (1956)
9. Pimentel, G. C., and McClellan, A. L., *The Hydrogen Bond* (W. H. Freeman & Co., San Francisco, 1960)
10. Andrews, L. J., *Chem. Revs.*, **54**, 713 (1954)
11. Orgel, L. E., *Quart. Revs. (London)*, **8**, 422 (1954)
12. Terenin, A. N., *Uspekhi Khim.*, **24**, 121 (1955)
13. McGlynn, S. P., *Chem. Revs.*, **58**, 1113 (1958)
14. Briegleb, G., and Czekalla, J., *Angew. Chem.*, **72**, 401 (1960)
15. McGlynn, S. P., *Radiation Research Suppl.*, **2**, 300 (1960)
16. Szent-Györgyi, A., *Introduction to a Submolecular Biology* (Academic Press, New York, 1960)
17. Booth, D., *Sci. Progr.*, **48**, 435 (1960)
18. Murrell, J. N., *Quart. Revs. (London)*, **15**, 191 (1961)
19. Tsubomura, H., and Kuboyama, A., *Kagaku to Kôgyô (Tokyo)*, **14**, 537 (1961)
20. Mason, S. F., *Quart. Revs. (London)*, **15**, 287 (1961)
21. Platt, J. R., *Ann. Rev. Phys. Chem.*, **10**, 349 (1959)
22. Price, W. C., *Ann. Rev. Phys. Chem.*, **11**, 133 (1960)
23. Ramsay, D. A., *Ann. Rev. Phys. Chem.*, **12**, 255 (1961)

23a. Andrews, L. J., and Keefer, R. M., *Advances Inorg. Chem. Radiochem.*, **3** (1961)

24. McConnell, H., Ham, J. S., and Platt, J. R., *J. Chem. Phys.*, **21**, 66 (1953)
25. Foster, R., *Tetrahedron*, **10**, 96 (1960)
26. Collin, J., *Z. Elektrochem.*, **64**, 936 (1960)
27. Yada, H., Tanaka, J., and Nagakura, S., *Bull. Chem. Soc. Japan*, **33**, 1660 (1960)
28. McCullough, J. D., and Zimmermann, I. C., *J. Phys. Chem.*, **65**, 888 (1961)
29. Drago, R. S., Carlson, R. L., Rose, N. J., and Wenz, D. A., *J. Am. Chem. Soc.*, **83**, 3572 (1961)
30. Good, M., Major, A., Nag-Chadhuri, J., and McGlynn, S. P., *J. Am. Chem. Soc.*, **83**, 4329 (1961)
31. Tamres, M., and Searles, S., Jr., *Am. Chem. Soc., Div. Phys. Chem.*, Paper No. 65 (Chicago, Sept. 3-8, 1961)
32. Tsubomura, H., and Lang, R., *J. Am. Chem. Soc.*, **83**, 2085 (1961)
33. De Maine, P. A. D., and Carapellucci, P., *J. Mol. Spectroscopy*, **7**, 83 (1961)
34. Orgel, L. E., and Mulliken, R. S., *J. Am. Chem. Soc.*, **79**, 4839 (1957)
35. Murrell, J. N., *J. Am. Chem. Soc.*, **81**, 5037 (1959)
36. Ferstandig, L. L., Toland, W. G., and Heaton, C. D., *J. Am. Chem. Soc.*, **83**, 1151 (1961)
37. Merrifield, R. E., and Phillips, W. D., *J. Am. Chem. Soc.*, **80**, 2778 (1958)
38. Chowdhury, M., *J. Phys. Chem.*, 65, 1899 (1961)
39. Chowdhury, M., and Basu, S., *Trans. Faraday Soc.*, **56**, 335 (1960)
40. Chowdhury, M., *Trans. Faraday Soc.*, 1482 (1961)
41. Czekalla, J., and Meyer, K.-O., *Z. physik. Chem. (Frankfurt)*, 27, 185 (1961)
42. Feldman, M., and Winstein, S., *J. Am. Chem. Soc.*, **83**, 3338 (1961)
43. See Evans, D. F., *J. Chem. Soc.*, 2753 1959, and earlier work given in this paper
44. Tsubomura, H., and Mulliken, R. S., *J. Am. Chem. Soc.*, **82**, 5966 (1960)
45. Evans, D. F., *J. Chem. Soc.*, 1987 (1961)
46. Jortner, J., and Sokolov, U., *J. Phys. Chem.*, **65**, 1633 (1961)
47. Rand, S. J., and Strong, R. L., *J. Am. Chem. Soc.*, **82**, 5 (1960)
48. Strong, R. L., Rand, S. J., and Britt, J. A., *J. Am. Chem. Soc.*, **82**, 5053 (1960)
49. Strong, R. L., and Perano, J., *J. Am. Chem. Soc.*, **83**, 2843 (1961)
50. Porter, G., and Smith, J. A., *Proc. Roy. Soc. (London)*, *A*, **261**, 28 (1961)
51. Gover, T. A., and Porter, G., *Proc. Roy. Soc. (London)*, *A*, **262**, 476 (1961)

52. Bridge, N. K., *J. Chem. Phys.*, **32,** 945 (1960)
53. Jortner, J., and Sokolov, U., *Nature,* **190,** 1003 (1961)
54. Bjorvatten, T., and Hassel, O., *Acta Chem. Scand.*, **13,** 1261 (1959)
55. Hassel, O., *Svensk Kem. Tidsskr,* **72:2,** 88 (1960)
56. Hassel, O., *Tidsskr. Kjemi Bergvesen Met.*, **3,** 60 (1961)
57. Hassel, O., *Nature,* **189,** 137 (1961)
58. Groth, P., and Hassel, O., *Proc. Chem. Soc.*, 343 (Sept. 1961)
59. Acker, D. S., Harder, R. J., Hertler, W. R., Mahler, W., Melby, L. R., Benson, R. E., and Mochel, W. E., *J. Am. Chem. Soc.*, **82,** 6408 (1960)
60. Kepler, R. G., Bierstedt, P. E., and Merrifield, R. E., *Phys. Rev. Letters,* **5,** 503 (1960)
61. Chesnut, D. B., Foster, H., and Phillips, W. D., *J. Chem. Phys.*, **34,** 684 (1961)
62. Cairns, T. L., *Am. Chem. Soc., Div. Org. Chem.*, Paper No. 75 (St. Louis, March 21–30, 1961)
63. Liptay, W., *Z. Elektrochem.*, **65,** 375 (1961)
64. Tamres, M., *J. Phys. Chem.*, **65,** 654 (1961)
65. Hassel, O., *Mol. Phys.*, **1,** 241 (1958)
65a. Reid, C., and Mulliken, R. S., *J. Chem. Phys.*, **76,** 3869 (1954)
66. Wallwork, S. C., *J. Chem. Soc.*, **494** (1961)
67. Hassel, O., and Hope, H., *Acta Chem. Scand.*, **15,** 407 (1961)
68. Hassel, O., Rømming, C., and Tufte, T., *Acta Chem. Scand.*, **15,** 967 (1961)
69. McCullough, J. D., Chao, G. Y., and Zuccare, D. E., *Acta Cryst.*, **12,** 815 (1959)
70. Chao, G. Y., and McCullough, J. D., *Acta Cryst.*, **13,** 727 (1960)
71. Chao, G. Y., and McCullough, J. D., *Acta Cryst.*, **14,** 940 (1961)
72. Ferguson, E. E., *J. Chem. Phys.*, **25,** 577 (1956)
73. Ferguson, E. E., *J. Chem. Phys.*, **26,** 1357 (1957)
74. Ferguson, E. E., *Spectrochim. Acta,* **10,** 123 (1957)
75. Collin, J., and D'Or, L., *J. Chem. Phys.*, **23,** 397 (1955)
76. Ferguson, E. E., and Matsen, F. A., *J. Chem. Phys.*, **29,** 105 (1958)
77. Aono, S., *Progr. Theoret. Phys. (Kyoto)*, **22,** 313 (1959)
78. Chang, I. Y., and Ferguson, E. E., *J. Chem. Phys.*, **34,** 628 (1961)
79. Popov, A. I., Marshall, J. C., Stute, F. B., and Person, W. B., *J. Am. Chem. Soc.*, **83,** 3586 (1961)
80. Person, W. B., Anderson, G. R., Fordemwalt, J. N., Stammreich, H., and Forneris, R., *J. Chem. Phys.*, **35,** 908 (1961)
81. Person, W. B., Humphrey, R. E., Deskin, W. A., and Popov, A. I., *J. Am. Chem. Soc.*, **80,** 2049 (1958)
82. Person, W. B., Humphrey, R. E., and Popov, A. I., *J. Am. Chem. Soc.*, **81,** 273 (1959)
83. Person, W. B., Erickson, R. E., and Buckles, R. E., *J. Am. Chem. Soc.*, **82,** 29 (1960)
84. Glusker, D. L., and Thompson, H. W., *J. Chem. Soc.*, 471 (1955)
85. Zingaro, R. A., and Tolberg, W. E., *J. Am. Chem. Soc.*, **81,** 1353 (1959)
86. Ferguson, E. E., and Matsen, F. A., *J. Am. Chem. Soc.*, **82,** 3268 (1960)
87. McKinney, P. C., and Barrow, G. M., *J. Chem. Phys.*, **31,** 294 (1959)
88. Tsubomura, H., *J. Chem. Phys.*, **24,** 927 (1956)
89. Jones, W. D., and Simpson, W. T., *J. Chem. Phys.*, **32,** 1747 (1960)
90. Friedrich, H. B. (Doctoral Thesis, University of Iowa, 1962)
91. Dewar, M. J. S., and Lepley, A. R., *J. Am. Chem. Soc.*, **83,** 4560 (1961)
92. Dewar, M. J. S., and Rodgers, H., *J. Am. Chem. Soc.*, **84,** 395 (1962)
93. Dewar, M. J. S., *Electronic Theory of Organic Chemistry* (Clarendon Press, Oxford, 1949)

Reprinted from THE JOURNAL OF CHEMICAL PHYSICS, Vol. 38, No. 1, 109–116, 1 January, 1963
Printed in U. S. A.

# Electron Affinities of Some Halogen Molecules and the Charge-Transfer Frequency

WILLIS B. PERSON*

*Laboratory of Molecular Structure and Spectroscopy, Department of Physics, University of Chicago, and, Department of Chemistry, University of Iowa, Iowa City, Iowa*

(Received 15 August 1962)

In an attempt to fix the value of the vertical electron affinity of the iodine molecule $E_{I_2}^v$ the potential curve for $I_2^-$ has been constructed from semiempirical considerations. From this curve the value of $E_{I_2}^v$ is estimated to be $1.7\pm0.5$ eV, in apparent contradiction to the value around 0 eV indicated by the electron-capture experiments. It is suggested that the electron-capture experiment may measure the vertical energy for capture to give an excited state of the $I_2^-$ ion, thus explaining the discrepancy.

Having fixed the value for $E_{I_2}^v$, the relationship predicting the charge transfer frequency for iodine complexes is reexamined critically. By fitting curves to the data for three different types of complexes ($\pi$ donors with $I_2$, amine donors with $I_2$, and complexes with I atoms), three sets of empirical constants required to force the fit are determined. An attempt is then made to calculate these sets of constants from *a priori* considerations. The success of this treatment supports the general theory, indicating especially that no major terms have been omitted from the estimation of the energy difference, $W_1-W_0$. Furthermore, the general argument supports the value of $E_{I_2}^v$ obtained in the earlier part of the paper as well as the reliability of the estimation of the other parameters. The analysis supports the perpendicular model for the geometry of the complexes between $I_2$ and $\pi$ donors.

Finally, similar analyses are presented for other halogens; $Br_2$, $Cl_2$, and ICl. Estimates of the vertical electron affinity for these halogens from the empirical potential curves agree reasonably well with estimates from the charge transfer frequency, in support of the results for $I_2$.

## INTRODUCTION

ONE of the more elusive properties of molecules has been the vertical electron affinity. This parameter is particularly important in the theory of charge transfer complexes because it is one which determines the energy separation between the "no-bond" state and the "dative" state. The frequency of the charge transfer band, for example, is given by [1,2]

$$h\nu_{CT}=I_D^v-E_A^v+G_1-G_0+X_1-X_0 \qquad (1)$$

Here $I_D^v$ is the vertical ionization potential of the free donor molecule, and is often fairly well known. $E_A^v$ is the vertical electron affinity of the acceptor, $G_1$ and $G_0$ are the interaction energy between donor and acceptor in the dative state and in the ground state, and $X_1-X_0$ is the resonance energy of interaction between the two states. All terms are evaluated at the equilibrium configuration of the complex.

If the value of $E_{I_2}^v$ were known definitely, and if $\Delta G$ and $\Delta X$ could be estimated with some confidence, it would be possible to test the validity of Eq. (1). Conversely, if reliable estimates could be made for $\Delta G$ and $\Delta X$, the observed frequency of the charge transfer band could be used to estimate the electron affinity of the acceptor molecule. Mulliken[3] originally estimated a value of 1.8 eV for $E_{I_2}^v$, and showed that this value was consistent with the charge transfer frequency with reasonable values of $\Delta G$ and $\Delta X$. In a recent re-evaluation of the data for $I_2$ complexes, Briegleb and Czekalla deduced a value of 0.8 eV for $E_{I_2}^v$.[4]

A slightly different approach has been proposed recently by Jortner and Sokolov.[5] In essence, they use the known electron affinity of the iodine atom[6] and the observed frequency for the charge-transfer complex between hexamethylbenzene and I atom[7] to evaluate the sum $\Delta G+\Delta X$ which they assumed constant. Under the assumption that this sum is the same for all complexes between hexamethylbenzene and the halogen acceptors (I atom, $I_2$, ICl, $Br_2$, and $Cl_2$), they use the observed frequencies of the charge-transfer bands in the complexes together with the value of $(\Delta G+\Delta X)$ determined from the I atom complex to evaluate the vertical electron affinities for these molecules. For $I_2$, they find $E_{I_2}^v=1.55$ eV.

From these analyses of the charge-transfer frequency, the value of the vertical electron affinity of $I_2$ is estimated to be $1.3\pm0.5$ eV. However, all these analyses are based on estimates of $\Delta G$ and $\Delta X$ (or on assump-

* Guggenheim Fellow, 1960–61. A portion of this paper was presented at the 141st meeting of the American Chemical Society, Division of Physical Chemistry, Washington, D. C., March 20–24, 1962.

[1] R. S. Mulliken, J. Am. Chem. Soc. **74**, 811 (1952).

[2] R. S. Mulliken and W. B. Person, Ann. Rev. Phys. Chem. **13**, 107 (1962).

[3] R. S. Mulliken, J. Am. Chem. Soc. **72**, 600 (1950).

[4] (a) G. Briegleb and J. Czekalla, Angew. Chem. **72**, 401 (1960); (b) See also, G. Briegleb, *Elektronen-Donator-Acceptor-Komplexe* (Springer-Verlag, Berlin, 1961).

[5] J. Jortner and U. Sokolov, Nature **190**, 1003 (1961).

[6] (a) O. H. Pritchard, Chem. Revs. **52**, 529 (1953). Since the preparation of the present manuscript, the author has learned of two new precise determinations of the electron affinities of halogen atoms, which suggest that the values listed by Pritchard are somewhat too high. (Thus, for the I atoms, $E_I=3.075\pm0.005$ eV, instead of 3.24 eV; for Cl, $E_{Cl}=3.620\pm0.007$ eV, instead of 3.75 eV.) These two studies are (b) B. Steiner, M. L. Seman, and L. M. Branscomb, J. Chem. Phys. **37**, 1200 (1962); and (c) R. S. Berry, C. W. Reimann and G. N. Spokes, *ibid.* **35**, 2237 (1961); Bull. Am. Phys. Soc. **7**, 69 (1962).

The effect of these changes on the present work is to raise the estimated potential curves for the $X_2^-$ negative ions by about 0.15 eV, and to decrease the estimated vertical electron affinities of the halogen molecules by about 0.1 eV.

[7] R. L. Strong and J. Perano, J. Am. Chem. Soc. **83**, 2843 (1961).

tions about their constancy) which may not be correct.

On the other hand, Biondi and Fox have measured the electron-capture cross section by $I_2$ as a function of the energy of the bombarding electrons.[8] In such experiments, the energy at which the cross section is a maximum should equal the value of the vertical electron affinity of $I_2$. Unfortunately, Biondi and Fox did not find a definite maximum in the electron cross section, although it was increasing rapidly as the energy of the bombarding electrons went to zero. Their results have been interpreted quite universally as indicating that the vertical electron affinity of $I_2$ is close to zero. However, Biondi and Fox themselves did not rule out the possibility that the maximum in the electron cross-section curve occurs for electrons with energy less than zero (a positive electron affinity, such as suggested in the analysis of charge-transfer bands reviewed above).

In an attempt to determine the best value for $E_{I_2}^v$, and to explain the apparent discrepancy between the values obtained from the charge-transfer spectra and that indicated by the electron-capture experiments, we have re-examined these various arguments in the following discussion. While it is still not possible to establish definitely the value of $E_{I_2}^v$, we hope that the situation will be clarified somewhat. The discussion is given in three parts: (1) The potential curve for $I_2^-$ is deduced from the principles of valence theory; from this the higher value of $E_{I_2}^v$ is indicated. An explanation is offered for the apparent discrepancy with the electron-capture experiments. (2) The relationship between the charge-transfer frequency and the ionization potential of the donor [Eq. (1)] is re-examined critically. (3) These arguments are extended to some other halogens.

## POTENTIAL CURVES FOR $I_2$

### Ground State

Let us consider the potential-energy curve for $I_2^-$. If it could be estimated reliably and then related to the curve for $I_2$, the electron affinity of $I_2$ would be known as a function of I–I distance. In order to determine the potential curve, we must estimate the dissociation energy $D_e$, the equilibrium distance $r_e$, and the frequency $\omega_e$, for the $I_2^-$ molecule–ion. The rest of the potential curve can then be constructed using a Morse potential function. The asymptote of the $I_2^-$ curve can be located with respect to the $I_2$ curve by the knowledge of the electron affinity of the I atom, since the energies of the dissociation products $[I(^2P_{\frac{3}{2}})+I(^2P_{\frac{3}{2}})$ for the $I_2$ molecule, and $I(^2P_{\frac{3}{2}})+I^-(^1S_0)$ for $I_2^-]$ differ by just that quantity.

First of all, we may expect that the ground state of $I_2^-$ is stable. This argument was first made by Mulliken[3] in terms of the number of bonding and antibonding electrons in $I_2^-$. Mulliken's argument is based on the electron configuration of $I_2^-$,

$$\cdots(\sigma_g)^2(\pi_u)^4(\pi_g)^4(\sigma_u).$$

Here only the valence shell $p$ electrons are shown. This configuration differs from $I_2$ only in the additional electron in the strongly antibonding $\sigma_u$ orbital. In $I_2$ the $\pi_u$ orbital is considered to be somewhat bonding, while the $\pi_g$ orbital is thought to be somewhat antibonding. Since they are both filled, their net effect is probably nearly nonbonding so that the bonding effect of the two $\sigma_g$ electrons is partially canceled by the one $\sigma_u$ electron. Therefore, we can expect, as a first approximation, that $D_e$ for $I_2^-$ will be about one-half that for $I_2$, or 0.7 eV.

In his earlier attempt to fix the internuclear distance, Mulliken[3] estimated that $r_e=3.5$ Å for $I_2^-$. There are several arguments which can be made to support this value or a lower value. For example, we can set the range of $r_e(I_2^-)$ between 4.3 Å (the sum of the van der Waals radii for two I atoms) and 2.67 Å ($r_e$ for $I_2$). A literal interpretation of the "half-bond" expected for $I_2^-$ might lead us to expect $r_e(I_2^-)$ to be half-way between these two values, $r_e(I_2^-)=3.5$ Å, or less.

The most useful and convincing way to estimate the properties of $I_2^-$ may be the use of the known properties of the $^3\Pi_{0u}{}^+$ excited electronic state of $I_2$. The electronic configuration of $I_2$ in this state is

$$\cdots(\sigma_g)^2(\pi_u)^4(\pi_g)^3(\sigma_u),\ ^3\Pi_{0u}{}^+.$$

This electronic configuration differs from that for $I_2^-$ only in that one weakly antibonding $\pi_g$ electron is missing. If the antibonding character of this oribtal is small enough, the properties of $I_2^-$ should be expected to be very similar to those of $I_2(^3\Pi_{0u}{}^+)$. For the latter, Herzberg[9] lists $\omega_e=128$ cm$^{-1}$, $r_e=3.02$ Å, and $D_e$ for the $^3\Pi_{0u}{}^+$ state is 0.54 eV.

Since the $\pi_g$ orbital should be somewhat antibonding for $I_2$,[10] it might be expected that $r_e(I_2^-)$ will be somewhat greater than $r_e$ for $I_2(^3\Pi_{0u}{}^+)$. Comparison of the properties of $Cl_2$ and $Cl_2^+$ suggests that the effect of an additional $\pi_g$ electron upon $r_e$ may be to cause an increase of about 0.1 to 0.2 Å. Thus, we conclude that the most likely value of $r_e$ for $I_2^-$ is $3.1_5\pm0.1$ A.

Other estimates, based upon Badger's rule and estimates of $\omega_e$[11] predict that $r_e$ should be between 3.1 and 3.5 Å. It is interesting to note that $r_e$ for the symmetrical $I_3^-$ ion was found to be 2.90 Å.[12] This value should be a lower limit for the bond length in $I_2^-$.

[8] M. A. Biondi and R. E. Fox, Phys. Rev. **109**, 2012 (1958); M. A. Biondi, *ibid.*, p. 2005; R. E. Fox, *ibid.* p. 2008.

[9] G. Herzberg, *Molecular Spectra and Molecular Structure. I. Spectra of Diatomic Molecules* (D. Van Nostrand Company, Inc., Princeton, New Jersey, 1950), 2nd ed.

[10] See reference 3, and also R. S. Mulliken, J. Am. Chem. Soc. **77**, 884 (1955).

[11] See reference 3 for this method of estimating $r_e$.

[12] R. C. L. Slater, Acta Cryst. **12**, 187 (1959).

We might expect by the same arguments that $\omega_e(I_2^-)$ may be slightly less than $\omega_e(I_2, {}^3\Pi_{0u}{}^+)$. We note that the force constant for $I_2$, ${}^3\Pi_{0u}{}^+$, is already only 0.36 of the value for $I_2$ in the ground state. It is possible that it should be even lower in $I_2^-$. Although it is difficult to be certain about $\omega_e$, values around 100 cm$^{-1}$ seem reasonable.

The value of $D_e$ from the excited state of $I_2$ (0.54 eV) is probably not helpful in estimating the properties of $I_2^-$. The rather close agreement between the predicted value (0.7 eV) and the observed value (0.54 eV) is no doubt due more to the result of the accident that the dissociation products of the ${}^3\Pi_{0u}{}^+$ state of $I_2$ include an I atom in the excited ${}^2P_{\frac{1}{2}}$ state than anything else.

To conclude, comparison of the ground-state electron configuration of $I_2^-$ with the ${}^3\Pi_{0u}{}^+$ excited state of $I_2$ suggests that $D_e=0.7\pm0.3$ eV, $r_e=3.15\pm0.1$ Å, and $\omega_e=115\pm30$ cm$^{-1}$ for $I_2^-$. Using the mean values of these parameters to construct a Morse potential function, and the electron affinity of the I atom (3.24 eV), we compute the potential curve for the ground state (${}^2\Sigma_u{}^+$) of $I_2^-$ shown in Fig. 1. For this curve we see that the vertical electron affinity of $I_2^-$ is 1.7 eV.

Because of the difficulty in predicting $D_e$, $r_e$, and $\omega_e$ exactly, we should consider the effect of variations in each of these parameters on the value of $E_{I_2}{}^v$. If we vary each separately, we find that a change in $\omega$ of as much as 50 cm$^{-1}$ has essentially no effect on $E_{I_2}{}^v$. The effect of changing $D_e$ by $\pm0.3$ eV is indicated in Fig. 1 by the shading. Changing $r_e$ merely shifts the potential curve for $I_2^-$ along the $r$ axis; thus, changes of $\pm0.1$ Å are also indicated by the shading. It is possible that both $r_e$ and $D_e$ might be in error, so that the true value of $E_{I_2}{}^v$ might be outside the limits suggested by the shaded region. However, we believe that the most probable value of $E_{I_2}{}^v$ is $1.7\pm0.5$ eV, as indicated in Fig. 1.

Thus, this attempt to deduce the shape of the potential curve for $I_2^-$, and hence $E_{I_2}{}^v$, ends by verifying the value originally deduced by Mulliken,[3] who reasoned along these same lines, even though we have revised substantially the estimates for some of the parameters involved. The results are summarized in Table III.

### Excited States

If the higher value of $E_{I_2}{}^v$ indicated above is really correct, then some explanation is needed for the apparent discrepancy with the results of the electron attachment experiments of Biondi and Fox.[8] A possible explanation is suggested immediately upon examination of the potential curves for the excited states of $I_2^-$ shown in Fig. 1. As shown there, it seems quite likely that both the ${}^2\Pi_{g\frac{1}{2}}$ and the ${}^2\Pi_{g(\frac{3}{2})}$ states of $I_2^-$ cross the potential curve for $I_2$ near its minimum. Thus, electron capture by $I_2$ to give $I_2^-$ in either of these excited states will occur with maximum probability for

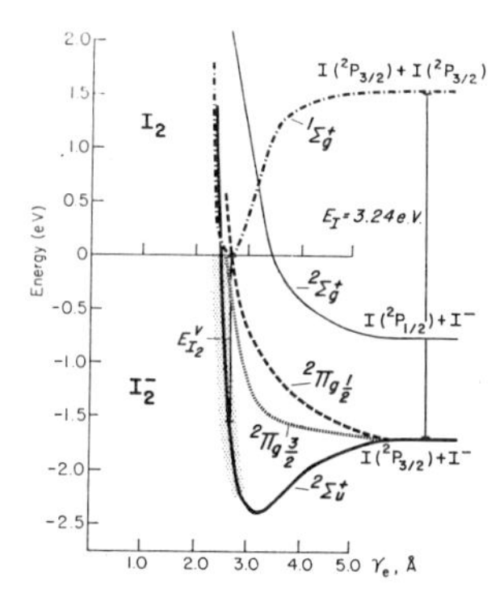


FIG. 1. Potential curves for the iodine molecule and for the $I_2^-$ molecule–ion. Also, see footnote 6.

electrons with energies near 0 eV. We suggest that it is this process which is responsible for the increase in electron cross section observed by Biondi and Fox,[8] rather than capture of an electron to give $I_2^-$ in its ground electronic state.

The potential curves for excited states of $I_2^-$ are deduced from the study of the absorption spectrum of $I_2^-$ trapped in a KI lattice by Delbecq, Hayes, and Yuster.[13] Their spectra and assignment give the vertical energy differences between the potential curves for the two ${}^2\Pi$ and the ${}^2\Sigma_g{}^+$ states and the potential minimum for the ground state. With these points and the knowedge that the excited states of $I_2^-$ are repulsive (since they all involve electron configurations with two electrons in the antibonding $\sigma_u$ orbital), the potential curves shown on Fig. 1 can be drawn. Although we cannot place too much confidence in the details of the resulting potential curves, there seems little doubt but that at least one of the excited-state potential curves of $I_2^-$ will cross the curve for $I_2$ at the minimum of the latter to give a maximum near 0 eV in the electron-capture experiments.

Recently Dunn[14] has determined the transition probabilities for electron capture to give different electronic states of the resulting molecule ion. This analysis states that the capture of an electron by $I_2$ to give the $\Pi_g$ states of the ion is forbidden. However, this result will not eliminate the mechanism proposed above, since the $\Pi_u$ states of the $I_2^-$ molecule ion should lie fairly close in energy to the $\Pi_g$ states. According to Dunn's analysis, there should be an interesting variation in the angular distribution of the negative ions, if the proposed mechanism is correct.[15]

[13] C. J. Delbecq, W. Hayes and P. H. Yuster, Phys. Rev. **121**, 1043 (1961).
[14] G. H. Dunn, Phys. Rev. Letters **8**, 62 (1962).
[15] The author is indebted to M. A. Biondi for drawing his attention to Dunn's work.

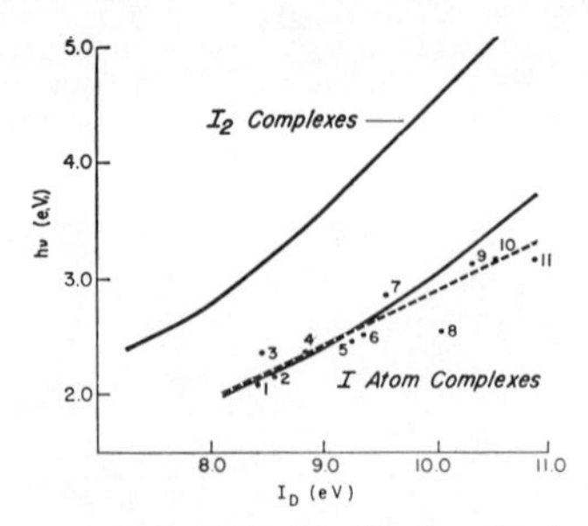


FIG. 2. A comparison of the plots of the charge-transfer energy ($h\nu_{CT}$) as a function of ionization potential of the donor, for iodine molecule complexes and for iodine atom complexes. The solid line through the data for the iodine atom complexes fits the equation

$$h\nu_{CT}=\left(\frac{I_D-8.0}{1-0.04}\right)\left\{1+4\frac{[1+0.2(I_D-8.0)]}{(I_D-8.0)^2}\right\}^{\frac{1}{2}}.$$

Key to donors: (1) mesitylene, (2) *o*-xylene, (3) *p*-xylene, (4) toluene, (5) benzene, (6) ethyl iodide, (7) diethyl ether, (8) carbon disulfide, (9) ethyl bromide, (10) ethyl alcohol, (11) methyl alcohol. The data for $I_D$ are from K. Watanabe, T. Nakayama, and J. Mottl, "Final Report on Ionization Potential of Molecules by a Photoionization Method", for Department of Army Project No. 5B99-01-004, Ordnance R & D No. TB2-0001 OOR No. 1624, Contract No. DA-04-200-ORD 480 and 737.

### $E_{I_2}{}^v$ AND THE CHARGE-TRANSFER FREQUENCY

Having shown in the discussion above that the higher value for $E_{I_2}{}^v(\sim 1.7$ eV) is probably correct, it seems appropriate now to re-examine critically the interpretation of the frequency of the charge-transfer band for complexes with $I_2$ and iodine atoms. One of the most useful implications of the charge-transfer theory lies in its potential ability to predict quantitatively the values of the parameters in Eq. (1), and hence predict the frequency of the charge-transfer band. However, very few attempts have been made either to predict the values of these parameters or to test the empirical values for the charge-transfer frequency for consistency since the first attempt by Mulliken.[1] One reason for this neglect has been the uncertainty in $E_{I_2}{}^v$, hence, we believe it is now worthwhile to make such an examination.

In order to use Eq. (1), it is necessary to evaluate the resonance term, $X_1-X_0$. If the energy of the pure dative state $\psi_1$ is $W_1$, and the energy of the pure "no-bond" state, $\psi_0$, is $W_0$, then

$$W_1-W_0=I_D{}^v-E_A{}^v+G_1-G_0=I_D{}^v-C_1. \quad (2)$$

Here the first part of this equation gives the energy difference in terms of the parameters of Eq. (1), and the second part states the usual assumption that for complexes with a given acceptor the term $E_A{}^v+\Delta G$ is a constant.[16] The evaluation of the resonance term by perturbation techniques leads to the familiar Eq. (3) below, if $W_1-W_0$ is large when compared to the resonance stabilization, $\Delta X$:

$$h\nu_{CT}=I_D{}^v-C_1+\frac{\beta_0{}^2+\beta_1{}^2}{I_D{}^v-C_1}. \quad (3)$$

Here the sum, $\beta_0{}^2+\beta_1{}^2$ is also usually assumed to be a constant $C_2$ for complexes with a given acceptor.

However, if $W_1-W_0$ is of the same order of magnitude as the resonance stabilization, the variation method must be used to evaluate $X_1-X_0$, and the charge-transfer frequency is given by the less familiar equation[2,16]

$$h\nu_{CT}=\left(\frac{W_1-W_0}{1-S^2}\right)\left[1+\frac{4\beta_1\beta_0}{(W_1-W_0)^2}\right]^{\frac{1}{2}}. \quad (4)$$

Here $S$ is the overlap integral between $\psi_1$ and $\psi_0$, and $\beta_1$ and $\beta_0$ are the resonance integrals $\beta_i=H_{01}-W_iS$; hence, $\beta_1=\beta_0-S(W_1-W_0)$.

### Empirical Constants

In order to evaluate empirically the parameters in Eq. (1), the data for the charge-transfer frequency for complexes between iodine, for example, and a series of chemically related donors are plotted against the ionization potential of the donor $I_D$. A curve of the form of Eq. (3) [or Eq. (4)] is then fitted to these points by adjusting the values of $C_1$ and $C_2$ [or of $C_1$, $\beta_0$, and $S$ in Eq. (4)]. In this way the constants of the general Eq. (4) have been evaluated for complexes between $I_2$ and $\pi$ donors and for those between $I_2$ and amine donors. The empirical values for the complexes with $\pi$ donors are $S=0.1$, $\beta_0=-0.6$ eV, and $C_1=5.2$ eV. For complexes with amine donors, the constants are $S=0.3$, $\beta_0=-2.5$ eV, and $C_1=6.9$ eV.

In order to follow the suggestion by Jortner and Sokolov[5] that the data for complexes with I atoms be used and thus eliminate the uncertainty in $E_A{}^v$, we must evaluate the parameters of Eqs. (3) or (4) for these complexes. In Fig. 2 the data[17] for the charge-transfer frequencies of several complexes are plotted against $I_D$. Comparison with the corresponding plot for $I_2$ complexes reveals the small slope for the data for I-atom complexes which has been interpreted[2] as being characteristic of a relatively large resonance interaction and the necessity of using Eq. (4) to interpret the data. The data for the I-atom complexes are fitted by the solid line in Fig. 2 when $S=0.2\pm0.1$, $\beta_0=-1.0\pm0.1$ eV, and $C_1=8.0\pm0.5$ eV. The fit to the data is about

[16] A complete discussion of the frequency of the charge-transfer band is given in reference 2. The present paper merely summarizes that analysis. In particular, we note that the assumption that $\Delta G$ is constant should hold only for complexes with chemically similar donors.

[17] T. A. Gover and G. Porter, Proc. Roy. Soc. (London) **A262**, 476 (1961).

as good as for the corresponding $I_2$ complexes. The deviations in Fig. 2 are worst for the complexes with *n* donors and probably represent failure of the assumption of constant values for $\Delta G$ and for $\beta_0$. The uncertainties in the constants indicate how much each parameter can be varied and still give a curve which fits the data fairly well.

### *A Priori* Evaluation of Parameters

Now let us attempt to deduce the values of the parameters of Eq. (4) from *a priori* considerations, and compare the resulting values with those found empirically above.

First of all, consider the overlap integral $S(\equiv\int\phi_A\phi_D dv$, where $\phi_A$ is the acceptor molecular orbital in $A^-$, and $\phi_D$ is the donor molecular orbital in $D^+$). In principle, this integral can be estimated using the tables of Mulliken, Rieke, Orloff, and Orloff.[18]

One merely notes that the acceptor orbital on the I atom will be a $5p$ atomic orbital; on $I_2$ molecules it will be the $\sigma_u$ antibonding orbital, approximated by a linear combination of $5p$ atomic orbitals. The donor orbitals are usually formed from linear combinations of $2p$ atomic orbitals on carbon or nitrogen atoms. Thus, $S$ can be approximated by evaluating $\int 2p_a\, 5p_b dv$ at appropriate distances using Slater or SCF atomic orbitals. In this way one can easily verify that $S$ is approximately 0.1 for the weakly interacting $\pi$-donors and approximately 0.3 for the stronger amine complexes.[19] In principle, if the geometry is known for a complex, one should be able to estimate $S$ fairly well. Since the frequency relation [Eq. (4)] is rather insensitive to $S$, it does not seem necessary at this point to refine the discussion further. Thus, the question of whether $S$ is determined from Slater orbitals or SCF orbitals will probably not change its value by more than $+0.1$, which is not significant in determining the frequency.

Next, let us consider the value of $\beta_0$. This parameter is one which is determined empirically fairly precisely, at least for the stronger complexes. On the other hand, $\beta_0$ is very difficult to estimate *a priori*, and is the most important parameter of Eq. (4) which cannot be so predicted.

However, it is still possible to define "reasonable" value of $\beta_0$. We may expect that $\beta_0$ will be larger for complexes in which the donor and acceptor approach each other closely. Thus, for the benzene–halogen complexes, it seems quite definite that the distance between the benzene and the halogen molecules in the complex is about equal to the van der Waals distance.[20]

On the other hand, $r_{DA}$ in the amine complexes is considerably less than the van der Waals distance ($r_{DA}$= 2.3 Å compared with 3.65 Å for the sum of van der Waals radii, and 2.03 for the sum of covalent radii).[18] The shorter distance in the amine complexes is consistent with the larger value of $\beta_0$ found empirically for this parameter. Furthermore, $\beta_0$ is closely related to the energy of formation of the complex and must also be made consistent with those data. This subject is discussed further below.

Finally, let us consider the value of $C_1$. As defined in Eq. (2), this constant is determined by $E_A{}^v$ and $\Delta G$. The previous discussion has indicated that $E_{I_2}{}^v$ should be about 1.7 eV. We must therefore deduce the values of $G_1$ and $G_0$ for the different types of complex. In addition, we should probably include a small correction term for the fact that the value $I_D$ used in the empirical evaluation of $C_1$ is the adiabatic ionization potential instead of the vertical value. This term will be expected to be zero except for the amine complexes, where we may expect a correction of about $-0.3$ eV.

The energy of stabilization of the dative structure $G_1$ is

$$-G_1 = V + e^2/r. \tag{5}$$

Here $V$ is the valence energy in the dative state of the stretched $D^+$–$A^-$ bond. Because of the stretching, it is expected to be small. At most, it could be only about 1.5 eV, the energy of a normal C–I or N–I bond. We may estimate $V$ as 0.2–0.5 eV for the weak $\pi$-complexes and about 1.0 eV for the stronger amine complexes.

The Coulomb energy in Eq. (5) can be estimated as the energy of attraction between two point charges at a distance $r$ apart. The negative charge is centered on the acceptor; the positive charge is on the donor. For $\pi$ donors, such as benzene, the positive charge will be at the center of the molecule; for amine donors, the positive charge will be expected to be centered on the N atom. If an I atom is the acceptor, the center of negative charge will be on the atom; for $I_2$, the center of negative charge will be at the midpoint of the I–I bond, since the $\sigma_u$ acceptor orbital is expected to be symmetric. Polarization effects will modify the electrostatic energy somewhat, but these effects are hard to separate from valence energy effects, and are probably already included in $V$. For strong complexes, we may expect the center of negative change to be slightly shifted from the center of the $I_2$ molecule away from the donor.

The only problem in estimating $e^2/r$ is thus to determine $r$, the distance between the charge centers as defined above. For the I-atom complexes, $r$ will be given approximately by the sum of the van der Waals radii of donor and acceptor. The van der Waals radius of the donor will vary somewhat for different donors, but not very much. For example, for $\pi$-donors, $r_D \simeq$ 1.7 Å, while for amine donors, $r_D = r_N \simeq 1.5$ Å. Thus,

[18] R. S. Mulliken, C. A. Rieke, D. Orloff, and H. Orloff, J. Chem. Phys. **17**, 1248 (1949).

[19] The major difference is that the donor and acceptor molecules approach each other much more closely in the amine complexes (see following text).

[20] O. Hassel and C. Rømming, Quart. Rev. **16**, 1 (1962).

TABLE I. Estimation of $G_0$ from the experimental heats of formation of some complexes with I atoms or $I_2$ molecules. (All units are eV).

| Complex | $X_0$[a] | Observed $\Delta H_f$ | $G_0$ |
|---|---|---|---|
| Benzene·$I_2$ | −0.48 | −0.09[b] | +0.39 |
| Benzene·$I_2$ | −0.09 | −0.07[c] | +0.02 |
| Durene·$I_2$ | −0.11 | −0.12[c] | −0.01 |
| $(CH_3)_3N·I_2$ | −1.60 | −0.53[c] | +1.1 |
| Pyridine·$I_2$ | −1.32 | −0.35[c] | +1.0 |
| $NH_3·I_2$ | −1.15 | −0.22[c] | +0.97 |

[a] Calculated from Eq. (7) or (8).
[b] G. Porter and J. A. Smith, Proc. Roy. Soc. (London) **A261**, 28 (1961).
[c] From Briegleb, reference 4(b).

we may choose $r=3.7$ Å for I-atom complexes, with confidence that it will not vary much for complexes with $\pi$-donors, amines, ethers, allyl halides, and alcohols.

For $I_2$ complexes with amine donors, there is little doubt that the geometry of the N···I–I group is linear, with N–I distance about 2.3 Å.[20] Thus $r=2.3+1.4$ (one-half the I–I distance), or $r=3.7$ Å, again. Even if the center of change is shifted slightly away from the donor, so that $r=3.9$ Å, for example, the effect on the energy will be small (about 0.1 eV decrease, for this example).

For $I_2$ complexes with $\pi$-donors in solution, the geometry is more controversial (see reference 2 for a full discussion). The two models which are generally considered are the "axial (or perpendicular) model" with the $I_2$ molecule perpendicular to the ring of the donor located on the symmetry axis, and the "resting (or parallel) model," with the halogen molecule parallel to the plane of the donor. For the present analysis, the difference between these two models lies in the difference in $r$. For the parallel model, $r$ is the same as for I-atom complexes, or $r=3.7$ Å. For the perpendicular model, $r$ is larger by 1.3 Å (one-half of the I–I distance) or $r=5.0$ Å.

Finally, consider the remaining terms which contribute to $C_1$; namely, $G_0$, the energy change when the D and A molecules are brought together into the configuration of the complex. This term is expected to be small and usually negative for weak complexes. For stronger complexes in which the D–A distance in the complex is much less than the van der Waals distance, we may expect $G_0$ to be large and positive because of steric repulsion.[2] While this parameter cannot be predicted independently, it is possible to limit it as follows.

The energy of formation of the complex from the isolated D and A molecules is[2]

$$\Delta E_f = G_0 + X_0. \tag{6}$$

The resonance stabilization of the ground state $X_0$ is readily shown to be[21]

$$X_0 = W_N - W_0$$

$$= \frac{W_1 - W_0 - 2\beta_0 S - [(W_1-W_0)^2 - 4\beta_0 S(W_1-W_0) + 4\beta_0^2]^{\frac{1}{2}}}{2(1-S^2)}.$$

When $W_1-W_0$ is large with respect to $\beta$, and when $S$ is negligible, then Eq. (7) reduces to the more familiar relation

$$X_0 = -\beta_0^2/W_1 - W_0. \tag{8}$$

As we have noted earlier, $\beta_0$ has been evaluated empirically, as well as $S$, and $W_1-W_0$ ($=I_D-C_1$). Thus, we can estimate $X_0$ from the empirical values of the parameters. It is then possible to use Eq. (6) to determine $G_0$, subject to the qualitative restrictions discussed earlier.

The experimental enthalpy of formation of the complex in solution in a nonpolar solvent from its component D and A solutions is not exactly $\Delta E_f$. However, it may be expected not to differ too much, since the solution energy of the complex should cancel pretty much the solution energies of the donor and acceptor, and since $\Delta E_0^0$ will not be too much different from $\Delta H_f$.

In Table I, we present the calculation of $G_0$ from the experimental heats of formation of the complex for several different complexes. In these calculations for $X_0$, the empirical values of the constants are used in Eqs. (7) or (8). We see from Table I that $G_0$ may indeed be quite large. We may now take the values of $G_0$ determined from Table I and calculate $C_1$, following the arguments given above. The results are summarized in Table II.

## Conclusions

Perhaps the most important result of the analysis given in this section is in the close agreement between the calculated and empirical values of $C_1$ shown in Table II. In all cases the agreement is within the experimental error in determining the empirical value. This agreement suggests that the values of the parameters selected in the preceding section are all approximately correct. It is true that $C_1$ is made up of several terms, and it is possible to stretch the value of each term a bit if necessary. However, this agreement lends further support to the value of $E_{I_2}{}^v$ deduced in the first part of the paper. Perhaps more important, the agreement for the I atom complexes, with known $E_I{}^v$, suggests that the evaluation of the other terms contributing to $C_1$ is approximately correct, and that no important terms in the energy separation $W_1-W_0$ have been omitted from consideration.

[21] Both this equation and Eq. (4) can be derived from the secular equation. [See, for example, equation II-19 of Briegleb, reference 4(b).]

TABLE II. Estimation of $C_1$ for complexes with I atoms and $I_2$ molecules (units are eV).

| Complex | $E_A{}^{v,a}$ | $V$ | $e^2/r$ | $-G_1$ | $G_0$ | $\Delta I$ | $C_1{}^b$(calc) | $C_1{}^c$(obs) |
|---|---|---|---|---|---|---|---|---|
| D·I | 3.2 | 0.3 | 3.9 | 4.2 | +0.4 | 0 | 7.8 | 8.0 |
| $\pi$-D·$I_2$, ⊥ model | 1.7 | 0.3 | 2.9 | 3.2 | 0 | 0 | 4.9 | 5.2 |
| $\pi$-D·$I_2$, ‖ model | 1.7 | 0.3 | 3.9 | 4.2 | 0 | 0 | 5.9 | 5.2 |
| Amine·$I_2$ | 1.7 | 1.0 | 3.9 | 4.9 | +1.0 | −0.3 | 7.3 | 6.9 |

[a] See footnote 6.
[b] $C_1=E_A{}^v-G_1+G_0+\Delta I$.
[c] Empirical values (see text).

We must conclude that attempts to use the above argument to confirm $E_{I_2}{}^v$ or to deduce the vertical electron affinities for other acceptors cannot be expected to yield values more accurate than about ±0.5 eV. Part of the reason for the uncertainty lies in the experimental uncertainty in $C_1$; the remainder lies in the difficulty in deducing the values of the other parameters with a total error less than this amount. In the case of the $I_2$ complexes the situation is complicated by the lack of certain knowledge of the geometry of the complexes with $\pi$ donors. If the parallel geometry were correct, it would suggest that a lower value of $E_{I_2}{}^v$ should be used. This suggestion would make it more difficult to obtain agreement between the $C_1$ values for the amine complexes, however.

Finally we may note that the results in Table II suggest that the perpendicular model for the geometry of the $\pi$ complexes of $I_2$ may be correct. This tentative conclusion, based on the better agreement between the $C_1$ values, is strengthened by consideration of the complexes between weak $n$ donors and $I_2$ (such as those with alkyl halides). If the geometry of these complexes is the same as for the strong $n$ donors (the amines) then we can be quite sure of the value of $r$, and hence $G_1$ The data for these weak complexes have always been correlated along with those for the $\pi$ donors, using Eq. (3) with the same empirical values of $C_1$ and $C_2$ [see Briegleb, reference 5(b), page 78]. This fact suggests that $G_1$ should be the same for those weak $n$ donors and for the weak $\pi$ donors. In turn, this suggests that $r$ for the $\pi$ complexes should be about the same as for the $n$ complexes. For the latter, assuming linear geometry, $r\simeq5.5$ Å. Since this is close to the value deduced for the perpendicular model of the $\pi$ complexes, we conclude again that the latter model may be correct.

TABLE III. Properties of some halogen molecules and halogen molecule ions.

| Halogen | $D_e$(eV) | $r_e$(Å) | $\omega_e$(cm$^{-1}$) | $E_A{}^b$(eV) |
|---|---|---|---|---|
| $I_2{}^{a,c}$ | 1.54 | 2.67 | 215 | $E_{I_2}{}^v=1.7\pm0.5$ |
| $I_2{}^a(^3\Pi_{0u}{}^+)$ | 0.54 | 3.02 | 128 | |
| $I_2^-$ | 0.7±0.3 | 3.1₅±0.1 | 115±50 | $E_{I_2}{}^{Ad}=2.4\pm0.3$ |
| $Br_2{}^{a,c}$ | 1.97 | 2.28 | 323 | $E_{Br_2}{}^v=1.2\pm0.5$ |
| $Br_2{}^a(^3\Pi_{0u}{}^+)$ | 0.46 | 2.66 | 170 | |
| $Br_2^-$ | 1.0±0.5 | 2.8±0.1 | 150±50 | $E_{Br_2}{}^{Ad}=2.6\pm0.3$ |
| $Cl_2{}^{a,c}$ | 2.48 | 1.99 | 564 | $E_{Cl_2}{}^v=1.3\pm0.4$ |
| $Cl_2{}^a(^3\Pi_{0u}{}^+)$ | 0.32 | 2.47 | 239 | |
| $Cl_2^-$ | 1.2±0.5 | 2.6±0.1 | 220±50 | $E_{Cl_2}{}^{Ad}=2.5\pm0.3$ |
| $ICl^{a,c}$ | 2.15 | 2.32 | 384 | $E_{ICl_2}{}^v=1.7\pm0.6$ |
| $ICl^a(^3\Pi_1)$ | 0.45 | 2.71 | 209 | |
| $ICl^-$ | 1.1±0.5 | 2.8±0.1 | 200±50 | $E_{ICl_2}{}^{Ad}=2.7\pm0.3$ |

[a] Data from Herzberg, reference 9.
[b] Estimated from the potential curves such as in Figs. 1 and 3. $E_A{}^v$ is the vertical electron affinity; $E_A{}^{Ad}$ is the adiabatic electron affinity (the difference between the *equilibrium* energies of the ion and molecule). Also, see footnote 6.
[c] The listed dissociation energy is $D_0$.

## ELECTRON AFFINITIES OF OTHER HALOGENS

It is of some interest to extend this line of reasoning to include other halogens. First let us deduce the po-

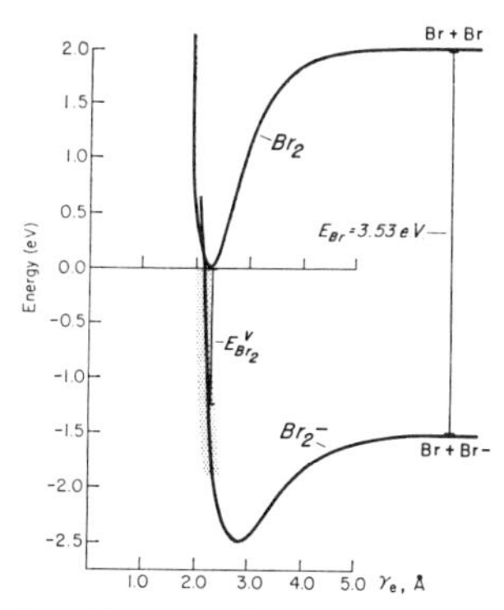


FIG. 3. Potential curves for the bromine molecule, and for the $Br_2^-$ molecule–ion. Also, see footnote 6.

TABLE IV. Electron affinities of some halogens from the charge transfer frequency (units are eV).

| Halogen and assumption | $e^2/r$ | $-G_1$[a] | $C_1$[b] | $E_A{}^{v,c}$ | $E_A{}^{v,d}$ |
|---|---|---|---|---|---|
| $I_2$, $\parallel$ | +3.9 | +4.2 | 5.2 | 1.0 | |
| $\perp$ | +2.9 | +3.2 | | 2.0 | 1.7±0.5 |
| $Br_2$ $\parallel$ | 4.0 | +4.3 | 5.1 | 0.8 | 1.2±0.5 |
| $\perp$ | 3.1 | +3.4 | | 1.7 | |
| $Cl_2$, $\parallel$ | 4.1 | +4.4 | 5.0 | 0.6 | 1.3±0.4 |
| $\perp$ | 3.2 | +3.5 | | 1.5 | |
| ICl, $\parallel$ | +3.9 | +4.2 | 4.9 | 0.7 | 1.7±0.6 |
| $\perp$ | +2.9 | +3.2 | | 1.7 | |

[a] $-G_1=\Delta G$, since $G_0$ is assumed to be equal to zero.
[b] Empirical value (see text).
[c] Computed from $E_A{}^v=C_1+\Delta G$.
[d] From the potential curves. See Table III. Also, see footnote 6.

tential curves for the ground states of the molecule-ions $X_2^-$. The properties of the ions can again be estimated from the properties of the $^3\Pi_{0u}{}^+$ states of the molecules. The dissociation energy of $X_2^-$ is taken in each case to be one-half of $D_e$ of the $X_2$ molecule. The equilibrium distance is taken to be that of the $^3\Pi_{0u}{}^+$ state, increased by 0.1 Å, and $\omega_e$ from the $^3\Pi_{0u}{}^+$ state is reduced by about 20 cm$^{-1}$. Such estimates may be somewhat in error for the smaller halogens, due to the increased antibonding character of the $\pi_g$ orbital as the internuclear distance increases. However, this effect is probably not very large except possibly for $Cl_2$. The properties estimated for the ions are tabulated in Table III. A typical Morse curve resulting from these parameters is shown in Fig. 3 for $Br_2$. The resulting values of the vertical and adiabatic electron affinities are given in column 5 of Table III. The uncertainties are estimated in the same way as for $I_2^-$.

Next, let us attempt to deduce the vertical electron affinities for these halogens from the spectra of their charge-transfer complexes. Data from these complexes are not as extensive as for $I_2$; however, the available data (mostly for $\pi$ complexes) can be plotted to give $h\nu_{CT}$ vs $I_D$. The resulting points almost fit the curve $f$ or $I_2$ complexes within the experimental error. A better interpretation, however, may be that the data for complexes with each of the other halogens fit a line parallel to that for $I_2$-$\pi$ complexes, but shifted slightly to higher frequencies. If so, the resonance energy, $X_1-X_0$, is the same for all the halogen complexes, but $C_1$ differs slightly. The resulting values of $C_1$ are; $I_2$ 5.2, $Br_2$ 5.1, ICl 5.0, and for $Cl_2$, 4.9 eV.

With these values of $C_1$, it is possible to evaluate $E_{X_2}{}^v$. For the same reason as for $I_2$ complexes with $\pi$ donors, the best assumption about the value of $G_0$ is about 0 eV. Similarly, we may expect the valence energy to be about the same as for $I_2$ complexes, or about 0.3 eV. Values of the Coulomb energy are tabulated in Table IV for these halogen complexes, assuming the parallel model and the perpendicular model for the geometry of the complex. The resulting values of $E_{X_2}{}^v$ are given in Table IV.

We see that the values of $E_{X_2}{}^v$ obtained from the analysis of the charge-transfer frequency data agree fairly well with those deduced from the potential curves, at least for the assumption of perpendicular geometry. The slight increase in $G_1$ and decrease in $C_1$ combine to give a decrease in $E_{X_2}{}^v$ as the size of the halogen decreases. It is not possible to rule out absolutely the lower values of $E_{X_2}{}^v$ which are required in Table II for the parallel model since the uncertainty in the estimate of $E_{X_2}{}^v$ from the potential curves in Table III is rather high. Since few data are available for complexes between $n$ donors and the halogen other than $I_2$, it is not possible to check the evaluation of $C_1$ and $E_{X_2}{}^v$ as for the $I_2$ complexes. Furthermore, it is not possible to say anything further about the question of geometry. However, the over-all consistency of the analysis is encouraging.

## ACKNOWLEDGMENTS

It is a pleasure to acknowledge the many helpful discussions with Professor R. S. Mulliken.

Partial financial assistance from the U. S. Army Research Office (Durham) and from the National Science Foundation is gratefully acknowledged.

Extrait du *Journal de Chimie Physique*, 1964, p. 20.

# THE INTERACTION OF ELECTRON DONORS AND ACCEPTORS,

by ROBERT S. MULLIKEN (*).

[*Laboratory of Molecular Structure and Spectra, Department of Physics, University of Chicago, Chicago 37, Illinois.*]

## SUMMARY

This paper reviews some aspects of the theory of the interaction of electron donors and acceptors, with particular reference to 1 : 1 interactions in solution, and also seeks to clarify and broaden some current viewpoints and applications of the theory ([1]). The classification of donors and acceptors, and the characteristics of various types of donor-acceptor interaction pairs (including ion-pairs, H-bonded complexes, 2-way charge-transfer complexes, reaction intermediates, contact pairs, intramolecular $\pi$ island interactions, etc.) are first discussed. Two sections are then devoted to the halogen complexes of $n$ donors and of $\pi$ donors. It is shown that the $n$ donor complexes conform well to predictions of the general theory, the main point of fresh emphasis (whose importance is indicated by Hassel's work on the geometry of solid complexes) being that some account must be taken of the participation of more than one dative resonance structure in stabilizing the normal state of most complexes. Interesting questions remain concerning the structure and charge-transfer spectra of complexes of the classical benzene-iodine type; probable answers are discussed. The last section deals, for the interaction of neutral-molecule even-electron donors and acceptors, with potential surfaces for the normal and charge-transfer excited states in each of three limiting cases ($W_1 > W_0$, $W_1 \approx W_0$, and $W_1 < W_0$, where $W_1$ and $W_0$ are the energies of the pure dative and pure no-bond structures respectively), and in each case for the two subcases of strong and weak interaction; the discussion in the third case is applicable also to compounds like NaCl, $CH_3NO_3$, $(CH_3)_3CCl$, etc. Photochemistry of contact pairs, the ionogenic effects of polar environments, and electron transfer between low-ionization-potential donors and high-electron-affinity acceptors (with possible relevance to biological systems and semi-conducting solid complexes) are briefly discussed. In discussing H-bonded complexes, new reasons favoring the importance of charge-transfer as against classical electrostatic forces are given (Sec. II).

(*) This work supported by NSF grant No. G-20375.

([1]) For a recent review, see R. S. MULLIKEN and W. B. PERSON, *Ann. Rev. Phys. Chem.*, 1962, **13**, 107-126. Both this and the present paper contain some new as well as review material, which we hope to include in a book now in preparation. A paper covering similar material was given by the present writer on the occasion of the presentation of the Peter DEBYE award of the American Chemical Society at the Society's Los Angeles meeting in April, 1963. For a much more detailed discussion of many topics, see G. BRIEGLEB's comprehensive book, Elektronen-Donator-Acceptor-Komplexe, Springer-Verlag, 1961.

## I. — Classification of donors and acceptors; quantum mechanical description of complexes.

This paper surveys some aspects of the theory of the interaction of electron donors (briefly, D) and acceptors (A), with particular reference to 1 : 1 complexes in solution. It includes revisions and improvements of some of the material presented in earlier papers. Table I shows a classification of donors and acceptors into major types, with a number of examples for each type. A donor of any type can in general form a compound or a complex with an acceptor of any type, but some types of DA pairs are more stable than others (see Sec. II). The wave function of the normal state (N) of any 1 : 1 DA combination can be written in the general form

$$\Psi_N = a\Psi_0(DA) + \Sigma b_i \Psi_{1i}(D^+A^-) + \Sigma c_j \Psi_{2j}(D^-A^+) + \cdots \quad (1)$$

In simple cases, only the first term and one or two terms of the first summation are important. In all cases, only those $b_i$'s and $c_j$'s can differ from zero for $\Psi_{1i}$'s and $\Psi_{2j}$'s of the same group-theoretical species as $\Psi_0$ under the actual geometry of the complex.

In a loose complex between an even-electron donor and an even-electron acceptor, the main term in Eq. 1 for $\Psi_N$ is $\Psi_0$. Among the excited states are some in which the main term corresponds to internal excitation of the donor or of the acceptor, while in other (CT, *i.e.* charge-transfer, states) the main term is usually one of the $\Psi_{1i}$ ($D^+A^-$). In a typical case, one might have for a CT state (here labelled V),

$$\Psi_V(DA) = a^*\Psi_{11}(D^+A^-) - b^*\Psi_0(DA) + d^*\Psi_{31}(D^*A) + \cdots, \quad (2)$$

with $a^*\Psi_{11}$ the main term. In Eq. 1, $\Psi_{31}$ refers to a donor-excited state. Absorption of light in a transition from state N to any state V gives a CT spectrum; such spectra are often very intense.

TABLE I

*Classification of Electron Donors and Acceptors.*
*Donor Types.*

| No. of Electrons | Functional Type | Structure Type | Examples (*) |
|---|---|---|---|
| Odd | Free Radical | R | Na, $C_2H_5$, H, NO, $NO_2$ |
| Even | Increvalent | *n* | $R_3N$,$R_3N$ oxide, Py (pyridine), Py N-oxide, $R_3P$, $R_2O$, dioxane, $R_2S$, RX; RCN, $R_2CO$, CO, $R_3PO_4$, $N_2O_4$, $RNO_2$; $X^-$Sl, $OR^-$Sl, $CN^-$Sl; — $NR_2$, — OR, — X (intramolecular $n\pi$ donor island groups). |
| | Sacrificial | σ | Al (aliphatic hydrocarbons) especially if cyclic (very weak); often RX, etc. |
| | | π | Ar (aromatic) and Un (unsaturated) hydrocarbons, especially if fortified by electron-releasing groups; — $C_6H_5$ etc. (intramolecular π donor island groups). |

(*) R may be H or an alkyl or other group; $R_2$ and $R_3$ may include two or three different R's. X, Y mean F, Cl, Br, or I. Sl means solvated.

*Acceptor Types.*

| No. of Electrons | Functional Type | Structure Type | Examples (*) |
|---|---|---|---|
| Odd | Free Radical | Q | X, OH, $NH_2$, H, $NO_2$. |
| Even | Increvalent | *v* | $BR_3$, $AlR_3$, $BX_3$, $AlX_3$, $SnCl_4$, etc.; $Ag^+ClO_4^-$, $Ag^+$Sl, etc.; $NO_2^+$Sl. |
| | Sacrificial | σ | $X_2$, XY, HX, HQ, RX, RQ, $CCl_4$, etc. |
| | | π | Ar and Un, especially if polycyclic or heterocyclic, or if fortified by electron-withdrawing groups (X, $NO_2$, COOR, CN) as e.g. in trinitrobenzene, maleic anhydride, tetracyanoethylene; quinones; — $C_6H_5$, — $NO_2$, — COR, — COOR (intramolecular π acceptor island groups). |

If D and A are odd free radicals (R and Q), their combination is an ordinary valence compound R — Q, usually more or less polar due to ionic resonance structures $D^+A^-$ and $D^-A^+$, with $b_1 \geqslant c_1$ ($b_1 = c_1$ only for cases like H — H or $H_3C$ — $CH_3$). Such molecules (e.g. HCl, $I_2$, $CH_3Br$) possess intramolecular (interatomic, or inter-radical) CT spectra, the familiar N,V spectra. However, the CT action is here more or less symmetrized or 2-way, because $\Psi(D^-A^+)$ as well as $\Psi(D^+A^-)$ resonance structures are important in Eqs. 1 and 2, except in highly polar molecules like those of NaCl vapor. In the latter, the CT action is nearly completely 1-way, just as in typical molecular complexes.

*Increvalent* donors are lone-pair (*n*) donors whose functioning in the $\Psi_{1i}$ terms of Eq. 1 can be described in terms of the donation of one electron from a lone pair located on a key atom [2] (e.g. N in $R_3N$), leaving $D^+$ with an odd electron which can form an additional valence bond (hence the term increvalent). Similarly, increvalent acceptors are vacant-orbit (*v*) acceptors

[2] This description is only approximate, because « lone » pairs are very often not entirely lone, and « vacant » orbitals are often not quite vacant. For example, the « lone » pair of electrons in $NR_3$ occupies an MO which is *mostly* on the N atom, but to an appreciable degree extends around the neighboring atoms; while in $BF_3$ the « vacant » orbital is a π MO which while mostly on the fluorines, is to a very appreciable extent (through intramolecular charge-transfer action, incidentally) on the boron. In $B(CH_3)_3$ and probably in $BCl_3$, the « vacant » orbital is more nearly though not entirely vacant.

whose functioning can be described in terms of the acceptance in the $\Psi_{1i}$ terms of an electron into a vacant orbital of a key atom (²) (e. g. B in $BR_3$) so that this atom can form an additional valence bond. Combinations between neutral $n$ donors and $v$ acceptors are the familiar dative compounds, with only $a$ and $b_{11}$ important in Eq. 1, and $b_{11} > a$. Here, also generally for 1 : 1 complexes of even donors and even acceptors, $\Psi_0$ is *a no-bond* structure (at least as respects *covalent* bonding), while the $\Psi_{1i}$'s and $\Psi_{2j}$'s contain a covalent bond between $D^+$ and $A^-$ or $D^-$ and $A^+$; when D and A are neutral molecules, the $\Psi_i$'s and $\Psi_j$'s are *dative* structures.

In *sacrificial* donors, donation is from a bonding molecular orbital, so that when such a donor enters into a complex the bonding *within* the donor is weakened, resulting in an increase in bond lengths and a decrease in corresponding vibration frequencies within the donor; however, a decrease in energy of course results for the complex as a whole. Similarly in sacrificial acceptors, acceptation of an electron is into an antibonding molecular orbital, with resulting weakening of bonding within the acceptor. Sacrificial donors and acceptors each fall into two classes, $\sigma$ and $\pi$, with respect to the type of molecular orbital involved in the donation or acceptation. Among sacrificial donors and acceptors, both $\sigma$ and $\pi$ acceptors are important, but $\sigma$ donors are weak and only $\pi$ donors are usually encountered.

There is a close analogy between *intermolecular* complex formation between $\pi$ donors and $\pi$ acceptors and *intramolecular* dative action (dative conjugation) between $\pi$ donor and $\pi$ acceptor *groups* within unsaturated (Un) and aromatic (Ar) molecules. For example, anisole can be represented by Eq. 1 using only the $a$ and $b_{11}$ terms; $\Psi_0$ then corresponds to $C_6H_5 — OCH_3$ with only a polar $\sigma$ bond connecting the two groups $C_6H_5$ and $OCH_3$ while the $\pi$ electrons form two « islands », one of six electrons in $C_6H_5$— and the other, in O of $— OCH_3$, of two electrons. The latter is a $\pi$ lone pair, hence this island belongs to an $n$, or say $n\ \pi$, donor type. (Most *intermolecular n* donors can be described as of $n\sigma$ type.) Donor-acceptor action from $— OCH_3$ as $n\pi$ increvalent donor to $C_6H_5$— as sacrificial $\pi$ acceptor yields the second term in Eq. 1, with coefficient $b_{11}$. In $C_6H_5NO_2$ on the other hand, the $C_6H_5 — \pi$ island acts as a sacrificial donor toward the four-electron $\pi$ island of $— NO_2$ acting as a sacrificial acceptor. Trigonal conjugation in $BF_3$ again illustrates $\pi,\pi$ intramolecular donor-acceptor action, (²) though now $n\pi$, $v\pi$, and 3 :1 instead of 1 : 1, in type.

Although it is convenient to speak of electron donors and acceptors, one should of course realize that these names do not refer to fixed absolute types of molecules, but rather to modes of functioning. The same atom or molecule may function sometimes as a donor, sometimes as an acceptor, depending on its partner.

Or not infrequently the same molecule may function simultaneously as donor and acceptor in a « 2-way charge transfer complex »; here at least one $c_j$ as well as one $b_i$ are important in Eq. 1 (³). Sometimes the same, sometimes different, parts of a molecule are involved in a mutually reinforcing 2-way D,A action. For example, in the $C_2H_4.Ag^+$ complex, $C_2H_4$ functions as a $\pi$ donor to $Ag^+$ as a $v$ acceptor, and simultaneously as a $\pi$ acceptor to $Ag^+$ (using a $d\pi$ electron) as an $n$ donor. On the other hand, pyridine ($C_6H_5N$) in 2-way complexes functions as an $n$ donor with the donor action focussed at the N atom and as a $\pi$ acceptor with the acceptor action spread over the entire ring. Intramolecular 2-way action can also occur. For example in $C_6H_5OCH_3$, the O atom is acceptor in its $\sigma$ bonds, but is $\pi$ donor to $C_6H_5$—, so that there is mutually reenforcing 2-way action in the $C_6H_5 — O$ link.

The classifications of donors and acceptors in Table I are valid not only for 1 : 1 but also for $n$ : 1 or other interactions. Thus the « ligands » in complex ions or other coordination compounds are either 1-way electron donors (e.g. $Cl^-$ in $PtCl_4^{=}$, or $NH_3$, when replacing $Cl^-$ in such a compound) or simultaneously donors and acceptors in 2-way action (examples, pyridine in complex ions or CO in carbonyls; CO can function as a simultaneous $n$ donor and $\pi$ acceptor like pyridine). In complex ions, the central metallic ion, e.g. $Pt^{++}$, is always a $v$ acceptor, but toward 2-way ligands it is also simultaneously an $n\pi$ donor.

The Eq. 1 type of description can be referred to as using the *resonance-structure method*. In this method, the *internal* structures of D and A, and of $D^+$ and $A^-$, and so on, are described using MO'S (molecular orbitals), — or of course AO's (atomic orbitals) in case D and A are atoms or atom-ions. But the overall structure and the DA interactions are expressed in Eq. 1 in terms of resonance between two or more wave functions (resonance structures) differing by the transfer of an electron one way or the other. The resonance-structure method is natural and convenient for 1 : 1 complexes, but its generalization for $n$ : 1 complexes, while entirely feasible, requires using a large number of resonance structures. The use of MO's for the complex as a whole is then greatly preferable, especially for dealing with spectra; the corresponding theory based on whole-complex MO's is « ligand field theory ».

The use of whole-complex MO's is a valid and not

(³) The idea of 2-way donor-acceptor action in complexes of cations containing $d$ electrons was first proposed for complexes by M. DEWAR in a discussion : *Bull. Soc. chim. de France*, 1951, 18, C 79.

inconvenient alternative to the resonance-structure method also for 1 : 1 complexes, but this method has not been much employed. For very loose 1 : 1 complexes the wave functions of the two methods, in an approximation where only two terms are used in Eq. 1, or only one MO configuration is used in the whole-complex MO method, become identical. A convenient example for comparing the two methods is that of the molecule $C_6H_5 - OCH_3$. As an alternative to the intramolecular resonance-structure description given above, the whole structure may be approximated in terms of a single MO electron configuration with $\pi$ MO's covering the whole molecule. A similar comparison can be made for an intermolecular $\pi,\pi$ complex. However, the planes of the donor and acceptor are now parallel, whereas in $C_6H_5OCH_3$ the intramolecular $\pi$ islands share a common plane. The whole-molecule or whole-complex MO's then must differ accordingly: in $C_6H_5OCH_3$ $\pi$ MO's based on a common plane are formed, but in a $\pi,\pi$ complex cross-linked MO's must be built as linear combinations of $\pi$ MO's of the donor and acceptor. In the former case both methods are convenient and instructive, in the latter the resonance-structure method seems preferable.

In both approaches donor-acceptor interactions are involved. The resonance-structure method exhibits charge distribution and donor-acceptor charge transfer explicitly, while in the MO method an LCAO — MO population analysis is required to obtain these. Intramolecular donor-acceptor action and CT spectra in $\pi$ systems (e.g. $\Phi OCH_3$, $\Phi NO_2$) have been usefully treated by both methods.

## II. — Types of 1 : 1 interactions.

Table II lists important classes of 1 : 1 donor-acceptor interactions, with specific examples in some cases. As already mentioned, the odd-odd R, Q combinations are mainly ordinary more or less polar molecules, and the even-even $n,v$ combinations if between neutral molecules are the familiar dative compounds.

Odd-even and even-odd combinations for the most part are not stable compounds, but are probably of frequent occurrence as short-lived complexes which function as reaction intermediates. Probable examples include loose complexes of aromatic hydrocarbons as even donors with iodine atoms as odd acceptors.

The interaction (type R, $v$ in Table II) of an alkali metal M, an odd donor of low ionization potential, with a $\pi$ acceptor of high electron affinity (e.g. tetracyanoethylene) results in essentially complete electron transfer to form an odd-even ion pair $A^-M^+$ ($a \sim 0$, $b_{11} = 1$ in Eq. 1). Donor-acceptor interactions which result in ion pairs (or, in sufficiently polar solvents, solvated ions), can conveniently be called ionogenic.

The most typical 1 : 1 complexes are those between even-electron donors and even-electron acceptors, especially if these are neutral molecules, but one or both may be ions. CT spectra are known especially well for complexes between even-electron neutral-molecule donors and acceptors. In some donor-acceptor interaction classes, CT spectra are not yet known, or are doubtfully known, but

TABLE II

*Typical 1 : 1 electron donor and acceptor combinations.*
(D and A classified by *Structure* Types).

| D \ A | Q | $v$ ($\sigma$) | $\sigma$ | $\pi$; $v\pi$ |
|---|---|---|---|---|
| R | Compound (*) | Reaction Intermediate; Ion-pairs ($TCNE^-Na^+$) | React. Interm. | Reaction Intermediate; Ionogenic pairs (Na, TCNE) |
| $n$ ($\sigma$) | Reaction Intermediate | Compound; Ion-pairs ($Q^-R^+$) | Complex ($n$, $X_2$); H-bond ($n$, HQ); (*) Contact ($n$, $CCl_4$) | 1-way and 2-way complexes; Ion-pairs ($I^-Tr^+$); Contact ($n$,$O_2$) |
| $\sigma$ | React. Int. | ($RX,AlX_3$) (*) | Contact ($\sigma$,$I_2$) | Contact ($\sigma$,$O_2$) |
| $\pi$; $n\pi$ | React. Int.; (Ar,I etc.) | 1-way and 2-way complexes | Complex ($n$, $X_2$); H-bond ($\pi$, HQ) (*) | Complex ($\pi$, $\pi$); (**) Intramolecular ($n\pi$,$v\pi$) Contact ($\pi$, $O_2$) |

(*) Often Even-Ionogenic in polar environment.
(**) Odd-Radical-Ionogenic in polar media if I — E small.

there is no reason to doubt that they exist for all D, A pairs.

In using Eq. 1 to describe a highly polar 1 : 1 combination, for which an LiF vapor molecule is a simple instructive example, one may often conveniently think of *either* odd-electron entities (here Li + F, of the types R and Q respectively) *or* even-electron ions (here $F^-$ and $Li^+$, of the respective types *n* and *v*) as the donor and acceptor. Using the odd-electron entities as starting point, *a* is small and $b_{11}$ large in Eq. 1; other resonance structures are unimportant. Using the ions as starting point, *a* is large and $b_{11}$ small; $\Psi_0$ of Eq. 1 now corresponds to the main structure. The second approach is especially instructive for 1 : 1 combinations which in state N are essentially ion-pairs in vapor (e. g. alkali halides) or in suitable solvents; for example, tropylium iodide ($I^-Tr^+$), $I^-Cr(NH_3)_6^{+++}$, or $I^-(C_6H_5NCH_3)^+$.

With the ion-pair approach to the state N description, one expects CT spectra in which the excited charge-transfer state V is nearly an atom-pair or a radical-pair. In Eq. 2, if D and A are respectively negative and positive ions, $D^+$ and $A^-$ are neutral radicals. In the examples above D is $F^-$ or $I^-$, $D^+$ is F or I, A is $Li^+$, $Tr^+$, etc., $A^-$ is the Li atom, tropylium radical, $Cr(NH_3)_6^{++}$ ion, or $C_6H_5NCH_3$ radical. In all the cases mentioned, a characteristic CT spectrum is known; these cases all fall under the *n*, *v* or *n*, $\pi$ ion-pair types of Table II.

For LiF and other alkali halides, several fairly low-energy charge transfer states, of more than one type, are expected. Since $F^-Li^+$ forms a simple prototype for even-even, especially ion-pair, complexes, a special discussion of its charge-transfer states seems worth while ([3a]). Omitting the 1 *s* and 2 *s* fluorine electrons, and the 1 *s* Li electrons, Eq. 1 now takes the special form

$$\Psi_N(F^-Li^+) = a\Psi_0(2p\sigma^2{}_{F^-}\, 2p\pi^4{}_{F^-}) + b_{11}\Psi_{11}(2p\sigma_F\, 2p\pi^4{}_F\, 2s_{Li}) + b_{12}\Psi_{12}(2p\sigma_F\, 2p\pi^4{}_F\, 2p\sigma_{Li}) + b_{13}\Psi_{13}(2p\sigma^2{}_F\, 2p\pi^3{}_F\, 2p\pi_{Li}) + \cdots, \quad (1')$$

where for $\Psi_{11}$, $\Psi_{12}$ and $\Psi_{13}$ only the $^1\Sigma^+$ state of the specified configuration must be used, since $\Psi_N$ is a $^1\Sigma^+$ state (as is $\Psi_0$). The no-covalent-bond resonance structure $\Psi_0$ is now somewhat stabilized by small amounts of the three $^1\Sigma^+$ dative structures $\Psi_{11}$, $\Psi_{12}$ and $\Psi_{13}$. (Self-consistent-field MO calculations indicate that all three of these are of appreciable importance). Charge-transfer spectral transitions from $\Psi_N$ to various essentially atom-pair CT states corresponding to $\Psi_{11}$, $\Psi_{12}$, $\Psi_{13}$ and numerons others can now occur. Following are the CT upper states for the allowed transitions of lowest frequency. (There are also many other CT states of the same electron configurations to which spectral transitions are forbidden or nearly so.) In probable order of increasing energy, these CT states are:

$$\left.\begin{array}{ll} \Psi(\sigma_F{}^2\pi_F{}^3 2s_{Li},\ {}^1\Pi) + \cdots; & \Psi_{11}(\sigma_F\pi_F{}^4 2s_{Li},\ {}^1\Sigma^+) \\ \Psi(\sigma_F{}^2\pi_F{}^3 2p\sigma_{Li},\ {}^1\Pi) + \cdots; & \Psi_{12}(\sigma_F\pi_F{}^4 2p\sigma_{Li},\ {}^1\Sigma^+) \\ \Psi(\sigma_F\pi_F{}^4 2p\pi_{Li},\ {}^1\Pi) + \cdots; & \Psi_{13}(\sigma_F{}^2\pi_F{}^3 2p\pi_{Li},\ {}^1\Sigma^+) \end{array}\right\} \quad (2')$$

followed by others in which $3s_{Li}$, $3p\sigma_{Li}$, $3p\pi_{Li}$, and so on are substituted for $2s_{Li}$, $2p\sigma_{Li}$, $2p\pi_{Li}$. The $^1\Sigma^+$ states here correspond to the V states of Eq. 2 (with $b^*$ and $d^*$ here small or negligible in the latter). Their main $\Psi$'s also appear as resonance structures helping to stabilize state N (cf. Eq. 1'); the $^1\Pi$ $\Psi$'s of course do *not*. Spectroscopic transitions from state N to the $^1\Sigma^+$ states are allowed with light polarized parallel to the line joining donor and acceptor; to the $^1\Pi$ CT states they are also allowed, but with light polarized perpendicular to this direction ([3a]). Analogues of both these types of transitions are expected also for intermolecular charge-transfer spectra (see Sec. III and IV).

In the odd-even ion-pairs $A^-M^+$ obtained from alkali metal donors and $\pi$ acceptors, electron spin resonance studies show a very small amount of back charge transfer from $A^-$ to $M^+$, analogous to the larger amounts in $F^-Li^+$ as expressed by Eq. 1'.

Leaving aside the stable R — Q and *n*, *v compounds*, the best known complexes are of the types (*n*, $\sigma$), and ($\pi$, $\pi$).

Complexes of the types $\pi$, *v* and *n*, $\pi$ are less familiar except where the donor and acceptor are such that a 2-way action can occur, in which case the complex (e.g. $C_2H_4.Ag^+$, see above, and $C_6H_6.Ag^+$) belongs simultaneously to the $\pi$, *v* and *n*, $\pi$ types. The apparently relatively low stability of 1-way $\pi$, *v* and *n*, $\pi$ types may perhaps be explained in terms of geometrical incompatibility leading to low formation energies, or perhaps by the easy occurrence of irreversible chemical reactions.

The most familiar $\pi$, $\sigma$ and *n*, $\sigma$ complexes are (*a*), complexes in which the acceptor is a halogen molecule (see Sec. III); (*b*), hydrogen-bonded complexes (the acceptor is of the type HQ with a more or less positive hydrogen atom; the type includes ordinary acids, alcohols, and even, weakly, $HCCl_3$). Halogen and HQ complexes, for both the *n*, $\sigma$ and $\pi$, $\sigma$ cases, show close parallelisms with respect to heats of formation, equilibrium constants, and the effects of complex formation on the infra-red X — X (or X — Y) or H — Q stretching frequency ([1]). Since classical electrostatic forces can hardly be held responsible for the $X_2$ complexes, these parallelisms strongly suggest that charge-transfer forces are

([3a]) Cf. R. S. Mulliken, *Phys. Rev.*, 1957, **51**, 310, for a discussion of the actually observed spectra of the alkali halide vapor molecules.

responsible in both cases, especially since it is not obvious that classical electrostatic forces could account for the often fairly strong H bonding observed in $\pi$, HQ complexes. An earlier objection of the present writer ([3b]) that the vertical electron affinities of HX molecules must be very strongly negative, is apparently not at all correct ([4]). However, it seems likely that at least most of the HQ complexes with $n$ donors are stabilized largely by classical electrostatic forces, though partly also by charge-transfer forces.

It is relevant here that at least some of the parallelisms mentioned may be explainable by a close parallelism between charge-transfer donor strength and donor electrostatic action, thus rendering more difficult the decision as to which property is the more responsible for H bonding. Thus in $n$ donors, for example $NR_3$, increasing donor strength (e.g. on substitution of $CH_3$ for H) is accompanied by decreasing ionization potential and increasing accumulation of negative charge on the lone-pair atom ([5]); this must cause increased electrostatic attraction for an HQ. On the other hand, this parallelism fails when one considers $n$ donors of the type RX, where electron donor strength certainly increases in the order $I > Br > Cl > F$, while accumulation of negative charge on the lone pair atom certainly increases in just the opposite order. Now as Schleyer and West have shown from spectroscopic evidence ([5a]), H-bonding strength (for MeOH and for $C_6H_5OH$) increases in the order $RI > RBr > RCl > RF$, strongly indicating that charge-transfer forces are predominant over electrostatic forces at least in H-bonded complexes of RX.

In the case of H bonding by $\pi$ donors, for example the methylated benzenes, each introduced methyl group increases donor strength and *also* increases the negative charge on the ring carbon atoms; and the predominance of electrostatic attraction of this ring of negative charge (already present of course in $C_6H_6$) for H of HQ with the H nuzzling into the middle of the ring possibly (although this seems doubtful) may be of major importance for the observed H bonding.

In evaluating electrostatic interaction energies, the detailed distribution of positive and negative charges in D and HQ must be taken into account; it is definitely NOT adequate to compute dipole-dipole or quadrupole-dipole interactions assuming plausibly located point dipoles with moments equal to observed over-all moments.

In seeking to evaluate the importance of CT forces in H bonding, a consideration of some details of the dative resonance structures involved is instructive. For definiteness, consider a complex D.HCl, where D might for example be an amine, or methyl ether (see Sec. V on the HCl complex of the latter), or a methylated benzene. Then

$$\Psi_N = a\Psi_0(D,\ HCl) + b\Psi_1(D^+ - [HCl]^-) + \cdots, \quad (3)$$

where the energy of $\Psi_0$ includes all classical electrostatic and van der Waals terms; $\Psi_1$ is built from molecular wave functions for $D^+$ and for $(HCl)^-$. The structure of $HCl^-$ is of particular interest. From valence-bond theory, it is clear that

$$\Psi(HCl^-) \approx \alpha\Psi(HCl^-) + \beta\Psi(H^-Cl) \quad (4)$$

with $HCl^-$ strongly predominant ($\alpha^2 \gg \beta^2$) except perhaps at small R values (less than $R_e$ of HCl). In MO theory, $\Psi(HCl^-)$ corresponds to an electron configuration... $\sigma^2\pi^4_{Cl}\sigma^*$, where (neglecting $s - p\sigma$ hybridization for simplicity) $\sigma$ is the H — Cl bonding and $\sigma^*$ a corresponding antibonding MO. Here $\sigma$ is polar in the direction $H^+Cl^-$, while (to make it orthogonal to $\sigma$) $\sigma^*$ has polarity $H^-Cl^+$. That is, roughly

$$\sigma \approx a\,3p\sigma_{Cl} + b\,1s_H; \qquad \sigma^* \approx a\,1s_H - b\,3p\sigma_{Cl}, \quad (5)$$

with $a^2 \gg b^2$.

A striking conclusion is that in order to agree with the unquestionable valence-bond result that the charge in $HCl^-$ is predominantly on the Cl, it is necessary that the inequality $a^2 > b^2$ shall be *greatly enhanced* in $HCl^-$ as compared with the corresponding inequality for the $\sigma$ MO in neutral HCl, because the $\sigma^*$ MO now has to have $H^-Cl^+$ polarity. That is, both the $\sigma$ and $\sigma^*$ MO's have to be extremely strongly polar, in opposite directions; the over-all polarity is now determined by *one* instead of both the $\sigma$ electrons, the other merely cancel-

([3b]) R. S. Mulliken, *J. Phys. Chem.* 1952, **56**. 801.

([4]) See D. C. Frost and C. A. McDowell, *J. chem. Phys.*, 1958, **29**, 503, who find for the *vertical* electron affinities (at the $R_e$ values of the neutral molecules), — 4,0 ± 0,2, — 0,77± 0,05, — 0,21 ± 0,5, and — 0,05 ± 0,05 ev for HF, HCl, HBr, and HI respectively. M. A. Biondi and R. E. Fox, *Phys. Rev.*, 1958, **109**, 2011, independently agree on the result for HI. Also, R. E. Fox, *J. chem. Phys.*, 1957, **26**, 1281, gets — 0,66 ev for HCl. At larger R, the HX electron affinities necessarily become strongly positive.

([5]) SCF — MO calculations indicate already in $NH_3$ an amazingly large negative charge, about 0,75 *e*, on the N atom (cf. D. Peters, *J. chem. Phys.*, 1962, **36**, 2743 ; this charge surely is increased on alkylation. Moreover, the lone pair MO in $NH_3$ is a hybrid wich protrudes out in the direction of donor action, for either $X_2$ or HQ.

([5a]) P. von R. Schleyer and R. West, *J. amer. chem. Soc.*, 1959, **81**, 3164. On the other hand. A. Allerhand and Schleyer, *Ibid.*, 1963, **85**, 1233, have shown from spectral shifts that very strong H bonding to $X^-$ ions occurs, with strength increasing in the order $Cl^- > F^- > Br^- > I^-$, indicating that here electrostatic forces (as would surely be expected) are predominant.

ling out the reverse polarity of $\sigma^*$; and *this* one electron has then to be fractionnally as much on the Cl as is the over-all charge. [A question may perhaps be raised here about whether the $\sigma$ MO's of different spin cannot be radically different, but the considerations involved are too complicated for discussion here.] Now since the $D^+ — [HCl]^-$ bond in $\Psi_1$ of Eq. 3 is essentially between an electron in the lone-pair MO of D and the odd electron in the $\sigma^*$ MO of $HCl^-$, and since also the CT resonance interaction matrix element is more or less proportional to the overlap integral between these two MO's, the foregoing discussion indicates that the $H^-Cl^+$ polarity of the acceptation MO in $\Psi_1$ of Eq. 3 should make HCl a *particularly strong* electron acceptor in CT interaction. Thus these considerations, though in some respects over-simplified, lend strong new support to the importance of CT forces in H-bonded complexes.

A further interesting aspect of the same considerations is seen by using the valence-bond Eq. 4 for $\Psi$ of $HCl^-$ in Eq. 3. Then

$$\Psi_N = a\Psi_0(D, HCl) + b[\alpha\Psi(D^+ — HCl^-) + \beta\Psi(D^+H^-Cl)] + \cdots, \quad (6)$$

in which the main dative structure contains a $D^+ — H$ bond which is favored by the fact that $\alpha^2 > \beta^2$ (probably $\alpha^2 \gg \beta^2$).

While both halogen and H-bonded complexes are definitely electron donor-acceptor complexes, the term charge-transfer complex is apparently of uncertain appropriateness for many or most H-bonded complexes. Similarly in certain other cases, for example $\pi$, $\pi$ complexes between donors and acceptors which contain strong local dipoles, it is possible that classical electrostatic forces may be comparable with or stronger than charge-transfer forces. Hence in general the term « donor-acceptor complexes » seems preferable to « charge-transfer complexes ».

Hydrogen-bonded and halogen complexes differ strikingly in *some* of their properties. While CT spectra are well known for halogen complexes of both $n, \sigma$ and $\pi, \sigma$ types, very little is known of such spectra for hydrogen-bonded complexes. However, Nagakura's paper presented at this conference discusses probable and possible examples of CT spectra of H-bonded complexes, and points out that the difficulty in observing others is to be explained by their expected short wave length locations, where they would usually be obscured by intra-donor or intra-acceptor bands.

Another important difference is that the stronger H-bonded complexes, namely those of the H acids, are very much more inclined than the halogen complexes to be even-ionogenic in polar environments, obtained either by dissolving D and HQ together in a polar solvent, or in the case of sufficiently acidic HQ, by polymerization of the D.HQ pair to form a crystal composed of $DH^+$ and $Q^-$ ions [6]. Then $D.HQ \rightarrow (D^+H)Q^-$ to form an ion-pair or an ionic crystal (or, in a sufficiently polar solvent, a pair of solvated dissociated ions $D^+HSl$ and $Q^-Sl$) in which the $D^+H$ and $Q^-$ ions each contain an even number of electrons. This reaction is essentially an enhanced donor-acceptor action in which an electron is transferred from the donor to the acceptor while *at the same time* the covalent bonding of the H atom is transferred from Q to D. However, as discussed above from the reverse viewpoint that the negative and positive ions of an ion-pair are weak donors and acceptors respectively, the electron tranfer from D to Q of HQ is not quite complete.

In earlier papers [6] this type of D, A interaction has been called dissociative charge transfer, and the two forms D.HQ (hydrogen-bonded, loose, complex) and $DH^+Q^-$ have been referred to as *outer* and *inner* complexes. The terminology in which D and HQ are referred to as proton acceptors and donors corresponds to a more special viewpoint than that in which they are thought of as electron donors and acceptors respectively, and is also open to objection because it is much more nearly true to say that an H *atom* (a proton *plus electron*), and its covalent bond, rather than a proton, have been tranferred from Q to D.

Among, $n$, $\sigma$ complexes, *symmetrized* complexes [6] like $I_3^-$ and $ICl_2^-$ are of interest. Formed from $I^-$ and $I_2$ or $Cl^-$ and ICl respectively, the two end atoms in the complex become more or less [6a] équivalent in a readjustment of the nuclear configuration. In a similar way, $BF_4^-$ can be regarded as a symmetrized $n,v$ complex formed from $F^-$ plus $BF_3$. $XeF_2$ if formed from $F^-$ plus $Xe^+ — F$ would be similar.

There are some compounds or complexes which can probably best be described as intermediate between $n$, $v$ and $\sigma$, $v$ in type. Thus MeX with $GaCl_3$ in MeX solution forms a 1 : 1 compound whose wave function can be written as

$$\Psi = a\Psi_0(MeX, GaCl_3) + b\Psi_1(MeX^+ — Ga^-Cl_3) + c\Psi_2(Me^+X — {}^-GaCl_3) + \cdots,$$

where in $\Psi_1$ MeX acts as an $n$ donor, in $\Psi_2$ as a dissociative $\sigma$ donor. Compounds of the same general type can be even-ionogenic in polar media, for example in MeF solution, $RF + BF_3 \rightarrow R^+SlBF_4^-Sl$, in which

(6) R. S. Mulliken, *J. amer. chem. Soc.*, 1952, **72**, 600; *J. Phys. chem.* 1952, **56**, 801.

(6a) E. H. Wiebenga, E. E. Havinga and K. H. Boswijk *Adbavances Inorg. Chem. Radiochem.*, 1961, **3**, 133.

case RF acts as a dissociative $\sigma$ donor in the same sense that HCl in water solution acts as a dissociative $\sigma$ acceptor ($H_2O + HCl \rightarrow H_3O^+Sl\ Cl^-Sl$).

In addition to the stable even-even compounds, the odd-even reaction intermediates and ion-pairs, the various types of complexes, and the intramolecular interactions between donor and acceptor $\pi$ islands, Table II contains reference to several kinds ($n\sigma$, $\sigma\sigma$, $n\pi$, $\sigma\pi$, and $\pi\pi$) of *contact pairs.* Here in state N the interaction between the donor and acceptor is so weak that there is no evidence of complex formation (though perhaps weak complexes would be formed at low temperatures), yet intense spectra are observed which apparently can be explained only as CT spectra. These can be understood if state N corresponds very nearly just to a donor *in contact* with an acceptor (almost only $\Psi_0$ in Eq. 1, with $a \approx 1$) and state V is represented by Eq. 2 with $b^* \approx 0$, though $d^*/a^*$ is probably of some importance.

The term « contact pair » is used here intentionally rather than « collision pair », because the theoretical interpretation (7) of the observed CT spectra does not require approach of the donor and acceptor closer than (or even as close as) their VAN DER WAALS contact distance. Even for such contacts, spectral absorption transition from N to V is allowed for $\Psi_{11} \leftarrow \Psi_0$, but its intensity is probably much increased, as suggested by MURRELL, by stealing from the typically intense $D^* \leftarrow D$ intra-donor transition, which can occur because of partial mixing of $\Psi_{31}$ with $\Psi_{11}$ in Eq. 2. The reason for this mixing is shown by the crossing of the $D^+A^-$ and $D^*A$ potential curves in the right-hand fig. 1, together with the fact that $\Psi_{11}$ and $\Psi_{31}$ are generally of the same group-theoretical species and are expected to interact strongly. Contact CT spectra have been observed for contact pairs which include a weak donor and/or a weak acceptor: donors of all types with $O_2$, amines with $CCl_4$ or similar compounds, aliphatic hydrocarbons with iodine, bromine, or tetranitromethane. It seems likely that many additional examples will be found.

Many of the *ion pairs* discussed above, for some of which CT spectra are known, can be regarded as being contact pairs rather than complexes in the sense that $\Psi_N$ is nearly pure $\Psi_0$, with $a$ nearly 1 in Eq. 1. This should be especially true where the ions are large, less true with small ions as in LiF. In the odd-even case of $Na^+(TCNE)^-$, back CT in $\Psi_N$ is very slight. However, such ion-pairs are of course held together strongly (in the absence of too polar environments) by electrostatic forces, and in that sense are complexes. Nevertheless, it is of interest to recall that the ionic radii of simple ions are approximately VAN DER WAALS radii, so that the ions are scarcely more closely in contact than are neutral molecules in contact pairs.

## III. — Halogen complexes with $n$ donors.

O. HASSEL and co-workers (8) have made many determinations of the crystal structures of *solid* complexes. These complexes fall roughly into two classes: (1), solids in which the crystal is made up of discrete 1 : 1 donor-acceptor pairs; (2), solids

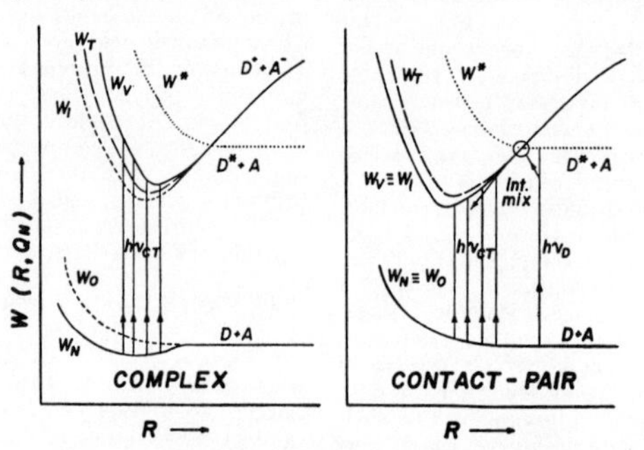


Fig. 1. — (R is some convenient internuclear distance; Q refers to all other nuclear coordinates; $Q_N$ refers to Q values to minimize $W_N$ at each R value).

Case I: Large $W_1 - W_0$

(7) R. S. MULLIKEN, *Rec. Trav. chim.*, 1956, **75**, 845; L. E. ORGEL and R.S. MULLIKEN, *J. amer. chem. Soc.*, 1957, **79**, 4839.

(8) O. HASSEL and Chr. ROMMING, *Qu. Rev. chem. Soc.*, 1962, **16**, 1.

in which no 1 : 1 associations exist, but donors and acceptors appear in alternating chains or sheets; these may be described as $\infty : \infty$ complexes. The two classes correspond respectively to cases where the 1 : 1 complexes in solution are especially tightly bound, or are weakly bound. The geometrical relations observed in the 1 : 1 units in the first class may probably safely be assumed to be nearly the same as for the corresponding 1 : 1 complexes in solution. For the $\infty : \infty$ class of solid complexes, one can obtain valuable indications of the directional characteristics of charge-transfer forces at work, but cannot necessarily derive reliable conclusions as to the geometry of the corresponding 1 : 1 complexes in solution.

The amine-halogen complexes afford examples of the first class. For the crystalline $Me_3N.I_2$ complex (Me = methyl) as an example, HASSEL finds that the two I atoms lie on a straight line passing through the N atom and the $Me_3N$ symmetry axis, with a N — I distance of 2,27 Å (much shorter than the N — I VAN DER WAALS distance), and an I — I distance of 2,83 Å, as compared with 2,67 Å in free $I_2$. The lengthened I — I distance is clear evidence of the sacrificial nature of the $I_2$ acceptor action, and the considerable magnitude of the lengthening corresponds to the fact the complex is a strong one.

For $Me_3N.I_2$, Eq. 1 becomes

$$\Psi_N = a\Psi_0(Me_3N, I_2) + b_{11}\Psi_{11}(Me_3N^+ - I_2^-) + \cdots \quad (7)$$

Just as in the case of the $HCl^-$ structure in $\Psi_{11}(D^+ - [HCl]^-)$ in Sec. II (cf. Eqs. 3-6), it is useful to look closely at the structure of $I_2^-$ in $\Psi_{11}$. In the valence-bond approximation,

$$\Psi(I_2^-) = \alpha\Psi(I\ I^-) + \beta\Psi(I^-I) + \cdots, \quad (8)$$

where for a free $I_2^-$ molecule, $\alpha = \beta$, but since the two I atoms are in non-equivalent locations in $Me_3N.I$, $\alpha \neq \beta$ is now expected; that is, the $I_2^-$ must be more or less polarized. On substituting $\Psi(I_2^-)$ from Eq. 8 into Eq. 7, we have

$$\Psi_N = a\Psi_0(Me_3N, I_2) + b_{11}\alpha\Psi(Me_3N^+ - I\ I^-) + b_{11}\beta\Psi(Me_3N^+I^-I) + \cdots \quad (9)$$

. . . . . . .

Examination of Eq. 9 suggests that $\alpha > \beta$; however, in view of the weakness of N — I bonds in general, and especially here where the N, I distance (2,27 Å) is considerably greater than a normal single bond distance (2,03 Å), also in view of the perhaps electrostatically more favorable [9] charge distribution in the third resonance structure in Eq. 9, the conclusion that $\alpha > \beta$ appears uncertain.

Expression of $\Psi(I_2^-)$ in $\Psi_{11}$ in the MO approximation throws further light on the question. In Eq. 7, we tacitly assumed that $I_2^-$ means *polarized* $I_2^-$. Eq. 7 can be rewritten as

$$\Psi_N = a\Psi_0(Me_3N, I_2) + b_{11}\Psi_{11}(Me_3N^+ - I_2^-) + b_{12}\Psi_{12}(Me_3N^+ - I_2^{-*}) + \cdots \quad (7')$$

where now $I_2^-$ means *unpolarized* $I_2^-$ in its normal state, with electron configuration $\ldots \sigma_g^2\pi_u^4\pi_g^4\sigma_u$, while $I_2^{-*}$ means unpolarized $I_2^-$ in the excited state of configuration $\ldots \sigma_g\pi_u^4\pi_g^4\sigma_u^2$; a linear combination of the two is equivalent to polarized $I_2^-$. It is seen that $b_{12} = 0$ in Eq. 7′ corresponds to $\alpha = \beta$ in Eq. 9, while $|b_{12}| \neq 0$ corresponds to $\alpha \neq \beta$.

Reference to Eq. 7′ in connection with the perturbation theory [6] resonance energy expression $(\mathcal{H}_{0,1i} - SW_0)^2/(W_{1i} - W_0)$ strongly indicates that $b_{12}^2$ is considerably or very likely much smaller than $b_{11}^2$, for the following reasons. First, $\Psi_{12}$ is known to be considerably higher in energy than $\Psi_{11}$ for $I_2^-$ in crystals, hence $(W_{12} - W_0) > (W_{11} - W_0)$. Further, $\mathcal{H}_{0,1i} - SW_0$, expected to be roughly proportional to the overlap intergal between the odd-electron MO in $Me_3N^+$ and that in $I_2^-$ or $I^-_2{}^*$ ($\sigma_u$ for $\Psi_{11}$, but $\sigma_g$ for $\Psi_{12}$), should be considerably larger for $\Psi_{11}$ than for $\Psi_{12}$, since the antibonding MO $\sigma_u$ should be considerably larger and extend out much more toward the N atom than the bonding MO $\sigma_g$.

Table III [10] gives some data for selected iodine complexes of amine and other *n* donors. Spectroscopic data on the CT band are given for the wavelength ($\lambda_{max}$) at maximum absorption coefficient, for the corresponding extinction coefficient $\varepsilon_{max}$, for the half-width $\Delta\nu_{\frac{1}{2}}$ in $cm^{-1}$, and for the oscillator strength $f$ (proportional to $\int \varepsilon_\nu d\nu$, or roughly to $\varepsilon_{max}\Delta\nu_{\frac{1}{2}}$). It should be especially noted that $f$, which is a good measure of intensity for the band, cannot be accurately gauged by $\varepsilon_{max}$ alone, since $\Delta\nu_{\frac{1}{2}}$ shows considerable variation. (Many authors give only $\varepsilon_{max}$). The interpretation of the size of $\Delta\nu_{\frac{1}{2}}$ is a matter of much interest, since it must be related to the forms of the potential surfaces (see below) of the N and V states. Hitherto little attention has been paid to $\Delta\nu_{\frac{1}{2}}$ or to the also interesting further details (asymmetry, etc.) of the CT band shape (but see BRIEGLEB) [1].

Table III also gives data for the intra-$I_2$ absorption band which appears in the spectra of all $I_2$ complexes,

[9] However, machine calculations (cf. D. PETERS, *J. chem. Phys.* 1962, **36**, 2743) on $NH_3$ indicate that the H — N bonds are strongly $H^+N^-$ polar, so that the charge on the N atom in $(H_3N)^+$ may be much less than 1, or even slightly negative; all the more so for $Me_3N$.

[10] Mostly from H. TSUBOMURA and R. P. LANG, *J. amer. chem. Soc.*, 1961, **83**, 2095.

TABLE III

*Iodine complexes with n donors (*).*

| | | | Charge-Transfer Band | | | | Vis. Band | | | |
|---|---|---|---|---|---|---|---|---|---|---|
| | $K_c$ (20 °C) (l/mole) | $-\Delta H$ (kcal/mole) | $\lambda_{max}$ (mμ) | $\varepsilon_{max}$ | $\Delta\nu_{\frac{1}{2}}$ (cm$^{-1}$) | $f$ | $\lambda_{max}$ (mμ) | $\varepsilon_{max}$ | $\Delta\nu_{\frac{1}{2}}$ (cm$^{-1}$) | $f$ |
| Free $I_2$ .... | | | | | | | 520 | 900 | 3 200 | ,012 |
| $Et_2O$ ..... | 1,16 | 4,2 | 252 | 5,650 | 6 900 | ,17 | 462 | 920 | 4 100 | ,017 |
| $Et_2S_2$...... | 5,62 | 4,6 | 304 | 15,000 | 7 200 | ,47 | 460 | 1 370 | 4 100 | ,024 |
| $H_3N$ ...... | 67 | 4,8 | 229 | 23,400 | 3 200 | ,32 | 430 | | | |
| $Et_2S$ ...... | 210 | 7,8 | 302 | 29,800 | 5 400 | ,69 | 435 | 1 960 | 4 200 | ,035 |
| $C_6H_5N$ .... | 269 | 7,8 | 235 | 50,000 | 5 200 | 1,12 | 422 | 1 320 | 4 300 | ,024 |
| $Et_3N$ ..... | 6 460 | 12,0 | 278 | 25,600 | 7 700 | ,85 | 414 | 2 030 | 5 700 | ,050 |

(*) All in heptane solution. Et = ethyl.

but shifted, broadened, and intensified. The data for « free $I_2$ » are those for iodine in heptane, which are essentially the same as for free $I_2$.

The rough parallelism seen in Table III between the equilibrium constants $K_c$, heats of formation $-\Delta H$, and intensities of the CT bands, is in agreement with theoretical expectations for complexes in which the CT theory in its simplest form applies, namely, where

$$\Psi_N = a\Psi_0(D, A) + b_{11}\Psi_{11}(D^+ - A^-) + \cdots;$$
$$\Psi_V = a^*_{11}\Psi_{11}(D^+ - A^-) - b^*\Psi_0(D, A) + \cdots \quad (10)$$

In this situation, the observed CT band should be polarized in the direction of the electron transfer required to obtain $\Psi_{11}$ from $\Psi_0$. A non-negligible term $a^*\Psi(D^*, I_2)$, — cf. Eq. 2, — also a term $e^*\Psi(D, I_2^*)$, should be added for $\Psi_V$ of Eq. 10; the presence of these terms should make some contribution to the intensity, but cannot possibly account for the high $f$ values of the CT bands in Table III. These can and must be explained essentially by the simple CT theory, with $b_{11}/a$ large in Eq. 7′, and best if $b_{12}|$ is small. For the amine-iodine complexes, a higher-frequency CT band is to be expected with a V state in which $\Psi_{12}$ of Eq. 7′ is substituted for $\Psi_{11}$ in Eq. 10, but this band may lie below λ 2 000, where it would be obscured by strong intra-$I_2$ and perhaps also intra-donor bands.

The shift of the visible intra-$I_2$ absorption band toward shorter wave lengths in increasingly tight complexes has been attributed mainly [7] to increasing crowding of the excited electron in the excited $I_2$ state as the intimacy of contact with the $n$ donor in the complex increases. In this intra-$I_2$ band, $\Delta\nu_{\frac{1}{2}}$ should increase if the I — I potential curve, at an I — I distance equal to that in state N of the complex, get steeper. Actually, $\Delta\nu_{\frac{1}{2}}$ of the intra-$I_2$ band *is* larger in the $n$-donor complexes than for free $I_2$, and is largest for $Et_3N.I_2$ among the complexes in Table III. This behavior is reasonably understandable as a result of crowding of one I atom against the donor, causing a steepening of the excited-state I — I potential curve [10].

In connection with the alcohol and ether complexes in Table III, HASSEL's results on the crystal structures of solid complexes of halogens with ethers, ketones, and the like are relevant. Although these are $\infty : \infty$ complexes, and so by no means conclusive indicators (see above) for the geometries of 1 : 1 complexes, they are at least suggestive for the latter. As an example, the dioxane-bromine solid complex shows chains of alternating dioxane and bromine molecules, with a linear arrangement of atoms O — Br — Br — O with the Br — Br linking O atoms of two different dioxane molecules. The striking feature is that the O — Br — Br — O line makes a considerable angle with the C — O — C plane of each dioxane molecule. This means that in Eq. 1 as generalized to the $\infty : \infty$ case, two dative resonance structures $\Psi_{11}$ and $\Psi_{12}$ are present for each O — Br link, one corresponding to each of the two lone pairs of the O atom. Let the C — O — C plane at one end of a dioxane molecule be taken as $yz$ plane; let the $z$ axis bisect the C — O — C angle and the $x$ axis pass through the O nucleus perpendicular to the C — O — C plane. The electrons of the most easily ionized lone pair of an ether, just as for $H_2O$ or an alcohol, occupy a $2\,p_x$ oxygen AO. If *only* this lone pair were used, in a dative resonance structure $\Psi_{11}(\text{ether}^+ — Br_2^-)$, maximum overlap would be obtained if the O — Br — Br line lay along the $x$ axis on one side of the C — O — C plane (or else the $Br_2$ should lie with its axis parallel to the $x$ axis and its center on the $z$ axis outside the O atom, a possibility once proposed [6] for the 1 : 1 complex), while if only the less easily ionized (largely $2\,s$) O atom « lone pair » [10a] were used (dative resonance structure $\Psi_{12}$), the O — Br — Br line should lie along the $z$ axis. The observed intermediate orientation shows that a combination of both dative structures ($\Psi_{11}$ and $\Psi_{12}$) is used. This is equivalent to saying that a *hybrid* dative structure $\Psi'_{11}$, in which the donor electron has come from a hybrid lone pair, is used.

Another example is that of the acetone-bromine complex. Here Hassel reports a structure consisting (aside from the H atoms) of parallel planes each containing $(CH_3)_2CO$ molecules linked by $Br_2$ molecules, with linear O — Br — Br — O arrangements again. Let the $\begin{matrix}C\\ \\C\end{matrix}\!\!>\!C = O$ plane of one acetone molecule be taken as $yz$ plane, with the $z$ axis along the C = O line and the $x$ axis passing through the O nucleus. Here the more easily ionized lone pair is largely [10a] $2p_y$ and the more deeply bound one largely [10a] $2s$ of the O atom. Hassel finds the O atom linked to *two* $Br_2$ molecules; that is, two O — Br — Br — O linkages start from one acetone oxygen (and end on those of two other acetone molecules). The two linkages are symmetrically disposed on opposite sides of the $z$ axis, with à 110° Br — O — Br angle. The orientation of the two linkages correspond to the use for each of them of one of two hybrid dative structures $\Psi'_{11}$ and $\Psi''_{11}$ corresponding to two hybrid « lone pairs » whose orbitals can be constructed as two equivalent linear combinations of the largely [10a] $2s_O$ and largely [10a] $(2p_y)_O$ MO's of the free acetone molecule.

Similar hybridization of lone pairs in halogen complexes of $n$ donors in general, except in those of the $R_3N$, MeCN, and similar types where only one « lone pair » [10b] is available, may well occur, with corresponding orientations of halogen molecules. An analogous comment applies to H-bonded complexes if predominantly stabilized by CT forces; the structure of ice is in accordance with this possibility.

## IV. — Halogen complexes of $\pi$ donors.

The classical cases of the benzene-iodine and methylated benzene-iodine complexes present some interesting problems which have not yet been conclusively solved. (See E. E. Ferguson's paper for some important contributions and a review.) The writer's first proposal [6] for the geometry of the benzene-iodine complex (« resting model ») placed the $I_2$ on one side of the benzene plane with its axis parallel to that plane and its center on the six-fold axis of the benzene. This is the model expected theoretically if Eq. 1 for state N of the complex is well approximated by the simple two-term expression

$$\Psi_N(Bz\,.\,I_2) = a\Psi_0(Bz,\, I_2) + b_{11}\Psi_{11}(Bz^+ — I_2^-) + \cdots,$$

with $Bz^+$ and $I_2^-$ *in their lowest states* in the $Bz^+ — I_2^-$ resonance structure.

The *axial model*, in which the I — I axis is coincident with the benzene six-fold axis, was at first dismissed as improbable [6] because it would require $b_{12}\Psi_{12}(Bz^{+*} — I_2^-)$ instead of $b_{11}\Psi_{11}$, with $Bz^+$ in an *excited* state corresponding to the removal of an electron from the deeper instead of the surface $\pi$MO. However, the axial model could be the stable one if the energy matrix element $\mathcal{H}_{0,12} — SW_0$ between $\Psi_0$ and $\Psi_{12}$ were sufficiently larger than the element $\mathcal{H}_{0,11} — SW_0$ to sufficiently overcompensate the less favorable energy denominator in the second-order perturbation [6] expression $(\mathcal{H}_{0,1i} — SW_0)/(W_{1i}—W_0)$ for $b_{1i}/a$. Arguments for the axial model were urged by Aono [10c]. This model is strongly suggested by Hassel's results on the crystal structure of solid benzene-halogen complexes, according to which benzene and halogen molecules alternate in endless chains with the halogen axes perpendicular (or nearly so) to the planes of both benzene neighbors; in *adjacent* chains, each benzene plane stands per-

([10a]) Like most « lone pair » orbitals in molecules (see second foot note in this paper), the MO here involved extends somewhat over other atoms in the molecule. Roughly, however, it can be treated as an AO localized on the O atom. The $2\,p_x$ lone pair in ethers is, however, very nearly a true lone pair of the O atom.

([10b]) Here again the « lone pair » is not quite lone, but extends to some degree outside the N atom [2]. In pyridine and related molecules, it is commonly assumed that the most easily ionized $\sigma$ MO is essentially an $s$, $p$ trigonal hybrid N atom AO occupied by a lone pair. But actually this MO must to *some* extent, and not unlikely to an *important* extent, extend around all the ring atoms. Conversely, all the other $\sigma$ MO's of appropriate symmetry must to some extent extend around the N atom. However, when pyridine functions as an $n$ donor, it could use a *hybrid* (i.e. linear combination) of these various $\sigma$ MO's which might be very nearly a localized N atom lone pair $\sigma$ AO like that usually assumed.

([10c]) S. Aono, *Prog. Theoret. Phys.*, 1959, **22**, 313.

pendicular to the axes of neighboring $I_2$ molecules (edgewise relation). As has been pointed out at the beginning of Sec. III, these crystal results are for $\infty : \infty$ complexes, and do not necessarily reflect the geometry of 1 : 1 complexes. However, they do show that the interaction of $\Psi_0$ and $\Psi_{12}$ must be dominant over that of $\Psi_0$ and $\Psi_{11}$ in the $\infty : \infty$ case, and it seems probable that this relation should carry over to the 1 : 1 case, unless edgewise interactions between neighboring linear chains are important in stabilizing the crystal in the $\infty : \infty$ case ([11]). It should be noted that, just as in the axially symmetric $R_3N.I_2$ complexes, the $I^{2-}$ in the dative structure in the 1 : 1 axial-model complex (but *not* in the crystal) must be somewhat polarized. Then

$$\Psi_N(Bz.I_2) = a\Psi_0(Bz, I_2) + b_{12}\Psi_{12}(Bz^{+*} - I_2^-) + b_{13}\Psi_{13}(Bz^{+*} - I_2^{-*}) + \cdots \quad (11)$$

Besides the resting and axial models for the 1 : 1 complex, another likely possibility is that of an intermediate form, the *oblique model.* As Ferguson shows in his paper, the infrared spectrum of the complex indicates that this is in effect the correct model. However, he suggests that the *equilibrium* state may be that of the axial model, but that the low-frequency high amplitude twistings and turnings expected for this loose complex cause the momentary geometry, which is determinative for the infrared benzene spectrum in the complex, to be that of the oblique model. A slightly oblique nearly-axial equilibrium model also seems not unlikely.

For the axial model, the familiar CT absorption spectrum, first observed by BENESI and HILDEBRAND, would be polarized *perpendicular* to the I — I axis, and parallel to the benzene plane. The relevant CT state would be given by Eq. 2 with $b^* = 0$ but $d^* \neq 0$. That is,

$$\Psi_V(Bz.I_2) = a^*\Psi_{11}(Bz^+ - I_2^-) + d^*\Psi_{31}(Bz^*, I_2) + \cdots \quad (12)$$

In $\Psi_{31}(Bz^*, I_2)$, $Bz^*$ would be benzene in its well-known $^1E_{1u}$ state. The rather high intensity of the CT band would then be due in part to intensity stolen from the very strong intra-benzene transition $^1E_{1u} \leftarrow {}^1A_{1g}$; however, the transition would be allowed and very likely fairly strong without this contribution. It is of interest that the perpendicularly-polarized CT band with which the observed CT band must be identified if the axial model should be correct is analogous to the perpendicularly polarized $F^-Li^+$ transition leading from state N to the first of the $^1\Pi$ states listed in Eq. (2′).

([11]) In the $\infty : \infty$ case, each $I_2$ interacts with two Bz, and each Bz with two $I_2$ in the same linear chain. There are also edgewise CT interactions with molecules of adjacent chains, which may be not negligible.

If $\Psi_V$ for the familiar CT spectrum is as in Eq. 12, there should occur at shorter wave lengths parallel-polarized bands with $\Psi_{12}$ and $\Psi_{13}$ as the leading terms in their CT upper states; but these bands would very likely be obscured by strong bands of Bz and $I_2$.

If the effective model is oblique rather than axial, the foregoing considerations are partially modified : in Eq. 12 for $\Psi_V$, $b^*$ is no longer zero, $d^*$ becomes smaller, some $\Psi_{12}$ enters although $\Psi_{11}$ must still be the predominant term, and the polarization of $\Psi_V \leftarrow \Psi_N$ becomes partly parallel, partly perpendicular. The various foregoing changes from the early simple interpretation of $\Psi_V$ as

$$a^*\Psi_{11}(Bz^+ - I_2^-) - b^*\Psi_0(Bz.I_2) + \cdots$$

may perhaps open new possibilities for an understanding of the previously puzzling apparent decrease of intensity of $\Psi_V \leftarrow \Psi_N$ in complexes of iodine with methylated benzenes as compared with that for benzene itself, and of the frequency shifts of the CT and visible bands at low temperatures in glassy solution ([12]).

It may be stressed again ([1]) here that the early discussion of the geometry of 1 : 1 complexes was based on a maximally simplified application of the resonance-structure theory, in which it was assumed plausibly and hopefully that the lowest-energy dative state would usually be strongly predominant over others of higher energy in $\Psi_N$. HASSEL's and other work, however, now indicate that it is usually essential for an adequate understanding to include one or two other dative states as resonance structures. Also, the need of including donor-excited or acceptor-excited states to some extent in $\Psi_V$ for the understanding of the intensities of CT bands has been emphasized (possibly over-emphasized, though perhaps not) by MURRELL ([12a]). These considerations do not show that the resonance-structure theory is unsatisfactory, but only that its use in maximally simplified form is usually inadequate.

## V. — Potential surfaces and CT spectra of complexes.

In understanding donor-acceptor complexes and their CT spectra, it is helpful to draw potential curves like those which are used for the electronic states of diatomic molecules. The energy (electron energy

([12]) J. S. HAM, *J. amer. chem. Soc.*, 1954, **76**, 3875. Perhaps the resting model is stabilized at low temperatures for benzene and toluene, but the axial model for mesitylene and hexamethylbenzene.

([12a]) J. N. MURRELL, *Qu. Rev. chem. Soc.*, 1961, **15**, 191.

plus nuclear repulsion energy) is plotted as a function of a coordinate $R$ chosen in some sensible though in general arbitrary way. Thus for amine-iodine complexes, the distance from the N to the nearer halogen atom is suitable; for $\pi,\pi$ complexes, the obvious choice is the perpendicular distance between the planes of the donor and acceptor, which are believed to be parallel. The resulting curves are, however, really only cross sections of many-dimensional potential surfaces which depend on a large number of other nuclear coordinates (collectively, Q) besides R.

Figures 1-4 show schematic potential curves for the case that D and A are *neutral even-electron* molecules. They illustrate *for this case* the effects on N states and CT spectra of the relative energies $W_0$ and $W_1$ of the pure states $\Psi_0$ and $\Psi_{11}$ of Eq. 1. *For simplicity, only one dative resonance structure* $\Psi_{11}$ is considered; this is enough to permit making several important points and to serve as a rough guide for more realistic complicated situations. For each of the cases $W_1 - W_0$ positive (case I, fig. 1, 2), nearly zero (case II, fig. 3), or negative (case III, fig. 4) *at R values $R_e$ corresponding to equilibrium in state N* of the complex, examples of the two sub-cases of strong and weak interactions, i.e. large or small

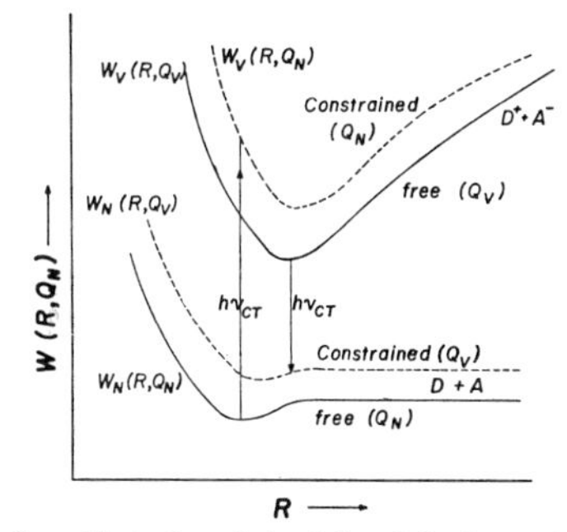


Fig. 2. — Illustrating effects of Franck-Condon constraints in Absorption, Photochemistry, and Fluorescence (effects especially large for contact pairs).

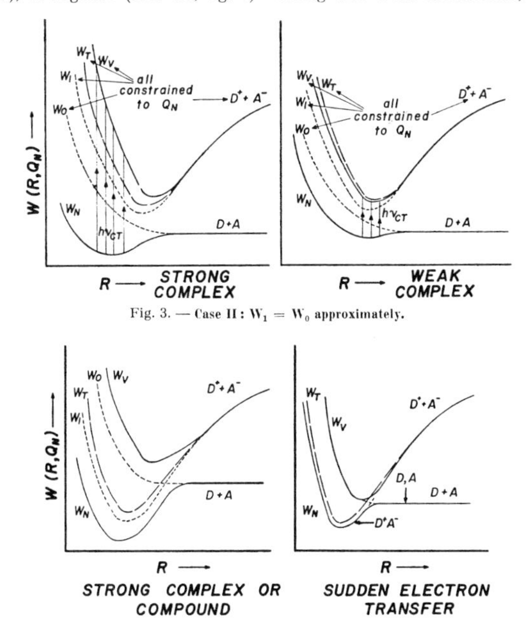


Fig. 3. — Case II: $W_1 = W_0$ approximately.

Fig. 4. — Case III: $W_1 < W_0$. Examples, $R_3N \rightarrow BCl_3$; $RBr \rightarrow GaBr_3$; $Na \rightarrow Cl$; low-I high-E $\pi,\pi$ pairs. Ionizing solvents lower $W_1$; they may also cause ionic dissociation.

$\mathcal{H}_{01} - SW_0)/(W_1 - W_0)$, are displayed. $W_1 - W_0$ is equal to $I_D - E_A - G_1 - G_0$, where $I_D$ = donor ionization energy, $E_A$ = acceptor electron affinity, $G_1$ = attraction energy between $D^+$ and $A^-$, $G_0$ = repulsion or attraction energy between D and A (for attraction, $G_0$ is negative) (*).

Case I with weak or moderately strong interactions is encountered in most complexes. The strong thioether and amine complexes of Table III probably belong to case II. The *n,v*, dative compounds belong to case III with strong interaction; Eq. 1 then reduces essentially to $\Psi_N = a\Psi_0 + b\Psi_{11}$ with $b > a$ or $b \gg a$. The same equation and potential curves also represent the major characteristics of highly polar odd-odd compounds such as alkali halide vapor molecules.

The right-hand case I curves (fig. 1) represent the situation for typical *contact pairs*. Here $R$ at contact is larger than $R_e$ for stable complexes. For both complexes and contact pairs, the Franck-Condon principle requires that the maximum of the observed CT band shall correspond to an approximately « vertical » transition (not much change of R, or of the other coordinates Q, from their initial values before absorption). Figure 1 (both left and right) has been drawn to show $R_e$ of state V (curve $W_V$) to be larger than $R_e$ of state N for a stable complex, but smaller than the VAN DER WAALS R of a contact pair. The latter condition is obviously necessary, and seems to account for the fairly large $\Delta\nu_{\frac{1}{2}}$ observed for contact CT spectra. (A steep upper-state potential curve in the given $R$ range should give a broad $\Delta\nu_{\frac{1}{2}}$.) The fact that $\Delta\nu_{\frac{1}{2}}$ is also fairly large for the CT spectra of complexes (of e.g. the alcohols and ethers in Table 3) probably indicates that $R_e$ of state V is different than in state N, and perhaps larger. However, further study of this interesting question is needed; in such a study, the various coordinates Q should not be forgotten.

Figure 1 further illustrates how the potential curve of a typical intra-donor excited state should intersect the $W_1$ curve so that its wave function, if of proper symmetry, should mix to some extent into $\Psi_V$, as in the term $d^*\Psi_{31}(D^*, A)$ of Eq. 2. This mixing, as mentioned earlier, may be largely responsible for the fairly high intensities of contact CT spectra.

For $\lambda_{max}$ in an absorption charge-transfer band, the Franck-Condon verticality principle of course applies to all the coordinates Q, not just to R. In figures 1, 3 and 4, the $W_N$ curve as drawn assumes that *all* the coordinates Q other than $R$ are adjusting themselves so as to make the energy of state N, at each $R$, as low as possible. The particular set of $Q$ values which thus minimizes $W_N$ at any $R$ is called $Q_N$; obviously $Q_N$ varies with $R$. Now when spectral absorption by a complex takes place, the coordinates $Q$ and $R$ retain, at $\lambda_{max}$, nearly the values $Q_N$ and $R_N$ also in the upper electronic state. However, especially in a loose complex, large-amplitude vibrations and internal rotations must occur in state N about the equilibrium value $R_e(N)$ and the corresponding equilibrium $Q_N$. The verticality principle must be applied with this fact in mind. In all the figures, vertical lines marked $h\nu_{CT}$ are shown for a *range* of $R$ values to take it into account.

After vertical excitation, for example to the V state, $Q_N$ no longer corresponds to equilibrium. Figures 1, 3 and 4 show for the V state a curve $W_V(R)$ appropriate to the *normal-state* values $Q_N$ of Q. Although they show an $R_e(V)$, this assumes $Q = Q_N$, so it, and the whole $W_V(R)$ curve, fail to correspond to a potential surface which minimizes the energy for state V. To minimize the energy one must have another set of values $Q = Q_V$, giving obviously a curve of lower energy. This is shown as $W_V(R, Q_V)$ in figure 2. In CT *fluorescence*, transitions down from $W_V(R, Q_V)$ with R near $R_e$ of $W_V(R_V, Q)$ are expected. Using the Franck-Condon verticality principle, $\lambda_{max}$ now corresponds to transitions to a curve $W_N(R, Q_V)$ which necessarily lies *above* $W_N(R, Q_N)$. Thus $\lambda_{max}$ of $h\nu_{CT}$ in emission may be at considerably longer wave lengths than in absorption.

The contrast between $Q_N$ and $Q_V$ should be at a maximum for contact-pairs, where $Q_N$ corresponds to the nuclear configurations of free D and A in van der WAALS contact in any of a variety of orientations. (Some orientations may of course be more favorable than others for CT band absorption). On the other hand, $Q_V$ corresponds to the equilibrium configuration for an ion pair with a covalent bond between $D^+$ and $A^-$, where in some cases formation of a bond may be accompanied by transfer of an atom or group from $A^-$ to $D^+$, or vice versa.

An interesting example in which CT absorption by a contact pair initiates a photochemical reaction has been studied by STEVENSON and COPPINGER ([13]). The contact pair is the donor $(C_3H_7)_3N$ with $CCl_4$ (or other $XCCl_3$ compound) as acceptor. With $CCl_4$, the observed reaction products are $HCCl_3$ and $(C_3H_7)_2(CH_3 - CH_2 = CH)NH^+Cl^-$. State V of the contact pair is $(C_3H_7)_3N^+(CCl_4)^-$ which at equilibrium (i.e. for $Q = Q_V$) may plausibly have the structure $[(C_3H_7)_3N^+ - CCl_3]Cl^-$. Thereafter an intramolecular (or perhaps intermolecular) reaction between

(*) For further details, see foot note 1.

([13]) D. P. STEVENSON and G. M. COPPINGER, *J. amer. chem. Soc.*, 1962, 84, 149.

one of the *n*-propyl groups and the $\rangle N^+ - CCl_3Cl^-$ could lead to the observed products.

In contrast to this extreme example of a very weak case I interaction, $Q_N$ and $Q_V$ may not differ very much in a strong case II complex (see fig. 3, left), where the amounts of dative character in States N and V should be more or less nearly equal ($a \approx b$ and $a^* \approx b^*$ in Eq. 4). The strongest amine complexes, e.g. $(C_2H_5)_3N.I_2$, very likely approach this case (NAGAKURA *et al*) ([1]). Nevertheless this very strong complex in heptane solution is not ionogenic, though it reacts to form ionic crystalline compounds if allowed time or if the concentration is increased. The fact that the CT band lies in the ultraviolet is consistent with a small $W_1 - W_0$ if the interaction between $\Psi_0$ and $\Psi_{11}$ is very strong (see left side of fig. 3). (We are here assuming that $b_{12}$ in Eq. 7′ can be neglected. This assumption is especially reasonable if $W_1 - W_0$ is indeeed small, where $W_1$ here refers to W of $\Psi_{11}$). A steep $W_V(R, Q_N)$ curve with its minimum at $R$ greater than $R_e(N)$ seems reasonable, and would account for a large $\Delta\nu_{\frac{1}{2}}$ as observed (see Table 3). For a *weak* complex with $W_1$ and $W_0$ nearly equal, the CT band should be at much longer wave lengths and might well have a much smaller $\Delta\nu_{\frac{1}{2}}$ (see right side of fig. 3).

In case III (fig. 4), $W_1 - W_0$ is positive at large $R$ values but becomes negative at smaller $R$ values so that at $R_e$ the predominant resonance structure is the dative function $\Psi_{11}$ (Eq. 10 with $b_{11} > a$, and $b^* > a^*_{11}$ at $R_e(N)$ with $Q = Q_N$). Several interesting variations of this case occur, depending on the strength of interaction of $\Psi_0$ and $\Psi_{11}$ and on the $R$ value, say $R_c$, at which the $W_1$ and $W_0$ curves intersect. The result of the interaction of $\Psi_0$ and $\Psi_{11}$ is of course an avoided crossing of the $\Psi_0$ and $\Psi_{11}$ curves in such a way that $\Psi_N$ is predominantly $\Psi_{11}$ for $R < R_c$, and predominantly $\Psi_0$ for $R > R_c$, with the reverse behavior for $\Psi_V$. The resulting $W_N(R)$ curve may or may not have a hump (activation barrier) depending on the strengths of interaction of $\Psi_0$ and $\Psi_{11}$ at the $R$ value at which the $W_1$ and $W_0$ curves intersect. For at least some of the dative compounds of the $n$, $v$ type, there is little or no activation barrier, in spite of the considerarable changes in $Q_N$ required as $R$ is decreased from $\infty$ to $R_e$ ([14]).

Interesting in the same connection are the hydrogen-bonded ($n,\sigma$ and $\pi,\sigma$) complexes. Consider for example $(CH_3)_2O$ plus HCl, which when mixed in the gaseous state show a drop of pressure indicating that a considerable fraction of molecules are paired as $(CH_3)_2O.HCl$ complexes ([15]). To understand this case a modification of figure 4, in the direction of the weak complex case of figure 3, is needed. Suppose that $W_1$ is below $W_0$ at small $R$ as in figure 4 (left side), but that its minimum is above the asymptotic energy of D + A. In the H-bonded complex, $W_N(R)$ has another minimum at relatively large R, where $W_1$ is still far above $W_0$ so that the expected CT band could be at short wave lengths. On decreasing R, $W_N(R)$, which is still not far below $W_0$, rises, because of exchange repulsions involved for $W_0$ at small $R$. At still smaller $R$, however, $W_1$ intersects $W_0$, and their interaction causes an avoided crossing; this creates an activation barrier in $W_N(R)$. The continuing drop of $W_1$ causes at smaller $R$ the inner minimum in $W_N(R)$ which no doubt corresponds to a dissociative-CT ion-pair complex (inner complex) with predominant structure $(CH_3)_2OH^+Cl^-$. Here $b_{11} > a$(or $\gg a$) in Eq. 9, but because $W_N$ at this minimum is higher than for the H-bonded minimum (outer complex), only the results of the latter have been observed ([16]). However, in certain other cases (e.g. aliphatic amines with H acids in non-polar solvents) where $I_D$ is smaller, hence $W_1$ is lower, the ion-pair minimum is lower and the expected presence of two minima has been established. Further, as already discussed in Sec. II, immersion in a polar solvent, or polymerization to form a crystal, lowers the inner minimum, if it is above the outer minimum in the isolated complex, so that it comes below the latter.

An important feature of case III for cases where the interactions of $W_0$ and $W_1$ are not very strong is that $Q_N$ is approximately $Q_O$ and $Q_V$ approximately $Q_1$ for $R > R_C$, but that $Q_N$ and $Q_V$ respectively approximate $Q_1$ and $Q_0$ for $R < R_c$. [Here $Q_0$ and $Q_1$ refer to $Q$ values which, at any $R$, would minimize the energies $W_0(R)$ and $W_1(R)$ respectively]. Thus in contemplating the avoided crossing of $W_0(R)$ and $W_1(R)$ to give rise to $W_N(R)$ and $W_V(R)$, one must for $W_N(R)$ work at all $R$ values with $Q_N$, but this must shift from being closer to $Q_0$ to being closer to $Q_1$ as one goes from $R > R_e$ to $R < R_c$; while $W_V(R)$ must show a converse behavior.

An interesting feature of the above-discussed complexes with two minima is the fact that the

([14]) See G. B. KISTIAKOWSKY and C. F. KLOTS, *J. chem. Phys.*, 1961, **34**, 712, 715, and forthcoming papers by S. H. BAUER *et al.*

([15]) O. MAAS and D. M. MORRISON, *J. amer. chem. Soc.*, 1923, **45**, 1675; J. SHIDEI, *Mem. Coll. Sci. Univ. Kyoto*, 1952, **9 A**, 97; A. GLADYSHEV and YA. H. SYRKIN, *Compt. R. Acad. Sci. U.R.S.S.*, 1938, **20**, 145. Studies by G. L. VIDALE and R. S. TAYLOR, *J. amer. chem. Soc.*, 1956, **78**, 294 on solutions of $(CH_3)_2O$ in liquid HCl indicates existence at various concentrations of complexes $(CH_3)_2O.HCl$, $(CH_3)_2OH^+Cl^-$, and $(CH_3)_2OH^+HCl_2^-$.

([16]) For further discussion, see ([6]).

predicted CT bands for the outer and inner complexes correspond to charge transfer in opposite directions, because of the reversal of energy position of the predominantly no-bond and predominantly dative states in the two cases. For both the outer and the inner complex, the maximum of CT absorption involves $W_V(R, Q_N)$, but whereas $Q_N$ more like $Q_0$ in the former case, it is more like $Q_1$ in the latter. But in *both* cases, $h\nu_{CT}$ is larger (cf. fig. 2) than if absorption went to $W_V(R, Q_V)$.

Although little is known about such CT spectra for simple cases such as $R_3N.HQ$ or $R_2O.HQ$, CT spectra such as those of $I^-(C_6H_5NCH_3)^+$ discussed in Sec. II are examples for the outer-complex case ([17]). Here, to be sure, there is no H bonding to stabilize an outer complex, but in principle such a complex might exist, and at least can occur as an $n$, $\sigma$ contact pair $C_6H_5N.CH_3I$ which could give rise to a CT spectrum (compare the rather similar case of $R_3N.CCl_4$ discussed above). Some of the solid complexes studied by HASSEL *et al*, e.g. dithiane, $S_8$, or quinoline with iodoform, may perhaps belong in a similar classification.

In the examples discussed in the two preceding paragraphs, the formation of an ion-pair inner complex corresponds to an even-ionogenic interaction of an $n$ donor with a $\sigma$ acceptor of type HQ or RQ. The same sort of interaction also occurs with $\pi$ donors. As was noted, the action is favored by a low ionization potential of the donor, since $W_1$ in figure 4 is then lower, and is also greatly assisted by a polar environment; it can be further assisted by formation of a complex anion, as e.g. in

$$C_6H_5CH_3 + HF + BF_3 \rightarrow C_6H_6CH_3^+BF_4^-$$

in HF solution. With halogens as $\sigma$ acceptors, even-ionogenic tendencies are much weaker, but for $\pi$ donors of sufficiently low ionization potential, there is evidence of ionogenic, probably even-ionogenic, action in the solid state, indicated by the occurrence of semi-conductivity; the complexes of perylene and pyrene with $I_2$ are examples. Another example is the carotene. $2I_2$ solid complex, perhaps of $(\text{carotene}^+ — I)I_3^-$ structure.

Most $\pi,\pi$ complexes appear to be no more than moderately strong [see Briegleb's book ([1])]. While at least most of them clearly belong to case I ($W_1 > W_0$ at all $R$ values), those having an unusually low $I_D$ and high $E_A$, which makes $W_1$ unusually low, show strong *odd-ionogenic* tendencies: in polar solvents they dissociate to odd ions (radical-ions, $D^+$ and $A^-$), and they form semi-conducting solids which are doubtless more or less ionic, perhaps often but not always containing $D^+$ and/or $A^-$ ions. Examples are furnished by paraphenylenediamine as $\pi$ donor and chloranil or tetracyanoethylene as $\pi$ acceptor.

The behavior of the low-$I_D$ and/or high-$E_A$ $\pi,\pi$ complexes suggests that in polar environments they may belong to case III, somewhat like the ion-pair inner complexes of the H-bonded $n,\sigma$ complexes. However, it seems much more probable that (except perhaps in extreme cases) they belong at least in non-polar solvents to the weak-interaction variety of case II with $W_1 > W_0$ as illustrated at the right side of figure 3. Their dissociation to odd ions in polar solvents may then be related not to $W_1 < W_0$ in the undissociated complex even under the influence of the solvent, but rather to a slow process in which solvation separates and stabilizes the odd ions.

The fact that $\pi,\pi$ complexes are not even-ionogenic is readily understandable. In $n,\sigma$ and $\pi,\sigma$ inner complexes, $\Psi_N$ is a predominantly dative structure but in the course of its formation an atom or radical *with its electron* has been transferred from the $A^-$ to the $D^+$, giving even-electron ions, for example

$$\Psi_{11}[(H_3N)^+ — (HCl)^-] \rightarrow \Psi_{11}(H_4N^+Cl^-);$$
$$\Psi_{11}[R_3N)^+ — (RQ)^-] \rightarrow (R_4N^+Q^-).$$

In $\pi,\pi$ complexes, there is no atom or radical which can be transferred, hence $\Psi_{11}(D^+ — A^-)$ can only break up, in polar solvents, into (solvated) odd ions $D^+$ and $A^-$.

Figures 1-4 assume that donor and acceptor, hence state N of the complex, are in singlet electronic states. This requires that of the two possible states for a structure $D^+A^-$ with odd-electron ions, the singlet structure, with a bond (however weak) between $D^+$ and $A^-$ be chosen for $\Psi_{11}$ or other dative $\Psi$'s entering into $\Psi_N$. But for every singlet dative function there is necessarily a corresponding triplet dative function $\Psi(D^+ \ldots A^-)$ with antibonding between $D^+$ and $A^-$. Just as in the interaction of two H atoms, the energy $W_T$ of the structure with antibonding should be higher than that, $W_1$, with bonding. Thus in every case there must be a triplet CT state whose $W(R, Q_N)$ curve must lie higher than $W_1(R, Q_N)$ but which in cases I and II (see figures 1 and 3) may be either below or above that of $W_V(R, Q_N)$ depending on how strong is the resonance interaction of $W_0$ with $W_1$ which pushes $W_V$ up above $W_1$. [Of course $W_T(R, Q_T)$ is below $W_T(R, Q_N)$, but may then be below or above $W_V(R, Q_V)$ in figure 2.] In case III (fig. 4) since now $W_1$ is below $W_0$ (for small R), $W_T$ is still above $W_1$ and thus necessarily above $W_N$, the more so the stronger the resonance interaction which now pushes $W_N$ below $W_1$.

([17]) Another example is found for $(n — Bu)_4N^+I^-$ in $CCl_4$: T. R. GRIFFITHS and M. C. R. SYMONS, *Mol. Physics*, 1960, **3**, 90. These authors conclude that in polar solvents, solvated ion pairs and solvated ions give additional and different spectra.

The right-hand diagram in figure 4 shows a hypothetical case in which the resonance interaction between $W_0$ and $W_1$ is extremely weak, so that $W_N \approx W_1$ is $\Psi_1(D^+ — A^-)$ and $W_V \approx W_0$ is $\Psi_0(D, A)$, while $W_T$ is $\Psi'_1(D^+ . . . A^-)$. Here $W_N$ and $W_T$ differ by having bonding or repulsion respectively between the radical-ions $D^+$ and $A^-$, and accordingly one might expect $W_T$ to be well above $W_N$. The paramagnetic behavior of $\pi,\pi$ complexes like paraphenylenediamine.chloranil in the solid and their dissociation in polar solvents suggest a situation like that in the right side of figure 4: ([18]) extremely weak resonance interaction so that $W_N \approx W_1$, and *also* extremely weak bonding in $W_1$ and extremely weak repulsion in $W_T$ so that it lies close above $W_N$. Extreme weakness of all these interactions is very improbable, although if one accepts it, one has a mechanism for rather sudden electron transfer from D to A at a critical distance of approach when a suitable $\pi$ donor and acceptor come together ([18]). Such a mechanism might then be operative for electron transfers in biological systems. However, a mechanism in which solvation or electrostatic effects in a solid separate a case II complex (or in extreme cases perhaps a case III complex) with $W_T$ well above $W_N$ into odd ions seems more likely.

## Acknowledgement.

The writer is much indebted to Professor W. B. PERSON for discussions and suggestions.

## DISCUSSION

**A. R. Ubbelohde.** — The attractive simplicity of Mulliken's theory of charge transfer bonds, for an isolated donor plus acceptor molecule pair, leaves unanswered the very interesting question of *cooperative charge transfer effects*, which needs to be considered in condensed systems generally and which becomes particularly urgent with crystalline systems.

One way in which the problem arises experimentally is in relation to the remarkable semi-conductor effects, which are found for example in aromatic hydrocarbons when these form solid solutions with electron donors such as the alkali-metals, Li, Na, K, etc.., or with electron acceptors such as halogen molecules. At low concentrations of these additives to the solid hydrocarbon, the marked lowering in activation energy for electrical conduction in these solids can be interpreted on one theory, in terms of charge transfer formation permitting easier generation of carriers than in the crystal free from such impurity. But as the concentration of electron donor or acceptor is progressively increased, the possibility of cooperative charge transfer effects may lead to the formation of chains of such bonds joining one aromatic molecule to the next, and eventually leading to transport of charge in « metallic » fashion.

$$\sim A \sim D \sim A \sim D \sim A \sim D \sim A \sim D \sim$$

We have obtained some experimental evidence of such cooperative charge transfer effects but need theoretical treatment of some of the electromagnetic and quantum mechanical features involved.

**R. S. Mulliken.** — A donor with two equally distant acceptor neighbors can certainly interact with both:

$$\Psi = \Psi(ADA) + \lambda\Psi(AD^+A^-) + \lambda\Psi(A^-D^+A^-) + \cdots$$

but terms with simultaneous charge transfer to both neighbors are presumably unimportant. Similarly for an acceptor molecule between two donor molecules. Although the existence of 2 : 1 complexes in solution has been established in some cases, they are probably generally disfavored by a saturation effect.

In a case such as for example an $I_2$ molecule surrounded by benzene molecules, the iodine should doubtless interact (in the manner suggested above) to some extent with its neighbors, but it seems likely that it tends to associate with a single partner as a 1 : 1 complex (with only small further interactions with other less near neighbors).

In a crystal Ar. . A. . Ar. . A. . Ar. . A. . (A = acceptor, e.g. $I_2$) certainly there are cooperative effects (an $\infty : \infty$ instead of à 1 : 1 complex). Semi-conductivity appears only for very large Ar (e.g. perylene), which have relatively low ionization potentials. Another similar example is the carotene-iodine complex of composition 2 carotene. 3 $I_2$, I think. One might suppose that the solid consists here of carotene positive and $I_3^-$ negative ions, or at least partly of these, introducing possibilities of hopping of electrons giving semi-conductivity. Although typical complexes in solution are not ionic, in the case of donors with unusually low ionization potential (e.g. the larger aromatic molecules, or alkali-metals) and/or high electron affinity (e. g. chlorazil); there is a strong tendency in polar solvents and in the solid state to form ions (often radical-ions). It is understandable in a general way that this situation, in solids, should permit cooperative phenomena and electron-hopping, thus semi-conductivity. But undoubtedly there is a variety of cases for different donor and acceptor pairs, and also perhaps irregular mixing of varying local structures or constituents in some solids. Thus a single unique theoretical treatment may not be very possible. However, for the moment, I myself am finding the theoretical understanding of 1 : 1 complexes in solution sufficiently complicated, with many varied cases, so that I have not given much attention to the solid semi-conducting complexes.

**B. Pullman.** — Tout le monde sait l'important apport de la théorie du Professeur MULLIKEN sur les complexes de transfert de charges dans le domaine de la chimie. Je voudrais signaler ici que les mécanismes mettant en jeu des transferts électroniques jouent également un rôle très important en biochimie. Beaucoup des travaux ont été consacrés récemment au rôle possible des complexes de transfert de charges en biologie moléculaire et également en pharmacologie. Ces différents travaux sont résumés dans notre livre récent, écrit en collaboration avec Mme PULLMAN, paru sous le titre « Quantum Biochemistry » (Wiley's Interscience Division, New York, 1963).

**J. H. Hildebrand.** — BENESI and I determined the equilibrium constant for the 1 : 1 complex of iodine and benzene in carbon tetrachloride. When we extrapolated to the amount of the complex in pure benzene saturated with iodine, we

([18]) The right side of figure 4 shows no activation barrier. This situation is possible if $R_e$ is large (comparable with van der WAALS distances, or somewhat less if there are classical electrostatic attractions); otherwise a barrier is expected.

accounted rather well for the enhancement of the solubility of iodine over what would be its « regular solution », non-complexed value. This may be regarded as evidence, not very strong perhaps, that the complex in pure benzene is still 1 : 1.

Gilbert LEWIS, in describing his theory of generalized acid-base interaction, included the factor of rapid, reversible interaction, with, therefore, virtually no activation energy. There are, however, reactions with considerable activation energy, that may be classed as oxidation-reduction reactions. I wonder what Prof. MULLIKEN thinks of the importance of such a distinction ?

**R. S. Mulliken.** — Regarding the benzene-iodine complex, I agree that Prof. HILDEBRAND's observation gives evidence for the existence of localized 1 : 1 complexes even for $I_2$ in pure benzene. But, if so, the $I_2$ molecules must change partners frequently, and at présent it does not seem clear to what extent the concept of 1 : 1 complexes is applicable for $I_2$ in pure benzene or its derivatives, although such complexes are well established for solutions of benzene plus iodine in inert solvents. Regarding rapid reversibility, this certainly is typical for DA complexes, but would not seem to be essential.

**A. Berlin.** — In which way are the electrons localized in the species formed by the dissolution of metals in their own salts, at concentration at which the conductance of the system is not yet electronic ?

**R. S. Mulliken.** — I am not sufficiently familiar with such systems to give a helpful answer.

**A. R. Ubbelohde.** — The problem I have already mentioned of the formation of *cooperative charge transfer bonds* can be presented in another way, which may help to clarify the issues involved. If one considers the electrostatic polarisation of a molecule B by a system (charge or dipole) A, then the juxtaposition of a second system A on the opposite side of B ABA will modify the effect of the first system A; according to circumstances, there may be enhanced or reduced interaction in this simple cooperative system. To quote specific dipole molecules, we are familiar with the enhancement of dielectric moment in clusters of $H_2O$ molecules, or nitrobenzene molecules in the liquids; and in the limit with very strong enhancement in ferro-electric polarisation in crystals.

What we need to know is to what extent a similar enhancement (or anti-effect) can arise when the polarisations are only « virtual » in the quantum mechanical description of dispersion forces, or charge transfer bonds. In the triad ABA can the donor system B simultaneously form charge transfer bonds with both acceptor systems, provided some kind of synchronisation is maintained? In the simplest problem of Van der WAALS dispersive attractions, the assumption would certainly not be justified or be made for multiple electrostatic interactions on any particular molecule. However, we need a theory of cooperative effects when the interactions are virtual.

**R. S. Mulliken.** — I believe my response to Prof. UBBELOHDE's first question largely covers this one also. But also, there is real charge transfer in charge-transfer complexes, and this must lead to new electrostatic forces if complexes are formed into a solid; and, in extreme cases, as remarked below, solid complexes become ionic in character.

**S. Leach.** — In the case of complexes between benzene and halogen molecules, HASSEL, by crystal structure determinations, found that the internuclear axis of the halogen molecule was aligned along the $C_6$ symmetry axis of the benzene molecule i.e. perpendicular to the ring plane. MULLIKEN, however, predicted, on theoretical grounds, an off the $C_6$ axis orientation for the halogen molecule. Since HASSEL's work refers to the crystal whereas MULLIKEN's predictions were for an essentially isolated 1 : 1 molecular complex, could it be that the difference between experiment and theory is due to crystal packing forces which cause the halogen molecule to prefer the $C_6$ axis alignment?

This might be checked by comparing the charge-transfer properties (spectra, etc..) of the liquid phase benzene-halogen solutions with those of the crystalline phase. Are there sufficient data to make this comparison ?

**R. S. Mulliken.** — In a general way one might say that crystal packing forces can cause altered geometrical arrangements in solid complexes as compared to 1 : 1 complexes in solution, especially if the charge-transfer forces are relatively weak, as is the benzene-halogen cases. However, I would rather say that the solid benzene-halogen complexes are $\infty : \infty$ instead of 1 : 1 complexes. In the cases of strong 1 : 1 complexes (e.g. pyridine $I_2$), the 1 : 1 units are still seen in the crystal. Nevertheless, it may be true that the alignment of the $I_2$ axis along the $C_6$ axis represents the equilibrium arrangement in the 1 : 1 as well as in the $\infty : \infty$ benzene-halogen complex. Dr FERGUSON's paper gives some evidence consistent with this possibility.

My early predictions of the geometry of 1 : 1 complexes were based on a maximally simplified application of the theory, and were statements of what seemed to me most plausible, rather than being rigorous predictions. The general trend of HASSEL's work on solid complexes indicates that it is usually necessary not to simplify too much in applying the theory. To be more specific, it appears that in general the normal state of a complex is stabilized by *more than one* dative resonance structure (or, one may say, by a hybrid dative structure) and that sometimes the *lowest-energy* dative structure may not have the predominant effect in causing stabilization.

As yet relativy little has been done in the spectroscopic study of solid complexes in comparison with 1 : 1 complexes in solution.

**W. Gerrard.** — In my view, it is helpful to look upon electron acceptor or donor function as a property of a particular part of a molecule. At a particular part of a molecule there may be acceptor function, and at another part, donor function, even, conceivably simultaneously. A donor part can donate to the acceptor part of another molecule, or to the acceptor part of the same molecule, and the matter of nomenclature arises. An alcohol is sometimes referred to as proton acceptor, sometimes as a proton donor !

**R. S. Mulliken.** — I agree. Some molecules include both donor and acceptor sites. Others may function either as electron donors or acceptors depending on the partner molecule. In still other cases, the same molecule may function simultaneously as donor and acceptor, often even at the same site. Thus ethylene with $Ag^+$ functions as donor to $Ag^+$ as acceptor, but at the same time as acceptor to $Ag^+$ as donor (two-way charge transfer). Again, pyridine as ligand in a complex ion (e.g. attached to $Pt^{++}$) functions as a $\sigma$ donor (at the N atom) but also as a $\pi$ acceptor (in the aromatic ring). Strictly speaking, one should speak rather of electron donor or acceptor *functioning* of molecules, rather than categorically of donors and acceptors. With respect to « proton acceptors » and « proton donors » I would rather call these electron donors and electron acceptors respectively, since I believe these terms describe the major function which is active, the motion of the H atom being a secondary result (which, besides, usually does not occur without the assistance of a polar environment).

**R. Buvet.** — I would ask to Prof. MULLIKEN if he thinks that it is possible in the present state to establish a qualitative or quantitative correspondence between :

— charge-transfer complex properties between two compounds in solution, for example, as they are indicated by spectroscopy,

— and electronic conductance of the same charge-transfer complex in the solid state.

Is it possible to think to such a correspondence in a given series of complexes, for example in the case of aromatic hydrocarbonhalogen complexes?

**R. S. Mulliken.** — My answer to Prof. UBBELOHDE's questions perhaps fairly well covers Dr BUVET's questions.

**A. Witkowski.** — Comment concerning the sudden electron transfer case : the potential energy curves of ionic and covalent states cross for some value of intermolecular distance. But in the environment of the crossing the BORN-OPPENHEIMER approximation is not valid, especially in the solid state, where the selective positions of donor and acceptor are fixed. I would like to know what is the experimental evidence for the possibility of separation of electronic and nuclear motions in the region of crossing, or if is no such evidence available, what is the evidence for influence of vibronic coupling for the electron transfer.

**R. S. Mulliken.** — It is true that we would be in some trouble with the BORN-OPPENHEIMER approximation in the ideal case where the potential energy surfaces cross with very little interaction (« sudden electron transfer »). But probably this extreme case is not actually realized, so that passage from the covalent to the ionic state is adiabatic, thus not completely sudden but only moderately rapid.

**W. Gerrard.** — With reference to the definition of acid-base interaction, and the rapidity of such interaction, involving little or no activation energy, the matter of definition again arises. The primary interaction of ethanol and hydrogen chloride involves the formation of the hydrogen bonded system

$$RO\begin{smallmatrix}\diagup H\\ \diagdown HCl\end{smallmatrix}$$

which is considered by many as related to acid-base reaction; quick and involving little or no activation energy. Nevertheless, there are others who believe the acid-base function operates only when ions are being formed, and this conceivably would involve activation energy. As a matter of fact, the « chlorine » *does* in time find its way from hydrogen to carbon, and some process as follows could be involved :

$$R-O\begin{smallmatrix}\diagup H\\ \diagdown HCl\end{smallmatrix} \rightleftharpoons ROH_2^+Cl^- \rightleftharpoons Cl^- + ROH_2^+$$
$$\rightleftharpoons RCl + H_2O.$$

Reprinted from THE JOURNAL OF CHEMICAL PHYSICS, Vol. 44, No. 5, 2161–2170, 1 March 1966
Printed in U. S. A.

# Infrared Spectra of Charge-Transfer Complexes. VI. Theory*

H. BRUCE FRIEDRICH† AND WILLIS B. PERSON
*Department of Chemistry, University of Iowa, Iowa City, Iowa*

(Received 26 July 1965)

The general theory of molecular vibrational transition intensities is discussed with emphasis on electronic reorientation contributions to the intensities. The wavefunctions used by Mulliken to represent the electronic states of donor–acceptor complexes are written to include an explicit dependence on the vibrational coordinates. These functions and the general theory are applied to the intensities of halogen vibrations in donor–acceptor complexes. Specific application to actual complexes requires the estimation of the derivative of the vertical electron affinity of the halogen molecule with respect to its internuclear distance, the electronic transition moment of the charge-transfer band, the coefficients in the donor–acceptor ground-state wavefunctions and the difference between the energies of the dative- and no-bond states. Evaluation of each of these parameters is discussed for a number of complexes of halogens and relationships between the wavefunction coefficients and the infrared frequency shifts are described The calculations indicate that all of the intensity enhancement of the halogen–halogen stretching vibration may be due to electronic reorientation during the vibration. Partly as a result of this conclusion, it is argued that no information about the geometry of the benzene–halogen complexes may be deduced form the infrared spectrum, at least in any simple way. The argument is extended qualitatively to hydrogen-bonded systems to indicate the probable similarity in explanation for the enhancement of the X–H stretching vibrations. In conclusion, a number of generalizations are presented regarding the spectra of complexes.

## INTRODUCTION

WHEN the infrared spectra of charge-transfer complexes are compared with those of the isolated molecules which form the complexes, three types of changes are found to occur: (1) the vibrational frequencies in donor or acceptor (or both) may be shifted, (2) the intensities of the bands may be changed considerably, and (3) new low-frequency bands appear due to the vibrations of one molecule in the complex against the other. For example, changes in the spectra of the electron acceptor have been illustrated in some of the earlier studies in this series[1–3]; examples of the third change are provided by the identification of the N–I stretching vibration in the strong amine–halogen complexes.[4–6] In this paper we discuss only the first two changes, and we mostly discuss only the changes observed[1–3] in the X–Y stretching frequency of an electron acceptor X–Y–Z, where Z represents an unreactive radical. Extension of the theory to vibrations of the donor and to other vibrations of the acceptor is reasonably straightforward, however.

The characteristic changes observed in the infrared spectrum of the X–Y bond stretching vibration on complex formation are a dramatic decrease in its frequency and an increase in intensity whose magnitudes depend upon the strength of the interaction between the electron donor and acceptor. Originally the observation[7] of an infrared absorption band due to the Cl–Cl stretching vibration of $Cl_2$ dissolved in benzene, for example, was interpreted as evidence favoring a complex with the $Cl_2$ molecule in an unsymmetrical location. Ferguson and Matsen have pointed out, however, that intensity enhancement can occur through an "electron vibration" mechanism even if the $Cl_2$ molecule is in a symmetrical environment.[8,9] According to their mechanism, the electron affinity of the acceptor (or ionization potential of the donor) changes during the vibration; as a result, the energy difference between the "no-bond" and "dative" states changes, so that the extent of mixing of these two wavefunctions changes during the vibration. Thus, the electron transferred from donor to acceptor in forming the dative structure may be thought to be vibrating back and forth between donor and acceptor with the frequency of the X–Y vibration. The resulting rather large change in dipole moment is responsible for the observed intensity enhancement.

We see that this mechanism is a "vibronic interaction" and may thus be described in that formalism.[10] As noted elsewhere[11] this description has been discussed

* Presented in part at the International Symposium on Molecular Structure and Spectroscopy, Tokyo, Japan, 1962, as Paper D110; abstracted from the Ph.D. thesis of H. B. Friedrich (1963). Paper V in this series is J. Am. Chem. Soc. **82**, 1850 (1960), and Paper IV is J. Chem. Phys. **35**, 908 (1961).

† NSF Cooperative Graduate Fellow, 1960–1962. Present address: Department of Chemistry, Gustavus Adolphus College, St. Peter, Minnesota.

[1] W. B. Person, R. E. Humphrey, W. A. Deskin, and A. I. Popov, J. Am. Chem. Soc. **80**, 2049 (1958).

[2] W. B. Person, R. E. Humphrey, and A. I. Popov, J. Am. Chem. Soc. **81**, 273 (1959).

[3] W. B. Person, R. E. Erickson, and R. E. Buckles, J. Am. Chem. Soc. **82**, 29 (1960).

[4] H. Yada, J. Tanaka, and S. Nagakura, J. Mol. Spectry. **9**, 461 (1962).

[5] F. Watari and S. Kinumaki (private communication).

[6] A. G. Maki (private communication).

[7] J. Collin and L. D'Or, J. Chem. Phys. **23**, 397 (1955).

[8] E. E. Ferguson and F. A. Matsen, J. Chem. Phys. **29**, 105 (1958).

[9] E. E. Ferguson and F. A. Matsen, J. Am. Chem. Soc. **82**, 3268 (1960).

[10] See, for example, A. D. Liehr, Z. Naturforsch. **13a**, 311 (1958) or R. Daudel and S. Bratož, Cahiers Phys. **75**, 39 (1956).

[11] R. S. Mulliken and W. B. Person, Ann. Rev. Phys. Chem. **13**, 107 (1962).

also by Jones and Simpson,[12] who considered the contribution to infrared intensities due to the "delocalization moment" and is essentially identical to the model proposed by Tsubomura[13] and by McKinney and Barrow[14] to explain analogous changes in the infrared spectra of the O–H stretching vibration upon hydrogen-bond formation.

In this paper we use the vibronic formalism of Liehr[10] and of Jones and Simpson[12] to derive an expression for the intensity enhancement, which reduces to that given by Ferguson and Matsen[8,9] with certain approximations. We have completely re-evaluated the parameters needed to predict the intensification of the X–Y stretching vibrations in the acceptor molecules. Although the details are thus changed considerably from the treatment given by Ferguson and Matsen,[9] we agree with their major conclusion that the observed intensification is indeed due to this mechanism.

Finally, we should like to point out again that this mechanism applies also to vibrations of isolated molecules, which change the relative weights of polar wavefunctions contributing to the ground-state structures of the molecule. We believe that these vibronic interactions are undoubtedly part of the correct explanation for the well-known failure of the bond-moment hypothesis to explain infrared intensities.[15] Although the parameters needed to predict quantitatively the "delocalization moments" are not known for most molecules, we believe the theory presented here can form the basis for an empirical, semiquantitative understanding of the deviations from the simple bond-moment hypothesis.[16]

## VIBRATIONAL TRANSITION MOMENTS

The intensity of a transition from Vibrational State $k$ to State $l$, occurring for the $K$th electronic state, is related to the transition-moment vector $\mu_{Kl,Kk}$, defined by

$$\mu_{Kl,Kk}=\iint \Omega^*_{Kl}(q,Q)\,\boldsymbol{\mu}\Omega_{Kk}(q,Q)\,dq\,dQ. \qquad (1)$$

Here $\boldsymbol{\mu}$ is the dipole-moment operator; $\Omega_{Kk}(q,Q)$ is the vibronic wavefunction for the $K$th electronic state and $l$th vibrational state, and depends upon both the electronic coordinates $q$ and the nuclear coordinates $Q$. In the usual formulation[17] the vibronic function is factored into a product $[\Omega_{Kl}=\Psi_K(q)\Phi_l(Q)]$. In (1) the integration over the electronic coordinates $q$ then gives unity; for the vibrational transition, only the dependence of $\boldsymbol{\mu}$ on the vibrational coordinates, $Q$, is considered, and $\boldsymbol{\mu}$ is expanded in a Taylor's series in $Q$. Neglect of higher-order terms (electrical anharmonicity) leads to the well-known dependence of infrared intensity on the dipole derivative $(\partial\mu/\partial Q_l)$, evaluated at equilibrium,[17] since

$$\mu_{Kl,Kk}=\left(\frac{\partial\boldsymbol{\mu}}{\partial Q_l}\right)_0\int \Phi^*_l(Q)\cdot Q_l\Phi_k(Q)\,dQ.$$

We should now like to recognize that for some vibrations the vibronic wavefunction $\Omega$ cannot be factored in this simple fashion since the electrons rearrange during the vibration so that the wavefunction describing them changes. Obviously this effect is more pronounced in some vibrations than in others.

Following Liehr[10] the vibronic function $\Omega$ is expanded in a set of approximate "floating orbital" functions:

$$\Phi_k^K(Q)\Psi_K(q,Q)=\Omega_{Kk}(q,Q)=\Phi_k^K\sum C_{jK}\theta_j(q,Q). \qquad (2)$$

Here $\theta_j$ are the ordinary approximate functions, constructed by the rules of either valence bond or molecular orbital theory, extended to include dependence on the vibrational coordinates explicitly. They are chosen to be orthonormal at the equilibrium configuration, but they are not, in general, orthogonal at other configurations. The function $\Phi_k^K$ is the vibrational wavefunction for the $k$th vibrational state of the $K$th electronic state.

Assuming that a vibration produces only small perturbations of the electronic functions and retaining only the first-order expansion terms in (2), Liehr[10] finds that

$$\Psi_k(q,Q)=\theta_K{}^0(q,0)+\sum_a[\nabla_{Q_a=0}\theta_j(q,Q)]\cdot Q_a+\sum_j\theta_j{}^0(q,0)\cdot C'_{jk}(Q),$$

where

$$C'_{jk}=\sum_a[\nabla_{Q_a=0}C_{jk}(Q)]\cdot Q_a. \qquad (3)$$

The vibrational transition moment for the $0\rightarrow1$ transition [Eq. (1)] now becomes

$$\mu_{K1,K0}=\iint \Psi_K{}^*(q,Q)\Phi_1{}^K(Q)\,\boldsymbol{\mu}\Psi_K(q,Q)\Phi_0{}^K(Q)\,dq\,dQ. \qquad (4)$$

Here $\Psi_K(q,Q)$ is defined in Eq. (3) and $\Phi_1{}^K$ is the vibrational wavefunction defined above [Eq. (2)].

[12] W. D. Jones and W. T. Simpson, J. Chem. Phys. **32**, 1747 (1960).
[13] H. Tsubomura, J. Chem. Phys. **24**, 927 (1956).
[14] P. C. McKinney and G. M. Barrow, J. Chem. Phys. **31**, 294 (1959).
[15] See, for example, I. M. Mills, Ann. Rept. Progr. Chem. (Chem. Soc. London) **55**, 55 (1958); C. A. Coulson, Spectrochim. Acta **14**, 161 (1959); J. Overend, in *Infrared Spectroscopy and Molecular Structure*, edited by M. Davies (Elsevier Publishing Company, Inc., New York, 1963), Chap. 10; T. L. Brown, J. Chem. Phys. **38**, 1049 (1963).
[16] (a) See W. B. Person and L. C. Hall, Spectrochim. Acta **20**, 771 (1964). (b) The most rigorous and satisfying application of vibronic theory to infrared intensities in molecules has been given for benzene by T. L. Brown, J. Chem. Phys. **43**, 2780 (1965).
[17] For example, E. B. Wilson, J. C. Decius, and P. C. Cross, *Molecular Vibrations* (McGraw-Hill Book Company, Inc., New York, 1955).

## APPLICATION TO CHARGE-TRANSFER COMPLEXES

First of all, we note that vibronic interactions in the isolated molecules may be changed when the two molecules are brought together to form the complex. However, we assume that such changes are small compared to the effect we wish to discuss.[18]

Mulliken[19] has written the wavefunctions for the ground and first excited states of the complex as

$$\theta_N(q,Q)=a(Q)\Psi_0(q,Q)+b(Q)\Psi_1(q,Q)$$

and

$$\theta_E(q,Q)=a^*(Q)\Psi_1(q,Q)-b^*(Q)\Psi_0(q,Q). \quad (5)$$

Here $\Psi_0$ is the wavefunction for the hypothetical "no-bond" structure $(D\cdots A)$ and $\Psi_1$ is the "dative function" corresponding to the state with complete transfer of the electron $(D^+\text{–}A^-)$. We have written Eq. (5) using the notation developed in the preceding section, showing explicitly the dependence on the nuclear coordinate. Here we assume that no further functions are needed in Eq. (2) and that $\theta_N(q,Q)$ and $\theta_E(q,Q)$ are "correct" in zero order.

Now consider the particular vibrational coordinate, $Q_l$. From Eq. (3) and the definitions above, it may be shown that

$$\theta_N(q,Q)\simeq a_0\Psi_0(q,0)+b_0\Psi_1(q,0)$$
$$+(\partial a/\partial Q_l)_{Q_l=0}Q_l\Psi_0(q,0)$$
$$+(\partial b/\partial Q_l)_{Q_l=0}Q_l\Psi_1(q,0). \quad (6)$$

The $\Psi$'s in Eq. (6) are equilibrium stationary functions. It has been assumed that the derivatives of the wavefunctions, $\partial\Psi_i/\partial Q_l$, are small compared with the derivatives of the coefficients; i.e., it is assumed that the principal contribution to $(\partial\theta/\partial Q_l)$ comes from a change in the relative weights of the two basis functions rather than from large distortions of the basis functions themselves during the vibration.

If Eq. (6) is inserted in Eq. (4), the transition moment for the $l$th vibration in the charge-transfer complex is found to be

$$\mu_{N1,N0}=\left\{\left(\frac{\partial\mu}{\partial Q_l}\right)_0+2\mu_{00}a_0\left(\frac{\partial a}{\partial Q_l}\right)_{Q_l=0}\right.$$
$$\left.+2\mu_{11}b_0\left(\frac{\partial b}{\partial Q_l}\right)_{Q_l=0}+\left[2a_0\left(\frac{\partial b}{\partial Q_l}\right)+2b_0\left(\frac{\partial a}{\partial Q_l}\right)\right]\mu_{01}\right\}$$
$$\times\int\Phi_1{}^0(Q)Q_l\Phi_0{}^0(Q)dQ_l$$
$$=M_{N1,N0}\int\Phi_1{}^N(Q)Q_l\Phi_0{}^N(Q)dQ_l. \quad (7)$$

Here the first term is the term found from an ordinary "stationary" orbital calculation[17] except that it includes the vibronic contribution discussed at the end of the introduction; the remaining three terms add to give the "delocalization moment" due to the complex formation and arises from electronic rearrangement during the vibration, under the assumption that it is the mixing coefficient and not the basis functions which depend strongly on the nuclear coordinates. The other terms are $\mu_{00}=\int\Psi_0{}^*\boldsymbol{\mu}\Psi_0d\tau$, the dipole moment of the complex for the "no-bond" structure; $\mu_{11}=\int\Psi_1{}^*\boldsymbol{\mu}\Psi_1d\tau$, the dipole moment of the "dative structure"; and $\mu_{01}=\int\Psi_1{}^*\boldsymbol{\mu}\Psi_0d\tau$, the overlap moment.

For many charge-transfer complexes (e.g., for benzene$\cdot I_2$) $\mu_{00}=0$. If $\mu_{01}$ is also neglected, then Eq. (7) reduces to

$$M_{N1,N0}=(\partial\mu/\partial Q_l)_0+2b\mu_{11}(\partial b/\partial Q_l). \quad (8)$$

(Here, and in the following, the zero subscript on $b$, indicating the equilibrium value, is omitted.) This equation is identical to that given by Ferguson and Matsen,[8,9] based on somewhat more intuitive arguments.

In the derivation given above, we have considered only one vibration associated with $Q_l$. If there are other vibrations which change the mixing coefficients $a$ and $b$ in $\theta_N$, we get an equation like Eq. (7) for each such vibration. We might expect the magnitude of the terms contributing to the delocalization moment to differ for each vibration (see below).

## NUMERICAL ESTIMATION OF INTENSITIES

### General Equations

It has been shown for weak complexes (those with $a\gg b$), that the perturbation charge-transfer theory gives[19]

$$b/a=-(H_{01}-W_0S)/(W_1-W_0)=-(\beta_0/\Delta W). \quad (9)$$

Here

$$H_{01}=\int\Psi_1{}^*\mathcal{H}\Psi_0dq; \qquad S=\int\Psi_1{}^*\Psi_0dq;$$

$W_1$, the energy of the dative state, is given by $W_1=\int\Psi_1{}^*\mathcal{H}\Psi_1dq$; and $W_0$, the energy of the no-bond state, is given by $W_0=\int\Psi_0{}^*\mathcal{H}\Psi_0dq$.

Differentiating Eq. (9) and making use of the normalization condition $(a^2+b^2+2abS=1)$, we find that

$$\left[1+\frac{b}{a}\left(\frac{b+aS}{a+bS}\right)\right]\left(\frac{\partial b}{\partial Q_l}\right)=-\frac{a}{\Delta W}\frac{\partial\beta_0}{\partial Q_l}-\frac{b}{\Delta W}\frac{\partial(\Delta W)}{\partial Q_l}. \quad (10)$$

Here

$$(\partial\beta_0/\partial Q_l)=(\partial H_{01}/\partial Q_l)-W_0(\partial S/\partial Q_l).$$

Thus, any vibration, which causes $\beta_0$ or $\Delta W$ to change, changes the mixing coefficient $b$ and hence has a "delocalization moment."

---

[18] M. Hanna, in a private communication, has suggested that this assumption may be questionable. However, it seems justifiable to assume this model here and investigate the consequences of it.

[19] R. S. Mulliken, J. Am. Chem. Soc. **74**, 811 (1952).

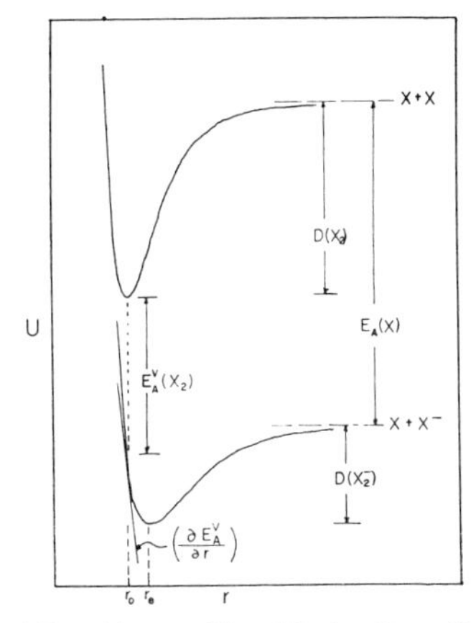


FIG. 1. Potential curves of $X_2$ and $X_2^-$ from Person (Ref. 20). $D(X_2)$ is the dissociation energy of $X_2$, $E_A{}^v(X_2)$ the vertical electron affinity of $X_2$ and $E_A(X)$ is the electron affinity of the X atom. The value of $(\partial E_A{}^v/\partial r)$ is equal to the slope of the tangent line labeled $(\partial E_A{}^v/\partial r)$ in this figure.

Since $\Delta W = I_D{}^v - E_A{}^v + \Delta G$, (where $I_D{}^v$ is the vertical ionization potential of the donor, $E_A{}^v$ is the vertical electron affinity of the acceptor, and $\Delta G$ is the difference in stabilization energies between the excited state and the ground state), we may now estimate which vibrations might change $\Delta W$. To a first approximation, the vibrations of either donor or acceptor do not change the distance between the centers of the molecules, so $\Delta G$ is relatively insensitive to these vibrational coordinates. However, as Ferguson and Matsen have noted,[8] the symmetric vibrations of the donor change its ionization potential, so $(\partial I_D{}^v/\partial Q_l)$ is different from zero for these vibrations. Examination of the figures given by Person[20] reveals that the vertical electron affinity of the halogen molecule is quite sensitive to the halogen–halogen bond distance. Hence, for the halogen–halogen stretching vibration of the acceptor,

$$\partial(\Delta W)/\partial Q_l = -(\partial E_A{}^v/\partial Q_l). \qquad (11)$$

Finally, we note, with Ferguson,[21] that vibrations which change the overlap integral between donor and acceptor orbitals also have a "delocalization moment."

### Enhancement of the Halogen Stretching Vibration

Now consider the X–Y stretching vibration of a diatomic halogen molecule in a weak complex ($b \ll a$, $S=0$). Then Eq. (10) reduces to the following, using Eq. (11):

$$(\partial b/\partial Q_l) = +(b/\Delta W)(\partial E_A{}^v/\partial Q_l). \qquad (11a)$$

[20] W. B. Person, J. Chem. Phys. **38**, 109 (1963).
[21] E. E. Ferguson, J. Chim. Phys. **61**, 257 (1964).

Substituting into Eq. (7) and using the definition of the electronic transition moment, $\mu_{EN}$, for the charge-transfer band $\mu_{EN} = \int \theta_E{}^0(q,\ 0)\, \boldsymbol{\mu} \theta_N{}^0(q,\ 0)\, dq = ab\mu_{11} - ab\mu_{00} + (a^2 - b^2)\mu_{01}$,[22a] together with the assumption that $\mu_{00} = S = 0$, leads to the following working equation:

$$M_{N1,N0} = \left(\frac{\partial \mu}{\partial Q_l}\right)_0 + \frac{2b(\partial E_A{}^v/\partial r)\mu_{EN}}{a(m_r)^{\frac{1}{2}}\Delta W}. \qquad (12)$$

Here we have also used the definition of the normal coordinate: $Q_l = (m_r)^{\frac{1}{2}} r$, where $m_r$ is the reduced mass of the halogen molecule and $r$ is the X–Y distance.

Now let us consider the derivative, $\partial E_A{}^v/\partial r$. If accurate potential curves were known for $I_2$ and for $I_2^-$, for example, we could evaluate this derivative almost exactly for weak complexes. Figure 1 shows the relationship between these curves, as estimated by Person.[20] The derivative is readily seen to be the slope of the $I_2^-$ potential curve at the equilibrium distance found in $I_2$ (which is assumed to be equal to $r_e$, the equilibrium distance of $I_2$ in the complex). This evaluation of $\partial E_A{}^v/\partial r$ differs considerably from the treatment of Ferguson and Matsen,[9] who neglected the slope of the $I_2^-$ curve completely.

If the $I_2^-$ potential curve is given by a Morse potential,[22b] the derivative evaluated at $r_e$ is

$$(\partial E_A{}^v/\partial r)_{r=r_e} = 2D(X_2^-)\beta\{1 - \exp[-\beta(r_e - r_0)]\} \times \exp[-\beta(r_e - r_0)]. \qquad (13)$$

Here $D(X_2^-)$ is the dissociation energy of $X_2^-$, $\beta$ is the Morse constant for $I_2^-$, $r_e$ is the $I_2$ equilibrium bond distance, and $r_0$ is the $I_2^-$ equilibrium bond distance. A corresponding equation holds, of course, for any halogen. These parameters in Eq. (13) have been estimated by Person for some of the halogens.[20] Values used to estimate $\partial E_A{}^v/\partial r$ for several halogens are shown in Table I. Several possible values are estimated for the parameters of $I_2^-$ in order to illustrate the sensitivity of $(\partial E_A{}^v/\partial r)$ to the values of the parameters. The rather large variation in $(\partial E_A{}^v/\partial r)$ with small changes in $r_0$ and $\omega_e$ for $I_2^-$ will make it difficult to establish that the calculations here are quantitatively correct, since the experimental values of $r_0$ and $\omega_e$ are not known.

Now consider the evaluation of $\Delta W$. For weak complexes, $\Delta W = W_1 - W_0$, and, as discussed earlier,[11,20] $\Delta W = I_D - C_1$. Actually this term is given more exactly by $\Delta W = W_E - W_N$ (which is approximately equal to $W_1 - W_0$ for weak complexes). Thus,

$$\Delta W = W_E - W_N = h\nu_{EN}. \qquad (14)$$

[22] (a) Equation (5) must give the exact representation of the true wavefunctions for this relationship to be true. Since we are using functions truncated after the second term of the basis set of functions, this representation of $\mu_{EN}$ is not likely to give the correct experimental intensity. See, for example, H. Hameka and L. Goodman, J. Chem. Phys. **42**, 2305 (1965). This representation of $\mu_{EN}$, however, does give an estimate for $\mu_{EN}$ which agrees quite well with the observed value, using reasonable values for $\mu_{11}$, $a$, $b$, etc. (b) G. Herzberg, *Spectra of Diatomic Molecules* (D. Van Nostrand Company. Inc., Princeton, New Jersey, 1950), 2nd ed., p. 101.

Here, $\nu_{EN}$ is the frequency of the charge-transfer band. Values of $\nu_{EN}$ are listed in Table II for the complexes to be considered here.

In principle, the electronic transition moment for the charge-transfer band, $\mu_{EN}$, required in Eq. (12), could be evaluated from spectroscopic studies of the charge-transfer absorption. Thus, the transition moment is related to the integrated absorbance as follows[11]:

$$\mu_{EN}^2=\left(0.92\times10^{-2}\int\epsilon d\nu\right)\Big/\nu_{\max}. \quad (15)$$

Unfortunately experimental difficulties and lack of interest conspire to make the necessary information rather difficult to find in the literature. In Table II we list the values for some complexes of interest which are available or which can be estimated from the approximate relationship[11] using the half-intensity width, $\Delta\nu_{\frac{1}{2}}$, and the maximum extinction coefficients, $\epsilon_m$:

$$\mu_{EN}\cong\{[(0.92\times10^{-2})/\nu_{\max}]\epsilon_m\Delta\nu_{\frac{1}{2}}\}^{\frac{1}{2}}. \quad (16)$$

Here we should include a shape factor correcting for the fact that the band shape is not triangular, but we consider error due to this neglect to be very small.

The final parameter which must be evaluated in Eq. (12) is $b$ (hence, $a$ can be calculated). There are several methods of approach which are possible here. The dipole moment of the complex, the intensity of the charge-transfer absorption, and the energy of formation of the complex may all be expected to depend on $b$.[23] Thus, it is possible—again, in principle—to obtain experimental estimates of $b$ from such measurements. However, the number of complexes for which such data are available is again very limited. Hence, we have used an alternate approach which depends on the relationship between $b$ and the change in frequency of the X–Y stretching vibration. This relationship is discussed in detail in the next section; the resulting values of $b$ are listed in Table III for the complexes considered here. We note that these values of $b$ are not shockingly different from previous estimates,[23] thus indicating a general consistency in the estimates. (See below.)

TABLE I. $X_2$ and $X_2^-$ potential parameters.[a]

| Halogen | $D_0$(eV) | $r^b$(Å) | $\omega_e$(cm$^{-1}$) | $\partial E_A^v/\partial r^c$ |
|---|---|---|---|---|
| $I_2$ | 1.54 | 2.67 | 215 | |
| $I_2^-$ | 0.70 | 3.15 | 115 | −4.25* |
| $I_2^-$ | 0.70 | 3.15 | 100 | −2.80 |
| $I_2^-$ | 0.70 | 3.05 | 115 | −2.39 |
| $I_2^-$ | 0.55 | 3.15 | 115 | −3.78 |
| $Br_2$ | 1.97 | 2.28 | 323 | |
| $Br_2^-$ | 1.00 | 2.80 | 150 | −4.35* |
| $Cl_2$ | 2.48 | 1.99 | 564 | |
| $Cl_2^-$ | 1.20 | 2.60 | 220 | −5.34* |
| ICl | 2.15 | 2.32 | 384 | |
| $ICl^-$ | 1.10 | 2.80 | 200 | −5.70* |

[a] See Ref. 20.
[b] $r$ is $r_0$ for $X_2$ and $r_e$ for $X_2^-$.
[c] Values expressed in electron volts per angstrom. Starred values were used to compute $M$ in Table IV.

[23] See Refs. 11, 19, and 20 together with G. Briegleb, *Elektronen-Donator-Acceptor-Komplexe* (Springer-Verlag, Berlin, 1961).

TABLE II. Estimated values of $\mu_{EN}$.

| Complex | $\nu_{EN}\times10^{-4}$ (cm$^{-1}$) | $\epsilon_m\times10^{-4}$ | $\Delta\nu_{\frac{1}{2}}\times10^{-3}$ (cm$^{-1}$) | $\mu_{EN}$ |
|---|---|---|---|---|
| (1) Benzene–$I_2$ | 3.42[a] | 1.64 | | 4.8[b] |
| (2) Pyridine–$I_2$ | 4.25[c] | 5.0 | 5.2 | 7.5 |
| (3) Benzene–$Br_2$ | 3.42[d] | 1.34 | | 4.7[b] |
| (4) Chlorobenzene–$Br_2$ | 3.50[d] | 0.73 | 5.4 | 3.3 |
| (5) Benzene–$Cl_2$ | 3.60[e] | 0.91 | 5.4 | 3.5 |
| (6) Benzene–ICl | 3.55[a] | 0.81 | 5.5[f] | 3.4 |
| (7) Biphenyl–ICl | 3.42[a] | 0.82 | 5.5 | 3.5 |
| (8) Toluene–ICl | 3.48[a] | 0.80 | 5.5 | 3.4 |
| (9) *p*-Xylene–ICl | 3.42[e] | 0.65 | 5.5 | 3.1 |

[a] L. J. Andrews and R. M. Keefer, J. Am. Chem. Soc. **74**, 4500 (1952).
[b] N. M. Blake, H. Winston, and S. A. Patterson, J. Am. Chem. Soc. **73**, 4437 (1951).
[c] C. Reid and R. S. Mulliken, J. Am. Chem. Soc. **76**, 3869 (1954).
[d] R. M. Keefer and L. J. Andrews, J. Am. Chem. Soc. **72**, 4677 (1950).
[e] L. J. Andrews and R. M. Keefer, J. Am. Chem. Soc. **73**, 462 (1951).
[f] No published spectra available.

### Comparison with Experimental Intensities

Having evaluated the parameters, we are now able to utilize Eq. (12) to compute the transition moment factors, $M$, for the X–Y stretching vibration in halogen charge-transfer complexes. The second term of Eq. (12), $M_{\rm del}$, was calculated for each complex using the parameters listed in Tables I–III. The value of $(\partial\mu/\partial Q)_0$ for each halogen-molecule stretching vibration was calculated from the experimental intensities observed in inert solvents. Since $(\partial\mu/\partial Q)_0$ is presumably directed along the halogen-molecule bond and since $M_{\rm del}$ is directed from D to A, the sum of the two terms in Eq. (12), $(\partial\mu/\partial Q)_0+M_{\rm del}$, depends on the geometry of the complex. The values of $M_{N1,N0}$ listed in Table IV were calculated assuming the halogen molecule to be perpendicular to the aromatic ring except for the pyridine complexes, where the halogen molecule was assumed to be in line with the N atom of pyridine.

Comparing the calculated values of $M_{N1,N0}$ with the experimental results in Table IV, we see that in most cases the two numbers agree within a factor of 2. (For example, the predicted *intensity* is about a factor of 4 higher than the observed *intensity*.) Considering the uncertainty in evaluating the parameters, this agreement is as good as could be expected. Certainly there seems to be little reason to doubt that the "delocalization moment" can account for all of the observed intensification.

Assuming Eq. (12) to be correct (see below), the errors in the predicted values of $M_{N1,N0}$ may arise

TABLE III. Estimated values of $b$.[a]

| Complex | $\nu_0$[b] ($cm^{-1}$) | $\nu_c$[c] ($cm^{-1}$) | $b^2+abS$ | $a$ | $b$ |
|---|---|---|---|---|---|
| (1) Benzene–$I_2$ | 207 R[d] | 201 R[d] | 0.077 | 0.95 | 0.24*[e] |
| (2) Pyridine–$I_2$ | 207 R[d] | 183 ir[e] | 0.31 | 0.77 | 0.45* |
| (3) Benzene–$Br_2$ | 312 R[d] | 301 R[d] | 0.094 | 0.94 | 0.26* |
| | | 305 ir[f] | 0.060 | 0.96 | 0.20 |
| (4) Chlorobenzene–$Br_2$ | 312 R[d] | 306 ir[f] | 0.051 | 0.965 | 0.18* |
| (5) Benzene–$Cl_2$ | 541 R[d] | 525 R[d] | 0.079 | 0.95 | 0.24* |
| | | 526 ir[g] | | | |
| (6) Benzene–ICl | 375 ir[h] | 355 ir[h] | 0.142 | 0.91 | 0.34* |
| | | 352 R[d] | 0.164 | 0.895 | 0.36 |
| (7) Biphenyl–ICl | 375 ir[h] | 358 ir[f] | 0.121 | 0.92 | 0.31* |
| (8) Toluene–ICl | 375 ir[h] | 356 ir[h] | 0.135 | 0.91 | 0.33* |
| (9) $p$-Xylene–ICl | 375 ir[h] | 354 ir[h] | 0.149 | 0.90 | 0.35* |
| (10) $(CH_3)_3N$–$I_2$ | 207 R[d] | 185 ir[i] | 0.61[j] | | |

[a] Calculated using Eq. (21), except for $(CH_3)_3N$–$I_2$.
[b] From a spectrum of the halogen in an "inert" solvent ($CHCl_3$ or $CCl_4$). R=Raman; ir=infrared.
[c] Frequency of the halogen vibration in the complex.
[d] H. Stammreich, R. Forneris, and Y. Tavares, Spectrochim. Acta **17**, 1173 (1961).
[e] E. K. Plyler and R. S. Mulliken, J. Am. Chem. Soc. **81**, 823 (1959).
[f] W. B. Person *et al.*, Ref. 3.
[g] J. Collin and L. D'Or, Ref. 7.
[h] W. B. Person *et al.*, Ref. 3.
[i] H. Yada, J. Tanaka, and S. Nagakura, Ref. 4.
[j] $\Delta k/k$ calculated using force constants from Ref. 4.

from errors in calculating $(\partial\mu/\partial Q)_0$ or $M_{\text{del}}$, and from assuming the wrong geometry. We have assumed that the charge on the halogen ion, $(XY)^-$, in the dative state is symmetrically distributed over the molecule and that $(\partial\mu/\partial Q)_0$ for the halogen molecule is the same in the no-bond state as it is in an inert solvent. Although both of these assumptions may contribute to the error,[9] we consider the major source of error for the $X_2$ complexes to be the estimated value of $(\partial E_A{}^v/\partial r)$ used to compute $M_{\text{del}}$. To test the latter error, we consider the comparison of complexes with the same halogen molecule, $X_2$, but with different donor molecules. If $(\partial\mu/\partial Q)_0=0$, then for two complexes, A—X–X and B—X–X, we find using Eq. (12):

$$M_{\text{del},A}/M_{\text{del},B} = [b_A(\mu_{EN})a_B(\Delta W)_B/a_A(\mu_{EN})b_B(\Delta W)_A]. \quad (17)$$

TABLE IV. Calculated vibrational-transition moment factors, $M_{N1,N0}$.

| Complex | $(\partial\mu/\partial Q)_0$[a] | $\lvert M_{\text{del}}\rvert$[b] | $M$[c] (calc) | $M$[d] (obs) |
|---|---|---|---|---|
| (1) Benzene–$I_2$ | 0 | 0.27 | 0.27 | 0.17[e] |
| (2) Pyridine–$I_2$ | 0 | 1.99 | 1.99 | 0.80[f] |
| (3) Benzene–$Br_2$ | 0 | 0.42 | 0.42 | 0.32[g] |
| (4) Chlorobenzene–$Br_2$ | 0 | 0.19 | 0.19 | 0.19[g] |
| (5) Benzene–$Cl_2$ | 0 | 0.50 | 0.50 | 0.20[g] |
| (6) Benzene–ICl | 0.51 | 0.63 | 1.14 | 0.67[h] |
| (7) Biphenyl–ICl | 0.51 | 0.60 | 1.11 | 0.82[g] |
| (8) Toluene–ICl | 0.51 | 0.62 | 1.13 | 0.71[h] |
| (9) $p$-Xylene–ICl | 0.51 | 0.61 | 1.12 | 0.60[h] |

[a] See Eq. (7). Calculated assuming $(\partial\mu/\partial Q)_0=1.537\times10^{-3}(A)^{\frac{1}{2}}$, where $A$ is the observed intensity of ICl in an inert solvent.
[b] $M_{\text{del}}=[2b(\partial E_A{}^v/\partial r)\mu_{EN}/a(m_r)^{\frac{1}{2}}\Delta W]$. See Eq. (12). Units are debyes per angstrom.
[c] Transition-moment factor calculated using Eq. (12) assuming the halogen to be perpendicular to the aromatic ring (or in line with the N atom of pyridine).
[d] Computed from $M=1.537\times10^{-3}(A)^{\frac{1}{2}}$.
[e] E. K. Plyler and R. S. Mulliken, J. Am. Chem. Soc. **81**, 823 (1959).
[f] A preliminary estimate by A. G. Maki (private communication).
[g] W. B. Person *et al.*, Ref. 3.
[h] W. B. Person *et al.*, Ref. 1.

The calculated and observed values of the ratio in Eq. (17) are shown in Table V for some $X_2$ complexes. The agreement here is much improved, suggesting that a major part of the difference between the calculated and observed $M$ factors for $X_2$ molecule complexes is indeed due to the uncertainty in estimating $(\partial E_A{}^v/\partial r)$. Unfortunately the number of complexes for which data are available is limited, so that it is not possible, at this time, to test these relations further.

## FREQUENCY SHIFT OF THE HALOGEN STRETCHING VIBRATION

Let us consider briefly the theory of the frequency shift in the X–Y stretching frequency in halogen complexes (D···X–Y–Z). Again, this theory can be modified to consider any vibration of the donor or of the acceptor; we concentrate on the halogen stretching vibration simply because it shows a large effect and has been studied most completely.

In order to discuss the frequency shifts we consider

TABLE V. Calculated intensity ratios.

| Complexes | Ratio (calc)[a] | Ratio (obs)[b] |
|---|---|---|
| (Benzene–$I_2$/pyridine–$I_2$) | 0.14 | 0.21 |
| (Chlorobenzene–$Br_2$/benzene–$Br_2$) | 0.45 | 0.59 |

[a] See Eq. (17). Transition-moment ratios calculated using the calculated $M$ values in Table IV.
[b] Calculated using the observed $M$ values of Table IV.

the potential function for the X–Y bond as a function of $r_{X-Y}$ (or $Q_l$). For the no-bond state, this function can be loosely approximated by the potential function for the free halogen; in the dative state, it is given closely by the function for the free $(XY)^-$ ion.[24] The potential function for the X–Y bond in the complex is then given in good approximation by

$$W_N(X\text{–}Y) = (a^2+abS)W_0(X\text{–}Y) + (b^2+abS)W_1(X\text{–}Y). \quad (18)$$

Here the coefficients $(a^2+abS)$ and $(b^2+abS)$ are the weights of the no-bond and dative structures, respectively. The resulting potential curve for the X–Y bond in the complex is given in Fig. 2. The properties of the curve representing the X–Y bond in the complex are given at any point by the weighted average of the properties of the two limiting curves. In particular, the force constant for the X–Y bond in the complex, $k_N \equiv \partial^2 W_N/\partial Q_l^2$, is given by

$$k_N \equiv (a^2+abS)k_0 + (b^2+abS)k_1. \quad (19)$$

Here $k_0$ and $k_1$ are the X–Y stretching force constants in the free X–Y and $(X\text{–}Y)^-$ molecules, respectively.[25] Using the normalization condition $(a^2+2abS+b^2=1)$, we may rearrange Eq. (19) to obtain

$$(k_0-k_N)/k_0 = \Delta k/k = [1-(k_1/k_0)](b^2+abS). \quad (20)$$

We believe Eq. (20) is the basic equation giving the approximate relationship between the frequency shift of the X–Y stretching vibration and the weight of the dative state in the structure of the complex $(b^2+abS)$.

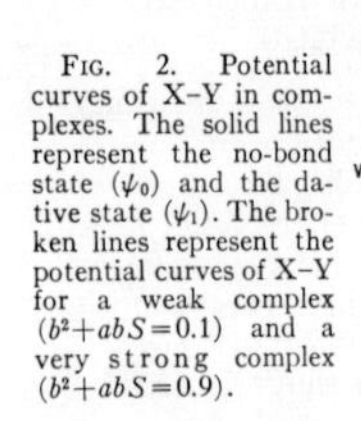

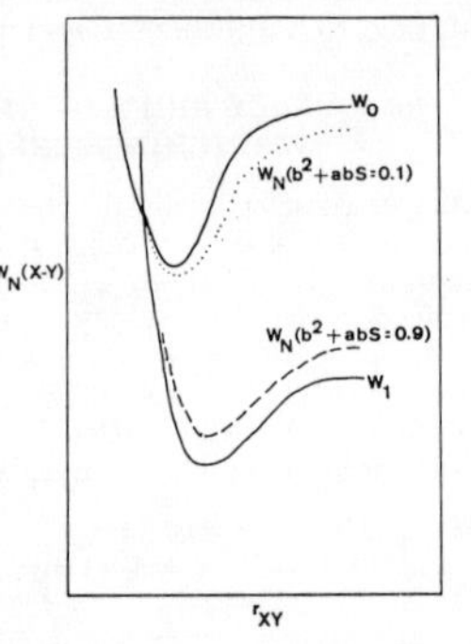


FIG. 2. Potential curves of X–Y in complexes. The solid lines represent the no-bond state ($\psi_0$) and the dative state ($\psi_1$). The broken lines represent the potential curves of X–Y for a weak complex ($b^2+abS=0.1$) and a very strong complex ($b^2+abS=0.9$).

[24] Note that these functions are approximations only to a part of the total potential surface for the complex.

[25] We note that this treatment ignores the dependence of $a$ and $b$ on $Q_l$, the normal coordinate describing the X–Y stretching vibration. We believe the terms which are ignored $[(\partial^2 a/\partial Q_l^2)$, $(\partial a/\partial Q_l)$, etc.$]$ are not important here. Also, in writing Eq. (18), we have assumed that $H_{01}=\frac{1}{2}S(W_1-W_0)$, which may be questionable.

TABLE VI. Comparison of values of $b$.

| Complex | $(\Delta\nu/\nu)$[a] | Dipole moment | $\Delta H$[f] |
|---|---|---|---|
| Benzene–$I_2$ | 0.24 | 0.17[b], 0.28[c] | 0.15 |
| Pyridine–$I_2$ | 0.45 | 0.5[d] | 0.6 |
| $(C_2H_5)_3N$–$I_2$ | ... | 0.6[e] | ... |
| $(CH_3)_3N$–$I_2$ | 0.61 | ... | 0.8 |

[a] The $b$ values are taken from Table III.
[b] R. S. Mulliken, Ref. 19.
[c] G. Kortum and H. Walz, Z. Elektrochem. **57**, 73 (1953).
[d] C. Reid and R. S. Mulliken, J. Am. Chem. Soc. **76**, 3869 (1954).
[e] Calculated from data of H. Tsubomura and S. Nagakura, J. Chem. Phys. **27**, 819 (1957). (Probably too large.)
[f] W. B. Person (to be published). It seems obvious that the values for the amine complexes here are too large.

We have already estimated frequencies for some $(X\text{–}Y)^-$ ions to be about one-half the frequencies of the free molecules.[20] Thus, the ratio $k_1/k_0 \cong 0.25$. As long as the interaction between the electron donor D and the X–Y bond is not so strong that we must treat the D···X–Y system as a triatomic molecule, the ratio $\Delta k/k$ is given in good approximation by $\Delta k/k \cong 2(\Delta\nu/\nu)$. Hence, for weak complexes, the weight of the dative structure may be evaluated approximately from the easy measurement of the simple relative frequency shift:

$$(b^2+abS) \simeq (\tfrac{8}{3})(\Delta\nu/\nu). \quad (21)$$

If we need to evaluate the coefficient $b$ alone, as in the previous section, then we may estimate the overlap integral $S$, between the no-bond and dative states and solve for $b$ by trial and error. Values of $S$ have been estimated for several types of complexes of interest.[20]

If both the X–Y stretching frequency and the D···X stretching frequency are known, as in $(CH_3)_3N\cdot I_2$, for example,[4] then the triatomic-molecule formula can be used to obtain $k_N$, the X–Y stretching force constant in the complex (assuming no interaction between the X–Y and D···X stretches). This value can be used in Eq. (20) to obtain estimates of the weight of the dative structure, which should be somewhat better than the values from Eq. (21).

As mentioned in the previous section, we have used the infrared data for the X–Y stretching frequency in several complexes to evaluate the weight of the dative structure and the coefficient $b$. These were tabulated in Table III. In order to test the analysis in this section, we have tried to compare values of $b$ estimated here with values obtained using other techniques.[23] Unfortunately, there are very few complexes for which such a comparison can be made at present. The results are shown in Table VI. We see that the agreement is phenomenally good; certainly it is much better than we expect from the approximations used not only in this method but also in the other methods of estimating $b$.

## GEOMETRY OF COMPLEXES

Because of the change in ideas from the earlier papers in this series, as well as from other discussions, it is perhaps worth emphasizing at this time that the infrared spectra yield essentially no information at all about geometry of the halogen complexes. As mentioned earlier in this paper, the original interpretation[1-3,7] of the appearance of the halogen–halogen stretching vibration in the infrared spectrum of complexed symmetrical halogens (as benzene–$Cl_2$ or benzene–$Br_2$, for example) was that the two halogen atoms must be in unsymmetrical locations; hence, the axial model or the oblique[26] model for the structure of the complex was indicated. As we see in this paper (and also in the papers by Ferguson and Matsen[8,9]) the enhancement of infrared intensity is adequately explained as a vibronic interaction, which can occur for any geometrical arrangement of the molecules in the complex. Hence, there is no information about geometry to be gained from the appearance of the halogen–halogen stretching vibration.

In addition to this discouraging conclusion, we must also admit that the changes in the infrared spectrum of the donor are not useful in determining geometry, at least for the interesting case of the benzene–halogen complexes. These changes were studied extensively for the benzene complexes by Ferguson,[27] who interpreted the changes to indicate $C_{6v}$ symmetry for the complex, hence, axial geometry for the complex.

However, in interpreting these results, it is important to remember that the benzene complexes are quite weak. The vibrations of the benzene against the halogen must occur at very low frequencies indeed. It would seem unlikely that the benzene–iodine stretching frequency (expected to be the highest frequency motion of these) could be higher than 75–100 $cm^{-1}$. Certainly the bending motions must be less than 50 $cm^{-1}$. (For the axial benzene–$I_2$ complex, we expect five new low-frequency vibrations of benzene against halogen: one benzene–halogen stretching motion, one doubly-degenerate bending vibration in which the benzene ring rotates about an axis in the plane of the ring, and one doubly degenerate pair of bending vibrations in which the benzene translates in a direction in the plane of the molecule.)

The excited states of these low-frequency vibrations are highly populated at room temperature; the complex cannot be studied in the "nonvibrating" state. The result is that the complex benzene molecule obeys $C_{6v}$ selection rules for *any* configuration of the complex in which the halogen molecule is on one side of the benzene. For less symmetrical equilibrium configurations (such as the "resting" model,[19] with $C_{2v}$ symmetry for the nonvibrating molecule) the low-frequency bending vibrations cause the configuration to change, so that the force field acting on the benzene molecule due to the halogen in the complex averages to $C_{6v}$ symmetry. Thus, we conclude that there is essentially no information about geometry of benzene–halogen complexes to be gained from the infrared spectra from experiments currently reported in the literature.

This conclusion and many of these arguments were also expressed by Ferguson in his recent review of this subject[21]; because of the importance of this question to the theory of complexes and because these arguments do reverse earlier ideas, which were commonly accepted, we thought it worthwhile to repeat them here.

[26] See Ref. 19 for discussion of the different models to be considered. Also see R. S. Mulliken, J. Chem. Phys. **23**, 397 (1955).
[27] E. E. Ferguson, J. Chem. Phys. **25**, 577 (1956); **26**, 1357 (1957); Spectrochim. Acta **10**, 123 (1957).

## FURTHER DISCUSSION

It has been shown empirically in earlier papers of this series[1-3] that there is a linear relationship between $\epsilon_a$ (the added effective charge) and $\Delta k/k$. If delocalization provides for all of the intensity enhancement for the halogen (or O–H) stretching vibration and if $\Delta k/k \cong b^2$, it follows from Eq. (12) that

$$\epsilon_a \cong \left[\frac{2(\partial E_A{}^v/\partial r)\mu_{EN}}{abm_r{}^{\frac{1}{2}}\Delta W}\right]\left(\frac{\Delta k}{k}\right) \cong \left[\frac{2(\partial E_A{}^v/\partial r)\mu_1}{m_r{}^{\frac{1}{2}}\Delta W}\right]\left(\frac{\Delta k}{k}\right). \quad (22)$$

Evaluation of the term in brackets using the same parameters we used to evaluate Eq. (12) does give a nearly constant value of this term for these complexes (which, however, is about 2–3 times larger than the experimental value). Considerable scatter is to be expected in a plot of $\epsilon_a$ vs $\Delta k/k$ and the empirical observation of a straight line is probably somewhat accidental.

Coulson[28] has described hydrogen-bonded systems in terms very similar to those employed here. Tsubomura,[13] using a similar model, estimated the intensification of the O–H bond stretching vibration and reached the conclusion that charge transfer (in a vibronic contribution) is the most important factor determining the increase in $\epsilon_a$. McKinney and Barrow[14] have used a one-dimensional model of the H bond and reached the same conclusion. Our results also suggest that this mechanism is correct, since a strong variation of $E_A{}^v$ with respect to the internal coordinate is also to be expected for the X–H bond, just as for halogen complexes.

In Fig. 3 we show the potential curves for HBr and for $(HBr)^-$ ion constructed[29] from information from the electron-capture process.[30] We see that a rather high value of $(\partial E_A{}^v/\partial r)$ is to be expected here, as well. We believe these curves are typical of the situation to be expected for all X–H bonds.

Thus, we conclude that the explanation for the intensification of the X–H stretching vibration upon

[28] See C. A. Coulson, in *Hydrogen Bonding*, edited by D. Hadzi (Pergamon Press, Inc., New York, 1959), pp. 339ff.
[29] R. S. Mulliken (private communication).
[30] D. C. Frost and C. A. McDowell, J. Chem. Phys. **29**, 503 (1958).

hydrogen-bond formation is probably exactly the same as that given here for the halogen stretching vibration on complex formation. Due to the lack of information for X–H bonds, however, it is much more difficult to treat these systems even as quantitatively as we have done here for the halogen vibration.

## CONCLUSION

We conclude that the enhanced infrared intensity of the halogen–halogen stretching vibration observed in halogen donor–acceptor complexes is explained by electronic reorientation during the vibration, as first suggested by Ferguson and Matsen.[8,9] Partly as a result of this conclusion, we note that the infrared spectra do not give any information about the geometry of the benzene–halogen complexes at least in any simple way. We have attempted to relate the infrared frequency shift of this vibration to the weight of the dative structure of the complex.

Finally, we must admit that our lack of exact knowledge about the structure of the dative state ($D^+$–$A^-$) prevents us from making a quantitative test of these ideas. Furthermore, the general lack of quantitative information about complexes (dipole moments in the ground state to get $b$, values of $\mu_{EN}$, inaccurate intensity measurements of the X–Y stretching vibrations, etc.) is a further bar to a quantitative test of these ideas. In addition, the use of second-order perturbation theory [Eq. (9), in particular] may be questioned for complexes in which $b$ values are as large as those reported in Table III. Certainly the equations presented here must be used with caution and the results are expected to be only approximately correct, even with accurate values for the parameters. Finally we must note that the wavefunctions of Eq. (5) are perhaps much too simple.[31] However, the picture seems to be quite consistent with generally accepted ideas about complexes and the semiquantitative results found here should be useful in predicting general features of the spectral changes expected when a weak complex is formed.

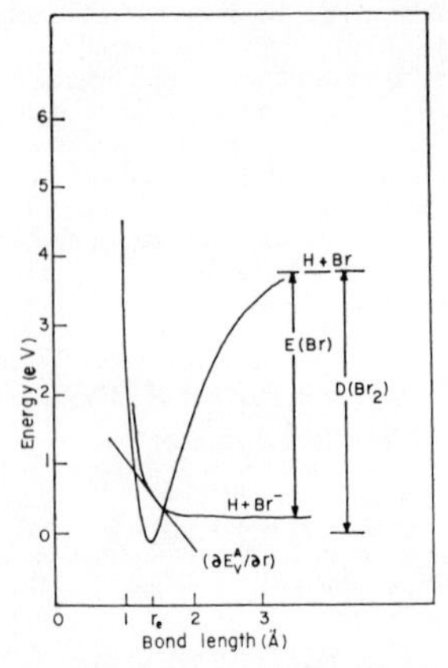


Fig. 3. Potential curves of HBr and $HBr^-$. $E$(Br) is the electron affinity of the Br atom and $E_A{}^v$ the vertical electron affinity of HBr. The slope of the tangent line gives $\partial E_A{}^v/\partial r$.

In considering the general problem of interpreting the infrared spectra of weak complexes, it seems to us that the approach used here is applicable. We assumed that the true structure of the complex could be described in terms of two limiting, idealized structures: the no-bond structure ($\psi_0$) and the dative structure ($\psi_1$), as suggested by Mulliken.[19] We then predict the vibrational spectrum of the complex from the (well-known) spectrum of the no-bond structure (i.e., the spectra of the isolated molecules) and from the (predicted) spectrum of the dative state. Here we considered only the X–Y stretching vibration, so we had to guess only that vibration in the dative state, which we assumed to be approximately the $(X\text{–}Y)^-$ ion. However, we should be able to predict the changes in spectrum of the donor molecule on complexing if we knew also the spectrum of that molecule in the dative state (i.e., if we knew the spectrum of the $D^+$ ion).

We see that we should not expect any very great changes in the spectrum of $D^+$ from that of D; the loss of one bonding electron is spread over several bonds (in aromatic donors) or the electron comes from a nonbonding orbital (in the case of nitrogen- or oxygen-containing donors). There might be intensity changes due to the change in electronic charge on D (to $D^+$) and also due to the change in the symmetry from the free D molecule to the complex D—X–Y–Z. Frequency changes should be small, however.

For acceptors, which do not accept the electron into a localized orbital as the halogens do, we should again expect the $A^-$ spectrum to be similar to the free A molecule spectrum. Furthermore, if the acceptor does maintain the added electron in a localized orbital involving the X–Y bond, then the changes in the Y–Z vibrations should be quite small.

Finally, we may predict that rather large intensity enhancements occur only for vibrations which change the weighting of the wavefunctions $\psi_0$ and $\psi_1$. We should expect the changes due to the change in total electronic charge and due to the change in symmetry to be relatively small. (Possibly, a change by a factor of 2 in bands already allowed, or by allowing gas-phase forbidden bands to appear weakly.)

As mentioned above, electronic rearrangements occur for vibrations of the donor or acceptor molecules, (1) which change the ionization potential of the donor (only the totally symmetric vibrations of the free donor do this), (2) which change the vertical electron affinity

[31] R. S. Mulliken, J. Chim. Phys. **61**, 20 (1964).

of the acceptor, or (3) which change the overlap between the no-bond and dative structures (effectively, this means vibrations which change the overlap between the donor atomic orbitals and the acceptor atomic orbitals). These larger intensity changes, then, should not occur for very many of the vibrations of the molecules in the complex. For benzene–$I_2$, for example, these changes should occur only for the two $a_{1g}$ and the $e_{1g}$ vibrations of benzene and for I–I stretching vibration. There is some question remaining about the magnitude expected for these "larger" changes. More data are needed before semiquantitative estimates can be made for other vibrations in the same way as attempted here for the halogen vibrations. However, a preliminary estimate by Ferguson[21] suggests that these ideas are reasonable for the $e_{1g}$ vibration of benzene.

## ACKNOWLEDGMENTS

Financial support from U.S. Army Research Office (Durham) and from Public Health Service Research Grant No. GM–10168 is gratefully acknowledged. We are grateful to Dr. E. E. Ferguson and to Professor R. S. Mulliken for many stimulating discussions. Comments by Professor D. A. Dows and by Professor M. Hanna have been most helpful.

# *General References*

[GR1] G. Briegleb, *Elektronen-Donor-Acceptor-Komplexe* (Springer Verlag, Berlin, 1961).
[GR2] L. J. Andrews and R. M. Keefer, *Molecular Complexes in Organic Chemistry* (Holden-Day, San Francisco, 1964).
[GR3] E. M. Kosower, in *Physical Organic Chemistry* (Wiley, New York, 1968)
[GR4] J. Rose, *Molecular Complexes* (Pergamon, Oxford, 1967).
[GR5] R. S. Mulliken and W. B. Person in *Physical Chemistry,* Volume III, Chapter 10, H. Eyring, D. Henderson, and W. Jost, Eds. (Academic, New York, 1968).

# *Partial Glossary of Symbols*

## A. SYMBOLS USED IN DESCRIBING WAVEFUNCTIONS

### 1. Symbols that represent wavefunctions

| *Symbol* | *Definition* | *Page** |
|---|---|---|
| $\Psi_N$ | Wavefunction for the ground state of the complex | 5 |
| $\Psi_V$ | Wavefunction for the charge transfer state | 6 |
| $\Psi_0\ (D, A)$ | Wavefunction for the no-bond structure | 5 |
| $\Psi_1\ (D^+ - A^-)$ | Wavefunction for the dative structure | 5 |
| $\Psi_{\mathrm{I}}, \Psi_{\mathrm{II}}$ | Abbreviation for components of the dative function $\Psi_1$ | 19 |
| $\Psi_T$ | Wavefunction for the triplet CT state | 21 |
| $\Psi_m^v$ | A vibronic wavefunction for a molecule in vibrational state $v$ and electronic state $m$ | 23 |
| $\Psi_m^0$ | Zero-order wavefunction for the $m$th electronic state of a complex | 164 |
| $\alpha(j), \beta(j)$ | Spin functions for electron $j$ | 19 |
| $\lambda_i\,(j)$ | The $i$th spinorbital, occupied by electron $j$ | 19 |

*The page on which the symbol first appears.

| | | |
|---|---|---|
| $S_0(1, 2)$ | Singlet spin function for electrons 1 and 2 | 177 |
| $T_{0\pm 1}(1, 2)$ | Triplet spin functions | 183 |
| $\phi_{a^-}$ | The orbital on the acceptor ion into which the electron is accepted | 20 |
| $\phi_d$ | The orbital on the donor from which the electron is donated | 20 |
| $\phi_{da}$ | The bonding bridging MO in the complex in the whole-complex-MO method | 181 |
| $\phi_{da}^*$ | The corresponding antibonding bridging MO | 183 |
| $\chi_v^m$ | A vibrational wavefunction for vibrational state $v$ and electronic state $m$ | 24 |

## 2. Coefficients Appearing in Wavefunctions

| | | |
|---|---|---|
| $a$ | Coefficient of $\Psi_0$ in $\Psi_N$ | 5 |
| $a^*$ | Coefficient of $\Psi_1$ in $\Psi_V$ | 6 |
| $b$ | Coefficient of $\Psi_1$ in $\Psi_N$ | 5 |
| $-b^*$ | Coefficient of $\Psi_0$ in $\Psi_V$ | 6 |
| $\mathfrak{N}, \mathfrak{N}''$ | Normalization factors | 178 |
| $\rho\,(\rho^*)$ | Ratio of $b$ to $a$ (or $b^*$ to $a^*$) in $\Psi_N$ (or $\Psi_V$) | 12 |
| $\rho_{mn}$ | Coefficient of $\Psi_n^0$ in the wavefunction for state $m$ in generalized resonance structure method | 164 |

## 3. Other Symbols Used in Connection with Wavefunctions

| | | |
|---|---|---|
| $\mathcal{A}, \mathcal{A}'$ | Antisymmetrizer operator and supplementary antisymmetrizer operator | 17, 18 |
| $\mathcal{H}$ | The exact Hamiltonian operator | 9 |
| $F_0$ | Weight of the no-bond structure in $\Psi_N$ | 5 |
| $F_1$ or $F_{1N}$ | Weight of the dative structure in $\Psi_N$ | 5 |

# B. ABBREVIATIONS USED IN DESCRIBING WAVEFUNCTIONS

| | | |
|---|---|---|
| AO | Atomic orbital | 35 |
| AS·MSOP | Abbreviation for antisymmetrized MSO product | 17, 20 |
| LCMO | Abbreviation for linear combinations of MOs | 181 |
| MO | Abbreviation for molecular orbital | 17 |
| MSO | Abbreviation for molecular spinorbitals | 19 |
| SCF | Abbreviation for self consistent-field | 17 |
| STF | Abbreviation for Slater-type-function | 181 |

## C. SYMBOLS USED TO DESCRIBE COMPLEXES

### 1. General

| | | |
|---|---|---|
| A, $A^-$, $A^*$ | Electron acceptor; its negative ion; A in an excited state | 1 |
| $\mathring{A}$ | The negative radical ion of A | 260 |
| D, $D^+$, $D^*$ | Electron donor; its positive ion; D in an excited state | 1 |
| $D^{\ddagger}$ | A positive radical ion | 232, 262 |
| D·A | A complex between D and A | 3 |
| $Q^-$ | A symbol representing the anion of an acid, e.g., $C_6H_5O^-$ | 231 |
| EDA | Electron-Donor-Acceptor (complex) | 301 |

### 2. Symbols Used in Classifying Complexes by Their Functioning

| | | |
|---|---|---|
| $a\sigma$ | Antibonding $\sigma$ acceptor | 4 |
| $a\pi$ | Antibonding $\pi$ acceptor | 4 |
| $b\pi$ | Bonding $\pi$ donor | 4 |
| $n$ | Non-bonding lone-pair donor | 4 |
| Q | Radical acceptor | 33 |
| R | Radical donor | 33 |
| $v$ | Vacant orbital acceptor | 4 |

## D. ABBREVIATIONS USED FOR CHEMICAL NAMES

| | | |
|---|---|---|
| Ar | Aromatic hydrocarbon (e.g., benzene) | 89 |
| Bz | Abbreviation for benzene | 34 |
| Chl | Abbreviation for chloranil | 89 |
| $R_3N$ | Alkyl (R)-substituted amines | 3 |
| Py | Abbreviation for pyridine | 47 |
| $Py^+$–R or R–$Py^+$ | N-alkyl pyridinium ion | 206 |
| TCNE | Abbreviation for tetracyanoethylene | 3 |
| TCNQ | Abbreviation for tetracyanoquinodimethane | 34 |
| TMPPD | Abbreviation for tetramethylparaphenylene-diamine | 253 |
| TNB | Abbreviation for trinitrobenzene | 55 |
| Un | Unsaturated compound (e.g., ethylene) | 273 |
| X–Y | General symbol designating a halogen molecule, such as I–Cl, or I–I | 46 |

## E. SYMBOLS USED IN CONNECTION WITH THE ABSORPTION OF LIGHT

### 1. Symbols Describing the Shape of the Band

| Symbol | Description | Page |
|---|---|---|
| $I_\nu^0$ | The intensity of the incident light at frequency $\nu$ | 25 |
| $I_\nu$ | The intensity of the transmitted light at frequency $\nu$ | 25 |
| $\epsilon_\nu$ | The extinction coefficient or molar absorptivity | 25 |
| $\epsilon_{max}$ | Molar absorptivity at $\lambda_{max}$ | 25 |
| $\lambda_{max}$ | Wavelength of maximum absorption | 25 |
| $\nu_{CT}$ | Frequency (in $sec^{-1}$) of the charge transfer band | 12 |
| $\nu_H$ | The higher frequency at which $\epsilon = \frac{1}{2}\epsilon_{max}$ for a CT band | 25 |
| $\nu_L$ | The lower frequency at which $\epsilon = \frac{1}{2}\epsilon_{max}$ for a CT band | 25 |
| $\Delta\nu_{1/2}$ | Half-intensity width in $cm^{-1}$ | 25 |

### 2. Symbols Related to the Absorbance

| Symbol | Description | Page |
|---|---|---|
| $A_A$, $A_C$, etc. | Absorbance of species $A$, $C$, etc. | 82 |
| $A_T$ | Total absorbance of a solution | 83 |
| $B$ | The integrated molar intensity (in vibrational spectroscopy) $\left(= \frac{1}{cl} \int \ln (I_\nu^0/I_\nu\, d\nu)\right)$ in cm $mmole^{-1}$ | 74 |

### 3. Symbols Referring to the Intensity of Absorption and to the Transition Dipole

| Symbol | Description | Page |
|---|---|---|
| $\mu_{VN}$ | The transition dipole for the charge transfer band | 16 |
| $\mu_{VN}^{el}$, etc. | More precise designation of $\mu_{VN}$ | 24 |
| $\mu_{mn}$ | General symbol for transition moment between electronic states $m$ and $n$ | 24 |
| $\mu_{mn}^{v_m, v_n}$ | Transition dipole between vibronic state $v_m$, $m$ and vibronic state $v_n$, $n$ | 24 |
| $f$ | Oscillator strength | 25 |

## F. SYMBOLS REFERRING TO OVERLAP INTEGRALS

| Symbol | Description | Page |
|---|---|---|
| $S_{01}$ | Overlap integral between $\Psi_0$ and $\Psi_1$ | 5 |

| | | |
|---|---|---|
| $S_{da-}$ | The overlap integral between $\phi_d$ and $\phi_{a-}$ | 20 |
| $S_{I,II}$ | The overlap integral between $\Psi_I$ and $\Psi_{II}$ | 19 |

## G. SYMBOLS REFERRING TO THE DIPOLE MOMENT

| | | |
|---|---|---|
| $\mu_{op}$ | The dipole moment operator | 15 |
| $\mu_N$ | The dipole moment of the complex in its ground state | 15 |
| $\mu_0$ | The dipole moment of the no-bond structure | 15 |
| $\mu_1$ | The dipole moment of the dative structure | 15 |
| $\mu_{01}$ | The dipole moment due to the overlap charge | 15 |

## H. SYMBOLS OCCURRING IN CONNECTION WITH EQUILIBRIUM CONSTANTS AND OTHER THERMODYNAMIC PROPERTIES OF SOLUTIONS

| | | |
|---|---|---|
| $c_i$; $c_i^0$ | Concentration of $i$th species (sometimes just called $c$; total concentration in solution of the $i$th species | 25 |
| $K$ | Equilibrium constant for complex formation (general) | 3 |
| $K_c$ | Equilibrium constant for complex formation (concentrations in moles liter$^{-1}$) | 81 |
| $K_x$ | Equilibrium constant for complex formation (concentrations in mole fraction units) | 81 |
| $K_a$ | Equilibrium constant in terms of activities | 91 |
| $\gamma_C$, $\gamma_A$, etc. | Activity coefficients of complex, of $a$, etc. | 91 |
| $\Gamma_i$ | Equilibrium product of activity coefficients ($i = x$ or $c$) | 91 |
| $\delta_i$ | Solubility parameter of species $i$ | 94 |
| $\Delta F^0$ | Standard free energy change on formation of complex | 91 |
| $\Delta H^0$ | Standard enthalpy of formation | 91 |
| $\Delta H_g^0$ | Enthalpy of formation of complex in gas phase | 94 |
| $\Delta H_v(D)$, etc. | Enthalpy of vaporization of liquid $D$, etc. | 94 |
| $\Delta H_{solv}(D)$ | Enthalpy of solvation of $D$, etc. | 94 |
| $\Delta S_i^0$ | Standard entropy of formation ($i = x$ or $c$) | 91 |
| $\overline{V}_S$, $\overline{V}_D$, etc. | The molar volume of solvent ($S$), $D$, etc. | 90 |
| $\Delta W_f$ | Change in internal energy on formation of complex | 95 |

## J. SYMBOLS REFERRING TO THE ENERGY OF THE COMPLEX

| | | |
|---|---|---|
| CT | Abbreviation for charge transfer | 6 |
| $C$ | Coulomb energy contributing to $G_1$ | 116 |
| $C_1, C_2$ | Empirical constants in the relationship between $h\nu_{CT}$ and $I_D$ | 124 |
| $E_A$ | Electron affinity of an acceptor (A) | 15 |
| $E_A^v$ | The vertical electron affinity of an acceptor (A) | 72, 116 |
| $G_0$ | The energy stabilizing $\Psi_0$ not including resonance | 72, 115 |
| $G_1$ | The energy stabilizing $\Psi_1$ not including resonance | 72, 115 |
| $H_{ij} = W_{ij}$ | $\int \Psi_i^\dagger H \Psi_j \, dv$ (if $i = j$, $W_{ij} = W_i$) | 11 |
| $I_D$ | Ionization potential of a donor (D) | 15 |
| $I_D^v$ | The vertical ionization potential of a donor (D) | 72, 116 |
| $V$ | Valence energy contributing to $G_1$ | 120 |
| $W_N$ | The energy of the complex in its ground state | 12 |
| $W_V$ | The energy of the complex in its charge transfer state | 12 |
| $W_0$ | The energy of the "no-bond" structure of the complex | 12 |
| $W_1$ | The energy of the "dative" structure | 12 |
| $W_{SN}, W_{SV}$ | Solvation energy of the complex in N or V states, respectively | 117 |
| $W'_V$ | Energy of a dative structure that does not have proper symmetry to mix with $\Psi_N$ | 166 |
| $X_0, X_1$ | The CT resonance energies ($= -\beta_0^2/\Delta$ or $\beta_1^2/\Delta$, respectively) | 118 |
| $\Delta$ | The energy difference between dative and no-bond structures | 12 |
| $\beta_i$ | $= W_{01} - W_i S_{01}$; $i = 0$ or 1 | 12 |
| $\beta_m^n$ | Resonance integral between structures $m$ and $n$ | 164 |
| $\Delta_{mn}$ | Energy difference between structures $n$ and $m$ | 164 |

## K. SYMBOLS RELATING TO NUCLEAR CONFIGURATION IN THE COMPLEX

| | | |
|---|---|---|
| $\mathbf{r}_i$ | Vector giving the coordinates of electron $i$ | 24 |

| | | |
|---|---|---|
| $R$ | A coordinate describing nuclear configuration | 23 |
| $R_{\mathrm{DA}}$ | The distance between the center of the donor molecule and the center of the acceptor | 68 |
| $R_{\mathrm{D^+A^-}}$ | Distance separating the charges on $D^+$ and A | 120 |
| $R_c$ | The point at which $W_0$ and $W_1$ potential curves cross | 223 |
| $R_e''$ | The equilibrium value of $R_{\mathrm{DA}}$ in ground state, $N$ | 113 |
| $R_e'$ | The equilibrium value of $R_{\mathrm{DA}}$ in CT state, $V$ | 113 |
| $Q$ | General symbol for a configuration coordinate of complex (as distinguished from $R_{\mathrm{DA}}$) | 129 |
| $Q_N$ | The general configuration coordinate in the ground electronic state of complex such as to minimize $W_N$ as $R_{\mathrm{DA}}$ varies | 130 |

## L. SYMBOLS RELATING TO VIBRATIONAL SPECTRA

| | | |
|---|---|---|
| $H_v(\sqrt{a}R)$ | Hermite polynomial of $v$th degree | 102 |
| $k_i^N$ | The force constant for stretching the $i$th bond associated with wavefunction $\Psi_N$ | 64 |
| $L_{ij}^{-1}$ | The normal coordinate transformation coefficient relating the $Q_i$ to $\Delta R_j$ | 78 |
| $\mathbf{M}_{0,1}$ | The total vibronic moment derivative | 71 |
| $m_r$ | The reduced mass for a diatomic molecule | 74 |
| $M_d$ | The delocalization moment (in terms of $Q_i$) | 74 |
| $M_d'$ | The delocalization moment (in terms of $R_i$) | 76 |
| $v$, or $v_i$ | Vibrational quantum number | 23 |

# *Author Index*

# *Subject Index*

The subject index refers both to the text and to the reprint section. Text numbers are given in roman type, page numbers referring to the reprint section in italics. In the indexing of specific complexes, the indexing of the reprints has been less complete than that of the text.

Hard and soft donors and acceptors, 276, *339*